# MATERIALS AND MANUFACTURING TECHNOLOGY

# MATERIALS AND MANUFACTURING TECHNOLOGY

**ROY A. LINDBERG**

Professor of Production Engineering
Mechanical Engineering Department
University of Wisconsin

ALLYN AND BACON, INC.
BOSTON • LONDON • SYDNEY • TORONTO

LIBRARY OF CONGRESS CATALOG CARD NUMBER 68-20571.

PRINTED IN THE UNITED STATES OF AMERICA.

ISBN: 0-205-01947-1

*Twelfth printing . . . August, 1979*

# PREFACE

Material science has opened up vast areas of knowledge in recent years. Many points that were formerly theory have now been proven, or disproven, by studies made with the electron-beam microscope. This advancement of knowledge is in turn of great benefit in understanding the relationships between atomic structure of materials and their reaction to various processes.

A basic knowledge of materials not only helps the student understand what is happening during the various processing operations but also why it is happening. In addition, it enables him to predict the results with a reasonable degree of accuracy.

While a basic knowledge of the properties of materials is essential, the student should also be familiar with the common code designations that are the everyday language of the production engineer.

Design and manufacturing are closely allied subjects. To be of maximum benefit, the student must have some knowledge of the limitations and capabilities of machine tools. It is not possible in a book of this type to discuss all types of manufacturing equipment; therefore, the metal-working processes have been grouped together, with an emphasis on common functions. For example, Chapter 15 includes all the processes used in making holes to the desired accuracy and finish. This treatment allows the student to more easily make meaningful relationships that will form a background for process and equipment selection.

The book has been so arranged that it can be used for one or two courses, depending upon the time alloted. The first nine chapters can be used as a basis for a course in materials and hot working processes. A second course would utilize the last half of the book and would include plastics, powdered metals, cold forming, metal cutting, and metrology.

The questions and problems at the end of each chapter are designed to stimulate the student to think through the material covered and to make engineering applications of the knowledge gained.

My sincere thanks and appreciation are extended to my colleagues for their encouragement, suggestions, and constructive criticism. Gratitude is also expressed to the publishers for their patience and cooperation in preparing this text.

R.A.L.

# CONTENTS

PREFACE v
1 SCIENCE, ENGINEERING, AND TECHNOLOGY 1
2 PROPERTIES OF METALS 10
3 FABRICATING CHARACTERISTICS OF METALS 54
4 SAND CASTING AND CASTING DESIGN 83
5 OTHER MOLDING PROCESSES 121
6 FORGING AND HOT WORKING 151
7 WELDING PRINCIPLES AND ARC WELDING 178
8 GAS WELDING, CUTTING, AND NONDESTRUCTIVE TESTING 233
9 NEWER WELDING PROCESSES 263
10 PLASTICS AND ADHESIVES 283
11 POWDER METALLURGY 320
12 COLD FORMING AND STAMPING 341
13 CUTTING-TOOL PRINCIPLES AND MACHINABILITY 380
14 TURNING AND RELATED OPERATIONS 406
15 HOLE MAKING AND FINISHING 439
16 STRAIGHT AND CONTOUR CUTTING 465

17 GRINDING AND RELATED ABRASIVE-FINISHING PROCESSES **493**
18 METROLOGY AND QUALITY CONTROL **524**
19 AUTOMATION AND NUMERICAL CONTROL **575**
20 NEWER METHODS OF MANUFACTURE **609**
APPENDIX **641**
INDEX **649**

1

# SCIENCE, ENGINEERING, AND TECHNOLOGY

TODAY'S RAPID ADVANCES in all phases of manufacture have been the result of both separate and combined efforts in science, engineering, and technology.

Webster says that *Science* is "a branch of study which is concerned with the observation and classification of facts, especially with the establishment . . . of verifiable laws." *Engineering* is the art and science by which the properties of matter and sources of power in nature are made useful to man in structures, machines, and manufactured products." *Technology* is an "applied science as contrasted with pure science."

No one-sentence definitions for terms as broad as these will be satisfactory. However, we know that modern engineering requires an integration and mastery of the facts (science), first-hand experience (technology), and a systematic method (experimentation and analysis). There can be no sharp dividing lines between one field and the other. Each is an essential element of a team needed in the development of a strong economy.

In the early development of engineering, some things were accomplished by the empirical approach—try a lot of ways of doing something until you find one that works best. This, however, is neither science nor engineering.

Today both science and engineering show a strong trend away from the empirical and descriptive toward the analytical and quantitative.

An economy based on trial-and-error technology must necessarily progress slowly. Industry has reached the point of diminishing returns as far as trial-and-error engineering is concerned. The production challenges of the next 25 years—the development of real innovations that can cause industry to run, rather than crawl, toward a truly significant increase in productivity and efficiency—must be met through the application of science and sound engineering, thus attaining manufacture on a broad scale.

Many engineers sincerely believe that almost any manufacturing problem can be solved by what they refer to as "common sense." And yet, beyond a certain point, common sense—which usually involves taking a practical, empirical look at the problem—has severe limitations. For example, the fact that all matter is composed ultimately of subatomic particles defies common sense, yet studies of those particles are fundamental to a real understanding of engineering materials.

## ROADBLOCKS TO MANUFACTURING PROGRESS

There are several roadblocks to manufacturing progress through science. One of these is the relatively small number of engineers who have applied their knowledge and skill to the manufacturing field. In the past, young people with analytic or scientific minds have not been attracted in sufficient numbers to the manufacturing field. Perhaps this is because manufacturing has traditionally been looked upon as "practical" rather than scientific. Fortunately, the challenges of today's manufacturing have caused a change in this attitude.

Neither industry itself nor educational institutions have shown a great amount of interest in manufacturing engineering. This type of thinking has not been limited to the processing of materials. An excellent example can be found in the electronics field. Except for some work on the electronic properties of vacuum tubes in physics departments, electronics was largely ignored by most universities prior to World War II. Before this time, improved electronic circuits were developed, often by radio amateurs, as a result of experiments, but few formal efforts were made to develop scientific theories in the field of electronics. Progress was slow.

When professors in engineering colleges began to apply mathematical principles to the operation of electronic circuits, they could explain, by analysis, some of the "strange" responses that were observed in the field. Analysis provided understanding, and understanding made it possible to develop more efficient circuits and circuits that could do entirely new things. The frequencies in use increased from a few million cycles to 10,000 megacycles, during World War II, as a consequence of new theories developed in engineering schools. During World War II, frequencies in use increased to 200,000 megacycles, and tomorrow the gap will be closed between today's frequencies and the visible spectrum.

Success in this field was made possible because university-trained scholars saw in the problems a real challenge and a chance to make a major contribution to engineering science. A similar historical sketch could be given for the mining and petroleum sciences. In each case, once the scholar was convinced that the problems were within his province, significant and accelerated progress was made. Something similar can be expected to happen in the manufacturing field.

It was not until after World War II that university scholars turned their attention to metal-cutting research. The increased horsepower and improved cutting-tool materials attracted their attention. The problems that became apparent could not all be solved by the empirical approach. More scientific instruments were brought to focus on metal-cutting forces and chip formation. Grinding, welding, and forming processes were also investigated from a scientific point of view.

European universities have been ahead of American universities in formally recognizing the need for courses related to manufacturing technology. All German technical universities have chairs and institutes for machine-tool engineering and for production engineering. Some of the English universities have similar chairs, but they are still rare in American universities.

European universities are engaged in teaching and research in the field of machine-tool design and operation. Courses are taught in the theory of metal cutting. These include the study of machinability of materials, tool wear, lubrication, cooling, and servomechanisms. The influence of friction and temperature is studied. Certain aspects of solid state, as well as the metallography and crystallography, of workpiece materials are investigated.

Some educators maintain that extremely broad engineering training, perhaps in mechanical engineering, best qualifies a man to enter the manufacturing field. Certainly the range of manufacturing problems is broad, and a wide acquaintance with various engineering disciplines is helpful. At the same time, it should be recognized that manufacturing engineering is a highly specialized field having the same relationship to the engineering sciences as, say, medicine has to the biological sciences. Carrying the parallel a step further, a supply of fully qualified tool and manufacturing engineers is as important to the future economic health of America as a supply of trained physicians is to the physical health of its inhabitants.

## NEED FOR RESEARCH

Ultimately, the degree of innovation—or manufacturing progress—depends on basic scientific research in the manufacturing area. Traditionally, the relationship of manufacturing industries to science has been a somewhat parasitic one. Although industry has applied the findings of science, it has not, to any great degree, sponsored basic scientific research that is applicable to manufacturing.

The scientist has contributed to today's advances in solid state devices which, in turn, allow the engineer to design more compact units. Computer technology, however, is an outgrowth of this work and is used by both the scientist and the engineer.

Today new fields of engineering and science are emerging. One of these, as described in the welding section of this book, is the field of plasmas. Another, also described, is that of lasers.

Despite the fact that science and technology have produced so much change and so much growth of knowledge, there is still the expectation that great breakthroughs will continue to come. Much accelerated development has been brought about by the space programs. Whole new industries can be attributed to the "spinoff" products and knowledge gained. This, along with several other factors, will provide pressure for still greater scientific advancements in industry.

## PRESSURES FOR SCIENTIFIC ADVANCEMENT IN MANUFACTURING

Several factors are at work in fostering the growth of this modern technological and scientific revolution.

The first of these is the upsurge in population (Fig. 1.1). This, coupled

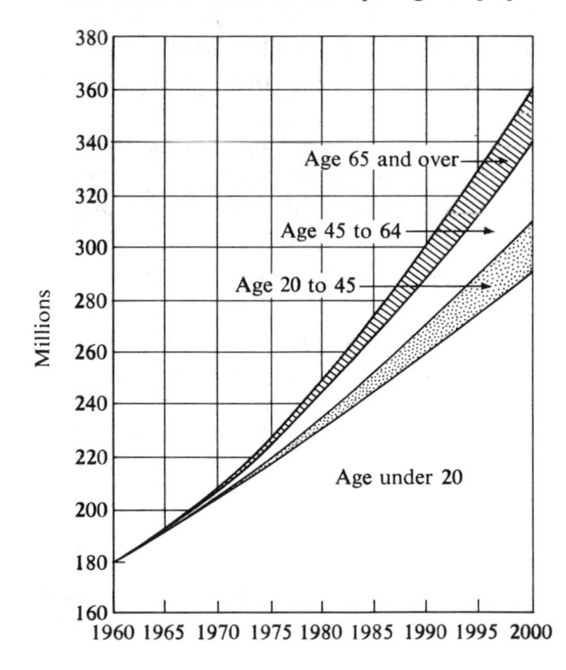


**Fig. 1.1.** United States population has grown explosively with about 30 million added in the 60s. It is estimated that 40 million will be added in the 70s, 50 million in the 80s, and 60 million in the 90s, to close to 360 million by the year 2000.

with the demand for a higher standard of living, has created huge markets into which working capital could be invested. It is estimated that 70 billion dollars for new tools will be needed annually in the next 10 years.

The rate of growth, and particularly the improvement of production in terms of labor and facility dollars invested, depends on the type and training of the personnel in the manufacturing industries. If they are aware of the technological advances that are available and if they contribute to this development, the rate of growth will be explosive.

One factor that has inhibited industrial growth in the past and may impede future manufacturing progress is the time lag between the development of a new manufacturing process, new machine, or new type of manufacturing equipment or control and its general adoption by industry. It is doubtful whether this time lag will be acceptable in the rapidly expanding economy of the future.

With few exceptions, recent breakthroughs in manufacturing technology, breakthroughs that can be exploited by most of industry, have been made as a result of research sponsored by the military services. Numerical control (Fig. 1.2), possibly the leading breakthrough of recent years, was, from its inception, developed and applied with Air Force funds.

Explosive forming and ultrahigh-velocity machining, two other recent breakthroughs, are primarily the result of Air Force sponsored research

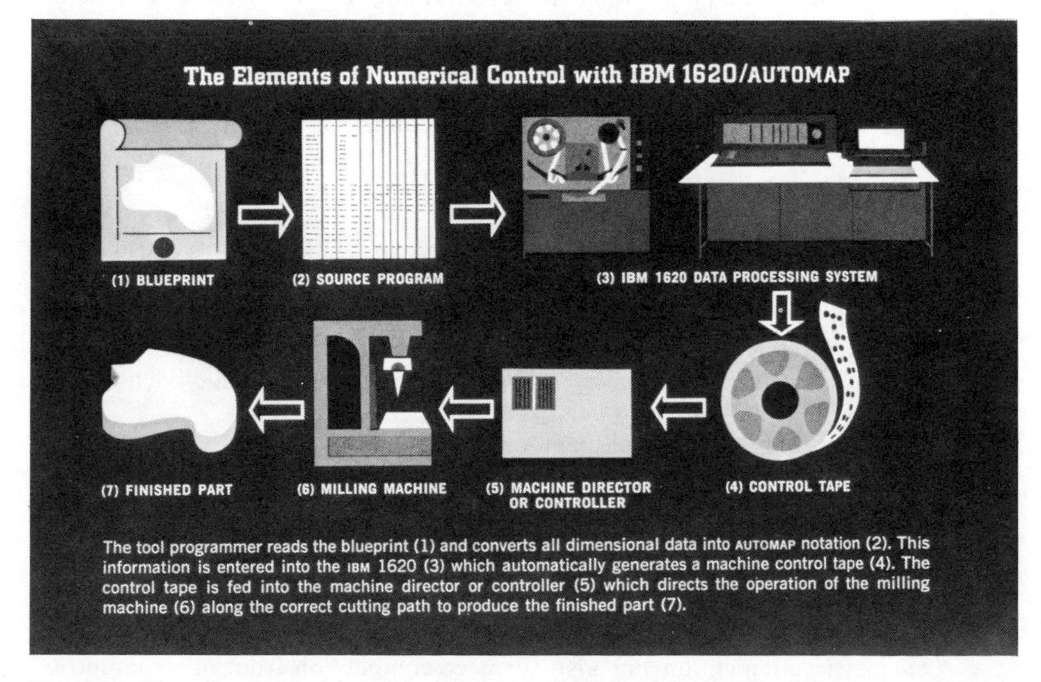


**Fig. 1.2.** The elements of numerical control. Courtesy International Business Machines Corp.

and development. Much of the research and development associated with grinding and shaping of the super alloys and, for that matter, the development of the super alloys themselves, has been financed by the government.

Military necessity was the impetus behind this work, pushing these new developments along at a fast rate, but it is entirely possible that they would have been conceived and put into action years earlier if engineers in private industries had turned their attention to basic research both in their own plants and through the medium of industry-wide programs.

Military-sponsored manufacturing research has had an effect that goes beyond its direct contribution to technology. It has shown what can be accomplished when sufficient funds and scientific and engineering talent are put to work on specific problems. It is doubtful, however, if industry can advance along a broad front on the basis of military-sponsored research alone.

Small companies are able to benefit from research performed by larger companies. A new machining process developed by a machine-tool builder, for example, or a new control developed by a control maker, can be applied by any company, large or small. With respect to basic research, the smaller company may be at a disadvantage. Still, it is surprising how many innovations in manufacturing could be, and sometimes are, performed with a minimum amount of investment and manpower. The basic equipment for explosive-forming research is, after all, merely a concrete-lined tank of water and a stick of dynamite. The raw material of research is ideas. Large companies or the military have no monopoly on ideas. Basic research, whether it is in the manufacturing field or any other, is an investment in qualified personnel.

## TODAY'S CHALLENGES

Today, with the advent of the space age, industry is faced with the challenge of developing new engineering materials. In order to do this, scientists are eagerly trying to understand the nature of materials. As yet, little is known as to what mechanism within a piece of material provides the properties of strength. Several theories have been offered, and the testing of them has begun (Fig. 1.3).

Not only is the prospect of creating new materials fascinating to engineers, but equally challenging is the processing of these new materials.

It is imperative to the economy of a country to develop better materials and more efficient production methods. It will be necessary to produce more reliable parts and assemblies, to closer tolerances and with lower labor time per unit. This will be necessary not only to meet domestic competition but also to meet competition from other countries.

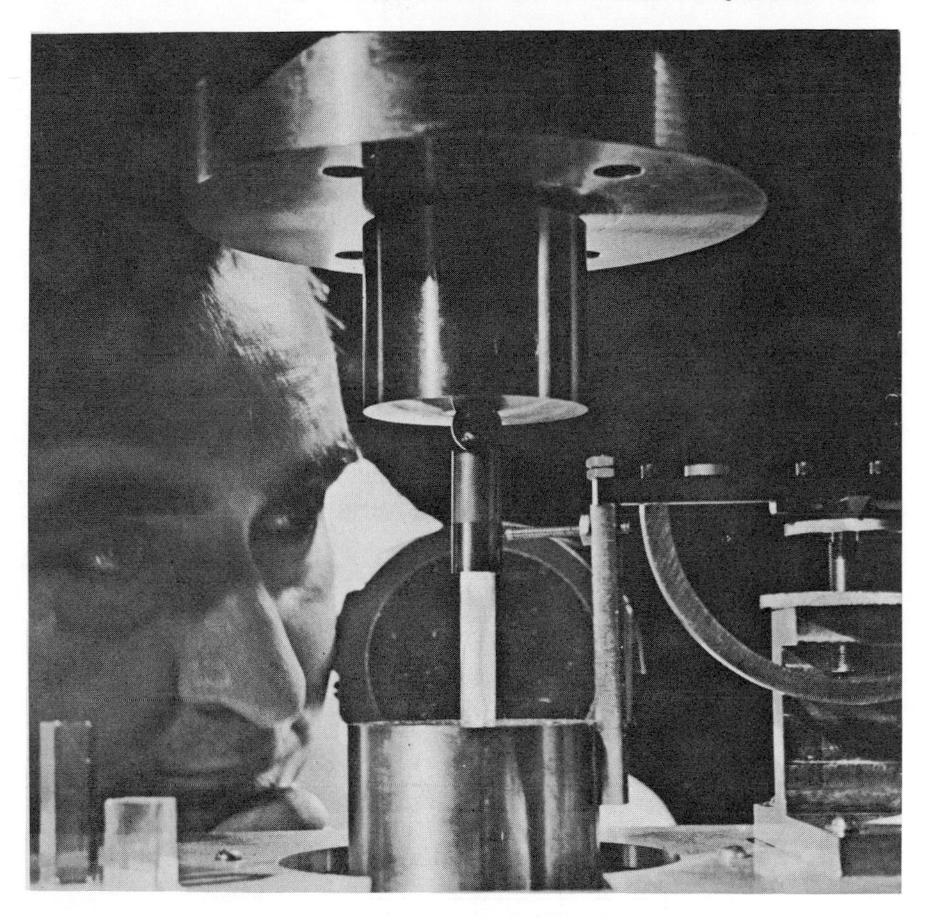

**Fig. 1.3.** Many aspects of material failure are still cloaked in mystery. No one yet knows exactly how or why fracture occurs. Considerable research work, such as this photoelastic compression test at NASA's Lewis Research Center, is directed toward a more thorough understanding of strain and fracture. Fundamental studies yield little information that can be used immediately but eventually could lead to better control of many physical and mechanical properties.

## FUTURE MANPOWER REQUIREMENTS

The highly automated and mechanized plant of the future will require fewer production workers per unit of output than does today's plant. There will be a higher proportion of setup and maintenance personnel who will have a greater degree of skill and training than is normally the case today. Also, a broader range of specialized skills—electronics, hydraulics, electrical—will be required. With a heavy investment in automated equipment, downtime will be costly, and emphasis will be placed on getting idle machines back into production in a hurry.

The manufacturing engineer, the man who plans the production line, will find his task increased in scope. His job will demand a higher degree of knowledge and skill. The basic items of tool engineering, tool design, fixture design, and so on, will still be there, but, in view of the extensive use of automated machines and automation, a somewhat broader knowledge will be required. A solid academic background, with emphasis on mathematics and manufacturing science, will be mandatory. Continuous study, coupled with experience in the field, will be necessary to keep abreast of the rapidly advancing technology.

Many of the technical personnel needed to fill manufacturing positions will not be graduate engineers; they will come from among the graduates of two-year technological courses. They will be people who have acquired a technical background that qualifies them for upgrading. With the fast rate of growth in science and technology, the upgrading of trained personnel will help fill the consequent demand for more engineers and engineering technicians.

It is expected that more and more manufacturing engineers, released from some of the time-consuming tasks by the aid of computers and technicians, will enter top management. These engineers will tend to encourage technical innovation, an attitude that will be the strongest impetus to real progress in engineering.

The organization chart (Fig. 1.4) shows how manufacturing processes and the process engineer fit into the overall operation of the plant. In this type of organization, all engineering is concerned with the end product, but the process or manufacturing engineer is more intimately associated with seeing it materialize. This chart lists some of the duties

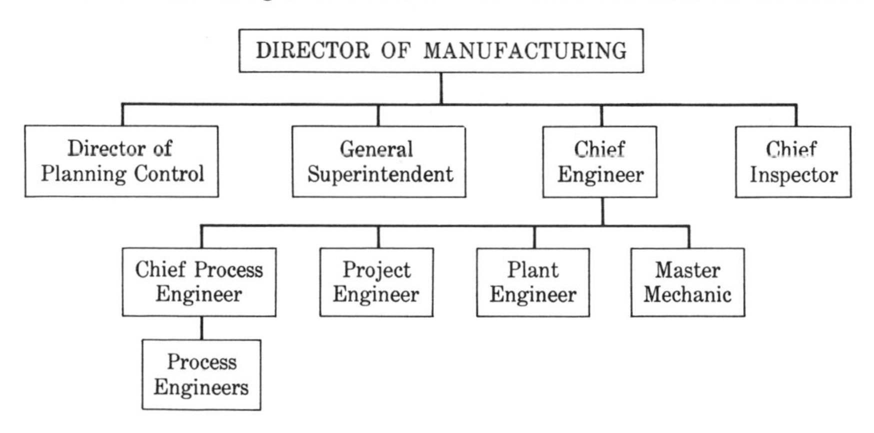


**Fig. 1.4.** Part of a manufacturing-plant organization chart. Some duties of the process or manufacturing engineer: (*a*) decide on manufacturing feasibility, (*b*) prepare manfacturing specifications, (*c*) decide whether to make or buy, (*d*) prepare tooling estimates, (*e*) prepare process drawings, (*f*) prepare machine load charts, (*g*) specify gaging required.

that are ordinarily assigned to the process engineer. Duties vary from company to company, depending upon the size and scope of the operation.

The manufacturing engineer must determine feasibility not only of the product designs the company produces, but even those purchased by the company. He must be able to determine whether parts should be purchased or manufactured. Oftentimes, specialty manufacturers can produce parts on a volume basis, thus reducing costs.

In judging the feasibility of manufacture, many things are taken into consideration. The process of manufacture, for example, could involve a choice for a given part between forging, casting, cold forming, or a custom extrusion and weldment. The manufacturing engineer should be able to give detailed information to the purchasing department concerning the selection of products as well as machines and equipment.

The manufacturing engineer must be able to work along with and advise product designers. This requires close cooperation, but it can result in fewer design changes and faster production.

Another area of responsibility is that of following up the use of the specified equipment and tooling for a job to determine if it is being used in accordance with the original plan. If this is not done, it is likely that neither the volume nor the quality of the parts will be as anticipated.

Many items require special tooling in the way of jigs, fixtures, die designs, templates, gages, etc. The tool and manufacturing engineer must be able not only to design them but also to prepare cost estimates.

This brief introduction to the role of the manufacturing engineer and the challenges that lie ahead is not intended to be comprehensive; rather, it is intended to give an idea of the area of work covered. As you study this book, you will see that many new manufacturing processes are constantly being introduced. The manufacturing engineer will be on the alert for methods that can be adapted to help meet the overall objective of *a better product at a lower cost.*

## QUESTIONS

**1.1.** State in your own words the relationship of science, engineering, and technology.

**1.2.** Contrast what you think to be the difference between a scientific approach and an empirical approach to a problem.

**1.3.** What factors make for rapid technological advance?

**1.4.** What are some types of work a manufacturing engineer will deal with?

**1.5.** What changes in duties can the manufacturing engineer expect in the years ahead?

# 2

# PROPERTIES OF METALS

THE PROPERTIES of metals have been studied ever since man found he could change the hardness of steel by heating it to a bright cherry red and then quenching it in water or other suitable media. What happened to the metal structure was at first theorized, but then, as instruments for study improved, facts replaced theory. The modern-day solid-state physicist is still deeply involved in the complexities of the problem. Some appreciation of the problem can be had when we realize that a piece of matter the size of a pinhead contains roughly 100 billion atoms. Fortunately, nature helps resolve this problem by making the atomic structure of a given solid into a definite pattern. Thus, when atoms are brought together, they tend to arrange themselves in infinitesimal cubes, prisms, and other symmetrical shapes. These geometrical units, joined to each other like perfectly fitted blocks, are embryos of the larger structure known as crystals or grains.

In this chapter we shall first examine the basic crystal structure of the metal, then ways in which the structure is changed to develop certain desirable properties, and finally how these properties can be tested.

## THE CRYSTAL STRUCTURE OF METAL

The metal structure is determined to a large extent by its content, but its structural details are controlled by its history. The history of the crystal structure begins when the metal starts to cool from the liquid state. As the molten metal cools, it solidifies and crystallization takes place in two stages, first *nucleation* and then *growth.* The growth of the nucleus takes place by the addition of atoms, one at a time. As crystals grow they join with other crystals and form a complicated structure resembling a pine-tree arrangement. The spacing of the pine-tree branches, or *dendritic* arms (Fig. 2.1), depends on how fast the crystals grow, which in turn is dependent on how fast the heat is transferred from the metal.

Growth continues in all directions until the crystal, or as it is usually referred to in metals, the *grain,* runs into interference from other grains. If two grains form alongside each other and have the same atomic arrangement, they will join together to form a larger grain. However, if the forming occurs on different axes, the last atoms to solidify must

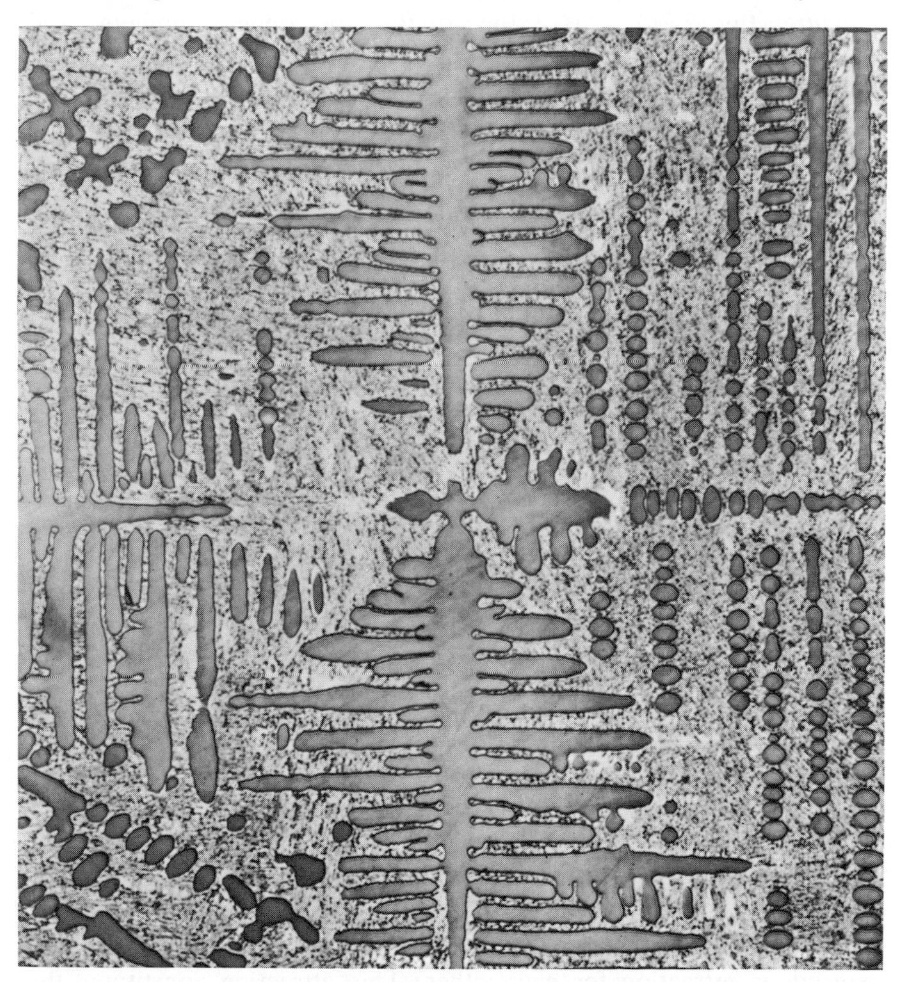

**Fig. 2.1.** The dendritic crystal solidification pattern of cast iron, 500×.

assume compromise positions. These misplaced atoms make up the grain boundaries. The disordered regions increase the deformation resistance of the metal.

Therefore fine-grained metals are generally stronger and harder than coarse-grained metals of the same composition.

Each grain consists of millions of tiny unit cells made up of atoms arranged in a definite geometric pattern. Each unit cell may take the form of an imaginary cube, with an atom in each corner and one in the center. This is called a *body-centered cubic crystal structure* and is the structure of iron at normal temperatures (Fig. 2.2*a*).

If, however, the center of the cube is vacant, and a single atom is contained in the center of each face, it is called a *face-centered cubic structure* (Fig. 2.2*b*). This is the structure of copper, aluminum, and nickel. It is also the structure of iron at elevated temperatures.

When the unit cell takes the form of an imaginary hexagonal prism, having an atom in each corner, another at each of the top and bottom hexagons, and three atoms equally spaced in the center of the prism (Fig. 2.2*c*), it is known as a *close-packed hexagonal structure.* This is the structure of magnesium, zinc, and titanium.

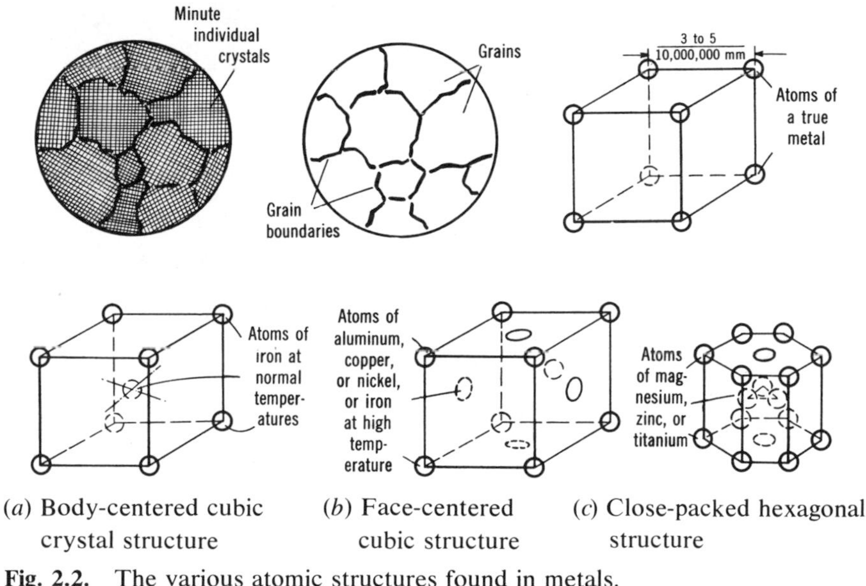


(*a*) Body-centered cubic crystal structure (*b*) Face-centered cubic structure (*c*) Close-packed hexagonal structure

**Fig. 2.2.** The various atomic structures found in metals.

The distance between the atoms is extremely small, sometimes only 3 ten-millionths of a millimeter and seldom more than 5 ten-millionths, as determined by x-ray diffraction. These closely spaced atoms have a tremendous attraction for each other. This attraction constitutes the force that resists any attempt to tear the metal apart.

## CRYSTAL STRUCTURE CHANGES DUE TO STRESS

Metals in everyday use are subjected to tremendous stresses and strains. When the metal is deformed or cut, certain rows of crystals slip or flow in fixed directions and in one or more parallel planes. This is why pure aluminum, copper, and nickel flow so readily under stress. The ability of the slipping crystals to hold together makes these metals extremely ductile.

Close-packed hexagonal structures (Fig. 2.2*c*) have only one set of densely packed slip planes, but they have three directions of ready slip in each plane. Generally, the close-packed hexagonal structures are not so ductile as the other two structures.

## CRYSTAL LATTICE IMPERFECTIONS AND DEFORMATION

One of the mysteries surrounding the enormous discrepancy between the theoretical and the observed shear strength of crystal structures was clarified with the aid of the electron microscope. It was shown that imperfections in the crystal structure, particularly those termed *dislocations,* provide a mechanism by which shear is accomplished at the critical stress values that correspond to those observed. These imperfections are sources of shear weakness and slip. *Slip* as a mode of deformation is based on the fact that a symmetric translation or movement of one unit lattice distance is periodic and therefore preserves the crystal structure on the planes that contain the defect and in a degree proportionate to the imperfection. The imperfections vary dimensionally and are termed *point type, line type,* and *plane type.*

**Point Defects.** The point defects in a crystal structure may show up as missing atoms or *vacancy*, as extra atoms or *interstitial*, or as displaced ions or *Frenkel defects*. In the case of a missing atom, a vacancy exists in the lattice structure, as shown in Fig. 2.3*a*. The vacancy may be the result of imperfect packing during the time crystallization occurred, or it may be due to increased thermal vibrations of the atoms brought about by elevated temperature. The individual atom may jump out of its position of lowest energy.

An extra or interstitial atom may become lodged in the crystal structure (Fig. 2.3*b*) without distorting or changing its shape. Atomic distortion may result, however, if the interstitial atom is larger than the rest of the atoms in the structure.

An ion displaced from its lattice into an interstitial site is referred to as a Frenkel defect (Fig. 2.3*c*).

An example of the interstitial solid solution is a mixture of iron and carbon. The carbon atom is smaller than the iron, so it does not fit

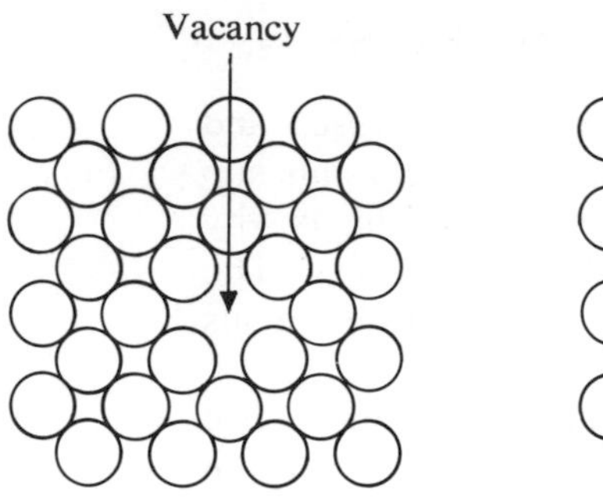


(*a*) A vacancy type of defect.

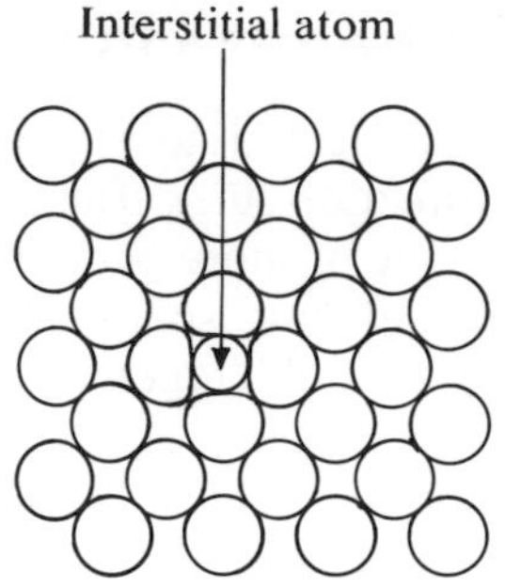


(*b*) An interstitial type of defect.

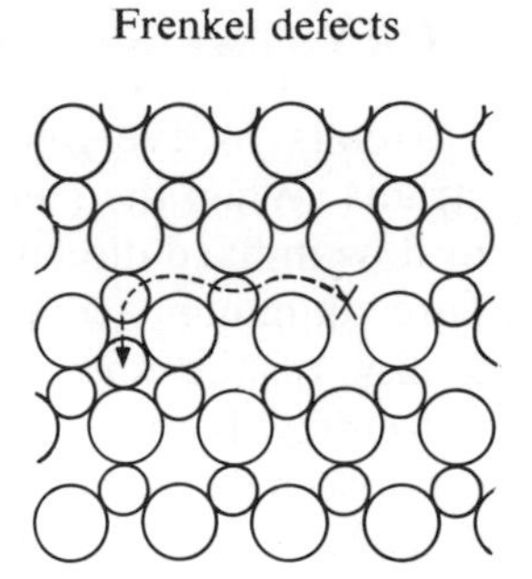


(*c*) Frenkel defects (ion displacement).

**Fig. 2.3.** Point defects in crystalline structure.

perfectly in the interstices of the iron crystal. As a result, wherever a carbon atom is located, it represents a strained condition which in turn severely limits the solubility of carbon in iron. At elevated temperatures, iron has a *face-centered* cubic form which, although it has more atoms per unit volume, also has larger interstitial spaces for maximum solubility of carbon. Therefore, above the upper critical temperature, steel is referred to as a *solid solution* of iron and carbon.

**Line Defects.** Line defects are better known as *dislocations*. They exist in the crystal structure as *edge dislocations, screw dislocations,* and *mixed dislocations*. The concept of dislocations was introduced in 1934 by Taylor, Orowan, and Polanyi, each working independently.

**Edge Dislocations.** Edge dislocations consist of an extra row or plane of atoms in the crystal structure (Fig. 2.4). The imperfection may extend in a straight line all the way through the crystal or it may follow an irregular path. It may also be short, extending only a small distance into the crystal. The passage of a single dislocation along a slip plane

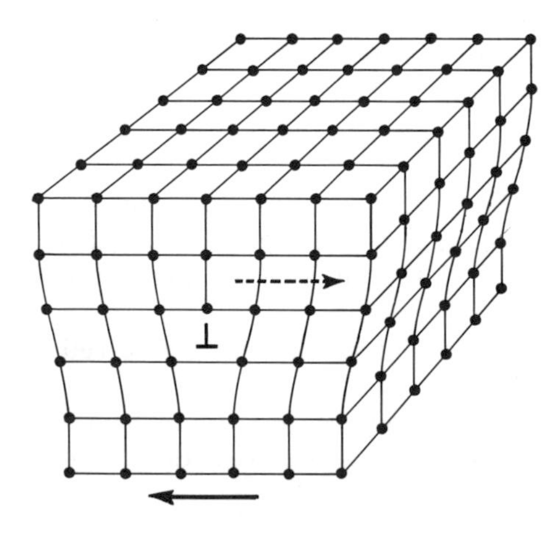


**Fig. 2.4.** An example of an edge dislocation. It is characterized by a compressive internal-stress field above the slip plane in equilibrium and tensile internal stresses below the slip plane. The dislocation is perpendicular to the slip direction and is represented by the symbol ⊥, where the vertical leg represents the extra plane of atoms.

through a crystal causes slip of one atomic distance along the glide plane (Fig. 2.5*b*). The slip of one active plane is ordinarily of the order of 1000 atomic distances and, to produce significant yielding, slip on many such planes is required.

**Screw Dislocations.** Screw dislocations can be produced by a tearing of the crystal (Fig. 2.5*a*) *parallel* to the slip direction. If a screw dislocation is followed all the way around a complete circuit it would show a slip pattern similar to that of a screw thread. The pattern may be either left or right handed. This requires that some of the atomic bonds are re-formed continuously so that the crystal has almost the same form after yielding that it had before.

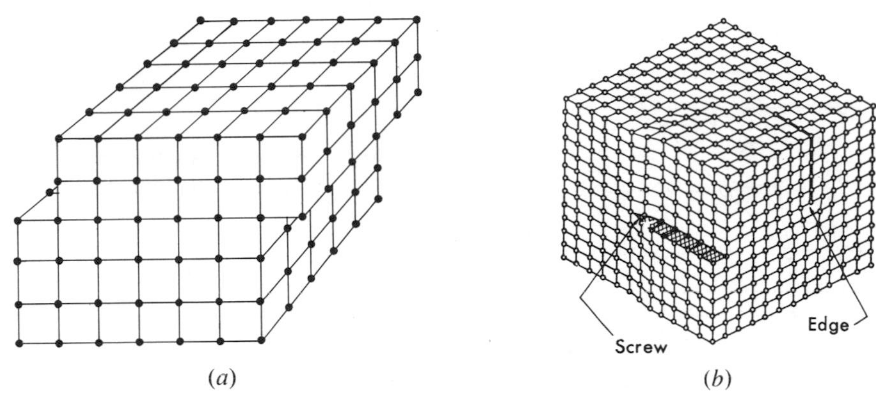


**Fig. 2.5.** Screw dislocations take place as a partial tearing of the crystal plane. Edge dislocations occur at the end of an extra half-plane of atoms.

During distortion of the crystal, there is constant thermal agitation among the atoms. This thermal energy allows the adjacent bond to be broken more easily; hence the continuation of the screw pattern.

**Mixed Dislocations.** The orientation of dislocations may vary from pure edge to pure screw. At some intermediate point they may possess both edge and screw characteristics.

The importance of dislocations is based on the ease at which they can move through crystals. As stated at the beginning of this discussion on crystal lattice slip, the amount of force required to cause shear stress on a crystal is much less than it should be theoretically. This can be accounted for by the fact that in edge dislocations there is a counterbalancing of interatomic forces on either side of the extra half-plane that makes for equilibrium. Only a small force is needed to displace the atoms from the equilibrium configuration. Even though the force is small, the movement of an edge dislocation is dependent upon the arrival or dispersion of *vacancies* or *interstitials*.

Although this condition is enhanced by the application of stresses, they are in essence temperature controlled.

Screw dislocations, unlike the edge dislocation, are not dependent on an extra half-plane inserted into the lattice. The result is that the screw dislocation can glide in any direction. A great many dislocations are necessary in order to develop macroscopic strains. Experimentally observed they are of the order of $10^5$ or $10^6$ $cm^2$ for annealed metals of average purity. They increase considerably to about $10^{10}$ or $10^{12}$ $cm^2$ when strained.

**Twinning.** A deformed material can take on a new orientation but still be a mirror image of the undeformed material in its original orientation (Fig. 2.6); hence the name "twin." The plane at $AA'$ is the *twin junction.*

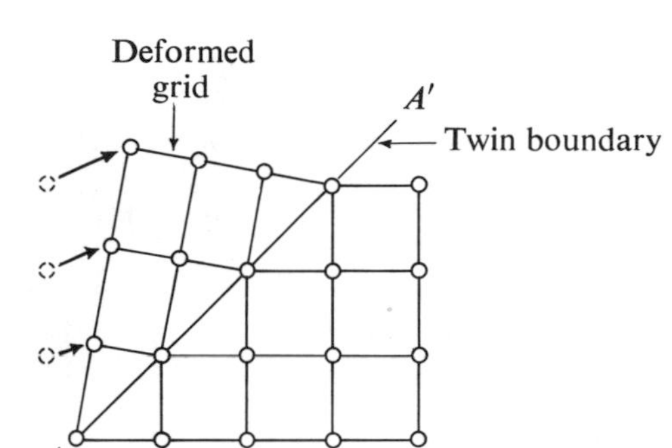


**Fig. 2.6.** An example of *twinning* of two body-centered cubic lattices.

The process of rebonding is required for both slip and twinning so only those materials that exhibit good bond mobility, that is, form new bonds easily, are susceptible to large amounts of plastic deformation before fracture.

## CRYSTAL STRUCTURE CONTROL

The structure of the crystal can be controlled to a limited extent by the rate at which it cools. It may also be modified by heat treatments, plastic deformation, and irradiation.

In plastic deformation, some changes in the shape of the crystals are made by reorientation and dislocations. Shown in Fig. 2.7, the dark "jagged" lines show the misalignment of atoms or dislocation caused by cold working the metal. As the metal is worked, these dislocations move through the material, concentrating in certain regions. The broad dark lines show areas of high concentration of dislocations.

Polycrystalline iron rolled at room temperature is shown in Fig. 2.8. The structure has been strained at three levels: 5, 9, and 30 percent. The 5 percent strain (*a*) shows no apparent pattern to the dislocations. In the 9 percent reduction (*b*), a pattern of high and low density is shown which also accounts for small (3 micron) embryonic cells' starting to form.

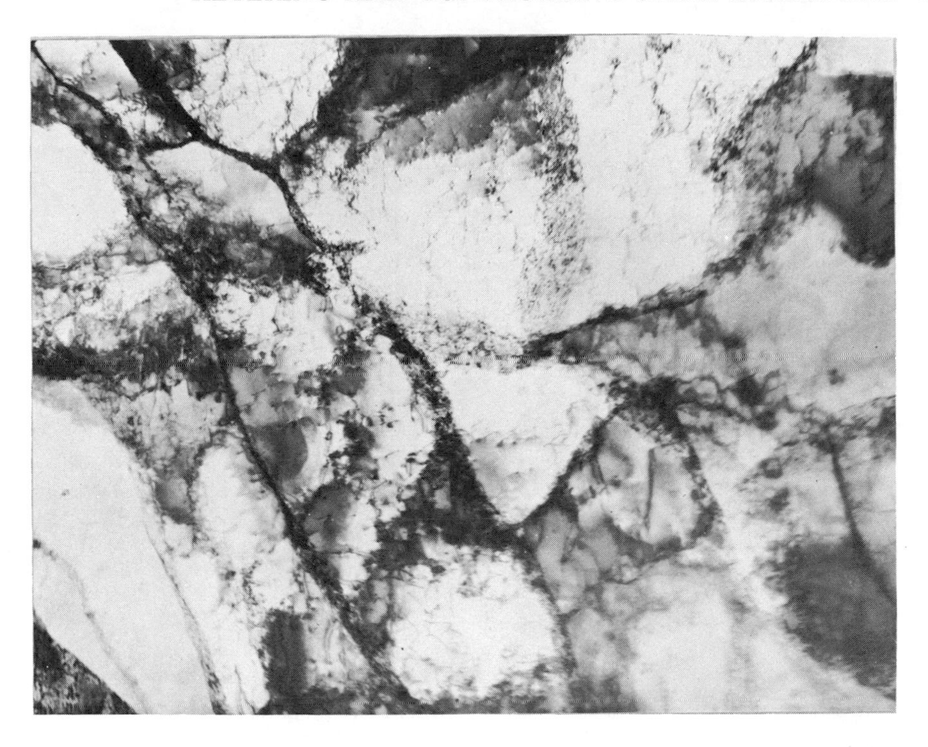

**Fig. 2.7.** The dark jagged lines show the misalignment of atoms or dislocation caused by cold working. Note that only a few dislocations exist within the outlined cell. The photo was made through transmission techniques, 12,000×. Courtesy United States Steel Corporation.

These cells are much smaller than the unworked grains. At 30 percent (*c*), very few dislocations exist in the matrix, but the average dislocation density has increased because of the extreme concentration of them in the cell walls. The cell diameter is now down to about one micron.

The average dislocation density (in iron tested at room temperature) rises with increasing strain. However, this average density is independent of deformation temperature as shown in Fig. 2.9. Both −95°F and 480°F fall on the curve.

## ALTERING AND CONTROLLING STEEL STRUCTURE

The most important single element in steel is carbon. As carbon is added to iron, the iron becomes steel, and as still more carbon is added, it becomes cast iron. The effect of carbon in varying percentages is best understood by studying an iron-carbon equilibrium diagram.

**The Iron-Carbon Equilibrium Diagram.** An iron-carbon equilibrium diagram (Fig. 2.10) shows both carbon percentages and the effect of

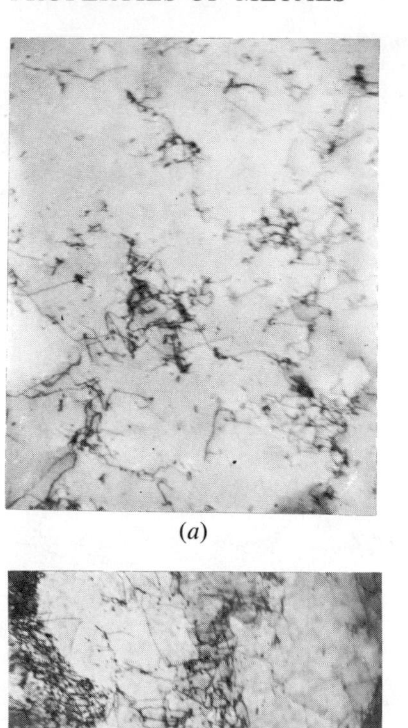

(*a*)

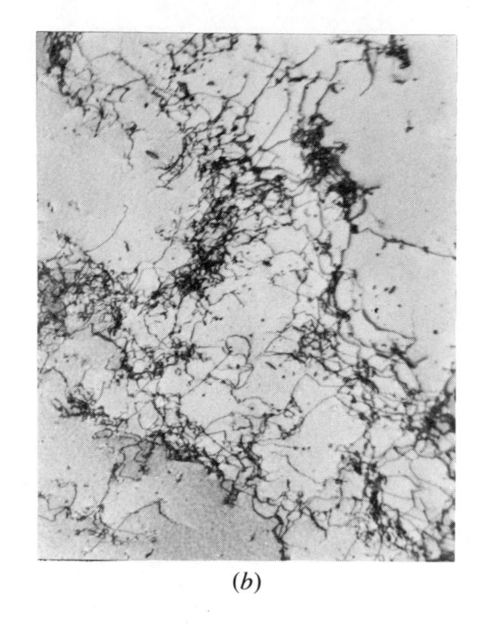

(*b*)

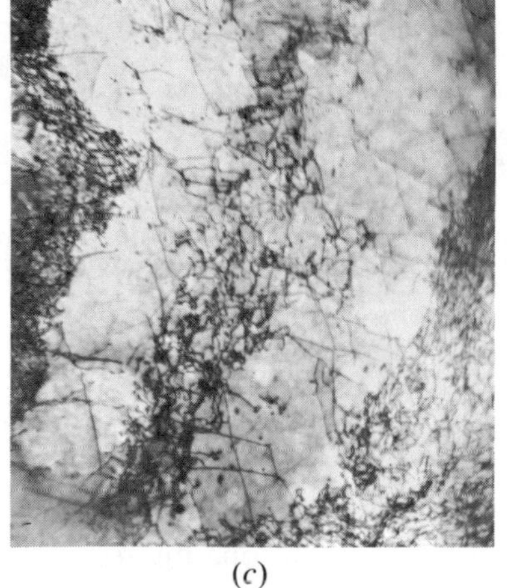

(*c*)

**Fig. 2.8.** Polycrystalline iron strained at 5% (*a*), 9% (*b*) and 30% (*c*). Note the dislocation density change. Progressive straining concentrates dislocations in embryonic cell borders until definite subgrains appear (*c*). Original magnification 30,000×.

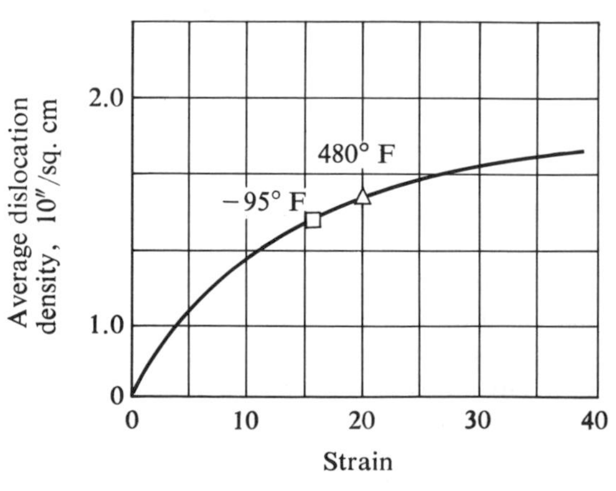


**Fig. 2.9.** Dislocation density rises with strain, increasing rapidly at first and then leveling off. Since the material was strained at both low and high temperature with little difference, the formation is apparently independent of temperature. Courtesy United States Steel Corporation.

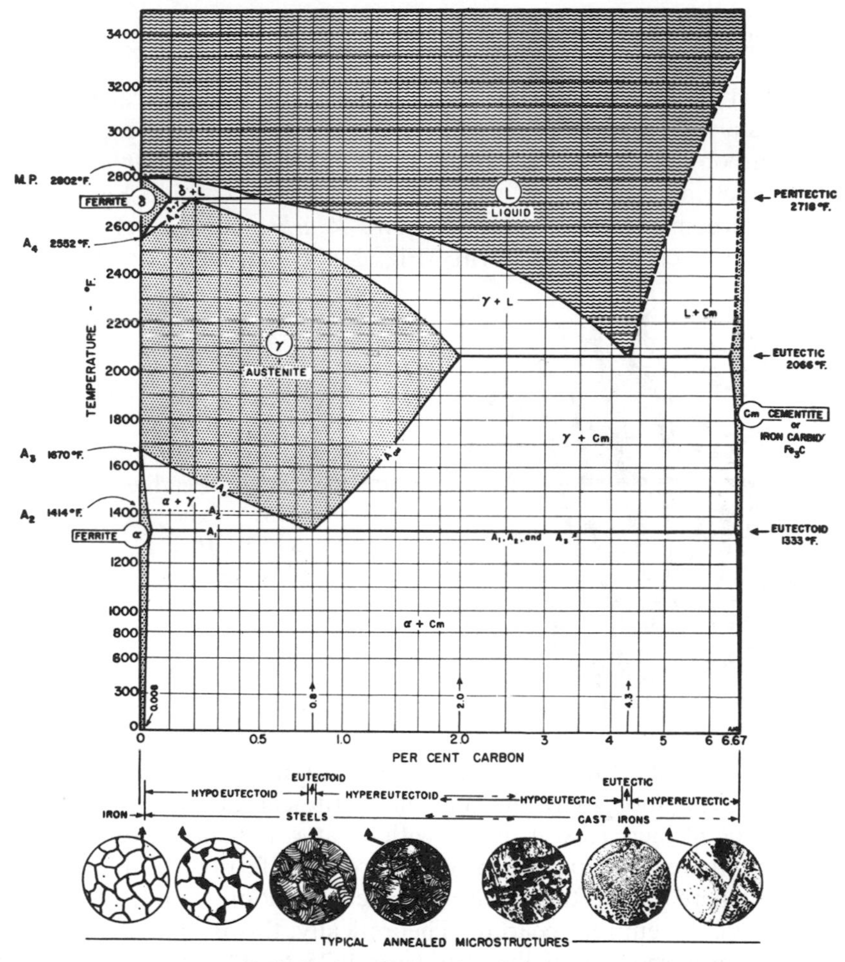


(Revised by Anton Brasunas from diagram in ASM Metals Handbook, 1948 ed.)

**Iron–Cementite Phase Diagram.**

**The critical points for pure iron are shown on the left; the changes in temperatures appear as lines progressing to the right as carbon is added.**

**Fig. 2.10.** The iron-carbon equilibrium diagram. Courtesy Metals Engineering Institute.

temperature. This diagram has been very carefully worked out and forms a basic guide for understanding critical temperatures.

For example, the first small shaded area from the ordinate represents pure iron. The diagram indicates a stable alpha iron (BCC lattice) up to 1670°F. Then it abruptly changes to gamma iron (FCC lattice) up to 2550°F and finally to delta structure above 2550°F to melting. These same changes in structure occur upon cooling at near the same temperatures.

As steel is cooled from an elevated temperature, several types of crystals start to form. They may be ferrite, cementite, pearlite, or others, depending on the carbon content of the steel and the cooling rate. *Ferrite* is practically pure iron; *cementite* is the iron carbide $Fe_3C$ (6.68 percent carbon and 93.32 percent iron). *Pearlite* consists of alternate layers of iron carbide and ferrite (Fig. 2.11). Normally, as the metal

**Fig. 2.11.** Pearlite structure. Courtesy The International Nickel Co., Inc.

cools slowly from a temperature above the upper critical, there is an automatic separation of ferrite and the ferrite-carbide mixture. As the carbon content increases, it unites with greater amounts of ferrite, thus causing an increase in pearlite and a decrease in ferrite. At the point of increase where all of the ferrite is in combination with carbon, the structure will be entirely pearlite. Theoretically, this is at approximately 0.80 percent carbon; actually, it is from 0.75 to 0.85 percent in plain carbon steels. This is called the *eutectoid point*. Eutectoid, taken from the Greek, means "most fusible." The excess carbon above 0.80 percent, as in *hypereutectoid* steels, is rejected by the austenite crystals and forms at the grain boundaries. Less than 0.80 percent carbon is a *hypoeutectoid* steel, whose structure, if cooled slowly, would be pearlite with excess ferrite. Pearlitic steel is generally defined in terms of the eutectoid amount of carbon and the slow cooling rate. However, owing to alloying elements and cooling rates, wide variations of pearlite structure exist.

If a specimen of hypoeutectoid steel is heated uniformly, it will be found that at approximately 1333°F the temperature of the steel will stop rising even though the heat is still being applied. After a short time, the temperature will continue to rise again. Ordinarily, metals expand as they are heated, but it is found that at 1333°F a slight contraction takes place and then, after the pause, expansion again takes place. This indi-

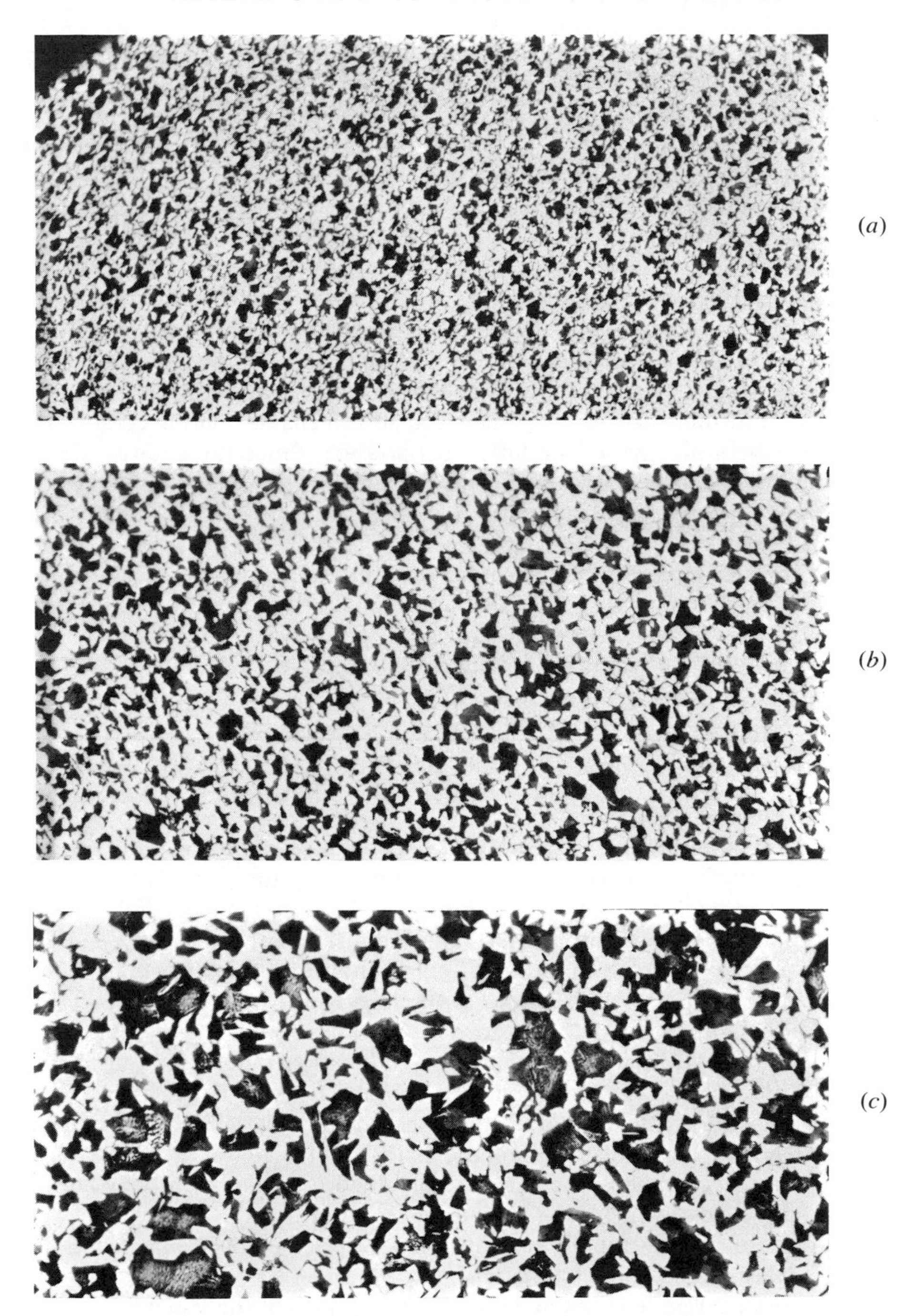


**Fig. 2.12.** Heating 1040 steel to progressively higher temperatures produces larger grain size, (*a*) 1500°F, (*b*) 1700°F, (*c*) 2100°F. Original magnification 200×.

cates the lower critical temperature of the steel, shown on the iron-carbon diagram as $A_1$. At this point, the structure changes from a BCC to an FCC lattice, or from alpha to gamma iron.

As the heating continues, another pause will be noted at $A_3$, at which time there is a transformation of ferrite to austenite. This varies with the carbon content; as shown in the diagram, and is known as the upper critical temperature. At this point the carbon goes completely into solution with the iron, so that it is now evenly distributed.

As the heating continues there is no further important change except that the grains grow larger until the melting point is reached (Fig. 2.12).

The final austenite grain size will depend upon the temperature, above the critical, to which the steel was heated. The grain size of austenite has a marked influence both on transformation behavior during cooling and upon the grain size of the final steel product. Since the grain size is so important in the behavior and quality of steel, it is necessary that the temperature be carefully controlled in relation to $A_3$ or the upper critical temperature.

**Grain Size and Crystal Structure Recovery.** Metal grains recrystallize at various temperatures depending upon the amount of cold working they have been subjected to. Shown in Fig. 2.13 is a specimen that was strained 16 percent at room temperature and then heated to 1025°F for various periods of time. In the first picture (*a*) the deformation of the grain structure is quite apparent. After being heated for 15 min at 1025°F (*b*), the dislocation density becomes less, and, after 16 hr (*c*), recovery or recrystallization is essentially complete with well defined subgrains.

This same process is also shown schematically in Fig. 2.14. On the left, normalized or annealed steel is brought up to the grain refinement zone and cooled to obtain fine grain size. Also shown is the grain growth that takes place when the grain refinement temperature is exceeded and the effect of time at an elevated temperature for grain growth. The diagram on the right represents work-hardened steel with a lower recrystallization temperature.

Recent evidence shows that recrystallization takes place through a process of subgrain coalescence or merging to form larger grains. That is, dislocations that make up the subgrain boundaries move out to the intersecting or connecting boundaries (Fig. 2.15).

Most steels when processed are involved in rolling or forging at high temperatures, which deforms the austenite structure; however, recrystallization occurs rapidly and little if any trace of the elongated grains are left at room temperature.

The finest grain is obtained by careful control of the initial crystallization temperature and then heating and deforming at the lowest temperature at which austenite is stable.

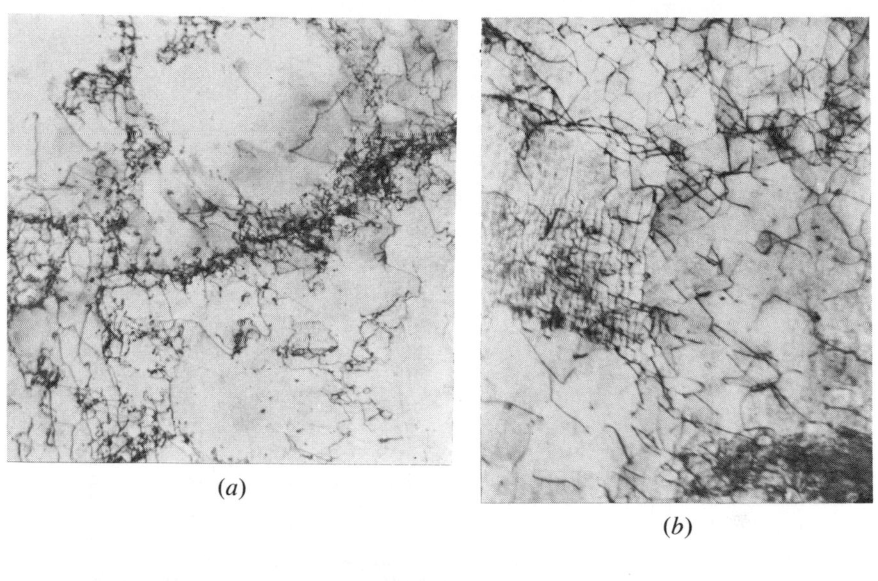

(*a*)

(*b*)

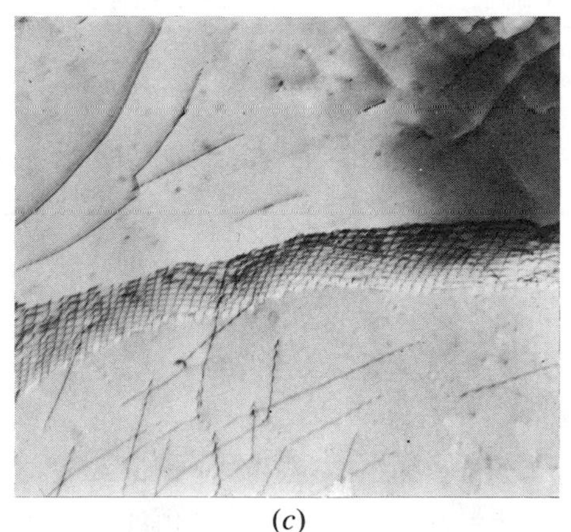

(*c*)

**Fig. 2.13.** Pure iron strained 16 percent at room temperature (*a*) shows many jagged dislocation lines. Heating at 1025°F for 15 min (*b*) straightens and redistributes these dislocations. Recovery is essentially complete after 16 hr (*c*).

**Isothermal Transformation Diagrams.** Another diagram that is essential in understanding the effect of temperature on steel structure is the isothermal (constant temperature) transformation diagram (Fig. 2.16). This diagram is also referred to as a TTT diagram because it plots *Time, Temperature,* and *Transformation*. It may also be referred to as *S*- or *C-curves* because of the line contours.

As can be seen on the diagram, the transformation of austenite involves a period of time or an incubation period in which no transformation takes place. Thus we have a relatively slow transformation at the beginning and again when it nears completion with a much more rapid

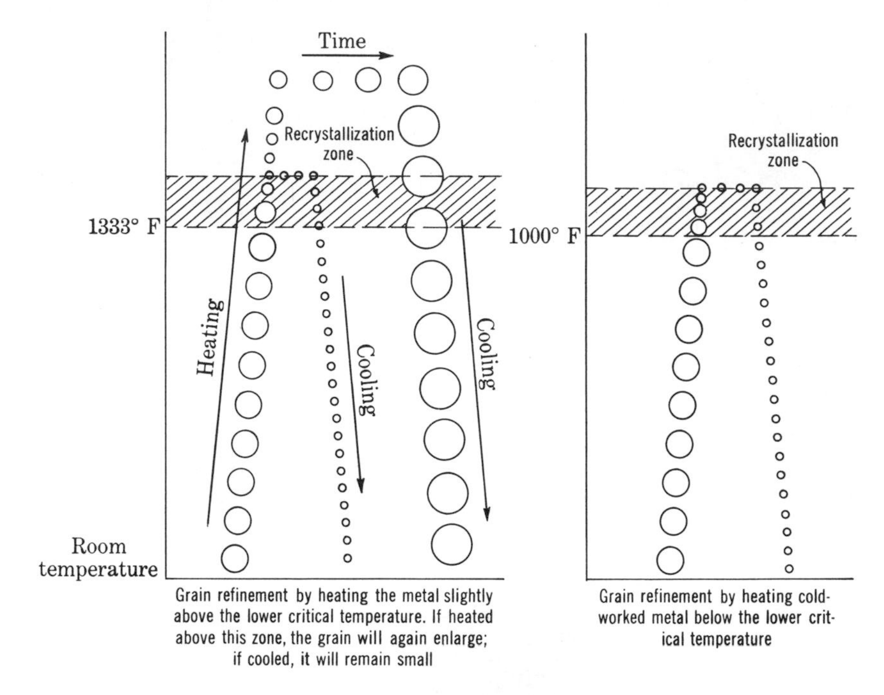


**Fig. 2.14.** Schematic diagram of grain growth and refinement.

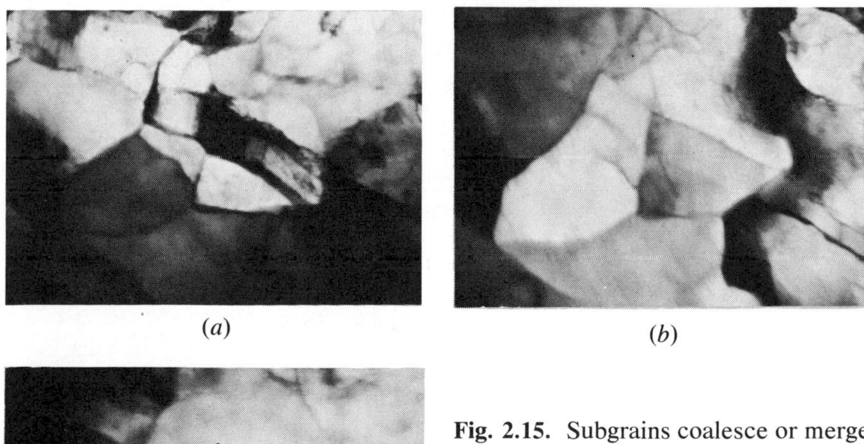

(*a*) (*b*) (*c*)

**Fig. 2.15.** Subgrains coalesce or merge to form larger grains during the annealing process. Here cold-worked iron is heated to 1250°F. for 30 sec (*a*), 2 min (*b*) and 3 min (*c*). Original magnification 40,000×. Courtesy United States Steel Corporation.

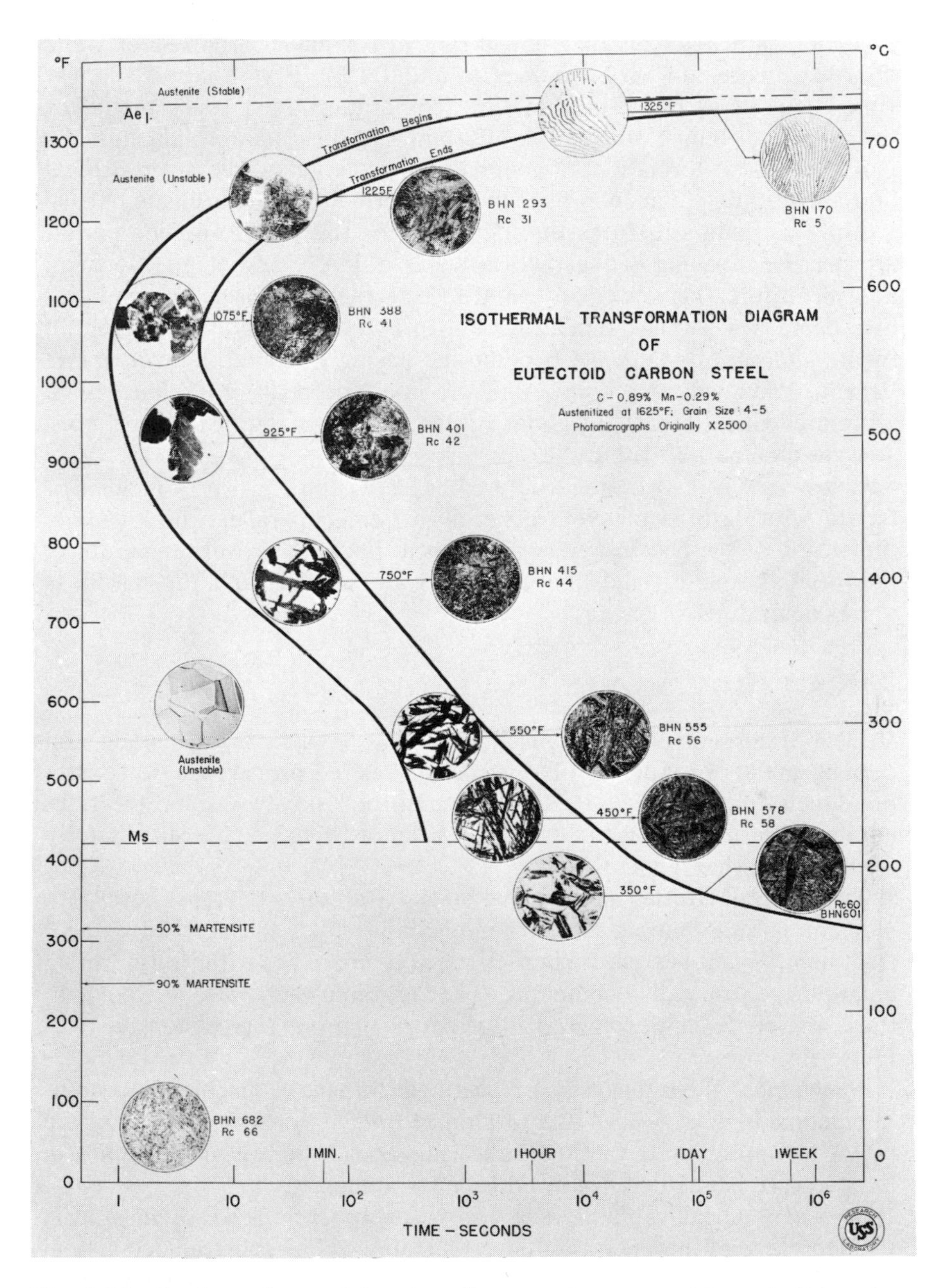


**Fig. 2.16.** Isothermal transformation diagram of a eutectoid steel. Courtesy United States Steel Corporation.

rate in between when 25 to 75 percent of the austenite transforms. The behavior shown in this diagram is typical for plain carbon steels. The shortest incubation period occurs at about 1000°F with a much longer time required as the transformation temperature approaches the $A_1$ or lower critical temperature at equilibrium. As the diagram indicates, the isothermal transformation behavior takes place below the lower critical and above the $M_s$ line, or where the transformation to martensite begins.

The two main structures that occur during this transformation period are pearlite, forming between $A_1$ and above 950°F, and bainite, forming at temperatures between 950°F and $M_s$. As the transformation temperature decreases within the pearlite transformation temperature range, the ferrite and carbide lamellae become increasingly smaller, known as *fine pearlite*. You will also notice that the finer the pearlite the harder the structure which is brought about by the close spacing of the hard constituent (cementite) within the soft ferrite matrix.

Below 950°F the bainite structure occurs, which consists of needles of ferrite with carbide platelets generally oriented parallel with the long dimension of the needles. As with the pearlite, at the lower temperature, the needles and platelets become finer or smaller, and the hardness varies accordingly.

## HEAT TREATMENT

Heat treatment is a term used to denote a process of heating and cooling metals in order to obtain certain desired properties. For example, steels that are to be used for cutting tools must be given the right combination of hardness and toughness in order to be able successfully to machine other metals.

**The Effect of Heating and Cooling Steel.** Carbon is the most important element in determining the ultimate condition of the steel after heat treatment. When heated, carbon chemically combines with iron, forming a carbide of iron called cementite. The carbon in each piece of steel is in the form of cementite, mixed in a matrix of iron, representing a two-phase structure.

**Annealing.** Steel that has been formed, forged, or machined accumulates considerable stress. A *stress-relief anneal* consists of heating the metal to approximately 500°F, allowing a soak period until a uniform temperature is attained, followed by air cooling.

Another annealing method is known as *spheroidized anneal,* and is used in place of the operation described above. The spheroidized anneal is used primarily to improve cold-working operations and machinability of the higher-carbon alloy steels. The process consists of heating the metal high enough (slightly above or slightly below the lower critical temperature) to permit the iron carbide layers to curl up in spherical forms. This allows better slippage of the iron crystals with less force needed to form or machine the metal.

A *full anneal* consists of heating the metal slightly above the $A_3$ line (Fig. 2.10), which allows the grains to recrystallize. It is soaked at this heat until there is a uniform temperature throughout the metal, and then it is furnace cooled. This allows the metal to break down into ferrite and pearlite structure. The full-anneal process ties up the furnace for a comparatively long time. Therefore, *cyclic annealing* is often used. The cooling phase is done in a molten-salt bath or controlled furnace at a constant temperature within the ferrite-pearlite range until full transformation takes place. Steels treated by this process include alloys of chromium-molybdenum within a carbon range of 0.12 to 0.25 percent.

**Normalizing.** Normalizing is a modified annealing process, the main difference being that the parts are allowed to cool in still air at room temperature. The metal is heated 50° to 100°F above the $A_3$ temperature, but in no case high enough or long enough to permit the grains to grow to any considerable size. Forgings that are to be machined are often normalized to restore uniform cutting conditions in the previously distorted crystals. It also allows the metal to recrystallize and form smaller grains. Normalizing produces a harder, less uniform section than does a full anneal. Heavy sections cool slower than thin sections, and therefore they will be softer and have less strength.

Normalizing is often done after machining or hot working to give uniform grain structure. If the parts are to be hardened, there will be less strain and less likelihood of cracking.

## HARDENING AND HARDENABILITY

The two terms hardening and hardenability are often confused. Hardenability refers to the *depth* to which a certain hardness level exists. Hardenability is constant for a given material but the hardness may vary, for it is dependent on the cooling rate. The cooling rate in turn is determined by (*a*) the amount of heat in the part being cooled, (*b*) the surface to volume ratio of the part, (*c*) the efficiency of the heat transfer from the part to the quenching media, and (*d*) the cooling capacity of the quenching media.

**Hardening.** The optimum combinations of strength and toughness can be obtained by first hardening the steel and then tempering it.

**Martensite.** A very hard interlaced needlelike structure (Fig. 2.17) can be formed by cooling the steel from the austenitizing temperature at a rate fast enough to miss the pearlite knee and bainite nose of the continuous cooling curve for the steel involved (Fig. 2.18). This usually means it will be quenched in water or oil. However, some tool steels become martensitic by quenching in air whereas other steels require

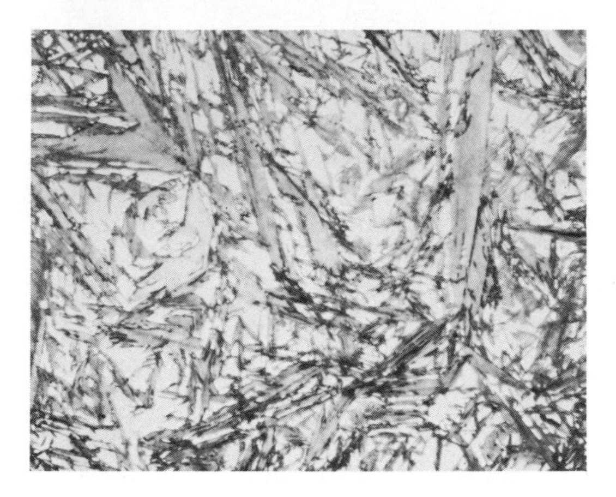

**Fig. 2.17.** Martensite structure.

such rapid quenching that it is practically impossible to prevent some transformation to ferrite, dense pearlite, or bainite, even in thin sections, because the alloy content is insufficient to form carbides. The time, temperature, transformation (TTT) diagram (Fig. 2.19) shows that, even with an extremely fast quench, only a small amount of martensite could be formed.

**Hardenability Tests.** Hardenability tests are made on sample rounds one inch in diameter and 4 inches long of the material in question. The most common method is that developed by Jominy and Boegehold. The sample is first normalized and then machined square on one end. After

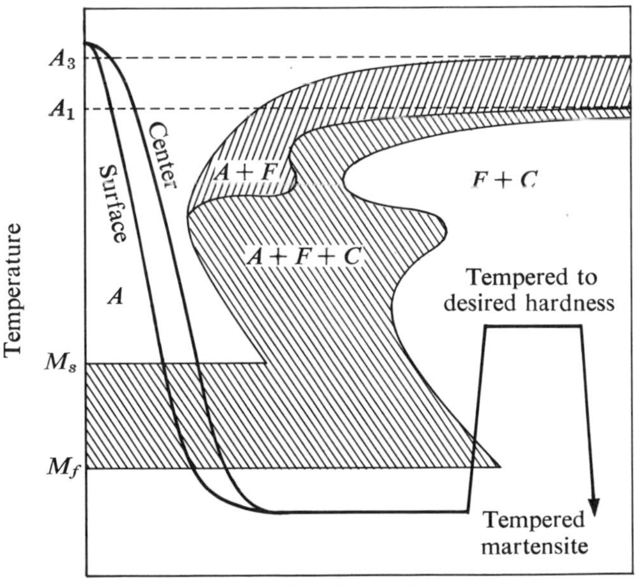


**Fig. 2.18.** Schematic cooling curves for hardening with the conventional quench and temper method.

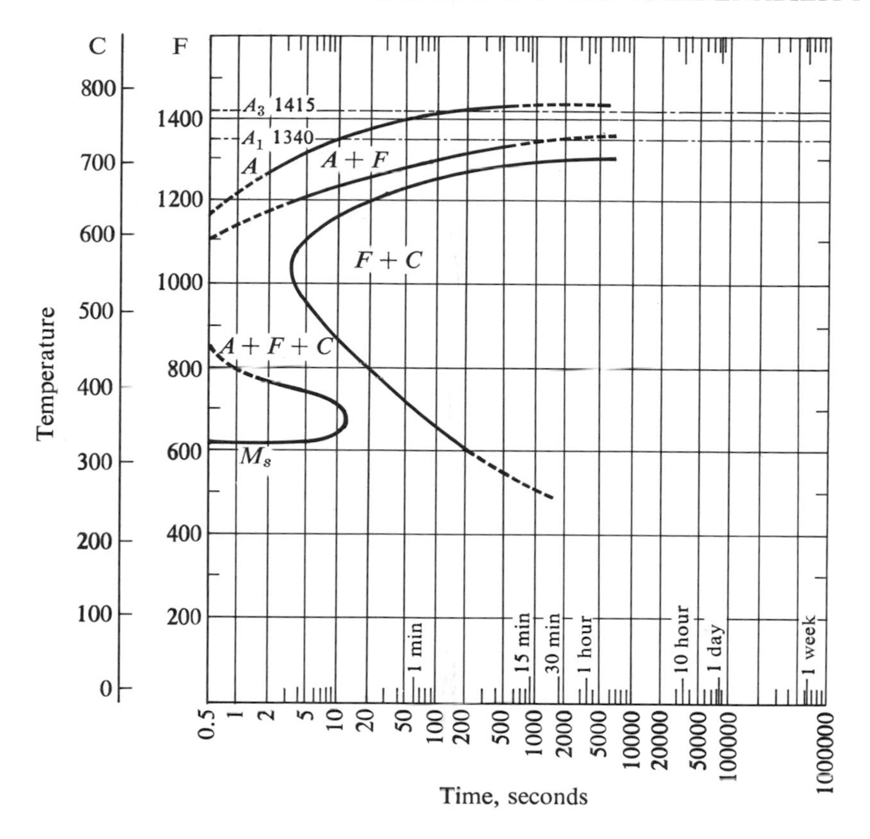


**Fig. 2.19.** TTT diagram for a plain carbon 1050 steel. Note that even with extremely rapid cooling rates some transformation of austenite in the 800° to 1150°F temperature range occurs.

being heated above the upper critical temperature, the bar is supported in a vertical position 1/2 in. above a 1/2 in. dia nozzle (Fig. 2.20) for end quenching for 10 min, after which it is quenched to room temperature. Two parallel longitudinal flats are ground on opposite sides of the bar on which the Rockwell C hardness tests are made at 1/16 in. intervals for a distance of 1/2 in. from the quenched end, 1/8 in. intervals to 20/16 in., and then at 1/4 in. intervals until the hardness falls to less than $R_c$ 20 or until a distance of 32/16 in. is reached.

The correlations between the center hardness of 1, 2, 3 and 4 in. rounds are superimposed over their equivalent end-quench curve position (Fig. 2.21).

Steel manufacturers are now listing H-steels or steels of given hardenability limits from which can be calculated definite responses to heat treatment. Thus the user can determine what quenching medium he will have to use to obtain a given hardness.

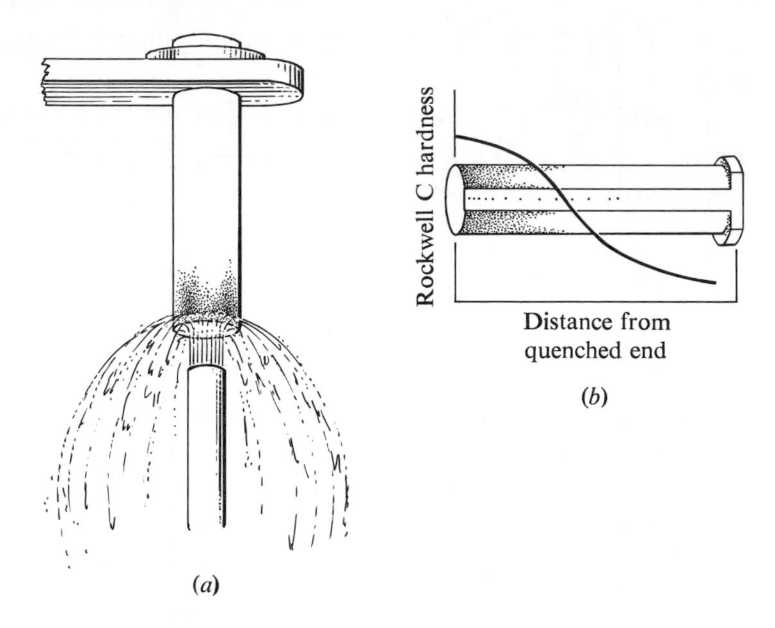


**Fig. 2.20.** End-quenching and method of hardness testing and end-quench hardenability specimen: (*a*) shows specimen being water quenched; (*b*) shows finished quenched specimen after grinding and checking Rockwell C hardness.

The correlation between the end-quench hardenability of eutectoid steel and the TTT diagram is shown in Fig. 2.22. At the top of the chart, the measured hardness curve is superimposed over a sketch of the end-quench bar. Four different locations on the bar, represented by *A*, *B*, *C*, and *D*, show the effect of the various cooling rates. *A*, for example, has a high hardness since the cooling rate was fast enough to entirely miss the pearlite zone or the "knee" of the transformation curve. Point *B* will be softer because the cooling curve intersects the pearlitic zone. The austenite is transformed, in part, to fine pearlite with some bainite, and the rest, as it continues to cool, will become martensite. *C* and *D* cool slowly enough to allow complete transformation to pearlite. The pearlite at *D* will be coarser and softer than at *C*. The diagram as shown is not intended to be highly accurate but rather to represent an approximation of typical heat treatment which involves transformation as it occurs during continuous cooling.

**Tempering.** Tempering is accomplished either in a furnace, heavy oil, molten lead, or molten salts. The usual tempering temperature ranges from 300°F to 1200°F, depending on the type of steel and the hardness required. For most steels the speed of cooling after tempering is of little consequence. In practice, the cooling after tempering is done in air.

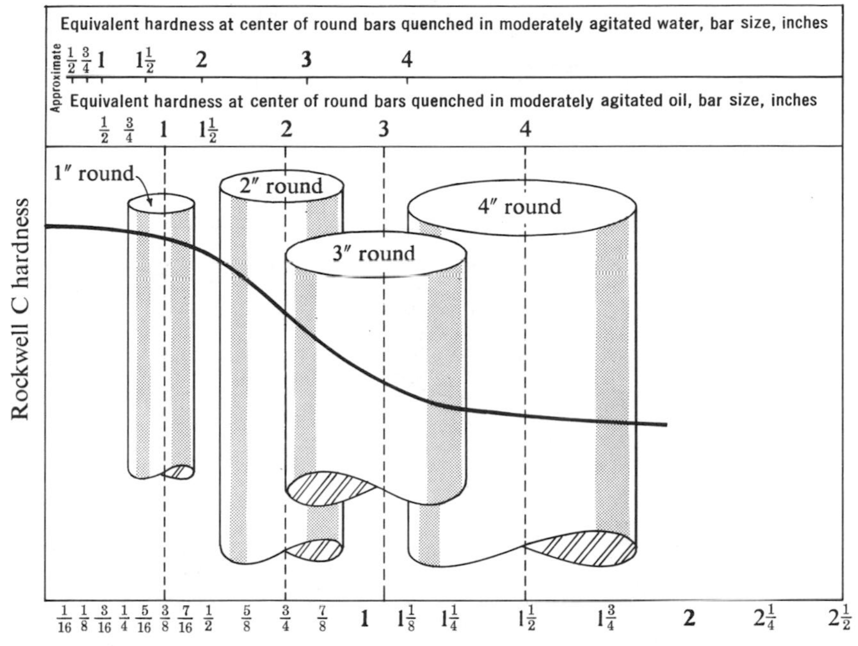


**Fig. 2.21.** Center hardnesses of various rounds correlated to their equivalent end-quench hardenability positions. Courtesy Republic Steel.

Some higher-alloy steels are double tempered to ensure stable structures. During the first tempering, new martensite is formed from some of the retained austenite. The second tempering operation helps temper the newly formed martensite.

**Martempering.** To minimize the stress and distortion in quenching, a pause in cooling can be made for the outside and inside of the part to equalize in temperature (Fig. 2.23). A molten-salt bath is usually used as an intermediate quenching medium for temperature equalization. If the equalization time is long, it may not be possible to avoid forming some bainite.

**Case Hardening.** In order to harden low-carbon or mild steels, it is necessary to add carbon to the outer surface and harden a thin outer case. The process involves two separate operations. The first is a carburizing process, where carbon is added to the outer skin, and the second is to heat-treat the carburized parts so that the outer surface becomes hard. The inside of the part, or the core, does not change materially in structure. The term "case hardening" includes the entire process of carburizing and hardening.

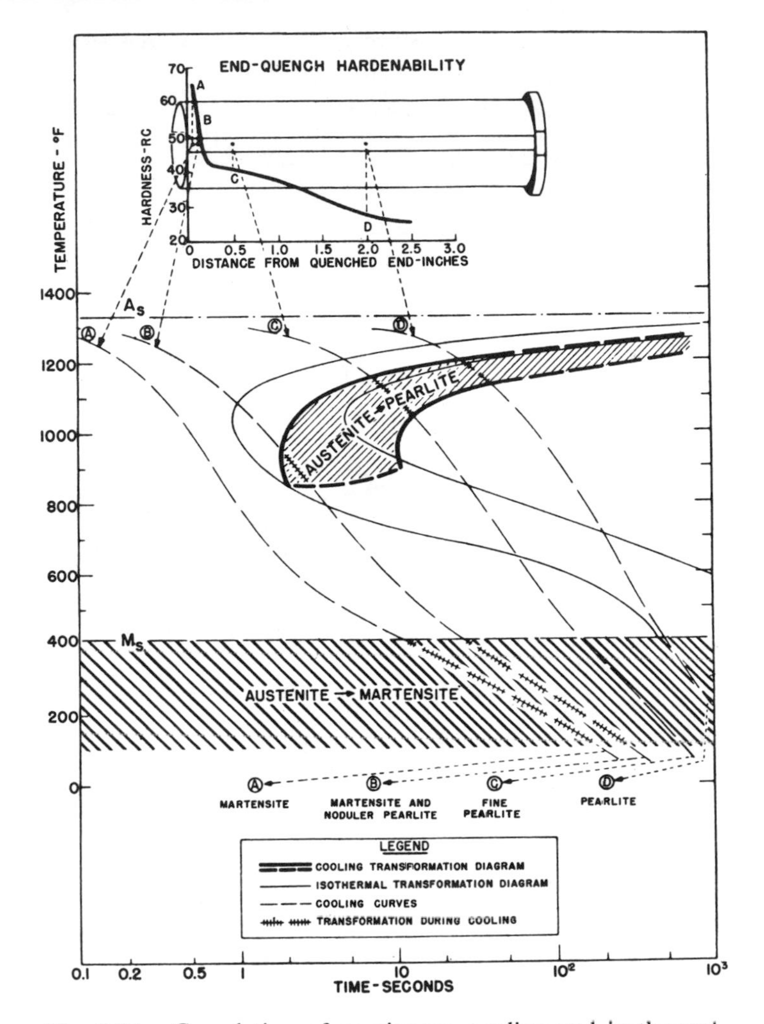


**Fig. 2.22.** Correlation of continuous cooling and isothermal transformation diagrams with end-quench hardenability test data for eutectoid carbon steel. Courtesy United States Steel.

CARBURIZING. The carburizing process consists of heating low-carbon steel in contact with a carbonaceous compound to a temperature that may vary from 1650° to 1700°F. This temperature is used because, at this heat, the steel is austenitic and absorbs carbon readily. The length of time that the material is kept in contact with the carbonaceous material depends upon the extent of carburizing action desired. The carbon from the carburizing material is absorbed by the steel, and the low-carbon surface is converted to a high-carbon steel. The result is a piece of steel with a low-carbon core and a high-carbon surface. After the ma-

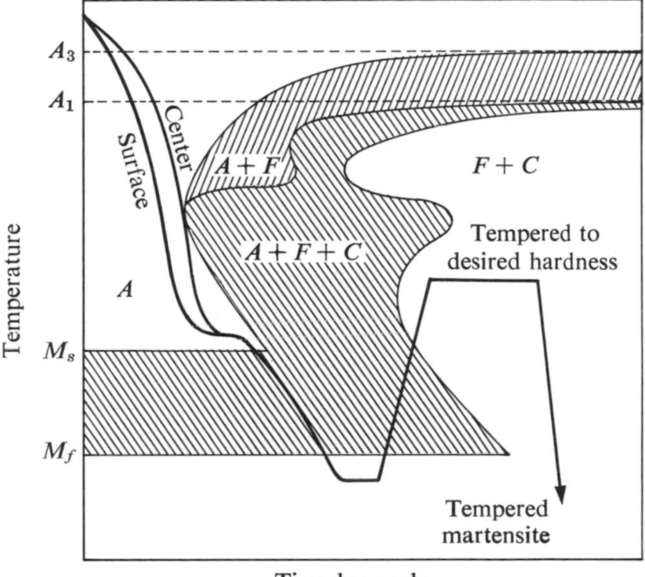


**Fig. 2.23.** Schematic diagram for the martempering method of hardening.

terial is quenched, the outer surface will be hardened while the inner core remains soft. This gives an ideal structure for many applications.

Many different carburizers may be used. These include wood charcoal, animal charcoal, coal, beans, nuts, bone, leather, or combinations of these materials. The thickness of the cases may be from 0.002 to 0.050 in., depending upon the length of time the steel is heated while in contact with the carbon.

CYANIDE HARDENING. Cyanide may also be used to case-harden steel. It is used to give a very thin but hard outer case. Sodium cyanide or potassium cyanide may be used as the hardening medium. The cyanide is heated until it becomes liquid. When the steel is placed in the cyanide bath, both carbon and nitrogen are added to the outer surface, resulting in a surface harder than that produced by the carburizing process.

NITRIDING. Nitriding is a case-hardening process in which nitrogen, instead of carbon, is added to the skin of the steel. This is used usually only on certain alloys which are susceptible to the formation of chemical nitrides. This forms a very hard case, which has a small amount of distortion, as the entire process is carried out at a relatively low temperature (950°F). No scaling occurs, since the steel is not exposed to air at elevated temperatures, and thus pieces can be finished to size before heat treatment. Ammonia gas is used as the nitrogen-producing material. The case that is produced is very thin (0.001 to 0.005 in. thick).

**Surface Hardening of Medium- and High-Carbon Steels.** Flame and induction hardening are performed on carbon steels containing 0.35 percent carbon or more. This process is also used on cast and malleable irons of suitable composition.

FLAME HARDENING. Flame hardening consists of moving an oxy-acetylene flame over the part, followed by a quenching spray. The rate at which the flame or *burner* is moved over the work will determine the depth of hardness for a given material. Burners can be obtained for different contours, as shown in Fig. 2.24. Quenching can be built into the burner head or be done as a separate operation after flame heating. The various types of burners vary from the standard torch welding tip to flat, contoured, and ring burners.

In flame hardening there is no sharp line of demarcation between the hardened surface zone and the core, so there is little likelihood that it will chip out or break during service.

The flame is kept at a reasonable distance from sharp corners to prevent overheating, and drilled or tapped holes are normally protected by filling them with wet asbestos or carbonaceous material.

Some *advantages* of flame hardening are:

(1) Large machined parts can be surface hardened economically.

(2) Surfaces can be selectively hardened with minimum warping and with freedom from quench cracking.

(3) Scaling is superficial because of the relatively short heating cycle.

(4) Electronically controlled equipment provides precise control of case properties.

The *disadvantages* include:

(1) To obtain optimum results, a technique must be established for each design.

(2) Overheating can cause cracking or, where thin sections are involved, excessive distortion.

INDUCTION HARDENING. Induction hardening is done by placing the part in a high-frequency alternating magnetic field. Heat is generated by the rapid reversals of polarity. The primary current is carried by a water-cooled copper tube and is induced into the surface layers of the workpiece.

The most commonly used sources of high-frequency current are:

(1) Motor-generators with frequencies of 1,000 to 10,000 cycles per second and capacities to 10,000 kw.

(2) Spark-gap oscillators with frequencies of 100,000 to 400,000 cycles per sec and capacities to 25 kw.

(3) Vacuum-tube oscillators operating at 500,000 cycles per sec with output capacities of 20 to 50 kw.

The depth of penetration of electrical energy decreases as the frequency increases; for example, the approximate minimum hardness for

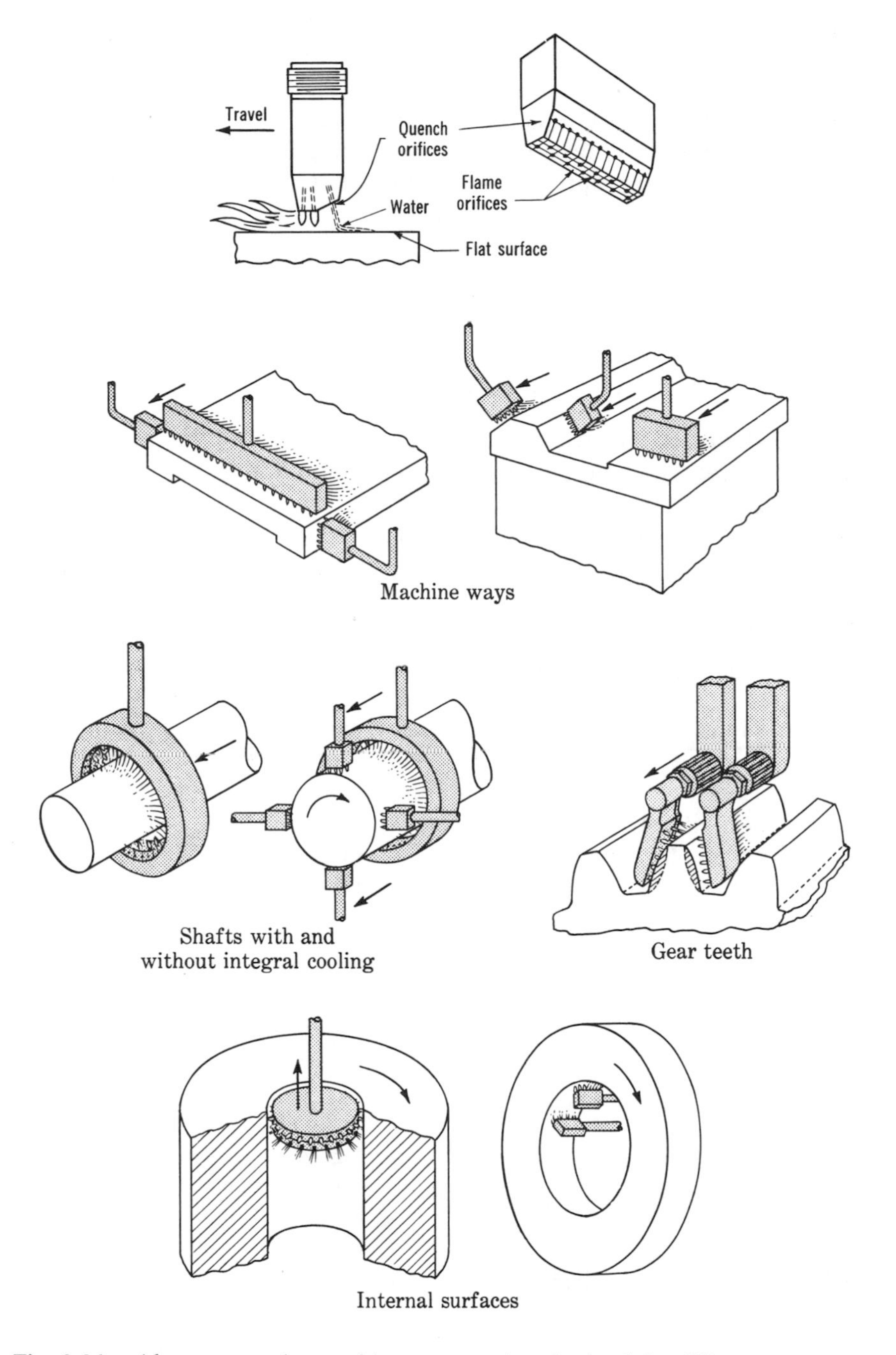

**Fig. 2.24.** Almost any shape of burner may be obtained for different contours.

3,000 cycles per sec is 0.060 in. and for 500,000 cycles per sec is 0.020 in. For this reason, thin-walled sections require high frequencies, and thicker sections must have low frequencies for adequate penetration.

Some *advantages* of induction hardening are:

(1) The operation is fast; comparatively large parts can be processed in a minimum time on automatic machines. Large truck crankshafts can be brought to the proper heat and spray-quenched in 5 sec.

(2) It can easily be applied to both external and internal surfaces.

(3) A minimum of distortion or oxidation is encountered because of the short cycle time.

## TESTING METAL PROPERTIES

A knowledge of the structure of the crystal is of little interest to us in technology unless it helps us to determine and control crystalline properties and obtain optimum results.

Oftentimes we are led into conflicting areas, that is, we want both strength and ductility. But anything done to increase strength reduces ductility and vice versa.

## TENSILE PROPERTIES

What is tensile strength? Simply, it is a measure of the maximum force a material will stand without being pulled apart. More practically for design purposes, it is the maximum force a material will sustain without yielding or without undergoing plastic deformation to more than a nominally acceptable standard. These properties are usually measured by means of a tensile test machine in which a specimen (Fig. 2.25) is

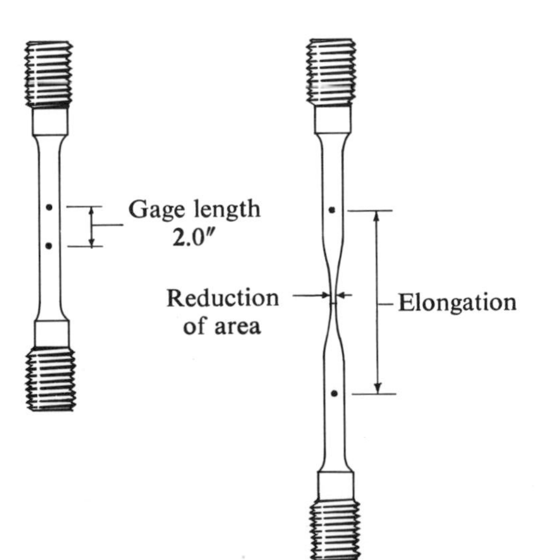


**Fig. 2.25.** A standard-type tensile specimen showing elongation and reduction of area.

placed and a load is applied at a constant rate. The amount of elongation of the specimen is measured accurately with an *extensometer*. The *engineering stress* (lb/in.²) is obtained by dividing the total load by the original cross section of the test specimen. The strain is the total elongation recorded by the extensometer divided by the gage length (in./in.). These relationships may be expressed as:

$$\text{stress} = S = \frac{L}{A_0} \qquad \text{and} \qquad \text{strain} = n = \frac{\Delta l}{l} = \frac{l_f - l_0}{l_0}$$

where $L$ = load or lbs force; $A_0$ = original cross-sectional area in in.$^2$; $l_f$ = final length; and $l_0$ = original length.

For example, a steel test specimen with a diameter of 0.505 in. is strained so that the two-inch gage-length marks are 2.125 in. apart. The load recorded is 20,000 lbs. Calculate the stress and strain.

$$S = \frac{20{,}000}{(\pi/4)(0.505)^2} = 100{,}000 \text{ psi}$$

$$n = \frac{2.125 - 2.00}{2} = 0.0625 \text{ in./in.}$$

**Stress-Strain Curves.** The data from the tensile test is used to plot the stress-strain curve (Fig. 2.26) in which the ordinate values are stress and the abscissa values are strain. The curve has two main regions. *AB* is the elastic region and *BC* the plastic. *B* represents yield and *D* the greatest force the sample will withstand. From *D* onwards, the force decreases

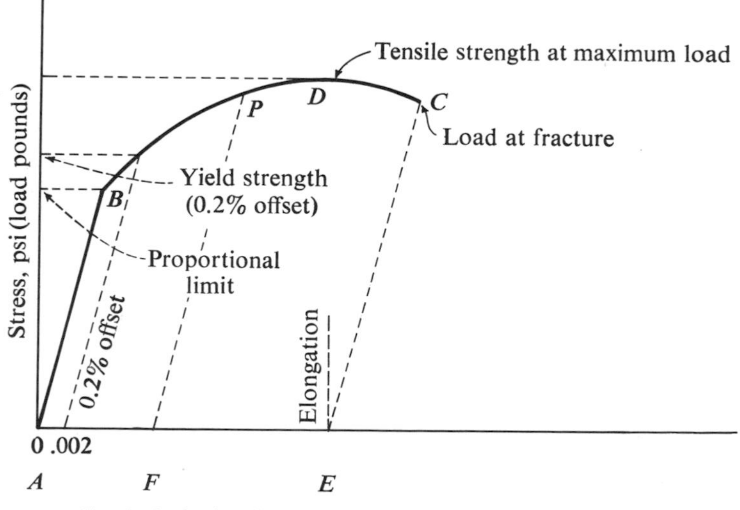


**Fig. 2.26.** A typical tensile-test plot for a ductile material.

because the test piece has begun to thin down locally, a process that quickly leads to fracture at *C*. From *B* to *D*, the material is being strain hardened due to the intersection of dislocations and the formation of pile-ups at the grain boundaries.

**Proportional Limit.** Proportional limit is the highest point on the stress-strain curve in which the stress is exactly proportional to the strain. It is also the highest point on the "curve" for which no "offset" tangent can be measured (Fig. 2.26). Theoretically it can be stated as being the highest stress at which the "curve" on the stress-strain diagram is a straight line. Proportional limit and elastic limit of many metals are considered to be approximately equal.

**Yield Point and Yield Strength.** The yield point of a material is equal to the stress at which a material undergoes a marked increase in strain without a corresponding increase in stress. On a stress-strain curve, the yield point may be marked by a sharp "knee" portion as shown by the dotted line curve in Fig. 2.27.

Some materials do not exhibit the yield-point phenomenon. For these materials the yield strength is the closest comparable property. On a stress-strain curve, the yield strength is usually determined by one of two common methods: "offset" or "strain under load." The offset method is based on a line drawn parallel to the straight-line portion of the curve which intersects the stress axis at a strain equal to the offset (Fig. 2.26).

Where the stress-strain behavior of a material is known, yield strength can be given as the stress corresponding to a specified strain as obtained by direct measurement. The specified deformation may be 0.5 percent

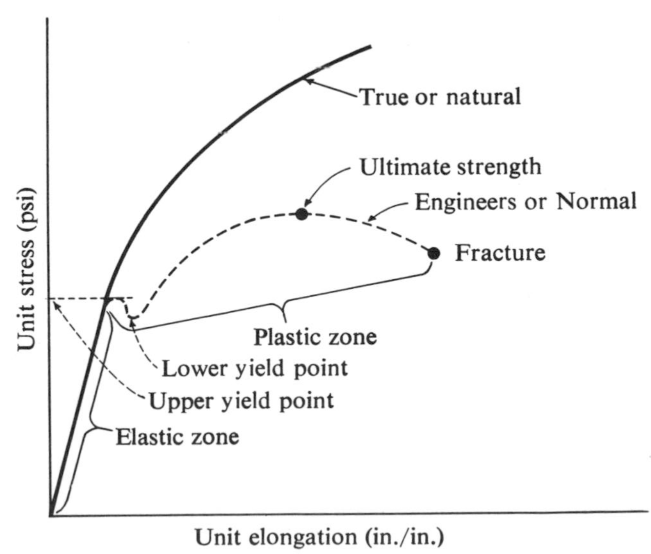


**Fig. 2.27.** Engineer's and true-stress, true-strain diagram.

extension under load, which corresponds to an offset of about 0.35 percent. This method is used primarily in determining tensile yield strength of copper and copper alloys.

The yield strength of a material may be determined according to the type of loading: compressive, flexural, shear, tensional or torsional. It is, however, generally assumed, unless otherwise specified, to refer to tensile yield strength.

Had the material been deformed by rolling, for example, before it was tested, the structure would have then been similar to the curve *FPDC* of Fig. 2.26. The yield point would be higher, *P* instead of *B*. The tensile strength would also be higher because the same maximum force would then be sustained by a deformed sample that has a smaller area. The elongation of the deformed material would be much smaller (*FE* instead of *AE*).

The yield strength can be determined by dividing the load at yield by the original area. For example, the yield load for a tensile specimen with a 0.505 in. dia was 15,000 lb. The yield strength can be calculated as follows:

$$Y = \frac{15{,}000}{(\pi/4)(0.505)^2} = 75{,}000 \text{ psi.}$$

**True Stress and True Strain.** The relationship just described is based on the original specimen cross-sectional area. If each time the load was instantly recorded and the calculations also were instantaneous, based on the diameter or cross-sectional area of the specimen, the stress calculated would then be the *true stress.*

*True strain* also may be defined as the instantaneous change per unit length compared to the original length.

True or natural stress-strain relationships are designated as follows:

$$\text{natural stress} = \sigma = \frac{L}{A_i}$$

$$\text{natural strain} = \epsilon = \frac{\Delta l_1}{l_0} + \frac{\Delta l_2}{l_0 + \Delta l_1} + \frac{\Delta l_3}{l_0 + \Delta l_1 + \Delta l_2} + \cdots$$

$$= \sum_{l_0}^{l_f} \frac{\Delta l}{l}$$

and by integration

$$\int_{l_0}^{l_f} \frac{dl}{l} = \ln \frac{l_f}{l_0}$$

where $L$ = load; $A_i$ = instantaneous area; $l_0$ = original length; $l_f$ = final length; and ln = natural log.

The difference between the engineering stress-strain curve and the true or natural stress-strain curve is shown in Fig. 2.27. As can be seen, the area affected is in the plastic-flow region. True stress continues to increase throughout the test. The maximum stress occurs at fracture. The engineer's stress-strain curve is easier to prepare and, in most cases, more useful for design purposes. The engineer is most concerned with the maximum load or maximum strain that a member can support without deformation.

If the same example is used as given previously for the engineering stress-strain relationship, we find that when, instead of using the original diameter of 0.505, the instantaneous diameter of 0.405 is used, the true or natural stress and strain would be:

$$\sigma = \frac{20{,}000}{(\pi/4)(0.405)^2} = 152{,}500 \text{ psi}$$

$$\epsilon = ln\frac{2.125}{2.00} = 0.0612 \text{ in./in.}$$

**Percent Elongation.** The total extension at fracture, as recorded by the extensometer divided by the original length, is the *elongation* of the specimen.

$$\text{Percent elongation} = \frac{\Delta l}{l_0} \times 100$$

where $\Delta l$ = change in length at fracture and $l_0$ = original length.

**Reduction of Area.** The ratio of change in cross-sectional area at fracture to the original cross-sectional area is the reduction in area.

$$\%\text{RA} = \frac{A_0 - A_f}{A_0} \times 100$$

where $A_0$ = original cross-sectional area and $A_f$ = final cross-sectional area.

**Modulus of Elasticity.** The modulus of elasticity, or Young's modulus, is the ratio of stress to the corresponding strain below the proportional limit. It is also commonly known as Hooke's Law and is expressed as:

$$E = \frac{\text{Stress}}{\text{Strain}} = \frac{S}{n} \text{ or given in units of psi} = \frac{\text{lbs/in.}^2}{\text{in./in.}}.$$

The approximate modulus of elasticity can be obtained from standard mechanical-properties tables, such as Table 3.6. With this information, load, and the length and cross-sectional area, both the strain and elongation can be determined. For example, cold-rolled steel has a modulus of elasticity of approximately 30 × $10^6$. If a rod 0.500 in. in diameter and 5 feet long is loaded in tension with a weight of 5,000 lb, what would the strain and elongation be?

$$30{,}000{,}000 = \frac{5{,}000/(\pi/4)(0.5)^2}{\text{strain}}$$

$$\text{Strain} = 0.00084/\text{in./in.}$$

$$\text{Elongation} = (0.00084)(60) = 0.050 \text{ in.}$$

**Ductility.** Ductility may be defined as the ability of a material to withstand plastic deformation without rupture. It is generally determined in a tensile test using a standard-sized test specimen with 2″ gage length. The test specimen is loaded in tension to rupture. It is then assembled and measured for length and diameter at the fracture. The increase in length is expressed as percent elongation, and the diameter as percent reduction in area, as given previously. These terms express the measure of ductility, since they are comparative.

Ductility may also be thought of in terms of bendability and crushability. Cup draw and twisting tests are considered to be indicators of ductility. The lack of ductility is often termed *brittleness*. Usually, if two materials have the same strength and hardness, the one that has the higher ductility is the more desirable.

## HARDNESS

In selecting a metal to withstand wear or erosion, there are three properties to consider—its ductility (already defined), toughness, and hardness. The most important of these is usually hardness.

Hardness is that property of a material that enables it to resist plastic deformation, penetration, indentation, or scratching.

Hardness is important from an engineering standpoint. Resistance to wear by either friction or erosion by steam, oil, and water generally increases with hardness.

**Hardness Testing.** Several methods have been developed for hardness testing. Those most often used are Brinell, Rockwell, Vickers, and scleroscope. The first three are based on indentation tests and the fourth on the rebound height of a diamond-tipped metallic hammer.

BRINELL. In Brinell hardness tests, a tungsten carbide or hardened steel ball, usually 10 mm in diameter, is employed as the indenter (Fig. 2.28). Ordinarily loads of 3,000, 1,500 or 500 kg are used. The load is applied for at least 15 sec on ferrous metals and for at least 30 sec on many nonferrous metals and for at least 2 min on magnesium. A definite, not a minimum, time of load application should be specified for soft metals. When the load is released, the diameter of the spherical impression is measured with the aid of a Brinell microscope. The Brinell hardness number is derived from the area of the indentation. However, conversion charts based on the diameter of the indentation in millimeters are used to give the hardness numbers directly.

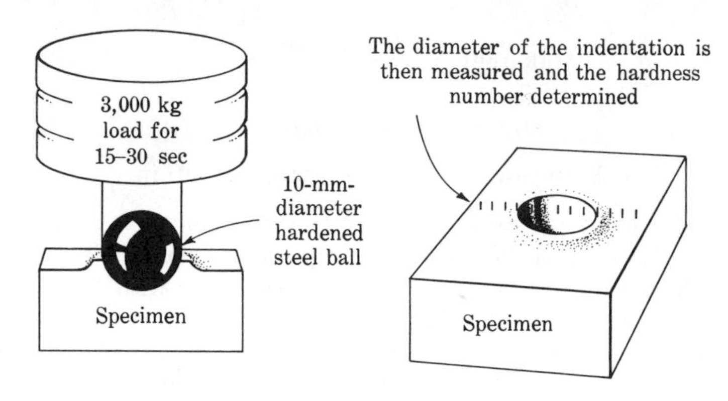


**Fig. 2.28.** Principles of Brinell hardness testing.

Brinell testing is most useful for larger workpieces such as castings and forgings having section thicknesses of 1/4 in. or more. The relatively large indentation is especially good for materials that are coarse grained, nonhomogeneous, or nonuniform in structure. The indentations produced give a better average reading over a wide variation in hardness.

Brinell testing machines range in size from small portable and bench models to ones weighing a ton or more.

ROCKWELL. The Rockwell hardness tester uses the principle of measuring the difference in penetration between a minor and major load as it is applied to the penetrator. The minor load is 10 kg, and the major load varies with the material being tested. If the material is known to be relatively soft, a 1/16-in.-diameter ball is used with 100 kg as the major load. This is known as the "B scale." If the material is relatively hard, a spheroconical diamond Brale penetrator is used with a 150-kg load (Fig. 2.29). The results are read on the C scale ($R_c$). Other scales are available and are useful for extremely hard surfaces. Also, very thin sections may be tested by using very small loads on a Rockwell superficial hardness tester. Carefully polished surfaces are required on all parts to be tested.

VICKERS. The Vickers hardness tester is very much like the Brinell hardness tester. The main differences are the size and shape of the penetrator and the magnitude of the applied loads. The penetrator used is a square-based diamond pyramid making a point angle of 136 deg. The hardness number is determined by measuring the diagonal across the square-surface impression made by the penetrator. Hardness numbers are identical with those of the Brinell scale up to 300; above this point, Brinell numbers are progressively lower. The Vickers hardness tester is used for both hard and soft materials and for precision testing of thin sheets.

SCLEROSCOPE. The scleroscope presents one of the fastest and most

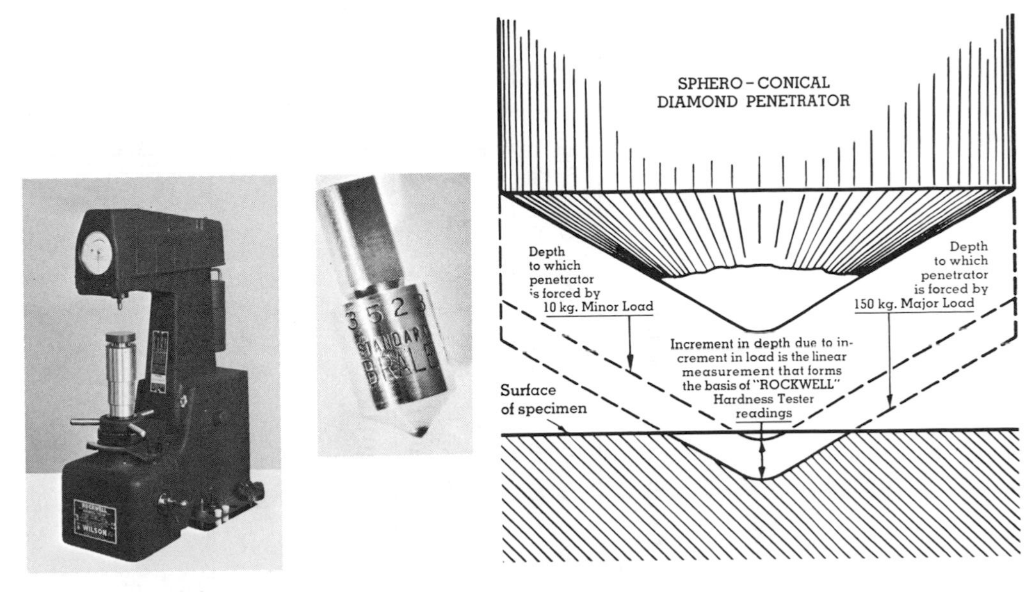


**Fig. 2.29.** Rockwell hardness tester with Brale penetrator used on hard metals. Courtesy Wilson Mechanical Instrument Div., American Chain and Cable Co.

portable means of checking hardness. The hardness number is based on the height of rebound of a diamond-tipped metallic hammer. The hammer falls free from a given height. The amount of rebound is observed on a scale in the background. The harder the material, the higher the rebound, and vice versa. Thin materials may be checked if a sufficient number of layers are packed together to prevent the hammer from penetrating in the metal to an extent where the rebound is influenced by the steel anvil. This is known as *anvil effect*.

HARDNESS CONVERSION CHARTS. In order to relate one method of testing hardness with another, hardness conversion charts have been worked out (Table 2.1). However, it should be borne in mind that these charts are only approximations. Since most hardness testing is based on indentation or severe work hardening in a localized area, much is dependent on the work-hardening characteristics of the metal. Tensile strengths are often listed on hardness conversion charts also. Although relationships exist between hardness, tensile strength, and yield strength, there are chances for error. It is always better, if possible, to use a tensile-testing machine and obtain the values directly.

## TOUGHNESS

It is not enough to know only how strong a metal is in tensile strength or how ductile it is; information as to how it reacts under sudden impact

**Table 2.1**
**A Rockwell-to-Brinnell Hardness Conversion Chart with Corresponding Ultimate Tensile Strengths**

| For Hardened Steel and Alloys | | For Unhardened Steel, Steel of Soft Temper, Gray and Malleable Cast Iron and Most Nonferrous Metal | | |
|---|---|---|---|---|
| Rockwell C 150 kg load "Brale" | Brinell 3000 kg load 10 mm ball | Rockwell B 100 kg load 1/16″ dia ball | Brinell 500 kg load 10 mm ball | Brinell 3000 kg load 10 mm ball |
| 60 | 614 | 100 | 201 | 240 |
| 59 | 600 | 99 | 195 | 234 |
| 58 | 587 | 98 | 189 | 228 |
| 57 | 573 | 97 | 184 | 222 |
| 56 | 560 | 96 | 179 | 216 |
| 55 | 547 | 95 | 175 | 210 |
| 54 | 534 | 94 | 171 | 205 |
| 53 | 522 | 93 | 167 | 200 |
| 52 | 509 | 92 | 163 | 195 |
| 51 | 496 | 91 | 160 | 190 |

| | | | | |
|---|---|---|---|---|
| 50 | 484 | 90 | 157 | 185 |
| 49 | 472 | 89 | 154 | 180 |
| 48 | 460 | 88 | 151 | 176 |
| 47 | 448 | 87 | 148 | 172 |
| 46 | 437 | 86 | 145 | 169 |
| 45 | 426 | 85 | 142 | 165 |
| 44 | 415 | 84 | 140 | 162 |
| 42 | 393 | 83 | 137 | 159 |
| 40 | 372 | 82 | 135 | 156 |
| 38 | 352 | 81 | 133 | 153 |
| 36 | 332 | 80 | 130 | 150 |
| 34 | 313 | 79 | 128 | 147 |
| 32 | 297 | 78 | 126 | 144 |
| 30 | 283 | 77 | 124 | 141 |
| 28 | 270 | 76 | 122 | 139 |
| 26 | 260 | 75 | 120 | 137 |
| 24 | 250 | 74 | 118 | 135 |
| 22 | 240 | 72 | 114 | 130 |
| 20 | 230 | 70 | 110 | 125 |
| | | 68 | 107 | 121 |

**Table 2.1 (Continued)**
**A Rockwell-to-Brinnell Hardness Conversion Chart with Corresponding Ultimate Tensile Strengths**

| For Hardened Steel and Alloys | | For Unhardened Steel, Steel of Soft Temper, Gray and Malleable Cast Iron and Most Nonferrous Metal | | |
|---|---|---|---|---|
| | | Rockwell B 100 kg load 1/16″ dia ball | Brinell 500 kg load 10 mm ball | Brinell 3000 kg load 10 mm ball |
| Brinell 3000 kg To Ultimate Tensile Strength For Steels | | 66 | 104 | 117 |
| | | 64 | 101 | 114 |
| | | 62 | 98 | 110 |
| | | 60 | 95 | 107 |
| | | 58 | 92 | 104 |
| Brinell 3000 kg load 10 mm ball | Ultimate Tensile Strength, psi | 56 | 90 | 101 |
| | | 54 | 87 | – |
| | | 52 | 85 | – |
| | | 50 | 83 | – |
| | | 48 | 81 | – |
| 200 | 100,000 | 46 | 79 | |
| 225 | 108,000 | 44 | 78 | |
| 250 | 122,000 | 42 | 76 | |
| 275 | 141,000 | 40 | 74 | |
| | | 38 | 73 | |

| | | | |
|---|---|---|---|
| 300 | 158,000 | 36 | 71 |
| 325 | 174,000 | 34 | 70 |
| 350 | 188,000 | 32 | 68 |
| 375 | 202,000 | 30 | 67 |
| | | 28 | 66 |
| 400 | 215,000 | 24 | 64 |
| 425 | 227,000 | 20 | 62 |
| 450 | 238,000 | 16 | 60 |
| 475 | 249,000 | 12 | 58 |
| | | 8 | 56 |
| 500 | 258,000 | 4 | 55 |
| 525 | 267,000 | 0 | 53 |
| 550 | 282,000 | | |
| 575 | 295,000 | | |
| 600 | 308,000 | | |

also is of prime importance. This quality is known as toughness. It is measured by the Charpy test or the Izod test.

Both of these tests use a notched specimen. The location and shape of the notch are standard. The points of support of the specimen, as well as the impact of the hammer, must bear a constant relationship to the location of the notch.

The tests are conducted by mounting the specimens as shown in Fig. 2.30 and then allowing a pendulum of a given weight to fall from a given height. The maximum energy developed by the hammer in the Izod test is 120 ft-lb and in the Charpy test 240 ft-lb.

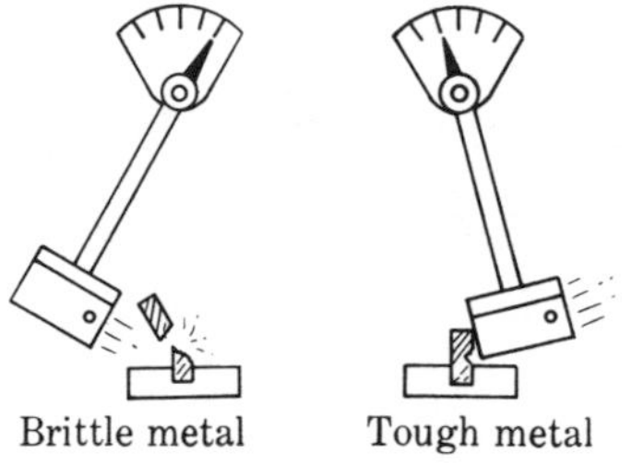


**Fig. 2.30.** The principle of the Izod impact test.

By properly calibrating the machine, the energy absorbed by the specimen may be measured from the upward swing of the pendulum. The greater the amount of energy absorbed by the specimen, the less will be the upward swing of the pendulum.

## SHEAR STRENGTH

Shear strength is expressed in the number of pounds per square inch required to produce a fracture when impressed vertically upon the cross section of a material. Methods of testing both single and double shear strength are shown in Fig. 2.31.

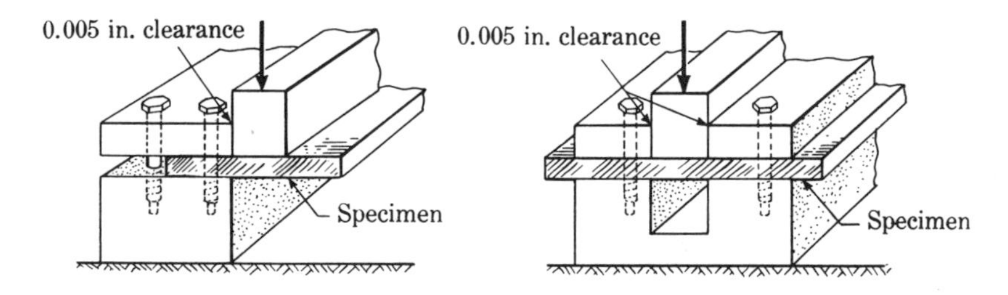


**Fig. 2.31.** Methods used in testing single and double shear.

Shear strength may be calculated as the amount of force needed to make the shear over a given cross-sectional area. For example, in single shear, assuming a required load of 12,300 lb for a bar $1/4 \times 1$ in. (area = 0.25 sq in.), the shear strength would be

$$\text{Shear strength} = \frac{\text{applied load}}{\text{area in shear}} \quad \text{or} \quad \frac{12{,}300}{0.25} = 49{,}200 \text{ psi}$$

In double shear, the area is doubled, resulting in twice the shear strength in pounds.

The shear strength of mild steels compared to ultimate tensile strength ranges from about 60 to 80 percent. The lower values are for the harder materials.

Tests made on 3/8-in.-dia annealed rods of Monel, steel, naval brass, and bronze containing 1.15 percent tin are shown in Table 2.2.

**Table 2.2**
**Shear Strength of Soft, 3/8-in.-Diameter Rods***

| Material | Shear strength (psi)† |
|---|---|
| Monel . . . . . . . . . . . . | 58,000 |
| Low-carbon steel . . . . . | 44,500 |
| Naval brass . . . . . . . . . | 37,000 |
| Bronze (1.15% tin) . . . . . | 35,500 |

*Courtesy The International Nickel Co., Inc.
†Taken in double shear.
Tests conducted by Columbia University, New York.

## CREEP

Creep is expressed as the plastic behavior of metal or plastics under constant load and at constant temperature. There are three stages of creep. In the first one the material elongates rapidly but at a decreasing rate. In the second stage, ordinarily of long duration, the rate of elongation is constant. In the third stage the rate of elongation increases rapidly until the material fails.

The design engineer is most concerned with second-stage creep, where elongation takes place at a constant specific rate. The percent of elongation and time required are dictated by the requirements of the particular application. An example would be 0.1 percent elongation in 10,000 hr. In rapidly rotating structural members such as rotors and blading of steam and gas turbines, the clearances are extremely small and critical. The designer will be satisfied with nothing short of experimentally determined stress of 1 creep rate unit (CRU) or 1 percent in 100,000 hr.

## METALS OF THE FUTURE

**Single Crystal Filaments or Whiskers.** The strongest solids yet known are single-crystal filaments that grow from the surface of metal under certain conditions (Fig. 2.32). Even though some are too small to be seen by the naked eye, they have tensile strengths up to several million psi and preserve it up to 90 percent of their melting point.

A promising approach is the impregnation of metals with the high-strength whiskers where they will act much the same as steel rods in concrete or glass fibers in plastics.

The filaments are usually single crystals with an extremely high length-to-diameter ratio with exceptionally high strength approaching that of pure atomic or ionic bonding. No one is really certain what makes the crystals grow, but it is postulated that the whiskers are caused by the

(*a*)

(*b*)

(*c*)

**Fig. 2.32.** Single crystal iron whiskers (*a*) grown by hydrogen reduction of ferrous sulfate; (*b*) magnified image, originally 50×; (*c*) iron oxide ($\alpha$ $Fe_2O_3$) whiskers growing on top of a thick oxide scale, originally 45,000×. Courtesy General Motors Research Laboratories.

difficulty the metal has in nucleating under certain conditions. Iron and copper whiskers are produced by a hydrogen reduction of the respective halide, almost always a chloride of the metal. The orientation of the fibers in the composite has a great effect on its strength. Most experimental composites have had random orientation, which gives the metal a medium strength as compared to having the fibers all one way and applying a load parallel to the orientation.

It is difficult to estimate the time that will be required to produce the quantity and quality of whiskers needed and to overcome the method of sorting and orienting in the composite. However the work has stimulated research in high-strength materials.

The need for structural materials to be used at 4000°F is focusing interest on ceramics and graphite. However, tests at these temperatures are difficult to make, and the data are difficult to interpret.

Another interesting development has been one that intersperses extremely fine, hard particles within and between metal grains. The result is a product that is different from the parent metal. The first commercial product of this kind is nickel with thorium oxide particles in its grain structure, which is able to retain its strength for long periods at 2400°F.

Explosive cladding of dissimilar metals may also lead to a range of engineering materials having properties not yet attainable. Most metals or alloys can be joined by this process.

## QUESTIONS

**2.1.** Explain with the help of a sketch how dendritic growth takes place in a metal crystal.

**2.2.** Name and sketch various atomic structures found in metals. Name the one with the greatest number of atoms in it.

**2.3.** Explain why certain metals flow more readily under stress than others do.

**2.4.** What do you understand by dislocation in a metal? Name the two types of dislocations.

**2.5.** What information about the properties of a material can be obtained from a tensile test?

**2.6.** How is a stress-strain curve plotted? How are values of modulus of elasticity, proportional limit, and ultimate tensile strength obtained from such a curve?

**2.7.** What is meant by strain hardening?

**2.8.** Describe in brief each of the four methods of testing hardness of a material. Convert 40 Rockwell C into a Brinell hardness number.

**2.9.** What is a unit of toughness?

**2.10.** What is creep?

**2.11.** What are interstitial and substitutional solid solutions? How do they differ from each other?

**2.12.** What is the effect of carbon on the ductility and toughness of steel?

**2.13.** Draw an iron-carbon equilibrium diagram. Show the following on the diagram: eutectoid point, hypoeutectoid steels, medium-carbon steels, high-carbon steels, cast irons.

**2.14.** What is the upper critical temperature for 0.8 percent carbon steel?

**2.15.** Distinguish between the following: annealing, normalizing, hardening, tempering.

**2.16.** How does carbon influence hardenability of steel?

**2.17.** Differentiate between cyaniding and nitriding.

**2.18.** Can low-carbon steels be flame hardened? Give reasons.

**2.19.** Lead when deformed at room temperature undergoes hot working. Explain.

**2.20.** What are single-crystal filaments? How are they formed?

## PROBLEMS

**2.1.** The data obtained from a tensile test conducted on a test piece of 0.505 in. dia and 2 in. gage length were as shown in Table P2.1.

**Table P2.1**

| | | | | | | | |
|---|---|---|---|---|---|---|---|
| Load (tons) | 3.66 | 4.17 | 4.81 | 5.08 | 5.22 | 5.46 | |
| Length (inches) | 2.05 | 2.10 | 2.18 | 2.24 | 2.30 | 2.37 | |
| Load (tons) | 5.60 | 5.68 | 5.71 | 5.67 | 5.40 | 4.84 | 4.02 |
| Length (inches) | 2.45 | 2.57 | 2.68 | 2.76 | 2.83 | 2.85 | 2.86 |

Plot the stress-strain curve. Find the following from the curve: (*a*) modulus of elasticity; (*b*) yield point; (*c*) ultimate tensile strength; (*d*) percentage of elongation; (*e*) yield point at 0.02% offset.

**2.2.** A 0.505-in. dia and 2-in. gage length specimen, when tensile tested, could stand a maximum load of 37,200 lb. The gage length at fracture measured 2.76 in. Determine the tensile strength of the material and percentage elongation.

**2.3.** A 3 1/2-in. dia hole is to be punched in a 10-gage low-carbon sheet. What is the shear force required for this job?

**2.4.** A brass shear pin is to be used to transmit a load. What should be the diameter of the pin if the transmitted load is not to exceed 10,000 lb?

**2.5.** If a turbine rotor is 20 ft (+000 or −0.005 in.) in dia at installation, how much larger could it be after 100,000 hr of use and still be within the specified creep rate?

## REFERENCES

Argon, A. S., S. Backer, F. A. McClintock, et al, *Mechanical Behavior of Materials,* Addison-Wesley Publishing Company, Inc., Reading, Mass., 1966.

Austin, J. B., "New Thoughts on Crystallization," *Metal Progress,* May 1963.

Chalmers, B., "Metallurgy Today," *International Science and Technology,* May 1966.

Accountins, O. E., "Whiskered Metals Reach for 1,000,000 psi," *Machine Design,* May 9, 1963.

Khol, R., "Materials Revolution at a Snail's Pace," *Machine Design,* May 12, 1966.

Black, P. H., *Theory of Metal Cutting,* McGraw-Hill Book Company, Inc., New York, 1961.

Dash, W. C., and A. G. Tweet, "Observing Dislocations in Crystals," *Scientific American,* October 1961.

Spencer, L. F., "Surface Hardening of Steels," *Machine Design,* January 1961.

VanVlack, L. H., *Elements of Material Science,* ed 2, Addison-Wesley Publishing Company, Inc., Reading, Mass., 1964.

# 3

# FABRICATING CHARACTERISTICS OF METALS

In THE PRECEDING CHAPTER an explanation was given as to how the various properties of metals, such as hardness, ductility, etc., are developed. This chapter deals with the characteristics that should be observed when fabricating some of the more important metals. By *fabricating characteristics* is meant how the metal reacts to machining, forming, welding, and casting.

Before considering the fabrication of a metal, it is important to know how it is classified, so that references can be made to a specific type of metal. Therefore, the method of classification will be given for each of the metals before the fabricating characteristics are described.

The more common metals will be discussed in this chapter. Not all metals are equally applicable to fabrication by each method. Therefore only those methods that are most applicable will be discussed.

## FABRICATING CHARACTERISTICS

**Machinability.** Machinability is an involved term with many ramifications. However, simply stated, it is the ease with which metal can be removed in such operations as turning, drilling, reaming, threading, sawing, etc.

Ease of metal removal implies, among other things, that the forces acting against the cutting tool will be relatively low, that the chips will be easily broken up, that a good finish will result, and that the tool will last a reasonable period of time before it has to be replaced or resharpened. Another way of expressing this is to give each material a *machinability rating*. This has been done for most ferrous metals, using the American Iron and Steel Institute (AISI) B1112 as the basis of 100 percent machinability. Thus another metal may be said to have a machinability rating of 60 percent, as in the case of one type of stainless steel, or 240 percent for an aluminum alloy.

**Formability.** The ability of a metal to be formed is based on the ductility of the metal, which, in turn, is based on its crystal structure. As mentioned before, the metal that has a face-centered cubic crystal has the greatest opportunity for slip—four distinct nonparallel planes and three directions of slip in each plane.

Other factors that govern, to a large extent, the flowability or ductility of a material are strain rate or dislocation formations, as discussed previously, by both hot and cold working, alloying elements, and heat treatments such as annealing and normalizing.

Grain size is also an important factor in formability. Fine grain size increases strength and produces a better finish. Large grain size materials are more ductile and are used for tough jobs like deep drawing of heavy strip.

Formability indicies have been developed for various materials subjected to different types of strain such as in joggling development, brake forming, and stretch forming and may be found in the *Metals Engineering Quarterly*, Volume 5, Number 1.

**Weldability.** It may be said that all metals are weldable by one process or another. However, the real criterion in deciding on the weldability of a metal is *weld quality and the ease with which it can be obtained.*

In deciding on the weldability of a metal, the characteristics commonly considered are the heating and cooling effects on the metal, oxidation, and gas vaporization and solubility.

The heating and cooling effects govern the hardness of the weld deposit. Metals that oxidize rapidly, such as aluminum, interfere with the welding process. Gases form in the welding of some metals causing porosity and weakening of the weld deposit.

**Castability.** The castability of a metal is judged to a large extent on the following factors: solidification rate, shrinkage, segregation, gas porosity, and hot strength.

Some metals, such as gray iron, are very fluid and have a slower solidification rate, which makes it possible to pour very thin sections and complex castings.

*Shrinkage* refers to the amount of contraction when the metal goes from a molten to a solid state. Shrinkage for gray iron is about half that for steel.

As the metal solidifies, forming a dendritic or pine-tree pattern along the mold edges, some alloying elements are excluded. This is termed *segregation.* Thus the surface of the casting may not have the same quality as that in the center.

Some molten metals have a high affinity for oxygen and nitrogen. As these gases become trapped, they cause voids or pinholes known as *gas porosity.*

Hot metals are comparatively low in strength, especially the nonferrous types. Care must be taken in the pattern and mold design to avoid stress concentrations that will tend to tear the metal as it solidifies. Metals that are particularly weak when hot, such as aluminum, are said to be *hot short.*

## STEEL

Steels are a broad group of iron-base alloys containing small amounts of carbon as their principal alloying element. They range from practically no carbon up to cast irons which have substantially more than 2 percent carbon. Most commercial steels fall in the range from 0.03 to 1.2 percent carbon. Manganese and silicon are also added to almost all steels to improve their properties.

Steels can be divided into two broad categories: *carbon steels* and *alloy steels.* Carbon steels make up about 90 percent of our present usage.

Alloy steels are generally used to obtain superior mechanical properties through heat treatment. For example, a new family of "extra high strength" *quenched and tempered alloy steels* have come into prominence because they have yield strengths over 100,000 psi compared to 30,000 psi for a typical carbon steel. They also show good toughness and weldability.

Still another group of steels known as *ultrahigh strength steels* are coming into prominence due to their high yield strengths that exceed 180,000 psi and may go as high as 600,000 psi. Included in this group are some stainless steels, some standard AISI alloy steels and maraging steels.

**AISI and SAE Steel Classifications.** The first code-numbering system for steel was made by the Society of Automotive Engineers (SAE). It consists of a four or five-digit system; the first two digits refer to the alloy content and the last two (or three) to the carbon content in *points* carbon, where one point is equal to 0.01 percent. A more recent classification is the AISI Code that makes use of the same SAE system but adds suffixes and prefixes for further identification, as shown at the bottom of Table 3.1.

Steel may be further classified as to carbon content as in Table 3.2.

**Table 3.1**
**AISI Standard Steels**
(XXs indicate nominal carbon content within range)

| AISI Series | Nominal Composition or Range |
|---|---|
| **Carbon Steels** | |
| 10XX Series | Non-resulphurized carbon steels with 44 compositions ranging from 1008 to 1095. Manganese ranges from 0.30 to 1.65%; if specified, silicon is 0.10 max. to 0.30 max., each depending on grade. Phosphorus is 0.040 max., sulphur is 0.050 max. |
| 11XX Series | Resulphurized carbon steels with 15 standard compositions. Sulphur may range up to 0.33%, depending on grade. |
| B11XX Series | Acid Bessemer resulphurized carbon steels with 3 compositions. Phosphorus generally is higher than 11XX series. |
| 12XX Series | Rephosphorized and resulphurized carbon steels with 5 standard compositions. Phosphorus may range up to 0.12% and sulphur up to 0.35%, depending on grade. |
| **Alloy Steels** | |
| 13XX | Manganese, 1.75%. Four compositions from 1330 to 1345. |
| 40XX | Molybdenum, 0.20 or 0.25%. Seven compositions from 4012 to 4047. |
| 41XX | Chromium, to 0.95%, molybdenum to 0.30%. Nine compositions from 4118 to 4161. |
| 43XX | Nickel, 1.83%, chromium to 0.80%, molybdenum, 0.25%. Three from 4320 to E 4340. |
| 44XX | Molybdenum, 0.53%. One composition 4419. |
| 46XX | Nickel to 1.83%, molybdenum to 0.25%. Four compositions from 4615 to 4626. |
| 47XX | Nickel, 1.05%, chromium, 0.45%, molybdenum to 0.35%. Two compositions, 4718 and 4720. |
| 48XX | Nickel, 3.50%, molybdenum, 0.25%. Three compositions from 4815 to 4820. |
| 50XX | Chromium, 0.40%. One composition, 5015. |
| 51XX | Chromium to 1.00%. Ten compositions from 5120 to 5160. |
| 5XXXX | Carbon, 1.04%, chromium to 1.45%. Two compositions, 51100 and 52100. |
| 61XX | Chromium to 0.95%, vanadium to 0.15% min. Two compositions, 6118 and 6150. |

**Table 3.1 (Continued)**
**AISI Standard Steels**
(XXs indicate nominal carbon content within range)

| AISI Series | Nominal Composition or Range |
|---|---|
| **Alloy Steels** | |
| 86XX | Nickel, 0.55%, chromium, 0.50%, molybdenum, 0.20%. Twelve compositions from 8615 to 8655. |
| 87XX | Nickel, 0.55%, chromium, 0.50%, molybdenum, 0.25%. Two compositions, 8720 and 8740. |
| 88XX | Nickel, 0.55%, chromium, 0.50%, molybdenum, 0.35%. One composition 8822. |
| 92XX | Silicon, 2.00%. Two compositions, 9255 and 9260. |
| 50BXX | Chromium to 0.50%, also containing boron. Four compositions from 50B44 to 50B60. |
| 51BXX | Chromium to 0.80%, also containing boron. One composition, 51B60. |
| 81BXX | Nickel, 0.30%, chromium, 0.45%, molybdenum, 0.12%, also containing boron. One composition, 81B45. |
| 94BXX | Nickel, 0.45%, chromium, 0.40%, molybdenum, 0.12%, also containing boron. Two compositions, 94B17 and 94B30. |

Additional notes: When a carbon or alloy steel also contains the letter "L" in the code, it contains from 0.15 to 0.35% lead as a free-machining additive, i.e., 12L14 or 41L40. The prefix "E" before an alloy steel, such as E4340, indicates the steel is made only by electric furnace. The suffix "H" indicates an alloy steel made to more restrictive chemical composition than that of standard steels and produced to a measured and known hardenability requirement (i.e.) 8630 H or 94B30H.

**Table 3.2**
**Steel Classification as to Carbon Content**

| Designation | Meaning |
|---|---|
| Low | 0.05 to approximately 0.30 percent carbon |
| Medium | 0.30 to approximately 0.60 percent carbon |
| High | 0.60 to approximately 0.95 percent carbon |

## FABRICATING CHARACTERISTICS OF STEEL

**Machinability.** The machinability of low-carbon steels is approximately 55 to 60 percent that of B1112. These steels are soft and draggy, generating considerable heat. They have a tendency to build up on the cutting edge, which also makes for inefficient cutting.

The medium-carbon steels cut better, even though the cutting pressures are higher. Both the hot- and cold-rolled steels machine better than the annealed. The machinability ranges between 65 and 70 percent.

High-carbon steels are less machinable than the medium-carbon steels, being at the other extreme—too hard. However, where fine finish and dimensional accuracy are needed, the high-carbon steels are used.

The machining qualities of plain low-carbon steels can be improved by the addition of small amounts of elements such as sulfur and phosphorus. These resulfurized and rephosphorized steels facilitate machining by permitting faster cutting speeds, longer tool life, and better surface finish. Manganese content of 1.00 to 1.90 percent also produces a free-cutting steel in the lower-carbon grades.

Since World War II, a number of grades of lead-bearing steels have taken their place among the free-machining steels. These leaded steels are actually sulfur, phosphorus, or manganese free-cutting steels that have 0.15 to 0.35 percent lead added to improve machinability. The addition of lead does not affect the basic mechanical properties of the steel.

**Formability.** Low-carbon steels have good forming qualities, as there is less carbon to interfere with the planes of slip. Steels in this class can be given a class-2 bend, which is a 90-deg bend with a minimum radius of *t* or the thickness of the metal. Classes of bends have been defined by the American Society for the Testing of Metals (ASTM).

Medium-carbon cold-rolled steels are too low in ductility for any practical degree of cold forming.

The hot-rolled medium-carbon steels are more ductile and can be bent with a 1-*t* radius up to 0.0090 in. thick.

**Weldability.** Plain carbon steel is the most weldable of all metals. It is only as the carbon percentages increase that there is a tendency of the metal to harden and crack. Fortunately, 90 percent of the welding is done on low-carbon (0.15 percent) steels. This amount of carbon presents no particular difficulties in welding. Near the upper end of the low-carbon range (0.27 to 0.30 percent), there may be some formation of martensite when extremely rapid cooling is used.

Medium- and high-carbon steels will harden when welded if allowed to cool at speeds in excess of the critical cooling rates. Preheating to 500° or 600°F and postheating between 1100° and 1200°F will remove any of the brittle microstructures.

The extra-high-carbon steels, or tool steels, having a carbon range of 1.00 to 1.70 percent, are not recommended for high-temperature welding applications. These metals are usually joined by brazing with a low-temperature silver alloy. Because of the lower temperature of this process, it is possible to repair or fabricate tool-steel parts without affecting their heat-treated condition.

## TOOL STEELS

Tool steels are so named because they have the properties needed in making tools for fixtures; for cutting, shaping, forming, and blanking of materials, either hot or cold; and for precision gages.

Tool-steel manufacturers have their own brand names for their products. Since these vary with each manufacturer, to avoid confusion for the consumer, the Joint Industry Conference (JIC) has set up some simple but effective means of identifying tool steels by classes. The system has received wide acceptance.

The JIC symbols are based partly on chemistry, partly on the cooling system required for hardening, and partly on physical characteristics. Some of the symbols adopted are shown in Table 3.3.

**Table 3.3**
**Joint Industry Conference Symbols for Tool Steels**

| Symbol | Meaning |
|---|---|
| T | Tungsten high speed |
| M | Molybdenum high speed |
| D | High carbon, high chromium |
| A | Air hardening |
| O | Oil hardening |
| H | Hot work |
| S | Shock resisting |
| W | Water hardening |

These easily understood symbols are derived from long-standing trade expressions and designations. Numbers following the letters are also used to designate variations of each type. For example, W8 and W12 are water-hardening steels with 0.80 and 1.20 percent carbon, respectively. The AISI and SAE codes are quite similar to the JIC code.

## STEEL CASTINGS

**Classification.** The American Society for Testing Metals (ASTM) has formulated the code for carbon and low-alloy castings shown in Table 3.4.

In addition to the listed code numbers, class numbers are given. Class numbers are used to indicate the tensile-strength and yield-point stresses. A class 80–50 cast steel, for example, would be one with a tensile strength of 80,000 psi and a yield point of 50,000 psi.

**Fabricating Characteristics.** The mechanical properties are quite similar in cast steels and wrought steels that have the same composition and heat treatment. However, there are a few differences. One differ-

**Table 3.4**
**ASTM Code for Carbon and Low-Alloy Castings**

| Designation | Meaning |
|---|---|
| 27-55 | Medium-strength carbon-steel castings for general application |
| 148-55 | High-strength steel castings for structural purposes |
| 216-53T | Carbon-steel castings suitable for fusion welding for high-temperature service (tentative) |
| 217-55 | Alloy-steel castings for pressure-containing parts suitable for high-temperature service |
| 351-52T | Ferritic-steel castings for pressure-containing parts suitable for low-temperature service |
| 356-58 | Heavy-walled carbon and low-alloy-steel castings for steam turbines (tentative) |
| 389-59T | Alloy-steel castings specially heat-treated for pressure-containing parts suitable for high-temperature service |

ence is that wrought products are more ductile in the direction of working and less so in the transverse direction, whereas castings have uniform properties in all directions. The other difference is that there is likely to be more variation in castings than in wrought products, since the equipment, pattern, and even the steel analysis vary from foundry to foundry.

Steel castings stay below the cast-iron range in carbon content. About 60 percent of all steel castings are in the medium-carbon range. They are used extensively in railroads, machine tools, rolling mills, construction equipment, and other applications as shown in Fig. 3.1.

**Fig. 3.1.** Steel castings can be used to fabricate a wide variety of equipment, from small gears for a gearbox to large machine frames.

## CAST IRON

**Main Types.** In order to understand the fabricating characteristics of cast iron, it is necessary to become familiar with each of the main types and how they are obtained.

Cast iron is a rather complex mixture containing 91 to 94 percent metal iron and varying proportions of other elements, the most important of which are carbon, silicon, manganese, sulfur, and phosphorus. By varying the amount of these elements and the heat treatment, the foundryman is able to produce a multitude of different irons, each with properties adapted to different uses.

**Gray Iron.** Gray iron and cast iron are often considered to be synonymous terms. The name is derived from the grayish appearance of its fracture, the color being caused by the large amount of visible graphite or carbon. The composition is largely pearlite and flake graphite (Fig. 3.2). The coarse pearlite and coarse graphite are usually associated with

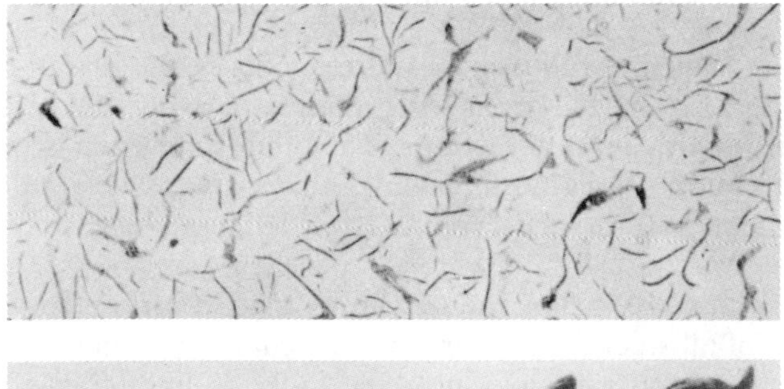

**Fig. 3.2.** The microstructure of gray cast iron and space models of flake graphite. Courtesy The International Nickel Co., Inc.

slow cooling of an iron that is not highly alloyed. Fine graphite and pearlite are usually the result of more rapid cooling, low silicon content, and such alloying elements as chromium and vanadium.

**Malleable Iron.** Malleable iron is made from white cast iron, which is a cast iron with virtually all of the carbon in the form of cementite (iron carbide, $Fe_3C$). It is called white iron because of the appearance of the fracture, which is brilliant white, caused by reflection from the facets of the iron carbide. White cast iron is made in the same way as gray iron, except that the graphitizing elements are modified and the cooling rate is considerably increased. It is the rapid cooling rate, often referred to as *chilling*, that produces the iron carbide structure. White cast iron has a hardness between 400 and 550 Brinell and is unmachinable. It is used in wear and abrasion applications such as liners on rock and ore crushers, rolling-mill rolls, plow points, and a variety of road machinery.

When white cast iron is properly annealed, it becomes malleable. The annealing is usually done at two temperatures, the first stage being at 1700°F. It is at this temperature that the coarse primary carbides become graphitized to flake nodules (Fig. 3.3). Then, as the temperature is

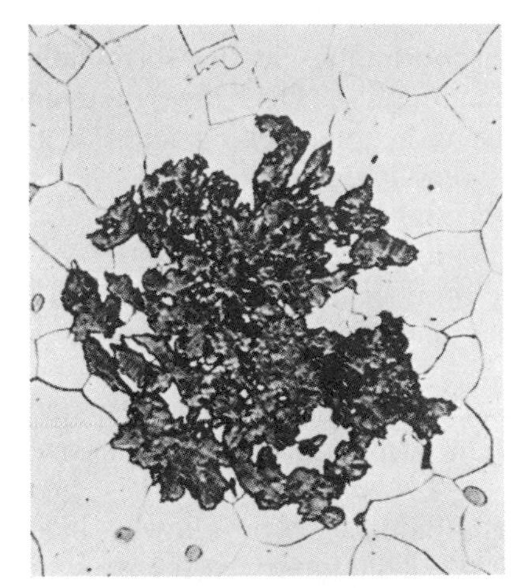

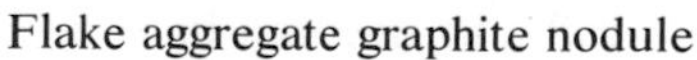

Flake aggregate graphite nodule

Spheroidal graphite nodule

**Fig. 3.3.** Photomicrographs showing the development of the graphite nodule in malleable iron. Courtesy The International Nickel Co., Inc.

lowered to about 1325°F, the iron carbide laminations of pearlite that have separated out are graphitized, leaving the final structure ferrite and graphite. The changes in the carbon are shown in Fig. 3.3.

The usual time needed for the two-stage annealing process is anywhere from 30 hr in modern furnaces to 150 hr in older batch-type furnaces.

**Ductile Iron.** For generations, foundrymen and metallurgists have searched for some way to transform brittle cast iron into a tough, strong material. Finally, it was discovered that magnesium could greatly change the properties of cast iron. A small amount of magnesium, about 1 lb/ton of iron, causes the flake graphite to take on a spheroidal shape (Fig. 3.4). Graphite in a spheroidal shape presents the minimum surface for a given

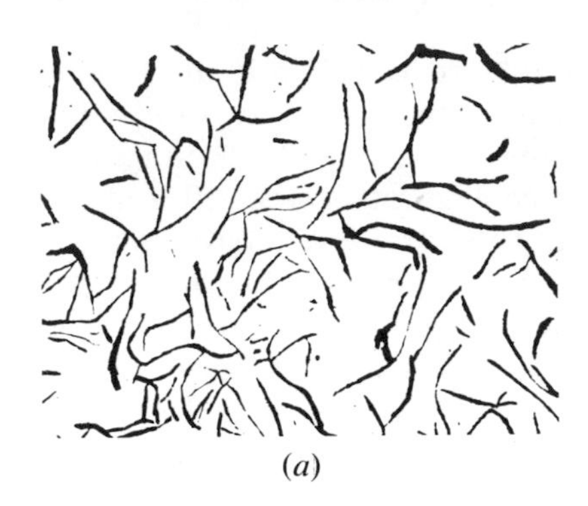

(*a*)

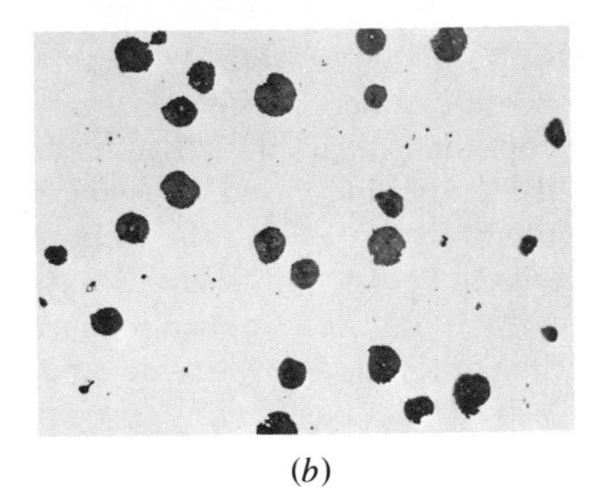

(*b*)

**Fig. 3.4.** Comparison of the microstructure of (*a*) gray cast iron and (*b*) ductile iron. Courtesy The International Nickel Co., Inc.

volume. Therefore, there are fewer discontinuities in the surrounding metal, giving it far more strength and ductility. Thus the processing advantages of steel, including high strength, toughness, ductility, and wear resistance, can be obtained in this readily cast material.

Ductile iron can be heat-treated in a manner similar to steel. Gray irons can too, but the risk of cracking is much greater.

There has been a rapid growth in the number of foundries licensed to produce this relatively new material that also goes by other names, such as nodular iron and spheroidal iron.

**Inoculated Irons.** Irons can be inoculated, while in the molten state, with certain materials that change their structure without materially changing their composition.

One of the early inoculants was ferrosilicon used on a lower-silicon iron. Addition of the missing silicon in the ladle improved the structure and reduced the tendency of the iron to chill and become hard around the edges. Other inoculants used with silicon are calcium, ferromanganese, and zirconium.

A familiar inoculated iron is Meehanite, which is produced by licensed foundries. It is a specially made white-cast-iron composition which is graphitized in the ladle with calcium silicide. It is produced in a number of grades, including a group for high-temperature service.

## CAST IRON CLASSIFICATION

**Gray Iron.** The most used classification for gray cast iron is that of the American Society for Testing Materials adopted in 1936. These specifications cover seven classes that are numbered in increments of 5, from 20 to 60. The class number refers to its tensile strength; for example, a class 40 cast iron refers to one having a minimum tensile strength of 40,000 psi. The carbon content becomes less as the strength requirement goes up. Alloys in increasing quantities are used to achieve strengths higher than 45,000 psi. Although high-strength cast irons have been developed, they are not widely used because they are still quite brittle.

**Malleable Iron.** About 90 percent of the malleable iron made is classified as 32510, which means that it has a yield point of 32,500 psi and an elongation of 10 percent. Grade 35018 has somewhat lower carbon content, but it has somewhat higher strength and better ductility. Grades 32510 and 35018 are known as standard types of malleable iron. There is an increasing demand for a newer type termed *pearlitic malleable*. Instead of having a ferrite matrix, it is pearlitic, which is stronger and harder than ferrite. These irons are obtained by interrupting the second-stage annealing process or by introducing larger quantities of manganese, which prevents graphitization of the pearlite, or by air cooling followed by tempering.

Pearlitic malleables have strengths up to 90,000 psi, but ductility is low. They are used where higher strengths and wear resistance are needed, along with less resistance to shock.

**Ductile Iron.** The general classification numbers for ductile irons are similar to those for gray cast iron, but a third figure is added which represents the minimum percent of elongation that can be expected. For example, an 80–60–03 has 80,000 psi minimum tensile strength, 60,000 psi minimum yield strength, with 3 percent minimum elongation. There are four regular types of ductile irons: 60–45–10, 80–60–03, 100–70–03, and 120–90–02. In addition, there are two special types, one of which is for heat resistance; the other, Ni-Resist (registered trademark of the International Nickel Company, Inc.), has both corrosion and heat resistance.

## FABRICATING CHARACTERISTICS OF CAST IRON

**Machinability.** GRAY IRON. The usual range of carbon is between 2.50 and 3.75 percent. The graphite flakes act to cause discontinuities in the ferrite. This helps the chips to break up easily. The graphite also furnishes lubricating qualities to the cutting action of the tool. Although gray cast iron is considered quite machinable, it varies considerably because of the microstructure. The ratings may range from 50 to 125 percent that of B1112.

One of the easiest checks used to determine the machinability of a given piece of material is the Brinell test. The normal hardness range of 130 to 230 Brinell will present no machining difficulties. Beyond this range, however, even a few points will have a large effect on the cutting speed that should be used. Hardnesses above 230 Brinell indicate that the structure contains free carbides, which greatly reduce tool life.

Tungsten carbide tools are usually used to machine cast irons because of the abrasive qualities of the material.

Machinability of gray cast irons can be improved through annealing. When this is done, machinability excels that of any other ferrous material.

MALLEABLE IRON. Malleable iron has a machinability rating of 120 percent and is considered one of the most readily machined ferrous metals. The reason for this good machinability is its uniform structure and the nodular form of the tempered carbon. The pearlitic malleables, which have a different annealing that leaves some of the carbon content in the form of combined carbides, have machinability ratings of 80 to 90 percent. (The term pearlitic is just a convenient term and does not mean that the microstructure is necessarily in pearlite form.)

DUCTILE IRON. The machinability of ductile iron is similar to that of gray cast iron for equivalent hardnesses and better than that of steel for equivalent strengths. Type 60-45-10 has both maximum machinability and toughness. Although the cutting action is good, the power factor is higher owing to the toughness of the material.

MACHINABILITY COMPARISON. By way of quick review, for rough approximations machinability ratings are given in Table 3.5.

**Table 3.5**
**Machinability Ratings for Types of Iron**

| Type of Iron | Machinability Index |
|---|---|
| Malleable (standard) | 110-120 |
| Gray iron (flake graphite) | 110 |
| Ductile iron | 90-110 |
| Meehanite | 76 |
| Gray iron (pearlitic) | 68 |

**Weldability.** GRAY IRON. Gray cast iron can be welded with the oxyacetylene torch or with the electric arc. However, owing to its low ductility, special precautions should be taken to avoid cracking in the weld area. Preheating is done, in the case of oxyacetylene welding, to make expansion and contraction more uniform. The opposite effect is used in arc welding, where short welds are made in order to keep the heat at a minimum. Slow cooling is essential to permit the carbon to separate out in the form of graphite flakes. Failure to do this will result in mottled or chilled areas.

Generally, the welding of gray cast iron is limited to repair work rather than fabrication. However, braze welding with bronze or nickel copper is frequently used in fabrication.

MALLEABLE IRONS. Malleable irons are not considered weldable, that is, in the same sense that gray cast iron is weldable. The reason for this is that the heat necessary to melt the edges of the break will completely destroy the malleable properties. Owing to the long-term anneal required to produce these properties originally, a simple annealing process would not restore them. There are times, however, where stresses are low or are in compression only; in this case, successful arc welding can be accomplished. A commercially pure nickel rod or a 10 percent aluminum bronze rod is used.

Since brazing can be done at a temperature of 1700°F or less, it is a preferred method of repair. Silver brazing is also used, since this procedure can be accomplished with even less heat.

DUCTILE IRONS. Ductile iron, being a high-carbon-content material, should be given special consideration, just as in welding gray cast iron. It can be welded by the carbon-arc process and most other fusion processes, either to itself or to other metals such as carbon steel, stainless steel, and nickel. The most easily welded types are 60 – 45 – 10 and the high-alloy variety. High-quality welds are made with cored electrodes having 60 percent nickel and 40 percent iron. Noncritical applications may be welded with ordinary steel electrodes.

This material can also be brazed with silver or copper brazing alloys. Crack-free overlays can be made with commercial hard-surfacing rods. This will provide special abrasion and corrosion resistance.

**Castability.** GRAY IRON. From the iron-carbon diagram (Fig. 2.10), it can be seen that carbon markedly lowers the melting point of iron. Many benefits accrue from this fact. The pouring temperature, for example, can be several hundred degrees higher than the melting point, making for excellent fluidity. This makes it possible to produce thin-section castings over a large area. Shrinkage is considerably less than that of steel because the rejection of graphite causes expansion during solidification. The amount of solidification shrinkage differs with the various casting alloys. Class 20 and 25 irons contain sufficient graphite so that there is virtually no shrinkage at all. The higher classes have one half to two thirds the solidification shrinkage of any ferrous alloys.

MALLEABLE IRONS. The molten white irons from which malleable irons are made have high fluidity, allowing complicated shapes to be cast. Good surface finish and close dimensional tolerances are possible. Since malleable irons require a prolonged heat treatment, internal stresses, that may have occurred as the casting solidified and cooled, are removed.

DUCTILE IRONS. Ductile iron is similar to gray cast iron when it comes to casting qualities. It has a low melting point and good fluidity in the molten state. Therefore, it can be poured in intricate shapes and thin sectional parts. The metal will flow several inches in sections as small as 1/16 in. in greensand molds at normal foundry pouring temperatures. Many parts are now being cast from ductile iron that formerly could not be cast, because gray iron did not possess adequate properties and steel could not be cast in such intricate shapes. Some of the advantages of ductile-steel castings are shown in the before-and-after design of a steel wheel (Fig. 3.5).

CASTABILITY COMPARISON. Some of the properties of the cast materials discussed are shown graphically in Fig. 3.6. You will note that,

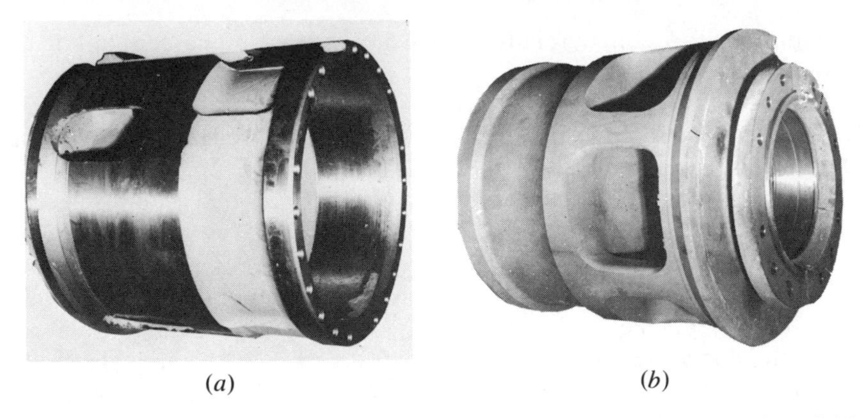

(a) (b)

**Fig. 3.5.** Some of the advantages of ductile iron are shown in the redesign of this heavy-duty wheel from (a) fabrication to (b) casting. Courtesy The International Nickel Co., Inc.

in general, yield strength, tensile strength, elongation, and modulus go progressively up from gray cast iron to malleable iron to cast steel to ductile iron, the exception being higher elongation and modulus for cast steels.

A summary of cast ferrous metals is given in Table 3.6. Important mechanical properties and fabricating qualities can easily be compared.

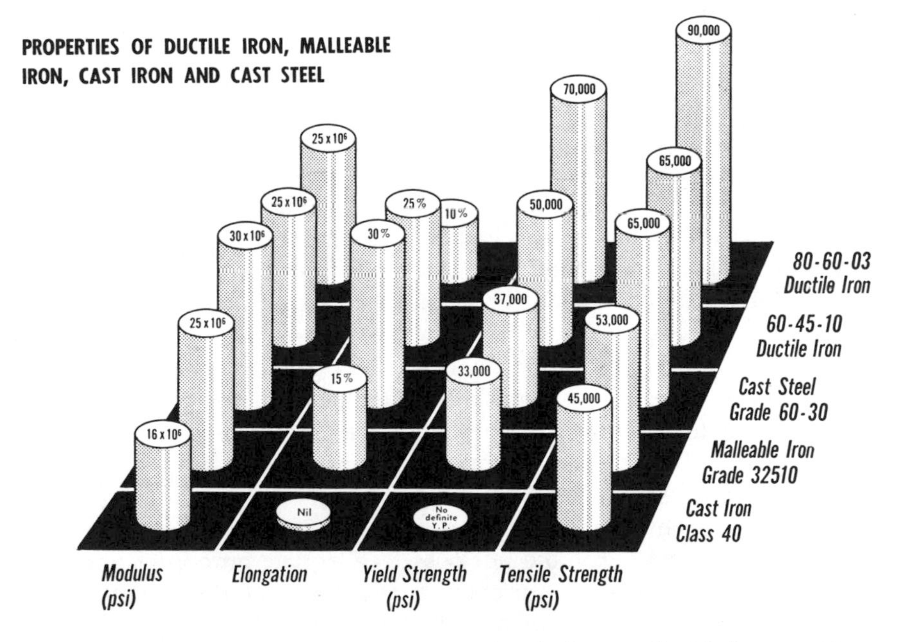


**Fig. 3.6.** Comparison of the properties of ductile iron, malleable iron, cast iron, and cast steel. Courtesy International Nickel Company, Inc.

**Table 3.6**
**General Summary of Cast Ferrous Metals**
(typical ranges)
Courtesy of *Machine Design*

| Metal | Tensile strength 1000 psi | Elongation percent | Hardness, Bhn | Modulus of elasticity ($10^6$ psi) | Maximum service Temp* (F) | Abrasive wear resistance | Damping capacity | Corrosion resistance | Weld-ability | Machin-ability | Cast-ability |
|---|---|---|---|---|---|---|---|---|---|---|---|
| Gray Iron | 20 to 60 | 0 to 1 | 140 to 300 | 10 to 19 | 700 | P to F | E | P to F | P | G | E |
| White Iron | 20 to 50 | 0 | 300 to 575 | 12 to 16 | 1100 | E | P | F | VP | VP | G |
| High-Alloy Irons | | | | | | | | | | | |
| Wear resistant | 23 to 90 | 0 to 0.5 | 250 to 700 | 10 to 17 | 1200 | E | P | F | VP | VP | G |
| Heat resistant | 20 to 90 | 0 to 1 | 100 to 500 | 11 to 19 | 1800 | P to G | P to G | F to E | P | F to P | G |
| Corrosion resistant | 13 to 90 | 0 to 1 | 100 to 500 | 10 to 19 | 1500 | P to G | P to G | G to E | P | F to P | G |
| Ductile Irons | 65 to 150 | 1 to 26 | 140 to 330 | 23 to 24 | 750 | P to F | F | P to F | P | E to F | E to G |
| High-Alloy Ductile Iron | | | | | | | | | | | |
| Austenitic | 54 to 69 | 8 to 40 | 130 to 180 | 23 to 25 | 1300 | P | F | G to E | P | F | G |
| Malleable Irons | | | | | | | | | | | |
| Standard | 50 to 56 | 10 to 25 | 110 to 156 | 25 | 750 | P | F | P to F | P | E | G |
| Pearlitic | 70 to 120 | 2 to 20 | 150 to 300 | 23 to 25 | 750 | F | F to P | P to F | P | G to F | G |
| Carbon Steels | | | | | | | | | | | |
| Low carbon | 50 to 75 | 27 to 38 | 108 to 150 | 29 to 30 | 900 | P | P | P | E | F | F |
| Medium carbon | 70 to 120 | 12 to 32 | 140 to 240 | 29 to 30 | 900 | F | P | P | G | F | F |
| High carbon | 100 to 200 | 2 to 23 | 200 to 400 | 29 to 30 | 900 | G | P | P | G | F to P | F |
| Low-Alloy Steels | 70 to 250 | 2 to 35 | 105 to 500 | 29 to 30 | 1000 | F to G | P | P to F | G to F | F | F |
| High-Alloy Steels | | | | | | | | | | | |
| Heat resistant | 65 to 200 | 2 to 35 | 130 to 400 | 27 to 29 | 2150 | P to G | P | G to E | G to F | F to P | F |
| Corrosion resistant | 70 to 200 | 8 to 60 | 140 to 400 | 27 to 29 | 1800 | P to G | P | G to E | G to F | F to P | F |
| Wear resistant | | | | | | | | | | | |
| Austenitic Mn | 80 to 200 | 15 to 55 | 180 to 550 | 27 | 500 | E | P | F | F | P | F |

E = excellent; G = good; F = fair; P = poor; VP = very poor. *In load-bearing applications.

## STAINLESS STEELS

**Classification.** The three main types of stainless steel are *austenitic, ferritic*, and *martensitic*. All of these groups have steels that contain at least 10.5 percent chromium. It is from this element that the excellent corrosion resistance is obtained. Current theory holds that a thin, transparent, and very tough film forms on the chrome surface. It is inert or passive and does not react with many other corrosive materials.

Stainless steels are often grouped according to number. One such system is that of The American Iron and Steel Institute which has the 200, 300, and 400 series. Another system specifies the steel by its chromium or nickel content, such as 18–8, meaning 18 percent chromium and 8 percent nickel. The SAE numbering system, in most cases, merely adds two digits to the front of the AISI numbers. For example, AISI stainless steel number 302 is classified as 30302.

A brief description of each of the three main types of stainless steels follows.

**Austenitic.** Austenitic stainless steels are known as the 300 series. The basic alloy in this series is type 302. It is the 18–8 stainless steel. It is called austenitic because the metal has an austenitic structure at room temperature. It has the highest corrosion resistance of all the stainless alloys. It also claims the greatest strength and scale resistance at elevated temperatures. It also retains ductility at temperatures approaching absolute zero.

The 300 series is generally selected for applications where strength, ductility, and good corrosion resistance are essential.

**Ferritic.** As the name implies, ferritic stainless has a low carbon-to-chromium ratio. This eliminates the effects of thermal transformation and prevents hardening by heat treatment. These steels are magnetic and have good ductility. They do not work harden to any appreciable degree.

Type 430 is the general-purpose ferritic. It is frequently used as an alternate for 302. It can be easily buffed to a mirror finish and, for this reason, is used extensively for trim molding and other decorative applications.

**Martensitic.** Because of the higher carbon-to-chromium ratio, martensitic steels are the only types hardenable by heat treatment. Hardness, ductility, and ability to hold an edge are characteristics of these alloys. These steels are magnetic in all conditions and possess the best thermal conductivity of the stainless types.

Type 410 is the most widely used steel in this group. It is tough and has good resistance to impact, and it attains a tensile strength of 300,000 psi when hardened.

## FABRICATING CHARACTERISTICS OF STAINLESS STEELS

**Machinability.** Ferritic stainless steels are the easiest to machine. The annealed martensitic are somewhat more difficult and the austenitic, because of their gumminess and work hardening qualities, are the most difficult.

Generally stainless steels require reduced cutting speeds (approximately one-half that used for carbon steels), finer feeds, and lighter cuts.

Rigid setups with sharp carbide tooling are recommended. Any rubbing action will work harden and glaze the surface.

Each class of stainless has a free machining type designated by the addition of sulphur or selenium as, for example, the austenitic grade 303 Se.

**Formability.** The low yield strength and high ductility which are characteristic of stainless steels permit the successful forming of complex shapes. However, stainless steels possess relatively high tensile strength, and they are subject to greater strain hardening than ductile materials are; therefore, forming pressures must be greater.

In selecting a stainless alloy for forming, the minimum tensile strength or the maximum ductility (elongation) should not be the main consideration. The alloy, rather, should be chosen by *the rate at which it work hardens*. For example, parts to be rolled out, such as angles and channels, should have a *high* rate of work hardening, as they are expected to have strength and stiffness when completed. On the other hand, items that are to be deep drawn must have a *low* rate of work hardening. The ferritic steels do not work harden appreciably, and they possess good ductility, making them the most suitable for forming.

Stainless steels can be obtained in various cold-rolled tempers, designated as 1/4, 1/2, 3/4, and full hard.

**Weldability.** When stainless steel is welded, it is helpful to compare it to plain carbon steel for important physical properties, as in Table 3.7.

**Table 3.7**
**Physical Properties of Stainless Steel and Carbon Steel**

| Property | Austenitic Stainless | Carbon Steel |
|---|---|---|
| Melting temperature (°F) | 2550–2590 | 2750–2775 |
| Thermal conductivity | 9.4 | 27 |
| Coefficient of thermal expansion | $9.6 \times 10^{-6}$ | $8.4 \times 10^{-6}$ |

The expansion and contraction of metals must be considered in all fabricating processes, but especially so in welding. The coefficients of expansion are given as a means of comparing one metal with another. The coefficient $8.4 \times 10^{-6}$ means that a piece of steel will expand 0.0000084 in. in length when its temperature is raised 1°F.

A number is also assigned to the coefficient of thermal conductivity. The number 27 for plain carbon steel means that it absorbs heat at 27 Btu/hr/sq ft/°F/ft. The coefficient of thermal conductivity is used in welding to estimate the amount of heat that will be required, based on the type of metal and thickness, and to determine what area will be heat affected. Metals with low thermal conductivity, such as the stainless steel listed, tend to localize the stress. The metal that is hot wants to expand but is hemmed in by the cold adjacent metal. When the metal cools and contracts, the cooler metal does not move, with the result that cracks form during solidification. Methods of dealing with these problems are described in the chapters on welding.

Although stainless steels are considered weldable by conventional processes, the best results are obtained when the molten metal is well protected from the atmosphere during welding.

## ALUMINUM

**Classification.** Aluminum is used extensively in both the wrought and cast forms. The wrought classification uses a four-digit system and the cast three digits, except for alloy 43.

**Wrought Classification.** The four-digit system put into effect in October 1954 by the Aluminum Association is as follows:

In Table 3.8, the first digit designates the alloy type. The second digit indicates alloy modification. The last two digits indicate the aluminum purity.

For alloys in use before this classification was devised, the last two

**Table 3.8**
**Sequence of Aluminum-Alloy Types in the Four-Digit System**

| Type of aluminum alloy | Number group |
|---|---|
| Aluminum—99.00% purity or greater | 1XXX |
| Copper | 2XXX |
| Manganese | 3XXX |
| Silicon | 4XXX |
| Magnesium | 5XXX |
| Magnesium and silicon | 6XXX |
| Zinc | 7XXX |
| Other element | 8XXX |
| Unused series | 9XXX |

numbers are the same as the old numbers; for example, 24S becomes 2024.

**Temper Designations.** Some aluminum alloys are strengthened by cold working, others by heat treatment. A temper designation of letters and numbers following the four digits and separated by a hyphen is used to indicate the condition or temper of the aluminum.

## FABRICATING CHARACTERISTICS OF ALUMINUM

**Machinability.** Pure grades of aluminum are too soft for good machinability. Work or strain hardening helps overcome some of the gumminess. The most easily machined alloy is 2011–T3. This metal is often referred to as the free-cutting aluminum alloy. It is used extensively for a broad range of screw-machine products. Small amounts of lead and bismuth help produce a better finish and easy-to-handle chips.

The softer alloys sometimes build up on the cutting edge of the tool. However, proper cutting oils and a highly polished tool surface will tend to minimize this.

Two important physical properties that should be kept in mind when machining aluminum are modulus of elasticity and thermal expansion. Table 3.9 gives a comparison between aluminum and mild steel.

**Table 3.9**
**Machining Comparison between Aluminum and Mild Steel**

| Property | Aluminum 2017 | Mild Steel |
|---|---|---|
| Modulus of elasticity (psi) | $10.5 \times 10^{6}$ | $30 \times 10^{6}$ |
| Thermal expansion | $13.1 \times 10^{-6}$ | $8.4 \times 10^{-6}$ |

Owing to the much lower modulus of elasticity (about one third that of steel), there will be much greater deflection under load. Care should be taken, in making heavy cuts and in clamping the work, to avoid distortion.

When dimensional accuracy is necessary, thermal expansion of the metal is an important consideration. Since the thermal expansion of aluminum is almost twice that of steel, care should be exercised in keeping heat down. Overheating can be reduced to a minimum by using sharp, well-designed tools, coolants, and feeds that are not excessive. It may be necessary, on larger parts, to make a thermal allowance in measuring.

**Formability.** Most aluminum alloys can be readily formed cold. However, the wide range of alloys and tempers makes for considerable variation in the amount of forming possible.

Generally, depending upon the alloy and temper used, aluminum requires less energy and horsepower than does the forming of heavier metals. Consequently, machine life is increased, power costs are lower, and maintenance is minimized.

**Weldability.** PHYSICAL PROPERTIES. The physical properties that must be considered in welding aluminum are given in Table 3.10.

**Table 3.10**
**Important Physical Properties for Welding Aluminum**

| Property | Aluminum | Mild steel |
|---|---|---|
| Melting point (°F) | Pure 1195 –1215 | 2750 –2775 |
| | Alloy 995 –1190 | |
| Thermal conductivity | Pure 135 | 27 |
| | Alloy 109.2 | |
| Thermal expansion | $13.1 \times 10^{-6}$ | $8.4 \times 10^{-6}$ |

By comparing some of the physical properties of aluminum and steel, it can be seen that the melting point of the aluminum is much lower and the thermal conductivity is muct higher. Therefore, more heat will be required to weld a corresponding thickness of aluminum.

The thermal expansion of aluminum is almost twice that of steel. This, coupled with the fact that aluminum weld metal shrinks about 6 percent in volume upon solidification, may put the weld in tension. Some restraint is often necessary to prevent distortion during the welding operation. However, excessive restraint on the component sections during cooling of the weld may result in weld cracking.

The speed at which the weld is made will be one of the determining factors in preventing distortion. A slow rate will cause greater area heating with more expansion and subsequent contraction. For this reason, arc- and resistance-welding methods that make use of highly concentrated heat are the most used.

PREHEATING. Preheating is necessary when welding heavy sections; otherwise the mass of the parent metal will conduct the heat away so fast that the welding arc will not be able to supply heat fast enough for adequate fusion. Preheating helps provide satisfactory fusion, reduces distortion or cracking in the weld, and increases welding speed.

Whenever aluminum alloys are heated above 900°F, they lose their heat-treated characteristics or tempers and revert to the annealed condition. Some annealing of the parent metal takes place when a weld is made. With the heat-treatable alloys there is also some loss of ductility in the joint. The desired temper can be restored in the fabricated part by heat treatment after welding.

OXIDE FILM. Aluminum and its alloys rapidly develop an oxide film when exposed to the air. This oxide has a melting point that is in excess of 3600°F, or about 2400°F above the melting point of pure aluminum. Therefore, the aluminum melts before the oxide coating does. When this happens there can be no fusion between the filler metal and the base metal. It is then necessary to remove the oxide film by suitable chemical cleaner, flux, mechanical abrasion, or action of the welding arc. Particles of oxide trapped in the weld will impair its ductility. Thorough brushing with a stainless-steel brush immediately before welding is a recommended procedure. Any grease or oil films remaining in the weld area will cause unsound welds.

GAS ABSORPTION. Molten aluminum readily absorbs available hydrogen from the air. When the weld pool freezes, most of the hydrogen is released because it is practically insoluble in solid aluminum. The released hydrogen may cause porosity in the weld which, in turn, may impair its strength and ductility. Porosity and hydrogen embrittlement can be reduced by having clean surfaces and moisture-free filler rods and by covering the weld area with an inert gas such as argon. Both TIG and MIG welding are common choices in welding aluminum. These methods provide a protective shield from the atmosphere, and there is no cleaning problem when finished.

**Castability.** The physical properties of aluminum that make it one of the most versatile of all foundry metals are low melting point, low specific gravity, and low density.

LOW MELTING POINT. The low melting point makes for minimum of sand burnout; that is, the sand can be used over and over again, only enough being added to take care of mechanical losses. Where metal or permanent molds are used, the low melting temperature of the aluminum promotes long life.

LOW SPECIFIC GRAVITY. The light weight of aluminum means that even moderate-sized castings can be efficiently handled manually. Only the largest castings require mechanically transported ladles or crucibles.

LOW DENSITY. Low density and low melting point combine to practically eliminate most problems of sand washes that occur when heavier metals are poured. Also, the lower pressures permit more lightly rammed molds with lighter molding equipment.

Special consideration must be given to casting shrinkage, hot shortness, and gas absorption when aluminum castings are made.

## COPPER AND COPPER-BASE ALLOYS

**Classification—Coppers.** Unlike other common metals, the various types of copper are better known by name than by code number.

Commercially pure copper is available in several grades, all of which have essentially the same mechanical properties: *Electrolytic tough-pitch copper* is susceptible to embrittlement when heated in a reducing atmosphere, but it has high electrical conductivity. *Deoxidized copper* has lower electrical conductivity but improved cold-working characteristics, and it is not subject to embrittlement. It has better welding and brazing characteristics than do other grades of copper. *Oxygen-free copper* has the same electrical conductivity as the tough-pitch copper and is not prone to embrittlement when heated in a reducing atmosphere.

Modified coppers include *tellurium copper,* which contains 0.5 percent tellurium for free-cutting characteristics (selenium or lead is also used for this purpose) and *tellurium-nickel copper,* an age-hardenable alloy that provides high strength, high conductivity, and excellent machinability.

**Classification—Copper-Base Alloys.** BRASSES. Brasses are principally alloys of copper and zinc. They are often referred to by the percentage of each, for example 70–30, meaning 70 percent copper and 30 percent zinc. Small amounts of other elements such as lead, tin, or aluminum are added to obtain the desired color, strength, ductility, machinability, corrosion resistance, or a combination of these properties. The main classifications of brasses are given in Table 3.11.

**Table 3.11**
**Classifications of Brasses**

| Alloy | Meaning |
|---|---|
| Alpha brasses | Contain less than 36% zinc and are single-phase alloys |
| Alpha-beta brasses | Contain more than 36% zinc and have a two-phase structure |
| Leaded brasses | Contain up to 88.5% copper and up to 3.25% lead |
| Tin brasses | Known as admiralty and naval brass when the percentage of tin is low, 0.75 to 1.0% |
| Nickel silvers | Brasses containing high percentages of nickel. (The designation 65–18 indicates approximately 65% copper, 18% nickel and the remainder zinc. Nickel is added primarily for its influence on color. When the percentage is high, a silvery-white color is obtained. Nickel also improves mechanical and physical properties.) |

**Machinability.** GENERAL SPEEDS AND FEEDS. For most copper alloys, it is generally good practice to use the highest practical cutting speed with a relatively light feed and moderate depth of cut. An exception is the machining of sand castings with high-speed tools, in which case low speeds and relatively coarse feeds are used to increase tool life.

After the scale has been removed, the higher speeds and lighter feeds can be resumed. Tool life can be considerably increased if the castings are sand blasted, pickled in acid, or tumbled to remove the extremely hard abrasive surface scale.

MACHINABILITY RATINGS. Steel and steel alloys have machinability ratings based on B1112 as 100 percent. Copper and copper-base alloys use either free-cutting brass (consisting of 61.50 percent copper, 35.25 percent zinc, and 3.25 percent lead) as the base material or B1112. The base must be designated.

**Formability.** The most-used alloys for forming are the 70–30 cartridge brasses. However, all brasses from 95–5 to 63–37 are used for pressworking operations. Trouble is encountered when less than 63 percent copper is used, because the beta phase is brittle and may produce fractures or waviness—a surface defect.

**Weldability.** All four of the commonly employed methods—gas, metal arc, carbon arc, and gas-shielded arc—are in general use for welding copper. It is also joined by brazing and a newer method, discussed later, ultrasonic welding. The gas method, because of the variety of atmospheres obtainable, finds wide use. Its disadvantages are cost and, in some cases, warpage with attendant high-conduction losses into the metal adjacent to the weld.

## MAGNESIUM

**Classification.** Magnesium is a strong, lightweight metal that is not used in its pure state for stressed applications but is alloyed with other metals such as aluminum, zinc, zirconium, manganese, thorium, and rare-earth metals to obtain the strength needed for structural uses. Its light weight is one of its best-known characteristics. Aluminum weighs 1 1/2 times more than manganese, iron and steel weigh 4 times more, and copper and nickel alloys weigh 5 times more.

The ASTM classification designates the principal alloying elements as in Table 3.12.

**Table 3.12**
**Alloying Elements Used with Magnesium**

| Symbol | Meaning |
|---|---|
| A | Aluminum |
| Z | Zinc |
| K | Zirconium |
| M | Manganese |
| H | Thorium |
| E | Rare-earth metals |

The numbers following the alloy types represent approximate percentages of each alloy, respectively. For example, the alloy designation AZ61A is an aluminum-zinc alloy with from 5.8 to 7.2 percent aluminum and 0.4 to 1.5 percent zinc. The letters A, B, or C may follow the number, indicating variations in composition or treatment.

Temper designations are similar to those used in aluminum sheet and plate.

Rolled magnesium alloys are supplied with either a chemical-treatment (chrome pickle) finish, an oil finish, or an acetic nitrate pickle. Plates over 1/2 in. thick are furnished with a mill finish and oiled, unless otherwise specified. The oil coating is superior to that given by the chrome pickle for normal shipment and storage.

## FABRICATING CHARACTERISTICS OF MAGNESIUM

**Machinability.** Magnesium and its alloys permit machining at extremely high speeds—usually at the maximum obtainable on modern machine tools. Using free-cutting brass as a base, it has a machinability rating of 500. Heavier depths of cut and higher rates of feed are also possible. Excellent surface finish is usually obtained because there is no tendency for the metal to tear or drag. Power requirements for magnesium alloys are about one sixth those for mild steel.

Either high-speed or carbide tools are used, but carbides usually result in better operating economy, especially at the high speeds.

Honing and lapping the tool surfaces ensure free cutting action and reduces the tendency of any particles to adhere to the tool tip.

Much has been said about the fire hazards in machining magnesium, but this becomes a problem only when making very fine cuts or when using dull tools.

Distortion of the machined part may be caused from (1) dull tools, (2) improper tool design, (3) taking cuts that are too fine.

**Formability.** The methods and equipment used for forming magnesium are the same types commonly employed on other metals, the principal difference being that magnesium is best formed at elevated temperatures, except in the case of mild deformations around generous radii. Working is usually done at temperatures between 300° and 700°F. Annealing can be avoided by carefully controlling both time and temperature if a strain-hardened condition is desired in the finished product. Magnesium has a hexagonal crystalline structure which offers fewer planes of slip than the cubic structure found in many other common metals.

**Weldability.** Magnesium can be joined by most of the fusion-welding processes. Since the development of TIG and MIG welding, they have become the most popular joining methods. Electric-resistance welding (spot, seam, and flash) is also used.

The shielded-arc method of welding magnesium is the most common. The blanket of helium or argon gas protects the molten metal from being oxidized. Flux or chemical cleaning is not necessary.

**Castability.** Magnesium can be sand cast, die cast, or cast in permanent molds. The three alloys used for sand casting are magnesium-aluminum-zinc, magnesium-rare earth, and magnesium-thorium.

Magnesium-aluminum-zinc alloys combine good strength and good casting characteristics with stable properties up to 200°F.

## TITANIUM

In its early development, titanium was heralded as the space-age metal. It possessed a high strength-to-weight ratio and good corrosion resistance even in salt water. Its properties lie between aluminum and steel in strength, density, elasticity, and serviceability at elevated temperatures. It is 60 percent heavier than aluminum but only 56 percent as heavy as stainless steel. Titanium also has excellent ductility. Despite all these desirable properties, some of the early enthusiasm for the metal subsided when the difficulties of production and fabrication were encountered.

In the molten state, titanium has an extreme affinity for oxygen and nitrogen, which act to harden and embrittle the metal. Melting must be done in a vacuum or an inert atmosphere. No method has been found to remove any contamination that may be acquired during processing. The special equipment needed and the relatively slow production rate have kept the priçe of titanium very high, being about 30 to 40 times that of stainless steel. However, extensive research and development are being continued, the goal being titanium alloys that are better and less costly than those in use at the present time.

**Classification.** Commercially pure titanium has a close-packed hexagonal (alpha) structure. Titanium alloys may have either close-packed hexagonal structures, body-centered cubic (beta) structures, or a combination of alpha-beta structures. Four groups are made from the foregoing, as follows:

COMMERCIALLY PURE TITANIUM. This group consists of 99 percent pure titanium and various percentages of carbon, oxygen, nitrogen, and iron. Designations, such as A–55, Ti–100A, R70, and others are used. The number approximates the tensile strength of the material in thousands of pounds per square inch. Standardization of designations has not progressed far as yet, however, it has been proposed that the letters A for alpha, B for beta, and C for combinations be used.

ALL-ALPHA WELDABLE ALLOYS. At the present time there is only one commercial alloy, A–110, in this group. It can be obtained in sheet, bar, and wire forms and may be welded with close to 100 percent joint efficiency.

ALPHA-BETA WELDABLE ALLOYS. This group comprises the majority of the titanium alloys. They can be heat treated and are available in bars, billets, and sheets. The designation may be by composition (for example, Ti–6Al–4V has 6 percent aluminum and 4 percent vanadium) or it may be simply C–120, indicating a combination-phase titanium with 120,000 to 130,000 psi tensile strength.

ALPHA-BETA NONWELDABLE ALLOYS. This group contains manganese, molybdenum, aluminum, and chromium as the main alloying elements. Some examples of code designations in this group are MST–4Al–4Mn, MST–3Al–5Cr, and C–110.

## FABRICATING CHARACTERISTICS OF TITANIUM

**Machinability.** Some of the problems associated with machining titanium are:

(1) The chip curls away from the tool at a sharp angle, producing high stress concentration and rapid rise in temperature. The temperature between the chip and tool may be as high as 2000°F.

(2) At elevated temperatures titanium has the tendency to dissolve everything in contact with it.

(3) Small particles of titanium weld themselves to the cutting tip, causing it to act dull, thus increasing cutting forces and heat.

(4) Forgings may have scale on the outside containing titanium carbides, nitrates, and oxides, which are extremely abrasive to the cutting tools. The scale can be removed with a solution of 10 percent hydrofluoric acid, 5 percent nitric acid, and 85 percent water. The soaking time required is 20 to 30 min.

(5) When the outside diameter of a bar on a lathe is machined, it is found that the slip planes of alpha titanium vary every 90 deg of work rotation. This results in changes in chip thickness and tends to develop pulsating pressures leading to chatter.

Machinability ratings vary with structure. The unalloyed alpha type has a rating of 38 percent of B1112, the alloyed alpha type drops to 29 percent, and combination alpha-beta is from 26 to 20 percent.

**Formability.** Cold forming of the close-packed hexagonal structures contained in titanium and magnesium is undesirable. Titanium is formed at 400° to 700°F, while the alloys require 800° to 1000°F. Forging is done at temperatures ranging from 1500° to 1850°F, depending on the section thickness and the alloy used.

Simple bending may be done cold, but the springback is about twice that of stainless steel.

**Weldability.** Resistance welds, because of the close proximity of the surfaces to be welded, can be made without resorting to inert atmosphere.

In fusion welding, air must be prevented from coming in contact with the molten metal. TIG welding is used with either argon or helium. Heavy welding requires a protection of inert gas on the bottom side of the weld also.

## QUESTIONS

**3.1.** What does a machinability rating mean?

**3.2.** What are four factors that govern the ductility of a metal?

**3.3.** Why may castings differ in content in the center from the content at the edges?

**3.4.** What explanation can be given for iron being less ductile than aluminum?

**3.5.** Identify each of the following metals: (*a*) B1112, (*b*) 1212, (*c*) 4140, (*d*) 12L14, (*e*) 8630, (*f*) 94B30H, (*g*) W12, (*h*) 80–50 cast steel, (*i*) 430 stainless, (*j*) 2011–T3, (*k*) AZ61A.

**3.6.** Why are thermal expansion and modulus of elasticity important considerations when machining aluminum?

**3.7.** What difficulties could be encountered in welding the higher alloy steels?

**3.8.** What are the properties of aluminum that make it more difficult to weld than plain carbon steel?

**3.9.** Why is malleable iron more ductile than gray cast iron?

**3.10.** What is the difference in structure between malleable iron and ductile iron?

**3.11.** What is the meaning of the following number classification of ductile iron: 60–45–10?

**3.12.** Why are malleable irons not considered weldable in the usual sense of the word?

**3.13.** Name several favorable factors for cast iron as a casting material.

**3.14.** (*a*) What type of stainless steel should be used for tools that must maintain a sharp cutting edge? (*b*) Why is this type the best?

**3.15.** What are the main considerations in choosing stainless steel for a forming operation?

**3.16.** What type of aluminum would 7075 be?

**3.17.** What aluminum is recommended for good machining qualities?

**3.18.** What is meant by an alpha-beta brass?

**3.19.** What do B1112 and a free-cutting brass have in common?

**3.20.** What explanation can be given for the fact that magnesium is not easily formed?

## PROBLEMS

**3.1.** What would be the minimum radius of a 90 deg bend for a 1/4 in. thick mild steel plate?

**3.2.** If B1112 steel can be machined at 300 sfpm (surface feet per minute) with a carbide tool, about what sfpm could be used on a 1040 hot-rolled steel?

**3.3.** Compare the tensile strength of a 2″ × 2″, 20–40 cast iron with an AISI 1020 steel of the same cross section.

**3.4.** Compare the amount of expansion that would take place in raising the temperature of 12″ × 18″ × 1/16″ sheet of aluminum from 70° to 800°F with that of the same size piece of steel raised the same number of degrees F.

**3.5.** Steel weighs 485 lb/cu ft. What is the approximate weight of the same size magnesium, aluminum, copper, and nickel alloys?

**3.6.** Compare the modulus of elasticity of a 2″ × 2″ mild steel bar and a similar bar of 2017 aluminum.

## REFERENCES

Black, P. H., *Theory of Metal Cutting,* McGraw-Hill Book Company, Inc., New York, 1961.

Dash, W. C., and A. G. Tweet, "Observing Dislocations in Crystals," *Scientific American,* October 1961.

Holmberg, M. E., "Precipitation-Hardening Stainless Steels," *Machine Design,* October 1959.

Paret, R. E., "How to Select the Right Stainless," *Machine Design,* June 1959.

Parker, C. M., "Steels for the CPI," *Chemical Engineering,* November 1966.

Wallace, J. F., and L. Lawrence, "Cast Irons and Steels," *Machine Design,* March 12, 1964.

4

# SAND CASTING AND CASTING DESIGN

A CASTING can be simply defined as a molten material that has been poured into a prepared cavity and allowed to solidify. Although the principle is simply stated, great skill and long experience are required to gain the knowledge and master the techniques to produce quality castings on difficult jobs.

Many casting processes have been developed over the years to fulfill specific needs of finish, accuracy, speed of production, etc. These processes include sand casting, shell-mold casting, plaster-mold casting, investment casting, permanent-mold casting, centrifugal casting, and die casting.

Sand casting, the most common of these processes, will be discussed in this chapter, along with design considerations. The terminology and basic concepts will carry over into other casting processes discussed in the next chapter.

## SAND CASTING

Sand casting is used primarily for steel and iron, but it can be used for brass, aluminum, bronze, copper, magnesium, and some zinc alloys. There are two principal methods of sand casting. One is known as greensand molding and the other as dry-sand molding.

**Greensand Molding.** This is the most widely used molding method. It utilizes a mold made of compressed moist sand. The term *green* tells us that there is moisture in the molding sand and that the mold is not dried out or baked. This method is generally the most expedient but usually is not suitable for large or very heavy castings. It is applicable for castings of rather intricate design, since the greensand does not exert as much resistance to the normal contraction of the casting when it solidifies as does the dry, baked sand. This reduces the formation of hot tears or cracks.

**Dry-Sand Casting.** Most large and very heavy castings are made in dry-sand molds. The mold surfaces are given a refractory coating and are dried before the mold is closed for pouring. This hardens the mold and provides the necessary strength to resist large amounts of metal, but it increases the manufacturing time.

Molds which are hardened by a carbon dioxide process, explained later, can also be considered in the dry-sand class.

**Casting Terms and Procedure.** A sand casting begins with a wood or metal pattern of the desired shape. It will be larger than the finished part, to allow for shrinkage of the metal, distortion, and machining. The pattern is placed on a *mold board,* which is a board a little larger than the open box, or *flask,* placed over it. Molding sand confined by the flask is then rammed around the pattern. Oftentimes, the sand next to the pattern is sifted or *riddled* to provide a smoother surface. This is called *facing sand,* and the remainder of sand needed to fill up the flask is called *backing sand.*

Patterns of simple design, with one or more flat surfaces, can be molded in one piece. Other patterns may be split into two or more parts to facilitate their removal from the sand. Two-part flasks are needed. The portion of the mold in the lower half of the flask is called the *drag,* and the upper portion is called the *cope* (Fig. 4.1).

Access to the cavity is provided by *sprues, runners,* and *gates,* as shown in the diagram. A *pouring basin* can be carved in the sand at the top of the sprue, or a *pour box,* which provides a larger opening, may be laid over the sprue to facilitate pouring.

After the metal is poured, it cools most rapidly where it contacts the sand. Thus the outer surface forms a shell which tends to hold its shape and pull the still molten metal from the center toward it. The last bit of metal near the center of the mass may not be of sufficient volume to fill this remaining space, so that a partial void or porous spot will be left in the center of the casting. These porous spots can be avoided by a reservoir or *riser* of molten metal near the cavity.

The size of the gates and runners is very important. It would appear that their sole function is to let the metal fill the cavity. However, if they are too large, the molten metal may flow too fast and erode the walls of the runner system. Also, turbulence may result in the mold cavity. If the gates and runners are not large enough, some of the metal may solidify

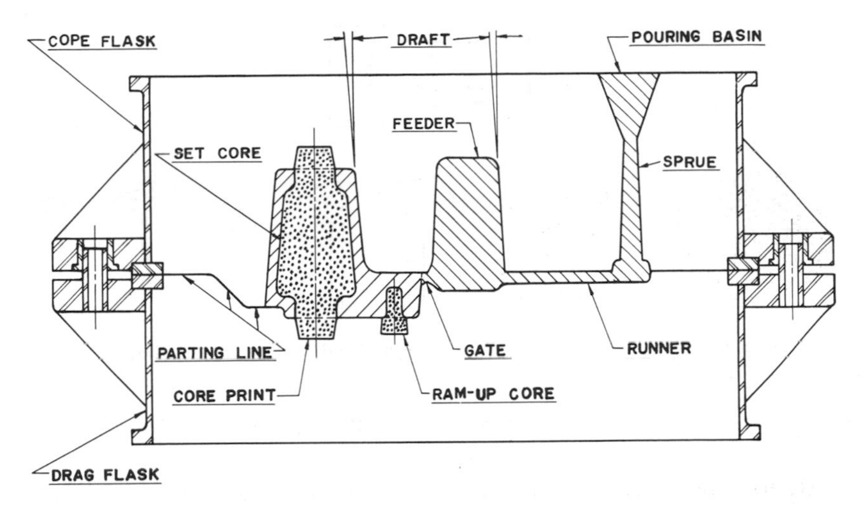


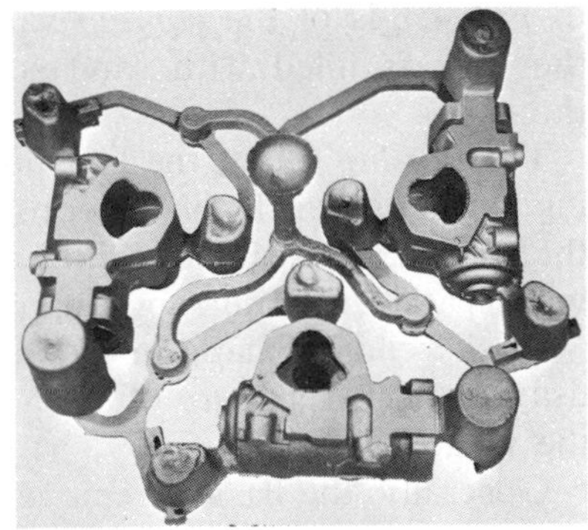

**Fig. 4.1.** Casting-mold terminology.

before the section is filled, making an incomplete part. The construction of proper sprues, gates, and risers is achieved by experience, study, and research.

The pattern must be tapered to permit easy removal from the sand. This taper is referred to as the *draft*. The draft for sand molding is generally 1 1/2 to 3 deg.

*Cores* are placed in molds whenever it is necessary to have a hole in the casting. As shown in Fig. 4.1, the core will be left in place after the pattern is removed so that the metal cannot occupy this space. Sometimes the core may be molded integrally with the greensand, and this is referred to as a *greensand core*. Generally, the core is made of sand bonded with core oil, some organic bonding materials, and water. The cores are placed in the mold and, after the pattern is removed, they are located and held in place by the *core prints*, as shown. When cores do not have much support, as shown in Fig. 4.2, support can be supplied by various types of *chaplets* that become part of the casting.

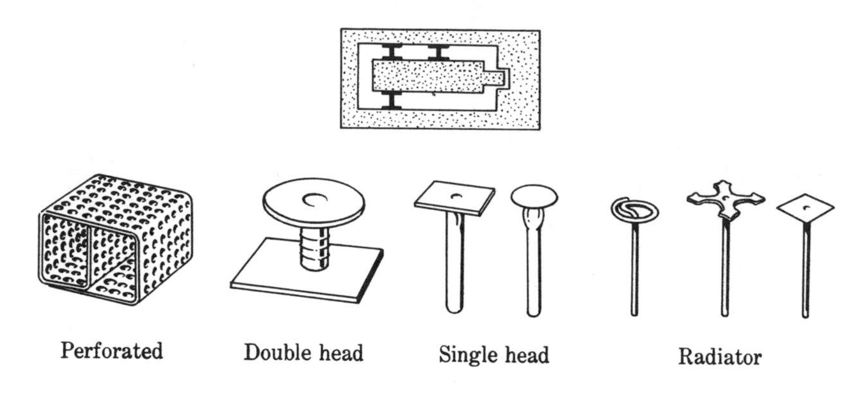


**Fig. 4.2.** Various types of chaplets used to support a molding core.

Exterior cores are used to facilitate molding. The small core shown in Fig. 4.1 is of the *ram-up* variety. It is placed in the pattern before the flask is filled with sand so that it rams up as an integral part of the mold.

The *parting line* is the line along which a pattern is divided for molding, or along which the sections of the mold separate. This surface and the pattern are sprinkled with parting dust so that the cope and the drag will separate without rupturing the sand.

Flasks can be made of wood, but they do not stand up under hard usage. Metal flasks are often made of aluminum or magnesium to keep the weight down.

Generally, the flask is removed before the metal is poured, to prevent it from becoming damaged by the hot metal and to speed up its reuse (Fig. 4.3). In its place is put a metal *slip jacket* which prevents the

**Fig. 4.3.** Metal flask and mold. Courtesy Hines Flask Co.

pressure of the molten metal from breaking the mold. Flasks are made in various ways to ease their removal from the mold. Some have a 5-deg inside taper, and others have hinged or quick-locking corners.

## TYPES OF PATTERNS

The selection of the type of pattern, of which there are several, is based largely on the complexity of the piece and the number of parts to be produced. Patterns can be classified as single, loose, solid patterns; split patterns; match-plate patterns; cope-and-drag patterns; loose-piece patterns; and composite patterns.

**Single, Loose, Solid Patterns.** This type of pattern is essentially a duplicate of the part to be cast except, of course, that it is larger, to provide for shrinkage and machining. It also has draft to facilitate its removal from the sand. This pattern is used only where a limited number of parts are to be produced. It contains no gates, sprues, or risers. These must be cut in by hand by the molder, which makes the process slow and comparatively expensive.

**Split Patterns.** Split patterns are made so that one part can be placed in the cope and the other in the drag. This facilitates not only the pattern removal but also the placement of the parting line. Tapered pins hold the two halves together during molding so that the cope half will be accurately located opposite the drag half.

**Match-Plate Patterns.** Match-plate patterns are usually made on an aluminum plate with the cope portion of the pattern on one side and the drag part of the pattern on the other. The cost of making this more elaborate pattern restricts it to cases where medium or large quantities of castings are to be produced. To justify its use, anywhere from a few hundred to a thousand parts should be cast per month. In use, the match plate fits between the cope and the drag, held in place by pins or guides that mate with those on the flask. The match plate is positioned between the two halves of the flask and, after each part has been filled with sand, the match plate is taken out. The mold cavity is then nearly complete, since runners, gates, and risers are included on the match plate (Fig. 4.4).

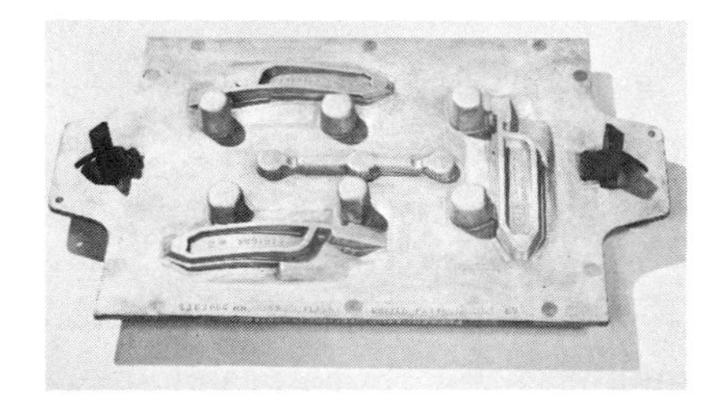

**Fig. 4.4.** Match-plate pattern. Courtesy Central Foundry.

Since runners and gates control the flow of metal to the cavity, having patterns for them mounted on the match plate with the part pattern assures that they will be the same in every mold produced, thus giving more uniform castings. Provision for proper location of the sprue is also provided.

**Cope-and-Drag Patterns.** To speed up operations so that both the cope and the drag can be "rammed up" at the same time, the match plate is made in two parts. This is sometimes also called a *double match-plate* pattern. In this procedure one workman can fill the cope side with sand while another fills the drag side, or they can both be filled by one man using mechanical equipment, described later.

**Loose-Piece Patterns.** Loose-piece patterns are needed when the part is such that the pattern cannot be removed as one piece, even though it is split and the parting line is made on more than one plane. In this case the main pattern is usually removed first. Then the separate pieces, which may have to be turned or moved before they can be taken out, are removed. Complicated patterns of this type usually require more maintenance and are slower to mold. However, their cost is sometimes justified by the end product.

**Composite Patterns.** Large and complicated castings, if produced in great quantity, are often made with *machine-equipment* patterns. They are similar to the cope-and-drag pattern, but different parts or a *composite* of parts can be made at one time. Each type of part is mounted on a half insert which is interchangeable with other half inserts (Fig. 4.5). This practice effects savings in equipment and cost, yet retains the low piece-price characteristic of high-volume equipment.

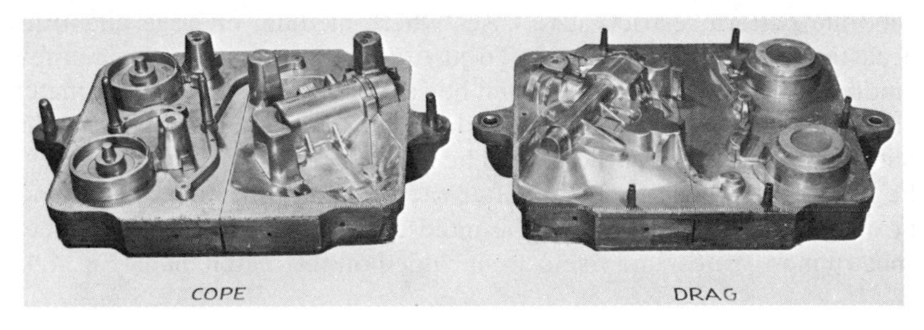


**Fig. 4.5.** Composite machine-equipment pattern. Courtesy Central Foundry.

## MOLDING SAND PREPARATION AND CONTROL

The selection and mixing of the molding sand constitute one of the main factors in controlling the quality of the castings. The two types of molding sand are natural and synthetic.

**Natural Molding Sands.** Deposits of natural molding sands are characterized by a clay content of 5 to 15 percent, distributed throughout the sand. Natural sands vary greatly and therefore must often be reconditioned to obtain certain desirable properties, as follows:

(1) The sand should present a smooth surface, which is governed by the *grain size* and *clay content.* The grain size can be determined by

running it through a series of sieves. The clay content is determined by washing a 50-gm sample through several cycles in a solution of sodium hydroxide. After the clay is removed, the sand is dried and weighed to see what percent it is of the original.

(2) After the sand has been tempered or moistened with water, its characteristics should be such that it may easily be worked in around the pattern. This is termed *flowability*. The moisture content of sand is determined by weighing the sand before and after drying it in an oven. It can also be found by a commercial "moisture teller," an instrument that registers the moisture in terms of electrical conductivity between two electrodes placed in thc sample.

(3) After the sand has been rammed in place around the pattern, it should not break out along the edges when the pattern is removed. This is known as *green strength*.

(4) The sand should be sufficiently open in structure to permit gas and steam, formed by the molten metal, to escape. This quality is known as *permeability*, and it can be checked with various instruments. One method is to use a standard rammed specimen and see how long it takes 2,000 cc of air to pass through it at a given air pressure. The molder frequently increases the permeability of the sand, after the cope half has been rammed, by inserting a *vent pin* several times to within 1/4 in. of the pattern. The vent pin is a 1/16-in. pointed steel wire.

(5) The sand must be sufficiently strong to withstand the flow of the molten metal but still weak enough to crumble as the casting starts to solidify and shrink. This quality is determined with a hardness checker on a standard rammed specimen. The hardness checker, when placed on the specimen, measures the penetration depth of a standard steel ball and registers the distance on a dial.

**Synthetic Sands.** Synthetic sands have as their base the purer silica sands, such as are found in dune sand. *Bentonite*, a clay derivative of volcanic ash, is added to give the right consistency to the sand.

**Sand Mixing.** To ensure proper mixing of sand, bonding clay, and water, a mechanical mixer, or *muller*, is used.

## CORES AND COREMAKING

As stated previously, a core is used to form internal passages in a casting. The characteristics of core materials are similar to those for molding sand. The core must withstand normal handling yet should not prevent the metal in contact with it from contracting normally. A good core will break down or crumble after it has performed its primary function of displacing metal.

**Core Construction.** Cores are made of fine sand mixed with core oil, some bonding material (usually natural or synthetic resins), and water.

After these materials are thoroughly blended, the mixture is placed in molds or coreboxes. Coremaking machines blow the core sand in the corebox. Vents allow the air to escape but trap the sand. By this process, cores can be made in a matter of seconds.

The cores are removed from the corebox and placed in metal coreboxes or on trays and baked at 350° to 450°F. After the cores are baked and cooled, those that consist of two or more parts are pasted together with a synthetic binder.

A newer method of curing cores is with carbon dioxide. When sand is mixed with sodium silicate binder, each grain of sand is coated with a thin film of this viscous liquid. After this coated sand has been packed in the corebox or mold, tiny liquid bridges of binder connect the grains of sand. When carbon dioxide gas is passed through the sand mass, the liquid binder thickens or sets so that the sand grains are bound together, making the sand rigid. The carbon dioxide reacts with the sodium oxide constituent of the binder to produce sodium carbonate. The silicic acid constituent deposits insoluble silica ($SiO_2$) which cements the sand grains together.

Small cores are easily gassed through a small rubber cup held over one end of the corebox. The other end must be vented or supported to allow free passage of gas through the core. Gassing coreboxes from the bottom allows the carbon dioxide to displace more easily the air in the core. Large cores and molds may be gassed by using manifold lids or covers on the coreboxes or mold flasks. In the case of very large cores or molds, direct injection of carbon dioxide gas with a lance is convenient. Cup, manifold, and lance gassing are shown in Fig. 4.6.

Gassing time will vary, depending upon the size and shape of the core or mold and the method of venting. Either under- or overgassing will cause the structure to be weak. However, undergassing is preferred to overgassing, because cores will air harden to some degree. Trial and error must be used to determine gassing times.

About 75 percent of the core sand hardened by carbon dioxide can be reclaimed after use by a pneumatic method. The broken-up core sand is mixed with a high-speed airstream and blown against an impact plate which breaks it down into the individual grains again. Reclaimed sand will require less binder than new sand.

This method of making cores has the advantage of being complete as soon as the cores are removed from the corebox. It also has less distortion as there is no curing time. Less draft is needed, since the cores develop high strength before they are removed from the corebox.

At high temperatures there is a partial fusion of the bond into silicate glass, causing the sand mass to become brittle and easily crushed to powder.

The carbon dioxide process has been extended so that molds can be

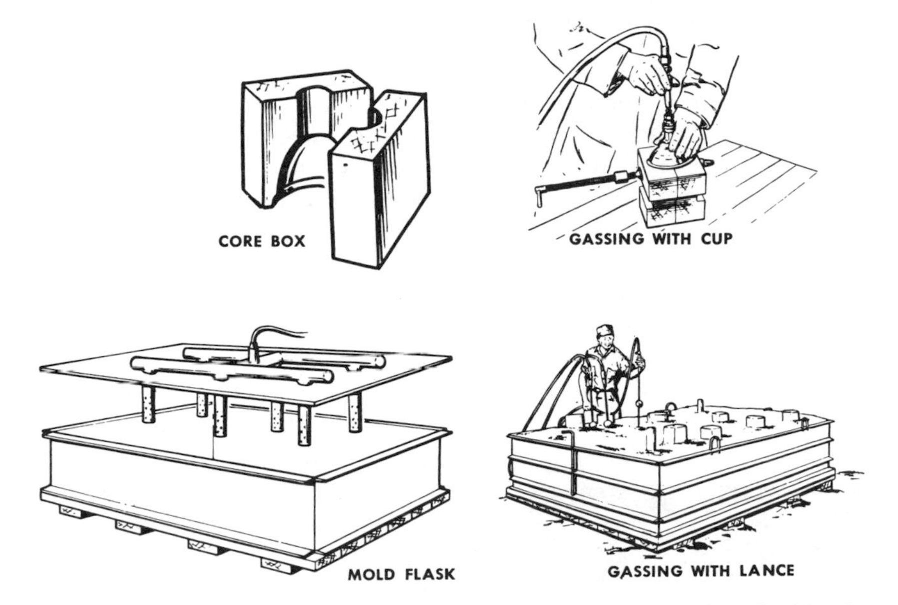


**Fig. 4.6.** Various methods of gassing cores and molds with carbon dioxide. Cup gassing (small core), manifold gassing (mold flask), and lance gassing (large mold) are shown. Courtesy Liquid Carbonic division of General Dynamics.

made by the same process. These molds are produced quickly and have sufficient strength to withstand pouring without an outer slip jacket.

## TYPES OF MOLDS AND SIZE CLASSIFICATION

**Bench Molds.** Small molds are conveniently made on a bench, hence the name.

**Floor Molds.** Floor molds are for the larger flasks that cannot be easily handled on a bench. This does not necessarily mean that the parts are large; it means that many similar small parts are made at the same time (Fig. 4.7).

**Fig. 4.7.** Pouring a large floor mold. Courtesy Allis Chalmers.

**Pit Molding.** Pit molding is done, as the name implies, in a pit or cavity cut in the floor to accommodate very large castings. The pit acts as the drag.

## MOLDING EQUIPMENT

Much of the handwork has been taken out of the modern foundry. Sand is distributed through pneumatic tubes to the molding stations, mechanical means are used to ram it in place, and conveyors transport the finished flask to the pouring floor. The mechanical operations used in filling a mold consist of pneumatic ramming, jolting, squeezing, and sand slinging.

**Pneumatic Rammers.** The simplest aid in making a mold is a pneumatic rammer. It is used in both bench- and floor-molding operations to eliminate hand ramming (Fig. 4.8).

**Jolt Machines.** Another way of packing the sand around the pattern is to give it a series of jolts. This is done by placing the flask with a match-plate pattern on the jolt machine (Fig. 4.9). The top half is filled with sand and subjected to a jolting action. This consists of raising the flask a short distance and allowing it to fall by gravity. The sudden action causes the sand to pack evenly around the pattern. The sand near the top, being less dense, gets very little packing. Sometimes weights are laid on top of the sand to add to the compacting action, thus eliminating any hand ramming.

**Fig. 4.8.** Sand ramming of cope with a pneumatic rammer. Courtesy Ingersoll-Rand.

**Jolt-Squeeze Machine.** Many moldmaking machines combine a squeezing action with a jolting action. The assembled flask is placed on the machine with a match-plate pattern between the cope and drag. The drag half is filled with sand and leveled off, and the jolting action rams the sand around the pattern. The flask is then turned over, and the cope is filled with sand and leveled off. A pressure board (Fig. 4.10) is placed on top of the cope, and squeezing action is brought about by the top platen of the machine. After the squeezing action, the platen is swung out of the way. The match-plate pattern is vibrated as the cope is lifted off and as it is drawn from the drag. With the pattern removed, cores can

**Fig. 4.9.** Large jolt machine used in packing sand around a match-plate pattern. Courtesy Osborn Manufacturing Co.

**Fig. 4.10.** Jolt-squeeze machine eliminates hand ramming of the sand. Courtesy Osborn Manufacturing Co.

be put in place and the cope returned to position. After stripping off the flask, the mold is moved to the pouring floor.

**Jolt-Squeeze Rollover Machine.** This machine is similar to the one just described. After the drag half of the flask is filled and jolted and a bottom board is placed over the drag, two arms lift the flask while the operator pivots it over. The cope is filled with sand and squeezed on the pattern plate. As the cope is lifted off, the pattern plate is vibrated to facilitate its withdrawal from the mold. Cores are then put in place, and the cope is returned to position. The flask is stripped off, and the finished mold, ready for pouring, is lowered to the floor by a jib crane (Fig. 4.11). Mold production rates are high with this type of equipment. Only 2 min are required to complete the entire operation.

**Diaphragm Molding Machines.** The automatic diaphragm molding machine uses air pressure on a diaphragm to squeeze the sand into the

(*a*)

(*b*)

**Fig. 4.11.** Jolt-squeeze rollover machine – courtesy Osborn Manufacturing Co. After the drag has been filled, jolted, and rammed, and the bottom board clamped in place, the operator hits a knee valve which causes arms to raise the flask so it can be turned over easily (*a*). The cope is then filled and rammed, after which the pattern plate is removed (*b*). The cope and drag are then reassembled for pouring.

mold. After the flask has been filled with a measured amount of sand, the diaphragm unit is moved into place. A measured amount of air is pumped into the chamber behind the pad, and the hydraulic cylinder raises the pattern to squeeze the mold. Pressures up to 450 psi are applied evenly to the mold, making possible uniform hardness ratings even in deep pockets (Fig. 4.12).

Because of its uniform pressure, this method has been used to make precision engine-block castings in greensand. A variation of less than 0.5 percent by weight in castings weighing 236 lb has been reported when the mold was made by this method.

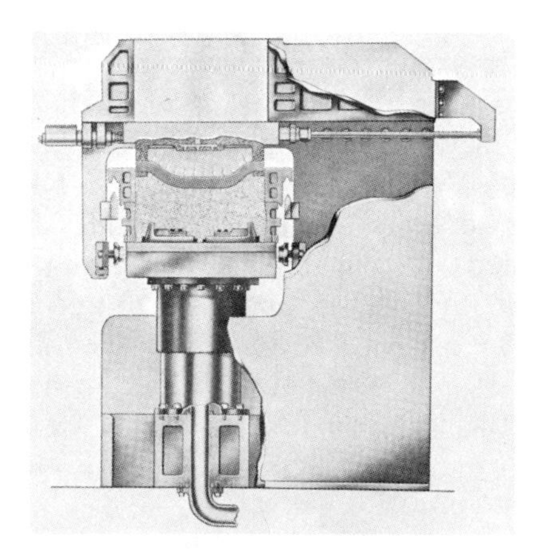

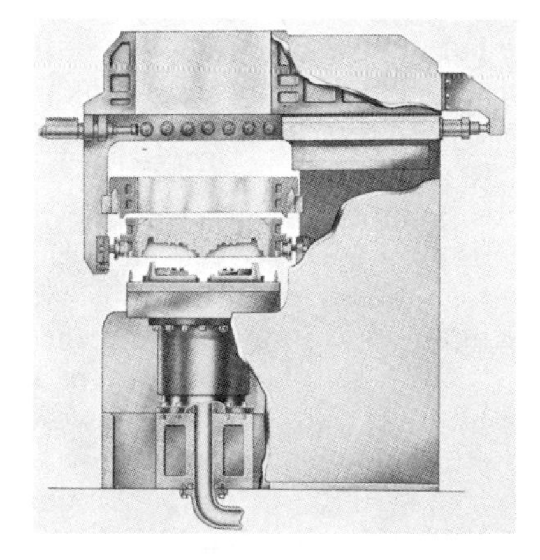

**Fig. 4.12.** Hydropneumatic high-density molding machine makes molds of uniformly high density throughout. Courtesy Taccone Corp.

**Sand Slingers.** Large floor and pit molds are filled with sand by overhead sand slingers (Fig. 4.13). The operator rides on a traveling

**Fig. 4.13.** A sand slinger can be used to fill a pit mold or a number of smaller molds in a relatively short time. Courtesy Gisholt Machine Co.

boom, directing the sand to the proper places. The sand is delivered to an impeller head by means of conveyor buckets. The impeller head throws the sand down at a rate ranging from 1,000 to 4,000 lb per min. The density of the sand can be controlled by the speed of the impeller head.

## MELTING THE METAL

The metal is melted in one of several types of furnaces, some of which are shown in Fig. 4.14. The choice of furnace is based on several factors listed briefly as follows:

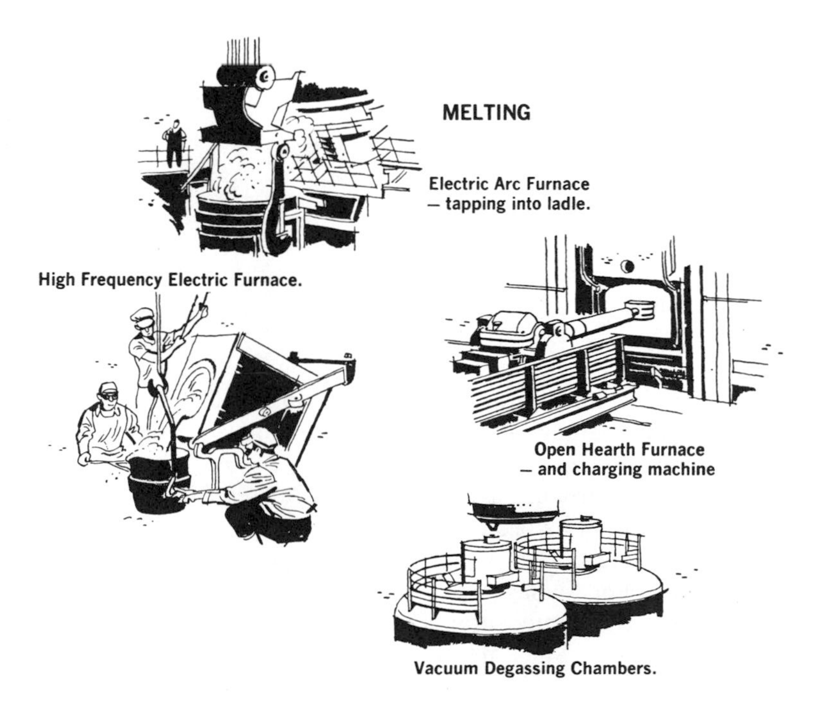


**Fig. 4.14.** Some furnaces used to melt steel.

(1) economy, including the cost of fuel per pound of melted metal, and the initial cost of the equipment, plus installation and maintenance expenses;
(2) temperatures required;
(3) quantity of metal required per shift and per hour;
(4) ability of the melting mediums to absorb impurities;
(5) method of pouring.

**Crucible, or Pot-Tilting, Furnaces.** Crucible furnaces, as the name indicates, consist simply of a crucible to hold the metal while it is being melted by a gas or oil flame. It is so arranged that it can be easily tilted when the metal is ready for pouring (Fig. 4.15). The capacity is generally limited to 1,000 lb. Also shown is a twin crucible furnace that utilizes the heat that leaves the firing chamber and passes over the heating crucible to the preheat crucible.

(*a*)

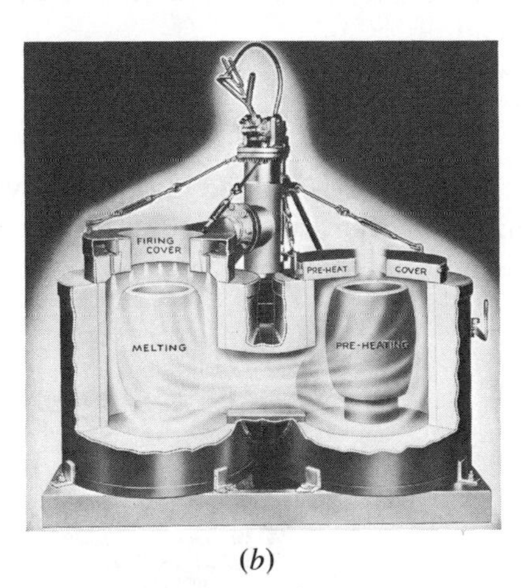


(*b*)

**Fig. 4.15.** (*a*) A crucible, or pot-tilting, furnace; (*b*) a twin-crucible furnace. Courtesy Randall Foundry Corp.

**Cupolas.** Cupolas are an outgrowth of the blast furnace. They are designed more specifically for cast-iron melting, and they do handle by far the greatest percentage of this metal. The cupola's wide usage stems from several advantages:

(1) Like a blast furnace, it can be tapped at regular intervals, as required, under continuous production.
(2) The operating cost is lower than that of other foundry furnaces producing equivalent tonnages.
(3) The chemical composition of the melt can be controlled even under continuous use. Since the molten metal comes in contact with the carbonaceous coke, the carbon content of the melt tends to be high (Fig. 4.16). It is sometimes necessary to transfer the molten metal to another

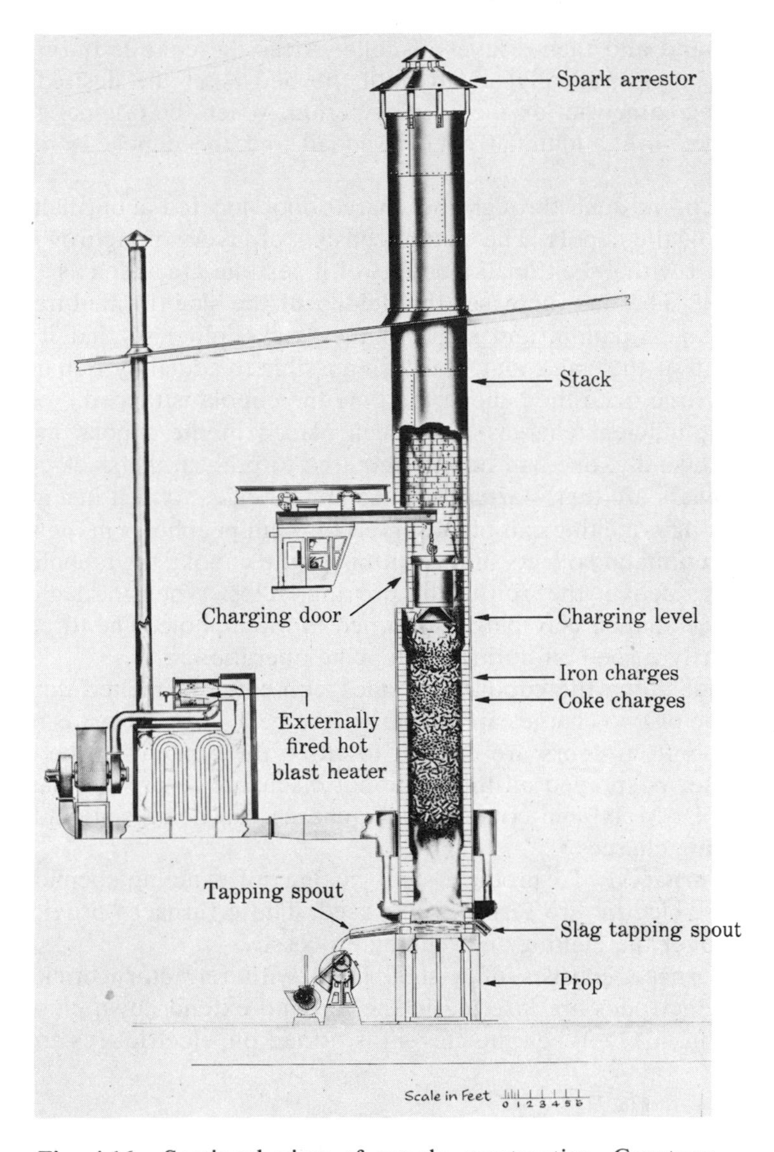


**Fig. 4.16.** Sectional view of cupola construction. Courtesy Whiting Co.

furnace, such as an air furnace, where some of the carbon can be burned out.

You will notice, in Fig. 4.16, that the cupola construction is essentially a steel shaft or cylinder lined with firebrick. A wind box and tuyeres are placed near the bottom to supply draft when connected to the blower. The bottom is covered with green foundry sand, rammed in place and tapered toward the taphole. Kindling wood is carefully placed on the sand and then a layer of coke. After the coke is burning well, more coke is gradually added until the bed is of the desired height, depending somewhat on the burning period. When the original coke bed is "burned in," additional coke is added and the cupola is ready for charging.

Charging is done through the charge door located about halfway up the side of the cupola. The charge consists of layers of pig iron or scrap alternated with coke. Small amounts of limestone are added as a flux for the metal. The flux increases the fluidity of the slag (formed from coke ash, metal oxidation, and some of the brick lining) so that it can be tapped off at the slag spout. It is also possible to add alloy-rich materials to the charge or to the ladle at the time the cupola is tapped.

When sufficient charges have been placed in the cupola, melting is started. Usually, one-half hour is required to preheat the stack contents. The blowers are then started, and the metal begins to melt in a matter of minutes. The melting can be observed through peepholes in the side. As the molten metal collects in the bottom of the cupola, the taphole can be opened to deliver the iron to the pouring ladles. When the ladle is full, the wedge-shaped clay plug is returned to the taphole. The forced air is temporarily turned off during the tapping operation.

To shut down the cupola, the stack contents are melted down until about one or two charges are left above the bed. The air blast is reduced, and the bottom doors are opened to allow the contents to be emptied out. Water is sprayed on the white-hot discharge to prevent damage to the cupola. Metal and coke from the charge can be reused gradually in succeeding charges.

**Arc Furnaces.** To produce steel castings of exacting chemical compositions, electric arc furnaces are used. These furnaces provide close control over the melting and refining processes.

The furnace consists of a shell lined with refractory brick. Three carbon electrodes are inserted in the top and extend down close to the metal (Fig. 4.17). When the current is turned on, electric arcs are struck

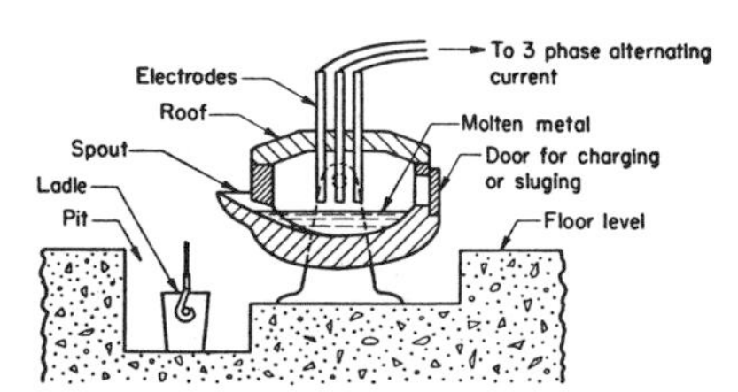


**Fig. 4.17.** Cutaway diagram of electric furnace for melting steel. Courtesy American Iron and Steel Institute.

between the electrodes and the charge of metal, providing the heat. Flux is added through a side door to form a protective cover for the metal. High temperatures, high melting rates, and close control can be maintained by these furnaces.

**Electric Induction Furnaces.** The high-frequency electric furnace shown in cross-sectional view (Fig. 4.18) is essentially an air transformer in which the primary is a coil of water-cooled copper tubing and the secondary is a metal charge. As the high-frequency electric current is

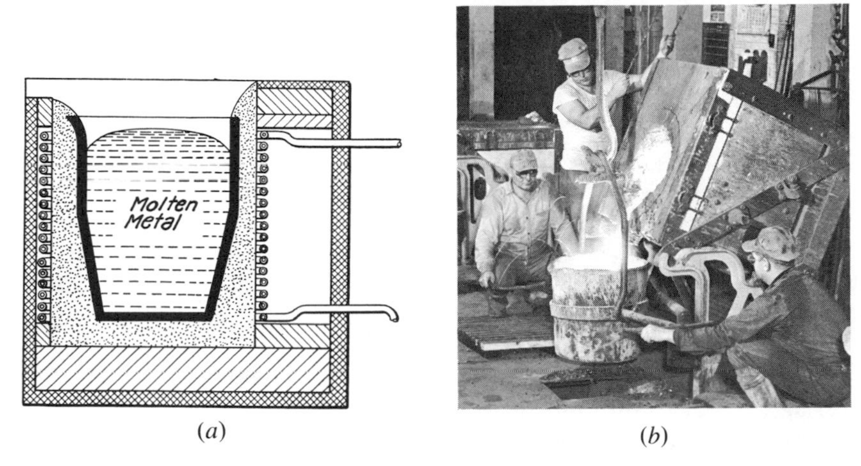


(*a*) (*b*)

**Fig. 4.18.** (*a*) Cross-sectional view of a high-frequency induction furnace. (*b*) Pouring steel from a high-frequency electric furnace. Courtesy Lebanon Steel Foundry.

passed through the primary coil, it induces a rapidly alternating magnetic field in the metal. This much heavier secondary current heats the metal very rapidly to the desired temperature. Cooling water in the copper tubing prevents the primary circuit from overheating.

Since melting conditions are obtained quickly, very little oxidation of the metal takes place, and no flux is used.

The induction furnace is proving valuable in foundries that do small-lot pouring of various alloys. It is also valuable in melting low-carbon steels, as there is no "pickup" of carbon, such as occurs in electric arc furnaces.

## POURING THE METAL

After the mold is made and the cores are set in place, the work is checked to see that all has been properly done. The mold is then closed and sent to the pouring floor. Several types of containers are used to move the molten metal from the furnace to the pouring area.

**Ladles.** Large castings of the floor and pit type are poured with a *bottom-pouring* ladle. When the casting is very large, two or three of these ladles are used at one time. For smaller or medium-sized molds, the *teapot ladle* is used (Fig. 4.19). The teapot ladle has a built-in spout

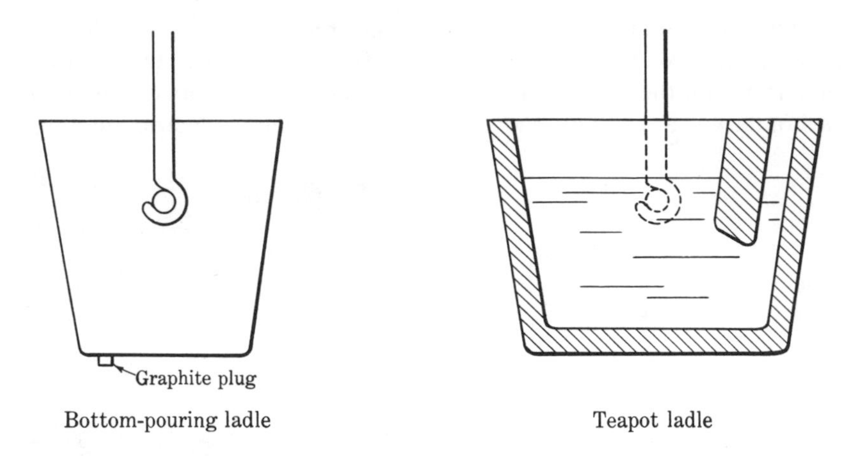


**Fig. 4.19.** Ladles used in pouring castings.

which allows the metal to be taken from the bottom and does not disturb the slag that forms on top.

Plain ladles are referred to as *lip-pouring* ladles. These receive the metal from the furnace and are placed on the pouring floor to fill the smaller ladles that are handled by monorail conveyors or by hand.

**Pouring.** Pouring the metal must be as carefully controlled as any part of the casting process. The temperature of the metal must be just right. If it is too hot, the hot gases will produce blowholes; if it is too cold, the metal will solidify prematurely and will not fill the entire cavity.

A newer pressure-pouring method now being used by some manufacturers has been developed to ensure sound castings. It consists of forcing molten metal through a refractory tube to the bottom of the cavity by means of compressed air.

**Automatic Equipment.** Foundry processes have become considerably automated. Shown in Fig. 4.20 is a semiautomated foundry process line using a hydropneumatic molding machine. Only the core setting and pouring need be done manually.

## FINISHING OPERATIONS

After the castings have had time to cool below the lower critical temperature, they are separated from the mold and placed on a shakeout table. Here the sand is separated from the casting. It is then taken to the finishing room where the gates and risers are removed. Small gates and risers can be knocked off with a hammer; larger ones require sawing, burning, or shearing. Unwanted metal protrusions, such as fins, bosses, and small portions of the gates and risers, need to be smoothed up to blend with the surface of the casting. Most of this work is done with a grinder, and the operation is known as *snag grinding*. On large castings it is easier to move the grinder than the work, so swing-type grinders are used (Fig. 4.21). Smaller castings are brought to stand- or bench-type grinders. Hand and pneumatic chisels are also used to trim castings. A more recent method of removing the excess metal from the casting is by means of a carbon-air torch. This consists of a carbon rod using high-amperage electric current with a stream of compressed air blowing at the base of it. This oxidizes and removes the metal as soon as it is molten. In many foundries this process has replaced all chipping and grinding operations.

Large castings may be taken to a special room where they are subjected to blasting operations of shot, sand, or water.

**Checking and Repairing.** The casting is cleaned and inspected for flaws. Surface defects can usually be repaired by welding. More recently, epoxy resins combined with steel or aluminum have been used to fill surface defects. These materials harden quickly and have very low shrinkage values.

Castings are tested both destructively and nondestructively. A destructive test consists of sawing the casting in sections to determine possible porosity locations. Other destructive tests are made by hammer blows and by loading to failure. Nondestructive tests made by x-ray, gamma rays, magnaflux, ultrasonics, etc., are discussed in Chap. 8.

**Heat Treatment of the Casting.** If heat treatment of the casting is necessary, it is usually done after the preliminary repair and cleaning. If the repair is made with epoxy resins, this would have to be done after heat treatment.

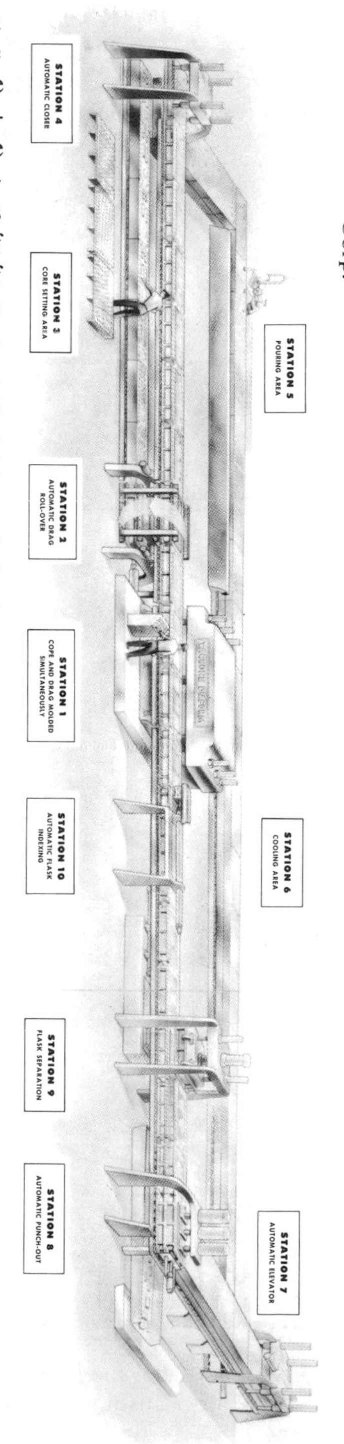


**Fig. 4.20.** Semiautomated molding line using a hydropneumatic molding machine. The only manual processes required are the core setting and the pouring. Courtesy Taccone Corp.

**Fig. 4.21.** A battery of swing grinders being used to smooth the casting surfaces where excess metal has been removed. Courtesy Lebanon Steel Foundry.

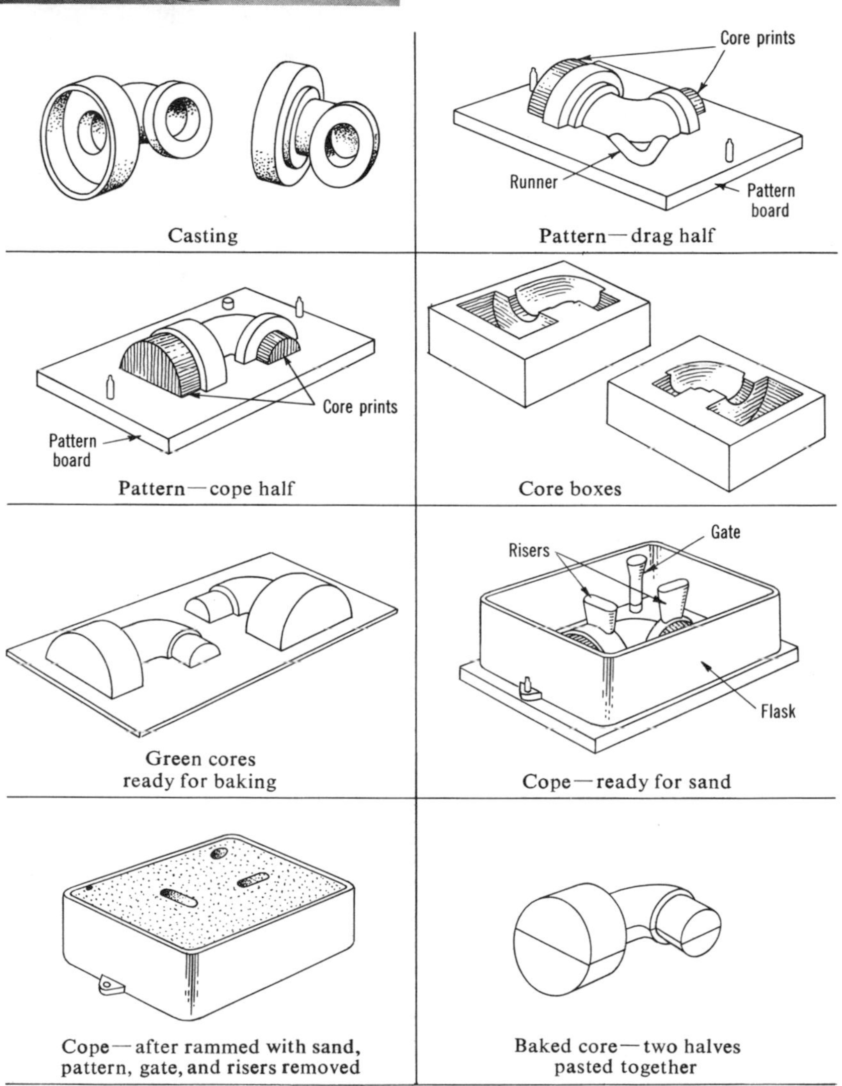


**Fig. 4.22.** A summary of the steps involved in producing a steel casting. Courtesy Adirondack Steel Casting Co., Inc. (Fig. 4.22 is continued on the next page.)

Various materials need different heat treatments. Malleable iron requires a prolonged anneal, steel castings may be quenched and tempered to develop maximum strength, and nonferrous castings may be normalized in preparation for welding or machining.

**Review of the Casting Process.** The casting process can become quite involved, especially in its initial presentation. Therefore, by way of brief summary, the important steps are reviewed with a series of line drawings (Fig. 4.22). These pictures will serve to tie the process and terms together.

## CASTING DESIGN

Good casting design usually results from the careful deliberations of the foundry engineer and the product-design engineer. As with almost all

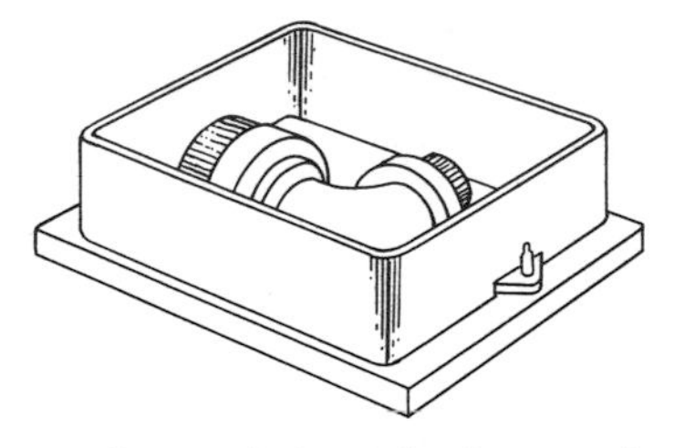

Drag ready for sand—after ramming with sand, bottom board is set on top of flask, flask inverted, and pattern removed

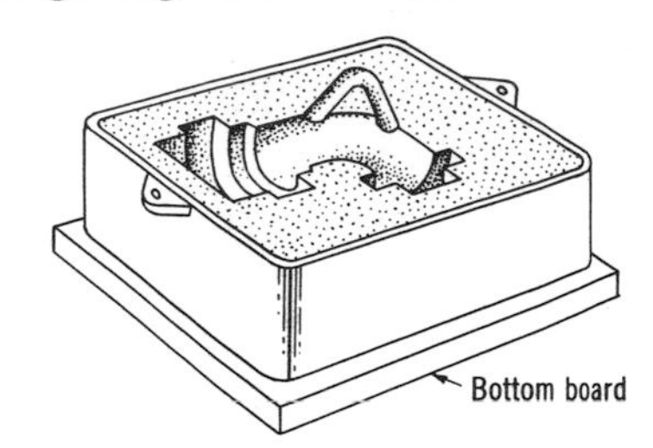


Drag—pattern removed

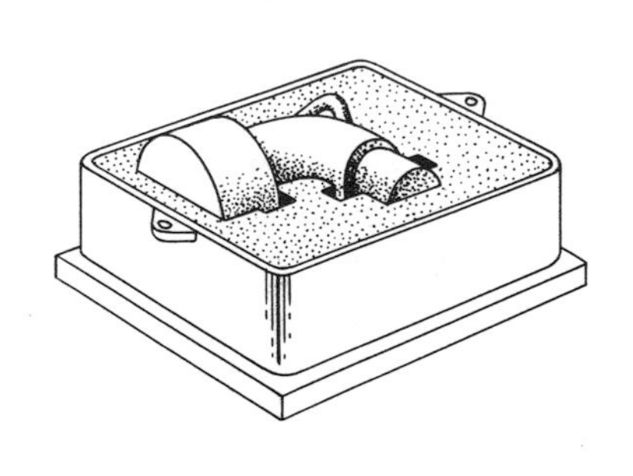

Drag with core set in place

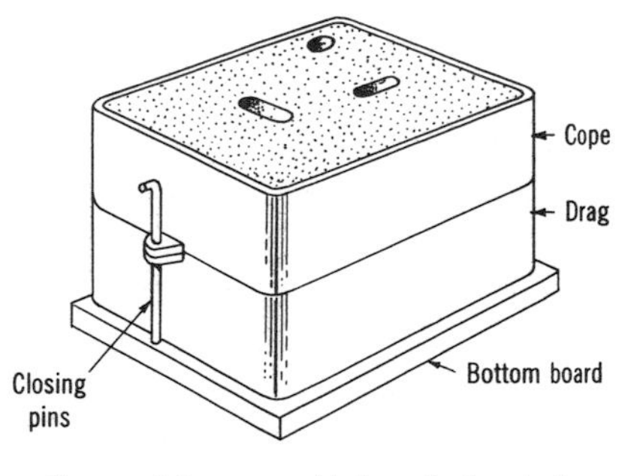


Cope and drag assembled, ready for steel

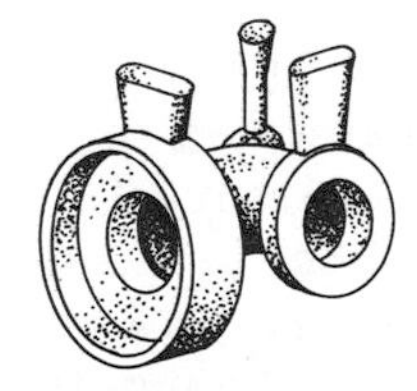

Casting as removed from sand—Risers and gate will be removed, casting chipped and ground where necessary, annealed, inspected, and ready for shipment

engineering, a *good* design is generally a compromise. Shown in Fig. 4.23 are two versions of the same fundamental design. The designer's ideal casting would be almost impossible to cast, and the foundryman's ideal would be extremely costly to machine.

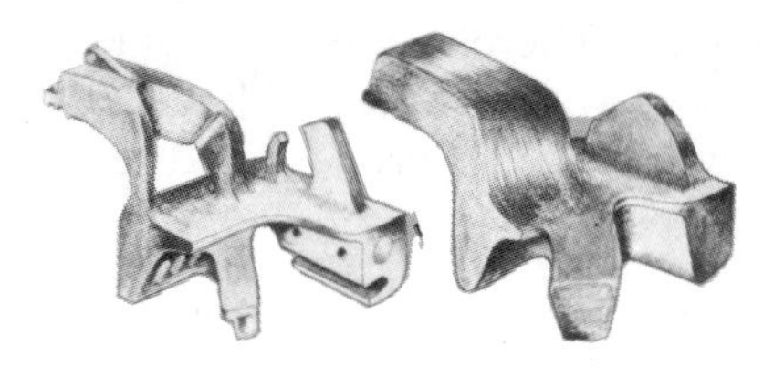

**Fig. 4.23.** The designer's and foundryman's views of an ideal casting. Courtesy *Machine Design.*

For a given part, there is usually an optimum point on the cost curve which represents the "best" design. The curves might look something like those in Fig. 4.24, where one curve represents an easy-to-cast, hard-to-machine casting and the other curve the difficult-to-cast, easy-to-machine casting. The lowest point of the total-cost curve represents the ultimate goal of the manufacturer.

## CASTING-DESIGN PRINCIPLES

Although cost reductions are often realized through good casting design, there are other important factors to keep in mind when designing a casting to obtain maximum strength. These include the solidification rate, effect of alloying elements, parting-line placement, avoidance of stress concentration, economic considerations, tolerances expected, and function.

**Solidification Rate.** The rate at which a given metal solidifies largely determines its microstructure. Therefore, in casting design it is important to consider carefully the effect of the size or volume of each cross-sectional area. As an example, normally gray cast iron, when poured in very thin sections, can become white cast iron, and normally white cast iron, poured in heavy sections, can become gray cast iron.

As noted on the iron-carbon equilibrium diagram in Chap. 2, there are two important cooling periods for cast iron. One is at the beginning of solidification, when iron carbides are starting to form and austenite makes its first appearance (about 2,065°F). The other is the transformation to pearlite at 1,333°F. The austenite will have rejected any carbon in excess of a eutectoid mixture (0.83 percent). The excess carbon rejected from the austenite may be dispersed throughout the austenite, or it may migrate and deposit itself on the primary carbide which separated out from the melt. The size and distribution of the graphite flakes are fairly well set at the time of solidification, so that subsequent heat treatment will not have any great effect on them.

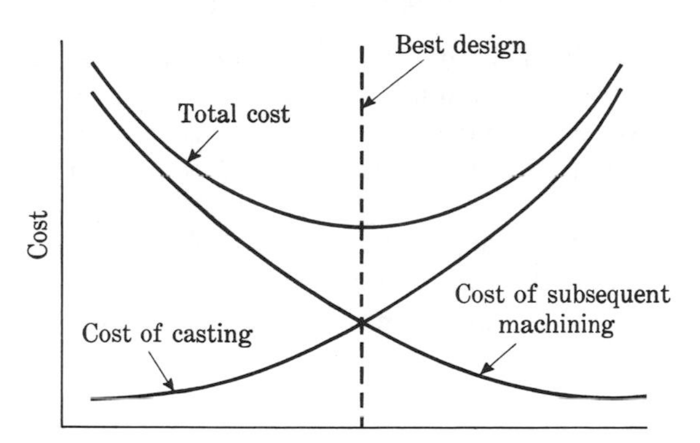


**Fig. 4.24.** The "best" casting design is represented by a summation of the cost of casting and the cost of machining.

To a great extent, the structure of the mold will govern the cooling rate of the metal. For example, a water-cooled mold will have more than twice the cooling action of a regular sand mold. Water cooling is seldom used unless white cast iron is desired.

THE USE OF CHILLS. Where localized cooling of a casting is desired, chills are used. Chills are usually composed of metals of various heat-dissipating capacities, placed in the mold to cause more rapid solidification in localized areas. They may be positioned at an intersection or joint that has a comparatively large volume of metal to cool, thus relieving a hot spot or maintaining a more uniform cooling rate and microstructure. They may also be located at points where it is desirable to have localized hardening, as in the case of bearing or wear surfaces.

In general, it will be well for the casting designer to be aware of the change in solidification rate that accompanies various cross-sectional areas. The sectional area of a casting largely determines its solidification rate, which, in turn, will help determine the microstructure and strength of the casting.

Heavy isolated sections should be avoided. Shown in Fig. 4.25, at the left, are three possible methods of making a small casting with a large central hub. If the hub is solid, as shown in the top section, porosity will occur in the center of the casting. If some of the metal is removed, as shown in the center section, there is some possibility that porosity can

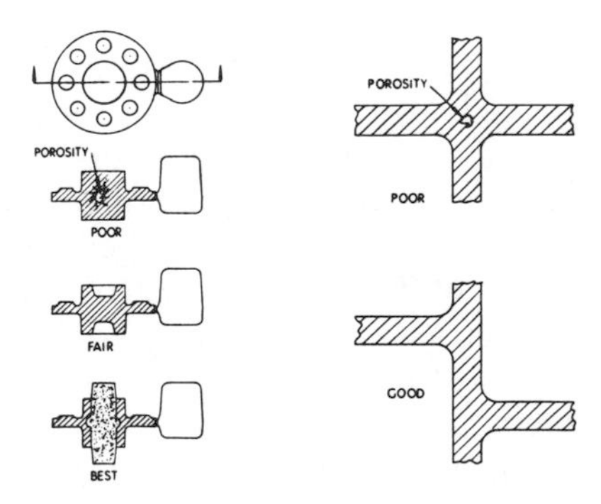


**Fig. 4.25.** Heavy, isolated sections cause porosity in the casting. Courtesy Central Foundry.

be avoided. If the hole is cored entirely through the hub, as indicated in the bottom section, porosity-free castings are assured.

Wherever possible, wall sections should be staggered to minimize the amount of metal concentrated at the junction, as shown at the right in Fig. 4.25.

Castings with heavy sections can be made if the heavy section is near the outside and is connected with the parting line. They can be gated directly, the metal entrance being increased by means of a gate pad (Fig. 4.26).

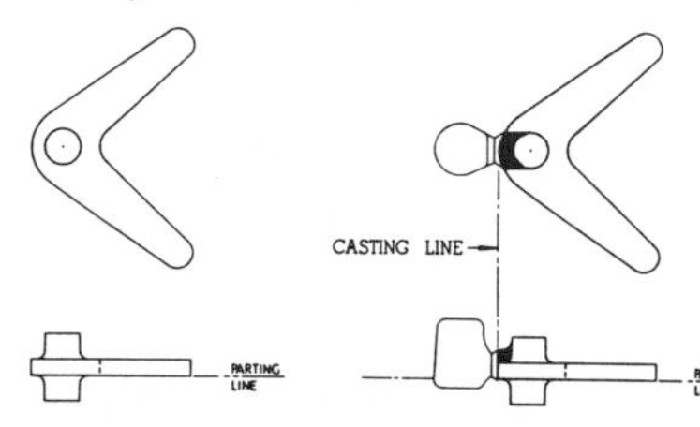


**Fig. 4.26.** The use of a gate pad to facilitate feeding the large hub section. Courtesy Central Foundry.

When specifying the desirable strength of a casting, references are often made to a test bar of a size equivalent to the casting cross section in question. One method is to pour a round bar that represents the cross-sectional area of the size that corresponds to the casting cross section. Of course, the cooling rate of a round bar will not be exactly the same as that of the casting area in question, so certain correlations have been worked out. One of these is that the center of a flat plate will cool at the same rate as a round bar whose diameter is twice the thickness of the plate. The edge, which cools at a much faster rate, will require a sample test bar the same diameter as the plate thickness. This variation in cooling rate is well known in the foundry and machine shops, where unmachinable chill is always found first on thin edges.

Even in fairly compact castings, differences in cooling rate will affect the final strength. For example, if a 3-in.-square bar were cast and various sizes of round-bar specimens were taken at the corners and center, as shown in Fig. 4.27, the results would be somewhat as indicated, even in a compact casting such as this. It will be noted that the hardness is higher at the more rapidly cooled corners than in the more slowly cooled center. Tensile strength corresponds closely to hardness and is higher for the harder specimens.

THE USE OF INSCRIBED CIRCLES. The effect of cooling on adjacent portions of a casting is not readily predictable on a useful quantitative basis, but the use of inscribed circles has been worked out by Harold T. Angus of The British Cast Iron Research Association. This method is a help in assessing relative cooling rates of castings at the design stage (Fig. 4.28). The circles are useful in determining chilling tendencies and likely mechanical properties.

When an inscribed circle is tangential to three major sides of a section,

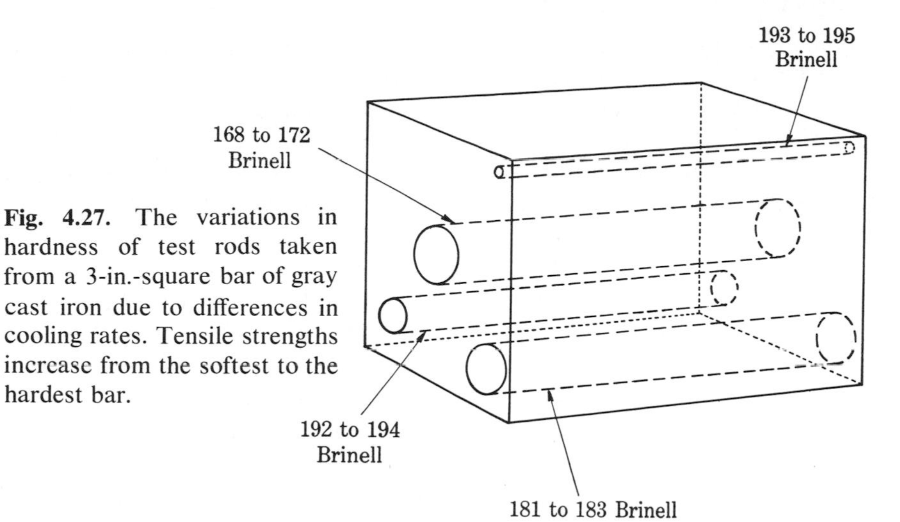


**Fig. 4.27.** The variations in hardness of test rods taken from a 3-in.-square bar of gray cast iron due to differences in cooling rates. Tensile strengths increase from the softest to the hardest bar.

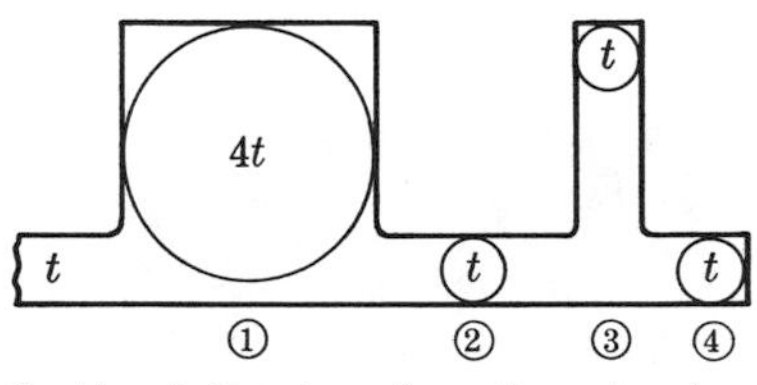


Sectional drawing of casting, showing inscribed circles

**Fig. 4.28.** Crosshatched areas show equivalent cooling rates of various casting sections.

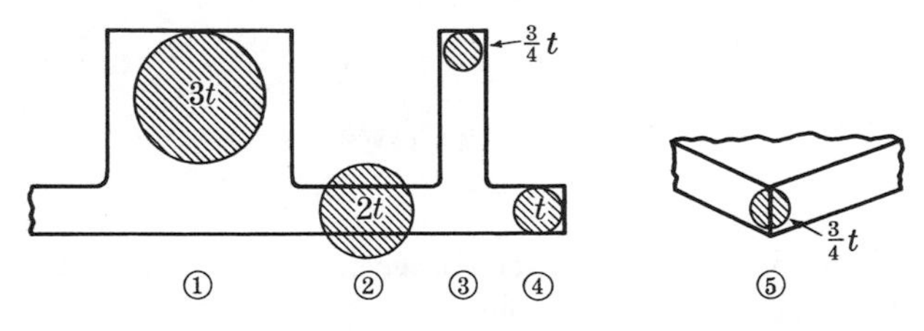


Equivalent diameter of bars which will cool at about the same rate as the particular position on the casting as in (1), (3), and (4) of Fig. 4.28, the cooling rate can be assumed to be equal to or faster than that of a bar of the same diameter, and the possibility of hard spots, depending on composition and cooling rate, must be considered.

It can be seen that the risk of obtaining hard spots would be greatest at points (3), (4), and (5). It would be necessary to specify iron that would not chill at these points.

Although appreciable variations occur in properties of apparently similar irons, it is possible to make a reasonable assessment of the likely properties of metal when poured into castings or test bars from the same analysis.

The structure and mechanical properties of a cast iron of a given composition depend primarily upon the cooling rate after pouring.

**Effect of Alloying Elements.** Alloys added to cast iron are used to enhance certain properties of a metal, as explained in Chap. 2. Here we are concerned with how alloys can be used to control the solidification rate and, consequently, the properties of the metal.

Silicon is the element normally used in castings to control chill. Nickel, copper, chromium, molybdenum, vanadium, and titanium are also used to affect the cooling rate. Where these elements are added in amounts greater than 0.2 percent, their effect must be considered. For example, 1 percent of nickel is roughly equivalent to 3 percent of silicon, and 1 percent of chromium roughly cancels the effect of 1.2 percent of silicon.

**Parting Line Placement.** Every new casting design should be thoroughly studied, tested, and evaluated before it is put into production. Shown in Fig. 4.29*a* is the original design of a bumper support. The parting line was matched with the flange, which required a large core on one side.

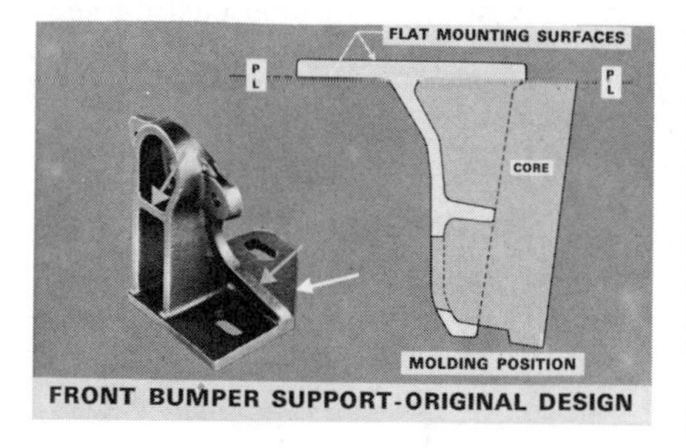


(*a*)

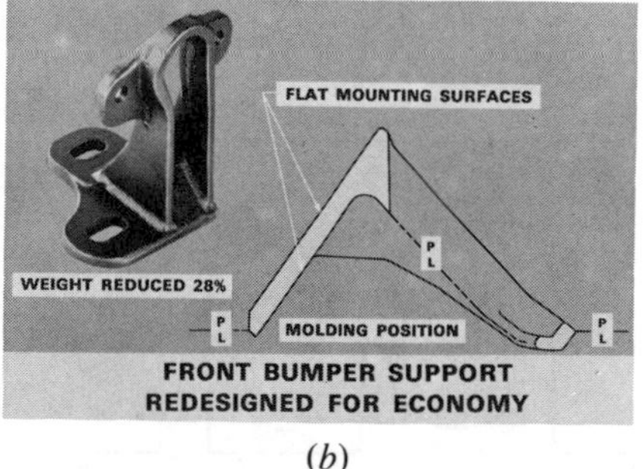


(*b*)

**Fig. 4.29.** (*a*) The original bumper design required a large core due to the center rib. (*b*) a study of the design showed the rib to be unnecessary. A savings resulted in casting weight, core material, and casting time. Courtesy Bradford-LaRiviere, Inc.

The stress analysis study of the large rib showed it was carrying no load. The bracket was redesigned without the rib, allowing a change in parting line as shown in Fig. 4.29*b*, resulting in 28 percent weight savings.

**Stress Concentrations.** Sharp corners should be avoided in casting design. Stress concentration is developed in any sharp inside corner during cooling, as shown in Fig. 4.30. If the stress exceeds the strength

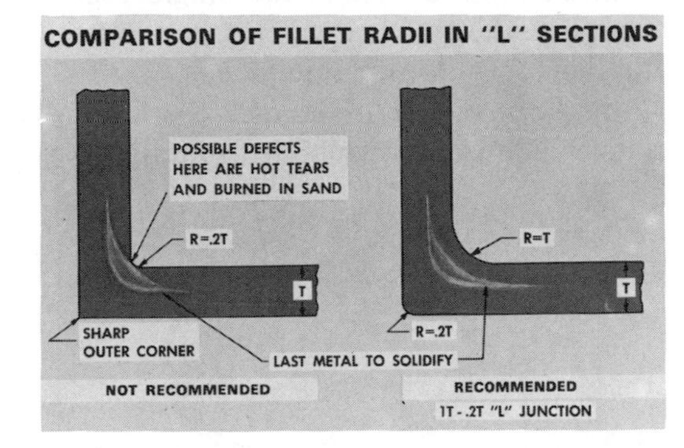


**Fig. 4.30.** Avoid sharp corners to minimize stress.

of the material (metal is weak immediately after solidifying), a *heat check* or *hot tear* may form. Very often, such flaws are small, unnoticed, and frequently not harmful, but, since the tendency for them to form is present in any sharp inside corner, better design is accomplished when a proper fillet is utilized. The illustration to the right in Fig. 4.30 shows how this stress is minimized. There is no spot for the concentration of cooling stresses to occur.

Whenever possible, castings should be designed with uniform thickness. Since this cannot always be done, accepted methods of varying sections are shown in Fig. 4.31. Sharp corners and small radii used to

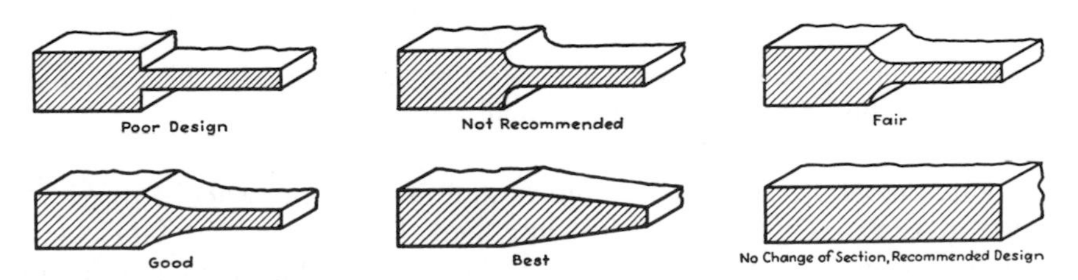


**Fig. 4.31.** Abrupt section changes are not recommended. Courtesy Steel Founders' Society of America.

change the thickness of a section are responsible for stress concentrations. A radius of 1 in., or a 15-deg taper, is an acceptable method of design when changes are made on both sides of the section.

**Economic Considerations.** Good casting design will incorporate savings wherever possible. The designer will consider such factors as multiple-cavity castings, fabrication, the use of cores, subsequent machining operations, and weight reduction.

MULTIPLE-CAVITY CASTING. Substantial savings can be realized by incorporating more than one finished part into a single casting. Molding and handling costs are greatly reduced, since only a fraction of the number of castings need be handled.

An example of this is the single casting of five V–8 engine-bearing caps (Fig. 4.32*a*). The casting is completely machined as a single piece and then cut into separate pieces.

Another example of a single casting for two parts is the front and rear stator-blade carrier which is separated during machining (Fig. 4.32*b*).

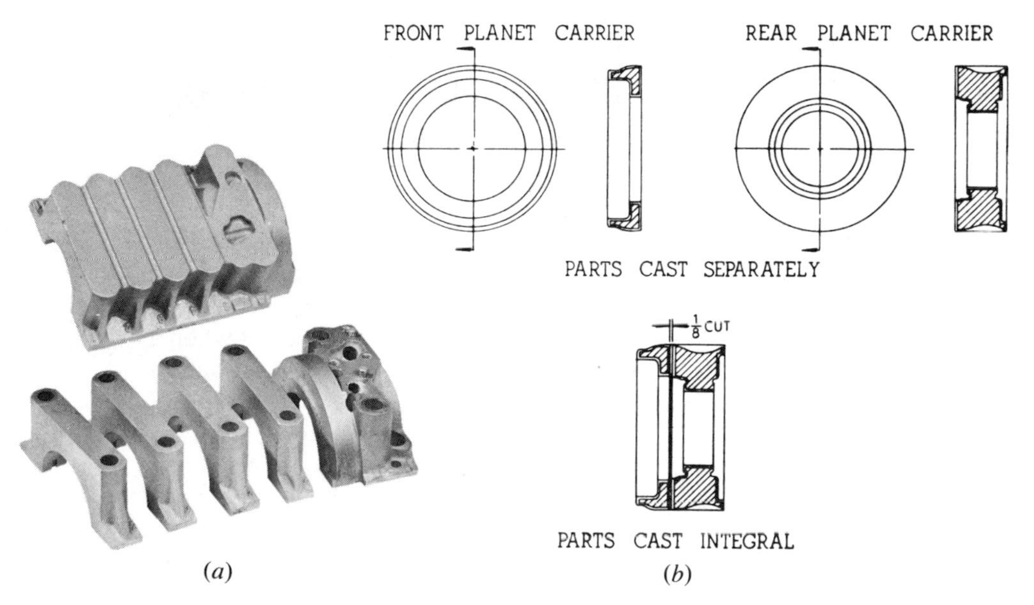


Fig. 4.32. Multiple and integral castings can result in cost reduction. Most of the machining is completed before separation. Courtesy Central Foundry.

FABRICATION. Fabricated designs, such as the rear-wheel truck hub shown in Fig. 4.33, can often be simplified. In this case the fabricated version was made by forging the main plate and welding a smaller flange in place. The casting on the right is lighter, stronger, and less costly than the fabricated design.

CORED HOLES. Engineers are often confronted with the problem of whether to specify cored holes or drilled holes. Sometimes the holes are required as oil passages; others are used to lighten the casting. Cored holes often require boring and reaming. The economics involved in cored holes versus drilled holes must be studied on an individual basis, but generally it is cheaper to core larger holes where appreciable metal

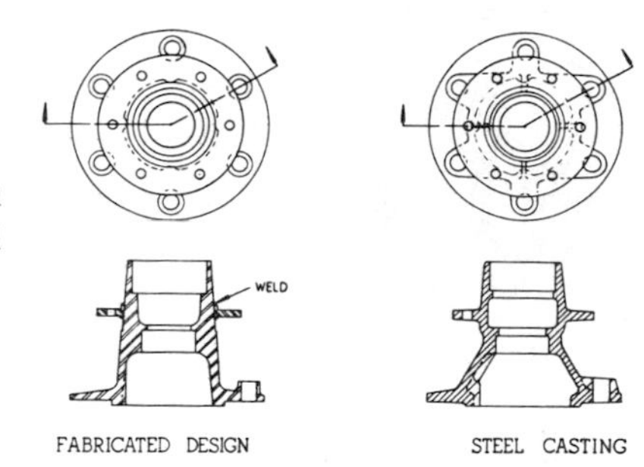


**Fig. 4.33.** Casting at right eliminates an added welding process. Courtesy Central Foundry.

can be saved and faster machining results. The universal joint yoke shown in Fig. 4.34 was originally a solid-steel forging. A rather slow drilling operation was required to make the shaft hole. In casting, this hole was cored, followed by a fast core-drilling operation, since most of the metal had already been removed.

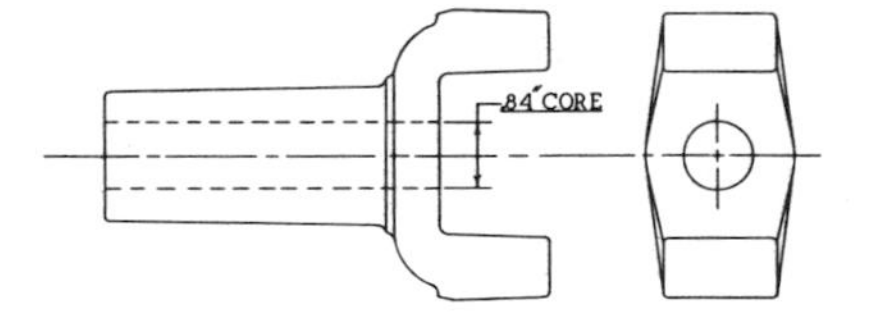


**Fig. 4.34.** Cored holes eliminate or reduce secondary operations of drilling and boring. Courtesy Central Foundry.

Many times, proper design can combine cores, resulting in a less expensive casting. Figure 4.35 shows the original and the improved designs of a truck wheel hub. The original casting could not be formed in greensand. A large *ring core* was necessary, together with a *body core* for the inner portion. The design at the right was considerably improved by eliminating the ribs and backdraft. This made it possible to make the mold in greensand, entirely eliminating the ring core.

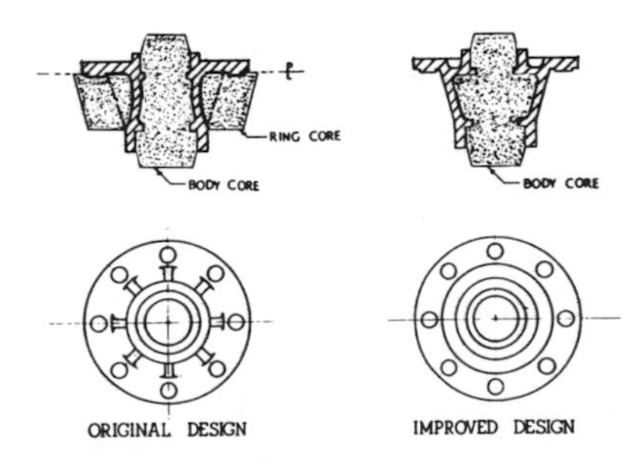


**Fig. 4.35.** Redesigning a truck wheel hub for one core instead of two. Courtesy Central Foundry.

SUBSEQUENT OPERATIONS. Good casting design takes into consideration how parts will be held for machining. The booster body shown in Fig. 4.36, used on several sizes of artillery shells, was originally made from brass bar stock on an automatic screw machine. After extensive testing, it was decided to use malleable iron castings. The manufacturer wished to use the same machines he had used on the brass bar stock.

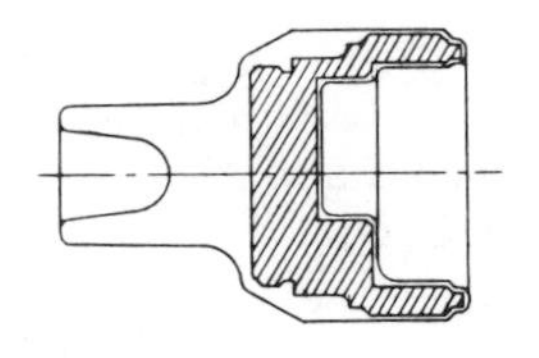

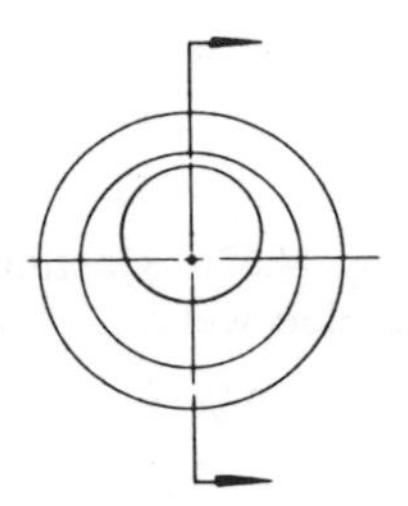

**Fig. 4.36.** A machining lug added to the booster body allows it to be held in a collet-type chuck for overall machining. The part was redesigned from brass bar stock to malleable-iron casting. Courtesy Central Foundry.

This problem was solved by adding a 1-in.-dia by 1.25-in.-long lug for holding it in a collet-type chuck. The booster body could then be machined completely during the single chucking and removed from the lug while machining the back side.

Casting designers should always watch for chances to combine two or more parts into a single casting. Shown in Fig. 4.37 are two brackets used in the support of a car air-conditioning suppressor. The single casting saved 13 percent by weight and 40 percent of the cost.

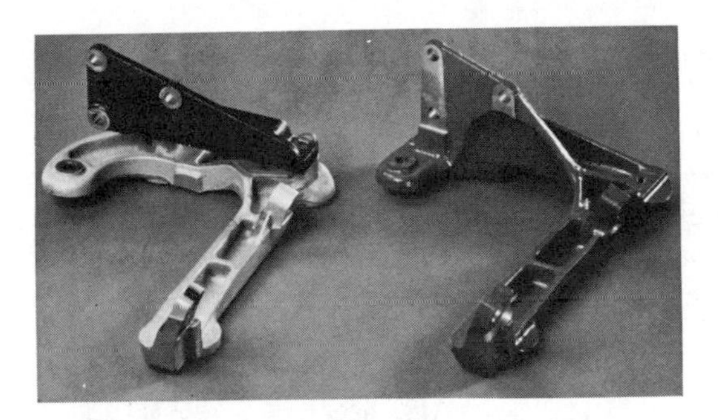

**Fig. 4.37.** Two support brackets combined into one casting.

Holes with diameters greater than the thickness of the metal can be punched in malleable cast iron. Two round holes and a square hole are punched simultaneously in an idler arm, as shown in Fig. 4.38.

WEIGHT REDUCTION. Weight reduction, formerly a problem primarily of aircraft and space engineers, is now also a problem for automotive engineers. Three main methods of reducing weight are:

(1) Reduce existing dimensions through stress analysis and related procedures.

(2) Convert existing ferrous materials to aluminum or magnesium. This usually results in increased costs, since lightweight metals are higher priced on a volume or strength basis.

(3) Convert existing parts from low- or medium-strength ferrous materials to a high-strength ferrous material. This permits reduction in physical size and weight. Often, the greater cost per pound of higher-

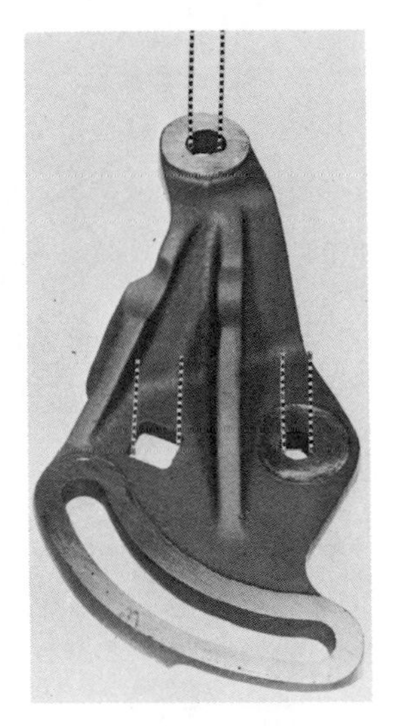

**Fig. 4.38.** Three holes are punched simultaneously in this malleable-iron casting, eliminating most of the secondary machining operations. Courtesy Malleable Castings Council.

strength material is offset, because less material is required. Shown in Fig. 4.39 are a gray-iron and a cast-steel differential carrier. By taking advantage of the higher-strength steel, it was possible to design the part 5 lb lighter and still maintain the original, or greater, strength.

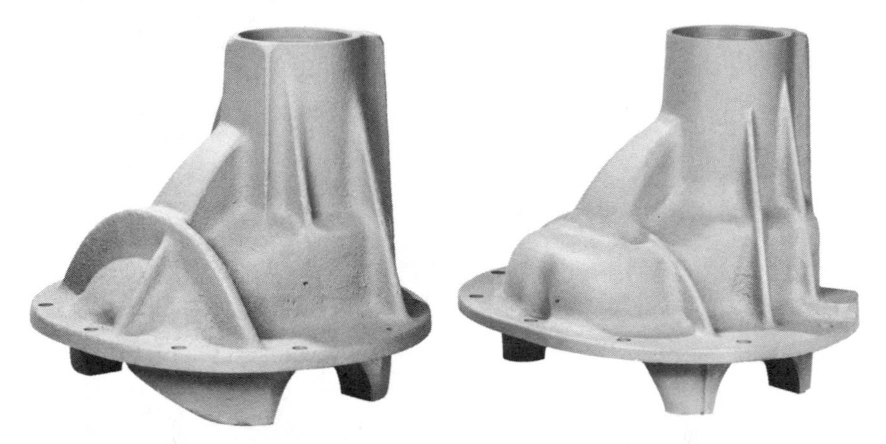

**Fig. 4.39.** The gray-iron casting at the left was redesigned for a pearlite malleable-iron casting, with a resultant weight saving of 5 lb. Courtesy Central Foundry.

The seven steps shown in Fig. 4.40 are recommended by the Malleable Founders Society as being basic in creating high-strength, lightweight castings.

**Sand-Casting Tolerances.** For ordinary greensand molds, the closest tolerances can be held in one part of the flask. The greatest error is introduced across the parting line. Several other reasons exist for tolerance build-up in greensand castings, as follows:

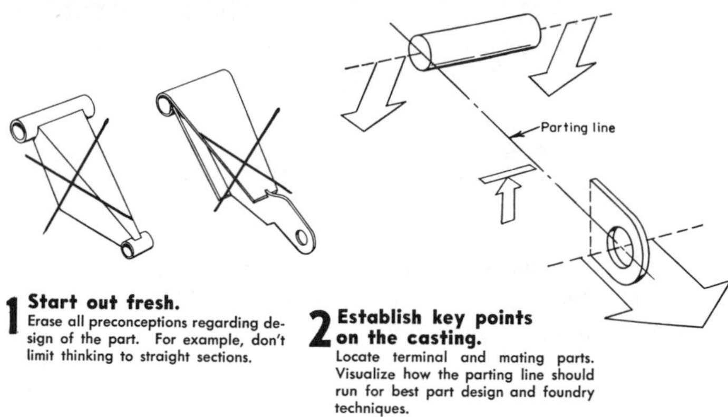


**1 Start out fresh.**
Erase all preconceptions regarding design of the part. For example, don't limit thinking to straight sections.

**2 Establish key points on the casting.**
Locate terminal and mating parts. Visualize how the parting line should run for best part design and foundry techniques.

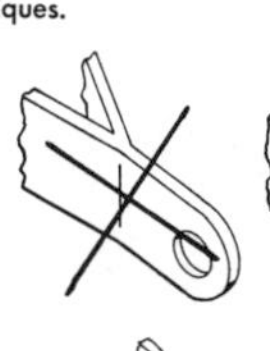

**3 Connect key points.**
Consider directions and magnitudes of service stresses when connecting the various terminal points.

**4 Check stresses at critical locations.**
Use experimental stress analysis techniques (brittle lacquer, for example) to refine the casting design.

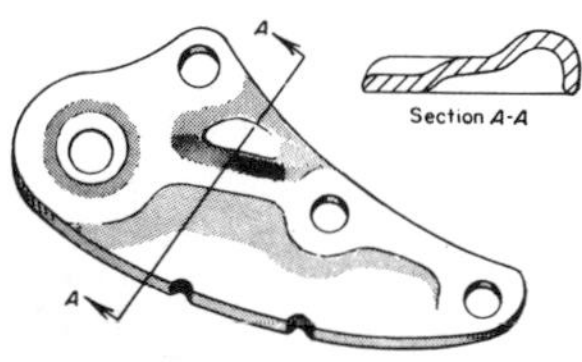


**5 Strengthen with ribs, corrugations, dimples.**
These simple shapes add greatly to strength, yet cost little in a cast part. Ribs should be used only in compression. Corrugations should be aligned along primary stress direction. Dimples should be oval in shape, with major axis of oval along the direction of stress.

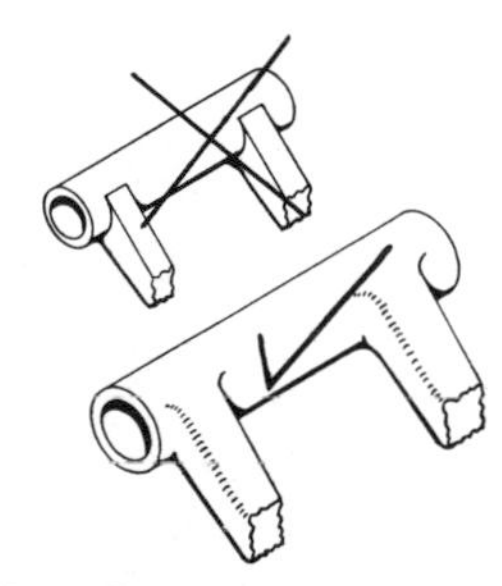

**6 Smooth out sharp corners.**
Use fillets and radii instead of sharp corners to promote metal flow and avoid stress concentration.

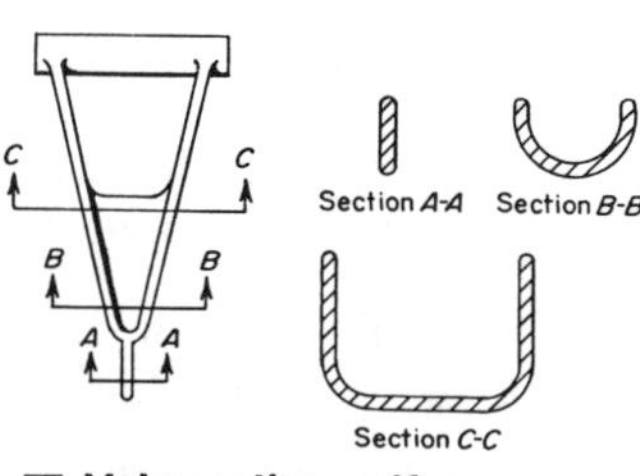
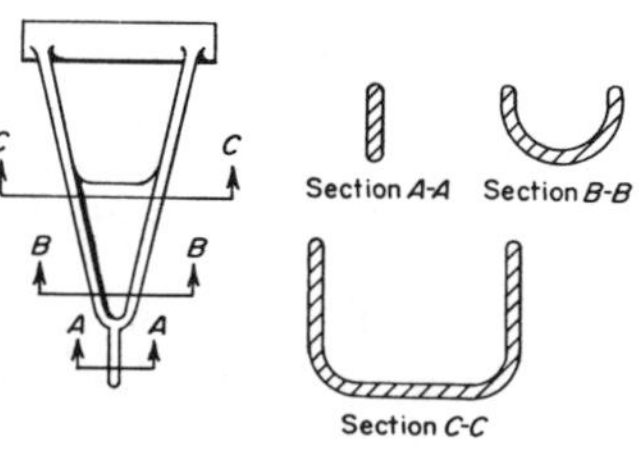


**7 Make sections uniform.**
Wall thicknesses in a malleable casting are usually from ¼ to 2 in. Sections should be designed to promote directional solidification toward the feeding head, thus insuring proper cooling and heat transfer.

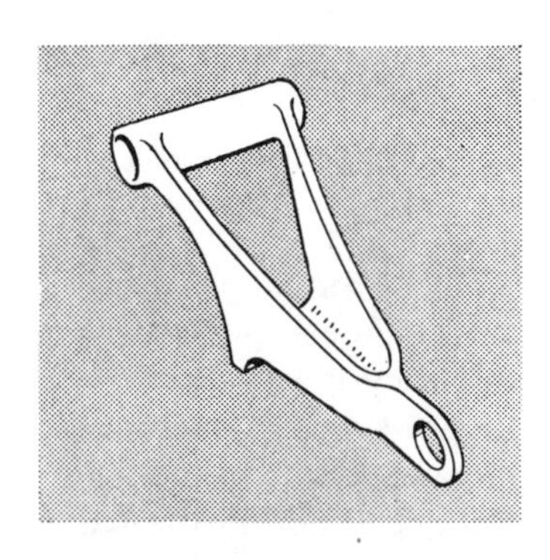

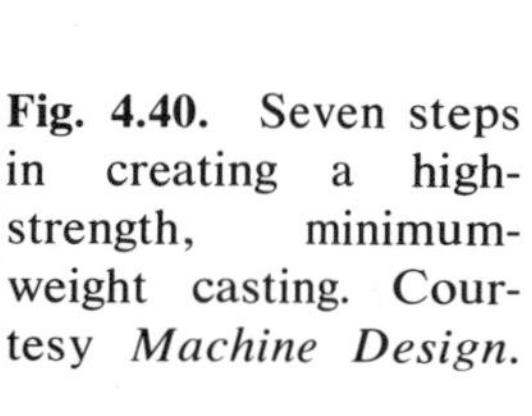

**Fig. 4.40.** Seven steps in creating a high-strength, minimum-weight casting. Courtesy *Machine Design*.

(1) The pattern may deviate from the drawing.

(2) Localized casting shrinkage was not fully compensated for by the patternmaker.

(3) Warpage may result, owing to restrained cooling of the casting. This is not always predictable, as it is difficult to say which portion of the casting will solidify first.

(4) Pattern rap, core shift, sand pressure, mold growth, and shake-out time may vary.

(5) Pouring temperature and rate of pouring will affect tolerance.

The first three items can be corrected by reworking the pattern or by modifying the molding practice. Where the quantity is large, trial-and-error corrections are entirely practical.

When extremely close tolerances are specified, there must be a correspondingly good finish. Since there is no standard specification for surface finish, this should be agreed on by the designer and the foundryman. The actual tolerances that can be supplied will vary in different foundries and according to the class of work, but Table 4.1 gives typical figures for sand castings. Closer casting tolerances can be obtained with hard molds, as discussed in Chap. 5.

**Table 4.1**
**Typical Sand-Casting Tolerances**

| Approximate size of casting (in.) | As-cast tolerance (in.) | Additional machining allowance per face (in.) |
|---|---|---|
| up to 8 | ±1/16 | 1/8 |
| 8 to 18 | ±1/8 | 3/16 |
| 18 to 30 | ±3/16 | 5/16 |
| above 30 | ±1/4 | 3/8 to 1/2 |

**Functional Design.** The overriding consideration in casting design is how functional it will be. Function is more important than cost, ease of manufacture, or appearance. The part must perform the job for which it is intended. A comparatively complicated casting should not be rejected on the basis of cost alone. It should be given a fair analysis and compared with other methods of producing the same item. A thorough analysis often leads to design improvements.

## NEWER DEVELOPMENTS

**Sand Casting.** The greatest recent development in the sand-casting process has been the use of expanded polystyrene beads or Styrofoam*

*Trade name of the Dow Chemical Co.

(*a*)

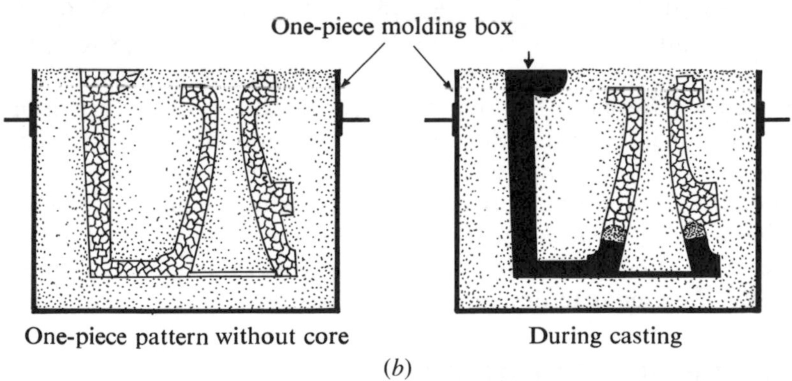


(*b*)

**Fig. 4.41.** (*a*) A large Styrofoam pattern can be easily transported. (*b*) The Styrofoam pattern vaporizes during the pouring process. Courtesy Full Mold Process, Inc.

patterns (Fig. 4.41*a*). The new method allows the pattern to be left in the mold, and it is vaporized by the molten metal as it rises in the mold during pouring (*b*).

Shapes can be cast now that were formerly impossible: undercuts are permitted, pattern draft is no longer required, coring is simplified, and loose pieces are done away with.

A new pattern must be made each time. If the pattern is relatively small and it is to be produced in quantity, the separate patterns can be made by steam expansion of the polystyrene beads in a metal mold as for the steam iron shown in Fig. 4.42. Large patterns are cut from the Styrofoam sheets with wood cutting tools or a hot wire.

Before molding, the pattern is coated with a zirconite wash in an alcohol vehicle. The wash produces a relatively tough skin separating the metal from the sand during pouring and cooling.

The molding sand is combined with a resin and a hardening agent which cause it to set at room temperature. Hard ramming of the mold is not required. Conventional foundry sand is used for backing up the mold.

No steam is produced and so no blow holes occur. Also, the pressure produced by the increased volume of the polystyrene as it vaporizes helps prevent any tendency of the mold walls to collapse.

ADVANTAGES. Patterns can be made within a relatively short time. Machining time is also greatly reduced because the part can be made closer to the desired dimensions with no draft, and undercuts can be put in where needed. Parting lines are no longer a problem.

LIMITATIONS. The repeatability of the process, when using the Styrofoam sheets, is dependent on the skill of those who cut and glue up the patterns.

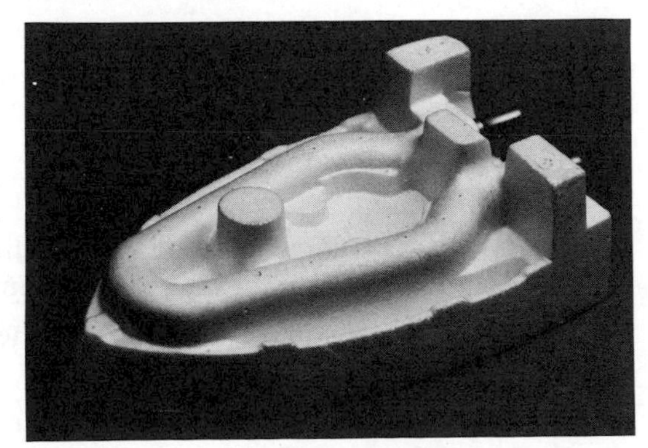

**Fig. 4.42.** A Styrofoam core for a steam iron made in a metal mold. Courtesy Full Mold Process, Inc.

## QUESTIONS

**4.1.** Explain the term *green* in greensand molding.

**4.2.** Why are dry-sand molds used?

**4.3.** Explain the terms (*a*) cope and (*b*) drag.

**4.4.** What is the function of a riser in a sand casting?

**4.5.** Why is it so important that the gates and runners be of the correct size?

**4.6.** (*a*) What is the function of the core in sand molding? (*b*) How may cores be held in place? (*c*) How may they be supported?

**4.7.** What are the advantages in having a match-plate pattern as compared to a split pattern?

**4.8.** How may the sand be tested for permeability?

**4.9.** What are the qualities of a good sand core?

**4.10.** What is the advantage of using carbon dioxide in making sand cores?

**4.11.** What advantages are claimed for the diaphragm molding machine?

**4.12.** Why would an induction-type furnace appeal to a small foundry dealing in alloy castings?

**4.13.** What steps are involved between the time a casting is poured and the time it has passed inspection?

**4.14.** What effect may the cross-sectional area have on the metallurgical structure of a casting?

**4.15.** Explain how localized cooling can be accomplished in a sand casting.

**4.16.** What is one method used to determine the cooling rate of various cross-sectional areas of the casting?

**4.17.** What general statement can be made regarding corners in casting design?

**4.18.** What are some methods of dealing with heavy sections in castings to avoid porosity and to obtain a uniform structure?

**4.19.** What provisions can be made in the casting design that will aid subsequent operations?

**4.20.** What are some principles to bear in mind when designing a casting for minimum weight?

## PROBLEMS

**4.1.** Calculate the amount of metal in pounds that would be required to make the casting shown in Fig. P4.1. Add 10 percent for sprues, runners, and risers and 5 percent for scrap loss.

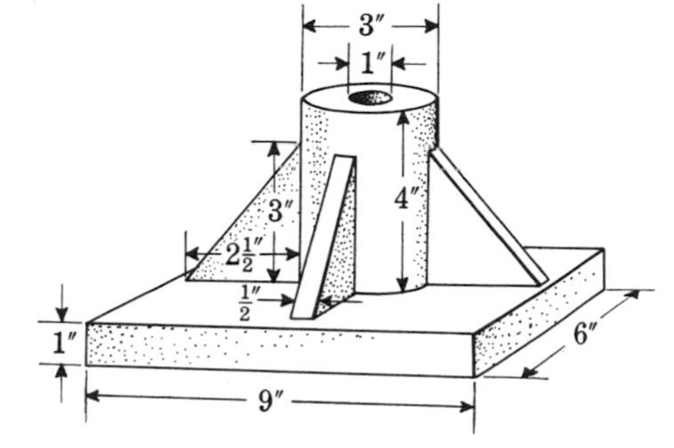


**Fig. P4.1.** Material: cast iron, 0.26 lb per cu in.

**4.2.** If 100 castings are made like the one shown in Fig. P4.1, find the approximate cost per casting. Time required to construct and finish the pattern, 4 hr. Time to make mold, 5 min. Core time, 1 min. Pouring time, 1 min 45 sec. Shakeout time, 3 min. Cutting sprues and risers, 2 min. Additional handling time, 8 min. Total labor and overhead charge is \$10.00 per hr. Cast iron cost is 17¢ per lb. A cooling time of 2 hr is charged at the overhead rate of \$5.00 per hr.

**4.3.** Phosphor bronze contains 85 percent copper, 10 percent tin, and 5 percent lead. Find the weight of each metal that would be required for a 300 lb charge.

**4.4.** A certain pig iron contains 2.5 percent silicon, and it is to be used with a scrap iron containing 1.8 percent silicon. In what proportions by weight should they be added to a 500 lb charge to obtain a final silicon content of 2 percent?

**4.5.** Show how the design of the casting in Fig. P4.5 can be easily improved to reduce casting defects and make the process easier.

**Fig. P4.5**

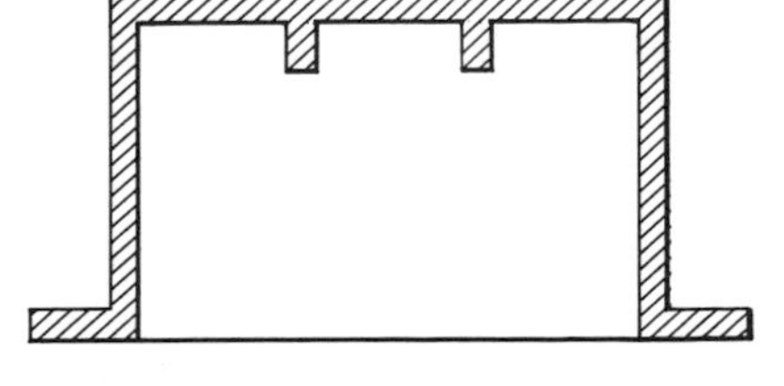

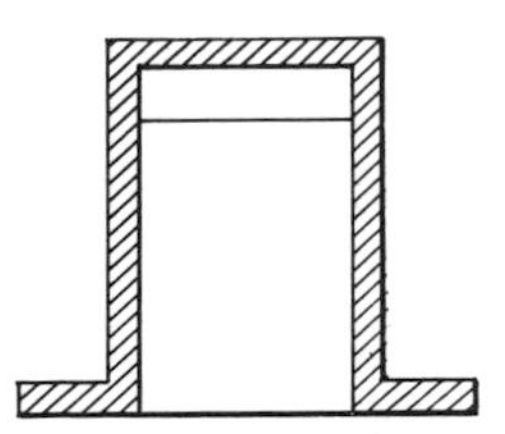

**4.6.** Make a sketch for the pattern used to make a rope pulley in Fig. P4.6. Show how it would appear in the flask with core and split pattern if necessary.

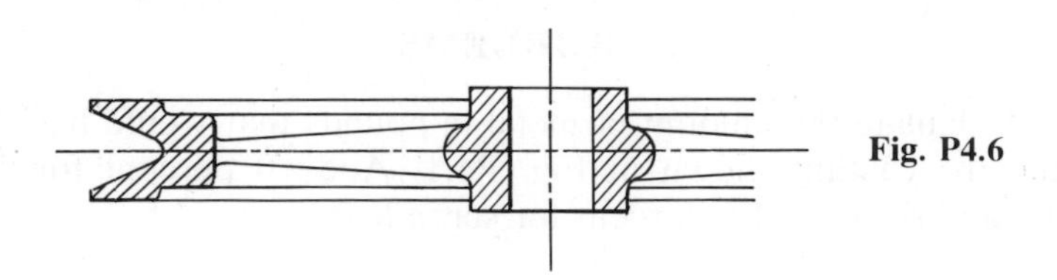

**Fig. P4.6**

## REFERENCES

Angus, H. T., *Physical and Engineering Properties of Cast Iron,* The British Cast Iron Research Association, Alvechurch, Birmingham, England, 1960.

Giordano, Felix M., "Disposable Patterns Facilitate Sand Casting," *The Tool and Manufacturing Engineer*, January 1965.

Casting Design Conference, Central Foundry Division, General Motors Corp., Saginaw, Michigan.

Malleable Founders Society, "Seven Steps in Designing Minimum-Weight Castings," *Machine Design,* November 1961.

Scharf, A., and Charles F. Walton, "Designing Gray Iron Castings," *Machine Design,* January 1957.

5

# OTHER MOLDING PROCESSES

SAND-CASTING METHODS have long been a favorite of the foundry because of their low cost and versatility. They may not, however, meet all the more exacting requirements of close tolerances with a minimum of machining and smooth surface finish. Other casting methods have been developed that meet these requirements. These processes are shell-mold casting, plaster-mold casting, investment casting, centrifugal casting, permanent molding, continuous casting, and extrusion casting.

## SHELL-MOLD CASTING

The shell-mold method of producing castings is basically a modification of the sand-mold process. It had its early development in Germany during World War II. Instead of using the regular foundry-sand mixture, which has clay and water as binders, a fine dry sand mixed with phenolic resin is applied to a pattern which is heated to approximately 450°F. The resin melts and flows in between the grains of sand, acting as a bond. This feature, plus curing on the pattern, produces a hard, smooth mold which is just as accurate as the pattern itself.

Jolting, squeezing, and ramming of the mold are eliminated in the shell-molding process. In place of the regular sand-molding equipment is an open-faced box mounted on trunnions. A resin-sand mixture is poured into the box to a depth of approximately 10 or 12 in. The pattern plate, which has previously been heated, is clamped face down on top of the box (Fig. 5.1). The box is inverted so that the sand-resin mixture drops directly on the hot pattern face.

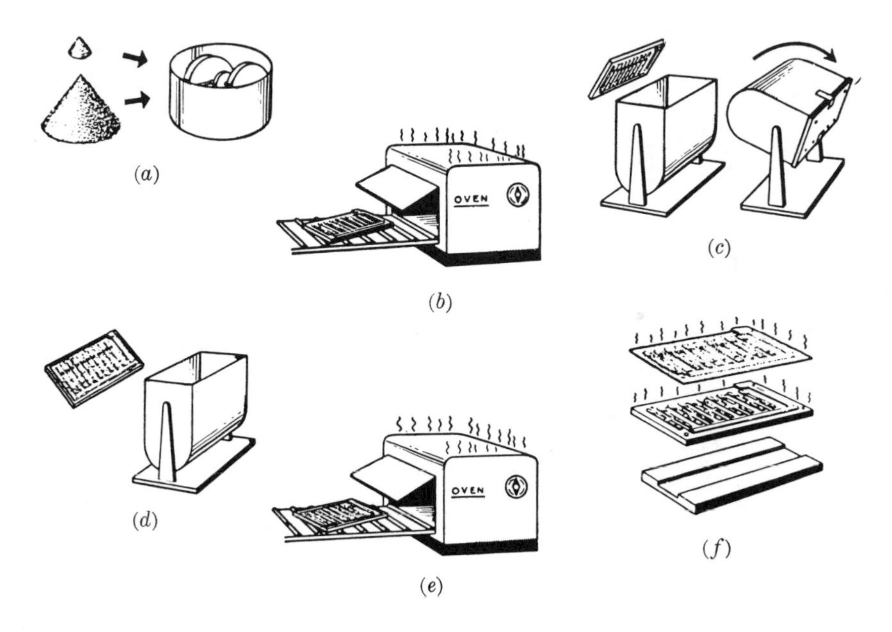


**Fig. 5.1.** The shell-molding process includes (*a*) mixing the resin and sand, (*b*) heating the pattern, (*c*) investing, (*d*) removing the invested pattern, (*e*) curing, and (*f*) stripping.

The length of time during which the mixture is in contact with the hot pattern determines the thickness of the shell. Usually, a 20- to 35-sec exposure will result in a shell 1/8 to 3/8 in. thick, which is adequate for most small to medium castings. The box is then returned to its original position, and the pattern, with the soft shell adhering to it, is removed and placed in an oven to cure. An oven temperature of approximately 450°F will require 40 to 60 sec to make the heat-hardening phenolic binder set.

Electric radiant-heat furnaces can be used effectively to direct the heat to the base and sides of projections, as shown in Fig. 5.2.

**Mold Ejection.** After the shell has had sufficient curing time, it is removed from the oven. The shell fits rather snugly on the pattern, so some means must be provided for rapid removal. This can be easily accomplished if knockout pins are incorporated into the pattern. These pins can be spring-activated, as shown in Fig. 5.3, or they can be part of a separately operated mechanical ejection system. In either case, the

Parabolic reflector
Focal point
Parabolic infrared radiant heat unit
Shell

**Fig. 5.2.** Radiant-heat furnaces are used effectively to cure shell molds.

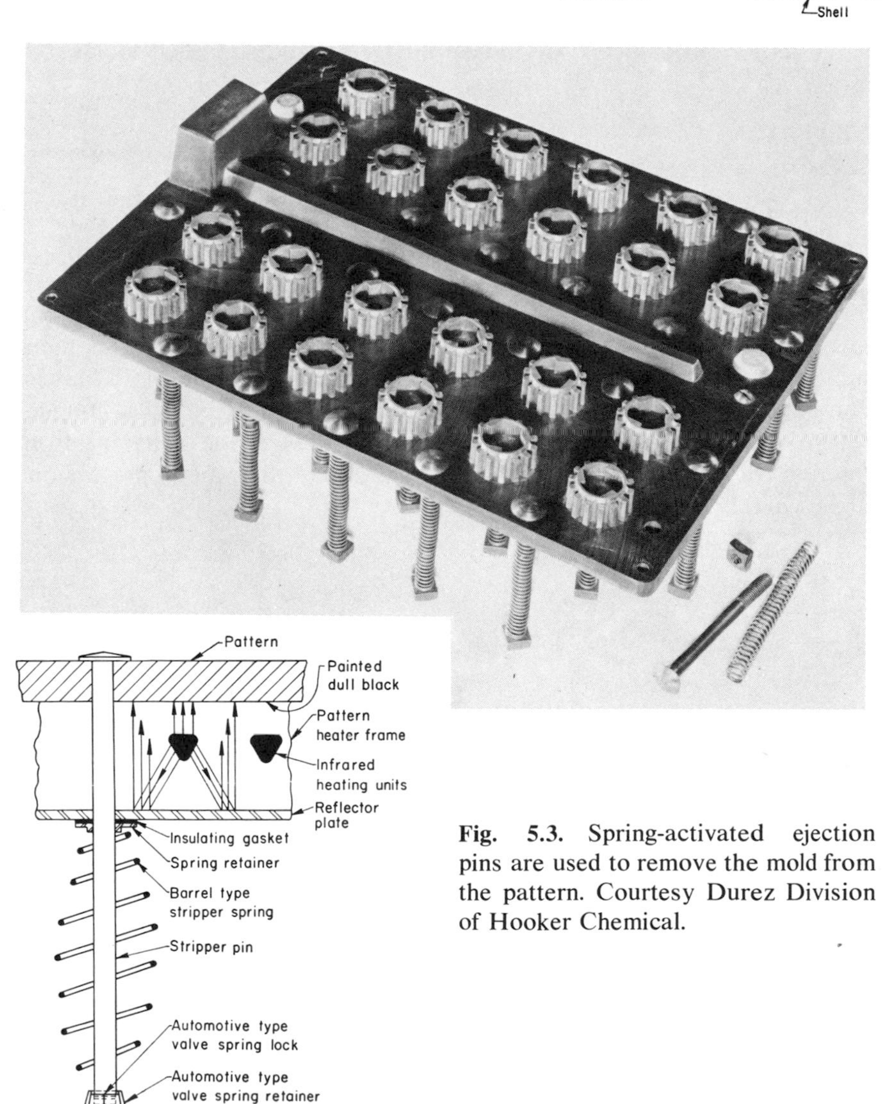


**Fig. 5.3.** Spring-activated ejection pins are used to remove the mold from the pattern. Courtesy Durez Division of Hooker Chemical.

method of removing the soft shell must be considered at the time the pattern is designed.

It is important to have the ejection pins work together uniformly. If uneven action takes place, broken or distorted molds are likely to result. The pattern also may be subjected to wear or damage.

**Draft.** The amount of draft necessary on shell molds is generally less than on sand molds, as shown in Table 5.1. Only a small amount is necessary to facilitate the removal of the mold from the pattern.

**Table 5.1**
**Comparison of Draft Allowances for Greensand and Shell Molds**

| | Greensand (deg) | Shell mold (deg) |
|---|---|---|
| Normal | 2 | 1 |
| Minimum | 1 | 1/4 |
| In pockets | 3 to 10 | 1 to 2 |

**Mold Assembly.** After the shell mold has been stripped from the pattern, it is ready for use, even though it is still warm. Various techniques are used to provide an intimate bond between the mold halves. These include clips, wire staples, tapes, and adhesives (Fig. 5.4). Tongue-and-groove arrangements at the joint help prevent the adhesive from entering the mold cavity and minimize *fins*, small thin metal protrusions at the parting line of the casting.

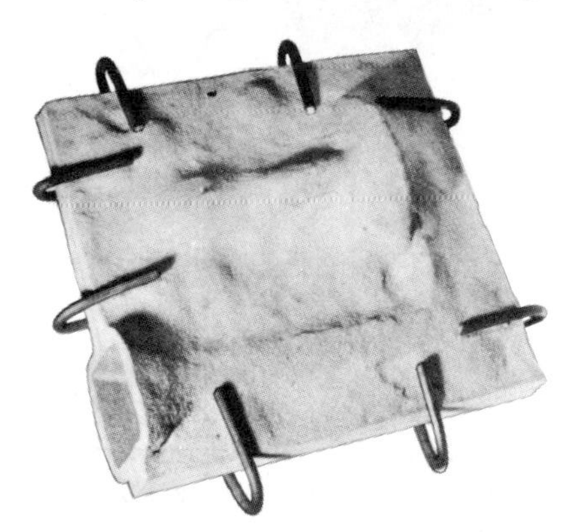

**Fig. 5.4.** The two halves of the shell mold can be assembled with clips as shown or, more commonly, with phenolic resin adhesives.

The most satisfactory method of sealing shell halves is with a thermosetting resin of the same general type as the shell-molding resin. The residual heat of the shell is enough to set the phenolic adhesive.

**Pouring.** When there is likely to be considerable metalostatic pressure on the mold, it should be supported with backup material such as metal shot, gravel, or molding sand.

**Shake-Out.** After the casting has solidified, regular shake-out techniques are used to free the casting from the shell. A series of screens may be used to separate the castings, chunks of mold, and backup materials.

**Sand Recovery.** Although it is possible to recover the sand from the sand-resin mixture of the mold, the economics of this procedure is questionable unless the quantity is large.

The resin has to be burned out at high temperature. When this is done, the sand is considered to be the same as new, and it may be used either for greensand molding or recoated with resin for shell molding.

**Patterns and Cores.** In order to withstand the heat of curing, the patterns used must be made of metal. Patterns made out of aluminum are satisfactory for short-run production, and gray cast iron, bronze, or steel may be used for high production. Although the pattern expands owing to the curing heat, this can be neglected, since the mold also shrinks after it is removed from the pattern.

Patterns for shell moldings are very accurate and are generally high-cost items. They are made by expert metal patternmakers or tool-and-die makers. They are used in a nearly polished condition.

Whenever possible, the cope and drag sections are placed on the same face of the pattern plate. This facilitates making a complete mold each cycle.

Cores are made by blowing the resin-coated sand into the corebox. The sand mixture is the same as that used for the mold. The corebox can be electrically heated or heated in an oven.

Hollow cores are made by filling the heated corebox with the coated-sand mix. After allowing a sufficient time to build up the proper wall thickness, the unfused sand can be dumped out. Completely hollow cores are made by dumping a predetermined amount of sand-resin mix into the hot corebox and then rotating it. Higher temperature are used in core production, so curing-cycle times range from 15 to 25 sec, depending upon size.

**Blowing Shell Molds.** A more modern approach to shell-mold making than the dump-box method is the shell-blowing method. In the dump-box method the sand is not always properly packed (Fig. 5.5). Places where the coated sand is loose or poorly packed lead to weak spots in the mold where the hot metal can break out.

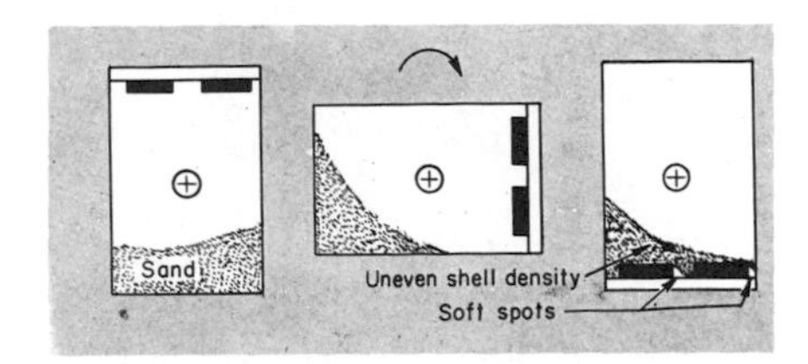


**Fig. 5.5.** The dump-box method of coating the pattern may leave soft spots and areas of unequal density in the mold.

Air disturbances and venthole problems can be eliminated by having the sand-resin mixture fall on the pattern from a height greater than 12 in. rather than by blowing it in. However, most modern shell-molding machines are built on the principle of blowing the sand in place.

**Shell-Mold Machines.** High-speed shell-mold-making machines are available that can handle from 30 to 200 full molds per hour, depending upon the pattern plate size and the number of stations available. These machines vary from the hand-operated single-station type to fully automatic multistation units that combine the investment and curing cycles into a standard mechanized operation.

**Costs.** Shell-mold costs are higher than for corresponding greensand molds. This is due primarily to the cost of the resin used. The latter, in turn, is governed by the market cost of the chemicals used to produce the resin. However, owing to the increased volume at which these chemicals are being manufactured, the price has steadily declined.

**Shell-Mold and Greensand Comparisons.** Shown in Fig. 5.6 are a greensand mold and a shell mold made to produce the same item. The shell mold weighs one tenth that of the greensand mold, but the cost of the shell mold is greater. In order to cut the ultimate cost of shell-molded parts, weight must be reduced, cores eliminated, and machining reduced or eliminated.

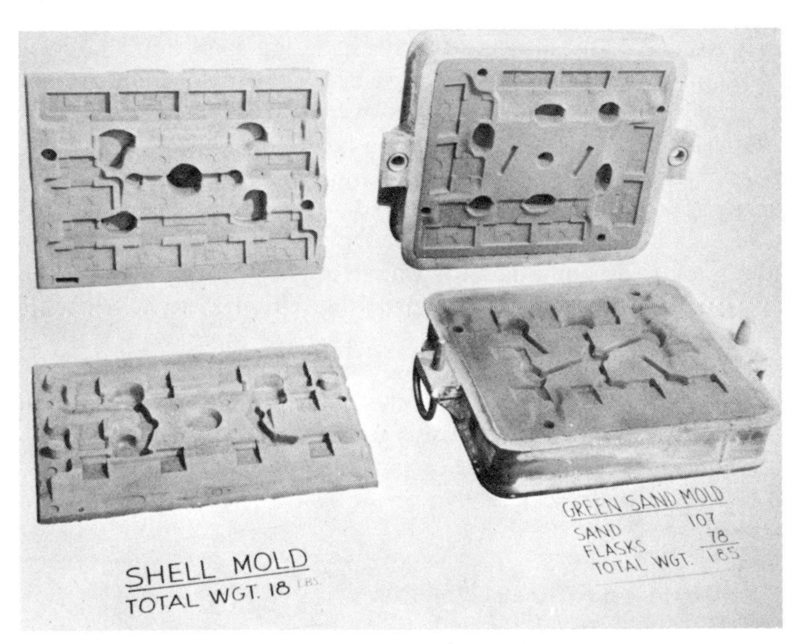


**Fig. 5.6.** Comparison of a greensand mold with a shell mold for the same item. The greensand mold has a total weight of 185 lb, the shell mold 18 lb. Courtesy Central Foundry.

The cost of the part shown at the left in Fig. 5.7 increased 40.4 percent when produced by the shell-molding process. The cost of the casting on the right increased 37.5 percent. The cost of the shell material alone amounted to over 2 ¢ per casting. In these examples, no advantage was gained by making the parts as shell-mold castings. No machining could be eliminated, and castings produced by either method performed satisfactorily in service. It is obvious that conversion to the shell process was impractical.

**Fig. 5.7.** The cost of the casting on the left increased by 40.4 percent when produced by shell molding, that of the casting on the right by 37.5 percent. Courtesy Central Foundry.

The transmission drum shown in Fig. 5.8 was redesigned from the greensand version, shown in the cross-sectional view, to a shell-mold casting. The results were better casting surfaces, longer tool life, less metal that had to be removed by machining. In addition, balance drilling was reduced. The savings in reduced machining alone were enough to justify the use of a shell mold instead of a greensand mold.

Some of the advantages of shell molding are summed up in the cross-sectional view of a hub, shown in Fig. 5.9 as cast in sand on the left, compared with the same part cast by the shell-mold process on the right. You will notice, on the greensand casting, that a considerable amount of finish stock must be machined from the inside diameter of the

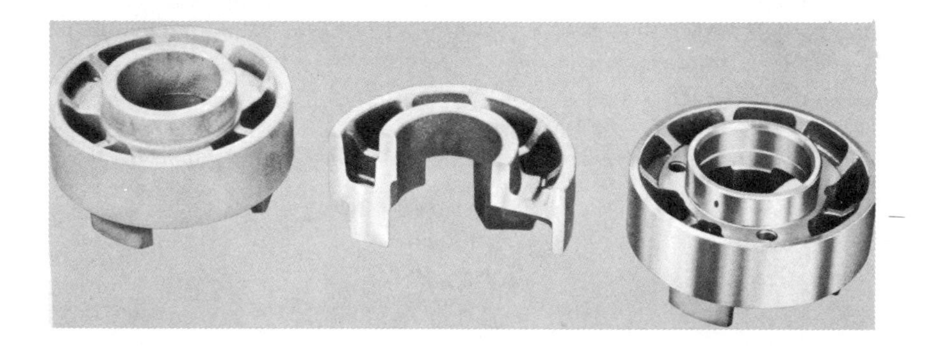

**Fig. 5.8.** This transmission drum is a good example of shell-mold casting. Note the deep pocket that can be made with only a small amount of draft.

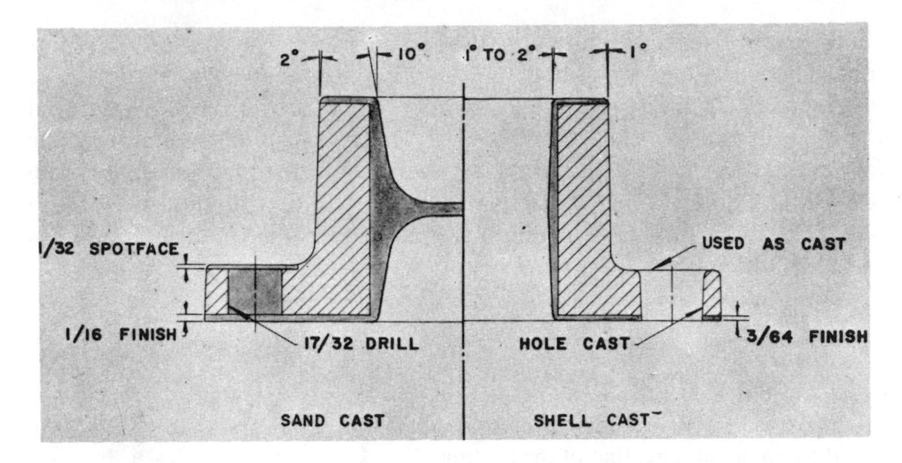


**Fig. 5.9.** Comparison of a greensand casting (left) with a shell-mold casting (right).

hub because the greensand molding requires a 10-deg draft and a 1/8-in. web in the hub. In the shell casting, however, only 1 to 2 deg of draft are needed, and no web is necessary. In the sand casting the flange would be cast solid, and spot-facing and drilling operations would be necessary on the 17/32-in.-dia holes. In the shell casting, these holes can be cast and the spot-face and drilling operations eliminated. The amount of finish stock allowed on the outside is also reduced. On the basis of these comparisons, it is up to the manufacturer to decide if the advantages gained will be sufficient to offset the increased cost of the shell casting.

The many variables that must be taken into account in casting make it difficult to set casting tolerances accurately. The tolerance chart, Table 5.2, gives a helpful comparison of what can be expected of each process. The figures refer to tolerances on one side of the parting line. It is necessary to add 0.004 in. to the values shown when the dimensions are across the parting line.

**Table 5.2**
**A Comparison of Greensand and Shell-Mold Casting Tolerances**

| | General tolerance chart | |
|---|---|---|
| Size of casting | Greensand | Shell casting |
| 0 in. –1 in. | ±0.023 | ±0.006 |
| 1 in. –2 in. | ±0.030 | ±0.012 |
| 2 in. –3 in. | ±0.030 | ±0.018 |
| 3 in. –8 in. | ±0.045 | ±0.030 |
| 8 in. –12 in. | ±0.060 | ±0.041 |

Although greensand is still the most widely used method of producing castings, Fig. 5.10 shows the growing use of the shell-mold process in the automotive field, owing to the increasing emphasis on both reduced machining time and reduced weight. Another area of growth of the shell-molding process is that of core making for rather intricate sand castings such as engine cylinder blocks, heads, etc.

**Fig. 5.10.** Typical automotive parts produced by the shell-molding process. Courtesy Central Foundry.

## PLASTER-MOLD CASTING

Plaster-mold casting is somewhat similar to sand casting in that only one casting is made and then the mold is destroyed. The advantages of plaster are the superior surface finish, good dimensional accuracy, improved metal characteristics, and finer detail.

**Material.** Metal-casting plaster is a specially formulated mold material for casting nonferrous alloys. Its main ingredients are 70 to 80 percent gypsum plaster and 20 to 30 percent fibrous strengthener. Water is added to make a creamy slurry.

**Process.** The plaster slurry is poured into the flask over the pattern, which is mounted on a match plate. When the plaster reaches its semiset state, the pattern is removed. After both the cope and the drag molds have been made, the match plate is removed, and the mold is put in an oven for drying. It is important that all the moisture be removed so that no gas or steam is formed when the molten metal is poured into it.

Cores, if needed, are made in the same way. Coreboxes are usually highly polished and are made from brass, plastics, white metal, aluminum, or magnesium. The process of pouring a plaster core and the assembly of its parts are shown in Fig. 5.11.

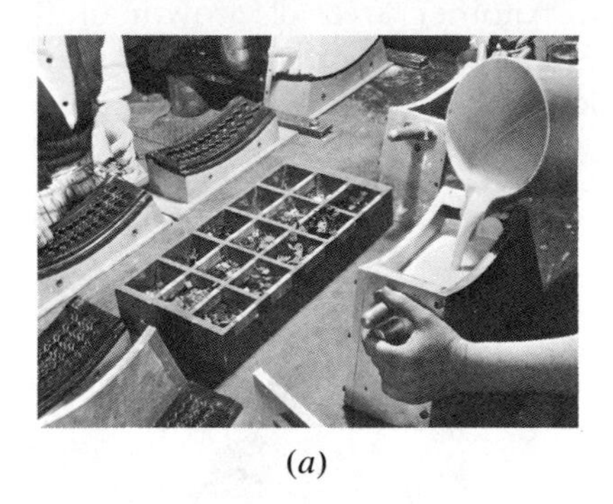

(*a*)

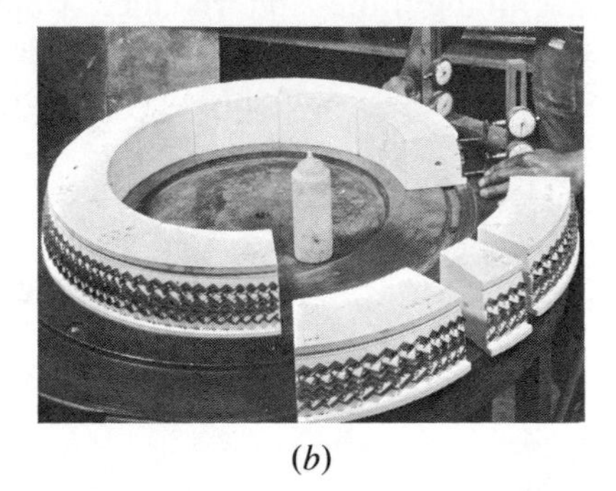

(*b*)

**Fig. 5.11.** (*a*) Pouring plaster into corebox to make cores for a tire-mold tread ring. (*b*) Assembling plaster cores into mold. Courtesy Aluminum Company of America.

Patterns are usually made out of lightly lacquered plaster, sealed wood, or polished metal. Flexible molded rubber and plastic composition materials may also be used for patterns when undercuts are encountered. The patterns are covered with a thin film of soap, lard oil, or paraffin oil to aid in removal from the plaster. Many commercially available waxes, when rubbed down with a soft cloth, make excellent parting compounds.

**Applications.** Plaster casting is generally limited to metals whose melting point does not exceed 2,100°F. It is particularly good in achieving fine finishes and close tolerances (32 to 125 microinches) in aluminum. Other metals regularly cast in plaster molds are beryllium, copper, brasses, and bronzes.

Owing to the fact that plaster has a low thermal conductivity, the metal remains in a fluid state longer and is able to fill very thin (0.040 in.) sections. Heat extraction can be increased by adding other ingredients to the plaster when needed.

Some examples of plaster-cast items in everyday use are a wide variety of gears, handles, levers, household hardware, toys, etc. Little or no machining is required on the finished part. Some plaster-cast parts are shown in Fig. 5.12.

**Expansion Plasters.** Plasters are available that expand when they set. The amount of expansion can be controlled by the plaster-to-water ratio. Values of 1/16 to 1/4 in. per ft can be obtained.

Expansion plasters are used in making patterns, match plates, and models. The plaster automatically compensates for the shrink allowance of the cast metal. For example, if a new metal pattern is to be made to

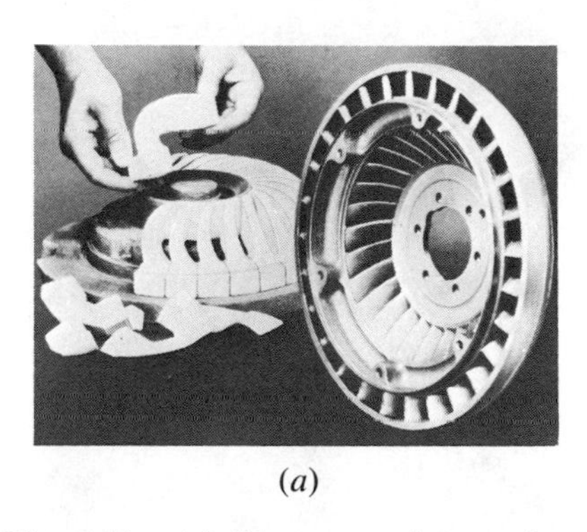

(*a*)

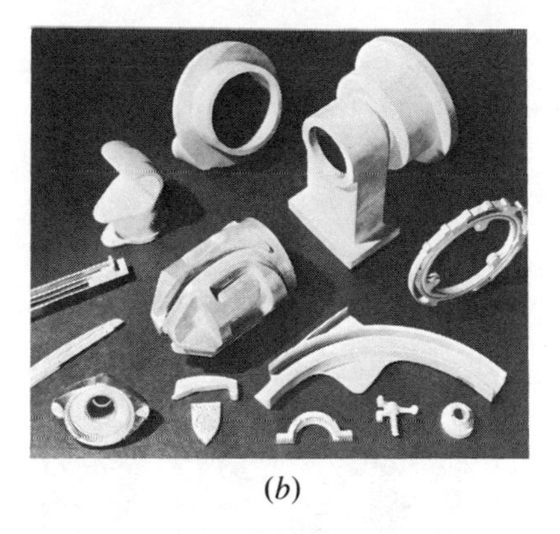

(*b*)

**Fig. 5.12.** (*a*) Plaster casting used to cast aluminum-alloy parts for automatic transmissions. (*b*) Typical casting applications of the plaster process. Courtesy Aluminum Company of America.

replace an old wooden one, the expansion plaster is cast over the wooden pattern. When it sets, the cavity will be larger than the wooden pattern. However, when the aluminum is cast in the cavity, it will shrink as it cools and be the same size as the original wooden pattern.

## INVESTMENT CASTING

**Lost-Wax Method.** Investment casting is a specialized process often called the *lost-wax method.* It was used centuries ago in China and Japan to produce beautiful statuary. Basically, it consists of pressing wax or plastic into a split metal mold. When the material is hard, it is taken out and the fins are removed. Other parts of the assembly may be fastened together at this time, or several assemblies may be gated together. The assembled gated clusters are then dipped in a very fine grained slurry of refractory powder and bonding material. This process is referred to as the *investment* of the pattern. After dipping, coarser sand is shaken over the cluster. This sand adheres to the wet slurry and builds up a thicker wall section. The built-up clusters are then placed in metal containers or flasks, and the final investment material is poured around them (Fig. 5.13).

A vibrating table equipped with a vacuum pump is used to eliminate all air from the investment. In some cases a partial vacuum is created around the flasks to assist in this process. After investment, the completed flasks are heated slowly to complete the drying of the mold and then to melt or burn out the wax, plastic, patterns, runners, and sprues. When the molds have reached a temperature of over 1000°F, they are ready for pouring. Vacuum is sometimes applied to the flasks during the pouring operation. In other cases, air pressure or centrifugal methods

(*a*)

(*b*)

**Fig. 5.13.** (*a*) The investment casting process starts with a wax pattern made by injecting a special type of hot wax into a die. (*b*) The wax pattern is covered with ceramic material, which now becomes the mold shell, later to be filled with metal.

may be employed to ensure complete filling of the mold cavities. Molds that are not difficult to fill may be poured by gravity, as in greensand molds.

When the metal has cooled, the investment material is removed by means of vibrating hammers or by tumbling. As with other castings, the gates and risers are cut off and ground down, and final inspection is performed.

**Frozen-Mercury Process.** Frozen mercury may be used for the pattern in place of the wax or plastic mentioned previously. This variation of the investment-casting process offers certain advantages, namely, excellent surface finish, intricate shapes, and high degree of accuracy. The pattern is made by pouring the mercury in precision steel dies and chilling at −100°F. In Fig. 5.14, the frozen-mercury pattern is being removed from the die.

When wax is heated for melting out, it expands with a volumetric increase of about 9 percent. Mercury has a volumetric expansion of 3.5 percent. Thus, the strains imparted to the mold walls are very low compared to the wax. Because of this, large investment castings can be made.

Unlike other investment-casting methods, in the frozen-mercury process the dipping is done in a slurry at −80°F. The shell mold is made by successively dipping in slurries of increasing viscosity until a thickness of about 1/8 in. or more is reached.

**Fig. 5.14.** Frozen mercury pattern after removal from steel die.

After the desired shell has been built up, the mercury is extracted by introducing the mercury at room temperature in the sprue or gate. After most of the mercury is washed out, the mold is allowed to come to room temperature and the remaining mercury is poured out. Recovery of the mercury is almost 100 percent. The remainder of the process is quite similar to the lost-wax method.

A frozen-mercury investment casting has been used to replace as many as 25 assembly operations. The frozen mercury can be welded into intricate shapes merely by bringing the two parts into intimate contact. Shown in Fig. 5.15 is a built-up assembly that has been redesigned for investment casting. In it 20 separate pieces have been replaced by 1.

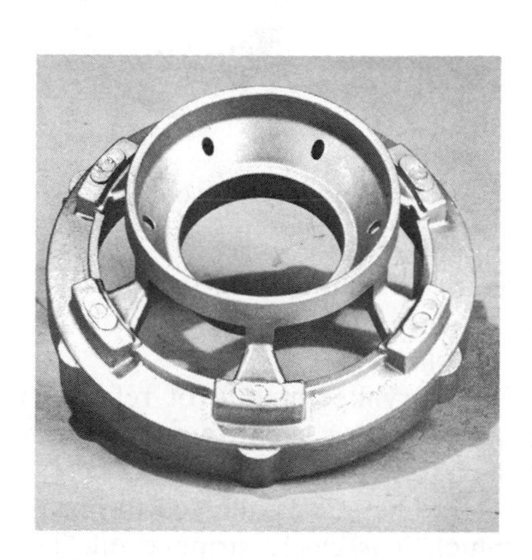

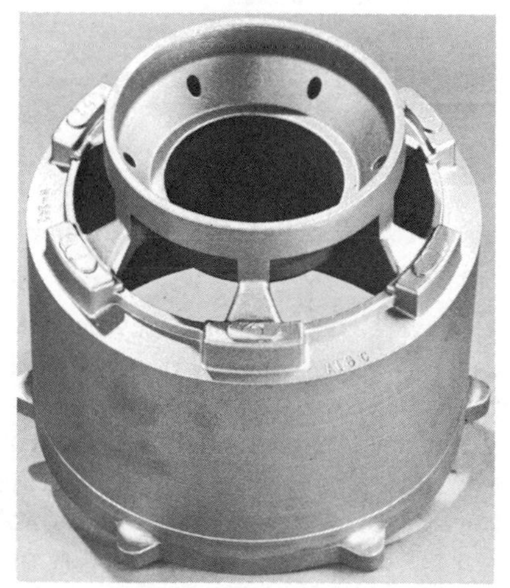

**Fig. 5.15.** A built-up assembly has been redesigned into a one-piece investment casting. Courtesy Kolcast Industries Inc.

**Applications.** Investment castings are used when complex parts are needed or when a complex part is made out of an alloy that is difficult to machine. The parts require no draft, since the pattern is melted out and the mold is destroyed when the part is removed. Investment castings are most appropriate when:

(1) One casting can be redesigned to take the place of several.

(2) The part will perform better if made out of tough alloy that may present machining difficulties.

(3) Welding, soldering, or mechanical fastening is undesirable.

(4) The quantity needed is not large.

(5) The design is subject to change after a few hundred pieces or before it is finalized into expensive long-run tooling.

(6) Accuracy and reproducibility are needed.

**Tolerance.** Tolerances as low as 0.005 in. per in. are obtainable by the investment method, but 0.005 in. or better will result in lower casting charges.

**Materials Cast.** Investment casting is applicable to both ferrous and nonferrous metals. It probably is most important for alloys which withstand high stresses, have good oxidation resistance, and are difficult to machine and weld. Various parts made by the investment process are shown in Fig. 5.16.

**Fig. 5.16.** Examples of investment-cast parts. Courtesy Arwood Precision Casting Corp.

**Shaw Process.** The newest of the investment-casting techniques is known as the Shaw process. By this process, castings can be made that range in size from a few ounces to 5,000 lb.

The process consists of pouring a slurrylike mixture of refractory aggregate, hydrolized ethyl silicate, and a jelling agent over a pattern. The pattern need not be expendable as in other investment processes. It can be made of wood, plaster, or metal. The mix is allowed to form a flexible jell over the pattern, after which it is easily stripped off. The mold is then ignited with a torch to burn off the volatile portion of the mix. After this, it is brought to a red heat in a furnace, which results in a rigid refractory mold.

Steel castings having 1/8-in. wall thickness and weighing 100 pounds have been cast in molds of this type. On small castings, tolerances may be held within 0.002 in. per in. but on larger pieces the tolerance is generally 0.010 in. per in. The surface finish varies from 80 to 120 $\mu$in. Parting lines are visible on the casting where the two mold halves are put back together. Pattern-to-finish casting time can be done in 2 hr.

The Shaw process originated in England, but has met with some variations in this country. By using the ceramic slurry to set on a chamotte-type sand mold, it can be baked out in 5 min. (A chamotte-type sand mold is made up of special molding sands containing lime, aluminum, and silicate.) Solid ceramic molds, often used in the Shaw process, require 4 to 5 hr to bake out the trapped gases. The thin ceramic facing allows the Shaw process to be automated. One machine is capable of turning out 120 molds per hour.

Both the Shaw process and the lost-wax method have been combined to gain the special advantages of each. The Shaw process has the advantage of low pattern and tooling costs. However, the pattern contours must be such that the flexible mold can be stripped off. This somewhat limits the angles or curvatures that can be made.

By setting wax-pattern components into the regular pattern or core-box, practically any configuration or complexity can be made. A closeup view of how the combined process operates is shown in Fig. 5.17. In (*a*), all but the curved portion of the blade in the impeller hub is a permanent type of metal pattern. In (*b*), the curved portion of the blade is added, with a consumable wax pattern. Shown in (*c*) is the cross section of the

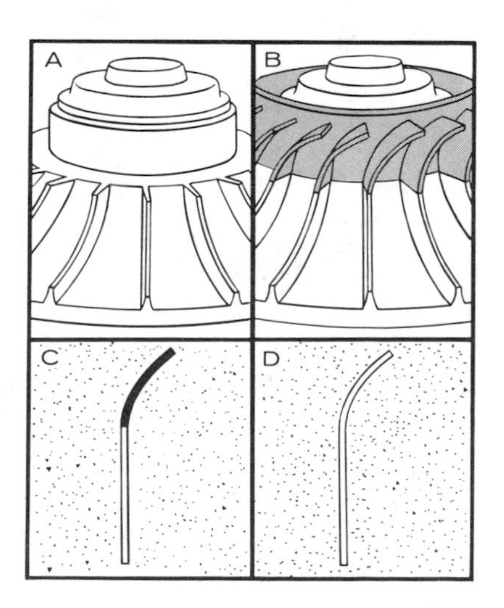


**Fig. 5.17.** A combination of the Shaw process and the lost-wax method was used to produce this impeller hub. (*a*) Metal pattern; (*b*) wax pattern added to metal pattern; (*c*) cross-sectional view after investment, metal pattern removed; (*d*) cross section after wax has been melted out. Courtesy Lebanon Steel Foundry.

blade after investment, with the metal portion of the pattern withdrawn and the wax insert still in place. At (*d*) the wax is melted out, leaving a clean, smooth contour.

## CENTRIFUGAL CASTING

Most molds are filled with metal simply by the force of gravity; however, methods have been developed that allow pressure to be exerted on the molten metal to provide greater density and more uniform structure. Various centrifugal methods are used to keep the metal under pressure as it is cast. The two principal ones are termed *centrifugal* and *centrifuge.*

**True Centrifugal Casting.** In true centrifugal casting, the molten metal is poured into a hollow cylindrical mold spinning about either a horizontal or a vertical axis at speeds sufficient to develop 60 to 75 g (gravities) of centrifugal force. This force causes the liquid metal to flow to the outside of the mold and to remain there in the shape of a hollow cylinder. Heavier components within the metal are thrown outward with greater force than the lighter particles. This helps eliminate light nonmetallic particles and impurities, which are congregated inward toward the axis of rotation through flotation (Fig. 5.18). These impurities can then be removed by a light machining operation.

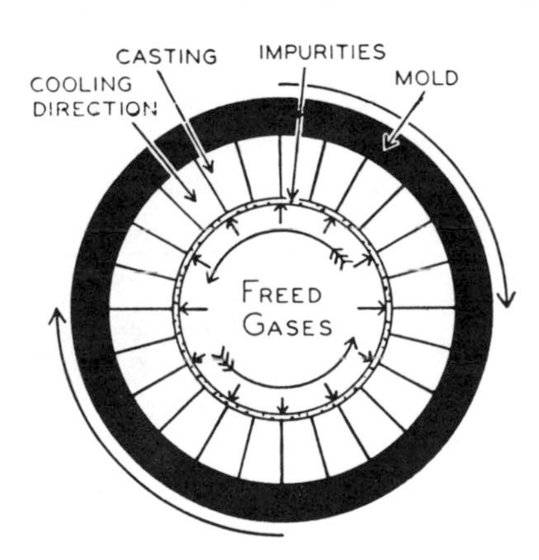


**Fig. 5.18.** The principle of centrifugal casting is to produce a dense high-grade metal by throwing the heavier metal outward and forcing the impurities to congregate inward. Courtesy Janney Cylinder Co.

The mold for centrifugal casting may be made by lining a cylindrical flask with 1 to 2 1/2 in. of sand. The pattern is placed in the center of the mold, and the sand is rammed between the flask wall and the pattern. The pattern is intermittently raised several inches at a time until the top is reached. The sand lining is washed with a mold dressing and baked in a core oven. Cast-iron pipe may be cast directly in cylindrical water-cooled molds with no sand lining.

After pouring the casting, spinning it, and allowing it to cool, it is shaken out, cleaned, and heat-treated to obtain various desired properties.

Some of the largest diameters in the greatest as-cast lengths are shown in Fig. 5.19. Even though only one special application is shown with specifications, the uses are many: rolls, sleeves, rings, bushings, cylinders, liners, tubes, shells, pressure vessels, retorts, pump rotors, and piping. Short lengths are centrifugally cast in diameters up to 62 in. or as small as 2 in.

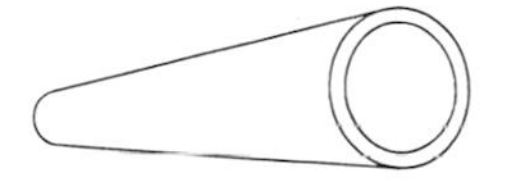


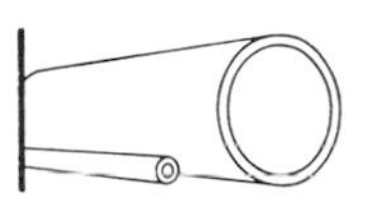


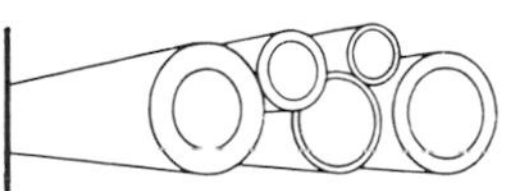


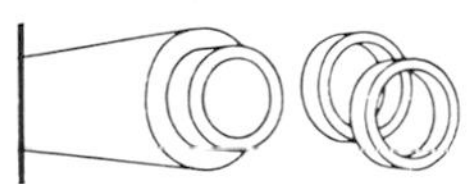


(*a*)

(*b*)

**Fig. 5.19.** (*a*) Some examples of centrifugal casting sizes that are available in stainless, carbon, and low-alloy steels. (*b*) A centrifugal casting used in a liquid-air converter machine having very close tolerance. Courtesy Sandusky Foundry and Machine Co.

Although the inner surface is usually round, the outside need not be. It can be any symmetrical shape—hexagonal, octagonal, or square. Occasionally, bosses and flanges are incorporated into the design.

**Centrifuge Casting.** The principal method employed in centrifuge casting consists of arranging the molds on a "wheel." The pouring sprue is at the hub, and the gates run down to each of the molds, similar to the spokes on the wheel. This method is particularly adaptable to small intricate castings that would ordinarily be difficult to gate. Risers are usually not necessary.

Molds used for centrifuge casting may be sand (green or dry), plaster, steel, cast iron, or graphite. Cores also may be used. The molds are made very close to tolerance, with little or no allowance for machining.

## PERMANENT MOLDING

One distinct disadvantage of the sand-casting processes is that the mold is destroyed each time it is used. It is natural that attempts would be made to make permanent metal molds. In the Middle Ages, iron molds were used to produce pewterware, such as cups, pitchers, and other utensils. Later, tin soldiers were made by pouring molten metal into iron molds that were hinged together. After a suitable time, the bulk of the still-liquid metal was poured out. The result was a thin-walled casting produced by what is termed *slush molding*.

Permanent molding includes two main types—die casting and permanent-mold casting.

**Die Casting.** The search for smoother, more accurate castings, produced at higher rates, led to applying pressure to the molten metal and the use of die-cut steel molds. By 1904, automobile bearings were being die cast. Die casting refers to both the process and the product. Today, die casting is used to produce unlimited quantities of parts of such uniformity and accuracy that machining costs are either greatly reduced or eliminated entirely. Die casting is generally considered a one-step process because molten metal is converted in a matter of seconds from a fluid into a finished or semifinished product.

The process consists of injecting molten metal at high pressures (100 to 100,000 psi) into a split metal die cavity. Within a fraction of a second, the fluid alloy fills the entire die, including all the minute details. Because of the low temperature of the die (it is water-cooled), the casting solidifies quickly, permitting the die halves to be separated and the casting ejected.

If the parts are small, several parts may be cast at one time in what is known as a *multiple-cavity* die.

**Machines.** Two main types of machines are used to produce die castings. The *hot-chamber* machine, shown in sectional view in Fig. 5.20, is used for casting zinc alloys. In this type of machine, a supply of molten metal is kept in a holding pot (*a*) so that, as the plunger (*b*) descends, the required amount of zinc alloy is automatically forced into the die (*c*). As the piston retracts, the cylinder is again filled with the right amount of molten metal.

The *cold-chamber* machine shown in Fig. 5.21 is used for aluminum, magnesium, and copper alloys. It derives its name from the fact that the metal is ladled into the cold chamber for each shot. This procedure is necessary in die-casting the higher-melting-point alloys, because these metals pick up iron if left in molten form in contact with an iron holding pot or plunger. This iron pickup not only would cause the plunger to freeze in the cylinder but would affect the properties of the die castings produced.

Castings are removed automatically from the die by ejector pins. Most of the castings will have flash where the two die halves come together. This may be removed with trimming dies or abrasive wheels on larger castings and by barrel finishing on the smaller sizes.

There are die-casting machines capable of handling over 60 lb of metal

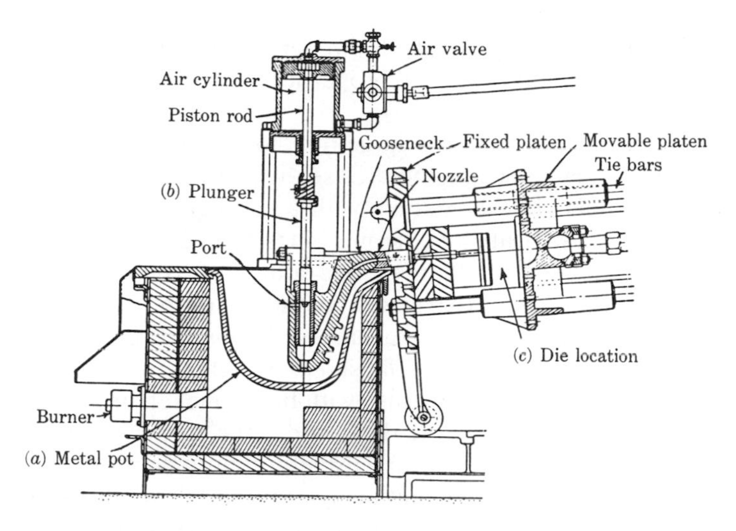


**Fig. 5.20.** The hot-chamber die-casting process. Courtesy New Jersey Zinc Company.

per shot. Automobile engine blocks are being made on machines of this type.

**Disadvantages.** The first cost of both the machine and the die is high. Machines cost upwards of $30,000, and dies may run from $1,000 to over $30,000. Die cost is high because of the accuracy to which they are made and the time required to put a high polish on all the cavity surfaces. Dies for zinc castings call for intermediate grades of die steels (heat treated for long production runs). The higher melting point alloys of aluminum, magnesium, and copper require special die steels properly heat treated.

The sharpest restriction in the use of die castings comes from the

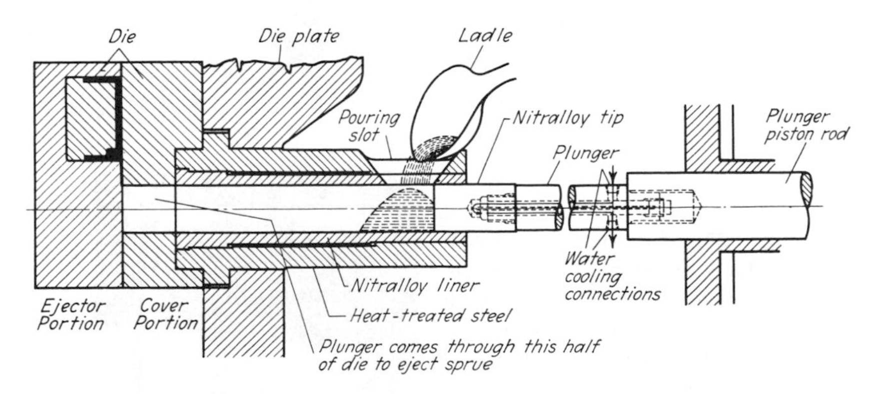


**Fig. 5.21.** Schematic view of the cold-chamber die-casting process. Courtesy American Zinc Institute Inc.

limited range of die-casting metals. However, through research and close quality control, maximum physical properties are being secured from the metals being used.

Some difficulty is encountered in getting consistently sound castings, particularly in the larger capacities. Gases tend to be entrapped, which results in small holes in the casting, with a subsequent loss in strength and the annoyance of leaks. Some machines are being modified to put the metal under vacuum to eliminate this source of trouble.

**Advantages.** Castings of extreme smoothness are produced with zinc alloys, and, if the die cavities are properly polished, castings can be plated with little or no additional finishing. This fact, together with the ease and speed of plating the castings with brass, nickel, or chromium, is the primary reason for using zinc die castings for hardware and decorative applications. They are used extensively in automobiles for decorative trim, fuel-system parts, ignition-system parts, window-operating mechanisms, chassis, and grille. The aircraft industry finds wide use for die castings, particularly those made from magnesium, because of the weight factor. Die castings, ranging in size from eight-cylinder engine blocks to tiny instrument parts, are used by almost every industry. A few examples are shown in Fig. 5.22.

Probably the greatest single advantage of die casting is the low end cost. Where production runs are long and an appreciable degree of complexity or precision is indicated, die castings are the most economi-

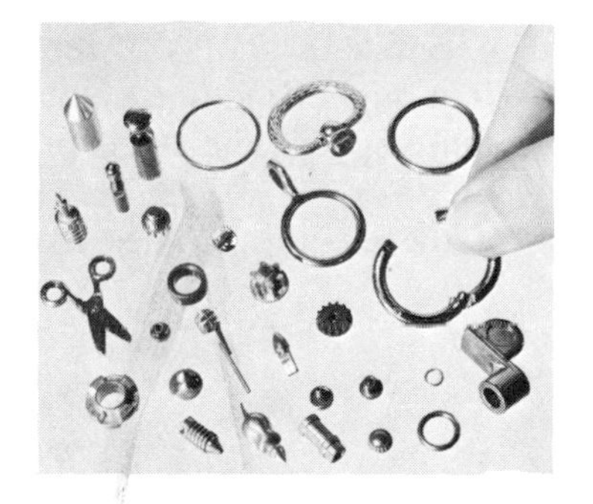

**Fig. 5.22.** A variety of die-cast products typifying some of the advantages of the process, such as clarity of detail, miniature size, good finish, close tolerance, and labor saving. Courtesy American Zinc Institute, Inc.

cal. The cost per item can be low because of the speed of production and the small amount of labor involved.

Compared with molded plastics, die castings have the advantages of greater strength and stability; they provide more secure anchorage for screws and other fasteners and are generally superior in durability.

Inserts of other materials can be placed in the die cavity and permanently assembled into the casting when the molten metal flows around it. These inserts are usually used to incorporate stiffening or bearing materials. The metallic parts of some zippers are die-cast around the cloth that supports them.

**Metals Used.** ZINC. Zinc alloys account for three fourths of the total tonnage of die castings produced, because they present the most favorable combination of low cost per casting, good physical properties, and ease of casting and finishing. The average melting point of zinc alloys is 716°F. Casting is commonly done at temperatures of 750° to 800°F. Speeds up to 500 casting cycles per hour can be obtained.

ALUMINUM. Aluminum alloys rank second in die-cast use. In some cases they cost no more than zinc castings of the same dimensions. The primary advantage is light weight. Die-casting rates are lower than for zinc, commonly from 80 to 200 die fillings per hour.

COPPER ALLOYS OR BRASS. Brass has the highest physical properties of the die-casting group, but, because of the high melting points of such alloys (ranging up to 1700°F), the die life is short, depending upon the die steels used and the mass volume of the casting sections. Copper alloys have a high specific gravity, and cost per casting is higher than for aluminum or zinc castings. However, this is offset by physical properties of strength, high hardness, and good corrosion resistance. About 150 cycles can be obtained per hour.

**Design Considerations.** As with other casting processes, it is desirable to have as nearly uniform sections as possible. Fillets and radii are preferred to sharp corners. Inserts may justify the higher cost of production due to slower operation, but they must be designed so that they can be positively located and locked in the die. Large flat surfaces should be avoided, especially where appearance in the final assembly is important.

**Table 5.3**
**Minimum Design Requirements**

MINIMUM DRAFT REQUIRED

| | |
|---|---|
| Outside walls: 1/2° minimum | Inside walls: 1° minimum |
| 1/8 to 1/4-in.-diameter cores: | 0.020 in. per in. in aluminum |
| | 0.016 in. per in. in zinc |
| 1/4 to 1-in.-diameter cores: | 0.016 in. per in. in aluminum |
| | 0.012 in. per in. in zinc |

**Table 5.3 (Continued)**
**Minimum Design Requirements**

MINIMUM TOLERANCES REQUIRED

Aluminum: ± 0.003 in. for first inch plus
± 0.001 in. per additional linear inch
Zinc: ± 0.002 in. for first inch plus
± 0.001 in. per additional linear inch

MACHINING ALLOWANCE. When machining is necessary, an allowance of 0.010 to 0.020 in. over the final dimension should be left to give a definite cut for the tool. Standard tools and cutters can be used quite satisfactorily, but greater speeds and better finishes can be obtained with larger rake and clearance angles.

The diagram in Fig. 5.23 shows the various stages in the design and production time of an average die casting.

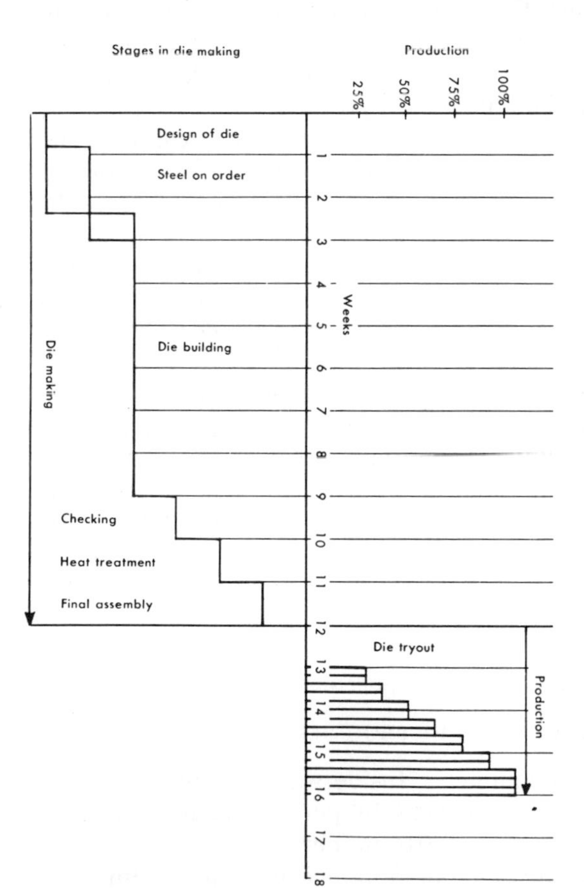


**Fig. 5.23.** Diagram showing the various stages in design and production of an average die casting. Courtesy The Hoover Company.

## PERMANENT-MOLD CASTING

The permanent-mold process is quite similar to the die-cast process in that the molds are made of metal—usually cast iron, die steels, graphite, copper, or aluminum. The cast mold is machined to obtain certain dimensional tolerances and to get proper draft angles. Vent plugs are inserted into the cavity to allow gases to escape when the molten metal is poured into the cavity.

**Pouring Permanent Molds.** Permanent molds may be poured individually or mounted on special turntables (Fig. 5.24). The turntable consists of an actuating mechanism for opening and closing the molds as they continually rotate around a central hub. The complete cycle of the machine can be made in 2 to 6 min, during which time the metal is poured, cooled, and ejected from the mold. Ordinarily, two operators are required, one to pour molten metal into the mold and the other to eject the castings after the molds are mechanically opened. The mold is automatically cleaned, coated with a refractory material, and closed to protect it from the heat of the molten metal preparatory to another cycle of operation. When cored castings are produced, a third operator is used to set the cores in the molds.

**Fig. 5.24.** A semiautomatic permanent-mold machine. Courtesy Eaton Mfg. Co.

**Disadvantages and Advantages.** The main disadvantage of permanent-mold casting is similar to that of permanent-mold die casting—high initial cost of tooling. The size of the castings is also limited by the mold-making equipment and the mold machine, usually not much over 100 lb.

The advantages of permanent-mold casting over sand casting include generally higher mechanical properties with smoother casting finish, good dimensional uniformity with less finish stock, improved pressure tightness, and ease of adaptability to automatic high production.

The higher mechanical properties stem from the rapid solidification, which improves soundness and microstructure. In the case of cast iron, the minute expansion caused by the conversion of combined carbon to graphite and ferrite is restricted owing to the rigidity of the mold. Thus internal shrinkage and porosity do not occur. The higher cooling rate of permanent-mold castings develops higher strength. For example, regular SAE 110, a fully annealed sand-cast gray iron, has a tensile strength of 20,000 psi, whereas a permanent-mold casting of the same material has a tensile strength of 57,000 psi.

The most frequently used casting processes are summarized in Table 5.4.

## NEWER DEVELOPMENTS

**Centrifugal Casting.** Ceram-Spun molds are a recent development in centrifugal casting. A predetermined weight (or volume) of semiliquid ceramic slurry is placed in the horizontal cylinder, which is then rotated at high speed. The ceramic materials are centrifuged against the flask wall to form the mold. Because the mold cavity is concentric with the flask, very high "g" values (up to 150 "gs") can be achieved. The high speed contributes to the soundness and density of the casting.

**Continuous Casting.** Continuous casting is a relatively new process of "quick freezing" poured parts formed in a graphite die and cut to length as it descends from a furnacc (Fig. 5.25).

The metal is melted in a rocking electric furnace and passed on to an enclosed casting furnace with a dry nitrogen atmosphere. The molten metal is tapped from the bottom of the furnace and is formed immediately in a water-jacketed graphite die. The large volume of metal above the die acts as a huge riser, feeding continuously and preventing shrinkage and cavities.

Below the die, a variable-speed roll draws the metal at a rate adjustable to the mass and cooling rate of the section. The movable cutoff saw can be introduced at any interval to cut the section to length. After cutting, the casting is swung from horizontal for checking and straightening.

Because the graphite dies are relatively inexpensive, the process can be used for almost any continuous shape in small or large lots. Dies usually cost between $25.00 and $100.00 and can be produced economically enough for orders as small as 500 lb.

USE. Continuous casting is applied to a broad range of copper-base alloys including those containing tin, lead, zinc, and nickel. The variety of parts in size and configuration is almost infinite; however, a few

**Table 5.4**
**Review of Principal Casting Processes**

| Process | Advantages | Limitations |
|---|---|---|
| Sand casting | Almost any metal can be used; almost no limit on size and shape of part; extreme complexity possible; low tool cost; most direct route from pattern to mold | Some machining always necessary; castings have rough surfaces; close tolerance difficult to achieve; long thin projections not practical; some alloys develop defects |
| Shell-mold casting | Rapid production rate; high dimensional accuracy; smoother surfaces; uniform grain structure; minimized finishing operations | Some metals cannot be cast; requires expensive patterns, equipment, and resin binder; size of part limited |
| Plaster-mold casting | High dimensional accuracy; smooth surfaces; almost unlimited intricacy; low porosity; plaster mold is easily machined if changes are needed | Limited to nonferrous metals and relatively small parts; mold-making time is relatively long |
| Investment casting | High dimensional accuracy; excellent surface finish; almost unlimited intricacy; almost any metal can be used | Size of part limited; requires expensive patterns and molds; high labor costs |
| Permanent-mold casting | Good surface finish and grain structure; high dimensional accuracy; repeated use of molds up to 25,000 times; rapid production rate; low scrap loss; low porosity | High initial mold costs; shape size and intricacy limited; high-melting-point metals such as steel unsuitable |
| Die casting | Extremely smooth surfaces; excellent dimensional accuracy; rapid production | High initial die costs; limited to nonferrous metals; size of part limited |
| Centrifugal and centrifuge casting | Centrifugal force helps fill mold completely; gases and impurities concentrated nearest center of rotation; solid good outer surface; gates and risers kept to a minimum | Alloys of separable compounds may not be evenly distributed; castings must be symmetrical; centrifuge, generally limited to small intricate castings |

**Fig. 5.25.** Continuous casting process components and typical parts. Courtesy American Melting and Refining Co.

general rules apply: (1) minimum inside dimension, 7/16 in. round and 1 in. for other shapes; (2) maximum outside dimension, 9 1/4 in.; (3) minimum section thickness, 5/32 in.; (4) tolerances on shapes other than round, approximately ± 1/2 percent of the outside dimension, but not less than ± 0.015 in.

The properties of the materials are considerably improved over sand-cast metals.

**Extrusion Casting.** The Russians were the first to publish any information on extrusion casting, which was in the form of a monograph in 1964. The process consists of extruding metal just as crystallization is complete or when the metal passes into a plastic state.

The essential features of the process are as follows: A mold is filled with a measured amount of metal (Fig. 5.26*a*). As the ram descends, the molten metal is forced upwards. When the forming process is finished, the punch pressure is transmitted to the top surface of the casting and then to the inner surface to consolidate the casting during crystallization. The metal remains under pressure until the casting has solidified.

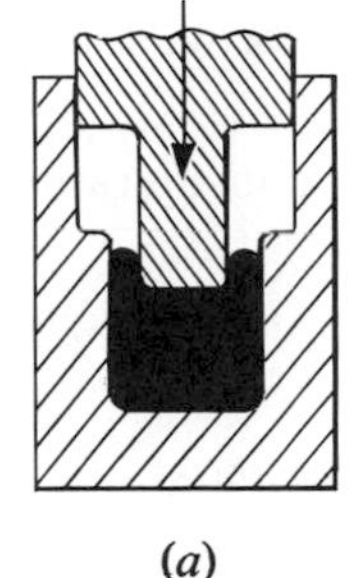

(*a*)

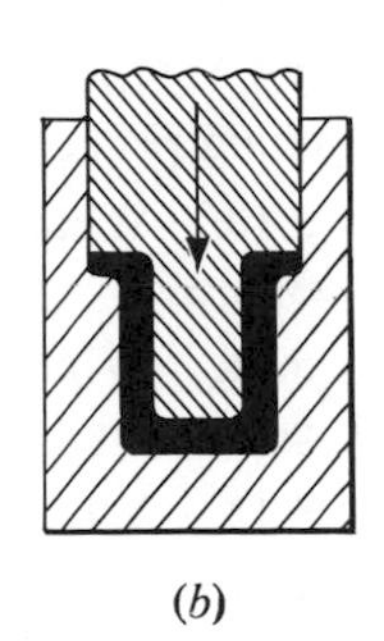

(*b*)

**Fig. 5.26.** Schematic of the extrusion casting process.

Although the process is somewhat similar to die casting, the following advantages are claimed:

(1) Both forging and casting alloys can be used.

(2) The metal has a shorter distance to travel so it loses very little fluidity and can fill the mold with less pressure and speed.

(3) No hydrodynamic pressures are involved as when the main stream of metal enters from the side, perpendicular to the core, without doing useful work and hindering the deep cavities from being filled.

(4) Deep, thin sections can be formed with complex ribs.

(5) No metal is lost on gate and riser system or flash.

The advantage claimed over hot forging is as follows: Copper alloy forgings have a restricted range and soon loose their plasticity. The last hammer blows may cause work hardening and cracks.

One disadvantage given is that, when components require a central through hole, a bottom web is left, 0.125 to 0.150 in. thick.

## QUESTIONS

**5.1.** Can the sand used in shell molds be economically recovered?

**5.2.** Why are shell-mold patterns generally more expensive than those used in greensand casting?

**5.3.** On what basis is it possible for shell molding to compete with sand casting?

**5.4.** What general tolerances can be expected in shell molding if taken on both sides of the parting line?

**5.5.** What are some advantages of plaster casting?

**5.6.** Explain what is meant by investment in the investment-casting process.

**5.7.** Why is frozen mercury a good pattern material to use in the investment-casting process?

**5.8.** Where does the investment-casting process find its best applications?

**5.9.** How does the Shaw process differ from conventional investment-casting procedures?

**5.10.** What are some considerations in the use of inserts in die castings?

**5.11.** What is the difference between centrifugal casting and centrifuge casting?

**5.12.** What materials are most often used for die casting? Why?

**5.13.** (*a*) What are some difficulties encountered in the die casting of comparatively large items such as automobile engine blocks? (*b*) What is one method of overcoming this difficulty?

**5.14.** Why are most die castings not made out of high-strength materials?

**5.15.** How do tolerances compare for die casting and investment casting?

**5.16.** What are some advantages and limitations of permanent-mold casting?

**5.17.** What factors lead to higher product strength in permanent-mold casting than in sand casting?

**5.18.** How does the amount of draft used on shell molds compare with that of greensand molds?

**5.19.** What are some possible causes of poor shell-mold castings?

**5.20.** What are some of the limitations of plaster casting?

**5.21.** What are the advantages and limitations of (*a*) continuous casting, (*b*) extrusion casting?

## PROBLEMS

**5.1.** The hub shown in Fig. 5.9 can be made by sand casting or shell molding. Assume the following savings per part are made by shell molding: weight reduction, 1/2 lb; machining time reduction, 5.2 min; tooling cost reduction, 1.7¢. Labor and overhead are calculated at $10.00 per hr. Cast iron cost is figured at 16¢ per lb. Assume the production rate is the same by either method. How much could be added to the shell-molding cost and still show a 10 percent cost reduction over sand casting?

**5.2.** Enlarge to full size the one-half scale sketches shown in Fig. P5.2, and indicate by adding shaded areas and dimensions how machining allowances and draft will differ between the sand-cast and the shell-molded part.

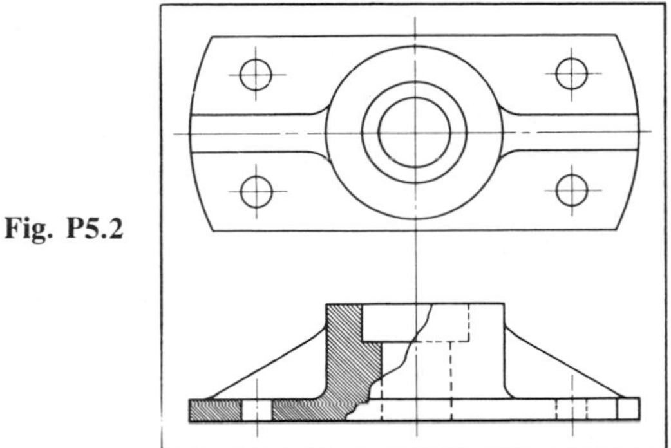

Fig. P5.2

**5.3.** How long after the design has been completed for a die casting can it be expected to be in final assembly? When in 50 percent production? When in 100 percent production?

**5.4.** (*a*) Calculate the amount of cast iron (density = 0.28 lb/cu in.) required in lbs. for the sand mold casting shown in Fig. P5.2; 100 pieces are required. Other considerations are: (1) 5% waste in metal due to spilling and over flowing; (2) 10% loss in metal in form of sprues, runners, and risers; and (3) 5% loss in slag.

(*b*) If approximate cost of pig iron is $0.04/lb (not considering lime, coke, etc.), find the total cost of the metal required.

(*c*) If approximate price of gray iron castings is $0.10/lb to $0.25/lb, calculate the price of castings shown in Fig. P5.2. What will the price of each unit be? (The lower cost figure, $0.10, is applied when quantity is very large and shape very simple. The higher cost figure, $0.25, applies when the quantity is small and the shape of the casting is complex.)

**5.5.** If 80 gray iron castings are made as shown in Fig. P5.2, the casting weight is 25 lb after cleaning and cutting off, but before machining. Find the approximate total cost with the following considerations:

Time for construction and finish of pattern, 8 hr.

Cost of material used in pattern, $5.00.

Time to make mold, 6 min.

Time to make core, 2 min.

Time for cleaning molds, placing core, and tying down cope and drag, 1 1/2 min.

Pouring time, 1 1/2 min.

Cooling time, 3 hr.

Shake-out time, 3 min.

Cutting sprues and risers and cleaning, 3 min.

Additional handling time, 10 min.

Labor and overhead charges at $10.00/hr. Cooling time is to be charged at overhead rate of $4.00/hr.

Cost of gray iron is to be taken as \$0.12/lb. Scrap, runner, riser, etc., waste to be taken as 10% of the weight of the casting.

(*a*) Calculate the approximate cost per casting. (*b*) Calculate the approximate *price* of gray iron casting in \$/lb.

**5.6.** How long do the following stages take in design and production of an average die casting? (*a*) die making as a whole? (*b*) to reach 100% production? (*c*) time for die making, after material has been received to final assembly? (*d*) time from the beginning of design to checking of the die? (*e*) time from checking to 50%, 75% production?

## REFERENCES

Broad, Edward M., "New Alloys Join the Ranks of Investment Casting," *Machine Design,* June 23, 1966.

Jones, Bruce M., "Centrifugal Casting at 150 'g's'," *The Tool and Manufacturing Engineer,* April 1966.

Bailey, R. W., "Continuous-Cast Copper Base Alloys," *Machine Design,* February 4, 1965.

Plyatskii, V. M., "Extrusion Casting," *Primary Sources,* Monument Press, New York, 1965.

# 6

# FORGING AND HOT WORKING

FORGINGS MAY be defined as metal that has been heated and hammered into shape. The process goes as far back as primitive man, who hammered and shaped his simple tools and weapons while they were still glowing hot from the fire. The men of ancient Greece and Cyprus had exquisite swords and daggers of forged steel. In the United States, forging played an important part in paving the way for interchangeable manufacture. In 1870 John Mahlon Marlin explored the then relatively new technique of the exact duplication of parts by forging the metal between closed dies. The result was the Ballard single-shot rifle which became world famous not only because of its performance but because replacement parts that actually fit could be obtained.

The advantages of forging parts for the Ballard rifle and other guns were greater strength, lighter weight, and better finish. These same factors are still important contributions of forging today.

In forging, greater strength is obtained largely through the kneading action on the hot metal as it is shaped between the dies. In this process of kneading the metal, a very beneficial grain flow takes place. The grain flow changes from a straight flow pattern into the exact repetition of the part contours (Fig. 6.1). The fact that the flow lines remain unbroken and that a tough, fibrous structure results is one of the greatest values of forging.

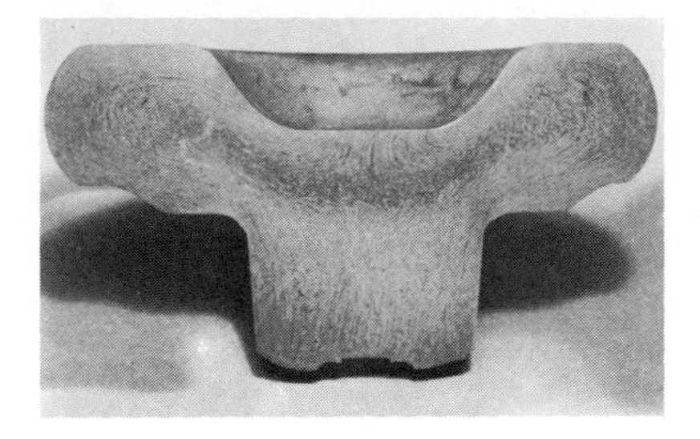

**Fig. 6.1.** The grain flow of the metal follows the part contour, contributing to the tough, fibrous structure of the forging. Courtesy The Ajax Manufacturing Co.

## FUNDAMENTAL CONSIDERATIONS

The effect of strain rate, temperature distribution, friction, and workpiece geometry are all fundamental considerations in understanding the forging process. These factors cannot be discussed in detail here. However, references at the end of the chapter provide resources for further study.

**Stress and Strain.** As hot metal is compressed between the forging dies there is a stress buildup because the metal suffers elastic deformation. As the yield stress is reached, plastic flow begins and continues under the same stress irrespective of the amount of strain (Fig. 6.2*a*). If deformation causes the material to harden, the stress required to maintain plastic flow will increase with increasing strain as shown by the broken line.

Stresses and strain hardening are relatively high in cold working of

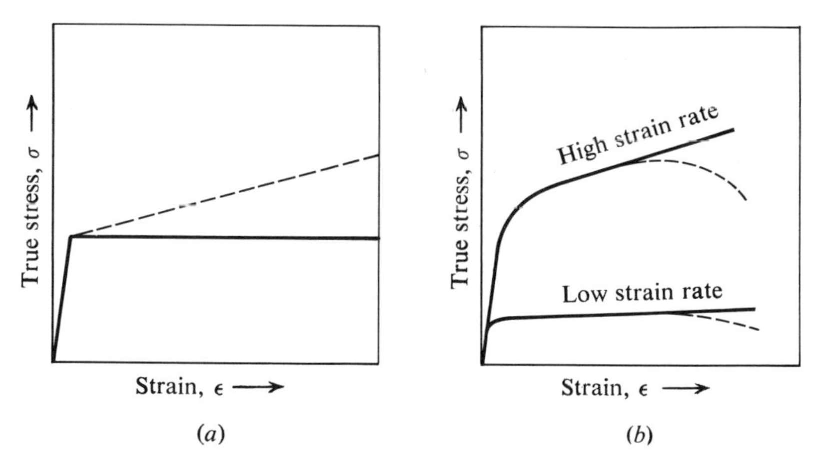


**Fig. 6.2.** (*a*) The stress builds up gradually on the metal in the forging die during the elastic deformation. As the yield stress is reached, plastic flow begins causing the stress to remain constant. If deformation causes the material to harden, there will be an increase in stress with increasing strain (broken line). (*b*) Hot working is characterized by little or no strain hardening, but the strain rate is significant.

metals. Hot working is usually characterized by little or no strain hardening but the strain rate is significant (Fig. 6.2*b*). Recrystallizations and transformations may lead to complex curves as indicated by the broken lines.

**Yield Stress.** Yield stress increases with decreasing temperature. At low strain rates, its value is 2.5 to 3 times as much at 1650°F as it is at 2200°F. The yield stress at forging temperature determines the ease with which it can be formed. It is not, however, the measure of its forgeability since a material of low yield strength may be totally unsuitable for forging due to premature cracking. One of the best tests for determining forgeability is the hot twist test. This method consists of a cylindrical test piece twisted on its own axis at various forging temperatures. The number of twists sustained without cracking is taken as a measure of its forgeability.

**Press Size Calculations.** Press size calculations for forgings from a theoretical point of view are very complex and still in their infancy. For this reason, tables or diagrams based on measured values in full scale production must be used.

## METHODS AND EQUIPMENT

The basic methods of forging hot metal are drop forging, automatic closed die forging, machine or upset forging, press forging, hand forging, and heavy forging.

**Drop Forging.** Drop forgings get their name from the fact that the upper half of a die is raised and allowed to drop on the heated metal placed over the lower half of the die. It is estimated that at least 90 percent of the total tonnage of forgings produced commercially are of the drop-forging type. Drop hammers are generally of either the gravity or the steam type.

GRAVITY HAMMERS. The most common kind of gravity hammer utilizes a hardwood board which is made to pass vertically between a pair of rotating rolls. The operator allows the rolls to press against the board, which raises it quickly to the top of its stroke where it is held by dogs. As the operator releases the holding dogs, the board, with the ram and die fastened to the end of it, falls directly on the hot metal.

Single-acting steam or air cylinders are also used to raise the ram in place of the board and rollers. Air-lift hammers allow variations in the length of the stroke, and consequently the force of the blow can be varied. This is particularly useful in the first few forming operations where the metal has to be rolled and gathered. After that, the full force of the hammer can be used for the finishing operations.

STEAM DROP HAMMERS. Steam drop hammers allow the operator to have constant control of the forging action. The motion of the ram is usually controlled by a foot treadle so that the upper die may be brought down with the force of a few pounds or of several hundred tons.

DROP FORGING STEPS. After the dies have been made and placed in the hammer, the metal must be selected and cut to the proper length. The proper length may be enough to complete one or several parts. It is then heated thoroughly in a furnace until it is in a plastic condition. Controlled-atmosphere furnaces will keep down decarburization and scaling of the metal. The heating furnaces are usually located in close proximity to the forging presses. It is important that the proper temperature for the particular metal and the specific job be maintained throughout the various forging stages.

The first forging operations consist of preliminary shaping of the metal, referred to as *fullering* and *edging* (Fig. 6.3). In these preliminary steps the dies are made to reduce the cross-sectional area in one place and gather it in another, which not only saves material but starts the fibrous grain flow of the metal.

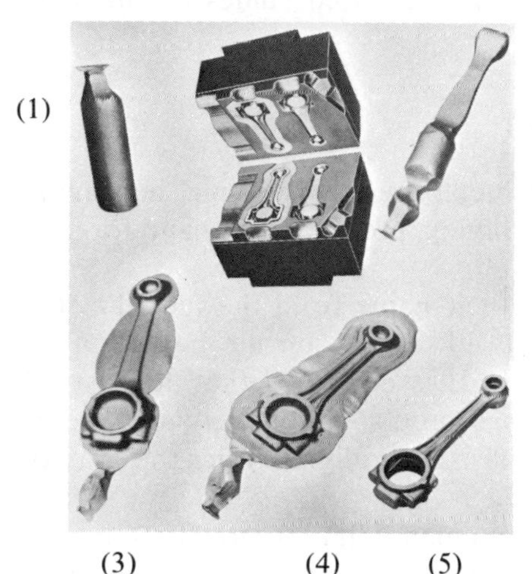


**Fig. 6.3.** The main forging steps are shown as fullering and edging (1, 2); blocking (3); finishing (4); and trimming (5). Courtesy The Drop Forging Association.

The shape of the connecting rod shown as an example of forging is further obtained by several successive blows of the hammer in the blocking cavity. In this operation the *flash,* or excess metal, begins to appear. Next, the metal is brought to the finishing impression dies. Here heavy blows compel the plastic metal to fill completely every part of the cavity.

Following the forming operation comes the trimming. This is usually done on a separate trimming press with special trimming dies. In this case, both holes of the connecting rod are punched out at the same time.

**Automatic Closed-Die Forging.** Automation in the forging industry is comparatively new. The difficulties involved in synchronizing feeds and hammer blows seemed to present a problem not easily solved. Now, however, machines have been built that do automatically all the operations connected with this closed-die forging.

The process consists of taking the precut parts from a hopper through a conveyor-type furnace and then through part orientation to the preform and forging operations. The parts are then hot-trimmed, inspected, heat-treated, coined, cleaned, inspected again, and made ready for shipment.

A schematic drawing of the type of machine used for automatic forging is shown in Fig. 6.4. You will note the two halves of the die are

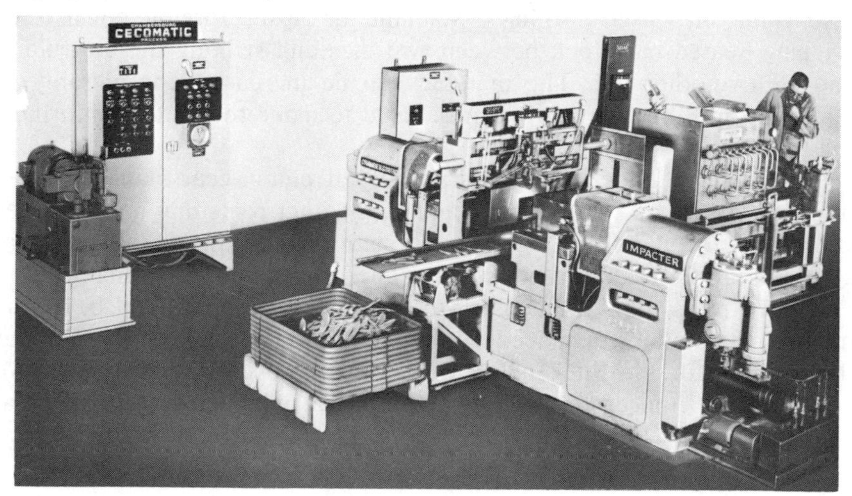


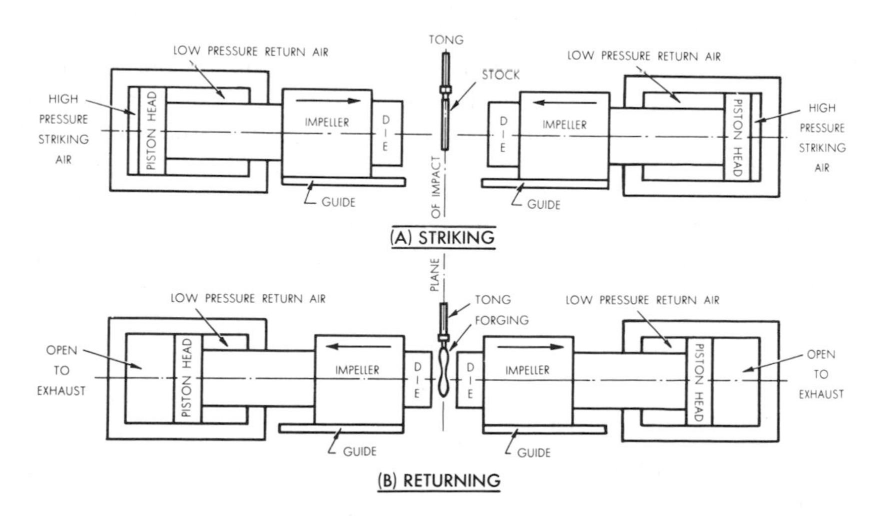


**Fig. 6.4.** The automatic forging machine and schematic diagram of operation. Courtesy Chambersburg Eng. Co.

mounted on air-actuated impellers. The plane of impact is maintained in the center of the machine by means of an electronic compensator employing sensing heads that detect blow eccentricity. The impacter, as it is called, is rated by a number which represents its maximum blow energy expressed in thousandths of foot-pounds.

Overall controls can incorporate automatic shutdown of stock movement in the event of any mechanical failures. Warning devices signaling such things as low stock supply or any malfunctioning can also be included.

**Machine or Upset Forging.** Machine or upset forging consists of gripping heated bar stock between two dies and striking the protruding end with another die. The material can be increased considerably in diameter, and the final blow can be used to make the metal conform to the die cavities (Fig. 6.5).

This process lends itself to the forging of pinion-gear blanks, flanges on axles, drawbars, valve stems, and many other parts that need a larger volume of metal on the end. Careful gathering of the metal results in controlled grain structure, with dense fiber for maximum strength.

Pierced parts may be made by the upset forging method also. In this method, the metal is displaced from the interior and made to flow around the outside, for the full length of the blank when necessary (Fig. 6.6).

The force required for upsetting can be calculated from the formula:

$$P = A_0 \frac{h_0}{h} C\left(\ln \frac{h_0}{h}\right)^m$$

$$A_0 h_0 = Ah \qquad \text{or} \qquad h_0/h = A/A_0 = D^2/D_0^2$$

Where: $A_0$ = original cross-sectional area; $h_0$ = original height; $h$ = final height; $A$ = final cross-sectional area; $C$ = average stress; $m$ = average strain exponent; $D_0$ = original diameter; $D$ = final diameter; and ln = natural log.

As an example, a 1 in. diameter mild-steel rod is to be cold headed to a 2 in. diameter. The length to be upset is 2 in. The strain hardening exponent is 0.5. The average stress on the metal is 50,000 psi. Calculate the upsetting force required.

$$P = \frac{\pi(1)^2}{4} \times \frac{2}{1} \times 50{,}000 \left(\ln \frac{2}{1}\right)0.5$$

$$= 27{,}000 \text{ lbs.}$$

The approximate forming force required to upset a 1 in. mild-steel bar under the conditions given is 13.5 tons.

**Press Forging.** In press forging, the plastic deformation of the metal is accomplished by a squeezing action. Both hydraulic and mechanical presses are used. The mechanical press is used for high-speed forging

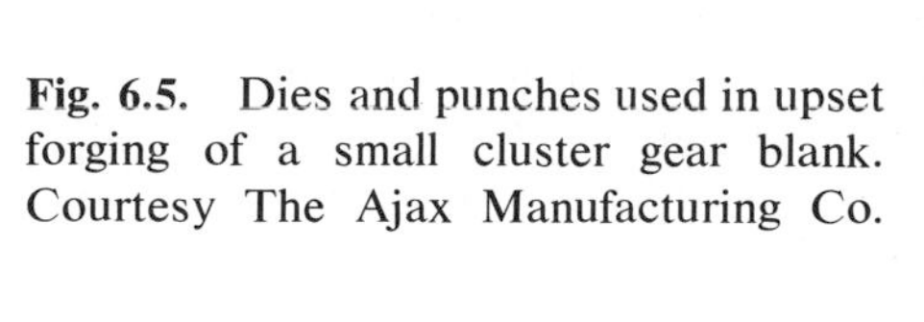

**Fig. 6.5.** Dies and punches used in upset forging of a small cluster gear blank. Courtesy The Ajax Manufacturing Co.

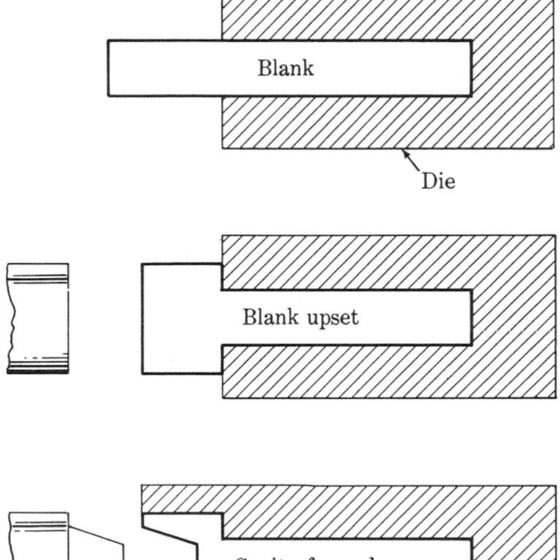


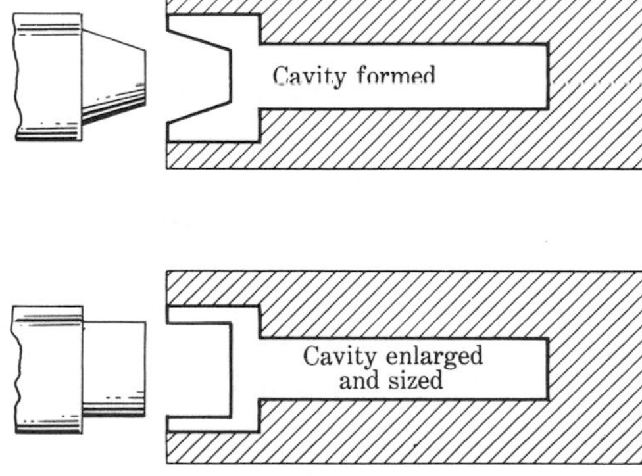


**Fig. 6.6.** Machine forging of a recessed cavity.

production. The hydraulic press is more versatile from the standpoint of varying the stroke length and the amount of pressure applied. Unlike forging with hammers using repeated impact blows, hydraulic presses operate entirely with squeezing force. It is applied continuously, with increasing intensity as the metal continues to flow.

Slugs of metal of the right size are prepared for the die cavity. As with drop forging, the metal may have to be shifted through a series of two or three dies before the finished part is produced.

Press forging is sometimes used in combination with drop-hammer forging. After the part has been forged in a drop-hammer press, it may be transferred to a mechanical press for sizing. A press forge may also be used for *coining*, which causes the metal to flow in a relatively small area by virtue of the heavy pressure exerted.

The structural characteristics of press forgings are generally equal to those of drop forgings. The process is generally limited to small parts and nonferrous alloy pieces under 30 lb. The largest presses are the hydraulic type used in press forging. Mechanical presses generally range from about 500 to 8,000 tons in capacity, and hydraulic presses from about 250 to 5,000 tons in capacity for drop forging. Recently, however, under government subsidy, the process was expanded to use presses of 50,000-ton capacity for press forging.

**Hand Forging.** Hand forging is a process for forming metal between flat dies. Since it is similar to the village blacksmith's methods, except that steam or hydraulic power is used, it is also called *smith forging*.

Hand forging requires a skilled operator to form metal parts without the aid of a die. The process is slow and is used only when the number of parts does not justify the use of a die. The size may range from less than 1 lb to parts weighing 200 tons or more. The larger sizes are handled by electric-hydraulic controls.

Die costs can be kept down on short runs by combining hand forging with press or drop forging. An example of this is shown in Fig. 6.7, where dies are used only for the last forming operation and for trimming. The part is started as a round bar, using flat dies to rough shape the ends. A hole is drilled through the center section and a saw cut made to separate the tines. A *bulldozer* is used to spread the tines and bend them to shape. (A bulldozer is a horizontal-type press especially adapted for bending and forming operations to prepare a forging for closed dies.) The completed blank is then put in the drop-forging die, after which it is trimmed to make the completed forging.

**Heavy Forging.** The biggest forgings are produced by steel mills. They include rotors for turbines and generators, propulsion shafts for big ships, anvil bases, and columns for big presses.

A heavy forging made to order for a particular job starts out as an outsize hot-top ingot. It may measure 110 in. in diameter and weigh 500,000 pounds. Working with cranes and heavy chains, experienced forging crews squeeze the heavy shape under a hydraulic press that exerts up to 50,000 lb pressure. With this tremendous pressure, the hot steel can be worked or subjected to a kneading action all the way to the center. This removes some internal stresses and gives a dense, tough structure.

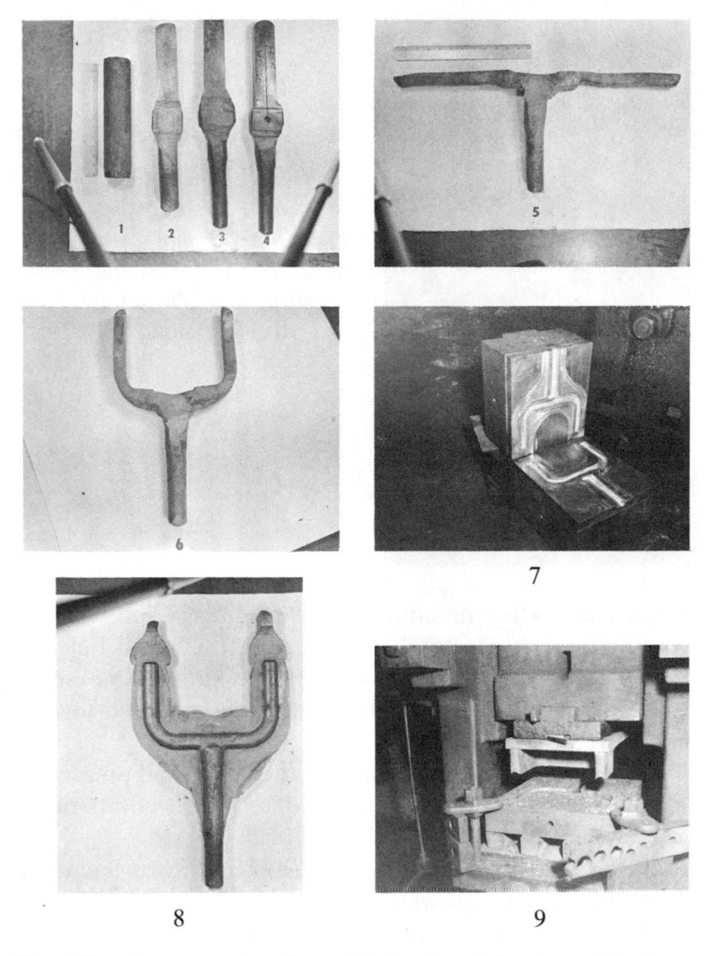


**Fig. 6.7.** An example of combining hand forging with drop forging to keep die costs down: (1) billet, (2) first forging, (3) second forging, (4) drilling and slotting, (5) bending to spread forks, (6) bending to fit yoke into die, (7) drop forging die, (8) untrimmed drop forging, (9) press with trimming die. Courtesy Merrill Brothers.

A turbine roller like that shown in Fig. 6.8 may take several weeks in the furnace and, with machining and testing, may require 3 to 4 months to complete.

After the forging crew get in their initial strikes, the ingot goes back to the reheating furnace in preparation for further forging to shape. The forging is alternately reheated 24 hr and forged 1 hr for about 1 week.

Before machining, the rotor is given a rough sonic test along its entire

**Fig. 6.8.** Forging a 100,000 lb steel ingot in a hydraulic forging press. Courtesy *Steelways*, published by American Iron and Steel Institute.

longitudinal axis. After machining, the forging is sent back to the furnace, where it is heated slowly to about 1,100°F and held at that temperature for 20 to 30 hr for stress relieving. The heat is then turned off, and the forging is left to cool. This process is repeated for from several days to one week.

The forging is again brought back to the lathe, and an asbestos oven is built around it. Rods are placed at predetermined points on the circumference to check for warpage.

After three constant readings at 1,000°F, the furnace ends are removed and the rotor is allowed to cool. More readings are taken, and, if the rotor has failed to come within 0.002 in. runout tolerance, it is scrapped.

**Auxiliary Forging Equipment.** In addition to the equipment already described, various other machines are needed for preliminary or small forging operations. These machines are helve hammers and forging rolls. The bulldozer, as previously described, is also considered an adjunct to the regular forging process.

HELVE HAMMERS. Helve hammers are high-speed hammers used in preparatory shaping of the metal or in *planishing*. Planishing is a term used to denote metalworking by short fast strokes for the purpose of smoothing the surface and obtaining greater overall accuracy.

FORGING ROLLS. Forging rolls are used primarily for reducing short thick sections of stock into long slender sections. The operator places the heated bar stock between semicylindrical rolls, where it is reduced in size by the half revolution of the roll. As the rolls reverse, the operator removes the stock and on the succeeding forward stroke inserts it between another set of rollers with smaller grooves. The process can be repeated between rolls with successively tighter clearance until the material is of the desired size and shape. Roll forging is used for making such parts as axle blanks, gear-shift levers, drive shafts, and leaf springs.

## FORGING PRACTICE

**Materials.** FORGING-QUALITY METALS. Almost all the commonly used metals and alloys can be used for forging. When forging-quality metals are specified, this term refers to the special care used in selecting the metals, thus eliminating defects. As shown in Fig. 6.9, the ingot that will be rolled out and used for forging operations is poured in a hot-top mold so that more of the gases escape during solidification. Also, a little more of the top and bottom of the ingot are cropped to ensure high quality.

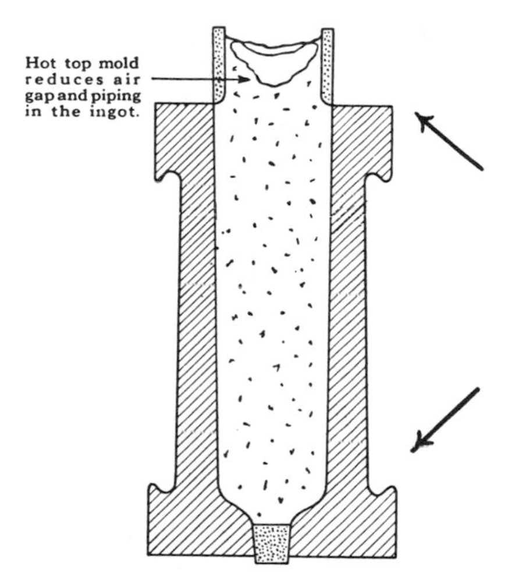


**Fig. 6.9.** Forging steels have a greater amount of cropping at both the top and the bottom of the ingot. In addition, a hot top may be placed on the ingot to keep the pipe or hollow area from forming in the metal that will be used for forging.

VACUUM-DEGASSED STEEL. Another method developed to ensure high-quality steel is vacuum degassing. Simply stated, the process consists of air-melted steel that is poured through a vacuum. During the process, much of the hydrogen, the major contributing factor to embrittlement and flaking in forgings, is removed. The steel produced is more ductile than air-poured steel.

Originally, the steel ladle was placed in an airtight chamber which was subjected to a vacuum. At the same time, helium gas was injected into the melt to increase the boiling action and expose more steel to the surface. A further protection was given the molten steel by pouring it into ingot molds filled with argon and protecting the pouring process with a blanket of argon.

VACUUM-ARC STEEL. Where exceptionally high-quality steel is required, vacuum-arc steel is made. The process consists of using air-melted steel to make an electrode. The electrode is then arc-melted into a water-cooled crucible under vacuum.

There is an increasing demand for large arc-melted ingots. As production facilities grow, it is expected that the price will drop. Most of the steel processed by vacuum has been of aircraft bearing quality. A major exception is the manufacture of hardened-steel rolls. Several of the major steel producers make such rolls at no premium to the user.

**Heat Treatment of Forgings.** After the forging has been formed, it may have large grains in one portion, where little or no forming is done, and fine grains where the metal is reduced. There is also some strain in the metal from the forging process. Therefore, *before* machining, the metal is generally normalized to produce uniform grain size and relieve some of the stresses (Fig. 6.10).

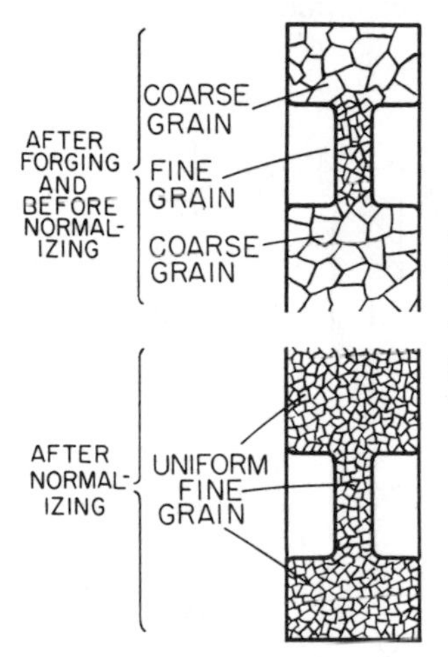


**Fig. 6.10.** In forged gear blanks, the crushing action reduces the grains in the web to a smaller size than the grains in the thicker sections. Normalizing refines this structure and produces grains of a more uniform size throughout the forging.

When steel forgings are to be hardened *after* machining, they are again normalized. This further refines the grain, takes out machining stresses, and helps eliminate cracking or distortion during quenching.

## QUANTITY AND COST

Parts that are to be made in closed dies generally require high production quantities to offset the high cost. Normally, these quantities will run from 10,000 to 100,000 pieces or more. As stated previously, this picture can be changed by using ingenuity in combining hand flat die forgings with closed dies. There are times, however, when limited quantities of forgings are produced simply because no other process can give the desired strength with the corresponding bulk.

In drop-hammer forging, as in casting, preliminary costs vary considerably, according to the complexity of the item. A rough approximation for small forgings of 1 lb or less would probably run from $1,000 to $2,000. Corresponding initial pattern costs for a casting would probably run between $100 and $200. However, differences to be considered will be the strength that can be obtained from a given weight of material and secondary machining and finishing costs.

## DESIGN CONSIDERATIONS

Most of the same factors that were discussed in sand-casting design apply also to forgings. Several factors which may call for slightly different design, include draft angles, fillets, corners, shrinkages, cavities, and dimensional tolerances. Many of these terms are clarified by referring to Fig. 6.11.

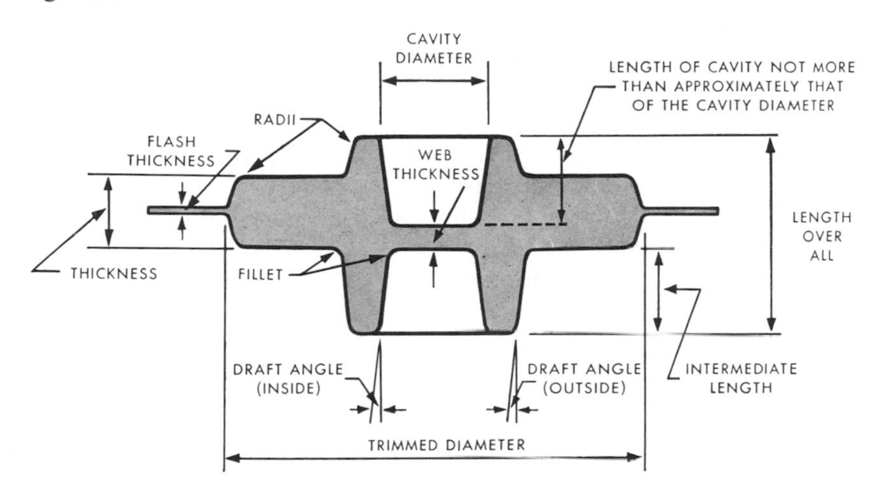


**Fig. 6.11.** The general design considerations shown are flash thickness, 0.040 in.; radii, a minimum of 1/16 in.; web thickness, 3/16 in. minimum; draft angles, 3 to 10 deg for drop forgings and 1 to 5 deg for press forgings. Where possible, the larger part of the impression should be in the upper half so that the scale drops out and is blown away.

**Draft Angles.** Conventional draft angles vary with the type and size of forging. Drop forgings have from 3 to 10 deg, press forgings from 1 to 5 deg, of draft. The deeper the cavity, the greater the draft.

New tools and techniques have been developed in the forging field that reduce draft and corresponding excess metal. This process, referred to as *precision forging,* can, in many cases, eliminate the draft angle entirely. Because of die-cost considerations, parts have been confined to relatively small sizes ranging in plan area from approximately 25 to 200 sq in.

Some of these forgings have been made in cold dies and others in heated dies. In the cold dies, ejection mechanisms are used. Also, a number of different protective coatings have been used on various materials to permit linear movement in the plastic condition without detracting from the surface finish.

Remarkable features have been produced in some precision forgings. Web thicknesses have been forged as thin as 0.094 in. with pocket depths of 1 1/4 in. with "no draft" walls.

**Fillets and Corners.** Sharp corners must be avoided because they not only tend to decrease the life of the forging dies, but they require excessive pressures to fill. This may cause stress concentrations in the work. Fillets and corner-radii tolerances are relative to the size of the forging. If they are not specifically designated, the die engineers alter them to fit conditions in production. In general, the minimum fillets and radii must not be less than 1/16 in., and preferably more.

**Shrinkage Factor.** Shrinkage factors are primarily considerations for the die designer. Thus, unless specifically stated, the die designers make the necessary allowances which take into consideration the type of material, forging method, and finishing temperatures.

**Cavities.** In general, cavities should not be specified that are deeper than their diameter.

**Tolerances.** Tolerances on dimensions vary according to the size of the part. For example, a 1-in. part may have a tolerance of ±0.008 to 0.015 in., depending upon the type of material used and the forging process. On parts 6 in. and over, the tolerance range is usually ±0.015 to 0.031 in. Closer tolerances are obtained by coining the forging. This consists of restriking the parts in a heavy-duty mechanical press. It is possible to obtain tolerances on thickness dimensions within ±0.005 and sometimes ±0.002 in. The process of coining often results in eliminating machining operations.

**Machining Allowance.** Machining allowance is generally a minimum of 1/32 in. per surface. This increases if the surface is large, because it becomes more difficult to hold it true.

**Web Thickness.** The minimum web thickness should be approximately 3/16 in. with a ±1/64-in. tolerance. The approximate flash thickness will generally range from 0.040 to 0.080 in.

**Finishing.** Removal of ferrous oxide scale formed during forging and heat treatment can be done in several ways—sandblasting, tumbling with various abrasive materials, or pickling in sulfuric acid solutions.

**Summary of Conventional Forging Design.** The drawing (Fig. 6.12) will serve to summarize some of the most important forging-design principles, as follows:

(1) Keep the parting line on one plane, if possible.

(2) Design all radii as large as possible. Small radii cause die checking and may be difficult to machine in the forging die.

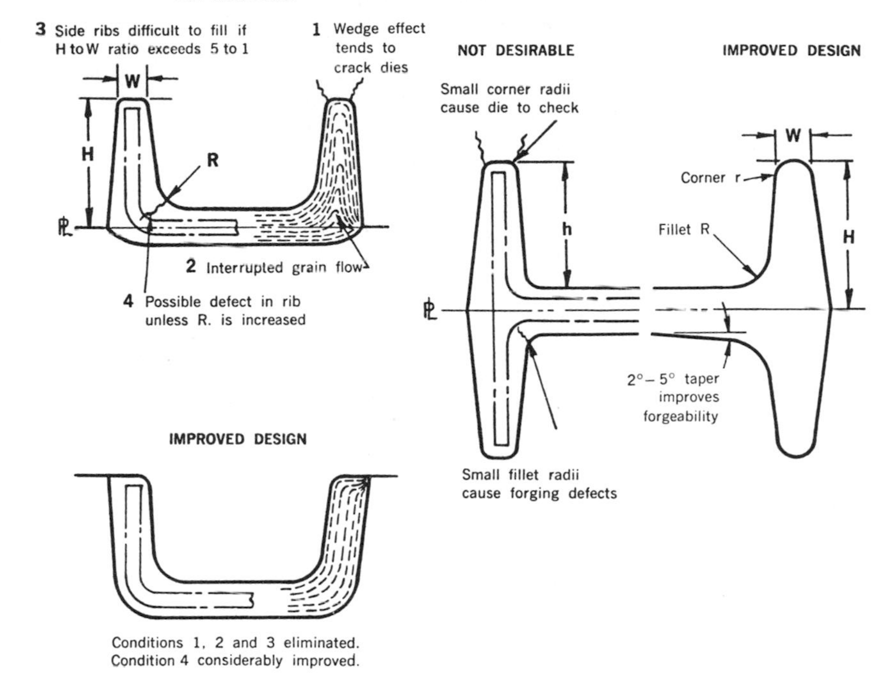


**Fig. 6.12.** Desirable and undesirable features in forging design. Courtesy *Metalworking*.

(3) Make generous fillets. Small fillets often cause forging defects.

(4) Make draft angles as large as possible. Small draft angles cause rapid die wear and may also cause the forging to stick in the die.

(5) For economy in large complex objects, forge in two or more parts and then combine the assembly by welding or other fastening methods. This is especially important for short-run jobs where the cost of tooling may be prohibitive.

Careful observance of these principles will result in better forgings at less expense.

## COMPARATIVE MECHANICAL PROPERTIES OF FORGED AND CAST MATERIALS

The designer, in addition to being familiar with general and specific requirements, must be able to specify materials that will assure the performance required. Mechanical properties of forged and cast materials are shown in separate graphs (Fig. 6.13).

Tensile properties are useful in determining steady loads in service. Fatigue tests are important in a product that will have ever changing stress cycles such as crankshafts, car axles, springs, gear teeth, etc.

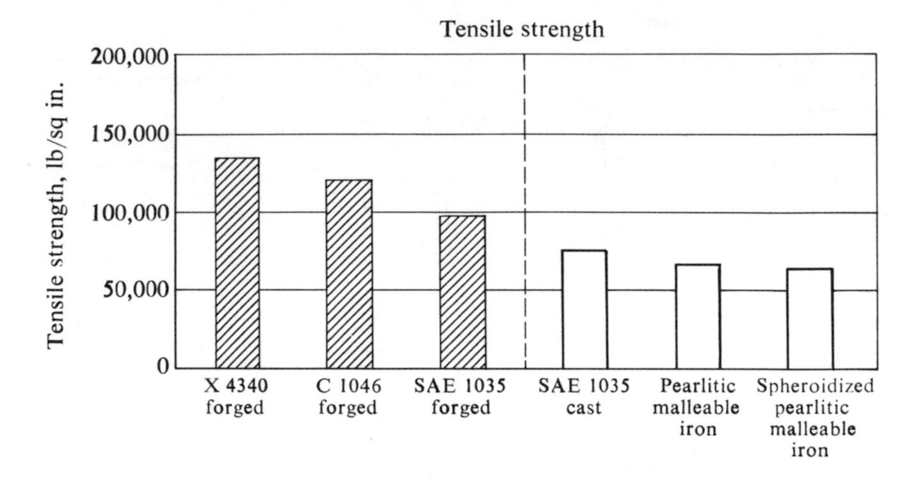
Tensile strength
Tensile strength, lb/sq in.
200,000
150,000
100,000
50,000
0
X 4340 forged
C 1046 forged
SAE 1035 forged
SAE 1035 cast
Pearlitic malleable iron
Spheroidized pearlitic malleable iron

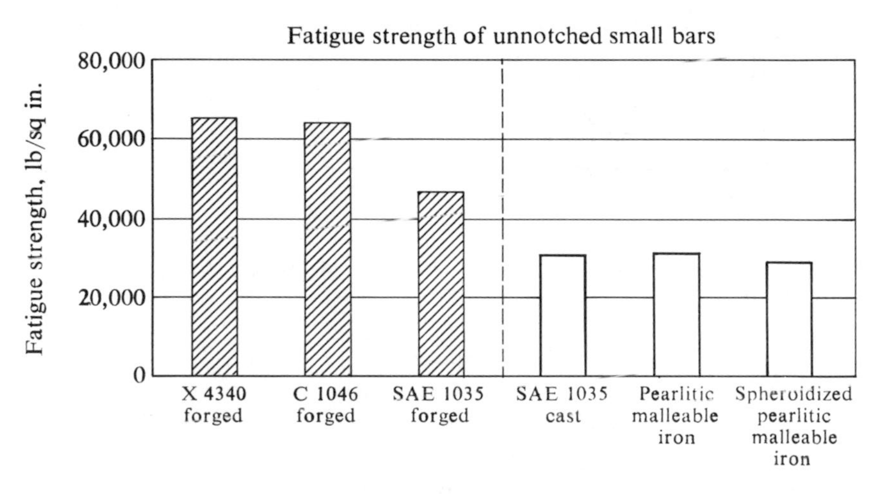
Fatigue strength of unnotched small bars
Fatigue strength, lb/sq in.
80,000
60,000
40,000
20,000
0
X 4340 forged
C 1046 forged
SAE 1035 forged
SAE 1035 cast
Pearlitic malleable iron
Spheroidized pearlitic malleable iron

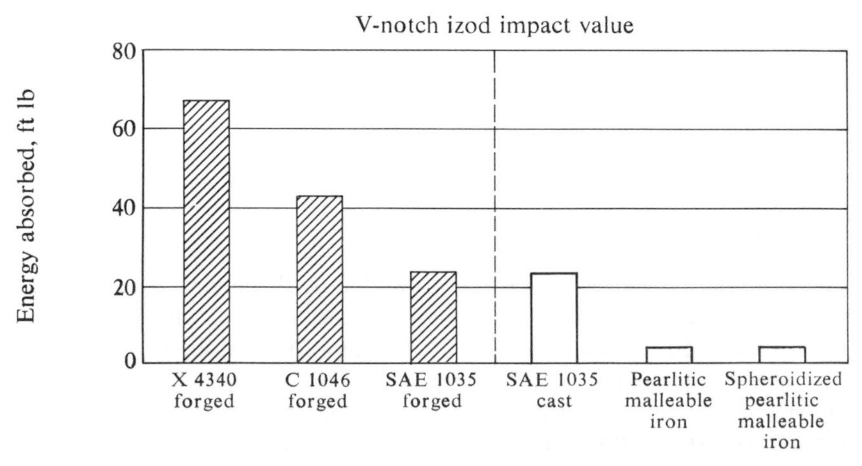
V-notch izod impact value
Energy absorbed, ft lb
80
60
40
20
0
X 4340 forged
C 1046 forged
SAE 1035 forged
SAE 1035 cast
Pearlitic malleable iron
Spheroidized pearlitic malleable iron

Impact values are helpful in selecting materials that may be subject to heavy accidental blows.

**Comparative Costs of Cast and Forged Parts.** It is difficult to generalize in comparing one process with another in that there are many variables to consider. However, the curves shown in Fig. 6.14 represent graphically a general basis of judging the relative costs of forgings and castings.

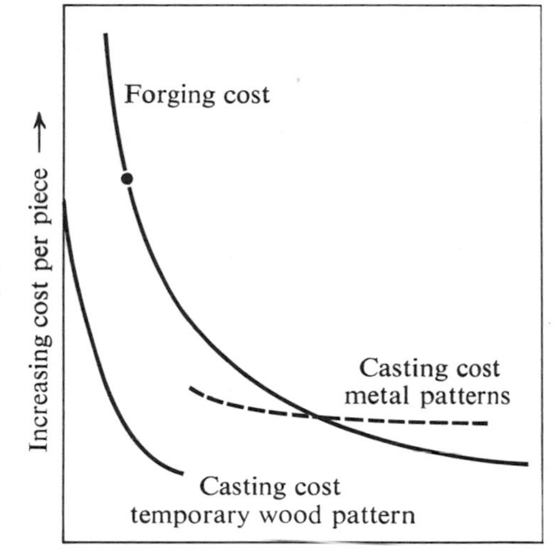


**Fig. 6.14.** Generalized cost comparison curves for quantities of forgings versus castings.

## HOT WORKING

**Hot Rolling.** Hot rolling consists of taking the hot ingot from a soaking pit, where it has been kept at an elevated temperature, and rolling it first into *blooms* (large oblong squares) and then through a series of other rollers into structural shapes, pipe, and tubing.

**Structural Shapes.** The familiar structural shapes such as channels, H-beams, angles, and rails are produced between various types of forming rolls (Fig. 6.15). The blooms are heated to 2200°F in a continuous-type reheating furnace before being rolled into structural-steel shapes. A single pass between the rolls may reduce the cross section as much as 25 percent. However, it may take as many as 26 passes to produce a structural shape such as an I-beam.

**Fig. 6.13.** Comparative mechanical properties of common forged steels and cast irons. The Brinell hardness values are X4340 – 276, C1046 – 248, 1035 forged – 196, 1035 cast – 198, pearlitic malleable – 190, spheroidized pearlitic malleable – 197.

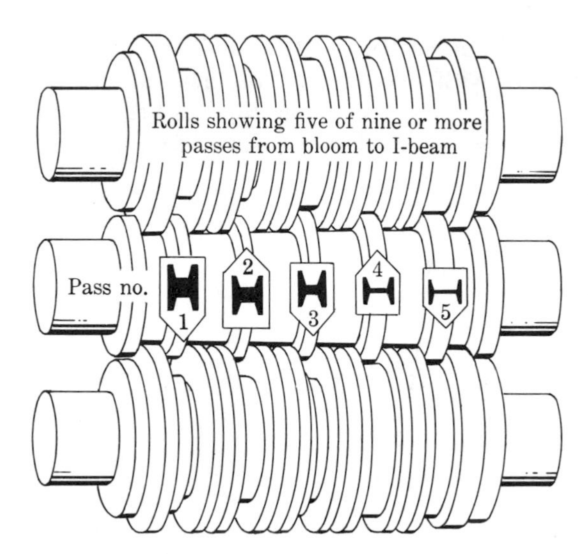


**Fig. 6.15.** Hot rolling structural shapes. Courtesy *Steelways*, published by American Iron and Steel Institute.

**Pipe and Tube Rolling.** Pipe and tubing are hot rolled by several different methods. They may be made out of flat sheet stock, in which case the seam is either made as a lap or a butt weld.

BUTT-WELDED PIPE. Pipe that is to be butt welded starts out as flat semifinished steel that is referred to as *skelp*. Grooved rollers bevel the edges so that, when they come together, they exactly match. The ends of the skelp are also cut so that they can easily enter the welding bell used to bend the edges of the heated (2600°F) skelp into a round form (Fig. 6.16).

Pipe size is designated by its nominal inside diameter up to 12 in. and its outside nominal diameter over 12 in. It is, however, the outside diameter of all sizes that is standardized so that lengths of any size will fit when they are coupled together.

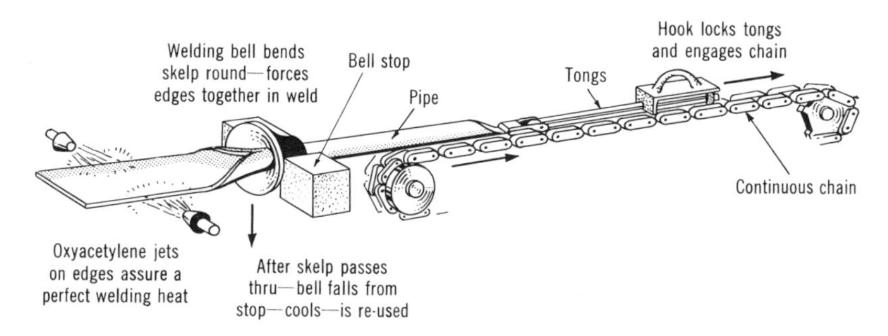


**Fig. 6.16.** The process of forming butt-welded pipe. Courtesy *Steelways*, published by American Iron and Steel Institute.

Pipe is purposely drawn oversize and reduced by sizing rolls. Then, while the pipe is still soft, the very hard scale on the surface is removed by additional rolling.

LAP-WELDED PIPE. The lap-welding process used in making pipe is quite similar to that used in making butt-welded pipe. The skelp is tapered at opposite edges so that they overlap where they are drawn through the die. The skelp is heated to a cherry-red color before entering the die and is reheated again to 2600°F after leaving the die. It is then rolled between forming rolls with a mandrel inside the pipe to press the lapped edges together, completing the weld. Additional rolling is done, as shown in Fig. 6.17, until the pipe is of the desired size and straightness.

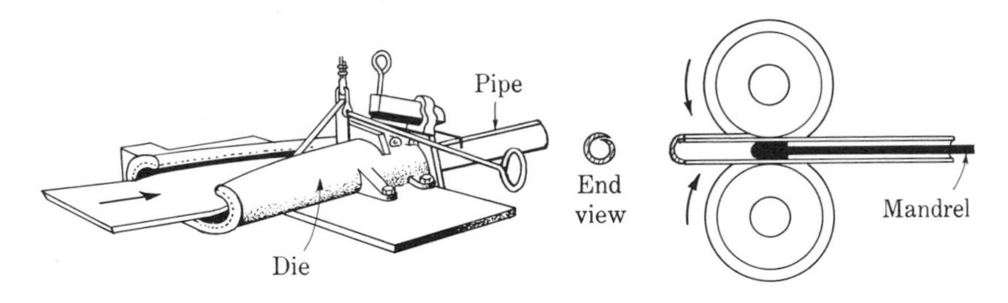


**Fig. 6.17.** The process of forming lap-welded pipe.

RESISTANCE BUTT-WELDED TUBING. Resistance butt-welded tubing is another type of tubing closely associated with butt- and lap-welded pipe. However, since heat is required only for the welding, this process will be discussed in Chap. 7.

SEAMLESS TUBING. Seamless tubing can be made by either piercing or drawing.

*Piercing.* The process of making hot-pierced tubing consists of passing a hot-rolled billet between two biconical rollers and over a plug held on a support bar, often referred to as a mandrel (Fig. 6.18).

The biconical rolls serve to spin the round heated billet and force it forward. The piercing action is actually started previous to placing it between the rolls, by drilling, punching, or piercing with oxygen. The

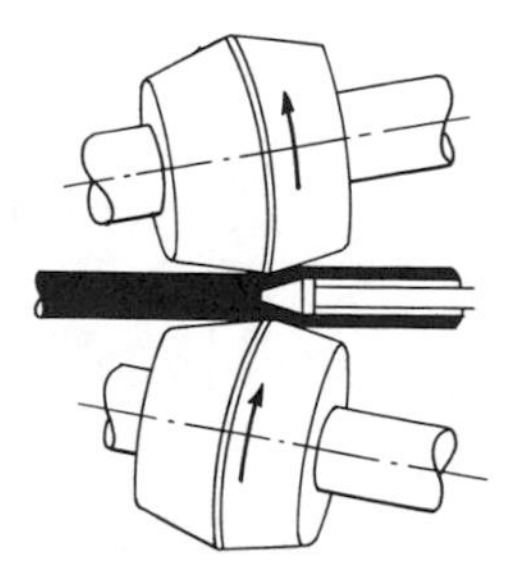

**Fig. 6.18.** Piercing of the hot billet is done between two biconical rollers and over a plug held on a support bar.

rolling, kneading action then moves the tube forward at high speed. The first pass makes a rather thick-walled tube, which is again rolled over a plug where it is converted into a longer tube with specified wall thickness. While it is still up to working temperature, it is passed onto a reeling machine which further straightens and sizes it (Fig. 6.19).

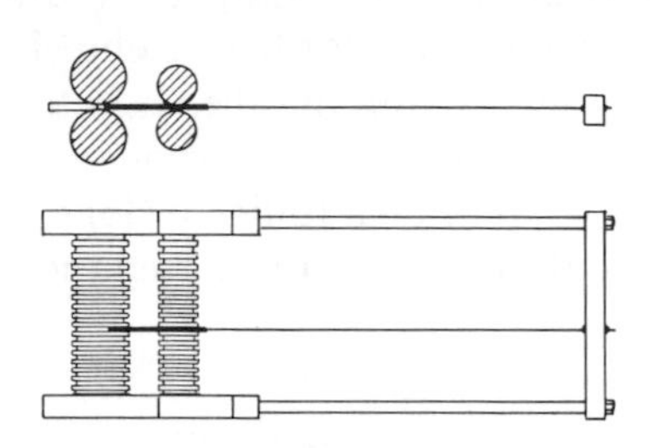

Rolling over a plug to thinner wall and greater length in an automatic rolling mill

Burnishing the surface in a reeling machine

**Fig. 6.19.** Specified wall thicknesses are obtained by further plug rolling and reeling.

If more accuracy and better finish are desired, the tube may be run through sizing dies. Smaller sizes are obtained by reheating the tubing in a furnace and running the tube through reducing rolls followed by straightening rolls. After cooling, the tubes are usually placed in a pickling bath of dilute sulfuric acid to remove the scale and oxide. If still greater accuracy and fine finish are desired, as is often the case in small-diameter tubing, the process is continued by cold drawing.

*Drawing.* Tubing and cylinders can be hot drawn from relatively thick plate stock. The heated blank is placed in position over the die or cavity. The punch descends and pushes the metal through the die to form a cup (Fig. 6.20).

The process may be continued through a series of successively smaller

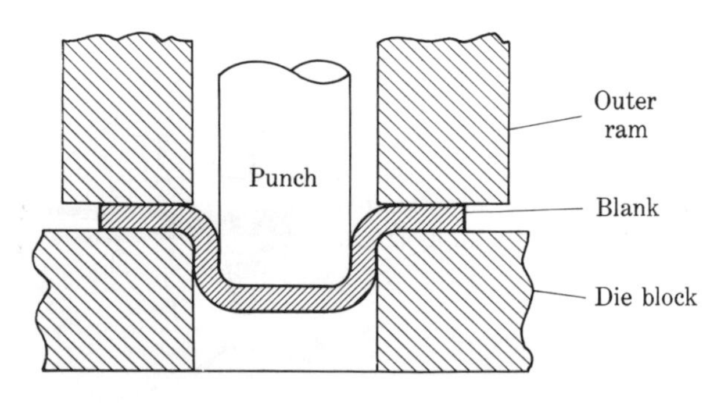


**Fig. 6.20.** Drawing a heavy plate into a cup shape. Successive draws can be made to make a deep cylinder or tubing.

dies and punches to obtain cylinders of the desired size and wall thickness. Oftentimes, a series of successively smaller dies is set up on a bench known as a *hot-draw bench.*

Seamless tubing and cylinders made in this way are used primarily for thick-walled cylindrical tanks.

## EXTRUSION

Extruding is a process in which a billet or slug of metal is pressed by movement of a ram until pressure inside the workpiece reaches the flow stress. On any further rise of pressure the material is squeezed out of an orifice and forms the extruded product (Fig. 6.21). The orifice can be anywhere in the container or the ram depending on the relative motion between the ram and the extrusion. Shown in Fig. 6.22, for solid stock, are forward (direct), radial, and reverse (indirect) extrusion methods and forward and reverse methods of producing hollow parts.

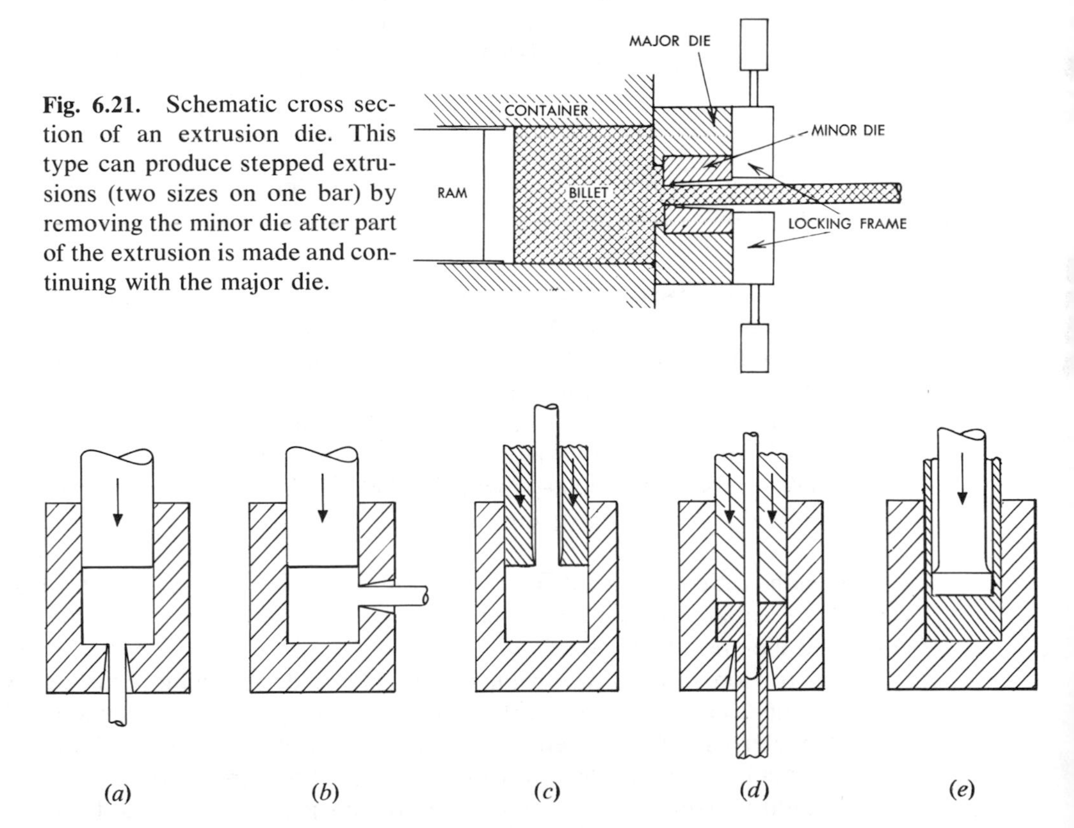


**Fig. 6.21.** Schematic cross section of an extrusion die. This type can produce stepped extrusions (two sizes on one bar) by removing the minor die after part of the extrusion is made and continuing with the major die.

**Fig. 6.22.** Extrusion of solid shapes (*a*) forward (direct), (*b*) radial, and (*c*) reverse (indirect). Extrusion of hollow shapes (*d*) forward (direct) and (*e*) reverse (indirect).

The reverse method gives smoother metal flow because of the absence of relative movement between the billet and the container. Forward extrusion is more complex since the outer metal flow is impeded by friction and the center tends to be in advance of the rest. A dead zone forms near the exit (Fig. 6.23). The dead zone is eliminated by using cone shaped entries into the die as shown in (*d*).

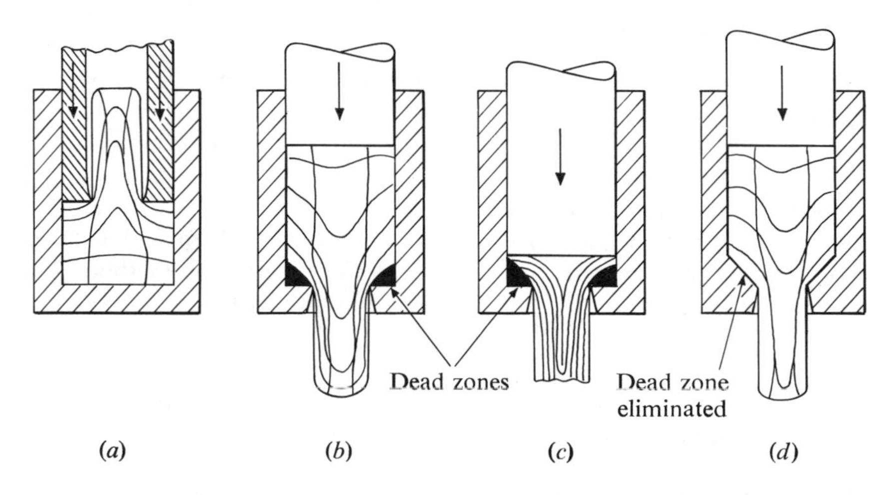


**Fig. 6.23.** Metal flow in extrusion: (*a*) indirect, (*b*) direct, (*c*) extrusion defect, (*d*) conical die entry, lubricated.

In the direct extrusion process, pressure reaches a maximum at the point of breakout at the die. When the extruding starts, the billet length decreases, reducing the contact area on the container wall. This continues until there is only a thin disk of the billet left at which time the pressure mounts rapidly due to the increased resistance to radial inflow of the disk at the die aperture (Fig. 6.24).

The most satisfactory billet length for the particular shape is determined mainly by (*a*) the extrusion ratio, (*b*) finished length, and (*c*) press capacity.

Extrusion ratio is based on the formula $R_e = A/a$ where $A$ = cross-sectional area of the container and $a$ = cross-sectional area of the combined die apertures.

The extrusion ratio plays an important role in determining pressure requirements. The ratio can be varied by changing the container and billet diameter used for a given section. The effect of the reduction ratio on pressures is shown in the curves in Fig. 6.25. Pressures normally are from 80,000 to 100,000 psi. Common extrusion press sizes range from 250 to 5,500 tons, but machines as large as 15,000 tons have been built.

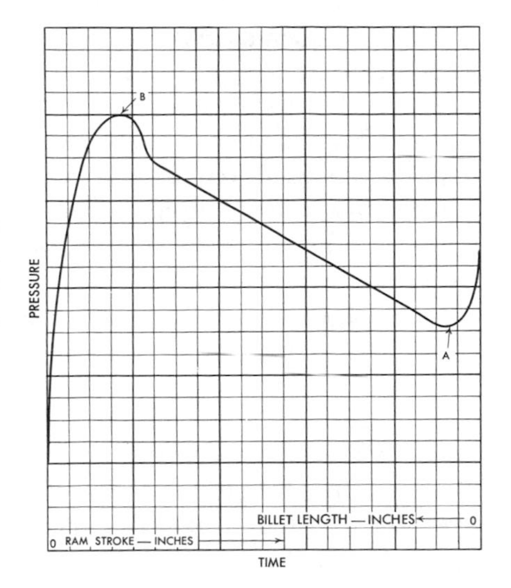


**Fig. 6.24.** The extrusion pressure graph curve shows high breakout pressure, then diminishes rapidly until only a thin disk is left when it again rises rapidly.

Commercial extruding is usually carried out with 5- to 7-in.-dia billets cut 12 to 27 in. long. The billets may be either hot rolled or cast. They usually have the outer surface removed or are peeled to eliminate surface defects in the billet.

Steel extrusions require a lubricant during the forming process. This is taken care of by carefully coating the hot billet with finely powdered glass. The glass becomes viscous when in contact with the hot billet and provides lubrication to the die and material.

**Materials.** Extrusions are most common among the nonferrous alloys of aluminum-, magnesium-, and copper-base alloys. However, various

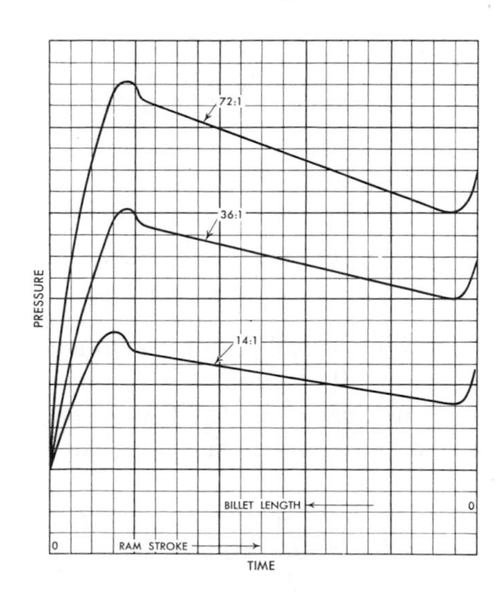


**Fig. 6.25.** The extrusion curves show the effect of reduction ratios 72:1, 36:1, and 14:1. All other factors were kept constant.

alloy steels such as AISI 4130, 4340, and 8360 are readily extrudable in rather intricate shapes. The regular 18–8 stainless steel also can be extruded, but tool life is somewhat shorter.

**Use.** Extruded shapes are exceptionally well suited to track and rail applications used extensively for materials handling. They are also used extensively in building hardware, automotive and aircraft parts, flooring strips, and many hardware items such as window sash, door trim, storm sash and doors, etc.

**Advantages.** Extrusions have to compete with similar shapes produced by rolling, casting, or forging. Therefore, they must offer certain advantages:

(1) Because of the high reduction ratio (the cross-sectional area of the billet to the cross-sectional area of the shape produced), the metal has excellent transverse flow lines. Structurally speaking, this condition not only adds to the strength of the part but makes easier any secondary operations that may be needed.

(2) Shapes can be extruded that are costly to produce by other means (Fig. 6.26).

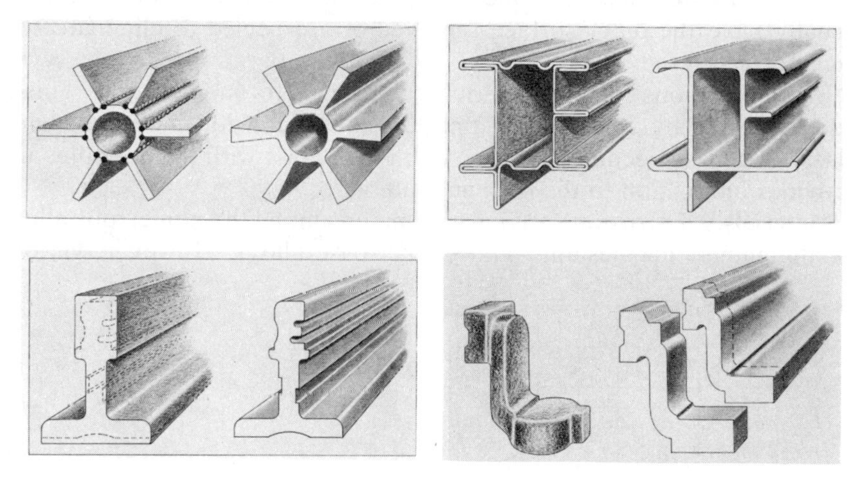

**Fig. 6.26.** Extrusions to facilitate design and eliminate steps in fabrication can be made in a wide variety of shapes. Courtesy Reynolds Metals Company.

(3) New experiments have indicated that thinner walls can be obtained if the forming pressure is increased. With 200,000 to 240,000 psi pressure, the present wall thicknesses may be decreased 30 to 60 percent. This will offer new design and application possibilities for aluminum extrusions.

(4) The cost of extrusion dies is relatively low, so that moderately short runs are practical. Dies are generally made from a chrome-molybdenum steel and heat treated to a Rockwell-C ($R_c$) of 47–50. More

intricate dies can be cut out of hardened steel or carbide by electric-spark or ultrasonic machining.

Extruded shapes can often replace weldments and members previously machined from bar stock. The tonnage figure must, of course, be considered. Where one or two special lengths are required, it is usually more economical to machine, weld, or fabricate by some other method. However, a small quantity such as 500 lb can sometimes be economically extruded.

A structural member, such as a special channel or angle, will carry a tooling charge of $180 to $200, and special shapes will rarely be more than $250. This is a small cost compared with mill roll charges. Also, the extrusion process allows low cost in-process redesign.

(5) There is more flexibility in design for adjacent thin and heavy sections as well as for difficult reentrant angles. Sharp corners, not practical in other processes, can readily be obtained by extrusion.

**Limitations.** There are some limitations to the extrusion process that should be considered:

(1) The size of the dies and presses that can be economically built is a limiting factor. The maximum size, at the present time, is about 60 in. wide.

(2) Extruding speed is slow compared to roll forming. Roll forming may run several thousand feet per minute, compared with about 700 fpm for extruding.

(3) Although the accuracy is good and entirely adequate for most applications, it is not so close as a machined part would be.

**Impact Extrusions.** Extrusions are also made by striking slugs of metal and forming them by high impact. Since this is essentially a cold-working operation, it is discussed in Chap. 12.

## QUESTIONS

**6.1.** Why are forgings considered as being better than cast or rolled steel products for critical situations?

**6.2.** Explain what happens to the relationship between true stress and strain when a forging is being made.

**6.3.** Why is there little or no strain hardening in hot working?

**6.4.** What is the relationship of yield stress to forgeability?

**6.5.** What methods are used to determine press size requirements for forging? Other than forging size, what are the variables?

**6.6.** What purposes do fullering and edging serve?

**6.7.** Why is it usually necessary to normalize a forging before machining it?

**6.8.** What accounts for a great share of the time used in preparing a large (200,000-lb) forging?

**6.9.** How does forging metal differ from ordinary hot-rolled metal of the same general composition?

**6.10.** (*a*) What is meant by vacuum-arc steel? (*b*) What is its principal use?

**6.11.** Name some considerations in deciding whether the forging process should or should not be used.

**6.12.** What factors have a bearing on the tolerance that can be achieved in forging?

**6.13.** How may costs often be reduced on large, complex forgings?

**6.14.** What is the difference in pipe-size designation for small and large diameters?

**6.15.** What methods are used to make seamless tubing?

**6.16.** What are some advantages that can be gained by using extrusions in fabrication?

**6.17.** How do extrusion dies compare in cost with forming rolls?

**6.18.** Why does the reverse extrusion method produce a better finish than the forward method?

**6.19.** Explain why the extrusion force curve is so high at the beginning.

**6.20.** What is the effect of extrusion ratio on the process?

**6.21.** Why can some extrusions compare favorably in price with the same item in rolled form?

## PROBLEMS

**6.1.** Express the following as ratios: (*a*) the tensile strength of forged C1046 to pearlitic malleable; (*b*) the fatigue strength of SAE 1035 to pearlitic malleable; and (*c*) the Izod impact value of X4340 to pearlitic malleable.

**6.2.** (*a*) A 4-in. extrusion channel die is made with 2-in. legs. The channel will be 1/4 in. thick. The container is 12 in. in dia. What is the extrusion ratio? How does it compare to the extrusion ratios shown in Fig. 6.25?

**6.3.** A forging similar to that shown in Fig. 6.6 is to be made from heated 2-in.-dia bar stock having a stress-strain curve of $\overline{\sigma} = 110{,}000$ $(\epsilon)^{0.23}$. The upsetting increases the diameter of the outer end to 3 1/2 in. in dia. How large an upset forging machine is needed to make this part?

**6.4.** If the extrusion shown in the lower right hand picture of Fig. 6.26 weighs 2.2 lb per ft, what is the minimum number of feet that would be needed before you could consider making this type of die?

# 7

# WELDING PRINCIPLES AND ARC WELDING

WELDING IS TERMED a fusion process for joining metals. Conventionally, heat is required to bring about the fusion process. By *fusion* is meant the intimate intergranular mixing of the two metals to be joined. Conceivably metals could be joined without heat provided the metal surfaces were free from oxidation and they were placed in intimate atomic contact. Since these conditions do not ordinarily exist except, to some degree, in outer space, fusion welding is classified as a thermal process. The heat is not only necessary to bring the metals to a molten or plastic state but also to help break down the oxides on the metal surfaces.

A solid exposed to the atmosphere absorbs molecules of air and water until a boundary film is formed that is denser than the surrounding atmosphere. The film has a thickness of several molecular layers on the order of $10^{-7}$ to $10^{-6}$ mm. X-ray defraction and cathode-ray studies show that these molecules are distributed in a definite order and represent something like a crystal lattice.

Successful fusion welds cannot be made without removing this gaseous oxide layer from the metal. So-called *cold* or diffusion-bond welding is done with a minimum amount of heat, less than 1,000°F, by applying pressures of 4,000 to 8,000 psi for an instantaneous period to 3 hr. Thus it is possible to make "cold" or nonthermal welds but not of the more commonly accepted fusion type.

The supervacuum of outer space evaporates the layers of absorbed gases on metal surfaces, which allows bare atomic contact. Under these conditions, and where unit pressures are greater, very little if any heat is needed to produce a weld. Experiments conducted by simulating space conditions and metals in intimate contact produced welds in copper in less than 10 min., welds that had a strength of 12,000 psi. Ordinarily a temperature of 1,980°F would have been required to make a weld in this material. What was a cause of concern for space flight may turn out to be quite useful not only commercially but also in the fabrication of space platforms.

Heat for welding metals is most often obtained by gas flame, electric arc, or electrical resistance. Additional metal for filling in and reinforcement is generally obtained from the electrode or filler wire available for that purpose. Some applications require no additional metal to make the weld.

In the previous chapters, metal properties were discussed. The evolution of metal from the crystal to the solid state was shown. How stresses come about and how they can be relieved were also described. In welding, we apply these fundamental principles. Most welding processes are, in effect, relatively small-scale, high-quality casting operations. Here we must deal not only with the high heat and metallurgical aspects of the cast metal but also with those of the parent metal, and particularly of the transition zone from the weld metal to the base metal. Thus the knowledge gained from the previous chapters will help in understanding the heat effects of welding and how they can be used in controlling grain size, microstructure, internal stresses, corrosion resistance, and contraction and expansion. After learning about the metallurgical structure of the weld and about stress and corrosion control, attention can be turned to the mechanical aspects of welding design.

## THERMAL EFFECTS OF WELDING

**Grain-Size and Hardness Control.** From the molten metal of the weld to the outer edge of the heat-affected zone, there will be a wide variation of temperature. Some of the metal has been heated far above the upper critical temperature, some just at critical, some not up to critical, and all the way down to the unaffected base metal. Therefore, as can be expected, the grain size of the weld will be rather large, becoming gradually smaller until the recrystallization temperature is reached. Here the grain size will be at a minimum and then will advance gradually larger again until it blends with the unaffected parent metal (Fig. 7.1).

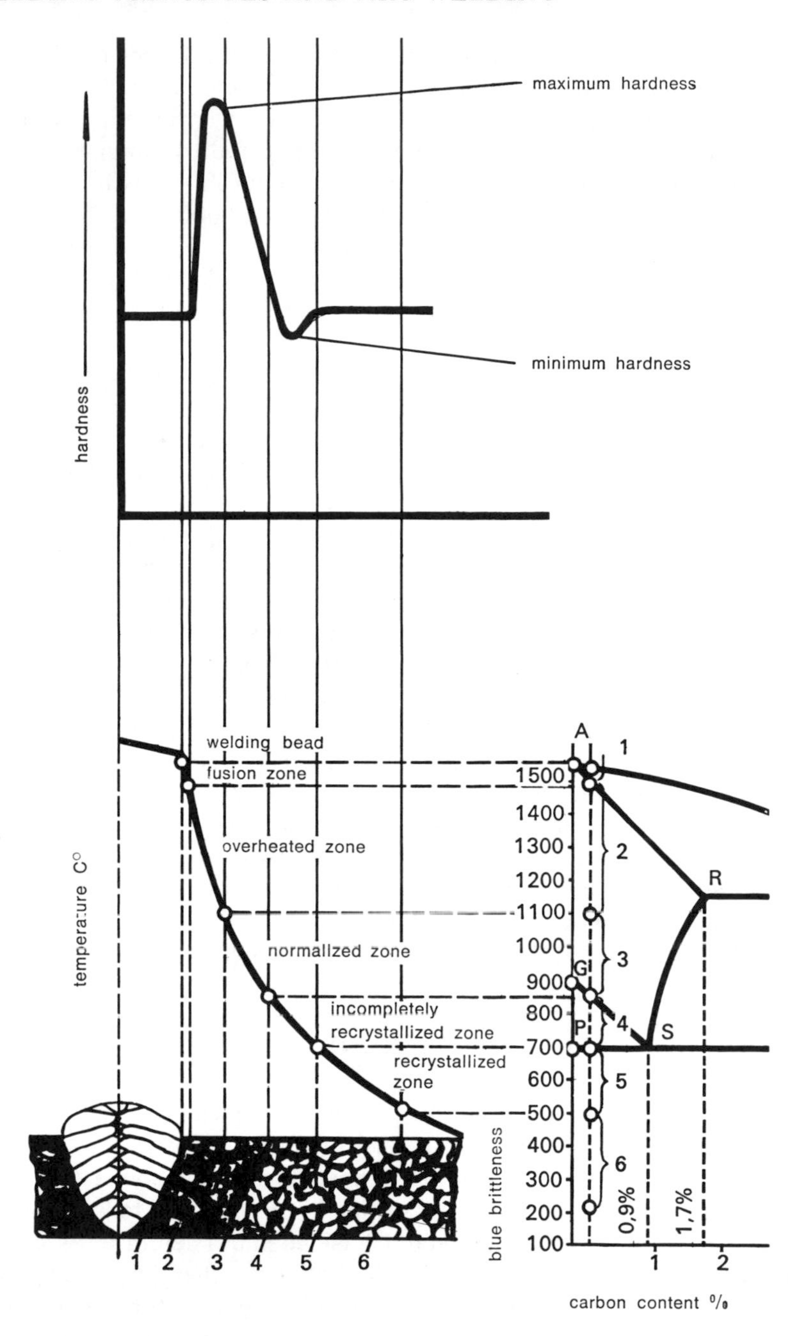


**Fig. 7.1.** The influence of welding heat on the distribution of hardness in the heat-affected zone of quenched and tempered structural steels (schematic). Courtesy Svetsaren.

The butt weld shown in Fig. 7.1 was made of high tensile strength steel. These steels are made by adding boron, aluminum, and vanadium for fine grain structure. The carbon content is limited to produce good weldability. These steels, in growing demand today, are delivered in their quenched and tempered condition.

The fact that these steels are hardenable to some extent can create problems. The heat of the weld in the transformation zone will be similar to annealing. Therefore the initial strength of the steel cannot be fully utilized. As shown schematically (Fig. 7.1), an increase in hardness appears in the zones that are heated above $AC_3$ due to the quenching effect or rapid heat transfer to the cold parent metal. The minimum hardness is found in areas that were heated just above the lower critical temperature.

Where and to what extent hardness peaks occur largely depends on the preheating of the parent metal and the heat input at the time the weld is made. By careful control of these factors, hardness in the softer part of the heat-affected zone can be kept above the guaranteed strength of the steel.

Quality welds require the proper proportion of heat input, preheat temperature, and welding speed. By applying certain heat flow and hardenability concepts to individual welding situations, these parameters can be approximated theoretically rather than by trial and error, resulting in considerable savings in time, effort, and expensive samples. For those interested in this topic of heat effects of welding, further information is given at the end of the chapter.

Highly cold-worked materials have locked-up stresses that are relieved by excessive heat from welding resulting in distortion and a decrease in strength.

**Internal-Stress Control.** The rapid heating and cooling of the weld can produce thermal stresses that are quite large. During heating there is rapid expansion of the base metal around the weld. If there is enough unaffected metal surrounding the weld zone, or if it is rigidly clamped, the expanding metal in the weld zone will be upset (Fig. 7.2). Upon

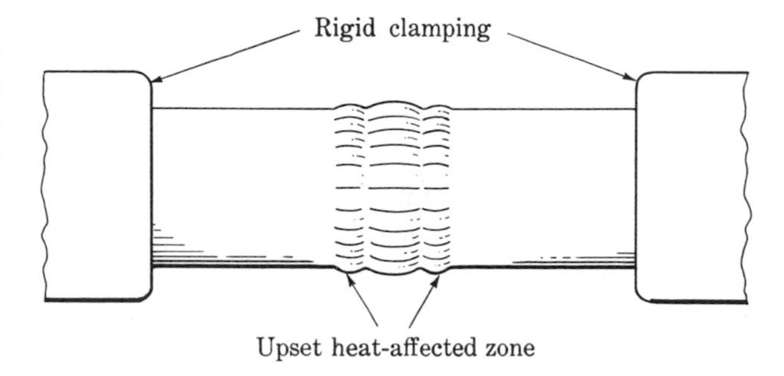


**Fig. 7.2.** The upset area adjacent to the weld will become a stress point, particularly in the less ductile metals that are rigidly clamped for welding.

cooling, the upset section will not return to its original dimensions, and internal stresses will be set up. Shrinkage of the solidifying weld metal will cause internal stresses in the same way. Therefore, in metals with low ductility, provision should be made for expansion and contraction during the welding operation.

Hot cracks are more apt to occur in the weld metal than in the base metal, the reason being that the weld metal is hotter than the base metal and cools last. A second reason is that weld metal is essentially a cast metal with columnar grains, as compared to the rolled and somewhat distorted grains of the base metal. Hot cracks may appear adjacent to the weld zone if the weld is made with ductile metal. For example, high-chromium steels may be welded with a combination chromium-nickel rod. The extra ductility furnished by the nickel content will make it possible to withstand the cooling stress. There is, however, a disadvantage in not using a rod of the same material as the base metal. In this case, chrome-nickel deposit will not have the same corrosion-resistant qualities as the base metal. Therefore, it may be necessary to go over the weld with a light deposit of high-chromium-content electrode.

**Corrosion Control.** Steels most often fabricated for corrosion-resistance purposes are the stainless steels. Shown in Fig. 7.3 are the various heat-affected zones of a chromium-nickel steel. The metal on the left side of the weld bead represents a modified chromium-nickel steel with columbium or titanium added to reduce harmful carbide precipitation. On the right side is shown the carbide precipitation zone. The longer this zone is held in the 1,500° to 800°F range, the more pronounced will be the carbide precipitation. Type 304 stainless, for example, is very corrosion resistant even by nitric acid, but, after welding, chromium carbides form at the grain boundaries, leaving the rest of the grain unprotected.

**Contraction and Expansion.** The rate at which metal contracts and expands is an important factor in knowing what to expect in the way of distortion. If metals that have a high rate of thermal expansion, such as aluminum (approximately twice that of steel), are clamped rigidly, there will be buckling in the weld area. Then, as cooling takes place, contraction results, putting the weld in tension. Many nonferrous metals are weak when hot (hot short), resulting in cracks as the metal shrinks and cools.

The speed at which the weld is made will determine, to a large extent, the amount of contraction and expansion. Slow speeds will cause greater area heating with greater expansion and, later, more contraction. Thermal expansion is associated with the increase in thermal vibration of atoms and lattice structures. The relative linear expansion of some common metals is shown in Fig. 7.4.

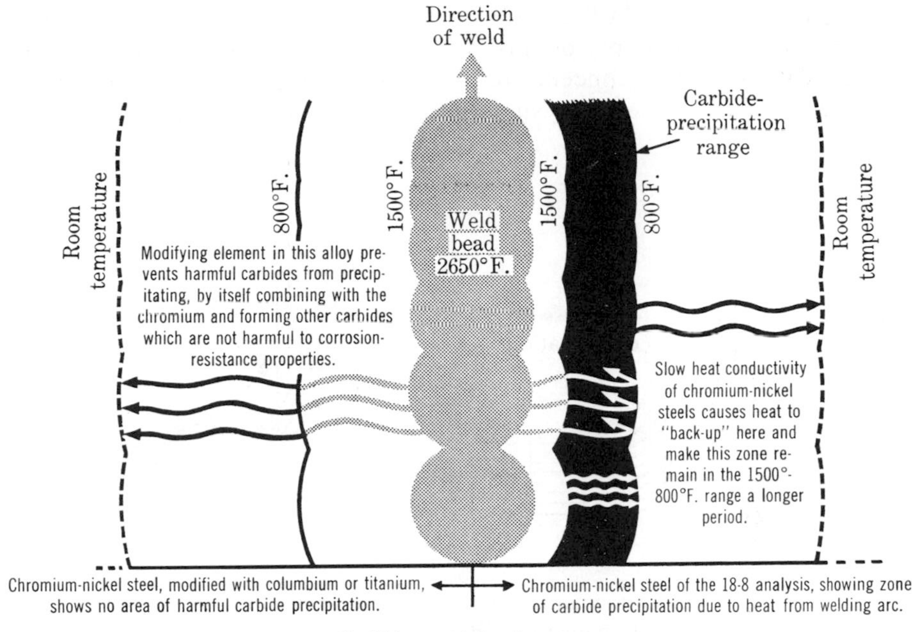


**Fig. 7.3.** The carbide precipitation zone of conventional 18-8 stainless steel compared with a modified chromium-nickel steel. Courtesy Allegheny Ludlum Steel Corp.

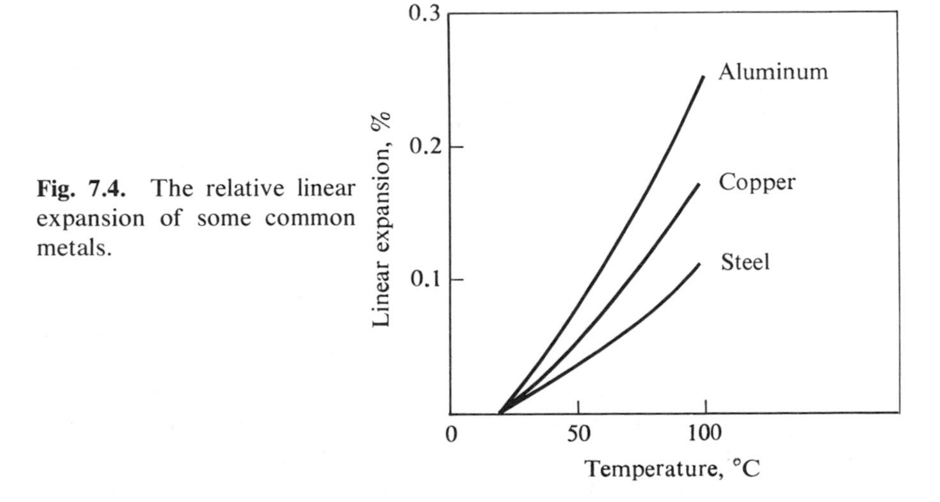


**Fig. 7.4.** The relative linear expansion of some common metals.

## STRESSES IN JOINT DESIGN

**Weld Contour.** Wherever possible, welded joints should be made with smooth-flowing lines that blend gradually with the parent metal.

Any abrupt change in the surface causes it to be a point of stress concentration. An example of this is a fillet weld shown in Fig. 7.5. The stress lines tend to concentrate at the toe of the weld when a convex bead is used, but they flow smoothly when a concave bead is used.

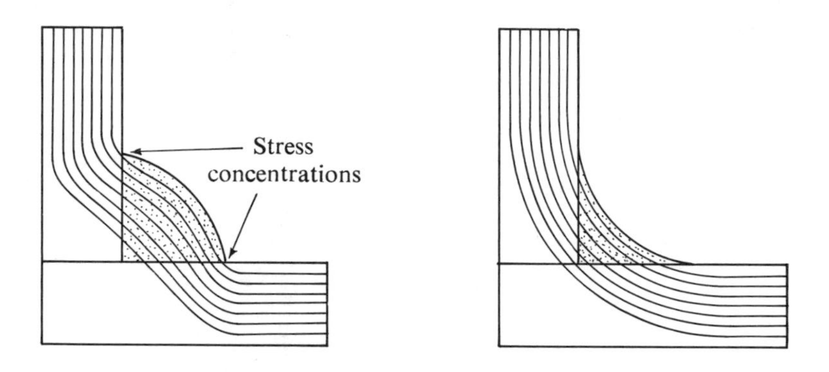


**Fig. 7.5.** Stress concentrations are caused in welded joints where there is an abrupt change in contour, as in the fillet weld on the left.

**Determining Stress Distribution.** Many engineered parts fail in service because of the stress concentrations in a particular area. In fact, most failures in high-duty equipment are caused by fatigue, which, in turn, is influenced by stress concentrations.

A recent aircraft manual cites over 50 failures which originated in regions of stress concentration, and most of them were fatigue failures. Stress concentrations are almost always traceable to improper design, fabrication, or maintenance.

Stress concentration tends to produce failure by fatigue even when the regions of stress concentration are very small. Several methods exist for the study of true stresses; these include mathematical theory of elasticity, strain gages, brittle lacquers, x-ray diffraction, and photoelasticity. The latter is still one of the most precise methods available in the analysis of two-dimensional stress systems.

PHOTOELASTICITY. In the photoelastic method, the joint contour is cut out of clear acetate plastic material. This model is placed in a field of polarized light produced by a polariscope. The interaction of the polarized light with the model produces stress patterns (Fig. 7.6*a*) from which the stress distribution can be determined.

The basic phenomenon of photoelasticity is that points of increasing stress produce cyclical variations in intensity of the observed light patterns. At zero stress, the image is black (*b*). Monochromatic light in the stressed transparent model produces alternate bright and dark bands

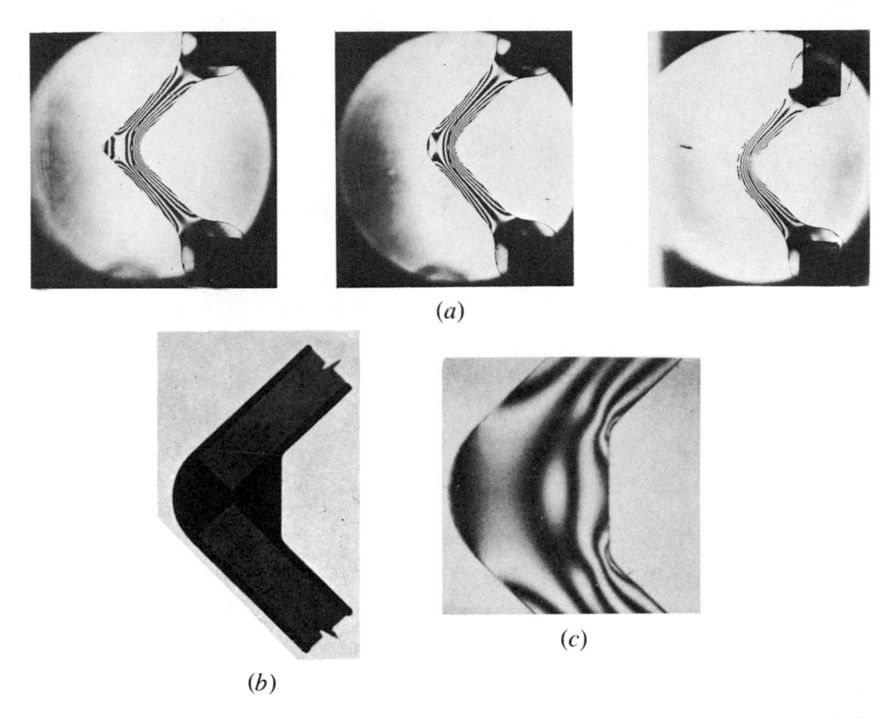


**Fig. 7.6.** Photoelastic stress patterns as determined on different corner-joint designs, but with the same loading, in upper group. Lower pictures show an unloaded and loaded model. Courtesy Steel Founders' Society of America.

called interference fringes. Examination of the photoelastic patterns developed shows a tendency of the fringe lines to group together in regions of stress concentration. The model can be gradually loaded so that the fringe lines can be distinguished and counted. The oval-shaped white dots in the knee of the specimen are known as *isotropic points* or points of equal stress in all directions. On this picture (Fig. 7.6*c*), there are six orders (white bars) and a half order (black spot) to the two stress-concentration points on the inside fillet. This gives a fringe order of $N = 6.5$ for a known load of 6.3 lb. The fringe order at the middle of the inside fillet is observed to be equal to $N = 4$. Therefore, the stress-concentration point is about 60 percent greater than in the region of the fillet. We therefore have a stress concentration factor of $6.5/4 = 1.625$.

Any discontinuity on the specimen-fillet surface which has a stress-concentration effect can initiate a fatigue crack in that region.

## WELDING PROCESSES

Table 7.1 shows all of the welding processes in popular use today and how they are classified. You will note that only five main methods of welding are given but with many types under each one. The major emphasis in this chapter will be on the two electrical methods, *arc* and *resistance* welding.

**Table 7.1**
**The Most Used Welding Processes**

| | | |
|---|---|---|
| **Brazing** | | Torch brazing<br>Twin carbon-arc brazing<br>Furnace brazing<br>Induction brazing<br>Resistance brazing<br>Dip brazing<br>Block brazing<br>Flow brazing |
| **Gas welding** | | Air acetylene welding<br>Oxyacetylene welding<br>Oxyhydrogen welding<br>Pressure gas welding |
| **Resistance welding** | | Resistance spot welding<br>Resistance seam welding<br>Projection welding<br>Flash welding<br>Upset welding<br>Percussion welding |
| **Arc welding** METAL ELECTRODE | Shielded | Arc spot welding<br>Arc seam welding<br>Shielded metal arc welding<br>Atomic hydrogen welding<br>Gas metal arc welding<br>Pulsed arc<br>Short-circuiting arc<br>Gas tungsten arc welding<br>Submerged arc welding<br>Gas shielded-stud welding |
| | Unshielded | Stud welding<br>Bare metal arc welding |

## ARC WELDING

The metallic arc welding process is termed a *nonpressure fusion process*. The heat is developed in an arc produced between a metal electrode or wire and the work to be welded. Under the intense heat developed by the arc, ranging from 5,000° to 10,000°F, a small part of the base metal, or work to be welded, is brought to the melting point. At the same time, the end of the metal electrode is melted, and tiny globules or drops of molten metal pass through the arc to the base metal, as shown by the high speed photographs in Fig. 7.7.

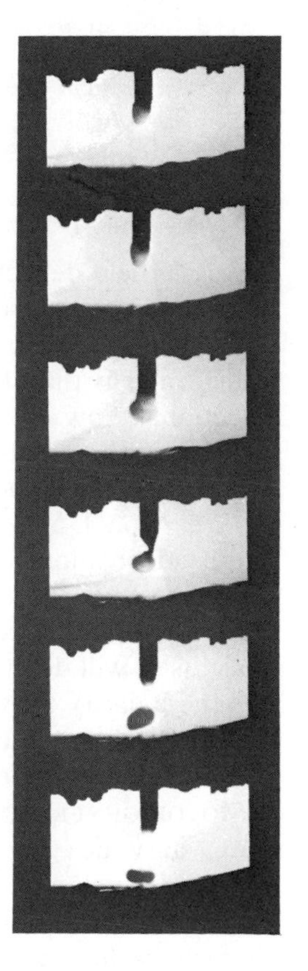

Fig. 7.7. High-speed photography is used to show how the metal is pinched off from the electrode as it passes through the arc in globular form. Photos were made at 3,600 frames per sec.

The heat liberated by the arc is a function of the power used. It may be converted as follows: 1 watt = 1 joule/sec or 1 joule = 1 watt/sec. In calculating the heat for a particular arc welding condition, the following formula can be used: $W = EIt$, where $W$ = heat in joules, $t$ = time in seconds, $E$ = volts, and $I$ = current in amperes.

## CURRENT SOURCES FOR MANUAL ARC WELDING

**Alternating-Current Transformers.** In the alternating current (ac) transformer, the primary coil is hooked directly to the power line. The secondary coil is either tapped at intervals to give different current settings, or the primary coil is moved in relation to the secondary or vice versa so as to vary the strength of the electrical field (Fig. 7.8). Small light-duty machines depend on a shunt that can be mechanically moved in or out of the central transformer core.

TRANSFORMER ADVANTAGES. The principal advantages of the transformer over the generator are lower initial cost—generally about 40 percent less—and lower maintenance costs. There are practically no moving parts in the transformer and, consequently, there is very little wear. It is also very quiet in operation. One particular operating advantage of the ac arc is that there is no *arc blow*. This is a phenomenon that sometimes occurs when welding with direct current. A distorted

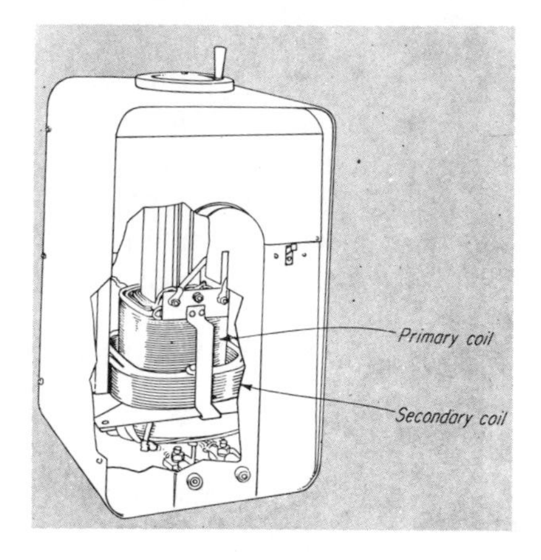


**Fig. 7.8.** The ac transformer used for arc welding.

magnetic field deflects the arc from its normal path, making it difficult to deposit the metal properly.

TRANSFORMER DISADVANTAGES. The main disadvantage of the transformer is that the *polarity* cannot be changed. Polarity refers to the directional flow of the current. In *straight polarity* the electrode or welding rod is negative, and in *reversed polarity* it is positive. Reversed polarity is recommended for nonferrous metals such as aluminum, bronze, Monel, and nickel, and also for making welds in the vertical and overhead positions. Reversed polarity generally gives a greater digging action, resulting in deeper penetration. Straight polarity is used in mild steel for greater speed and smoother beads. However, the type of electrode used will determine these properties to a much greater extent than will the polarity.

**Direct-Current Machines.** Direct current (dc) machines are of the motor-generator type or the ac–dc rectifier type.

MOTOR-GENERATOR SETS. The motor-generator set consists of a heavy-duty dc generator driven by a suitable motor or engine. The voltage of such a generator will usually range from 15 to 45 volts across the arc, although any setting is subject to constant variation owing to changes in the arc conditions. Current output will vary from 20 to 1,000 amp, depending upon the type of unit. Most generators are made of the variable voltage type, so that the voltage automatically responds to the demands of the arc. However, the machine shown in Fig. 7.9 is of the dual-control type in which the open circuit voltage can be preset. On machines having only one control setting, the voltage is automatically adjusted so that it is proportional to the amperage.

AC-DC RECTIFIERS. The transformer rectifier unit consists of a three-phase transformer that reduces line voltage to the proper level (Fig. 7.10*a*). The rectifiers that change the alternating current to direct current are usually made of silicon or selenium. However, silicon now

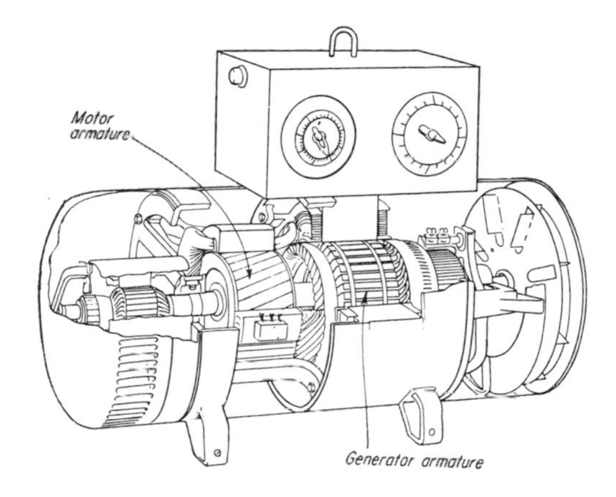


**Fig. 7.9.** A dual control motor-generator type of welding machine.

seems to be gaining favor. The cooling fins around the rectifier are made of steel with a layer of selenium sprayed on them, or they may be laminated with layers of silicon. Silicon diodes are also used.

ADVANTAGES OF AC–DC MACHINES. The ac–dc machines have the advantage of being able to supply either type of current as needed (Fig. 7.10*b*).

## WELDING CURRENT CHARACTERISTICS

Before any decision as to what type of equipment is best suited for a given welding operation, it is necessary to have a basic knowledge of how current affects the arc, what the power requirements are, what is meant by duty cycle, and the improved electrical controls available.

**Arc Characteristics.** A *drooping arc voltage* curve (Fig. 7.11) provides the highest voltage when the circuit is open and no current is flowing. This makes for easy starting of the arc. As the arc is struck, the current rises to maximum, and the voltage droops or goes back to normal. During the welding operation, voltage varies directly with the length of the arc.

The *constant arc voltage* (CAV) and *rising arc voltage* (RAV) are also shown in Fig. 7.11. As the name implies, constant arc voltage machines maintain a preset voltage regardless of current draw.

Rising arc voltages provide increasing amperage to maintain a constant arc length. Current delivered is a function of the wire feed rate.

Both stick electrode and tungsten arc welding require a drooping voltage.

**Characteristic Power Source.** A gas-shielded metal arc can use drooping, CAV, and RAV types. CAV has proved to be the most adaptable so far.

**Power Requirements.** Power requirements are determined by the type of work to be encountered throughout the life of the machine. If the

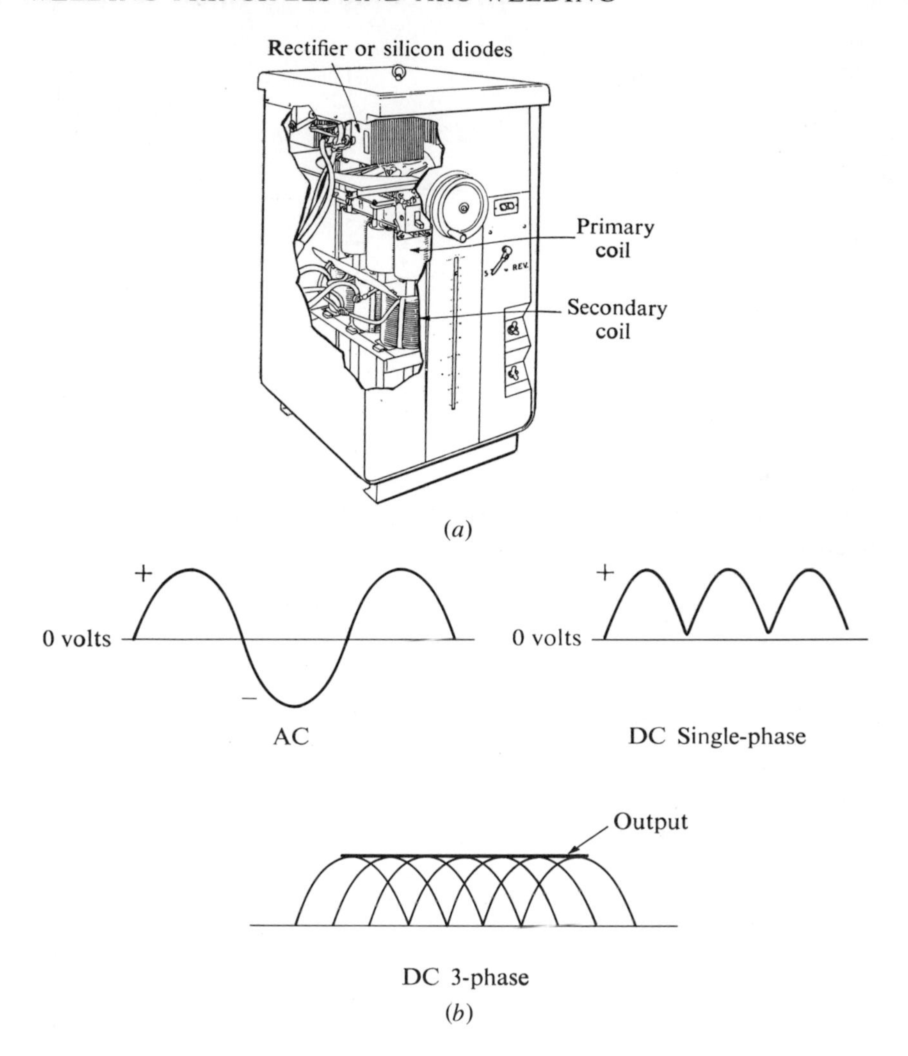


**Fig. 7.10.** (*a*) Direct current can be supplied for welding through a rectifier. (*b*) Current wave forms obtained from various power supplies.

actual work requires a 400-amp machine, it is usually wise to get a 500-amp welder. The additional capacity is often used.

Most welding machines are designed to operate on either 230- or 440-volt lines.

**Duty Cycle Requirements.** Duty cycle is defined as the ratio of load time to total time. For the purpose of standard rating, 10 min is the time period of one complete test cycle. For example, when a 60 percent duty cycle is used, the load is applied 6 min and shut off 4 min.

It is possible to draw more than the rated current from a power supply

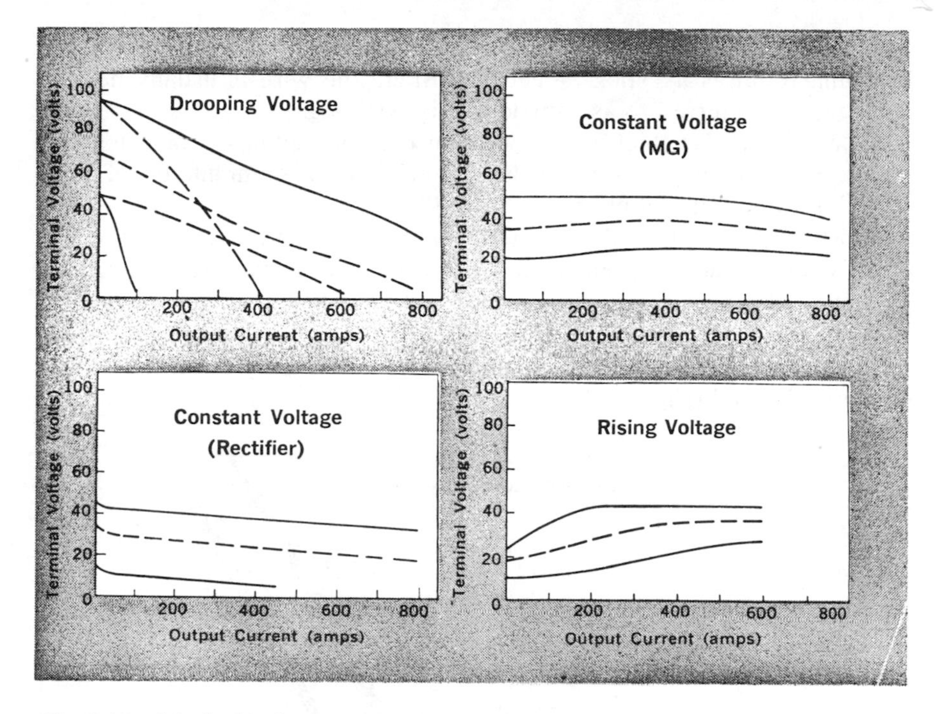


**Fig. 7.11.** Typical voltage-curve characteristics. Drooping arc voltages provide highest starting potential. Constant arc voltages maintain the preset voltage regardless of current draw. Rising arc voltages provide increasing arc voltage with increasing amperage. Courtesy *Metalworking.*

at reduced duty cycles. For example, a power supply rated to produce 100 amp on a 60 percent duty cycle (6 min out of 10) might be operated at 150 amp on a 27 percent duty cycle. Conversely it is possible to use a machine at less than 100 percent duty cycle for extended periods of time. For example, a machine rated to produce 100 amp on a 60 percent duty cycle could be used to produce 77 amp at 100 percent duty cycle. The percent duty cycle is a ratio of the square of the rated current to the square of the load current multiplied by rated duty cycle: % duty cycle = $(I_r)^2/(I_l)^2 \times$ (rated duty cycle) where $I_r$ = rated current and $I_l$ = load current.

## ELECTRODE TYPES AND SELECTION

**Electrode Types.** Electrodes can be broadly classified as consumable and nonconsumable. The consumable perform the dual function of supplying the filler material for the weld as well as maintaining the arc. The nonconsumable are tungsten wires, carbon rods, or copper electrodes.

The consumable electrodes may again be classified as to use, for manual welding or for automatic and semiautomatic machines. Manual welding requires a "stick" electrode. Since the wire is usually heavily coated, it is referred to as "shielded-arc" welding.

The coating for shielded-arc welding may be cellulosic, or it may be made of mineral or iron powder materials or a combination of both cellulosic and mineral materials.

**Shielded-Arc Stick Electrodes.** CELLULOSIC COATING. The cellulosic coatings derive their name from wood pulp, sawdust, cotton, and various other cellulosic compositions obtained in the manufacture of rayon. This coating forms a gaseous shield for protection around the arc stream during the welding operation and a light slag covering for the deposited metal (Fig. 7.12).

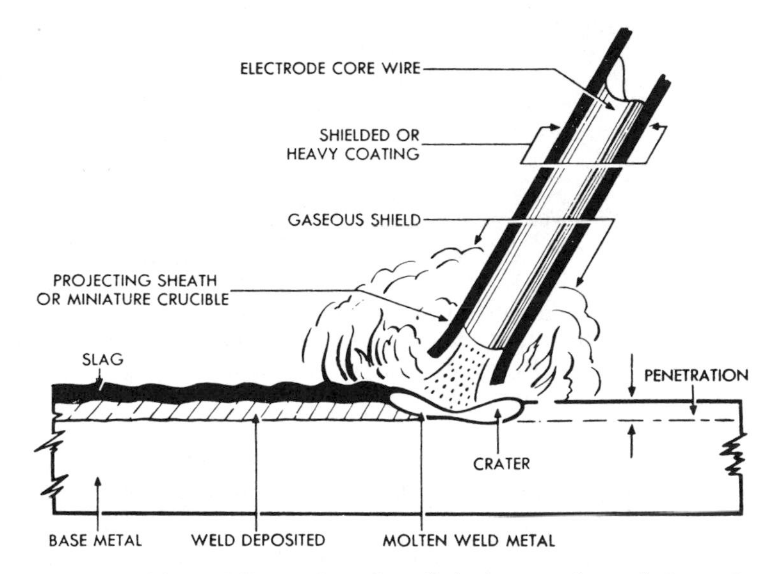


**Fig. 7.12.** The welding action of a cellulosic-coated metal electrode.

*Use.* The expanding gas furnished by the burning cellulosic material acts to give a forceful digging action to the weld. It can be used in any position but is particularly useful in vertical, horizontal, and overhead positions.

MINERAL COATING. Mineral coatings are manufactured from natural silicates such as asbestos and clay. By adding oxides of certain refractory metals such as titanium, the harsh digging action of the arc is modified to give an arc that is softer and less penetrating. This type of action is desirable in making welds where the fit-up is poor and on sheet metal parts where shallow penetration is desired. The mineral-coated rod depends on the heavy slag for protection and control of the chemistry of the deposited metal.

The heavy slag acts to slow the cooling rate of the metal, allowing gas to escape and slag particles to rise to the top. Cooling stresses are reduced, and a more homogeneous microstructure results.

*Use.* The burn-off rate of mineral-coated electrodes is much higher than for the cellulosic type, resulting in a larger molten pool and faster deposition rate. Because of this and the large amount of slag produced, they are more advantageously used for *downhand* welding. Downhand welding is used in making flat welds and those inclined up to 45 deg.

IRON-POWDER COATINGS. The use of iron powder in the electrode coating is comparatively new. The addition of iron powder brought about several desirable effects; particularly, the welding speed increased, and the appearance of the bead improved. The iron powder in the coating not only helps to form an effective crucible at the end of the rod, which is longer than that formed with other coatings, but it more effectively concentrates the heat and gives an automatically consistent arc length. Welding is made considerably easier since the electrode can be dragged upon the work without having to hold it a certain height above the work. Slag is often self-removing. There is less sticking of the electrode, which often occurs in the small sizes. Finally, it furnishes more inches of weld per electrode because the iron in the coating goes into the weld. Iron-powder electrodes require about 25 percent higher current setting than for equal-sized conventional rods. Also, because of the heavy coating, there are about half as many rods per pound.

LOW-HYDROGEN ELECTRODES. Low-hydrogen electrode is the name given to a mineral-type coating that was at first called "lime ferritic." It is now called "basic low hydrogen," since it is the basic or alkaline property of the slag and its low hydrogen content that provide the unique qualities for welding difficult-to-weld steels. Low-hydrogen electrodes developed from the idea that the basic calcium-carbonate, calcium-fluoride covering used on stainless-steel-core wire might work well on a mild-steel-core wire alloyed with manganese and molybdenum. It proved to be satisfactory in welding armor plate that had been so prone to cracking. However, the welding of heavy armor plate was still erratic, and it was found that this was because some hydrogen was contained in the coating. Consequently, the hydrogen was reduced to a minimum.

In 1956, iron powder was added to the coating. There are now three main types of low-hydrogen, iron-powder electrodes: conventional E7016; E7018, containing approximately 30 percent iron powder; and E7028, containing approximately 50 percent iron powder. The iron-powder addition increases the burn-off or deposition rate of the metal.

In use, there may be some porosity each time the arc is struck unless a special effort is made to keep the arc short. The shielding carbon dioxide gas is largely produced from the decomposition of the calcium carbonate, which constitutes the largest amount of any ingredient in the covering. Since it requires only a fraction of a second for the electrode covering to reach this decomposition temperature, a short arc must be maintained or the metal will not be adequately protected from the atmosphere.

Moisture in the coating of low-hydrogen electrodes has a detrimental effect on the quality of the weld if it is over 1.9 percent. A general recommendation is that these electrodes be stored where the temperature is at least 10°F above the outdoor temperature. This is not always adequate. The best results are obtained by keeping the electrodes in a dry-rod storage oven at 250° to 350°F.

ELECTRODE SIZE. A general rule in selecting electrode size is never to use one that is larger in diameter than the thickness of the metal to be welded. For vertical and overhead welding, 3/16 in. is the largest diameter you should use, regardless of the plate thickness. Larger electrodes make it too difficult to control the deposited metal.

**Electrode Classification.** The American Welding Society and the American Society for Testing Materials have jointly established a code for welding electrodes based on tensile strength, position, current supply, and application.

The code is based on an "E" prefix with four or five digits following it. The first two digits of four-digit numbers refer to minimum ultimate tensile strength. The third digit refers to the welding position in which the electrode can be used, as shown in Table 7.2. The fourth digit refers to current, and indirectly to coating, as shown in Table 7.3.

**Table 7.2**
**Meaning of Third Digit in the E Code**

| Digit | Meaning |
|---|---|
| 1 | All positions |
| 2 | Horizontal and flat positions |
| 3 | Flat position |

**Table 7.3**
**Meaning of Fourth Digit in the E Code**

| Digit | Meaning |
|---|---|
| 0 | Direct-current reverse when third digit is 1 |
| 0 | Direct-current reverse and alternating current when third digit is 2 or 3 |
| 1 | Alternating-current or direct-current reverse |
| 2 | Straight polarity or alternating current |
| 3 | Alternating- or direct-current straight |

**Table 7.3 (Continued)**
**Meaning of Fourth Digit in the E Code**

| Digit | Meaning |
|---|---|
| 5 | Direct-current reverse (lime or titania sodium, low hydrogen) |
| 6 | Alternating- or direct-current reverse (titania or lime potassium, low hydrogen) |
| 8 | Alternating- or direct-current reverse (titania, calcium carbonate, fluorspar, and iron powder) |

As an example, take the code number E6010. The "E" stands for "electrode." The first two digits, 60, indicate that the minimum ultimate tensile strength is 60,000 psi. The third digit, 1, indicates that it is possible to weld in all positions. The fourth digit, 0, indicates direct-current reverse polarity.

## ELECTRODES FOR AUTOMATIC AND SEMIAUTOMATIC WELDERS

Automatic and semiautomatic machines are made to handle the consumable electrode in either the solid or the cored-wire form.

Cored wire (Fig. 7.13) is easily used in automatic machines. It contains not only a flux and shielding for the molten metal deposit but also arc stabilizers, metallic deoxidizers, and alloys.

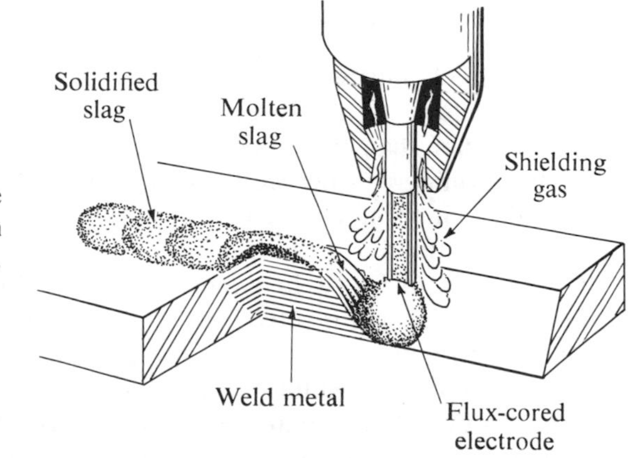


**Fig. 7.13.** The cored-wire electrode is used with shielding gas to weld steel.

Two types of cored wire are made, "single pass" and "multipass." The single pass was developed to weld over rusty and unclean plate by the use of metallic deoxidizers in the core. In multipass welds, dilution of the weld metal with the base metal is not so serious and is made for welding on clean plate.

To reduce this need for two types of cored wire, a nominally single-pass wire has been developed which is able to accept heavy rust and mill scale and yet be satisfactory for multipass work.

The cored wire is normally used with a shielding gas, usually $CO_2$; however, it has been used without a shielding gas in mass production work.

The cored-wire electrode can be used on CAV machines of high capacity, which makes for high deposition rates. It produces deep penetration and can often be used with simpler joint designs. In the manual electrode welding method, about one fifth of every stick is wasted. Also, there is more time lost in changing electrodes. The cost of $CO_2$ is about 15 percent of that of argon, and the flow is approximately half that required for helium and two thirds that required for argon.

The nonconsumable electrode used in automatic and semiautomatic welding is a tungsten or tungsten alloy rod. It is most commonly used with an inert gas to provide the shielding—thus, the name TIG for tungsten-inert-gas welding, which is described later.

**Basic Selection Factors.** Much of the information needed for selecting an electrode for a given job is contained in standard codes jointly established by The American Welding Society (AWS) and The American Society for Testing Materials (ASTM). Factors that must be considered are mechanical and physical properties required of the completed structure, current source, and position of the weld.

## WELD POSITIONING

The main advantage of being able to position a joint is to increase the speed of welding. Therefore, in selecting the electrode, if the weld is normally other than flat, repositioning it for welding should be considered. Positioning can be accomplished manually, but there are so many universal positioning devices on the market that manual positioning can seldom be justified.

To make weld beads in either the vertical or the overhead position requires fast-freezing electrodes with lower deposition rates and special techniques to help overcome the force of gravity. An example of the advantage of positioning a fillet weld from vertical to horizontal is shown in Fig. 7.14. You will note that the speed increased from 4.5 ipm to 19.2 ipm.

Good positioning is important not only in respect to the weld direction but also in respect to improving the *operating factor*. By operating factor is meant the percentage of time that the arc is actually welding. Here the convenience of the supplies, machine controls, and accessibility are important. A 50 percent factor means that only half the number of feet per hour have been welded that would be done with a 100 percent operating factor. (See Fig. 7.15).

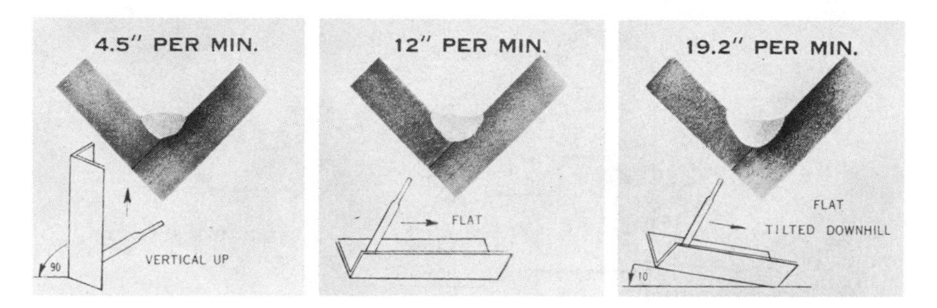


**Fig. 7.14.** Effect of position on speed of welding fillet welds in plate 3/8 in. or thinner. Courtesy Lincoln Electric Company.

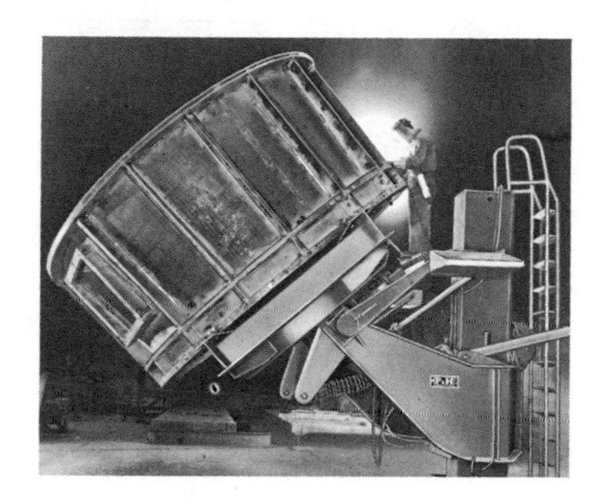

**Fig. 7.15.** This large welding positioner is able to put all welds on this $12\frac{1}{2}$-ton rotor housing in the downhand position. Courtesy Harnishfeger Corp.

## BASIC TYPES OF AUTOMATIC AND SEMIAUTOMATIC ARC WELDING

Automatic and semiautomatic arc welding processes usually involve wire feed systems. Shown in Fig. 7.16 is a simplified drawing of the major components required to make up a semi- or fully automatic system.

In addition to the basic components shown, automatic welding also requires holding fixtures, variable speed travel arrangements, specialized torches for directing the shielding, guiding devices, etc. The selection of semi- or fully automatic methods usually depends on the economics and quality requirements. Obviously, to justify the extra expense, repetitive parts are required for fully automatic processes.

There are four popular automatic arc welding systems: two with consumable electrodes, submerged arc and gas metal arc (GMA); two with nonconsumable electrodes, TIG as mentioned previously, and resistance welding.

**Submerged-Arc Welding.** Submerged-arc welding derives its name from the fact the entire welding action is submerged beneath a mineral material known as flux (Fig. 7.17). The arc is started by a short initial contact with the work, by a steel-wool fuse ball, or by a high-frequency spark. As the flow of current starts, the welding arc becomes submerged

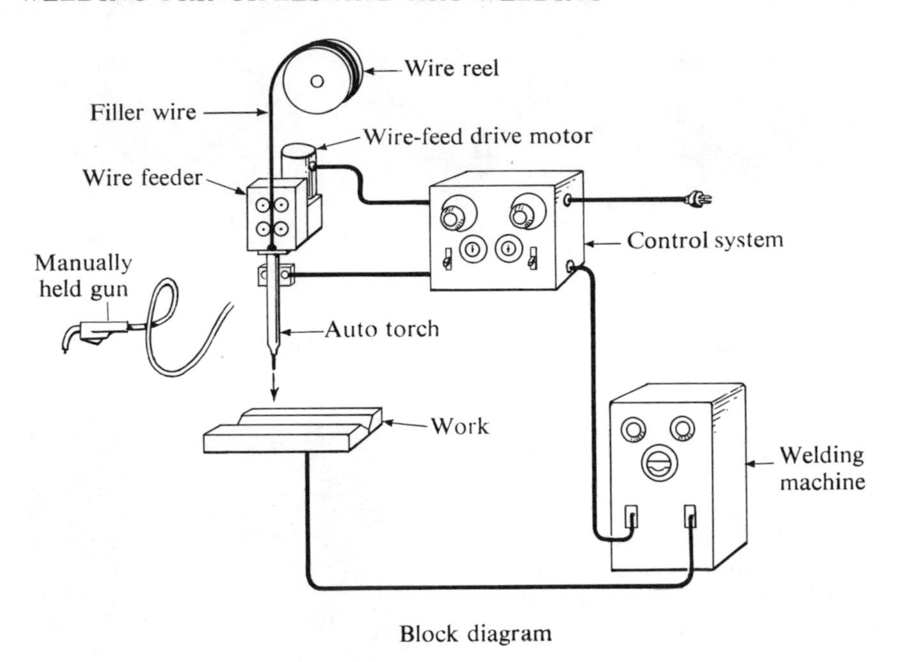


**Fig. 7.16.** The basic components required for automatic or semiautomatic welding.

in a sea of molten flux, which shields the arc and the molten metal from oxidation and covers the hot weld deposit, allowing it to cool more slowly.

The bare wire electrode, ranging in size from 5/64 to 1/4 in. and occasionally to 3/8 in., is held on a coiled reel and fed mechanically by

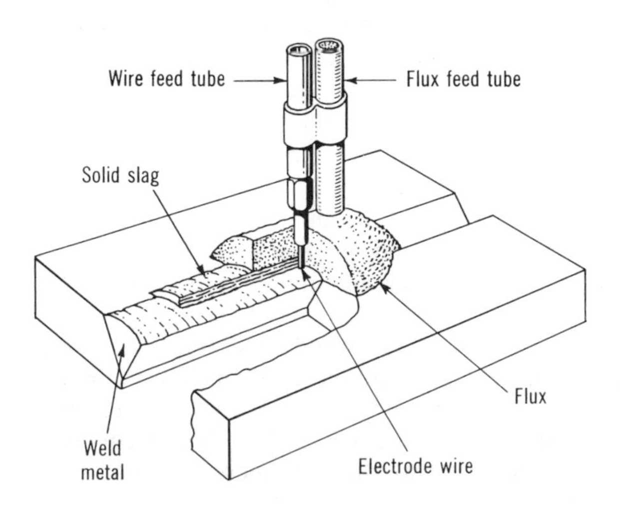


**Fig. 7.17.** Submerged-arc welding is used with either alternating or direct current for heavy welding and high deposition rates.

means of continuously rotating drive rolls. The speed is varied according to the travel of the machine and the needs of the job.

Automatic heads are available for either alternating or direct current. The ac transformers are better suited to heavy plate work and where continuous seams are used. The low current settings required for light-gage work make the ac arc difficult to control.

The dc type head on reversed polarity can be used to advantage on heavier-gage metal. (A minimum of 14-gage metal is recommended.) Direct current has superior arc-striking qualities and generally better arc control. However, it is still subject to arc blow. Welding voltage control used on this equipment helps make the process automatic.

ADVANTAGES. The submerged arc can handle extremely high current densities. For example, currents as high as 600 amp can be used with 5/64-in.-diameter electrode wire. This creates a current density of 10,000 amp per sq in., or 6 to 10 times that carried by a manual electrode of equivalent size. The advantage of high current density is, of course, the high *melt-off* rate that can be obtained. Speeds can be obtained that range from 15 ipm on heavy fillet welds in 1-in. plate to 150 ipm on 1/4-in. material.

The high speeds and high current densities result in deep penetration. Butt welds can be made in steel plate up to 5/8 in. thick without any edge preparation. Because of the deep penetration, the use of smaller V's is practical. Shown in Fig. 7.18 is an example of a weld made in heavy plate with only a 15-deg U-groove type of preparation. This weld was made by using two electrodes at once in tandem position, or one in back of the other. A side-by-side or transverse position is also used. By the use of multiple electrodes, speeds can be increased from 1 1/2 to 2 times that of the single electrode.

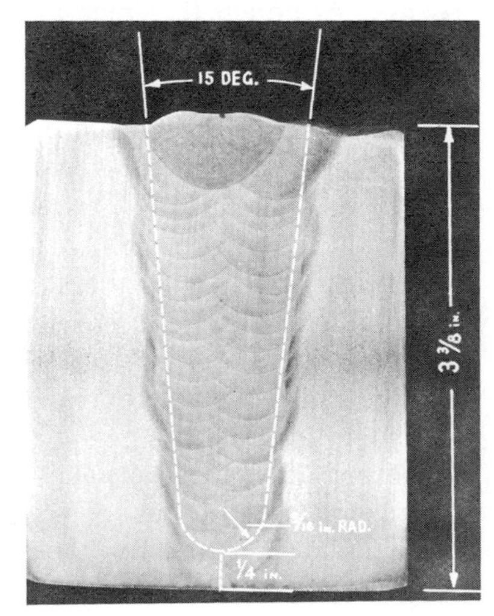


**Fig. 7.18.** A cross section of a multipass weld made with submerged arc. Courtesy Linde Co., division of Union Carbide Corp.

With the submerged arc, the quality of the weld is excellent and uniform. There is no spatter, and the fused flux on top of the weld pops off by itself, revealing a weld that needs little or no finishing.

USE. Metals welded by the submerged-arc process are low-carbon steels, low-alloy steels, stainless steels, and high-alloy steels. Special precautions, such as preheat and postheat, must be taken in welding the high-alloy steels. Common nonferrous metals are also welded by this process.

Because of the large amount of molten metal, most of the welding is limited to the flat position.

**Gas Metal Arc Welding.** The gas metal arc (GMA) welding process (Fig. 7.19) uses a continuously fed electrode wire with a shielding gas. The process was developed shortly after World War II. At first only inert gases were used so it was called metal inert gas welding or MIG. Now, however, the process has been expanded to use $CO_2$ gas, which is not inert, so the term gas metal arc is more inclusive.

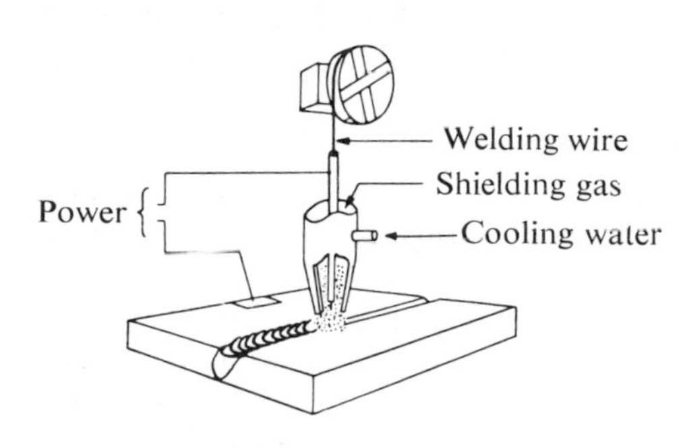


**Fig. 7.19.** Basic components of the gas metal arc process.

The use of $CO_2$ as shielding gas is comparatively new for the GMA process. Although the inert gases worked well on aluminum and other nonferrous materials, their use resulted in excessive costs for low and medium carbon steels.

A variation in the $CO_2$ process is the use of flux-cored wire as described previously in the section on electrodes. The gas shielding envelope is maintained to exclude oxygen and nitrogen from the weld. The flux in the electrode aids in the stabilization of the arc, provides for scavenging of the weld metal, provides for a thin slag covering of the deposited weld, and can provide some alloying elements.

ADVANTAGES. Welding costs are cut with the GMA process due to the high melt-off rate developed by the high welding current and comparatively low-cost $CO_2$ gas.

Another advantage, although not a typical characteristic, of the cored wire arrangement is the deep penetration and small weld size that can be obtained as compared to that made with the stick electrode (Fig. 7.20). As an example, for a 3/8-in. fillet weld, the travel speed for the flux-

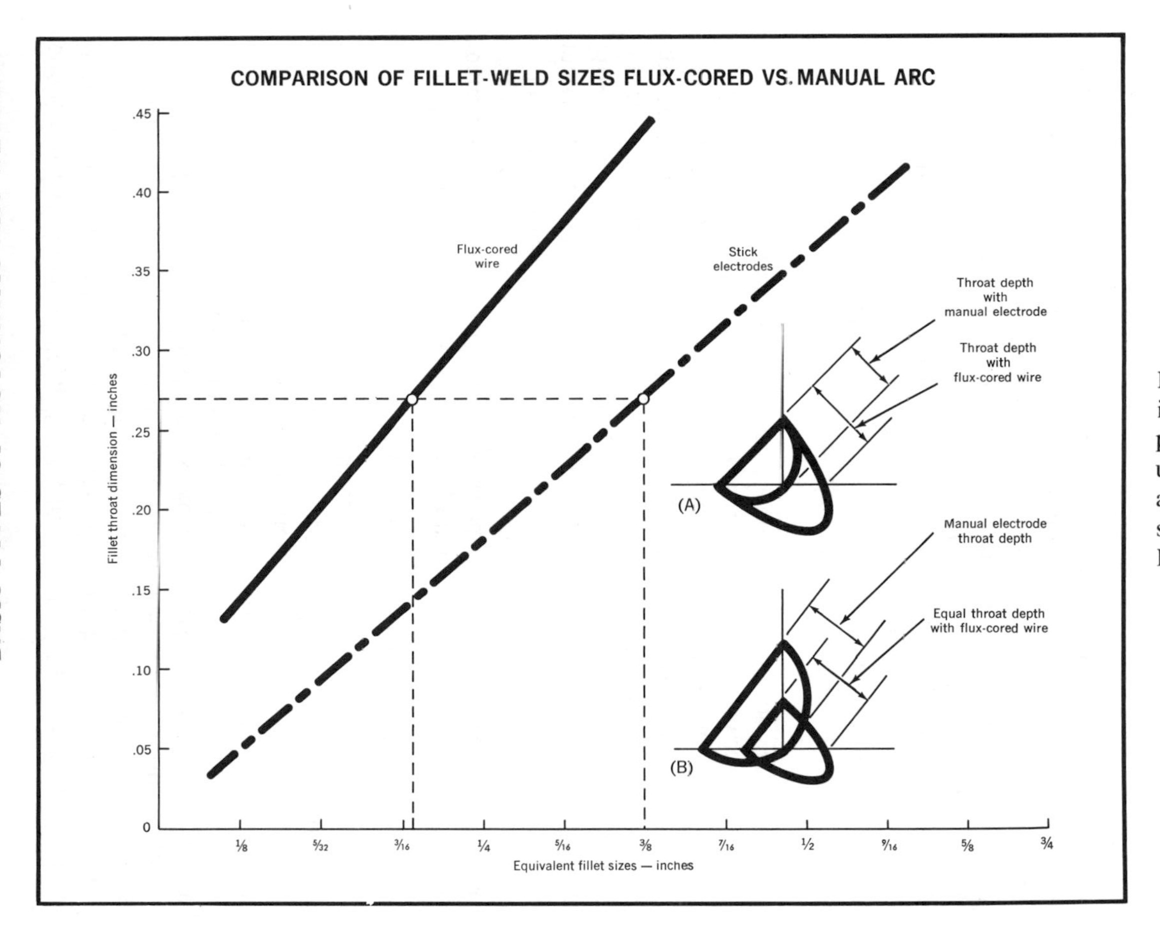


**Fig. 7.20.** A smaller weld is required for the $CO_2$ process than for the manual stick process to produce a given throat depth and strength. Courtesy Lincoln Electric Co.

cored process could be 16 in./min with a 3/32 wire at 500 amps, but, because the penetration is deeper, the fillet can be reduced to a 1/4-in. size for the same strength. Thus the same weld specifications can be met with 100 amp less current and a travel speed of 23 in./min. This is 14 in./min faster than with the stick electrode. Also, there is a saving of nearly 50 percent in weld metal deposit.

Weld penetration is a function of the voltage, current, travel speed, type of electrode, arc length, and position of weld. Some experimental work has been done to determine weld penetration mathematically. The equations derived vary, but one that can be used is as follows:

$$P = 2 \times 10^{-5} E^{0.56} I^{1.40} V^{-0.29} \text{ where } V = \text{travel speed in./min.}$$

USE. The GMA process is now primarily used for welding ferrous materials. A CAV power supply is usually used with fixed (but adjustable) wire feeds.

The $CO_2$ cored-wire process is limited at the present time to ferrous materials, primarily the low-carbon and low-alloy, high-strength steels. However, research is continuing to produce additional filler wires to match other metals. The GMA process is used extensively in tank construction and on structural steel and will handle thin materials as well as very heavy plate as used in ship and barge construction.

**Variations of the Gas Metal Arc Process.** There are several variations of the gas metal arc process due to arc types, current variation, and small (micro) wire size.

METAL TRANSFER TYPES. Three arc lengths are now in standard use (Fig. 7.21). Of particular interest is the short circuiting process, which consists of allowing the electrode to touch the metal and short circuit anywhere from 20 to 200 times/sec. Metal transfer occurs when the arc is out. The transfer cycle period is shown on the oscillogram (Fig. 7.22) as A through E, and the arcing period is shown as E through I. Because the arc melt is less than the feed rate, the end of the electrode advances to the workpiece and extinguishes the arc. The short-circuiting arc has been found to do an excellent job of welding stainless steel when used

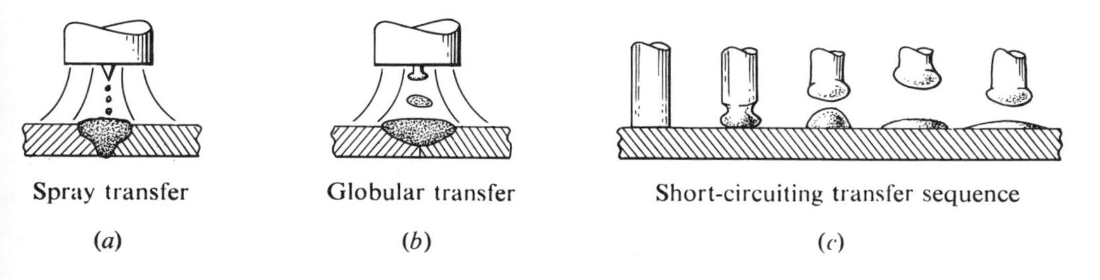


**Fig. 7.21.** The three types of arcs used in gas metal arc welding.

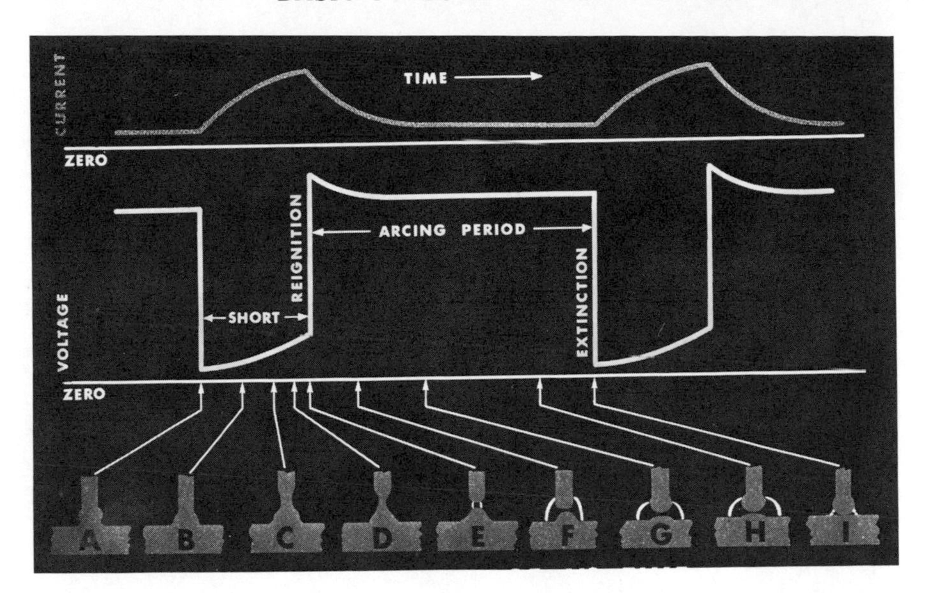


**Fig. 7.22.** Current and voltage versus time oscillograph of a typical short-circuiting metal-transfer cycle. Courtesy Union Carbide.

with a new mixture of 90 percent helium, 7.5 percent argon, and 2.5 percent oxygen.

The spray type of metal transfer is used on heavier materials because it gives the best penetration. The globular metal transfer is considered conventional for $CO_2$ welding of steel.

CURRENT VARIATION. The pulsed arc is based on current variation originally developed for gas metal arc welding of aluminum. Normally the minimum usable current for a 1/16 in. dia wire is 150 amp. If lower current settings are used, the metal transfer across the arc is ineffective. Higher currents are often too much. The pulsed arc allows the use of a *low average* or background current with a high pulse to spray the metal at the proper time (Fig. 7.23*a*). The three-phase background current

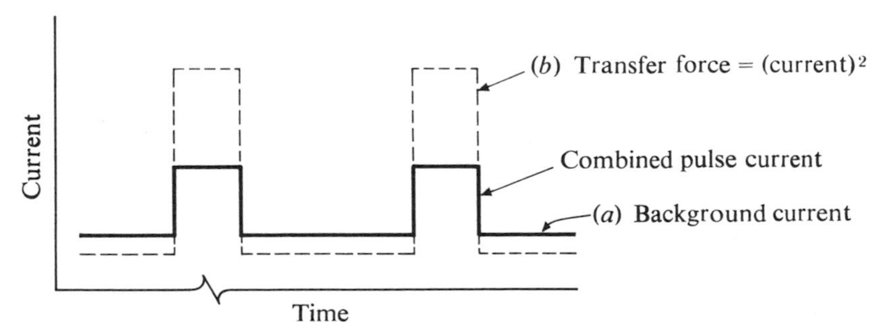


**Fig. 7.23.** Schematic of the pulsed current wave form (*a*).

heats the wire and weld joint and the pulse current, a half wave rectified ac supply, sprays the metal across the arc.

MICRO WIRE. The small (micro) wire $CO_2$ process, along with mixtures of $CO_2$ and argon, has made a major breakthrough for welding thinner materials, for pipe welding, and for welding out of position. This process has been widely accepted since it can be utilized anywhere that manual stick electrodes were formerly used. No design changes are required, and the work can usually be done in half the time.

ARC SPOT WELDING. Arc spot welding is another variation of the GMA process. The previously described wire feeders and power sources are used, but a timing device and special gun nozzle are added. Shown schematically in Fig. 7.24 are the steps required to make an arc spot weld. The arc actually penetrates the top sheet and welds into the bottom sheet, producing a weld nugget.

This type of spot welding has the advantage of being portable and of requiring access to only one side of the workpiece. Only a small amount of surface cleaning is required.

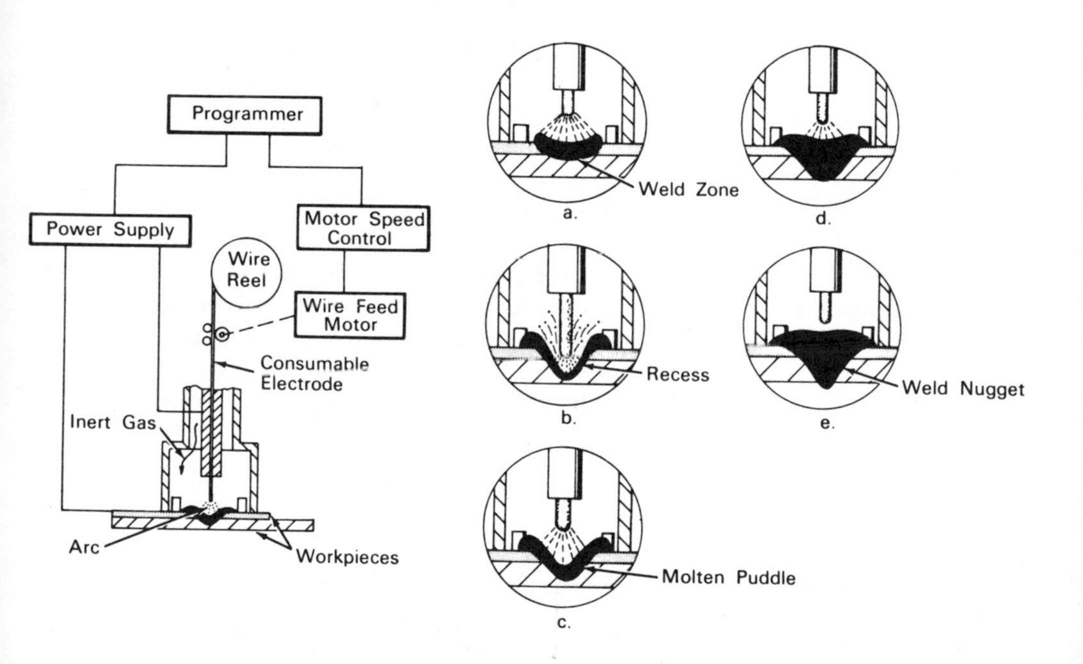


**Fig. 7.24.** After the arc is struck, the temperature and length of arc are varied by a predetermined program of (*a*) preheat, (*b*) blowing the metal out of the weld zone to make a recess, (*c*) melting the sides of the recess, (*d*) forming the weld nugget, and (*e*) interruption of the arc.

**Gas Tungsten Arc Welding (TIG).** The gas tungsten arc welding process was developed in the early 1940s so that the aircraft industries could weld nonferrous materials. It is a gas-shielded process in which the intense heat of the arc is maintained between a virtually nonconsumable tungsten electrode and the work (Fig. 7.25). The shielding consists of an inert gas or a mixture of inert gases. Filler metal may or may not be added, depending upon the requirements of the joint. The wire, if needed, may be added manually or automatically as a cold wire.

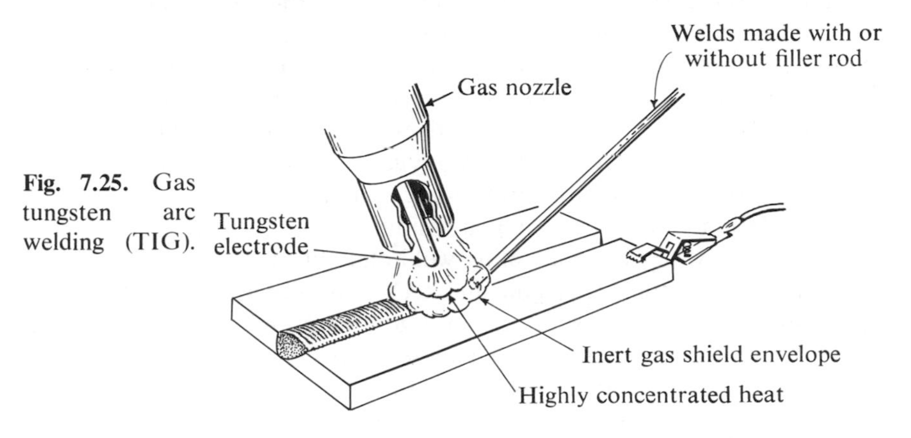


**Fig. 7.25.** Gas tungsten arc welding (TIG).

Advantages. There is practically no postweld cleaning required. The arc and weld pool are clearly visible to the welder. There is no metal transferred through the arc, and the process can be used in all positions.

Use. TIG welding is used primarily for joining nonferrous metals including aluminum, magnesium, silicon-bronze, copper and its alloys, nickel and its alloys, stainless steels, refractory, and precious metals. It can be used also to weld carbon steels and it is often used to make the root pass in joining alloy steel pipe. TIG is a highly preferred process for thinner materials where extreme quality is required. The main drawbacks are the heat input characteristics and relatively slow speed of operation.

## SHIELDING GAS SELECTION

There are several types of shielding gases that are used with the gas metal arc process. Argon and helium are chemically inert so will not combine with the products of the weld zone.

**Argon.** Argon has a low thermal conductivity restricting the arc plasma, which results in high arc densities. The bead made is relatively narrow with deep penetration at the center. The primary use of argon as a shielding gas is for nonferrous metals.

**Helium.** Helium is lighter than air and has a high thermal conductivity. The helium arc plasma expands under heat, reducing the density, which results in a broader weld bead with relatively shallow penetration.

**$CO_2$.** $CO_2$ is not an inert gas but a compound gas of carbon monoxide and oxygen. It is able to dissociate and recombine. It is this factor that allows more heat energy to be absorbed in the gas. It also uses the free oxygen in the arc area to superheat the weld metal transferring from the electrode to the work. It has a wider arc plasma than argon but less than helium. Depending upon the arc type used, the weld deposit cross section will show a medium narrow, deeply penetrating weld.

**Argon-Oxygen Mixtures.** Argon-oxygen mixtures (1 to 5 percent oxygen) permit better control in welding stainless steel.

**Argon-Helium Mixtures.** Argon-helium mixed (20 to 90 percent helium) are used to obtain exacting weld characteristics.

**Argon-$CO_2$ Mixtures.** Argon and $CO_2$ are sometimes used in mild steel welding to control spatter. An argon-helium-$CO_2$ mixture is used primarily for welding austenitic stainless steels. Shown in Fig. 7.26 are characteristic bead formations for various shielding gases with both straight and reverse polarity.

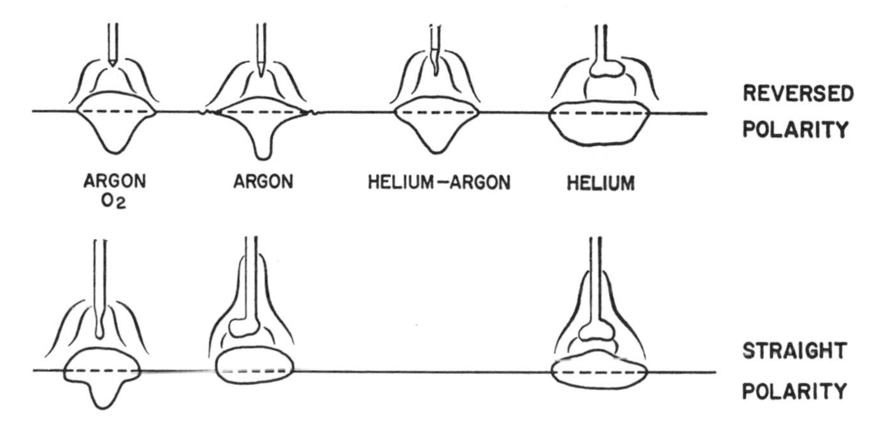


**Fig. 7.26.** The effect of various shielding gasses and polarity on weld bead contour and penetration.

## STUD WELDING

Stud welding is a means of arc welding a fastener to a workpiece. There are three main methods of accomplishing this: (*a*) conventional electric arc, (*b*) shielded arc, and (*c*) capacitor discharge stud welding.

**Conventional Stud Welding.** The conventional method of stud welding consists of loading a stud into a spring type chuck on the end of a welding gun. A ceramic ferrule is also placed in the end of the gun. When the trigger is pulled, the stud retracts from the weld plate about 1/16 in. (Fig. 7.27), making an arc. The arc melts a small portion of the end of the stud and the base metal. After a preset time interval, the current is automatically turned off, and the stud is instantly plunged into

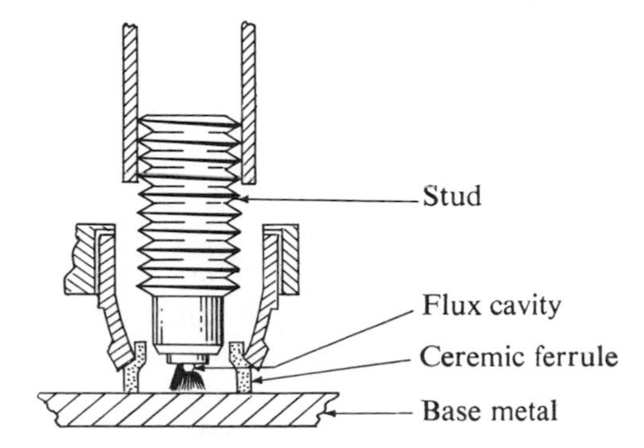


**Fig. 7.27.** As the stud is withdrawn from the metal, a pilot arc is started at the flux pocket as shown.

the pool of metal by the spring action of the gun (Fig. 7.28). The ceramic ferrule placed over the end of the stud confines the arc to a limited space, making a neat appearing fillet. At the end of the stud is a recess, filled and capped, that contains a flux which acts as a chemical cleaner for the surface before welding.

Another variation of the conventional method has been developed that uses a small aluminum disk pressed against the end of the stud, rather

**Fig. 7.28.** A stud welding gun used to weld studs to an oil tank opening for cover fastening. Courtesy KSM Products Inc.

than the flux pocket. The arc causes the aluminum to vaporize and clean the metal.

**Shielded Arc Stud Welding.** Because of the oxide coating that exists on aluminum, some means must be provided to properly prepare the surface and protect the weld at the instant it is made. A standard stud welding gun can be equipped with a special adapter and controls to apply a protective atmosphere around the weld when it is made. An inert gas such as argon or helium is used. The latter is preferred for aluminum studs larger than 3/16 in. dia because of the greater penetration.

**Capacitor Discharge Stud Welding.** A bank of capacitors, when discharged to make a weld, ionize the atmosphere between the tip and the work. This effectively cleans both surfaces and melts the full diameter of the stud and corresponding area in the workpiece. A hammer blow then exerted by the gun completes the weld, which is made without producing a fillet or marking the face of the workpiece.

The capacitor discharge method also has the advantage of being able to weld many combinations of ferrous and nonferrous metals without distortion or discernable heat. Thus studs can be welded in place without annealing or marring a painted or enameled workpiece face.

Multiple gun units are now available along with automatic hopper feeds and indexing fixtures to speed production. Handling rates are now as high as 4,000 welds/hr.

## RESISTANCE WELDING

Resistance welding is based on the well-known principle that heat is generated by the resistance offered to the flow of electrical current. The amount of heat generated in the workpiece depends on the magnitude of the welding current, the resistance of the current-conducting path, and the time the current is allowed to flow. This is expressed by the formula $H = I^2RT$, where $H$ = heat generated, in joules; $I$ = current, in rms amperes; $R$ = resistance, in ohms; $T$ = time of current flow, in seconds.

The heat generation is directly proportional to the resistance offered by any point in the circuit. Since the interface of two joining surfaces is the point of greatest resistance, it is also the point of greatest heat. In simple resistance welding a high-amperage, low-voltage current is passed from one adjoining plate to the other until the metal becomes heated to a temperature that is high enough to cause localized fusion and the formation of a weld nugget. Additional pressure is applied, after the current is turned off, to squeeze the two parts into a localized homogeneous mass.

The principle types of resistance welding can be divided into two groups: lap welds (including spot, seam, and projection) and butt welds (including upset butt and flash butt).

## SPOT WELDING

Spot welding is the basic type of resistance welding. Therefore, points covered in explaining it will not be repeated for the other processes. The fundamental circuit components are shown in Fig. 7.29.

Current, time, and pressure are recognized as the fundamental variables of spot welding. For welding most metals, these must be kept within very close limits. Therefore, they will be discussed from the standpoint of their relation to current control.

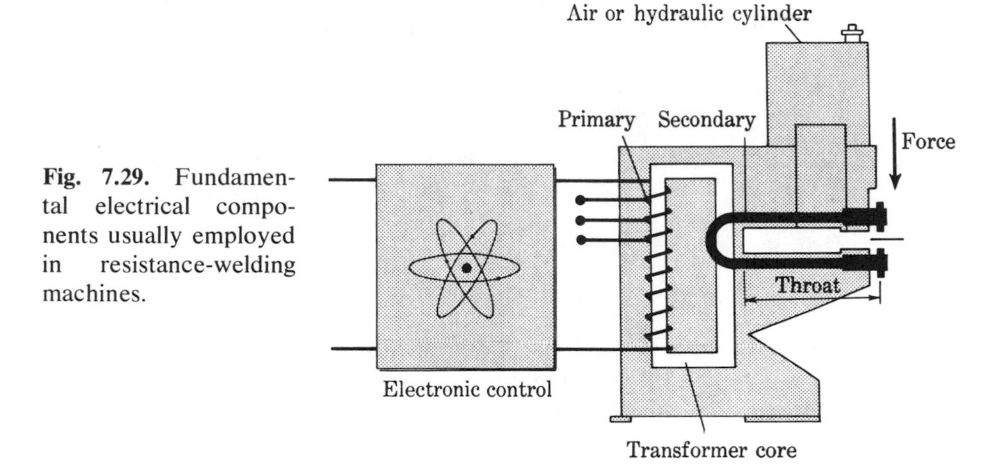


**Fig. 7.29.** Fundamental electrical components usually employed in resistance-welding machines.

**Current Control.** Electronic tubes are used as the switching devices for stopping and starting the primary circuit. Thyratron tubes are used for currents up to 40 amp, ignitrons for currents above 40 amp. Electronic contactors eliminate most of the noise and maintenance common to mechanical-type contactors. Solid state controls are now gradually replacing tube controls as they stay cool and require little maintenance.

Magnetic contactors can also be used as a switching device. They are made to open the power circuit when the ac wave approaches zero. Magnetic contactors have the advantage of low initial cost, but maintenance cost is higher, and they are not able to function consistently in rapid welding cycles.

**Time Control.** The time involved in spot welding is relatively short. Usually, the duration of current flow is a fraction of a second. For example, a spot weld can be made in two 1/16-in.-thick pieces of mild steel in 15 cycles or 1/4 sec when using 60-cycle current.

Low-carbon steels offer little or no metallurgical problems when spot welded. Once the nugget is formed, the metal passes from austenite back to pearlite. If medium- and high-carbon steels were to be treated in the same manner, the rapid cooling of the spot would result in brittle martensite. This does not mean that medium- or high-carbon steels cannot be resistance welded, but other processes, such as quench and postheat, are needed to permit the proper heat treatment of the spot immediately after it is made.

The normal timing sequence consists of *squeeze time*—the interval between application of the tip pressure to the work and application of welding current; *weld time*—the time for current flow; *hold time*—the interval after the current is shut off but during which the electrodes are held in place to forge the metal while it is cooling and to draw off the heat; and *off time*—the interval allowed for the work to be transferred to a new location before the cycle repeats (Fig. 7.30).

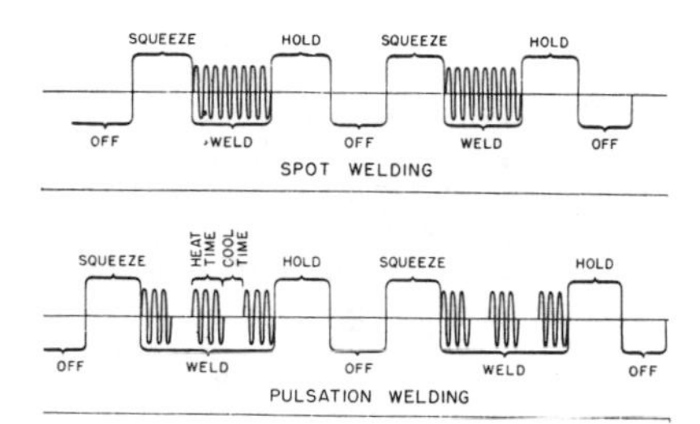


**Fig. 7.30.** The normal timing sequence for spot welding compared with pulsation welding.

Also shown in Fig. 7.30 is the *pulsation-welding* cycle. The main difference between this and the conventional cycle is that, instead of the weld period being one period of current flow, there is an intermittent flow with cooling no-current intervals between current-flow periods. This method is frequently used on multiple-layer welds, projection welds, and welds on two pieces of steel thicker than 1/8 in. It also has the advantage of increased electrode life.

When more precise current control is needed, as in welding aluminum or magnesium, a three-phase welder is often used. These machines can provide a slowly rising, rather than a rapidly rising, wave front. This is known as electronic *slope control*. Slope control is variable from steep rise (1 cycle) to gradual rise (15 cycles). It reduces spitting when the welding current comes on, and it gives the electrodes a chance to seat themselves. Also, a modulated delay of a secondary current can be obtained, thus eliminating the formation of cooling cracks (Fig. 7.31).

Timers are *synchronous* or *nonsynchronous*. The nonsynchronous ones are those that start and stop the flow of welding current at any time with respect to the voltage wave form. The opening and closing of the contactor is not necessarily synchronized with the line-voltage wave form. This can affect the ac frequency as much as ±1 cycle. There are many noncritical conditions that this small deviation will not affect.

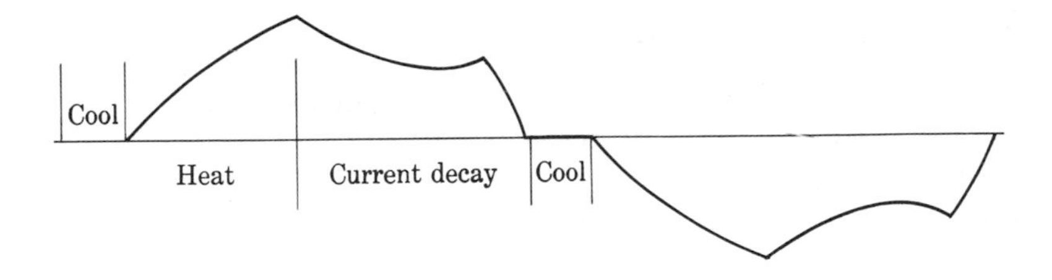


**Fig. 7.31.** Added current control obtainable with three-phase machines.

## SEAM WELDING

Seam welding is a continuous type of spot welding. Instead of using pointed electrodes that make one weld at a time, the work is passed between copper wheels or rollers which act as electrodes. Thyratron and ignitron tubes are used to "make and break" the circuit.

The appearance of the completed weld is that of a series of overlapping spot welds which resemble stitches, hence the name *stitch welding*.

Seam welding can be used to produce highly efficient water- and gastight joints. A variation of seam welding, called *roll welding*, is used to produce a series of intermittent spots (Fig. 7.32).

Most seam welding can be done on two types of standard machines. In one machine, the rotation of the wheels is at right angles to the throat; in the other, the rotation is parallel to the throat.

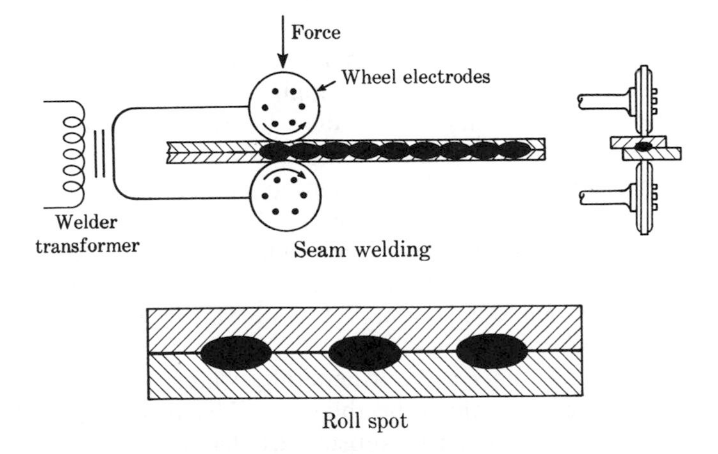


**Fig. 7.32.** Seam and roll spot welding. Seam welding can be done as shown, or one of the wheels can be replaced with a flat backing electrode that supports the work for the entire length of the seam.

## PROJECTION WELDING

Projection welding is another variation of spot welding. Small projections are raised on one side of the sheet or plate where it is to be welded to another. The projections act to localize the heat of the welding circuit. During the welding process, the projections collapse, owing to heat and pressure, and the parts to be joined are brought in close contact (Fig. 7.33).

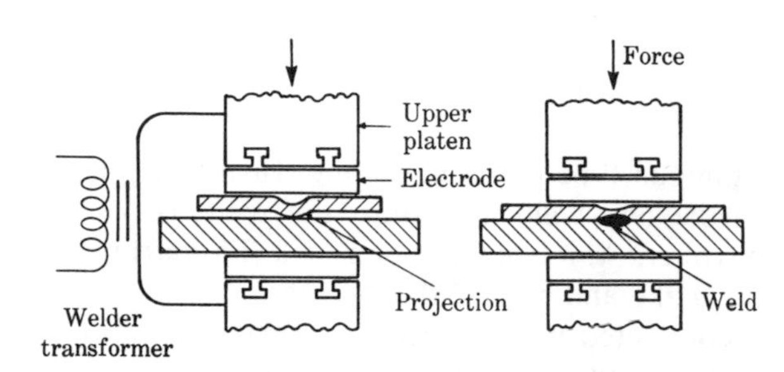


**Fig. 7.33.** In projection welding, the current is concentrated in the areas of the raised projections. During the welding process the projections are flattened by heat and pressure. Several projections can be welded at one time.

**Advantages.** This method of welding gives longer electrode life, since the electrodes can be made of harder material, with less wear and maintenance resulting from fusing and overheating. The prelocated spots permit welds that would be impractical by other resistance methods. Outer or top surfaces can be produced with no electrode marks, making it possible to paint or plate them without grinding or polishing.

**Limitations.** One of the main limitations of projection welding is the fact that it can be used only on a comparatively small group of metals and alloys. These are low-carbon steels, high-carbon and low-alloy steels, stainless and high-alloy steels, zinc die castings, terneplate, and some dissimilar and refractory metals. With brasses and coppers, the method has not been too satisfactory. Aluminum applications are rare, although they have been practical in special cases.

**Applications.** One of the most common applications of projection welding is for attaching small fasteners, nuts, special bolts, studs, and similar parts to larger components. A wide variety of these parts are available with performed projections.

Projection-welded joints are not generally water- or gastight but can be made so by sweating solder into the seam. This is usually satisfactory unless the parts are exposed to substances that will attack the solder.

## UPSET BUTT WELDING

The material to be welded is clamped in suitable electrode clamps. The ends to be welded touch each other as the current is turned on. The high resistance of the joint causes fusion at the interface. Just enough pressure is applied to keep the joint from arcing. As the metal becomes plastic, the force is enough to make a large, symmetrical upset that eliminates oxidized metal from the joint area. The excess metal is then removed by machining.

## FLASH BUTT WELDING

In flash butt welding, the ends of the stock are held in very light contact. As the current is turned on, it causes flashing and great heat. The metal burns away, and the pieces move together in an accelerated motion, maintaining uniform flashing action. When the inner faces reach the proper temperature, they are forced together under high pressure, and the current is cut off (Fig. 7.34).

Although quite similar in equipment to the flash method, upset butt welding uses a different control system.

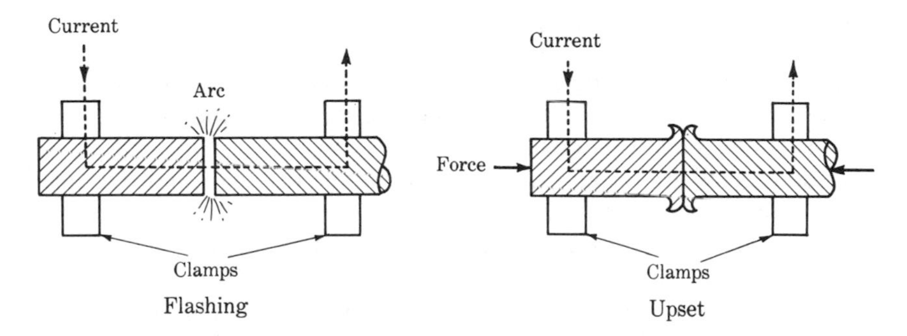


**Fig. 7.34.** Flash butt welding.

**Advantages and Use.** Flash butt welding offers strength factors up to 100 percent. No extra material, such as welding rod or flux, is required. Generally, no special preparation of the weld surface is required.

Dissimilar metals with varying melting temperatures can be flash butt welded. The size and shape of the parts should be similar, but a 15 percent variation in end dimensions is permissible for commercial use.

The process is regularly used for end joining of rods, tubes, bars, forgings, fittings, etc. Heavy forgings can sometimes be eliminated by welding small forgings to bar stock.

## PERCUSSION WELDING

In percussion welding, two workpieces are brought together at a rapid rate. Just before they meet, a flash of arc melts both of the colliding surfaces. The molten surfaces are then squeezed together by the collision, and some of the metal is forced out to the sides of the joint.

Percussion welding is particularly good for joining small-diameter wires, for example, welding 0.002- and 0.015-in.-dia wires in electronic applications, and for materials of widely differing properties. Wires can be joined by soldering or other processes, but the advantage of percussion welding is that it is almost instantaneous.

The welds are produced either by stored-energy type machines or by rapid dissipation of current from a standard 60-cycle-per-second ac source.

Parts other than fine wires have pinpoint-type projections that are formed or coined into the part. These localize the heat so that the small area is instantaneously vaporized upon contact.

Some metal combinations that have been welded by this process, with excellent results, are copper to Nichrome, copper wire to type 304 stainless plate, thorium to thorium, and thorium to Zircaloy.

## WELDING ECONOMICS

The engineer should be able to properly assess fabrication costs by various methods in order to choose the one that gives the most satisfactory results at the lowest cost.

Welding costs are based on three main items: (*a*) labor and overhead, (*b*) welding supplies, (*c*) power consumption.

As with most processes, there are certain operating variables that are difficult to standardize such as (*a*) the amount of filler metal required for a given joint, (*b*) the actual weld time compared to the overall time (operating factor), (*c*) allowance for handling and cleaning.

**Weld Deposit Cost.** To find the actual amount of metal used in making the weld, the cross-sectional area in sq in. can be multiplied by a factor of 3.4 which will give the lb per linear ft. To estimate the weight of stick electrode required, an additional 50 percent should be added. Submerged arc welding deposit in lb/ft of joint can be determined from the nomograph (Fig. 7.35). The example given is for 590 amp. with a 1/8 in. electrode at a travel speed of 30 ipm. The weight of the weld deposit is shown as 0.10 lb/ft of joint with dc positive polarity or 0.13 lb with dc negative polarity.

The weld metal weight in lb/linear ft for various type joints is given in Table 7.4.

Both too much welding and too little welding can be uneconomical. The engineer should be able to recognize both conditions for combined safety and economy.

A simple illustration will serve to point out the added cost of overwelding. If 5/16-in. fillet welds are being made instead of 1/4-in., as

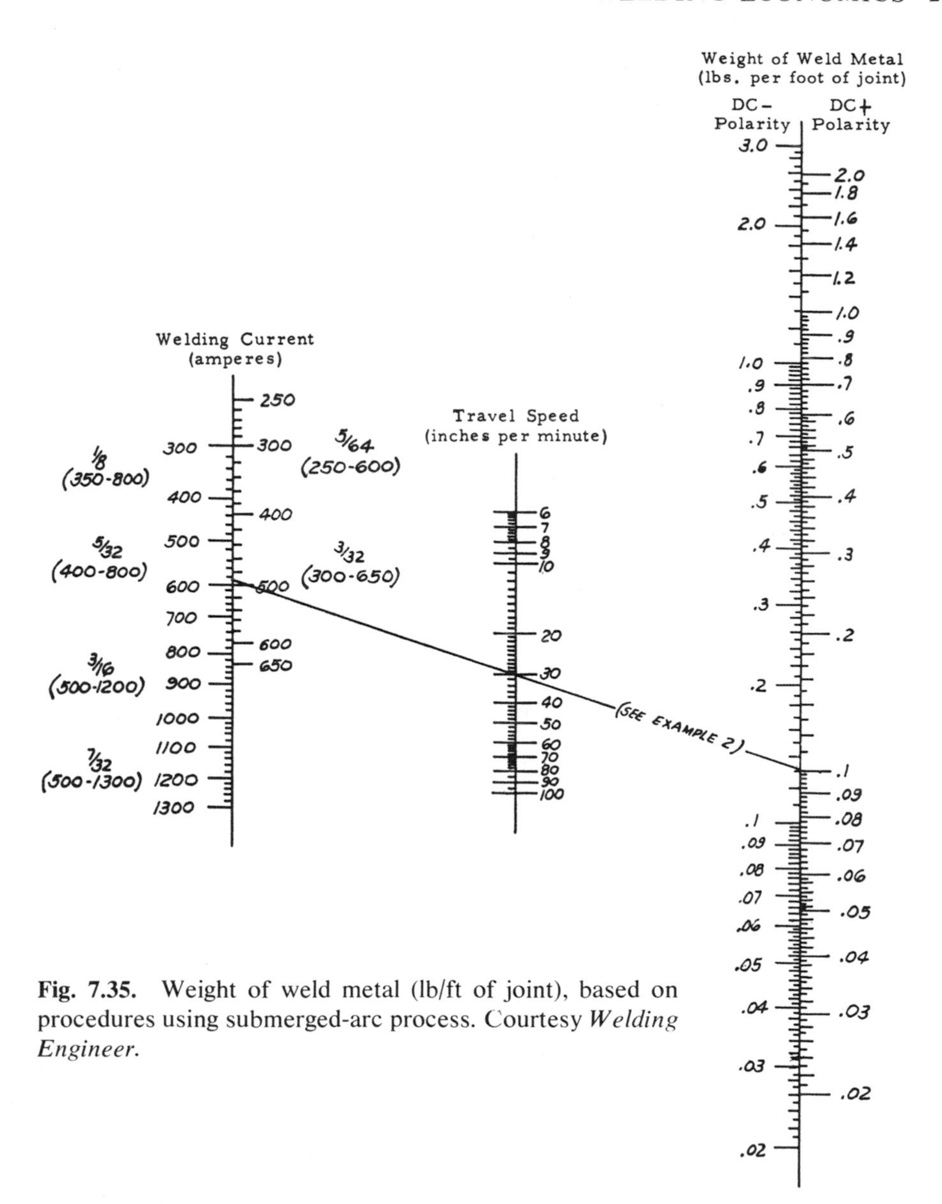


**Fig. 7.35.** Weight of weld metal (lb/ft of joint), based on procedures using submerged-arc process. Courtesy *Welding Engineer.*

called for, the overweld material amounts to 58 percent. Even more striking is the overweld that calls for 3/8 in. and is made 1/2 in. (Fig. 7.36). The excess metal amounts to 78.5 percent. The waste is not only in the weld deposit, but also in the labor and overhead of depositing it.

**Operating Factor.** The operating factor (OF) is the amount of time the weld is actually being made out of the total time. For example, if it takes an operator 2.06 min to use up one electrode and an additional

**Table 7.4**
**Weight of Weld Metal, lb/ft of Joint, Based on Joint Design**
(Courtesy of *Welding Engineer.*)

| "d" or "r" Dimension | Segment of Weld Bead | BUTT WELDS | | | | | | | | | | | | | | | | FILLET WELDS | | |
|---|---|---|---|---|---|---|---|---|---|---|---|---|---|---|---|---|---|---|---|---|
| | | "t" Dimension | | | | | | "t" Dimension | | | | r | Included Angle | | | | | Values below are for leg size 10% oversize, consistent with normal shop practices. (leg size increased 10%) | | |
| | | 1/16" | 1/8" | 3/16" | 1/4" | 3/8" | 1/2" | 1/16" | 1/8" | 3/16" | 1/4" | | 14° | 20° | 60° | 45° (1/2 of 90°) | 70° | flat | convex | concave |
| 1/16" | | | | | | | | | | | | .021 | | | | | | | | |
| 1/8" | | .027 | .053 | .080 | .106 | .159 | .212 | | | | | .083 | .0065 | .0094 | .031 | .027 | .037 | .032 | .039 | .037 |
| 3/16" | | .040 | .080 | .119 | .159 | .239 | .318 | .027 | | | | .188 | .0147 | .021 | .069 | .060 | .084 | .072 | .087 | .083 |
| 1/4" | | .053 | .106 | .159 | .212 | .318 | .425 | .035 | | | | .334 | .026 | .037 | .123 | .106 | .149 | .129 | .155 | .147 |
| 5/16" | | .066 | .133 | .199 | .265 | .390 | .531 | .044 | | | | .531 | .041 | .059 | .192 | .166 | .232 | .201 | .242 | .230 |
| 3/8" | | .080 | .159 | .239 | .318 | .478 | .637 | .053 | .106 | | | .750 | .059 | .084 | .276 | .239 | .334 | .289 | .349 | .331 |
| 7/16" | | .091 | .186 | .279 | .371 | .557 | .743 | .062 | .124 | .186 | | 1.02 | .080 | .115 | .376 | .326 | .456 | .394 | .475 | .451 |
| 1/2" | | .106 | .212 | .318 | .425 | .637 | .849 | .071 | .142 | .212 | | 1.33 | .104 | .150 | .491 | .425 | .595 | .514 | .620 | .589 |
| 9/16" | | .119 | .239 | .358 | .478 | .716 | .955 | .080 | .159 | .239 | | | .132 | .190 | .621 | .538 | .753 | .651 | .785 | .745 |
| 5/8" | | .133 | .265 | .398 | .531 | .796 | 1.06 | .089 | .177 | .266 | | | .163 | .234 | .766 | .664 | .930 | .804 | .970 | .920 |
| 11/16" | | .146 | .292 | .438 | .584 | .876 | 1.17 | .097 | .195 | .292 | .389 | | .197 | .283 | .927 | .804 | 1.13 | | | |
| 3/4" | | .159 | .318 | .478 | .637 | .955 | 1.27 | .111 | .212 | .318 | .424 | | .234 | .337 | 1.11 | .956 | 1.34 | 1.16 | 1.40 | 1.32 |
| 13/16" | | .172 | .345 | .517 | .690 | 1.04 | 1.38 | .114 | .230 | .345 | .460 | | .275 | .396 | 1.30 | 1.12 | 1.57 | | | |
| 7/8" | | .186 | .371 | .557 | .743 | 1.11 | 1.49 | .124 | .248 | .372 | .490 | | .319 | .459 | 1.50 | 1.30 | 1.82 | 1.58 | 1.90 | 1.80 |
| 15/16" | | .199 | .398 | .597 | .796 | 1.19 | 1.59 | .133 | .266 | .398 | .530 | | .367 | .527 | 1.73 | 1.50 | 2.07 | | | |
| 1" | | .212 | .425 | .627 | .849 | 1.25 | 1.70 | .142 | .282 | .418 | .566 | | .417 | .599 | 1.96 | 1.70 | 2.38 | 2.06 | 2.48 | 2.36 |
| 1-1/16" | | .226 | .451 | .677 | .902 | 1.35 | 1.80 | .150 | .301 | .451 | .602 | | .471 | .676 | 2.22 | 1.92 | 2.68 | | | |
| 1-1/8" | | .239 | .478 | .716 | .955 | 1.43 | 1.91 | .159 | .318 | .477 | .637 | | .528 | .758 | 2.48 | 2.15 | 3.02 | 2.60 | 3.14 | 2.98 |
| 1-3/16" | | .252 | .504 | .756 | 1.01 | 1.51 | 2.02 | .168 | .336 | .505 | .672 | | .588 | .845 | 2.77 | 2.40 | 3.36 | | | |
| 1-1/4" | | .265 | .531 | .796 | 1.06 | 1.59 | 2.12 | .177 | .354 | .531 | .706 | | .651 | .936 | 3.07 | 2.66 | 3.72 | 3.21 | 3.88 | 3.68 |
| 1-5/16" | | .279 | .557 | .836 | 1.11 | 1.67 | 2.23 | .186 | .372 | .557 | .743 | | .718 | 1.03 | 3.38 | 2.93 | 4.10 | | | |
| 1-3/8" | | .292 | .584 | .876 | 1.17 | 1.75 | 2.34 | .195 | .389 | .584 | .777 | | .789 | 1.13 | 3.71 | 3.21 | 4.50 | 3.89 | 4.69 | 4.45 |
| 1-7/16" | | .305 | .610 | .915 | 1.22 | 1.83 | 2.44 | .203 | .407 | .610 | .814 | | .836 | 1.24 | 4.05 | 3.51 | 4.91 | | | |
| 1-1/2" | | .318 | .637 | .955 | 1.27 | 1.91 | 2.55 | .212 | .425 | .636 | .849 | | .938 | 1.35 | 4.42 | 3.82 | 5.36 | 4.62 | 5.58 | 5.30 |
| 1-9/16" | | .332 | .664 | .995 | 1.33 | 1.99 | 2.65 | .221 | .442 | .664 | .884 | | 1.02 | 1.46 | 4.79 | 4.15 | 5.81 | | | |
| 1-5/8" | | .345 | .690 | 1.04 | 1.38 | 2.07 | 2.76 | .230 | .460 | .690 | .920 | | 1.10 | 1.58 | 5.18 | 4.49 | 6.29 | 5.43 | 6.55 | 6.22 |
| 1-11/16" | | .358 | .716 | 1.07 | 1.43 | 2.15 | 2.87 | .239 | .477 | .716 | .956 | | 1.19 | 1.71 | 5.59 | 4.84 | 6.80 | | | |
| 1-3/4" | | .371 | .743 | 1.11 | 1.49 | 2.23 | 2.97 | .249 | .495 | .743 | .990 | | 1.28 | 1.84 | 6.01 | 5.20 | 7.29 | 6.29 | 7.59 | 7.21 |
| 1-13/16" | | .385 | .769 | 1.15 | 1.54 | 2.31 | 3.08 | .257 | .513 | .770 | 1.03 | | 1.37 | 1.97 | 6.45 | 5.58 | 7.81 | | | |
| 1-7/8" | | .390 | .796 | 1.19 | 1.59 | 2.39 | 3.18 | .266 | .531 | .796 | 1.06 | | 1.47 | 2.10 | 6.90 | 5.97 | 8.36 | 7.23 | 8.72 | 8.28 |
| 1-15/16" | | .411 | .822 | 1.23 | 1.65 | 2.47 | 3.29 | .274 | .549 | .823 | 1.10 | | 1.56 | 2.25 | 7.36 | 6.38 | 8.94 | | | |
| 2" | | .425 | .849 | 1.27 | 1.70 | 2.55 | 3.40 | .283 | .566 | .849 | 1.13 | | 1.67 | 2.40 | 7.85 | 6.80 | 9.52 | 8.23 | 9.93 | 9.43 |

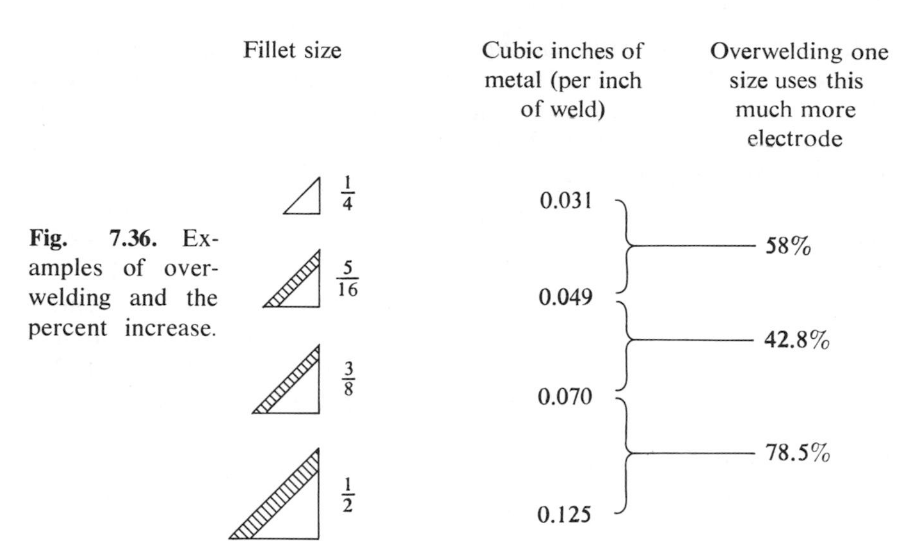


**Fig. 7.36.** Examples of overwelding and the percent increase.

2.06 min to clean the slag and insert a new rod into the holder, we have a 50 percent OF.

**Weld Time.** Actual weld time can be estimated in three ways: (1) determine the weight of the weld metal/linear ft and the deposition rate for the given welding current; (2) use standard welding data, which lists arc travel speeds in in./min, and apply this to the total length of the weld; (3) actually time the welder.

## HEAT FLOW AND METALLURGICAL WELD PARAMETERS

As mentioned previously in this chapter, additional information would be given for those interested in learning more about the heat effects of welding.

As weld control becomes more critical due to maintaining given strength requirements in alloy steels, certain parameters such as heat input, preheat temperature, and welding speed must be closely controlled. In the past, trial and error has often had to suffice for determining the proper proportion of these parameters and their magnitude. Now, however, by applying certain heat flow and hardenability concepts to individual welding situations, these parameters can be approximated theoretically, resulting in substantial savings of time, effort, and expensive sample preparation.

**Basic Concepts.** There are four basic metallurgical and welding concepts used to predict the approximate welding parameters that produce the desired weld. These concepts are: (*a*) the ideal critical diameter, (*b*) the Jominy and Boegehold end quench, (*c*) heat flow during welding relationships, (*d*) heat input versus Jominy end quench data.

**The Ideal Critical Diameter.** The ideal critical diameter can be defined as that diameter of bar that will form a 50 per cent martensite structure when quenched with $H = \infty$. It is calculated from the chemical composition and grain size of a given steel. Each alloy changes the hardenability factor. For example, by choosing a steel such as AISI 8740 with a number seven grain size, the hardenability factor can be determined with the aid of Table 7.5. Follow the percent alloy column to determine each factor such as 40 percent carbon of a number seven grain size = 0.329 and 85 percent manganese = 0.587, etc. The sum total of the hardenability factors, as shown in the upper right corner of Table 7.5, is 1.639. Below the example in Table 7.5 is the ideal critical diameter (DI), which in this case is 4.35 in.

**End Quench Hardenability.** The second concept makes use of the Jominy and Boegehold end quench as discussed in Chap. 2. The hardness readings are taken every 1/16 in. on a bar that is quenched on the end by a fountain of water. The hardnesses are shown in Fig. 7.37.

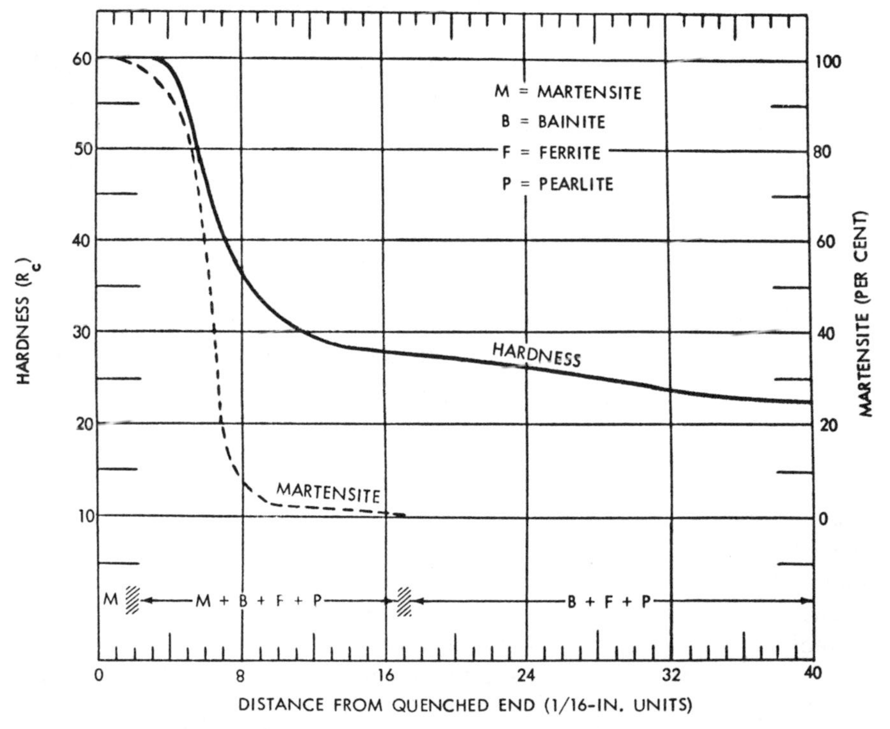


**Fig. 7.37.** Typical end quench hardenability curve. This plot shows Rockwell C hardness values and corresponding per cent martensite for a one-in. by four-in. standard cylindrical steel test bar in relation to the distance from the quenched end (DQE). Courtesy United States Steel Corporation.

**Heat Flow.** The calculation of heat flow during the welding process in relation to the plate thickness and radiation is the third concept. Heat flow may be described as either two or three dimensional. Two-dimensional flow is found in single pass butt welds or where there is no flow perpendicular to the plate. Three-dimensional flow is obtained in very thick plates. The plates are thick enough so that, even with increases in thickness, heat flow will have no effect on the cooling rate.

**Heat Input Versus End Quench Data.** The fourth concept is based on the volt-ampere factors obtained from a graph (Fig. 7.38), which gives the requirements for several welding speeds that produce a given Jominy distance from the quenched end (DQE). A change in the plate thickness or joint geometry, however, changes the data obtained from the graph. Hence it is necessary to multiply the data by a geometry factor as obtained from Table 7.6 for each type of joint.

**An Example of Applying Basic Heat Flow Concepts.** To illustrate how the basic concepts are applied, an example using a butt weld of 1/2-in. SAE 1340 steel plates will be considered. The grain size best suited will be no. 7, and the alloying constituents of 0.40 percent carbon, 0.90 percent

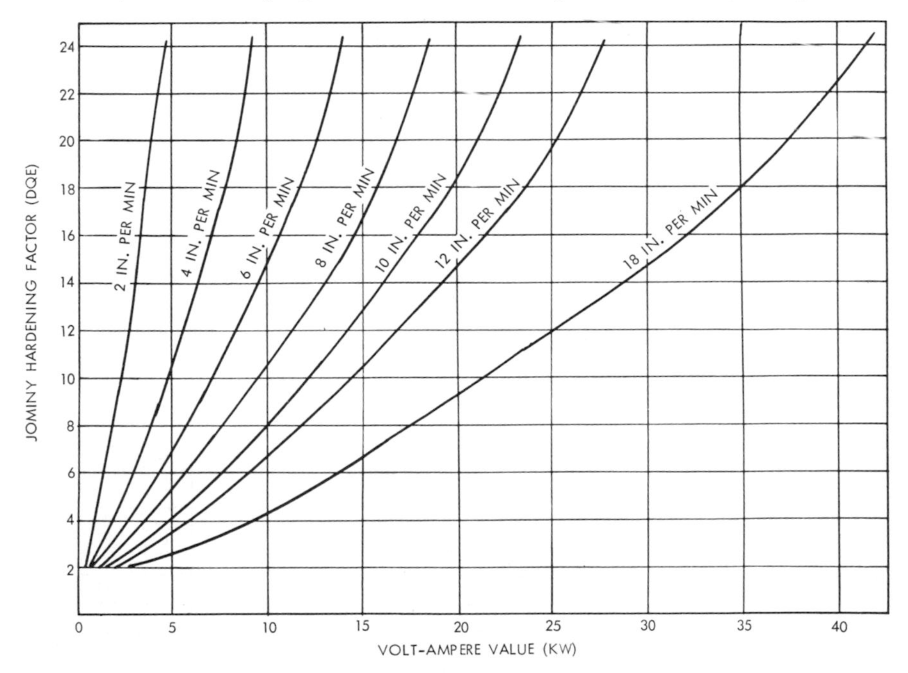


**Fig. 7.38.** Volt-ampere values for last pass butt weld in 1/2-in. steel plates. Knowing the distance from the quenched end (DQE) and welding travel speed, the proper volt-ampere value to produce the DQE can be obtained. Courtesy *Welding Journal*.

**Table 7.5. Calculation of Ideal Critical Diameter**
Courtesy of Republic Steel Corporation

| Base Hardenability Characteristic Due to Carbon and Grain Size | | | | | Hardenability Characteristics of Alloys | | | | | | | | | |
|---|---|---|---|---|---|---|---|---|---|---|---|---|---|---|
| Grain Size | | | | Percent of | | | | | | | | | | |
| No. 5 | No. 6 | No. 7 | No. 8 | Alloys | Mn | Si | Ni | Cr | Mo | V | P | S | Al | Ti |
| ......... | ......... | ......... | ......... | 0.01 | 0.014 | 0.003 | 0.002 | 0.009 | 0.013 | 0.061 | 0.011 | −0.003 | 0.006 | −0.008 |
| ......... | ......... | ......... | ......... | 0.02 | .028 | .006 | .003 | .018 | .025 | .097 | .022 | −0.006 | .012 | −0.018 |
| ......... | ......... | ......... | ......... | 0.03 | .041 | .009 | .005 | .027 | .037 | .137 | .033 | −0.009 | .017 | −0.025 |
| ......... | ......... | ......... | ......... | 0.04 | .054 | .012 | .006 | .036 | .049 | .146 | .044 | −0.011 | .022 | −0.034 |
| ......... | ......... | ......... | ......... | 0.05 | .067 | .015 | .008 | .045 | .061 | .146 | .054 | −0.014 | .028 | −0.043 |
| ......... | ......... | ......... | ......... | 0.06 | .079 | .018 | .009 | .053 | .072 | .140 | .064 | −0.018 | .033 | −0.053 |
| 0.021 | ......... | ......... | ......... | 0.07 | .091 | .021 | .011 | .061 | .083 | .137 | .073 | −0.020 | .039 | −0.062 |
| .050 | 0.012 | ......... | ......... | 0.08 | .103 | .024 | .012 | .069 | .094 | .124 | .083 | −0.024 | .044 | −0.072 |
| .076 | .038 | 0.005 | ......... | 0.09 | .114 | .027 | .014 | .077 | .104 | .111 | .092 | −0.027 | .049 | −0.081 |
| .101 | .062 | .029 | ......... | 0.10 | .125 | .029 | .015 | .085 | .114 | .097 | .101 | −0.032 | .054 | −0.092 |
| .120 | .084 | .052 | 0.017 | 0.11 | .136 | .032 | .017 | .093 | .124 | .086 | ........ | ........... | .059 | −0.099 |
| .138 | .103 | .071 | .037 | 0.12 | .146 | .035 | .019 | .101 | .134 | .072 | ........ | ........... | .064 | −0.112 |
| .155 | .121 | .088 | .056 | 0.13 | .156 | .038 | .020 | .108 | .143 | .061 | ........ | ........... | .069 | −0.123 |
| .170 | .136 | .104 | .070 | 0.14 | .166 | .041 | .022 | .115 | .152 | .037 | ........ | ........... | .074 | −0.134 |
| .184 | .150 | .119 | .084 | 0.15 | .176 | .043 | .023 | .122 | .161 | .025 | ........ | ........... | .079 | −0.146 |
| .198 | .164 | .133 | .097 | 0.16 | .186 | .046 | .024 | .129 | .170 | ........ | ........ | ........... | ........ | ........... |
| .211 | .176 | .146 | .110 | 0.17 | .195 | .049 | .026 | .136 | .179 | ........ | ........ | ........... | ........ | ........... |
| .224 | .188 | .158 | .122 | 0.18 | .204 | .052 | .028 | .143 | .188 | ........ | ........ | ........... | ........ | ........... |
| .236 | .199 | .169 | .134 | 0.19 | .213 | .054 | .029 | .149 | .196 | ........ | ........ | ........... | ........ | ........... |
| .247 | .210 | .179 | .146 | 0.20 | .222 | .057 | .030 | .156 | .204 | ........ | ........ | ........... | ........ | ........... |

**EXAMPLE**

To calculate the ideal critical diameter of AISI 8740 Steel having No. 7 grain size and of the following composition:

| Analysis | | Hardenability Characteristics |
|---|---|---|
| 0.40% | Carbon (No. 7 grain size) | 0.329 |
| 0.85% | Manganese | 0.584 |
| 0.30% | Silicon | 0.083 |
| 0.55% | Nickel | 0.079 |
| 0.50% | Chromium | 0.318 |
| 0.25% | Molybdenum | 0.244 |
| 0.010% | Phosphorus | 0.011 |
| 0.030% | Sulphur | 0.009 |
| | Sum of Hardenability Characteristics | 1.639 |

Ideal Critical Diameter
D.I. = 4.35 Inches

| Grain Size No. 5 | No. 6 | No. 7 | No. 8 | Percent of Alloys | Mn | Si | Ni | Cr | Mo |
|---|---|---|---|---|---|---|---|---|---|
| .258 | .221 | .188 | .156 | 0.21 | .231 | .060 | .032 | .163 | .212 |
| .268 | .231 | .198 | .166 | 0.22 | .239 | .062 | .033 | .169 | .220 |
| .278 | .241 | .208 | .176 | 0.23 | .247 | .064 | .035 | .175 | .228 |
| .288 | .250 | .217 | .184 | 0.24 | .255 | .067 | .037 | .181 | .236 |
| .297 | .260 | .225 | .193 | 0.25 | .263 | .070 | .038 | .187 | .244 |
| .306 | .269 | .233 | .201 | 0.26 | .271 | .073 | .040 | .193 | .251 |
| .314 | .277 | .241 | .209 | 0.27 | .279 | .075 | .041 | .199 | .258 |
| .322 | .285 | .250 | .216 | 0.28 | .287 | .078 | .042 | .205 | .265 |
| .330 | .292 | .259 | .223 | 0.29 | .294 | .080 | .044 | .211 | .272 |
| .337 | .299 | .267 | .230 | 0.30 | .301 | .083 | .045 | .217 | .279 |
| .343 | .306 | .274 | .238 | 0.31 | .308 | .085 | .047 | .222 | .286 |
| .350 | .313 | .281 | .246 | 0.32 | .315 | .088 | .048 | .228 | .293 |
| .356 | .320 | .288 | .253 | 0.33 | .322 | .090 | .049 | .234 | .299 |
| .362 | .327 | .295 | .260 | 0.34 | .329 | .093 | .051 | .239 | .306 |
| .368 | .333 | .301 | .266 | 0.35 | .336 | .095 | .052 | .244 | .312 |
| .374 | .339 | .306 | .272 | 0.36 | .343 | .098 | .053 | .249 | .318 |
| .380 | .345 | .312 | .278 | 0.37 | .349 | .100 | .055 | .255 | .324 |
| .386 | .351 | .318 | .284 | 0.38 | .355 | .102 | .057 | .260 | .330 |
| .392 | .357 | .324 | .290 | 0.39 | .362 | .105 | .058 | .265 | .336 |
| .398 | .362 | .329 | .296 | 0.40 | .368 | .107 | .059 | .270 | .342 |

Hardenability Characteristics (Continued from bottom of page 223)

| Percent of Alloys | Mn | Si | Ni | Cr |
|---|---|---|---|---|
| 0.91 | .606 | .214 | .124 | .472 |
| 0.92 | .609 | .216 | .125 | .475 |
| 0.93 | .613 | .218 | .126 | .478 |
| 0.94 | .616 | .220 | .128 | .481 |
| 0.95 | .620 | .221 | .129 | .485 |
| 0.96 | .623 | .223 | .130 | .488 |
| 0.97 | .627 | .225 | .131 | .491 |
| 0.98 | .630 | .227 | .132 | .494 |
| 0.99 | .633 | .229 | .134 | .497 |
| 1.00 | .637 | .230 | .135 | .500 |
| 1.02 | .643 | .234 | .137 | .506 |
| 1.04 | .650 | .238 | .139 | .511 |
| 1.06 | .656 | .241 | .142 | .517 |
| 1.08 | .662 | .245 | .144 | .522 |
| 1.10 | .669 | .248 | .146 | .528 |
| 1.12 | .675 | .251 | .148 | .534 |
| 1.14 | .681 | .255 | .150 | .539 |
| 1.16 | .687 | .258 | .153 | .545 |
| 1.18 | .694 | .262 | .155 | .550 |
| 1.20 | .702 | .265 | .157 | .555 |

Conversion of the sum of hardenability characteristics into the ideal critical diameter (D.I.).

| Sum | D.I. Inch | Sum | D.I. Inch |
|---|---|---|---|
| 0.740 | 0.55 | 1.550 | 3.55 |
| 0.778 | 0.60 | 1.556 | 3.60 |
| 0.813 | 0.65 | 1.562 | 3.65 |
| 0.845 | 0.70 | 1.568 | 3.70 |
| 0.875 | 0.75 | 1.574 | 3.75 |
| 0.903 | 0.80 | 1.580 | 3.80 |
| 0.929 | 0.85 | 1.585 | 3.85 |
| 0.954 | 0.90 | 1.591 | 3.90 |
| 0.978 | 0.95 | 1.597 | 3.95 |
| 1.000 | 1.00 | 1.602 | 4.00 |

**Table 7.5. Calculation of Ideal Critical Diameter (Continued)**
Courtesy of Republic Steel Corporation

| Base Hardenability Characteristic Due to Carbon and Grain Size | | | | Hardenability Characteristics of Alloys | | | | | | | | | | | Conversion of the sum of hardenability characteristics into the ideal critical diameter (D.I.). | | | |
|---|---|---|---|---|---|---|---|---|---|---|---|---|---|---|---|---|---|---|
| Grain Size | | | | Percent of | | | | | | Percent of | | | | | | D.I. | | D.I. |
| No. 5 | No. 6 | No. 7 | No. 8 | Alloys | Mn | Si | Ni | Cr | Mo | Alloys | Mn | Si | Ni | Cr | Sum | Inch | Sum | Inch |
| .403 | .368 | .334 | .301 | 0.41 | .374 | .110 | .061 | .275 | .348 | 1.22 | .710 | .268 | .159 | .561 | 1.021 | 1.05 | 1.607 | 4.05 |
| .408 | .373 | .339 | .306 | 0.42 | .380 | .112 | .062 | .280 | .354 | 1.24 | .718 | .271 | .161 | .566 | 1.041 | 1.10 | 1.613 | 4.10 |
| .413 | .378 | .344 | .310 | 0.43 | .386 | .114 | .063 | .285 | .360 | 1.26 | .725 | .275 | .164 | .571 | 1.060 | 1.15 | 1.618 | 4.15 |
| .418 | .383 | .349 | .315 | 0.44 | .392 | .117 | .064 | .290 | .365 | 1.28 | .733 | .278 | .166 | .576 | 1.079 | 1.20 | 1.623 | 4.20 |
| .423 | .387 | .354 | .320 | 0.45 | .398 | .119 | .066 | .295 | .371 | 1.30 | .741 | .281 | .168 | .581 | 1.097 | 1.25 | 1.628 | 4.25 |
| .428 | .392 | .358 | .325 | 0.46 | .404 | .121 | .067 | .300 | .377 | 1.32 | .749 | .284 | .170 | .586 | 1.114 | 1.30 | 1.633 | 4.30 |
| .433 | .397 | .362 | .330 | 0.47 | .409 | .124 | .069 | .304 | .382 | 1.34 | .757 | .287 | .172 | .590 | 1.130 | 1.35 | 1.638 | 4.35 |
| .438 | .402 | .366 | .334 | 0.48 | .415 | .126 | .070 | .309 | .387 | 1.36 | .765 | .290 | .175 | .595 | 1.146 | 1.40 | 1.643 | 4.40 |
| .443 | .407 | .372 | .338 | 0.49 | .420 | .128 | .072 | .313 | .393 | 1.38 | .772 | .294 | .177 | .600 | 1.161 | 1.45 | 1.648 | 4.45 |
| .448 | .412 | .377 | .343 | 0.50 | .426 | .130 | .073 | .318 | .398 | 1.40 | .780 | .297 | .179 | .605 | 1.176 | 1.50 | 1.653 | 4.50 |
| .452 | .417 | .382 | .348 | 0.51 | .431 | .133 | .074 | .323 | .403 | 1.42 | .787 | .300 | .181 | .609 | 1.190 | 1.55 | 1.658 | 4.55 |
| .456 | .422 | .387 | .352 | 0.52 | .437 | .135 | .076 | .327 | .408 | 1.44 | .794 | .303 | .183 | .614 | 1.204 | 1.60 | 1.663 | 4.60 |
| .461 | .427 | .391 | .356 | 0.53 | .442 | .137 | .077 | .331 | .413 | 1.46 | .801 | .306 | .185 | .618 | 1.217 | 1.65 | 1.667 | 4.65 |
| .465 | .431 | .396 | .360 | 0.54 | .447 | .139 | .078 | .336 | .418 | 1.48 | .808 | .309 | .187 | .623 | 1.230 | 1.70 | 1.672 | 4.70 |
| .469 | .435 | .400 | .364 | 0.55 | .452 | .141 | .079 | .340 | .423 | 1.50 | .815 | .312 | .190 | .627 | 1.243 | 1.75 | 1.677 | 4.75 |
| .473 | .439 | .404 | .367 | 0.56 | .457 | .144 | .081 | .344 | .428 | 1.52 | .822 | .315 | .193 | .632 | 1.255 | 1.80 | 1.681 | 4.80 |
| .477 | .443 | .408 | .371 | 0.57 | .462 | .146 | .082 | .349 | .433 | 1.54 | .828 | .318 | .196 | .636 | 1.267 | 1.85 | 1.686 | 4.85 |
| .481 | .447 | .412 | .375 | 0.58 | .467 | .148 | .084 | .353 | .438 | 1.56 | .835 | .321 | .198 | .641 | 1.279 | 1.90 | 1.690 | 4.90 |
| .485 | .450 | .416 | .378 | 0.59 | .472 | .150 | .085 | .357 | .442 | 1.58 | .841 | .323 | .200 | .645 | 1.290 | 1.95 | 1.695 | 4.95 |
| .489 | .454 | .419 | .382 | 0.60 | .477 | .152 | .086 | .361 | .447 | 1.60 | .848 | .326 | 203 | .650 | 1.301 | 2.00 | 1.699 | 5.00 |

| | | | | | | | | | | | | | | | | | |
|---|---|---|---|---|---|---|---|---|---|---|---|---|---|---|---|---|---|
| .493 | .458 | .423 | .386 | 0.61 | .482 | .154 | .087 | .365 | .452 | 1.62 | .854 | .329 | .205 | .653 | 1.312 | 2.05 | 1.703 | 5.05 |
| .497 | .461 | .427 | .389 | 0.62 | .487 | .157 | .088 | .369 | .456 | 1.64 | .860 | .332 | .208 | .657 | 1.322 | 2.10 | 1.708 | 5.10 |
| .500 | .464 | .430 | .393 | 0.63 | .492 | .159 | .090 | .373 | .461 | 1.66 | .866 | .335 | .210 | .661 | 1.332 | 2.15 | 1.712 | 5.15 |
| .504 | .467 | .433 | .396 | 0.64 | .496 | .161 | .091 | .377 | .465 | 1.68 | .872 | .338 | .212 | .665 | 1.342 | 2.20 | 1.716 | 5.20 |
| .507 | .470 | .436 | .400 | 0.65 | .501 | .163 | .092 | .381 | .470 | 1.70 | .878 | .340 | .215 | .670 | 1.352 | 2.25 | 1.720 | 5.25 |
| .510 | .473 | .439 | .403 | 0.66 | .505 | .165 | .094 | .385 | .474 | 1.72 | .884 | .343 | .217 | .673 | 1.362 | 2.30 | 1.724 | 5.30 |
| .513 | .476 | .442 | .407 | 0.67 | .510 | .167 | .095 | .389 | .479 | 1.74 | .890 | .346 | .219 | ......... | 1.371 | 2.35 | 1.728 | 5.35 |
| .517 | .479 | .446 | .410 | 0.68 | .514 | .169 | .096 | .393 | .483 | 1.76 | .896 | .349 | .222 | ......... | 1.380 | 2.40 | 1.732 | 5.40 |
| .520 | .482 | .449 | .413 | 0.69 | .519 | .171 | .097 | .396 | .487 | 1.78 | .902 | .351 | .225 | ......... | 1.389 | 2.45 | 1.736 | 5.45 |
| .523 | .485 | .452 | .415 | 0.70 | .523 | .173 | .099 | .400 | .491 | 1.80 | .908 | .354 | .228 | ......... | 1.398 | 2.50 | 1.740 | 5.50 |
| .526 | .488 | .455 | .418 | 0.71 | .527 | .175 | .100 | .404 | .496 | 1.82 | .914 | .357 | .231 | ......... | 1.407 | 2.55 | 1.744 | 5.55 |
| .530 | .491 | .458 | .422 | 0.72 | .531 | .177 | .101 | .407 | .500 | 1.84 | .920 | .359 | .234 | ......... | 1.415 | 2.60 | 1.748 | 5.60 |
| .533 | .494 | .461 | .425 | 0.73 | .536 | .179 | .102 | .411 | .504 | 1.86 | .925 | .362 | .237 | ......... | 1.423 | 2.65 | 1.752 | 5.65 |
| .536 | .497 | .464 | .428 | 0.74 | .540 | .181 | .104 | .415 | .508 | 1.88 | .930 | .365 | .240 | ......... | 1.431 | 2.70 | 1.756 | 5.70 |
| .539 | .500 | .467 | .431 | 0.75 | .544 | .183 | .105 | .418 | .512 | 1.90 | .936 | .367 | .243 | ......... | 1.439 | 2.75 | 1.760 | 5.75 |
| .542 | .502 | .470 | .433 | 0.76 | .548 | .185 | .106 | .422 | .516 | 1.92 | .941 | .370 | .245 | ......... | 1.447 | 2.80 | 1.763 | 5.80 |
| .544 | .505 | .473 | .436 | 0.77 | .552 | .187 | .107 | .425 | .520 | 1.94 | .946 | .373 | .246 | ......... | 1.455 | 2.85 | 1.767 | 5.85 |
| .547 | .508 | .476 | .439 | 0.78 | .556 | .189 | .109 | .429 | .524 | 1.96 | .951 | .375 | .247 | ......... | 1.462 | 2.90 | 1.771 | 5.90 |
| .549 | .511 | .479 | .441 | 0.79 | .560 | .191 | .110 | .432 | .528 | 1.98 | .955 | .378 | .249 | ......... | 1.470 | 2.95 | 1.775 | 5.95 |
| .551 | .513 | .481 | .444 | 0.80 | .564 | .193 | .111 | .436 | .531 | 2.00 | .960 | .381 | .250 | ......... | 1.477 | 3.00 | 1.778 | 6.00 |
| .554 | .516 | .484 | .447 | 0.81 | .568 | .195 | .112 | .439 | .535 | 2.10 | ......... | ......... | .262 | ......... | 1.484 | 3.05 | 1.785 | 6.10 |
| .556 | .519 | .487 | .450 | 0.82 | .572 | .197 | .113 | .443 | .539 | 2.20 | ......... | ......... | .275 | ......... | 1.491 | 3.10 | 1.792 | 6.20 |
| .559 | .521 | .490 | .453 | 0.83 | .576 | .199 | .114 | .446 | .543 | 2.30 | ......... | ......... | .288 | ......... | 1.498 | 3.15 | 1.799 | 6.30 |
| .561 | .524 | .492 | .456 | 0.84 | .580 | .201 | .116 | .449 | .547 | 2.40 | ......... | ......... | .303 | ......... | 1.505 | 3.20 | 1.806 | 6.40 |
| .563 | .526 | .494 | .458 | 0.85 | .584 | .203 | .117 | .453 | .550 | 2.50 | ......... | ......... | .318 | ......... | 1.512 | 3.25 | 1.813 | 6.50 |
| .566 | .529 | .497 | .461 | 0.86 | .588 | .205 | .118 | .456 | .554 | 2.60 | ......... | ......... | .333 | ......... | 1.519 | 3.30 | 1.820 | 6.60 |
| .568 | .531 | .500 | .464 | 0.87 | .592 | .207 | .120 | .459 | .558 | 2.70 | ......... | ......... | .351 | ......... | 1.525 | 3.35 | 1.826 | 6.70 |
| .571 | .534 | .502 | .467 | 0.88 | .596 | .208 | .121 | .462 | .561 | 2.80 | ......... | ......... | .369 | ......... | 1.531 | 3.40 | 1.833 | 6.80 |
| .573 | .537 | .504 | .469 | 0.89 | .599 | .210 | .122 | .466 | .565 | 2.90 | ......... | ......... | .387 | ......... | 1.538 | 3.45 | 1.839 | 6.90 |
| .574 | .539 | .507 | .471 | 0.90 | .602 | .212 | .123 | .469 | .568 | 3.00 | ......... | ......... | .405 | ......... | 1.544 | 3.50 | 1.845 | 7.00 |

**Table 7.6**
**Geometry Factors**

The proper factor is multiplied by the volt-ampere value obtained from Fig. 7.38 to provide the required volt-ampere value for a weld other than a last-pass butt weld or for plate thicknesses other than 1/2 inch. Courtesy of *Welding Journal.*

| Plate thickness (in.) | Bead on plate or last-pass butt | First-pass vee or U welds | First-pass fillet or lap welds | First-pass vee weld with backing strip |
|---|---|---|---|---|
| 1/8 | 0.25 | 0.25 | 0.38 | 0.3 |
| 1/4 | 0.5 | 0.48 | 0.75 | 0.6 |
| 3/8 | 0.75 | 0.73 | 1.1 | 0.8 |
| 1/2 | 1.0 | 0.95 | 1.5 | 1.1 |
| 5/8 | 1.25 | 1.2 | 1.9 | 1.3 |
| 3/4 | 1.5 | 1.4 | 2.25 | --- |
| 1 | 2.0 | 1.9 | 3.0 | --- |
| 1-1/4 | 2.5 | 2.3 | 3.7 | --- |
| 1-1/2 | 2.9 | 2.7 | 4.3 | --- |
| 2 | 3.5 | 3.3 | 5.4 | --- |

manganese, 0.04 percent phosphorus, 0.04 percent sulfur, and 0.20 percent silicon are specified.

CALCULATION OF IDEAL CRITICAL DIAMETER. See Table 7.5. The sum of the factors for each element = 1.349, which gives a DI of 2.20 in.

DETERMINATION OF MAXIMUM DQE COOLING RATE. A maximum heat affected zone hardness of $30R_c$ is desired. With a carbon content of 0.40 percent, the initial hardness (IH) is equal to 56 $R_c$, Table 7.7. The hardness ratio = 56/2.06 or $R_c$ 27.1. From Table 7.7, the DQE is taken as 3/4 in. or 12 as shown in Table 7.8. The lower the DQE value, the lower the hardening tendencies of a given steel. The cooling rate for 12 is shown to be 16.3°F/sec, Fig. 7.39.

HEAT INPUT CALCULATION. A simplified approach to the heat input problem of welding is based on the temperature and mass of metal deposited per second. The relationship is shown by the following formula:

$$\text{Heat input in cal/sec} = M(\triangle T H_s) + M(H_1)$$

where M = weld mass deposited in grams/sec, $\triangle T$ = temperature change °C, $H_s$ = specific heat of the metal in cal/grm/°C $H_1$ = latent heat of the metal in cal/grm.

The heat required in calories per second can be changed to kw as follows:

$$\text{kw} = \frac{\text{cal/sec} \times 3{,}600}{252 \text{ cal} \times 3{,}412 \text{ BTU}}.$$

It must be noted that this formula is reduced to basic conditions and does not take into account the heat lost by convection and radiation. The following example is used for clarification.

A 60 degree butt weld is made in 3/8 in. thick mild-steel plate at 24 in. per minute. A root gap of 1/16 in. is allowed between the plates, and a reinforcement of 1/8 in. is desired. What is the heat input in terms of kw if the temperature of the deposit must be raised 1,500°C?

M = 0.080 + 0.106 + 0.276 = 0.462 lbs/ft (Table 7.4)
= 0.0154 lbs/sec or 6.99 grams/sec
= 7 grms/sec × 0.107 cal grm/°C × 1500°C + 7 grms/sec × 65 cal/grm
= 1,120 cal/sec + 455 cal/sec
= 1,575 cal/sec

$$= \frac{1{,}575 \text{ cal/sec} \times 3{,}600 \text{ sec}}{252 \text{ cal} \times 3{,}412 \text{ BTU}}$$

Rate of heat input = 6.6 kw.

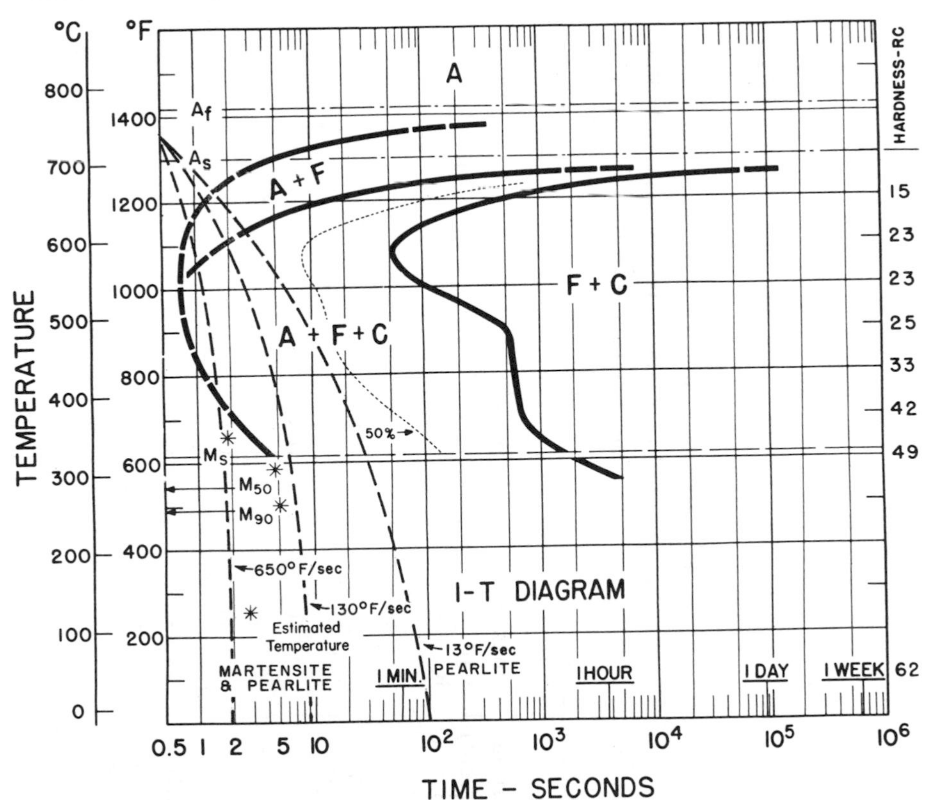


**Fig. 7.39.** Isothermal transformation diagram for AISI 1340 steel with continuous cooling rate curves superimposed.

**Table 7.7**
**Distance from Quenched End (DQE)**

Using the far left column, the initial hardness (IH) corresponds to the percent carbon present in the alloy. Then, finding the critical diameter in column three and moving to the right until the proper initial hardness to distance hardness (IH/DH) ratio is reached, the DQE is determined. Courtesy of Republic Steel Corporation.

| Carbon (percent) | Initial (maximum) hardness IH (Rockwell C) | Ideal critical diameter DI (in.) | Distance from quenched end (DQE) | | | | | | | |
|---|---|---|---|---|---|---|---|---|---|---|
| | | | 1/4 in. | 1/2 in. | 3/4 in. | 1 in. | 1 1/4 in. | 1 1/2 in. | 1 3/4 in. | 2 in. |
| 0.10 | 38 | 0.50 | 4.90 | | | | | | | |
| 0.11 | 39 | 0.55 | 4.42 | | | | | | | |
| 0.12 | 40 | 0.60 | 4.03 | | | | | | | |
| 0.13 | 40 | 0.65 | 3.70 | 6.00 | | | | | | |
| 0.14 | 41 | 0.70 | 3.47 | 5.15 | | | | | | |
| 0.15 | 41 | 0.75 | 3.25 | 4.50 | | | | | | |
| 0.16 | 42 | 0.80 | 3.07 | 4.18 | | | | | | |
| 0.17 | 42 | 0.85 | 2.90 | 3.88 | 6.00 | | | | | |
| 0.18 | 43 | 0.90 | 2.75 | 3.68 | 5.13 | | | | | |
| 0.19 | 44 | 0.95 | 2.61 | 3.50 | 4.70 | | | | | |
| 0.20 | 44 | 1.00 | 2.48 | 3.33 | 4.40 | | | | | |

$$\text{Ratio} = \frac{\text{Initial Hardness}}{\text{Distance Hardness}}$$

| | | | | | | | | | | |
|---|---|---|---|---|---|---|---|---|---|---|
| 0.21 | 45 | 1.05 | 2.33 | 3.20 | 4.13 | 5.28 | | | | |
| 0.22 | 45 | 1.10 | 2.17 | 3.08 | 3.93 | 4.75 | 5.70 | | | |
| 0.23 | 46 | 1.15 | 2.05 | 2.96 | 3.76 | 4.40 | 4.95 | 5.75 | | |
| 0.24 | 46 | 1.20 | 1.96 | 2.86 | 3.60 | 4.15 | 4.58 | 5.00 | 6.00 | |
| 0.25 | 47 | 1.25 | 1.88 | 2.76 | 3.45 | 3.95 | 4.32 | 4.65 | 5.15 | 6.00 |
| 0.26 | 48 | 1.30 | 1.80 | 2.66 | 3.32 | 3.78 | 4.13 | 4.40 | 4.72 | 5.25 |
| 0.27 | 49 | 1.35 | 1.73 | 2.57 | 3.21 | 3.65 | 3.95 | 4.18 | 4.45 | 4.83 |
| 0.28 | 49 | 1.40 | 1.67 | 2.49 | 3.10 | 3.53 | 3.77 | 4.02 | 4.26 | 4.53 |
| 0.29 | 50 | 1.45 | 1.62 | 2.42 | 3.01 | 3.41 | 3.65 | 3.87 | 4.08 | 4.29 |
| 0.30 | 50 | 1.50 | 1.57 | 2.34 | 2.93 | 3.30 | 3.53 | 3.73 | 3.91 | 4.10 |
| 0.31 | 51 | 1.55 | 1.53 | 2.27 | 2.84 | 3.20 | 3.43 | 3.61 | 3.78 | 3.96 |
| 0.32 | 51 | 1.60 | 1.49 | 2.21 | 2.75 | 3.10 | 3.33 | 3.51 | 3.67 | 3.83 |
| 0.33 | 52 | 1.65 | 1.46 | 2.16 | 2.66 | 3.01 | 3.24 | 3.42 | 3.57 | 3.71 |
| 0.34 | 53 | 1.70 | 1.43 | 2.11 | 2.59 | 2.93 | 3.16 | 3.33 | 3.47 | 3.59 |
| 0.35 | 53 | 1.75 | 1.40 | 2.06 | 2.52 | 2.86 | 3.08 | 3.25 | 3.38 | 3.49 |
| 0.36 | 54 | 1.80 | 1.38 | 2.01 | 2.45 | 2.80 | 3.00 | 3.17 | 3.29 | 3.40 |
| 0.37 | 55 | 1.85 | 1.36 | 1.96 | 2.38 | 2.74 | 2.94 | 3.10 | 3.21 | 3.32 |
| 0.38 | 55 | 1.90 | 1.34 | 1.91 | 2.33 | 2.68 | 2.88 | 3.04 | 3.14 | 3.25 |
| 0.39 | 56 | 1.95 | 1.32 | 1.87 | 2.27 | 2.63 | 2.83 | 2.97 | 3.08 | 3.18 |
| 0.40 | 56 | 2.00 | 1.30 | 1.83 | 2.23 | 2.58 | 2.78 | 2.92 | 3.02 | 3.11 |
| 0.41 | 57 | 2.10 | 1.26 | 1.75 | 2.13 | 2.50 | 2.69 | 2.82 | 2.91 | 3.00 |
| 0.42 | 57 | 2.20 | 1.24 | 1.69 | 2.06 | 2.42 | 2.61 | 2.73 | 2.83 | 2.91 |

**Table 7.8**
**Cooling Rate Versus DQE**

The cooling rate that produces a given DQE is obtained from this table. This cooling rate then can be used to calculate the required welding heat input. Courtesy of Republic Steel Corporation.

| Bar Position | Distance from Quenched End (in.) | Cooling rate (F per sec) |
|---|---|---|
| 1 | 1/16 | 490 |
| 2 | 1/8 | 305 |
| 3 | 3/16 | 195 |
| 4 | 1/4 | 125 |
| 5 | 5/16 | 77 |
| 6 | 3/8 | 56 |
| 7 | 7/16 | 42 |
| 8 | 1/2 | 33 |
| 10 | 5/8 | 21.4 |
| 12 | 3/4 | 16.3 |
| 14 | 7/8 | 12.4 |
| 16 | 1 | 10.0 |
| 20 | 1-1/4 | 7.0 |
| 24 | 1-1/2 | 5.1 |
| 32 | 2 | 3.5 |
| 40 | 2-1/2 | 3.1 |

## QUESTIONS

**7.1.** Why can welds be made out in space without the addition of heat?

**7.2.** Sketch the cross section of a welded butt joint to show the different metal structure from the center of the weld to the parent metal.

**7.3.** How can hardness of a high tensile strength joint be controlled during and after welding?

**7.4.** What happens to a butt weld during cooling if both parts are rigidly clamped?

**7.5.** (*a*) What is the main element in stainless steel that prevents corrosion? (*b*) Why do ordinary stainless steels lose their corrosion resistance during welding?

**7.6.** What is the principle involved in the process of checking weld joints by the photoelastic method?

**7.7.** When may a fillet weld be made with a convex contour?

**7.8.** Compare the advantages and disadvantages of the transformer type of welder with the motor generator type.

**7.9.** Explain the *use* of each of the following arc characteristics: (*a*) drooping arc voltage, (*b*) CAV, (*c*) RAV.

**7.10.** What is meant by duty cycle requirements for an arc welding machine?

**7.11.** Identify the following types of electrodes: E9015, E8030, and E6011.

**7.12.** Explain the difference between the action of a mineral- and a cellulosic-coated electrode.

**7.13.** In what way are low-hydrogen electrodes more effective than other mineral-coated electrodes?

**7.14.** Name several advantages of using cored wire over "the stick" electrode.

**7.15.** What are some applications for submerged arc welding?

**7.16.** Why is GMA more inclusive than the term MIG?

**7.17.** What are some of the main advantages of the $CO_2$-cored wire process?

**7.18.** Compare pulsed-arc and microwire welding as to use.

**7.19.** (*a*) What is the primary application of TIG welding? (*b*) Why is it not competitive with some of the previously mentioned processes?

**7.20.** Sketch what you think the cross section of a $CO_2$ gas-shielded butt weld should look like.

**7.21.** What are some of the important controls in spot welding?

**7.22.** What special controls are used for spot welding aluminum?

**7.23.** Make a sketch to show how a standard 3/4-in. square nut could be prepared for projection welding it to 1/2-in. thick steel plate.

## PROBLEMS

**7.1.** To obtain a 0.350 throat dimension in a fillet weld, approximately what size fillet would be required for (*a*) the stick electrode? (*b*) the flux-cored wire? (Use the next larger nearest standard size.)

**7.2.** (*a*) A 1/4-in. fillet weld is made with the stick electrode. Find the equivalent size made with the cored-wire electrode. (*b*) These welds are flat faced. What is the savings in pounds of deposited metal/100 ft with equivalent welds but with a convex face?

**7.3.** A mild steel strip 1/2″ × 5″ × 10″ is restrained in expanding in length as the temperature in the bar rises from 100°F to 1,500°F. Assume the coefficient of expansion to be $6.6 \times 10^{-6}$/°F and Young's Modulus to be 29,000,000 lb/in.$^2$, what will the nature of the stress be in the bar?

**7.4.** A 5-in. wide mild steel strip 1/2 in. thick is welded, as shown in Fig. P7.4, to a 1/2-in. thick plate. The AWS specifies an allowable working stress of 158,000 lb/in.$^2$ for the E70 series electrodes when used in *shear* loading. Determine the allowable tensile load, *P*. *Note:* A fillet weld loaded at right angles is 30% stronger than if it is loaded parallel. Use Table 7.4 for fillet size.

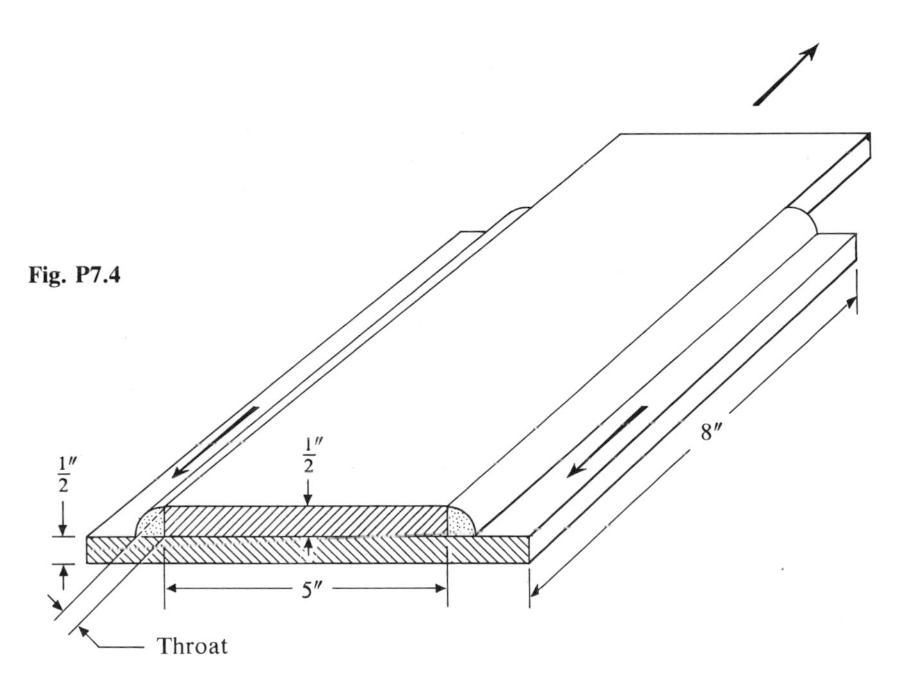

**Fig. P7.4**

**7.5.** (*a*) An arc resistance welding operation is being performed with a current of 265 amp, a circuit voltage of 24 volts, and an electrode travel speed of 14 in./min. How much heat is liberated by the arc to fuse the weld? Atmospheric heat losses reduce the efficiency to 80 percent. (*b*) What diameter of electrode do you think would be suitable to carry out the above welding operation on mild steel? (*c*) What amperage would you recommend for use with 3/16-in. rod to weld two 1/2-in. thick mild steel plates in a lap weld? (*d*) Does the 1/4 in. rod require three times as much amperage as the 1/8 in. rod under average conditions?

**7.6.** (*a*) What is the approximate temperature of the flame of an electric arc? 6,000°F, 11,000°F, or 17,000°F? (*b*) Moisture in the coating of low-hydrogen electrodes has a detrimental effect on the quality of the weld if it is over 2%, 5%, 10%? (*c*) The best welding results are obtained by keeping the electrodes in a dry-rod storage oven at 100°F, 300°F, or 600°F?

**7.7.** (*a*) Steel plates of 3/8 in. thickness are to be welded together using the square-butt welding technique. Electrodes 3/16 in. dia are used at a speed of 10 in./min at 25 volts and 250 amp. Calculate the penetration. (*b*) Calculate the actual cost in dollars/ft to make a 3/8-in. fillet weld, assuming a 5/32-in. electrode is being used in a flat-position, with the submerged-automatic process. A travel speed of 10 in./min using dc straight polarity at 700 amp is selected.

**7.8.** Compare the cost of three methods of making a 60° single V-butt weld between two 1/2-in. thick plates 6 ft long. Use Table 7.4 for weight of weld/ft of joint. Use 1/8 in. reinforcement. Labor and overhead are $10.00/hr. Current cost is 2¢/kwh. Additional information is given in Table P7.8.

**Table P7.8**

| | Oxyacetylene | Manual Arc | Submerged Arc |
|---|---|---|---|
| Speed | 5 in./min/pass | 10 in./min | 22 in./min |
| Passes | 3 | 3 | 1 |
| Cost of weld material | 15¢/lb | 11¢/lb | 22¢/lb |
| Joint preparation | 15 min | 15 min | none |
| Setup time | 10 min | 10 min | 5 min |
| Oxygen or kwh | 58 cu ft/hr | 0.02/kwh | 0.02/kwh |
| Acetylene | 56 cu ft/hr | | |
| Oxygen | 1.5¢/cu ft | | |
| Acetylene | .4¢/cu ft | | |
| Operating factor | 10% | 50% | 100% |

## REFERENCES

Abella, F. J., and R. P. Sullivan, "Gas Metal-Arc Welding of Stainless Steel Using Short-Circuiting Transfer," *Welding Journal.*

Albom, M. J., "Diffusion Welding," *Machine Design,* September 16, 1965.

"Basic Metallurgical and Welding Data Used to Approximate Weld Parameters," *General Motors Engineering Journal,* Second Quarter, 1965.

Blair, H. S., "Today's Welding Power Sources," *Metalworking*, January 1964.

Blodgett, O., "Estimating Welding Cost," *Welding Engineer*, July 1966.

Doar, G. E, and R. Stout, "A Tentative System for Preserving Ductility in Weldments," *Welding Journal*, vol 22, no 7, pp 278S–299S, July 1943.

Irons, G. S., and R. D. Regan, "For High Joining Speeds, Which Welding Process?" *Metalworking*, October 1964.

Jhaveri, P., W. Moffatt, and C. Adams, Jr., "The Effect of Plate Thickness and Radiation on Heat Flow in Welding and Cutting," *Welding Journal*, vol 41, no 1, pp 12S–16S, January 1962.

Norcross, J. E., "Cored Wire Welding Curve Is Climbing," *Welding Design and Fabrication*, March 1966.

Puschner, M., and R. Killing, "Investigation of the Weldability of High-Tensile, Killed, Fine-grained Steels," *Svetsaren*, no 1, 1966, Gotenborg, Sweden.

Smith, D. M., "Plasma Spraying of Refractory Materials," *General Motors Engineering Journal*, Second Quarter, 1963.

*Welding Aluminum*, American Welding Society, United Engineering Center, New York, N.Y., 1966.

Wooding, W. H., "Have You Tried $CO_2$ Weld Shielding?" *American Machinist/Metalworking Manufacturing*, November 1963.

8

# GAS WELDING, CUTTING, AND NONDESTRUCTIVE TESTING

GAS WELDING generally refers to heating metal for both fusion- and braze-welding purposes with an oxyacetylene flame.

Scientists knew as early as 1895 that a mixture of oxygen and acetylene would burn at a very high temperature, but there was no suitable means of mixing and controlling them. It was not until 1903 that a welding torch was developed which could mix and control oxygen and acetylene to produce a flame of about 6,000°F.

## OXYGEN

Oxygen can be obtained by either a liquid-air or an electrolytic process.

**Liquid-Air Process.** Most of the oxygen used for welding is obtained by the liquid-air process. Air is compressed and cooled to a point where the gases become liquid. Then, as the liquid air is allowed to rise in temperature, the various gases can be taken off separately. Nitrogen is given off first at −321°F. Oxygen is given off next at −297°F. These gases, having been thus separated, are further purified and compressed into cylinders for use as needed.

**Electrolytic Process.** In the electrolytic process, oxygen is obtained by passing a direct current through water to which an acid alkali has been added. The electric current breaks the water down into its chemical elements of hydrogen and oxygen. The oxygen collects at the positive terminal, while hydrogen is collected at the negative terminal.

## ACETYLENE

Acetylene is a fuel gas composed of carbon and hydrogen, $C_2H_2$. It is obtained from the chemical reaction of water and calcium carbide:

$$CaC_2 + 2H_2O \rightarrow C_2H_2 + Ca(OH)_2$$

The reaction provides acetylene gas and hydrated (or slaked) lime as a sludge. A special hopper for dropping the calcium carbide into a tank of water at controlled rates is referred to as an *acetylene generator*. A 300-lb generator is one into which 300 lb of calcium carbide are loaded and slowly dropped into 300 gal of water. Since calcium carbide generates 4.5 cu ft of acetylene per pound, a generator of this type could operate at full load for a period of 4.5 hr.

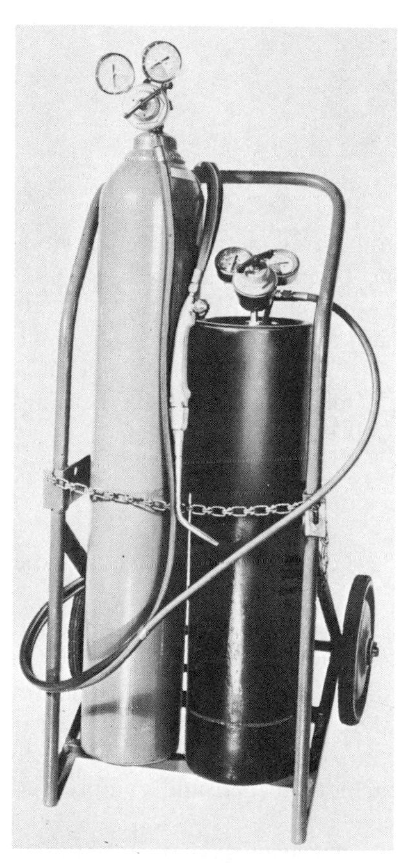

**Fig. 8.1.** A portable oxyacetylene welding outfit, complete with carrying cart. Courtesy Linde Co., division of Union Carbide Corp.

The principal items of equipment needed for gas welding are oxygen and acetylene cylinders, regulators, torches, and hose (Fig. 8.1).

**Advantages and Disadvantages of Gas Welding.** ADVANTAGES. The oxyacetylene flame is generally more easily controlled and not as piercing as metallic arc welding. Therefore, it is used extensively for sheet metal fabrication and repairs.

The oxyacetylene torch is versatile. It can be used for brazing, braze welding, soldering, preheating, postheating, heating for bending, heat treatment, metal cutting, metal cleaning, etc.

It is very portable; small-sized units are comparatively light, weighing only 150 lb complete with carrying truck. In this size, the oxygen cylinder contains 80 cu ft and the acetylene cylinder 60 cu ft. This, and even larger units, can be moved almost anywhere for needed repairs or fabrication.

DISADVANTAGES. The disadvantages of oxyacetylene welding are as follows:

(1) It takes considerably longer for the metal to heat up than in arc welding. There is almost no instantaneous pool.

(2) Harmful thermal effects are aggravated by prolonged heat, and there will, in most cases, be a larger heat-affected area. This often results in increased grain growth, more distortion, and, in some cases, loss of corrosion resistance.

(3) Oxyacetylene tanks are leased, and when they are kept beyond a specified period, demurrage charges are incurred.

(4) Oxygen and acetylene gases are rather expensive.

(5) There are safety problems involved in handling and storing the gases.

(6) Flux applications and the shielding provided by the oxyacetylene flame are not nearly so positive as those supplied by the inert gas in TIG, MIG, or carbon dioxide welding.

## BRAZING

Brazing differs from welding in that the metals to be joined are not melted and deeply fused with the filler metal. Instead, brazing employs a filler metal which melts at a temperature below the melting point of the metals to be joined and is drawn into the close-fitting joint by capillary action. The surfaces must be clean so that a *wetting* action can take place. *Wetting* refers to the attraction that occurs between the molecules of the alloy and the base metal on properly cleaned surfaces at the right temperature.

**Joint Design.** The two main types of joints used in brazing are the butt joint and lap joint. The scarf joint is a compromise between these two. The joint should be selected on the basis of service requirements, such as mechanical strength, electrical conductivity, and pressure tightness. Other considerations are the brazing process to be used, fabricating techniques, and production quantities.

LAP JOINTS. Lap joints should be used, in preference to other types of joints, where strength is a primary consideration. An overlap of three times the thickness of the thinnest member will usually give maximum efficiency (Fig. 8.2*a*). Overlaps greater than this lead to poor brazing owing to insufficient penetration, inclusions, etc. This joint is also recommended when leak tightness and good electrical conductivity are required.

BUTT JOINTS. The butt joint can provide a smooth joint of minimum thickness. Since the butt joint is more difficult to fit up and to preserve alignment and the necessary clearances, its strength will not generally be as reliable as the lap joint (Fig. 8.2*b*).

SCARF JOINTS. The scarf joint is an attempt to maintain the smooth contour of the butt joint and, at the same time, provide the large joint area of the lap joint (Fig. 8.2*c*). The scarf joint is stronger than the butt joint, but it requires considerable preparation. Both scarf and butt joints,

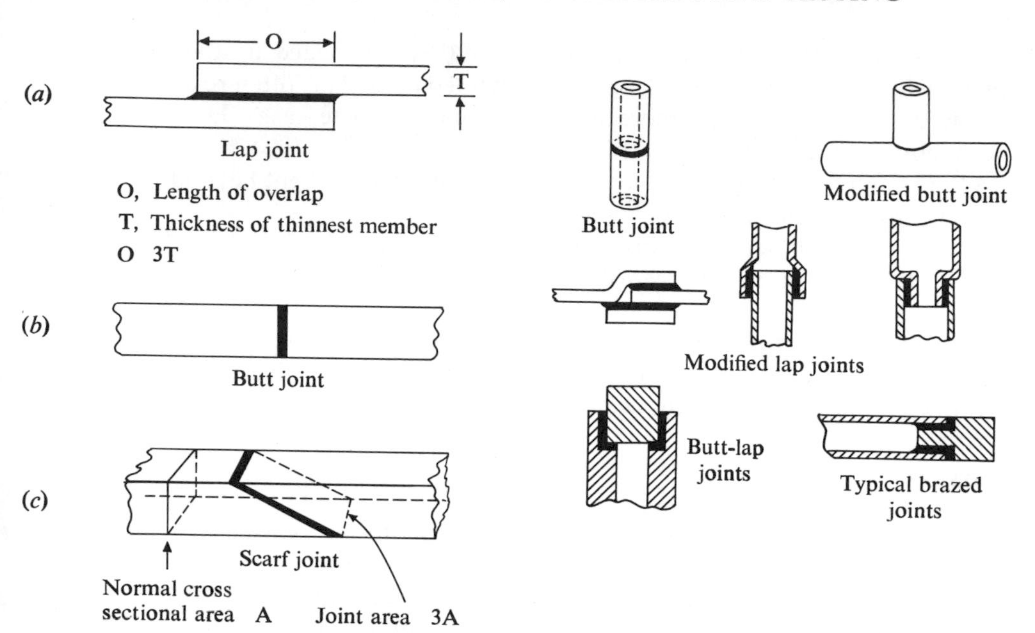


**Fig. 8.2.** Basic types of brazed joints. Courtesy Air Reduction Sales Co.

when properly made with silver-brazing alloys, are considerably stronger than the parent material.

STRESS DISTRIBUTION. The brazed joint should be designed to minimize stress concentrations that may tear the joint. Two approaches are used: one is to increase the flexibility of the joint; the other is the opposite—to stiffen the flexible members.

JOINT CLEARANCE. Joint clearance is the distance between the surfaces of the joint into which the brazing filler metal must flow. The clearance varies greatly with conditions. It is entirely different at room temperature from what it is at brazing temperature. In general practice, the clearance at room temperature is specified. For any given combination of base and filler metals there is a *best* joint clearance. Values below the minimum clearance are weak because the alloy does not flow into the joint, and clearances beyond maximum result in greatly lowered strength. The general range of joint clearance is from 0.002 to 0.010 in.

Sometimes, when one member surrounds another and the internal member has a greater expansion rate than the outside member, the clearance will be lost and capillary action will be prevented. In this case, clearance near maximum should be used. Another factor to consider is the situation when an internal member has greater contraction than the outer member. When this occurs, the braze may be fractured, but the right braze material—one with a long temperature range between solidus and liquidus and with a sluggish flow—may be able to bridge large gaps and preserve sufficient strength to resist cracking or cooling.

INTRODUCTION OF BRAZING ALLOYS INTO THE JOINT. Brazing alloys may be introduced into the joint by *face feeding* or by *preplacement.* Face feeding consists of holding the alloy in the hand and applying it when the joint has reached the proper temperature, as indicated by the flux becoming liquid. In preplacement the alloy is placed in or near the fluxed joint, and at brazing temperature it melts and flows into the joint. Preplaced filler metal may be in the form of rings, washers, shims, slugs, powder, or sprayed-on coatings.

In designing joints for preplaced alloy, it is best to arrange for the alloy to flow with the aid of gravity and to provide for inspection (Fig. 8.3). When a groove is used for the wire, as shown, it must be deducted from the joint area in determining the amount of overlap needed.

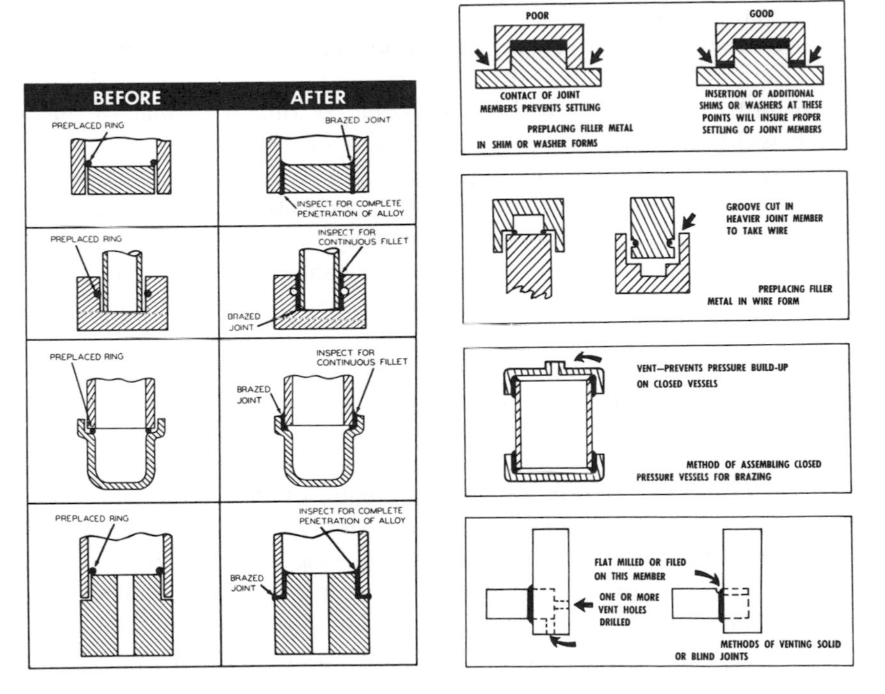


**Fig. 8.3.** Joints with preplaced filler material. Courtesy Air Reduction Sales Co.

Inert atmospheres are used to help prevent oxidation of the metal surface during the brazing process. Some atmospheres eliminate the need for flux; others do not. Controlled atmospheres are used primarily in furnace brazing but they may be used in induction and resistance brazing.

**Brazing Processes.** Although the AWS master chart of welding processes lists eight methods for the application of brazing alloys, only four are of extensive commercial importance; these are torch, induction, furnace, and dip.

## BRAZE WELDING

Braze welding is somewhat similar to brazing in that the base metals are not melted but are joined by an alloy of lower melting point. The main difference is that capillary action is not used to draw the filler material into the joint. A braze-welded joint is designed very much like a welded joint, but an effort should be made to avoid sharp corners that are easily overheated or ones that become points of stress concentration.

Braze welding is used extensively for repair work and some fabrication on such metals as cast iron, malleable iron, wrought iron, and steel. It is also used, although to a lesser extent, on copper, nickel, and high-melting-point brasses and bronzes. Some of the brasses and bronzes melt at a point so near that of the filler metal that fusion welding, rather than brazing, is done.

Braze welding is not recommended on parts that will be subjected to temperatures higher than 650°F.

**Strength.** The strength of a braze-welded joint is similar to that of fusion welding. It is, however, dependent on the quality of the bond between the filler metal and the base metal. The braze-welded joint depends also on the quality of the filler metal after it is deposited. It is important that the filler metal be free from blowholes, slag inclusions, and other physical defects. Some of the same forces used in brazing—namely, interalloying and intergranular penetration—are at work in braze welding.

**Interalloying.** The metal surface must be cleaned, usually by mechanical means or solvents, so that the filler metal will wet the surface. Commercial flux is used to remove the oxide film at the time of brazing. In a narrow zone at the interface of the two alloys—the base metal and the filler metal—there will be some diffusion of both materials. This is especially true in such metals as copper, zinc, and tin.

**Intergranular Penetration.** The brazing heat, usually dull red in the case of ferrous materials, is enough to open up the crystal grain structure to allow the brass filler material to penetrate along the grain boundaries. It is this action that gives braze welding its great strength.

## HARD SURFACING—METAL AND CERAMIC APPLICATIONS

Hard surfacing or hard facing generally consists of applying alloys or ceramic materials to a metal surface. The purposes of the application are to increase resistance to wear, abrasion, cavitation, corrosion, heat, or impact. In most cases the hard-facing alloys can be applied by means of the various electric-arc methods already discussed or by the use of oxyacetylene, oxypropane, or oxyhydrogen flames. Sometimes a combination of both electric arc and gas is used, as in the more recent *plasma* method.

Since the welding processes already discussed can be used for depositing hard-surfacing materials, the use of each will be pointed out briefly, and then an explanation of the newer flame-spray methods will be given.

**Oxyacetylene Process.** The oxyacetylene process is generally used to apply hard-facing materials where it is necessary to achieve a surface with the minimum of required finishing. It is especially good in applying crack-free applications of nonferrous alloys.

Tungsten carbide is often used as the hard-surfacing material. It is handled by having small particles of it in a mild-steel tube; as the tube is melted, the tungsten carbide particles become fused to the surface in a matrix of mild steel.

**Inert-Gas Process.** The inert-gas process is particularly adaptable where a flawless deposit is required, mainly on new construction rather than building up a worn surface for repair. Shown in Fig. 8.4 is a schematic view of how tungsten carbide particles are fed down from a vibratory hopper to the base metal, where they are fused on by the addition of a wire that flows in around the particles. This composite surface is very abrasive-resistant yet is not brittle.

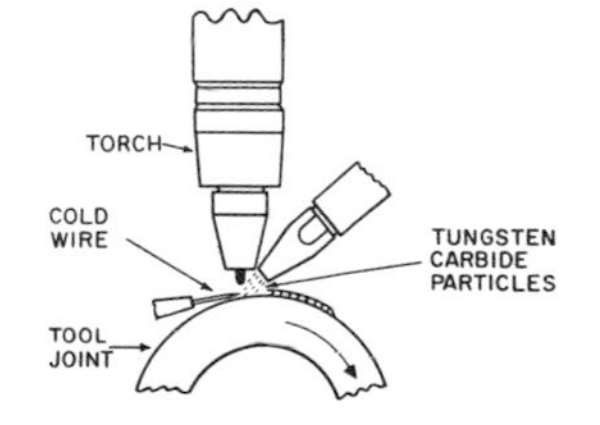


**Fig. 8.4.** Schematic view of an inert-gas tungsten-arc hard-surfacing machine used to apply tungsten carbide particles in a steel matrix. Courtesy Linde Co., division of Union Carbide Corp.

**Submerged-Arc Process.** The submerged-arc process is of most value in large runs of similar work. Multiple electrodes and oscillation techniques have made this method a very fast one. The welding head can be made to oscillate back and forth up to 3 in. A forward speed of 3 to 5 ipm can be maintained with a 2-in.-wide oscillation. The pellet rolls shown in Fig. 8.5 have been hard-surfaced using the oscillating technique. A mild-steel wire was used in conjunction with a submerged-arc hard-surfacing composition which produced a deposit hardness of $R_c$ 58. The main disadvantage is the high heat input that may cause cracking of the hardened overlays.

**Manual Metal Arc Process.** This process is the one most frequently used. It is usually more adaptable to position and location than are the other methods. It is quick and economical for short jobs, and a wider range of alloys may be used.

Hard-surfaced rods are divided into five groups, according to the alloy content. The groups are arranged in order of increasing hardness, starting with group I, which has less than 20 percent alloying element. The alloys used are mainly chromium, tungsten, manganese, silicon, and carbon. Group V consists of crushed tungsten carbides. These particles may be fused to strips of mild steel, embedded in a high-strength rod, or packed in a mild-steel tube.

**Fig. 8.5.** Pellet rolls before and after hard surfacing with the submerged-arc process and oscillation. Courtesy Linde Co., division of Union Carbide Corp.

## FLAME SPRAYING

The principle of flame spraying consists of feeding a suitable wire or powder through a gun, where it is heated and vaporized by an oxyacetylene or oxypropane flame and then propelled by compressed air, or other means, to become embedded on the surface of the workpiece. There are several variations of this method, and these will be discussed later.

**Surface Preparation.** Surfaces to be sprayed must be absolutely clean and sufficiently roughened to offer the maximum amount of mechanical anchorage for the coating. The methods employed are blasting, machining, or the use of a bond coat.

BLASTING. The use of clean, sharp, crushed steel grit or aluminum oxide blasted against the surface by compressed air will provide reentrant angles for mechanical bonding.

MACHINING. Surfaces that are to be machined after spraying or hard surfacing need an exceptionally strong bond. If a heavy coating or built-up surface is required, an undercut will first have to be made to provide room.

Grooves are made, to provide additional bonding surface, with a standard 1/8-in. cutoff tool ground down to 0.045 or 0.050 in. wide and rounded on the end. The grooves are cut about 0.025 in. deep and about 0.015 in. apart. If an exceptionally strong bond is not required, grooving at the ends may be sufficient.

A faster method is to cut threads on the undercut surface. The threads are also rolled down with the rotary tool until they are only partly open. This method is entirely satisfactory for applications that do not require exceptionally high bond strengths.

PREPARATION OF FLAT SURFACES. Shrinkage stresses that tend to lift the coating away from flat surfaces may be overcome either by spraying over the edge, to give a clamping action, or by cutting short slots, near the edge, that taper inward. The base metal may also be heated to 350°F to equalize cooling stresses.

It is important that these operations be done dry, as oil of any kind would impair the bond. The surface should not be handled until after metallizing. If this is not possible, the part should be wrapped in paper or a clean cloth before it is removed from the lathe. Any oil or grease that may come in contact with the work surface must be removed by vapor degreasing or other chemical cleaning methods.

BOND COAT. Molybdenum is used as bond coat because it can adhere to a smooth surface. When it is applied, surfaces that are not to be coated must be masked or oiled. Care must be taken that oil does not run into the undercut. The oil moisture can be boiled off by running the flame of the gun over the area. With the exception of copper or copper-base alloys, molybdenum bonds well to most metals.

**Methods of Flame Spraying.** There are three main methods of applying flame-sprayed metallic and ceramic coatings. These are the rod-and-wire method, powder method, and plasma method.

ROD-AND-WIRE METHOD. The rod-and-wire method of flame spraying is based on the theory that coating materials must become fully molten by means of a flame before they are released. The process uses compressed air to atomize fully the molten metal or oxides and project them against a prepared surface where they are embedded, assuring good mechanical adhesions (Fig. 8.6).

The compressed air helps cool the work parts, so the coatings may be applied successfully not only to metals but also to glass, wood, asbestos, and certain plastics.

POWDER METHOD. The powder-spray method uses an oxyacetylene welding torch with a modified tip which permits the powdered metal to be sprayed through the flame. A carrier gas—argon, helium, nitrogen, carbon dioxide, or compressed air—conveys the powdered metal to the torch tip. The fuel gas can be acetylene or hydrogen (Fig. 8.7). Modern units are capable of depositing high-melting-point materials, including oxides such as alumina and zirconia.

PLASMA METHOD. The plasma arc torch (Fig. 8.8) is a newer method of applying refractory materials in powder or wire form. The torch and

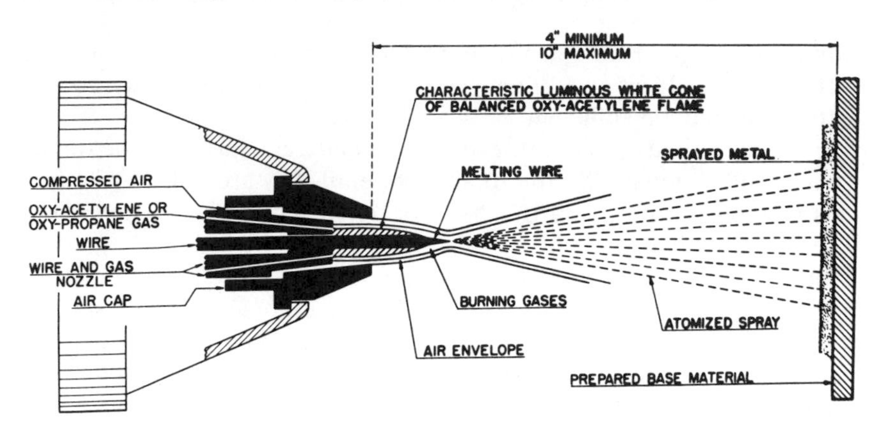


**Fig. 8.6.** This type of equipment is used for wire-type metallizing or rod-type application of ceramic coatings. Courtesy Metco Inc.

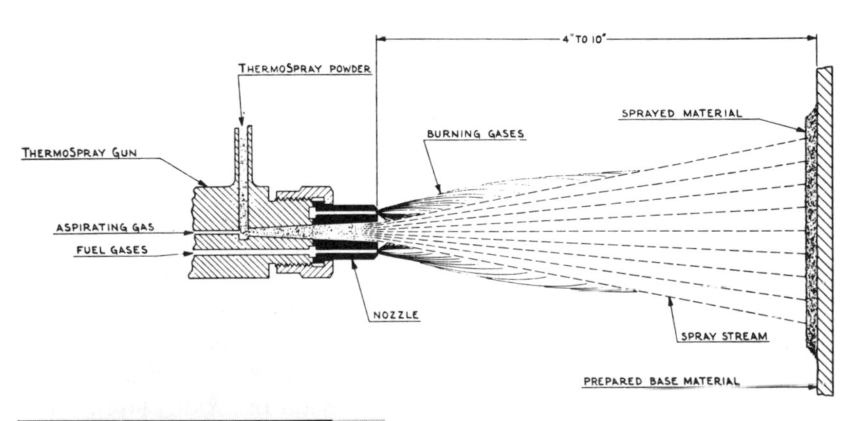


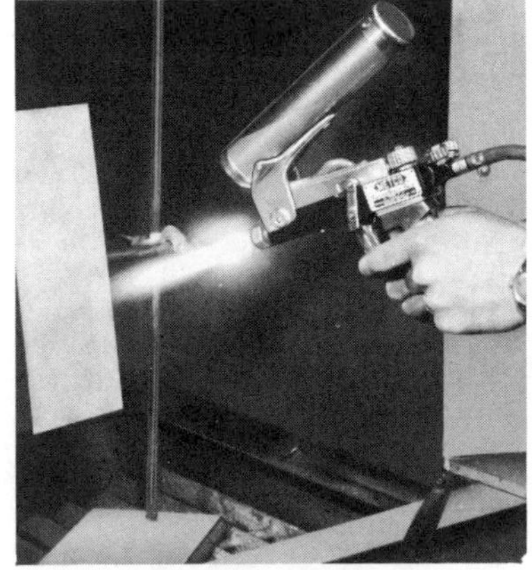

**Fig. 8.7.** Powder-spray methods. Courtesy Metco Inc.

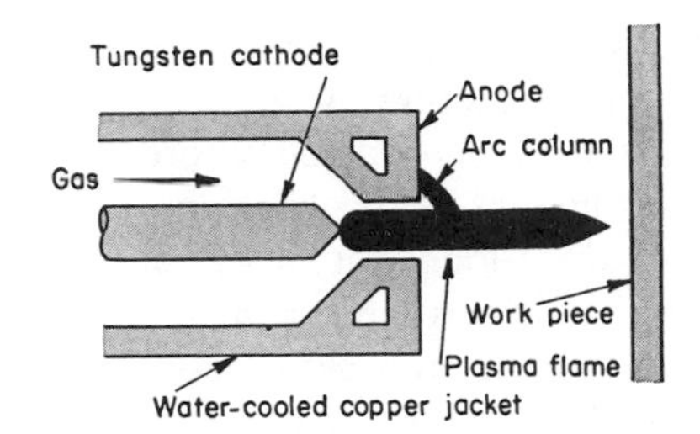


**Fig. 8.8.** Plasma method.

process will be described in more detail in this chapter under metal cutting. The main advantage of the plasma torch for spraying is its ability to reach high temperatures of 20,000°F to 30,000°F. When fine powder is used, the individual particles are melted and accelerated to velocities of the gas stream. They are propelled against the workpiece where they flatten and solidify to form a coating. Both refractory metal compounds and ceramics are applied by this method. Shown in Fig. 8.9 is an example of an alumina coating on steel.

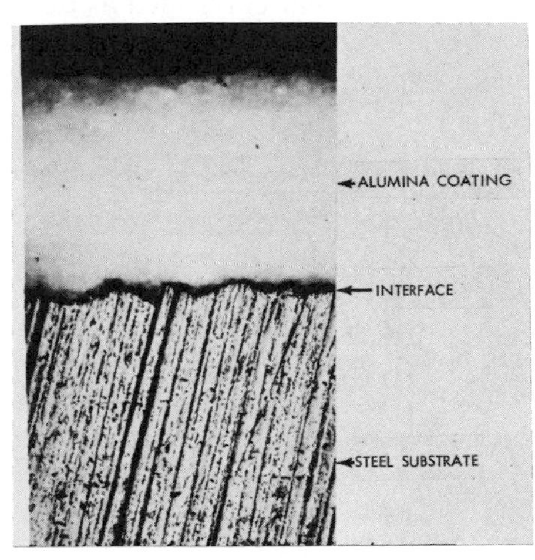


**Fig. 8.9.** This is a magnified cross section of an alumina coating deposited on a steel substrate by a plasma torch. Continuous-surface coatings can be developed. The acceleration of the particles leaving the torch results in a good bond at the interface and a homogenous coating. Courtesy *General Motors Engineering Journal.*

## FLAME AND ARC CUTTING

Flame cutting is a term used to describe the process of separating metals by the use of an oxy-fuel flame. It refers almost exclusively to ferrous metals. In fact, plain carbon steels are the only metals that oxidize before they melt. When steel is brought up to a temperature of between 1,400°F and 1,600°F, it reacts rapidly with oxygen to form iron oxides ($Fe_3O_4$). The melting point of the oxide is somewhat lower than that of the steel. The heat generated by the burning iron is sufficient to melt the iron oxide and some free iron which runs off as molten slag, exposing more iron to the oxygen jet.

Theoretically, 1 cu ft of oxygen is required to oxidize about 3/4 cu in.

of iron. Actually, the *kerf,* or cut, is not entirely oxidized, but 30 to 40 percent of the metal is washed out as metallic iron.

Once the cut is started, there should be enough heat to continue it without the use of preheating flames, using only the oxygen, but in practice, this does not work out. Excessive radiation at the surface, small pieces of dirt, and paint scale make it necessary that the oxygen be surrounded with preheating flames throughout the cutting operation.

Although acetylene is widely used as the fuel gas, other gases that can be used are hydrogen, natural gas, and propane—often at considerable savings.

## FLAME-CUTTING EQUIPMENT

**Manual Cutting.** The cutting torch somewhat resembles the welding torch, but it has provision for several neutral flames and a jet of high-pressure oxygen in the center (Fig. 8.10). The neutral preheating flames are adjusted and controlled as described previously in gas welding. The oxygen needed for cutting is controlled by a quick-acting trigger or lever-type valve.

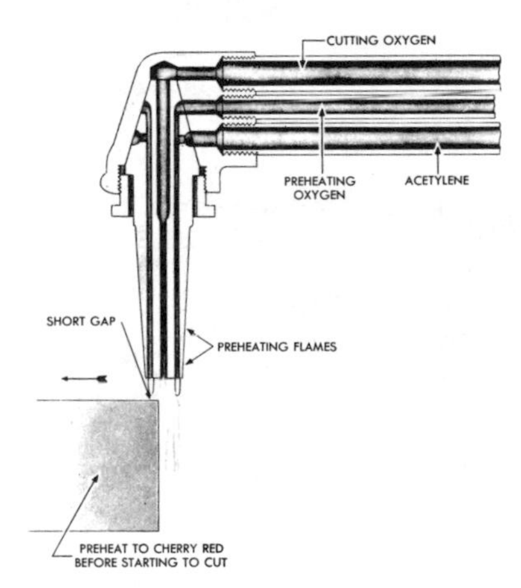


**Fig. 8.10.** The cutting torch has an oxygen jet surrounded by preheating flames.

The thickness and type of material to be cut will determine the tip size. Best results are obtained when the cutting oxygen pressure, cutting speed, tip size, and preheating flames are controlled to give a narrow, clean cut. Cuts that are improperly made will produce ragged and irregular edges, with slag adhering at the bottom of the plates.

**Machine Cutting.** Although manual flame cutting is entirely satisfactory for a wide range of cutting operations, machine flame cutting is finding increased application. Machine flame cutting provides greater speed, accuracy, and economy. The machines can be used for straight-line cutting, circle cutting, plate-edge preparation, and shape cutting.

STRAIGHT-LINE AND CIRCLE CUTTING. Machine flame cutting consists, in its simpler form, of mounting the cutting torch on a carriage driven by a variable-speed motor. This unit is mounted on a track for straight cutting. Mechanized flame cutting often replaces machine cutting. Where extreme accuracy demands machine-tool finishing, the cut can be made to remove all but a small finishing cut.

Most portable cutting machines are made to work both on a straight-line track or in a circular path. Various sized circles can be cut by the setting made on a radius-rod attachment.

SHAPE CUTTING. Shape cutting refers to any contour cutting of metal. It can be accomplished by freehand manipulation of the torch, but this is not generally satisfactory, except for very rough work. The newer developments in this field are the *optical tracer* and *numerical* controls. Other methods of template tracing are also used.

*Photoelectric tracers.* A photocell tracer (Fig. 8.11) is, in essence, an electric eye that guides the motorized unit used to control the path of the cutting torch. It is possible to trace simple pencil or ink sketches of intricate shapes. Kerf width is provided for so that it does not have to have an oversize drawing. Full or reduced-size drawings may be used. The optical tracer scanning a reduced-scale template has its signals transferred through multipliers. This system is generally referred to as ratio control.

**Fig. 8.11.** Signals to command the torch carriage drive are given by the optical tracer as it scans the drawing. Courtesy Air Reduction Sales Company.

*Numerical control.* Numerical control is now gaining momentum as a means of directing the cutting torches. Tapes are prepared either from a computer output or by manual programming. They are then fed into a director that is designed specifically for cutting machines. The director generates signals that command the machine for cutting, or rapid traverse, and all other necessary functions to produce an automatic operation (Fig. 8.12).

**Fig. 8.12.** A large-type numerically controlled multiple-torch cutting machine. The rectangular box to the right of each torch is an electromechanical sensor which maintains proper tip to work distance. Courtesy Air Reduction Sales Company.

*Template tracers.* Although not quite so easy to use as a photocell, templates are still widely employed as a means of guiding the cutting torch. Strip templates can be easily formed and used, with a special tracing head to guide the cutting torch. Magnetic-type tracer heads are used on solid ferrous-metal-type templates. Simple templates for hand guiding of the tracer head can be cut out of plywood.

STACK CUTTING. Considerable time can be saved in cutting a number of identical parts by stacking them and cutting them all in one pass.

## PHYSICAL AND METALLURGICAL EFFECTS OF TORCH CUTTING

Flame cutting of mild steel has very little effect on the metal adjacent to the cut. However, as the carbon content or alloys are increased, there will be some hardening of the edge due to the quenching action of the adjacent metal and the atmosphere. The hard edges may be difficult to machine, and the lowered ductility of this hard layer may cause cracking under load. The best method of avoiding this condition is to preheat the metal. Medium-carbon steels should be heated from 350°F to 700°F, and low-alloy high-tensile-strength steels from 600°F to 900°F.

Heavy plates do not warp when flame cut, but plates 1/2 in. or less in thickness may have to be clamped or the amount of cutting done at one time may have to be restricted. Judicious postheating with a torch can be effective in reducing or eliminating warpage.

**Cutting Chemically Oxidizing Metals.** Some metals are very difficult to cut with the oxyacetylene torch due to the chemically oxidizing flame that forms from the high-melting-point, refractory oxides in the cut surfaces. Common metals of this type are aluminum, stainless steel, and cast iron. Several methods have been used to overcome this problem, but the best solution now is the plasma torch.

## PLASMA TORCH PRINCIPLES

**Formation of Plasma.** Plasma is frequently referred to as the fourth state of matter since it exhibits properties not found in solids, liquids, or gases.

An electric-arc plasma is formed in a gas vapor when there is a strong potential gradient between two electrodes. Accelerating electrons emitted from the cathode collide with and excite atoms in the gas.

The excitation may take the form of complete ionization, orbital displacement of an electron (Fig. 8.13), or merely increased kinetic energy. Electrons freed by ionization are accelerated in turn, which causes more collisions and additional ionization. The number of charged

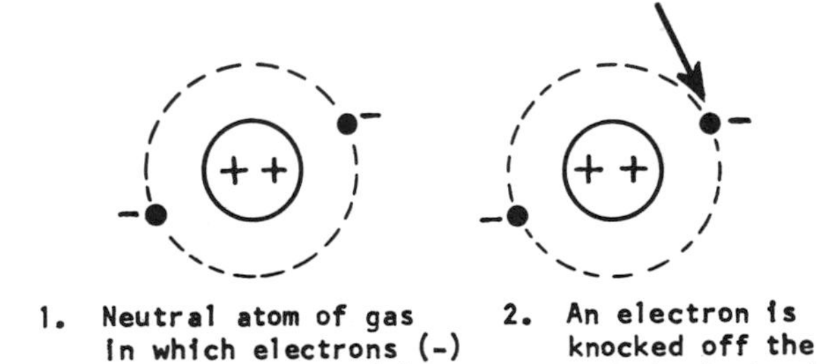


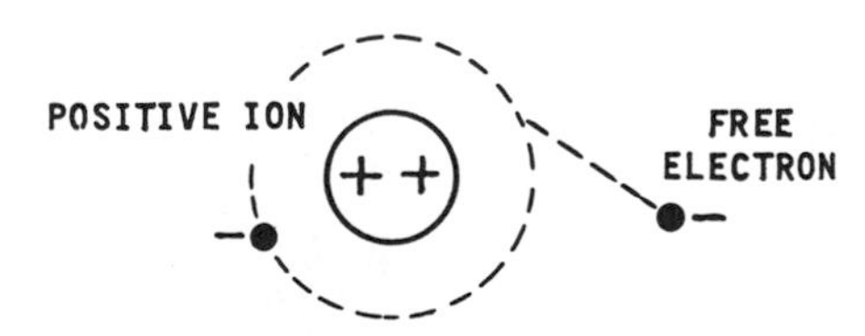


**Fig. 8.13.** Ionization of gas.

particles in the gas increases rapidly until it becomes conductive, allowing a spark to pass between the electrodes. The spark marks a tremendous increase in current flow and electron emission, which raises the temperature of the plasma, making still greater current flow. The process is limited only by the external circuit.

**Thermal Pinch.** The boundary layer of gas near the walls of the torch remains relatively cool with the discharge current concentrated at the warmer, more conductive, central region of the plasma. This effect is often referred to as thermal pinch (Fig. 8.14).

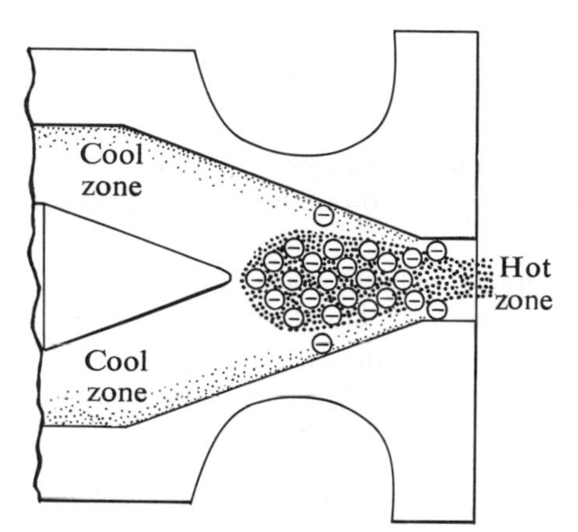


**Fig. 8.14.** The diagram at left illustrates the thermal pinch effect. Electrons move more freely in the hot zone, which is in the center of the plasma, than in the cool zone of boundary layer gases. Therefore, the majority of the electrons and the discharge current are concentrated in the hot zone.

**The Plasma Torch.** There are two main types of plasma torches: transferred-arc and nontransferred-arc. The nontransferred-arc has both electrodes in the torch as shown in Fig. 8.8 as it is used for metal spraying. The transferred-arc torch shown in Fig. 8.15 uses an external workpiece as one of the electrodes. This type is preferred for metal cutting as a greater portion of the energy is transferred directly to the workpiece.

**Plasma-Flame Cutting.** Plasma flame cutting is also referred to as constricted-tungsten-arc cutting.

Recently an improved nozzle has been developed for plasma-arc cutting torches (Fig. 8.16). The early torches for cutting steel had com-

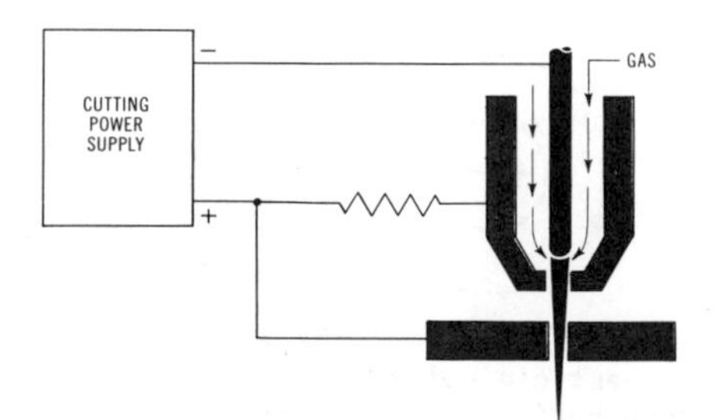


**Fig. 8.15.** Schematic view of the transferred-arc plasma torch. The work is subjected to both plasma heat and arc heat. A thoriated tungsten electrode is used with direct current, straight polarity.

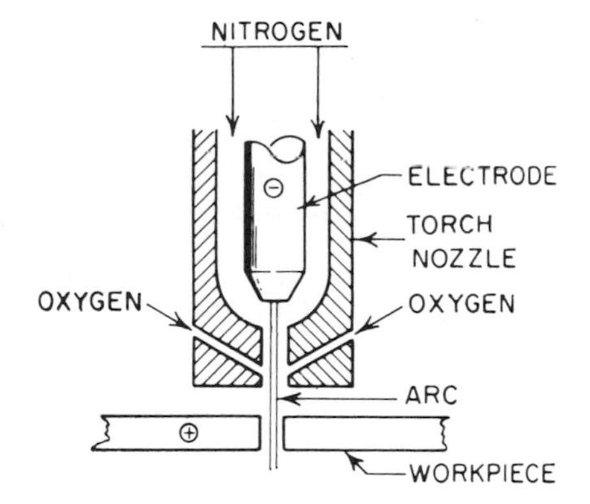


**Fig. 8.16.** The newer plasma cutting-torch design permits the addition of oxygen into the plasma below the electrode, which makes it last much longer.

pressed air in constant contact with the tungsten cathode. This was extremely errosive, and the cathode had to be replaced after every three or four hours of operating time.

Nitrogen surrounds the electrode in the new design and is the plasma-forming gas. Oxygen is added to the plasma below the electrode, providing a highly reactive cutting gas.

The plasma torch has several advantages over the fuel-oxygen torch:

(1) It can cut faster and leave a narrower kerf. Mild steel 3/4 in. thick can be cut at 70 ipm, compared to oxygen cutting of 20 ipm.

(2) Operating costs are less. Figure 8.17 shows comparative operating costs between the two methods. The ratio of savings in favor of the arc is now about 3:1. This figure is exclusive of labor rates.

(3) The plasma flame can be used to cut any metal, since it is primarily a melting process.

(4) Plasma-flame energy seems to be unlimited. The greater the power used, the greater the volume of kerf material that can be removed.

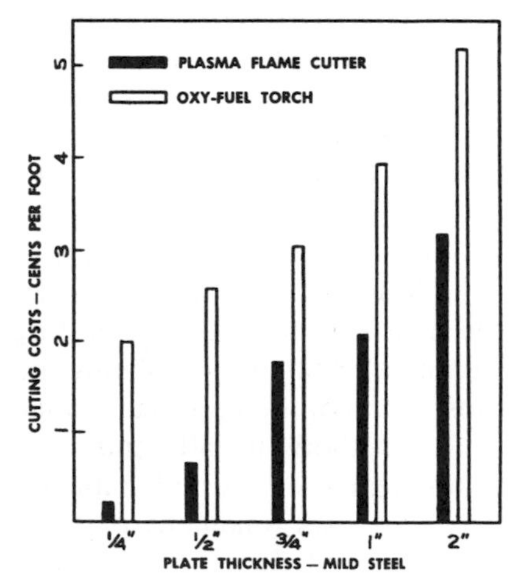


**Fig. 8.17.** Comparative cutting costs between the plasma flame cutter and the oxy-fuel torch. Courtesy Thermal Dynamics.

Oxy-fuel cutting is limited to the maximum temperature of the chemical reaction (burning).

The main disadvantage of the process is the high initial cost of the equipment. Thus only those who do quantity cutting will be able to amortize the cost in a reasonable time.

## LANCE CUTTING

An oxygen lance is a simple device consisting essentially of a length of steel pipe (usually 1/8- or 1/4-in. size), a length of hose, some couplings, a control valve, and an oxygen tank, complete with regulator.

With this equipment no preheating flame is provided, so an auxiliary torch is needed. After heating the metal to the kindling temperature, the lance is brought over to start the cut. Other methods are also used to obtain the heat necessary to start the cut, such as placing a red-hot piece of steel on the starting point or heating the end of the lance until it is red hot. When it is brought in contact with metal and the oxygen is turned on, the end of the pipe will burn brilliantly, furnishing enough heat to start the cut.

The oxygen lance is an excellent tool for piercing holes in steel. For example, a hole 2 1/2 in. in diameter can be cut in 1-ft-thick steel in a matter of 2 min. It is routinely used in tapping blast and open-hearth furnaces.

## NONDESTRUCTIVE TESTING

Nondestructive testing usually refers to an evaluation of quality characteristics and reliability that are beyond the scope of the visual inspection. The characteristics inspected are those that affect product performance and safety. Therefore, to avoid failure of the finished product, certain standards must be set for each type of test.

Nondestructive tests are those that do not impair the function of the part or the material. They are used to reveal imperfections or faults in the material or in the fabrication of the product. Imperfections are classified, according to origin, as: *inherent,* resulting from melting and solidification of the materials; *processing,* resulting from fabrication of the finished product, as in forging, drawing, welding, grinding, heat treatment, rolling, and plating; or *service,* cracks resulting from use, the most common being fatigue cracks.

The methods usually employed to reveal these imperfections include radiography (x-rays and gamma rays), sonic and ultrasonic imperfection, magnetic tests (magnetic particles and eddy currents), penetrant inspection (dye and fluorescent penetrants), and leak tests (oil and kerosine, soap bubbles, freon, halide, mass spectrometer).

## RADIOGRAPHY

Radiography is basically a method of taking pictures. Instead of using visible light rays, the radiographer uses invisible short-wavelength rays developed by x-ray machines, radioactive isotopes (gamma rays), and variations of these methods. These rays penetrate solid materials—metals, glass, wood, plastic, leather, and ceramics—and reveal their defects on a film or screen. Flaws show up on films as dark areas against a lighter background (Fig. 8.18).

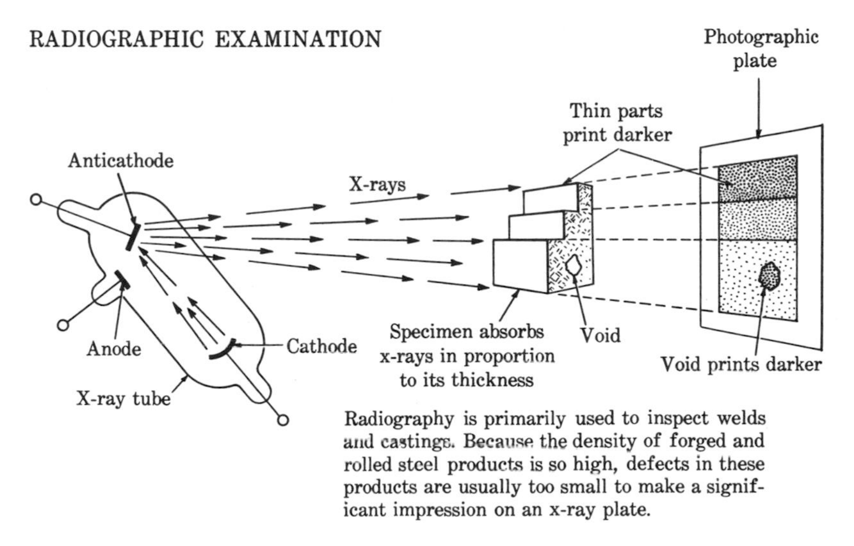


**Fig. 8.18.** Principles of industrial radiographic examination. Courtesy *Steelways*, published by American Iron and Steel Institute.

Four main types of film, based on grain size and speed, are used in radiographic recording. Fine grain offers high quality with excellent contrast. High-speed films are used for the examination of steel, brass, etc., with machines of limited kilovoltage.

**X-ray Machines.** The x-ray machine is used mainly to inspect the interior soundness of various materials. On welds, it is used to detect gas porosity, slag inclusions, incomplete penetrations, cracks, and burn-through. It is used, to a lesser extent, on castings, forgings, and cold-formed parts.

The unit consists basically of a high-voltage transformer, an x-ray tube, and controls. These machines are classified according to their maximum voltage (designated as kVp or kilovolt peak) which may range from 50 kVp (50,000 volts) to 15 to 24 MeV (15 to 24 million electron volts). They will handle up to 20 in. of steel or its equivalent.

Selection of the proper machine size will depend on the thickness and density of the material, its absorption characteristics, the time available for inspection, and the location of the parts to be inspected.

A 50-kVp unit can be used on very thin sections. A 150-kVp unit can be used on 5-in. aluminum, or its equivalent in light alloys, and on 1-in. steel, or its equivalent. A lead-foil intensifying screen is sometimes used with a fluorescent (calcium tungstate) intensifying screen. With this setup the strength of a 150-kVp unit is stepped up enough to check steel 1 1/2 in. thick. Other sizes include 250, 400, 1,000, and 2,000 kVp and 15 to 24 MeV, the largest being capable of penetrating 16-in.-thick steel with a lead-foil screen, and 20 in. with a fluorescent screen.

**Gamma Radiography.** Gamma rays are used for the same industrial purposes as x-rays, but there are two main advantages – portability and low cost. Gamma-ray machines are more portable because of their smaller size, and they do not require an outside source of power, or water or oil cooling. Thus they are available for smaller and more complex jobs. They can also cover a wider range of metal thickness with the same radiograph.

This does not mean that gamma rays are preferable to x-rays for every application. X-rays are generally better for the thinner materials. Also, the radiation source can be turned off when not in use. Trained technicians are needed to operate either type of equipment.

Up until 1942 the only source of gamma rays was radium; however, with the advent of nuclear fission, the door was opened to several types of man-made isotopes. Gamma rays have various wavelengths, which may be compared to x-rays generated at different voltages. The source of the radioactive rays is, however, constantly decaying. The period in which the energy of the isotope decays to half its original value is called its *half-life*. Radium, for example, has a half-life of 1,620 years.

There are five radioactive materials commonly used as gamma-ray sources. These are radium, cobalt 60, iridium 192, thulium 170, and cesium 137. Each of these materials has particular applications. Cobalt 60, for example, is equivalent to x-rays of about 1.3 mev, with a half-life of 5.3 years. Its capacity range on steel is from 1 to 6 in. Its advantages include high specific gravity, low cost, and a wide range of applications.

**Related X-ray Techniques.** Related x-ray techniques include fluoroscopy, x-ray television systems, and xeroradiography.

FLUOROSCOPY. Fluoroscopy differs from radiography in that the image is viewed directly on a fluorescent screen rather than on a film. This technique offers fast, low-cost, reasonably sensitive inspection. Because the part can be viewed from all directions, the overall quality level may be equal to that of radiography. The initial cost of the equipment may exceed the cost of a radiographic setup.

TELEVISED X-RAY. Televised x-ray is a direct-viewing, remote, instantaneous system. Materials to be inspected are x-rayed and viewed simultaneously on a remote monitor screen. No film is used. Images can be viewed in normal light. The image size can vary from one-half to three times the original object.

XERORADIOGRAPHY. Xeroradiography is a rapid method for dry processing x-ray images. It will frequently disclose discontinuities that cannot be seen on film. Images can be viewed in less than 1 min after exposure.

## SONIC AND ULTRASONIC INSPECTION

**Sonic Inspection.** Sonic inspection consists of tapping the specimen with a hammer and listening to the sound waves through a stethoscope. The blow by the hammer sets up both natural and forced vibrations. Sounds will be similar in good sections, but a high-pitched, reedy sound will occur if a defect is present. The disadvantage of this method is that the defective area is not located exactly, and the findings depend largely on the operator's skill.

**Ultrasonic Inspection.** Ultrasonic flaw-detection equipment makes it possible to locate very small checks, cracks, and voids too small to be seen with x-rays. Variations of the technique permit inspection of sheets, forgings, shafts, and even bars up to 42 ft long.

Many new applications of ultrasonic inspection are being developed. The object here is only to present some of the basic operating principles. There are two principal inspection techniques: *reflection testing,* in which the echoes of the pulse are used to detect discontinuities, and *through-transmission testing,* in which two transducers are used to indicate the presence and extent of discontinuities.

REFLECTION TESTING. The ultrasonic inspection unit consists of a pulse generator or transducer which sends out high-frequency electrical pulses to a crystal. The crystal sends a pulse and receives an echo that is then amplified to show up either as a straight line or as a pip on the cathode-ray oscilloscope screen. A pip indicates a defect (Fig. 8.19).

Ultrasonic waves are not transmitted through air, so a suitable coupler must be used to couple the searching unit to the work. Oil, grease, glycerin, water, or other similar liquids are suitable for this purpose.

Ultrasonic inspection is set up for continuous automatic operation, as shown schematically in Fig. 8.20. Units of this type have been used to inspect tubing that has a continuous weld. The unit not only detects the flaw but also marks it. Poor welds, caused by improper adjustment, laminations, etc., are brought to the attention of the operator by a signaling horn. Immediate detection makes possible prompt adjustment, reducing scrap loss. Instantaneous marking assures 100 percent inspection even at speeds up to 100 fpm.

THROUGH-TRANSMISSION TESTING. In through-transmission testing, the transducers are arranged coaxially facing each other on opposite

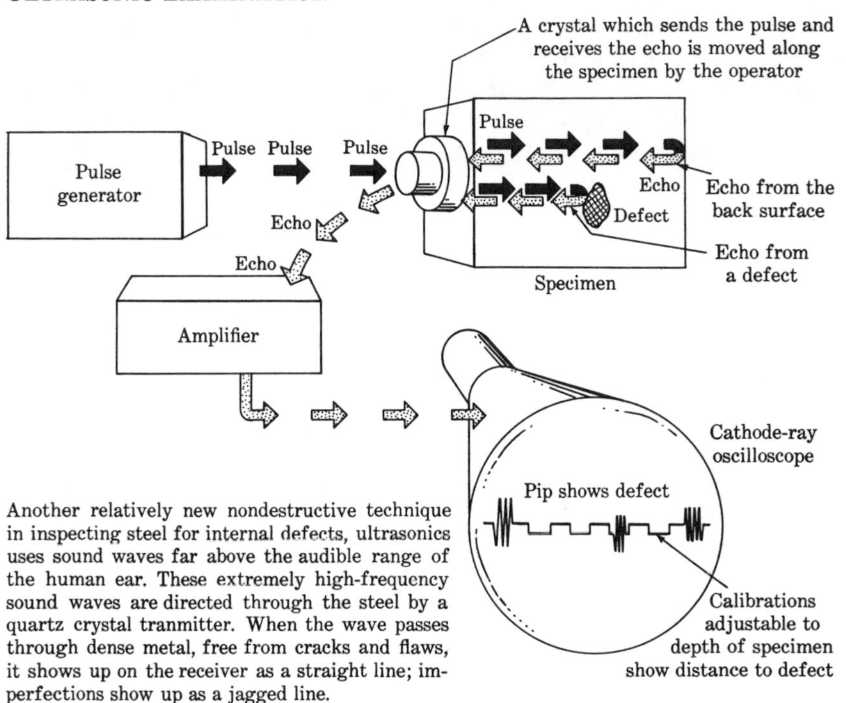


**Fig. 8.19.** The ultrasonic inspection unit consists of a pulse generator or transducer, a transmitter, amplifier, and cathode-ray oscilloscope. Courtesy *Steelways*, published by American Iron and Steel Institute.

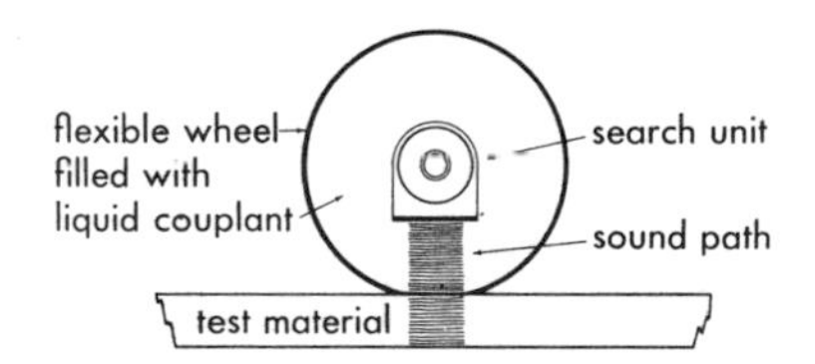


**Fig. 8.20.** A schematic view of continuous-type ultrasonic testing used on seam-welded pipe at rates up to 100 fpm.

sides of the test piece. Pulses are received by the second transducer. A reduction in amplitude of the received signal indicates the presence and size on discontinuities, but not their location in depth (Fig. 8.21).

A portable ultrasonic thickness tester is shown in Fig. 8.22. It measures the thickness from one side by determining the resonant frequency of vibration in the thickness direction.

A transducer probe, vibrating at ultrasonic frequencies, is placed in contact with one side of the material to be tested. The sound waves travel through the material and are reflected back by the opposite surface. When the transmitted and reflected waves are in phase, "resonance

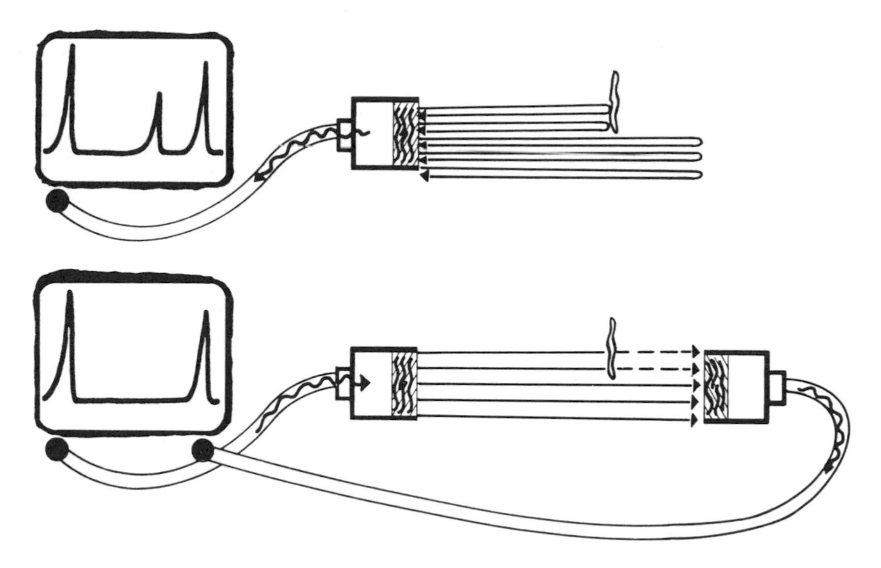

**Fig. 8.21.** Schematic diagrams to show both *reflected* (top diagram) and *through* ultrasonic-testing techniques. Courtesy Branson Ultrasonic Corp.

conditions" are established. Since the velocity is a known constant, the fundamental frequency required to produce resonance is an accurate and reliable measure of unknown thickness. The signal produced is heard on a headset, and on some models the thickness of the material is indicated directly in inches.

Reflections will occur at any discontinuity, and the resonance pattern will change when the audigage is tuned for the thickness of a laminated section. Voids may also be detected in bonded material by the same method.

In summary, ultrasonics can be used on a variety of materials, with only negligible loss in great thicknesses. Angular cracks can be readily

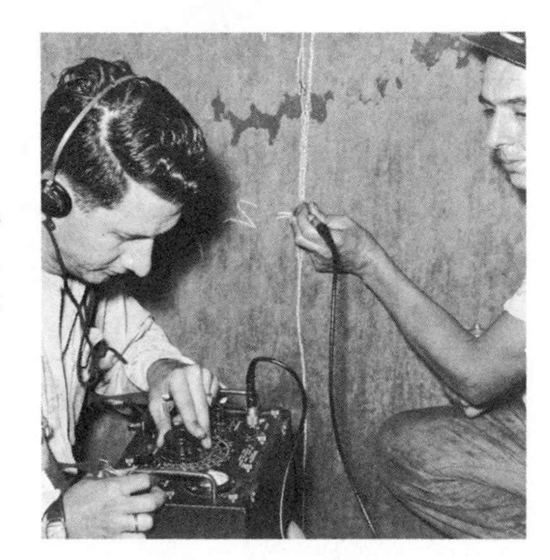

**Fig. 8.22.** An ultrasonic thickness tester used to inspect the extent of corrosion on a paper-pulp digester. Courtesy Branson Ultrasonic Corp.

found. The initial equipment is considered low in cost, as is the cost of necessary supplies.

## MAGNETIC TESTS

**Magnetic-Particle Inspection.** Magnetic-particle inspection is used to find nonvisible cracks at or near the surface on ferrous materials. The method utilizes magnetized particles (usually iron powder), in *dry* or *wet suspension,* applied to a workpiece in which an intense magnetic field has been formed. Breaks or flaws in the magnetized part set up local magnetic fields. These local fields, and the flaws that cause them, can be revealed by the way in which the iron powder is attracted to them.

The parts can be magnetized by either direct or alternating current. The former finds subsurface defects and is commonly used for inspecting castings and welds. Alternating current is usually used to check finished machined parts.

WET-SUSPENSION METHOD. The wet-suspension method uses a paste or iron powder and oil or water. The part is dipped into this magnetic-particle paste to show tiny surface defects. Fluorescent paste is often used with this method. An ultraviolet light causes the defect to stand out brightly. Figure 8.23 shows a part as it appears before inspection, after treatment with a wet suspension, and, lastly, with a fluorescent paste.

**Fig. 8.23.** A kingpin immediately after grinding, treated with a wet magnetic paste and a flourescent paste, and brought out with an ultraviolet light. Excessive heat during grinding had caused dangerous cracks. Courtesy Magnaflux Corp.

DRY-SUSPENSION METHOD. The dry method is more sensitive to subsurface defects. Magnetic powder is dusted or blown over the part to be inspected.

EQUIPMENT. Magnetic-particle inspection is done with either stationary or portable equipment. Stationary equipment is used on smaller parts for hand or automatic inspection. Portable equipment ranges from small hand-held magnetic yokes to large 6,000-amp heavy-duty power units. Some equipment is specially designed for automatic conveyorized production lines.

Magnetic-particle inspection does not show cracks that are parallel to the magnetic field, so magnetism in two directions is needed to show all discontinuities.

**Eddy-Current Testing.** Eddy currents are used to detect cracks and porous or embrittled areas and also to locate welds and sudden changes in both ferrous and nonferrous metals.

Simply stated, eddy currents are composed of free electrons, made to drift through metal under the influence of an induced electromagnetic field. The minute eddy currents explore the part and give back information as to cracks, porosity, or any discontinuity in the metal. The eddy currents act as compressible fluids; thus, when they encounter a discontinuity, they detour around it and, as a result, are compressed, delayed, and weakened (Fig. 8.24*a*). This causes a relatively large electrical reaction in the coil, which can be amplified and reflected on a cathode-ray tube or by other means.

Some instruments use a hand detector pickup (Fig. 8.24*b*) to determine electrical conductivity in absolute units, with a readout on a meter scale. Since other properties, such as hardness, alloy proportions, ther-

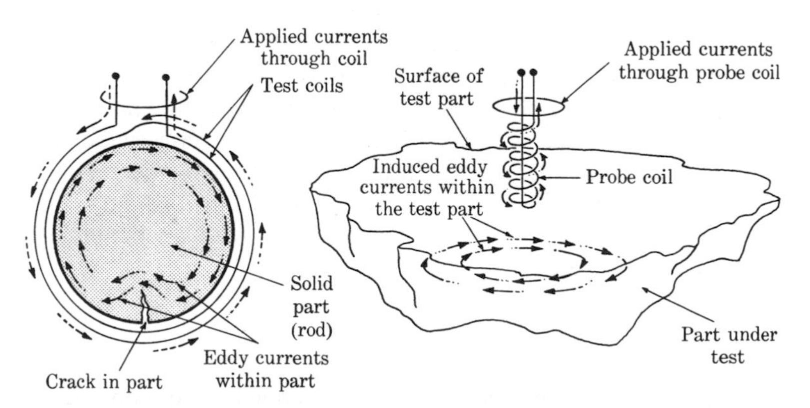


(a) A part is passed through a test coil. The applied current in the coil induces eddy currents within the part. These are affected by a crack, or other change in the part.

(b) Electric currents in the probe coil induce eddy currents within the test part. These react back on the applied current, for "readout."

**Fig. 8.24.** Two methods of using eddy currents to inspect a part for flaws or other characteristics. Courtesy Magnaflux Corp.

mal conductivity, etc., are related to electrical conductivity, these instruments have a wide variety of uses.

## PENETRANT INSPECTION

Penetrant inspection is usually used to examine nonporous materials for defects that are open to the surface.

Surface defects that can be located include all types of cracks (connected with welding, forging, grinding, shrinkage, and fatigue), porosity, seams, laps, cold shuts, and lack of bond between two metals. There are two kinds of penetrant inspection—*dye* and *fluorescent*. In both, the penetrant is applied by brushing, spraying, or dipping.

**Dye Penetrants.** In dye-penetrant inspection, the penetrant is applied to a clean, dry surface and allowed to soak for a while. The dye penetrates surface defects. Excess dye is wiped off, and a thin coating of developer chemical is applied. The developer acts like a blotter, drawing the dye to the surface. Contrast between the color of the developer (usually white) and the dye penetrant (usually red) outlines the surface flaws. The red area will indicate the extent of the crack or defect.

Commercial dye-penetrant kits are made up in pressurized cans. This makes it a very portable inspection tool that can be used anywhere in the plant or field.

Both dye and developers have been made in water-wash formulas. This aids in removal and reduces cost in moderate- to high-volume inspection.

**Fluorescent Penetrants.** In the fluorescent method the solution is applied just like the regular dye penetrant, but the surface is inspected under a black light. Cracks or flaws glow brightly under this light. A black light is one that is near the ultraviolet range in the spectrum.

The black-light source is normally a 100-watt mercury-vapor bulb of the sealed-reflector type, used with a suitable transformer. The transformer unit allows the black light to be plugged into ordinary 110-volt 60-cycle-per-second circuits. The complete kit, with an example of its use, is shown in Fig. 8.25.

## LEAK TESTS

A number of techniques are used for leak checking, depending on the degree to which the leaks are tolerated. Most methods involve the passage of a "tracer" medium from one side of a pressurized leak to the other and the subsequent detection of the medium on the latter side.

**Oil and Kerosine.** Oil and kerosine are often used to check leaks in cast iron. Inhibitors or fluorescent substances are placed in the oils to make any penetration show up brightly when examined with an ultraviolet light.

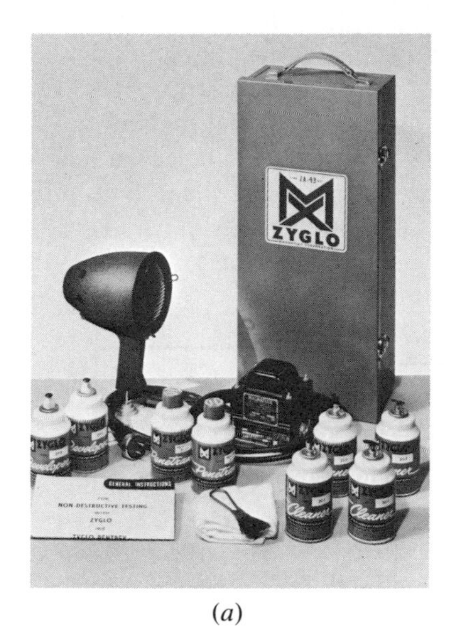


(*a*)

(*b*)

**Fig. 8.25.** (*a*) Fluorescent-penetrant inspection kit. (*b*) Crack in aircraft aluminum landing-gear cap. Courtesy Magnaflux Corporation.

**Soap Bubbles.** Small vessels may be pressurized and submerged to detect leaks. Larger vessels can be coated with a mixture of glycerin and soap in the area to be tested. Air pressure from the inside will cause bubbles to form in the area of the leak. Air testing can be very hazardous and should not be attempted without careful safety precautions.

**Freon.** Extremely small leaks are checked with an inert gas, usually Freon-12. The areas to be checked are "sniffed" or probed with a sensitive sampling device. The presence of a leak is indicated on a meter or by an audible alarm. It is possible to measure leaks quantitatively by using set standards.

**Halide.** A variation of the preceding method is to have the areas "sniffed" with a halide torch. The presence of a leak shows up in the change of color in the torch flame.

**Mass Spectrometer.** The most sensitive leak-test detection method devised is the use of the mass spectrometer. It measures the rate of tracer gas, usually helium, through leaks. This sensitive instrument is used to convert any helium gas coming through the leak into an electrical signal. The location and size of the leaks are determined on the basis of the signal received.

**Ohmmeter.** NASA has developed a simple leak test for tanks that relies on the law of electrical resistance.

A strip of water-soluble paper is placed outside of the tank on the weld to be tested. A strip of aluminum foil is placed over the paper. Both strips are fastened to the tank with adhesive tape. One ohmmeter contact is made to the tank wall and the other to the aluminum strip (Fig. 8.26).

An initial reading is taken on the ohmmeter. If a leak occurs during the hydrostatic test, the paper will absorb water, and the resistance in the circuit is lowered. The exact location of the fault is found by removing the foil strip and finding the water stain on the paper. The method is considered safe since no personnel are needed near the tank during the hydrostatic test.

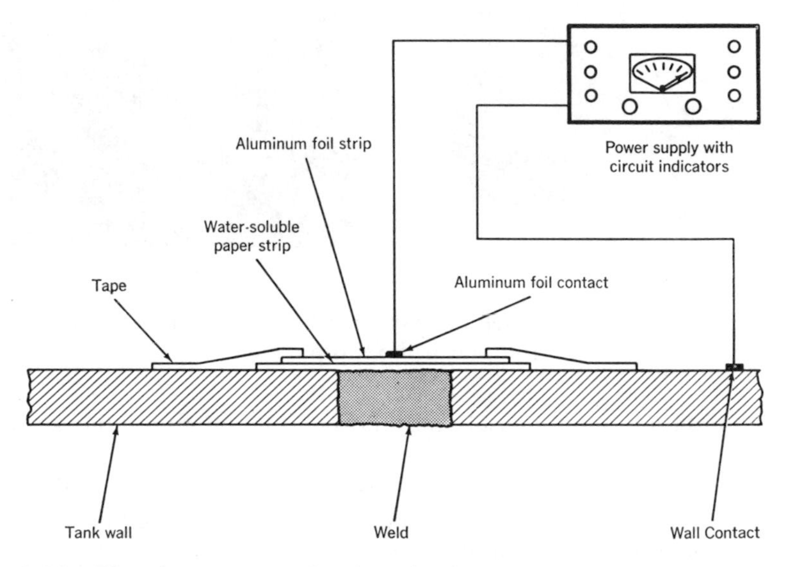


**Fig. 8.26.** The ohmmeter test for detecting leaks.

## QUESTIONS

**8.1.** How is acetylene obtained?

**8.2.** Give what you consider to be the main advantages and disadvantages of gas welding.

**8.3.** What is the main difference between welding and brazing?

**8.4.** Why is the joint clearance so significant in brazing?

**8.5.** How does braze welding differ from brazing?

**8.6.** How are tungsten carbide particles and other hard materials made to adhere to a metal surface?

**8.7.** What is the principle involved in preparing a surface for a flame sprayed metal coating?

**8.8.** Why can steel be cut quite easily with the proper oxyacetylene torch?

**8.9.** What are three methods used to make machine flame cutting accurate?

**8.10.** How can hard surfaces be avoided when medium carbon or alloy steels are cut?

**8.11.** Why are some nonferrous materials difficult to cut with a torch?

**8.12.** Explain what is meant by plasma.

**8.13.** What is meant by thermal pinch?

**8.14.** What improvement has been made on plasma cutting torches?

**8.15.** What do you consider to be the two outstanding advantages of the plasma torch as used in cutting?

**8.16.** Explain the difference between radiography and x-rays.

**8.17.** What is the principle of ultrasonic inspection?

**8.18.** How can the thickness of a part be checked from one side ultrasonically?

**8.19.** Explain the difference between the wet and dry suspension methods of surface inspection.

**8.20.** Explain the principle of penetrant inspection.

**8.21.** How can an ohmmeter be used to check for leaks?

## PROBLEMS

**8.1.** Flame machining can be very competitive to conventional machining, especially where the tolerance is greater than about 0.030 in. Find the cost comparison for cutting 1-in.-thick mild steel under the following conditions: (*a*) oxygen = 130 cu ft/hr; (*b*) acetylene = 12 cu ft/hr; (*c*) cut = 14 in./min; (*d*) preheat time = 0.5 min; (*e*) band sawing = 1½ in./min; (*f*) 3 cuts, each 30 in. long; (*g*) labor and overhead = $10/hr; (*h*) preheat flow = 1.25 oxygen to acetylene; (*i*) oxygen cost = 0.03/cu ft; (*j*) acetylene = 0.01/cu ft.

**8.2.** If the flame-cut surfaces in the previous problem were to be machined on the following basis, how much would the cost be? Original setup time = 30 min; additional setup time/part ≐ 10 min; milling approach and overtravel = 4 in; feedrate = 8 in/min.

**8.3.** A convenient method of estimating pressure drop through hose connectors is as follows:

$$Q = 1{,}300D^2\sqrt{P_2(P_1 - P_2)}$$

where $Q$ = ft³ of $O_2$/hr. at 1 atmosphere and 70°F; $P_2$ = absolute downstream pressure, psi; $P_1$ = absolute inlet pressure, psi; and $D$ = diameter.

Absolute pressure is obtained by adding 14.7 to gage pressure.

The formula is set up for $O_2$. Multiply $1.19 \times Q$ for acetylene, and multiply $0.85 \times Q$ for propane.

*Problem:* The pressures measured on each side of an oxygen hose connector are 90 psi on the inlet side and 86.7 psi on the downstream side while 0.125 inches = orifice diameter in the connector. What is the oxygen flow rate?

**8.4.** Using the data in Table P8.4, find the comparative (actual torch time) cost of making 10 cuts across a $4' \times 8'$ plate 1 in. thick. Each cut strip will be $3'' \times 1'' \times 4'$. Cutting speed = 14 in/min. Allow for a torch idle time of 30 sec/cut.

**Table P8.4**

| | Cu ft/hr | Cost/cu ft |
|---|---|---|
| Oxygen | 130 | 0.03 |
| Acetylene | 12 | 0.01 |
| Natural gas | 25 | 0.005 |
| Propane | 12 | 0.009 |

**8.5.** Compare the cost of cutting 200 mild steel blanks that are 3/4 in. thick and 3 ft. in dia by the plasma arc and the oxy-fuel methods.

## REFERENCES

Agnew, S. A., and E. H. Daggett, "Pulsed Spray Arc Welding," *Welding Journal,* April 1966.

Anthes, C. C., "Recent Developments in Oxy-Fuel-Gas Cutting," *Welding Journal*, April 1960.

Obrien, R. L., and R. J. Wickhan, "Advances in Plasma Arc Cutting," *Welding Journal,* April 1965.

*Welding Handbook,* Section One, Fifth Edition, American Welding Society, New York, N.Y., 1962.

*Welding Handbook,* Section Three, Fifth Edition, American Welding Society, New York, N.Y., 1964.

Worthington, J. C., "Analytical Study of Natural Gas–Oxygen Cutting, Theory and Application," *Welding Journal,* March 1960.

Worthington, J. C., and A. L. Cooper, "Close-Tolerance Flame Cutting," *American Machinist,* August 7, 1961.

9

# NEWER WELDING PROCESSES

THE NEWER DEVELOPMENTS in welding have been separated here from the more basic processes so that the reader will have a better picture of the technological advances made in the past few years. These have come in response to the increasing demand for speed in fabrication and for better methods of dealing with reactive materials. Some of the processes can be considered new; others are modifications of older ones.

## ELECTROSLAG WELDING

This unique process was first developed in Russia. Granular flux is placed in the gap between the plates being welded, and, as the current is turned on, submerged-arc welding takes place in a U-shaped starting block, tack-welded to the bottom of the joint. As the flux melts, a slag blanket from 1 to 1 1/2 in. thick is formed. At this point the arc goes out, and current is conducted directly from the electrode wire through the slag (Fig. 9.1). The high resistance of the slag causes most of the heating for the remainder of the weld. When the arc goes out, the operator switches the transformer from constant current to constant potential.

This process is used to weld metals from 1 1/2 to 15 in. thick, the maximum reported thickness being 40 in. Welding is done with the joint positioned vertically, starting at the bottom and progressing upward. The molten metal and slag are retained between the work parts by means of

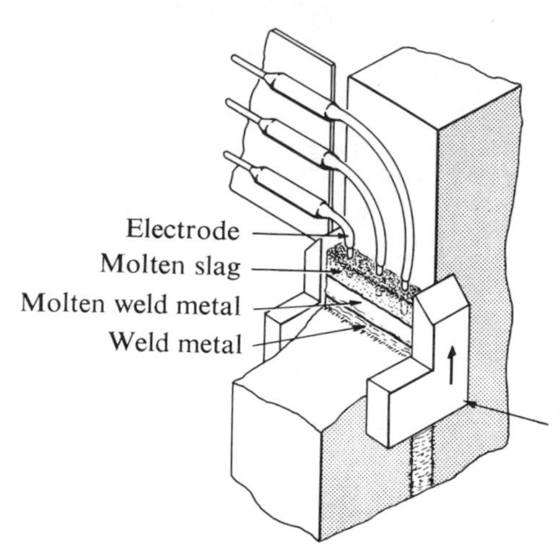


**Fig. 9.1.** Schematic view of the main elements of electroslag welding. More than one electrode can be used, as shown.

copper shoes. The water-cooled copper bar helps the metal at the bottom of the pool to solidify, thus forming the weld. Temperature sensors on the water-cooled copper shoes sense the build-up of weld metal and cause the whole mechanism to move upward as needed. Two other probes control the addition of granular flux from the hopper over the weld zone. As solidification occurs, the electrode wire is fed vertically into the slag pool at a rate somewhat less than it is used, thus maintaining the end of the wire at a constant distance from the molten metal pool. Single stationary electrodes are used for metals in the 2-in. range. By moving the electrode back and forth horizontally, welds can be made in metals 6 to 8 in. thick. When multiple electrodes are used, welds of almost unlimited thickness can be made.

**Applications.** As first developed, electroslag welding was used in the fabrication of heavy forgings and castings. Today Russian industry considers it to be the preferred method for automatic welding. It is also being used in Europe and the United States. Common types of joints for which electroslag welding is adapted are shown in Fig. 9.2. The minimum practical thickness of weld is 1 inch.

**Vertical Submerged-Arc.** Vertical submerged-arc welding is similar to electroslag welding except that the flux and alloying elements are added by a stationary consumable guide tube (Fig. 9.3). The steel tube has a 3/16 in. ID and a 5/8 in. OD on which a 1/16-in.-thick special flux coating is extruded. It serves as a current-carrying conductor and guides the electrode wire into the melt zone. A few ounces of submerged-arc composition are put in the bottom of the joint before the weld is started. This charge becomes molten and forms a thick protective layer over the melt zone. The process is used for joining plates from 5/8 in. to more than 6 in. thick. Preparation of the joint is simple, requiring only square edges spaced approximately 1 in. apart.

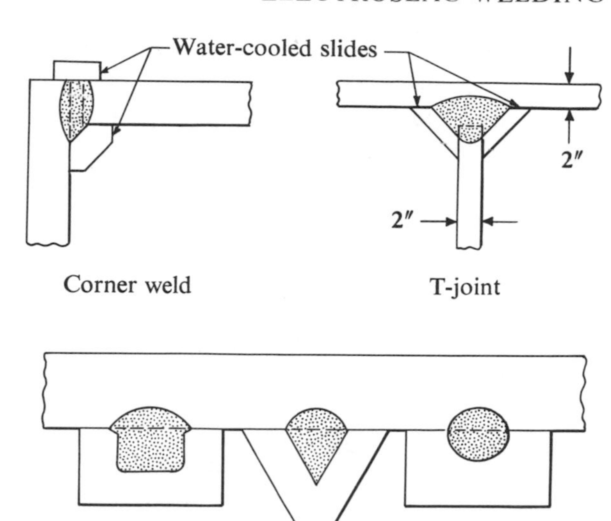


**Fig. 9.2.** Shaped slides are used in the electroslag process to make the joints shown.

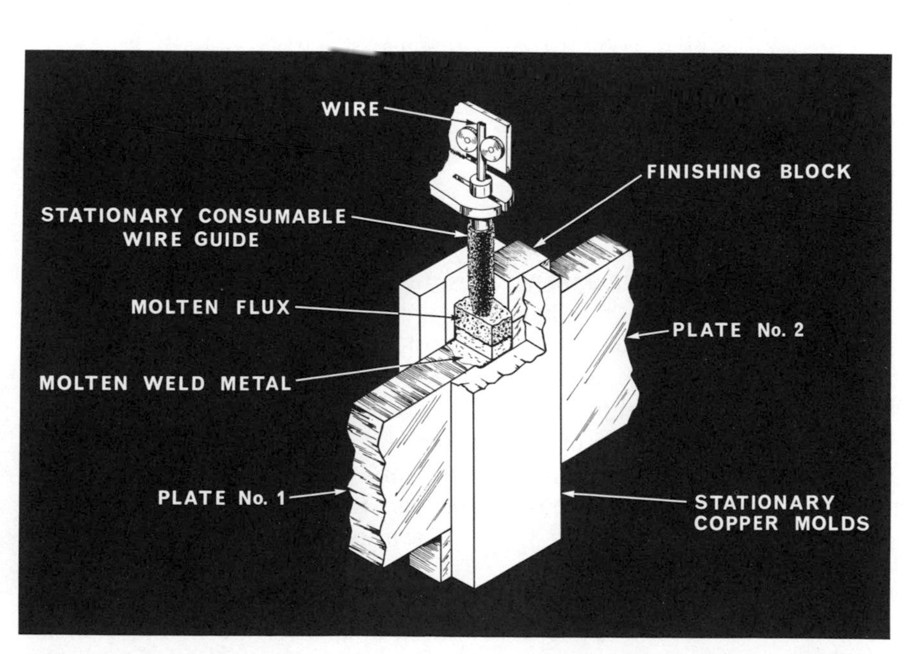


**Fig. 9.3.** Schematic view of the vertical submerged-arc process. Courtesy American Bridge.

**Electrogas.** The electrogas process is also an adaption of electroslag welding. The basic difference is that an open arc is maintained under a protective atmosphere of shielding gas (Fig. 9.4). It is used for single-pass welding of metal sections ranging from 1/2 in. to more than 2 in. thick.

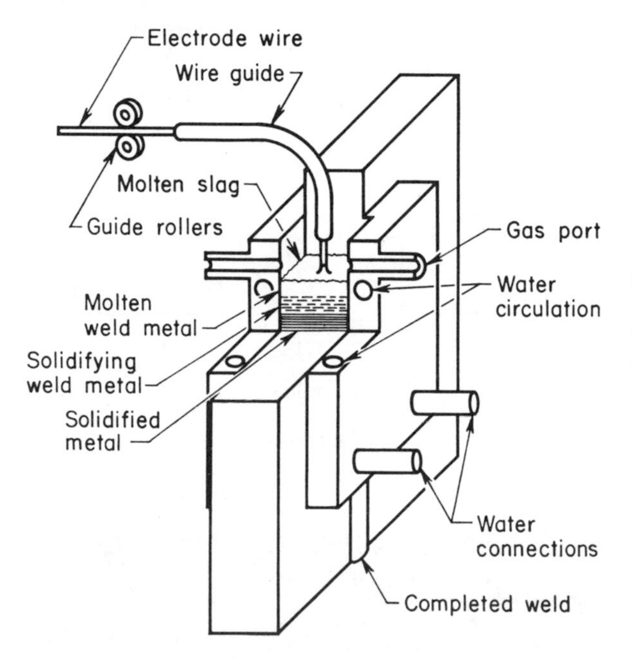


**Fig. 9.4.** The electrogas process is similar to the electroslag except that an open arc is maintained under a layer of shielding gas.

## ELECTRON-BEAM WELDING

Electron-beam welding is, perhaps, the most spectacular of the recent welding developments that have become commercially feasible. Heat to melt and fuse the metal is generated not by a flame or an electric arc but by a stream of highly accelerated electrons. These electrons are accelerated to about one tenth the speed of light, at which velocity they attain a potential temperature of millions of degrees. This energy does not appear as heat, however, until the electrons hit the work. The beam can be easily controlled in intensity and can be focused to a pinpoint or broadened to cover a wide track. Electromagnetic focusing is shown schematically in Fig. 9.5.

Either electrostatic or electromagnetic focusing is used. Electrostatic is simpler, but electromagnetic results in a narrower beam and a higher concentration of energy. Sharp focus is desirable for most welding. The out-of-focus broad circle can be extremely useful in scouring contaminants from the face of the work prior to welding. With the beam out of focus and the power reduced, the operator can track the weld seam

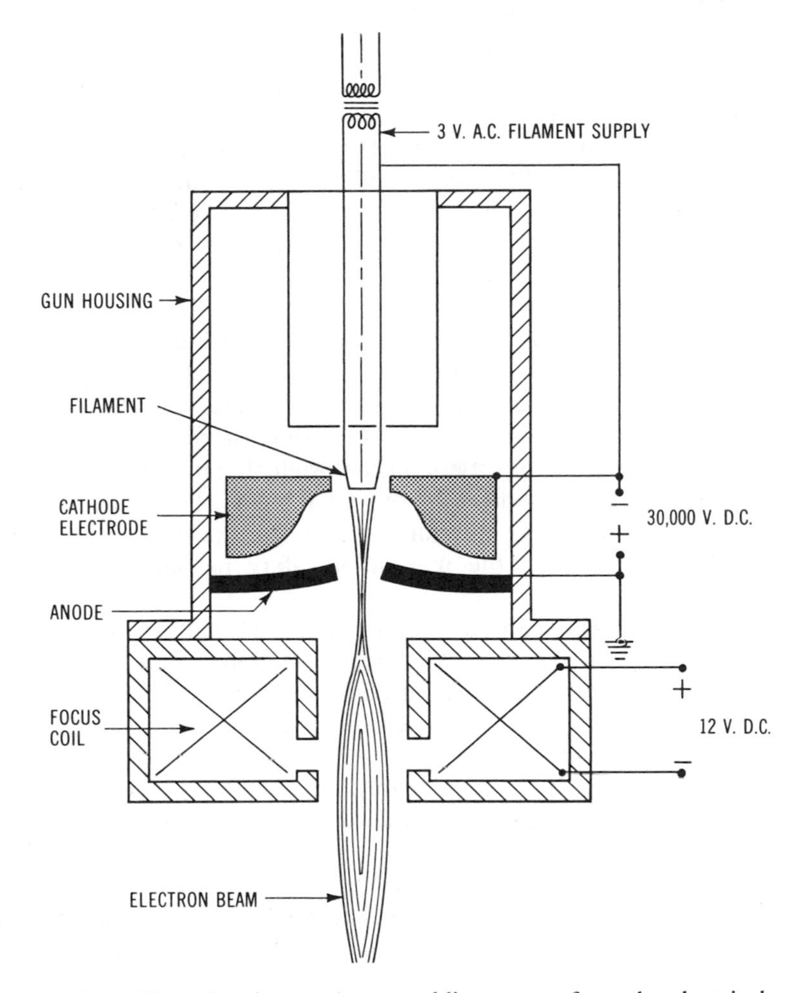


**Fig. 9.5.** Heat for electron-beam welding comes from the electrical differential between the anode and cathode. The fast moving stream of electrons is focused into a very small diameter and appears to hit the workpiece as a single column of energy. The kinetic energy is converted to thermal energy, melting and fusing any known metal. Courtesy Sciaky Bros., Inc.

beforehand to vaporize oxides and other surface impurities. These are drawn off, since the entire operation is done under a high vacuum.

The vacuum required is 1 ten-millionth of an atmosphere, or $0.1\mu$ (micron) Hg. In this vacuum, gaseous impurities total less than 1 ppm, as opposed to at least 20 ppm as the best level attainable with inert-gas-shielded methods.

**Electron Acceleration.** The source of energy for electron-beam welding is the tungsten cathode filament which, when heated to about

2,000°C, emits a cloud of electrons. The electrons are pulled or accelerated toward the workpiece by an accelerating anode. The cathode potential at the filament is fixed at 60,000 volts. The accelerating anode varies between 0 and 30,000 volts. It is normally set at about 15,000 volts to maintain a 30,000-volt potential between the cathode and the accelerating anode. The enclosure for the beam or "electron gun" is not water cooled because the beams are prevented from striking the gun walls by the field near the anode.

The x-rays produced by the 15,000 volts are soft and can be shielded by the steel and glass in the welding chamber. At higher voltages, lead shielding of the weld box and auxiliary equipment may be required.

**Applications.** The hard vacuum makes it possible to weld such highly reactive vacuum-melted materials as titanium, zirconium, and hafnium with the same control of purity as in the original material.

Owing to the close control of the heat input, the chance of burn-through is greatly reduced. Section thickness in the joint area need not be beefed up to provide the margin of safety needed with stick-arc welding.

Edge and butt welds can be made in metals as thin as 0.001 in. Also, small thin parts can be welded to heavy sections. Butt joints can be made in plates up to 1/4 in. without edge preparation.

The welded joint is usually cleaner and brighter than the parent metal. Surface cleanup is eliminated.

Electron-beam welds have a highly desirable depth-to-width ratio. Because of the small weld puddle, the heat-affected zone is substantially smaller than with TIG welds (Fig. 9.6).

Precise heat control is obtainable. When welding one-of-a-kind assemblies, the operator can gradually increase beam energy to obtain proper welding conditions. If fusion is incomplete the first time, an additional pass can be made at a slightly higher input.

More recently a non- or low-vacuum electron-beam welder has been developed. It is very much like the present welder except that the vacuum chamber in which the beam is formed is evacuated to a lower pressure. If contamination of the workpiece must be held to a minimum, a shielding gas, either argon or helium, is used.

The Plasma Electron Beam (PEB) gun is shown in Fig. 9.7. It consists of a hollow cylindrical cathode surrounded by an anode cylinder or shield. The cathode is water cooled, and the whole assembly is mounted on a vacuum flange. Plasma is formed within the cathode shield through the ionization of the gas by the high-voltage current.

Welds have been made in steel and aluminum alloys and in titanium. Single-pass penetration depths in excess of 1 in. have been made in steel. The penetration depth in aluminum is somewhat less due to the high thermal conductivity.

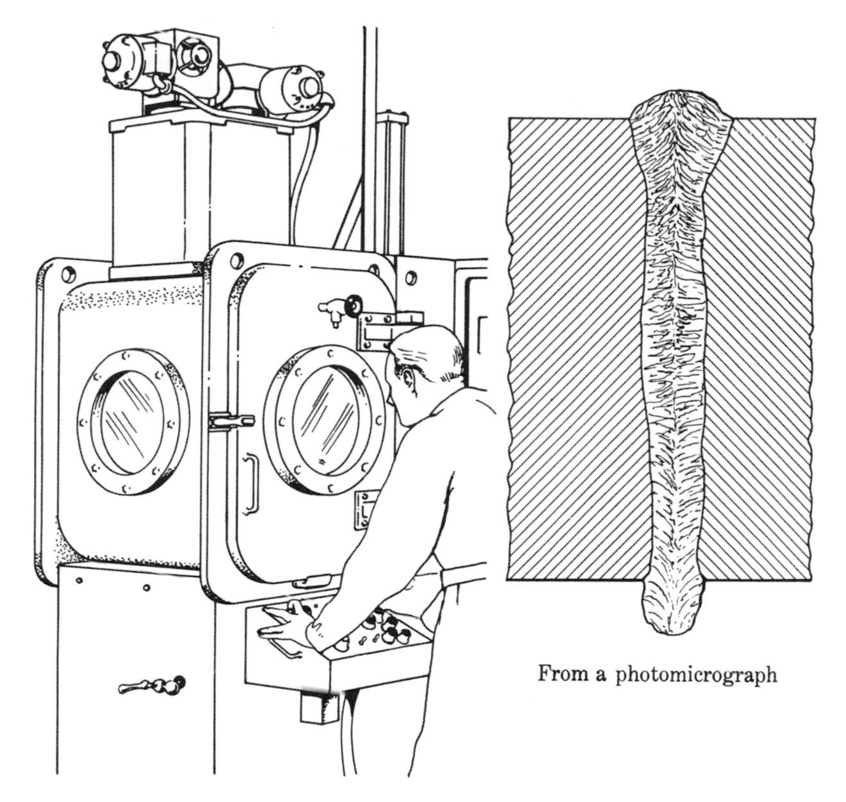


**Fig. 9.6.** Electron-beam welding equipment and a photomicrograph of a weld made by the process in 1/2-in.-thick 304 stainless steel.

The nonvacuum process is faster since it eliminates the need for evacuation of the area around the workpiece and the need for multiple-part tooling setups. Although the new process has advantages, it does not replace the conventional vacuum system where freedom from contamination is critical. Also the nonvacuum technique is limited to workpiece thicknesses of one inch.

The main disadvantage of the nonvacuum system is that the inert gases diffuse the electron beam. As a result, only part of the electron beam energy is available to produce the weld. For example, a beam passing through 3/8 in. of shielding gas at atmospheric pressure requires about three times the power of a beam on hard vacuum. Also the diffusion is an important factor in affecting the width of the weld melt zone. For this reason the beam is restricted to applications where the travel through the shielding gas is 1/2 in. or less. In the hard vacuum, the beam can travel efficiently over a distance of several inches.

An outgrowth of the hard-vacuum and the nonvacuum system has been the "soft-vacuum" system. This system, which consists of two

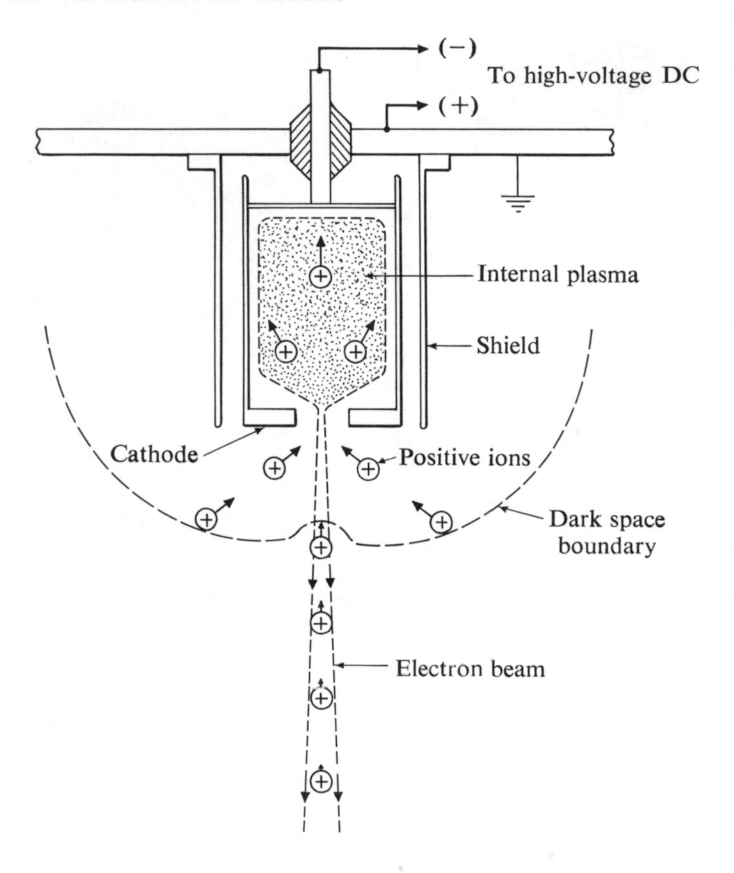


**Fig. 9.7.** Schematic of the Plasma Electron Beam gun, showing beam formation in shielded PEB cathode.

chambers (Fig. 9.8), has the advantage of welding at a considerable distance from the point where the beam leaves the gun. The gun housing is made as small as possible and is evacuated by both a mechanical pump and a diffusion pump. The welding chamber is evacuated only by a mechanical pump. Connecting the two chambers is a passage that can be closed off by a special valve. What makes the system work is that opening the valve between the two chambers does not cause the pressure to equalize. At 100 microns, the air molecules have a long free path so that only a few of them randomly find their way into the gun chamber and are disposed of by the diffusion pump.

After welding, the valve between the two chambers is closed so that the gun chamber can remain at 0.1 micron while the welded part is removed and a new part loaded.

Welding chambers are made as small as possible; however, chambers made to take both small and large assemblies can be used. When small

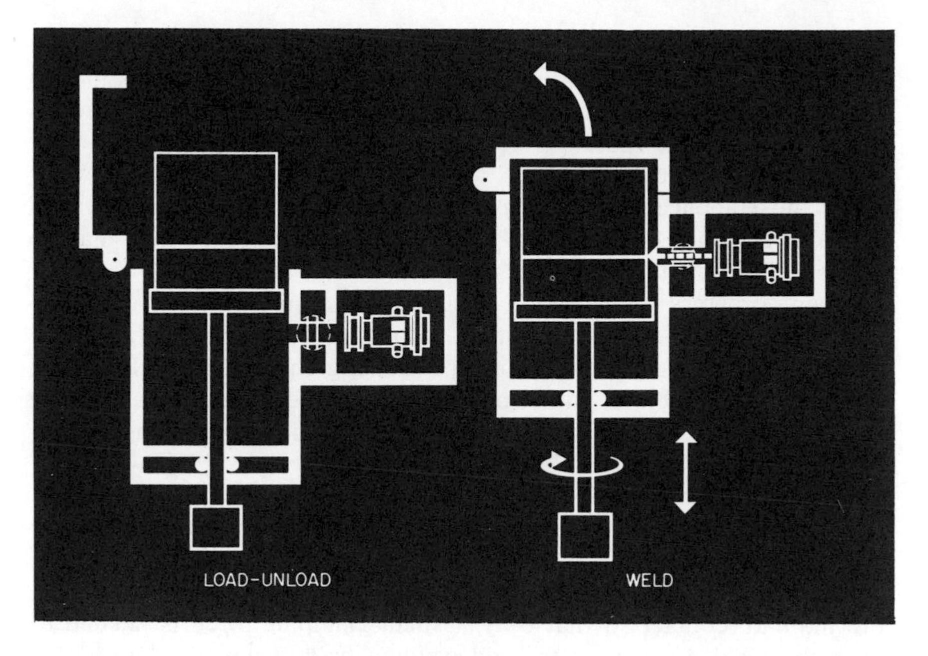


**Fig. 9.8.** The "soft-vacuum" two-chamber electron beam welding system. Courtesy *Tool and Manufacturing Engineer.*

assemblies are welded, the effective chamber volume can be reduced by filling the unused sections with metal blocks.

## ULTRASONIC WELDING

In ultrasonic welding, the surfaces to be joined are made to slide in contact under a compressive force (Fig. 9.9). Shear vibrations are introduced which produce what is termed a *solid-state metallurgical bond* over half the surface width. The remainder is a mechanical bond.

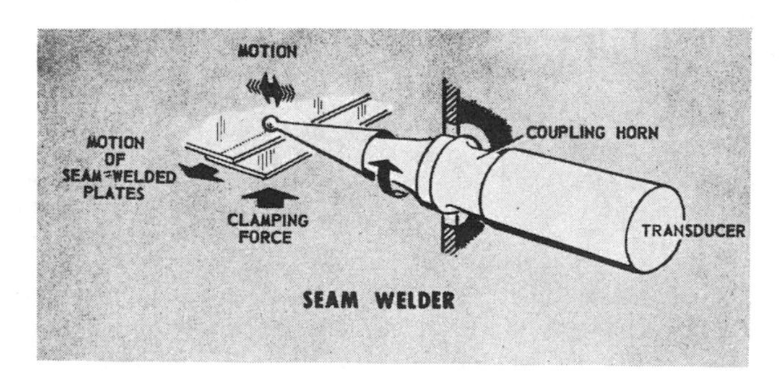


**Fig. 9.9.** A basic arrangement in ultrasonic welding. Courtesy *American Machinist.*

As shown schematically in Fig. 9.9, a solid metal conical horn serves as a mechanical step-up transformer to amplify small vibrations and to transfer them through a coupling system to a vibratory element which is in contact with the metal to be welded. The contact is at very light pressure – only 25 psi.

For most applications, the tips are either spherical or cylindrical in shape. They can be mounted on conventional machines, or they may be portable (Fig. 9.10).

Recent progress in ultrasonic welding has resulted in new and improved transducer-coupling systems which permit more energy to be delivered to the weld zone. Spot-type welding equipment, with capacities ranging up to 8,000 watts, has been developed, enabling aluminum sheets in excess of 0.090 in. thick to be spot-welded with strengths that exceed military specifications.

Perhaps the most interesting feature of ultrasonic welding is the true metallurgical bond formed in the solid state, with no melting of the metals being joined. The high-frequency current used in ultrasonic welding shatters the contaminating surfaces and brings the base metals into intimate contact so that only a small clamping force is needed to make the weld. Compared to cold welding or diffusion welding where the pressure may cause a 30 percent reduction in area, ultrasonic welding would produce only a 5 percent deformation.

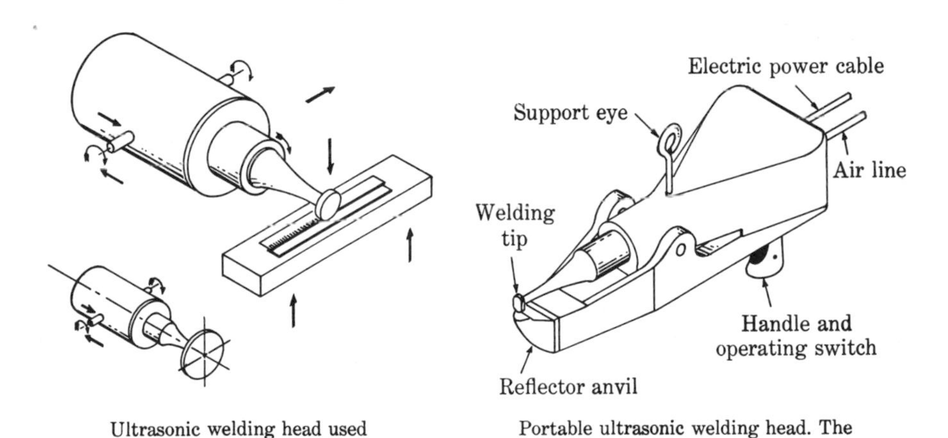


Ultrasonic welding head used in continuous-seam welder.

Portable ultrasonic welding head. The design can be varied to suit the geometry of the weldment.

**Fig. 9.10.** Ultrasonic welding tips, whether for continuous or for spot welding, are cylindrical or spherical in shape and can be portable or mounted in conventional spot and seam welders. Courtesy *The Tool and Manufacturing Engineer.*

## DIFFUSION WELDING

Metals can be welded by heating them and pressing them together. This type of welding is also referred to as pressure bonding and forge welding. Time, temperature, and pressure depend on the materials being joined and the joint design. No filler material is added and no melting occurs. Pressures as low as 10 psi have proven satisfactory for some welds. The joining surfaces must be free of oxides and make good contact so that atoms can migrate across the interface.

Each new application of diffusion welding requires rather high development costs. It is likely to be used mainly in the aerospace, nuclear, and electronic industries. Many high-temperature materials used in missiles and rockets are successfully diffusion welded. The process is often used to join dissimilar metals such as titanium to stainless or tungsten to molybdenum.

## PLASMA ARC WELDING

The principles of the plasma arc were explained in connection with metal spraying and cutting in Chap. 8. The process is presented here also since it is considered one of the newer ones in fusion welding.

The "transferred" arc, in which the workpiece is an electrode in the electrical circuit, is shown in Fig. 9.11.

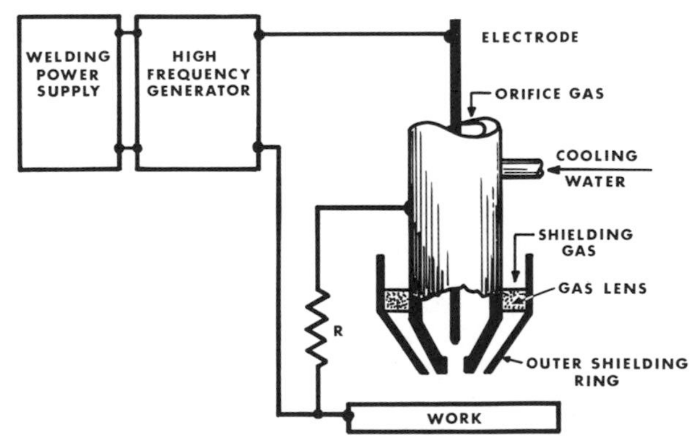


**Fig. 9.11.** Schematic of the "transferred" type of plasma torch used for welding. Courtesy *Welding Journal.*

The process operates in the same field as gas tungsten-arc welding but has some advantages in certain applications. The term "keyhole" is associated with plasma-arc welding and is used to describe the shape of the hole that is produced at the leading edge of the weld puddle. The plasma jet displaces the molten metal allowing the arc to pass completely through to the workpiece. This keyholing effect is the chief difference between the plasma-arc and the gas tungsten-arc processes.

A backing plate (Fig. 9.12), which consists of a channel supplied with inert gas, is used to protect the back side of the weld from the atmosphere when nonferrous metals are being welded.

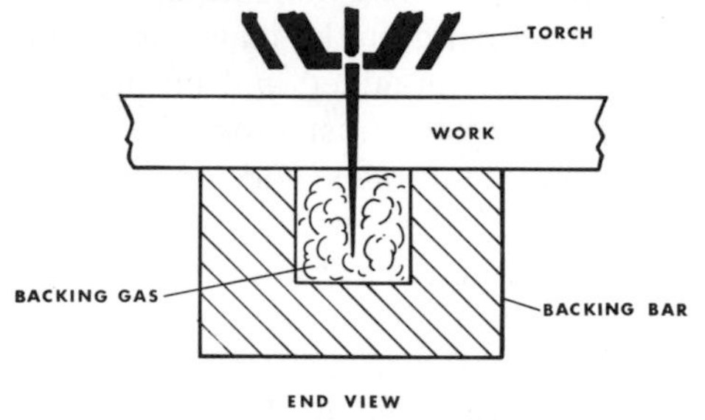


**Fig. 9.12.** Typical backing bar for plasma arc welding. Courtesy *Welding Journal.*

## FRICTION WELDING

Friction welding is a process that uses heat generated to bring the surfaces to be joined to a hot plastic state and then produces the weld by a forging and upsetting action. The parts to be joined are rotated against each other at high peripheral speeds. The localized rubbing soon causes the metal to reach the plastic state. Continued relative rotation causes the metals, now in a plastic state, to mix. The application of pressure produces a flash, or raised portion at the joint, which carries the surface oxides and other impurities away from the joint itself.

The relationship of three variables, pressure, speed of rotation, and time, makes the process relatively simple. High speed of the operation is essential to prevent a zone that is excessively affected by heat. The process can be diagramed as shown in Fig. 9.13.

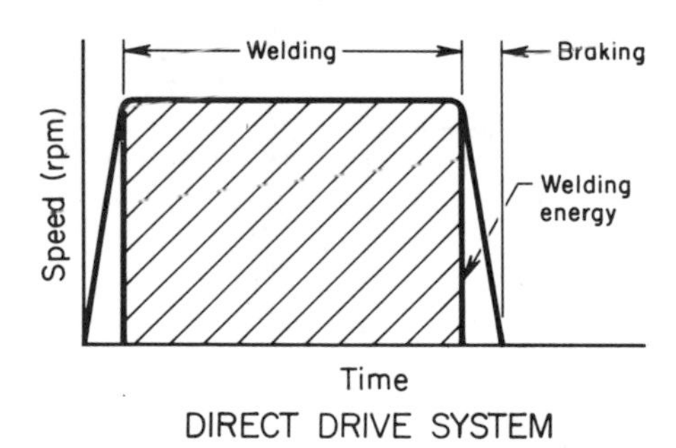


**Fig. 9.13.** The friction welding cycle.

**Application and Limitations.** The major advantages of friction welding are: speed, ability to weld dissimilar materials, and ease of operation or automation. A preset automatic cycle can be programed into the machine by previous study of the thermal conductivity of the material and the part geometry.

The major limitations are the differences in the plastic temperatures of the metals that are being welded, and, further, the process is limited by part geometry. If materials of widely varying plastic ranges are joined, the one having the lower temperature will tend to mushroom, wasting some material. Ideally the part geometry should be circular, but some deviations are acceptable. Small parts do not lend themselves to friction welding since they do not have the rigidity required for the axial pressure. Also, excessive speeds would be required to bring the metals to the plastic state.

## EXPLOSIVE WELDING

Explosive welding is done in two ways, by angular and parallel placement (Fig. 9.14). Of the two methods, the most common is parallel.

The process consists of placing the plates together but with a "standoff" space which can be arranged by dimples, filler rod, or natural roughness of the work.

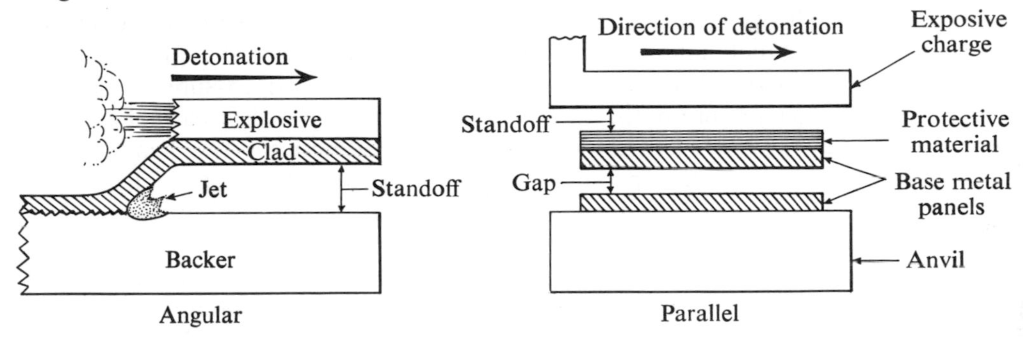


**Fig. 9.14.** Angular and parallel methods of explosive welding.

As the gelatin placed on one side of the work is detonated, the entrapped gas between the parts compresses and superheats the metal surfaces. Because there is little gas, there is little heat, but the reaction is so fast the work cannot carry the heat away fast enough to prevent shallow-melting or softening.

An interesting phenomenon that occurs in explosive welding is the eruption of microscopic swirls on the interface surfaces (Fig. 9.15).

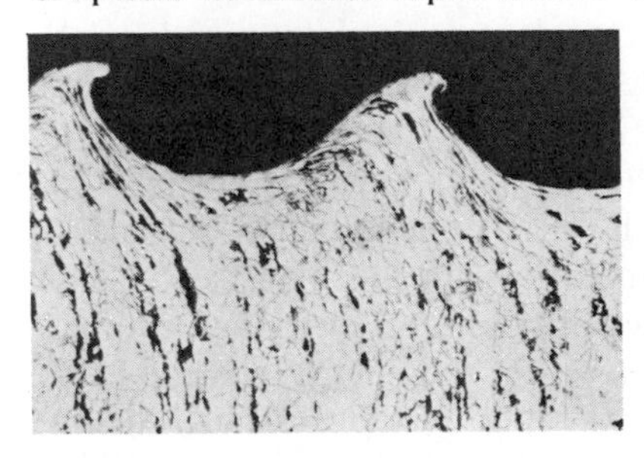

**Fig. 9.15.** Eruption of surface jets on a steel projectile. These increase the weld strength by interlocking with similar jets on the opposing surface. Courtesy *The Tool and Manufacturing Engineer.*

These are called jets, and they contribute greatly to the strength of the weld by their interlocking action. Surface jetting is dependent on the angle at which the plates come together. The various angles that have been reported are 1 to 15 deg for welding aluminum and anywhere from 7 to 50 deg for welding steel, provided both plates are covered by explosives.

Explosive welding can be considered a cold-welding process, since only the local interface region reaches welding temperature. The effects of heat treatment should not be lost.

Since this process is still in its early research stages, very little quantitative data are available. Enough is known, however, to indicate that the process is practical.

So far, most of the explosive welding has been done on aluminum and mild steel, but successful welds have been reported in aluminum to Inconel, aluminum to stainless steel, stainless steel to itself, and aluminum sandwiched between stainless steel.

**Explosive Cladding.** Explosive cladding makes use of the explosive welding technique to bond a plate of one material to that of another.

Explosive cladding is now used in production. An example of its use is the manufacture of the silverless coins. In this process, slabs of copper 3 in. thick, 2 ft wide, and 6 ft long are placed between two 1-in. layers of cupro-nickel. A separation of a few thousandths of an inch is maintained. Five hundred pounds of explosives are applied on each side of the "clad" which is exploded at the same time. The high impact of the three sheets causes a bonding of the metals. The clads are then cleaned and sent to the rolling mills to be reduced to the required thickness.

Almost any two wrought metals or alloys with enough ductility and impact strength can be joined by explosive bonding. Clad plates range from small sizes up to areas as large as 230 sq ft.

## LASERS

**Laser Action.** The newest of the high-energy sources to become a tool of the production engineer is the laser. The term laser stands for light amplification by stimulated emission of radiation. It goes back to Bohr's atomic theory and wave amplification theory. Electrons can exist only in definite shells. The energy of the electron is determined by the distance from the nucleus; for example, in Fig. 9.16 an electron in shell $E_2$ has more potential energy than one in shells $E_1$ or $E_0$. Energy is given off when the electron falls back to a lower shell. Thus the electron can absorb or give off energy as it changes levels. A phenomenon that makes the laser possible is that in certain materials an electron can be in a semistable level (Fig. 9.17) so that, when a photon of radiation of exactly the right frequency triggers the electron, the energy can cause the electron to fall to a lower level giving off a photon of radiation which is exactly in phase with the photon that caused the triggering action. Thus

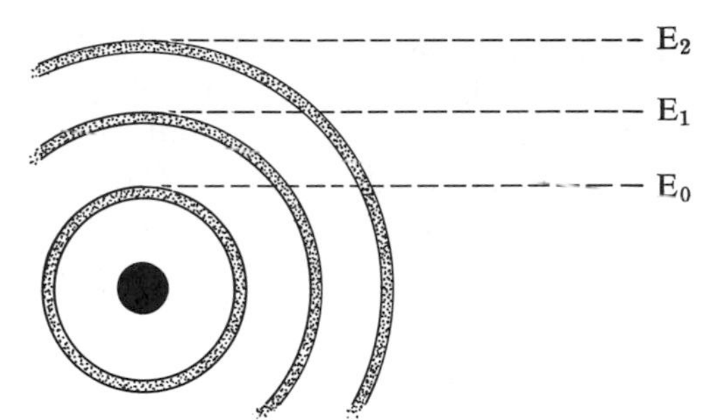


**Fig. 9.16.** Atomic model showing electron energy levels.

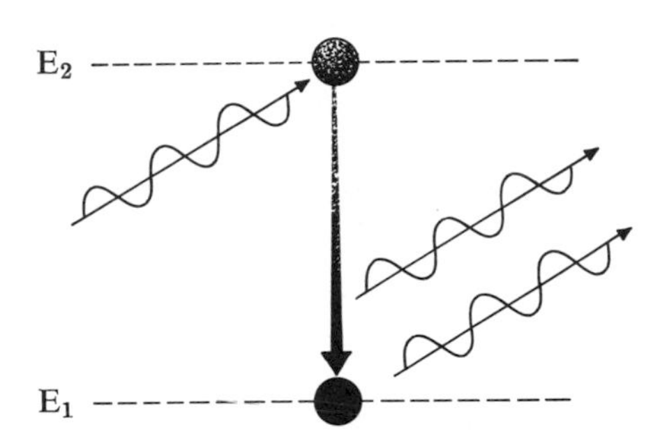


**Fig. 9.17.** Stimulated emission of radiation or doubling of radiant energy.

we have a doubling of radiant energy or amplification.

The first lasers were made using a solid ruby rod as an optical cavity. Chromium atoms "suspended" in the cavity were "pumped" by a discharge from a xenon flash tube (Fig. 9.18). By absorbing the light from

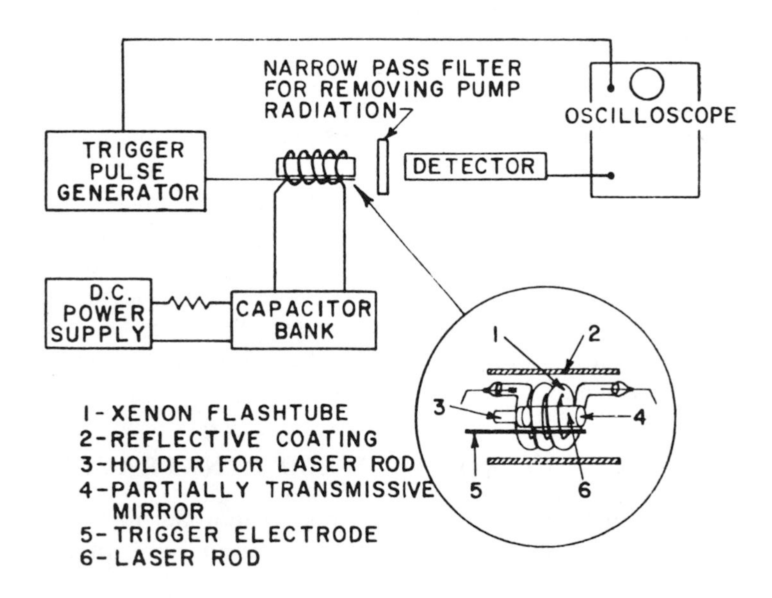


**Fig. 9.18.** Diagram of apparatus for flash excitation of solid-state laser.

the flash tube, the electrons are raised two levels above the ground state. As some of the atoms release their energy in the form of light, they return to an intermediate or ground state. The reflecting walls of the ruby rod or optical cavity are mirrors, one of which is partially transmitting so that some of the radiation can escape from the cavity as highly *coherent* or in phase light. It has a very narrow frequency band and is quite parallel.

**Laser Materials.** There are now four broad family types of lasers: solid light pumped (ruby or glass); solid injection or semiconductor (mainly gallium arsenide); gaseous (helium-neon or xenon); and liquid or plastic.

The glass rods convert energy more efficiently than ruby but have shorter life. The neodymium-doped calcium tungstate laser is reasonably efficient and can be used for continuous-wave laser action. The output, however, is only in milliwatts. The same is true for gas lasers. For large amounts of almost instantaneous energy, the ruby rod is still the most satisfactory.

Lasers are now available with outputs of from 50 to 1,500 joules. Most production applications require outputs of 1 to 10 joules at the rate of up to 1,200 pulses/hr.

**Efficiency.** Ruby-rod lasers have low energy conversion efficiency; only 1/2 to 1 percent of the pumping energy is converted into useful coherent light. Efficiency is improved by cooling the ruby rod with liquid nitrogen or water.

**Welding Applications.** The laser has the advantage of being able to operate in virtually any atmosphere. Any material that can be welded by conventional methods can be welded by the laser beam. Welding dissimilar metals poses no problem. Both materials are melted and subjected to infinite mixing. As the metals cool, grains grow across the former interface and result in a homogenous material. No flux is required.

A potentially large field for the laser welder is in microelectronics. Subminiature components can be welded and modified. Production laser welders are now being used in joining fine wires (0.015 in. to 0.030 in.) and in joining fine wires to foil. The weld can be concentrated in a pinpoint no bigger than 0.010 in. in diameter. The laser beam delivers energy only to light-absorbing surfaces and can be used to weld through transparent materials.

## AUTOMATIC CONTROL OF WELD PENETRATION

Although automation of welding does not come under new processes, it would be an incomplete picture not to mention some of its newer important aspects.

**Feedback Control.** Automatic welding has been with us for a long time but it is based on preset controls. If some variable changed outside the anticipated limits, a bad weld resulted. Now, however, work is being done on *adaptive controls*. That is, a sensing device is incorporated into the welding arc area to produce a signal that is used to make corrections before a bad weld occurs. The feedback system must be able to change the weld settings automatically when such conditions occur as: change in the gap between the metals, encountering tack welds, variation in heat sink of the base metal, etc.

There are two main types of adaptive controls: (1) arc penetration control system and (2) melt-through weld control penetration system.

ARC PENETRATION CONTROL SYSTEM. The arc penetration control has a penetration sensor on the same side as the arc (Fig. 9.19). In TIG the

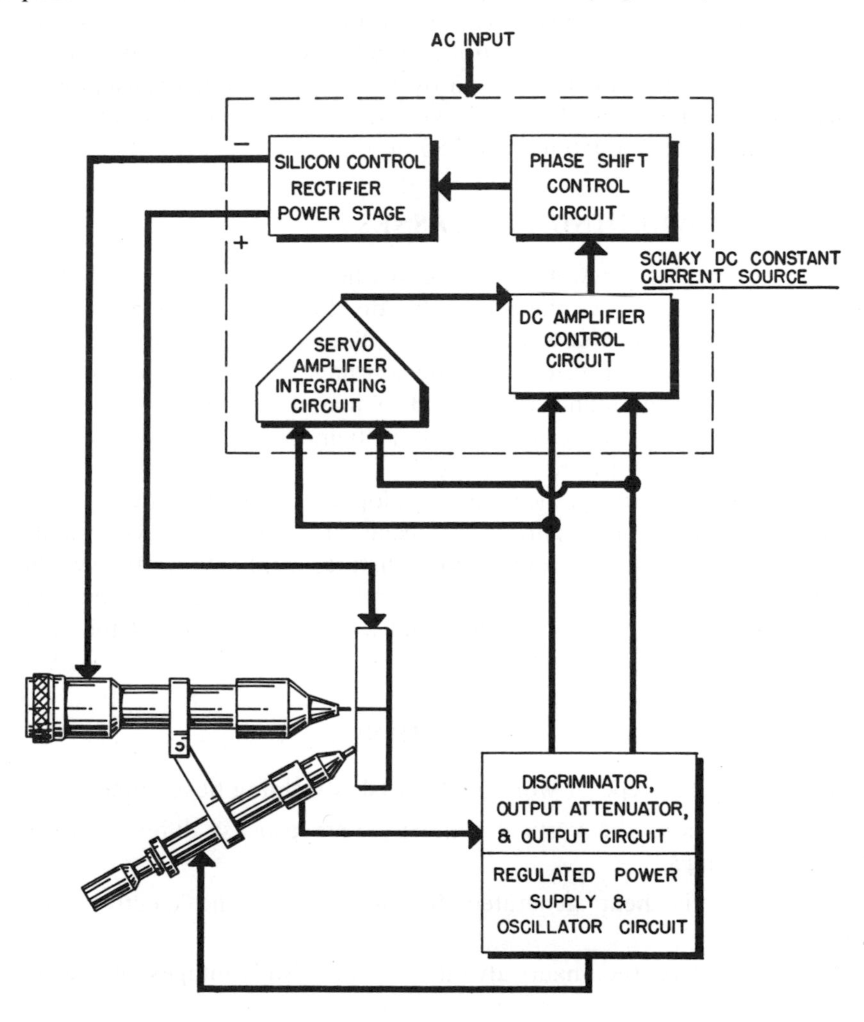


**Fig. 9.19.** Electrical block diagram of the arc penetration control system. Courtesy *Welding Engineer.*

tungsten electrode will advance into the work or withdraw from it due to variances in fitup or changes in the heat sink. Movement of the electrode with respect to the work is detected by a sensor, which produces a signal automatically increasing or decreasing welding current to maintain a constant electrode to work position. The result is constant penetration of the arc. Dual probes are also used to average out the mismatch in butt joints.

Another development in the use of arc penetration control is opposed-arc welding in which two TIG welding torches are used, one on each side of the butt joint. Experimental work has shown energy requirements may be reduced to 33 percent of that required to make the joint in the normal method. Furthermore, distortion is greatly reduced.

MELT-THROUGH WELD PENETRATION CONTROL SYSTEM. This system, although similar to the arc penetration control system, is based on a more direct measurement of penetration by having a photocell sensor on the back side of the weld. It has proven successful in making aluminum welded joints from 0.050 in. to 0.750 in. thick.

## FUTURE WELDING PROCESSES

Some of the newer welding processes have been described, but it is safe to say that, before the ink is dry on this page, other newer techniques will have been developed, descriptions of which will be found in current scientific literature and trade journals. This only points up the fact that nothing is static in the field of manufacturing processes. The reader should keep abreast of what is being developed by regularly checking current literature in the field.

Even at this writing, pending developments include low-hydrogen electrodes that will require no special storage precautions, and a square-wave a-c power source with distinct applications for welding aluminum. Other improvements are ion-beam welding, self-correcting resistance-welding control circuits, and automatic guiding of the arc in high-speed welding operations.

## QUESTIONS

**9.1.** For what type of work is electroslag welding best suited?

**9.2.** How does the vertical submerged-arc process differ from electroslag welding?

**9.3.** How is heat generated for welding in the electron beam process?

**9.4.** What are the main advantages and disadvantages of electron beam welding?

**9.5.** What is the principle involved in making ultrasonic welds?

**9.6.** Explain the significance of the "keyhole" in plasma arc welding.

**9.7.** What are some advantages plasma arc has over gas tungsten arc welding?

**9.8.** For what type of work geometry is friction welding best suited?

**9.9.** How is bonding brought about in explosive welding?

**9.10.** What is the main purpose of explosive cladding of metals?

**9.11.** What causes the amplification of energy in the laser?

**9.12.** What are some advantages and disadvantages of the *ruby* laser?

**9.13.** What is the main advantage and the main disadvantage of laser welding?

**9.14.** What is meant by "adaptive control" for welding?

**9.15.** What is the principle of the arc-penetration control system?

**9.16.** What are some limitations of ultrasonic welding?

## PROBLEMS

**9.1.** Refer to Chap. 7 and find the appropriate heat input ($H_i$) for a weld made under the following conditions:

Plate thickness, 3/4 in.
Initial temperature, 70°F.
Material, SAE 1030 steel.
Butt weld with backing plate.
Welding speed, 20 in./min.
Grain size, No. 6.
Max $R_c$ hardness, 27.
Alloy content, 80% Mn, 0.06% P, 0.06% S, 0.30% Si.
1/8 in. root, 1/8 in. reinforcement, 60°V.

**9.2.** Friction welding requirements are as given in Table P9.2. Assume each of the items is round and each pair of mating numbers equal in area. (*a*) What is the surface foot speed used on weld #1? (*b*) What is the ratio of weld area to weld time for each one? (*c*) What is the relationship of heating pressure to area?

**Table P9.2**

| Materials welded | Weld area | Rotational speed (rpm) | Total weld time (sec) | Heating pressure (psi) | Forging pressure (psi) |
|---|---|---|---|---|---|
| 1) SAE 1141 to 1020 | 0.55 | 4,400 | 4.0 | 12,000 | 17,000 |
| 2) SAE 1010 to 1141 | 1.96 | 4,400 | 13.0 | 8,400 | 28,000 |
| 3) Copper to copper | 1.77 | 5,400 | 8.5 | 6,400 | 10,000 |
| 4) Bronze to bronze | 0.14 | 1,700 | 7.0 | 11,000 | 35,000 |

**9.3.** In the case of mild steel friction welding about 10 hp/in.$^2$ of cross section must be delivered to the weld area. What would be the hp requirement for #2 weld of Problem 9.2?

**9.4.** Diffusion welds may be produced that are 2 to 2 1/2 times the normal brazed joint strength. Find the normal brazed butt joint strength in psi for beryllium copper and what it can attain by diffusion welding.

**9.5.** Electroslag welding can deposit up to 40 lb/hr/electrode. How long would it take to weld two 8-in.-thick mild steel plates 4 ft high with a 2-in. gap, exclusive of setup time, if 3 electrodes are used?

# 10

# PLASTICS AND ADHESIVES

THE WIDESPREAD USE of plastics covers the comparatively brief span of about 30 years. These materials are still in infancy when compared to other common ones such as steel, copper, and brass, which were known and used in ancient times. Nevertheless, plastics now rank second to steel as an industrial material. The cubic-foot production of plastics is 50 percent greater than that of all nonferrous metals combined.

Prior to World War II, plastics usage was restricted to areas where no other material was suitable. Today, by contrast, they are considered, in many cases, a superior replacement. As an example, the average modern car contains 70 lb of plastics which have replaced 400 lb of steel and other metals.

Plastics are not cheap; pound for pound, they cost more than most natural materials, but they are chosen because of their favorable weight ratios, ease of fabrication, service life, and lower end-product cost.

## COMPOSITION OF PLASTICS

Plastics have their origin in the chemical synthesis of materials from five different sources: agriculture, agriculture and petroleum, petroleum and coal, petroleum and mineral, and mineral. Of these, the coal-and-petroleum source is the most used.

## POLYMERS AND POLYMERIZATION

**Polymers.** The basis of understanding plastics lies in gaining a knowledge of their giant molecules. Each molecule has hundreds or thousands of atomic groups, or repeat units. The backbone of these repeat units is the carbon chain. Attached to it are one or more of seven other elements: hydrogen, oxygen, nitrogen, chlorine, fluorine, sulfur, and silicon.

When other elements join with the carbon chain, it usually acquires four other atoms. These smaller units or "building blocks" are called *mers,* from the Greek *meros,* meaning part.

**Polymerization.** When the smaller, *single* mers or *monomers* join together they are called polymers (Fig. 10.1*a*). Two or more different types of monomers may join on the same chain to form a copolymer (Fig. 10.1*b*).

Monomer ratios can be made to vary over a wide range to develop certain desirable properties. Polymers consist of huge numbers of mole-

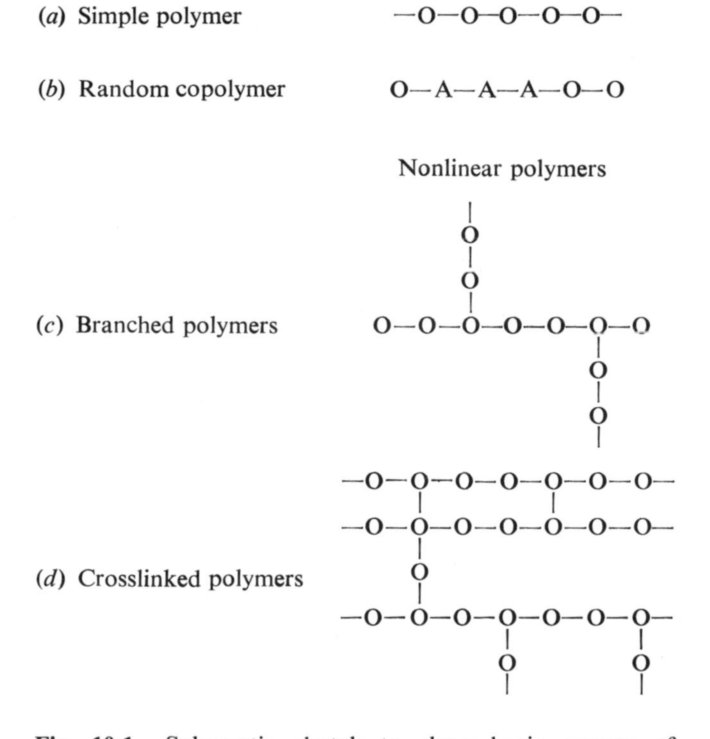


**Fig. 10.1.** Schematic sketch to show basic groups of polymers synthesized from nonidentical monomers O and A.

cules strung together so that dimensionwise the chain is hundreds of times longer than it is wide. If all the molecular chains could be arranged with their maximum dimension one way, the physical properties would also be directional. This orientation can be achieved on some polymers. In polyethylene, for example, the average molecule may be 300 times longer than it is wide.

**Molecular Orientation.** Under normal circumstances the molecule chains arrange themselves in a random manner. Orientation comes about by outside pressures. For example, the force required to make the plastic flow into an injection press will cause orientation. There will generally be more orientation towards the outside dimensions than in the center, due to flow conditions. This arrangement causes internal stresses and affects dimensional stability. It can also produce increased tensile strength and fatigue resistance. Internal stresses act to preload the part, reducing the temperature at which distortion occurs.

**Copolymerization.** It is oftentimes desirable to combine the properties of one polymer with that of another (Fig. 10.1*b*). An example of this is the ABS plastics, a combination of acrylonitrile, butadiene, and styrene. However, simply mixing the two polymers together may not work since they may not be miscible, or they may separate out in time. To overcome this, the two monomers must be polymerized together. Common copolymers are: styrene and acrylonitrile, vinyl chloride and vinyl acetate, isobutylene and isoprene.

The way two monomers add into a chain is dependent on their relative reactiveness with respect to the two possible free radical groups that appear on the growing end of the chain. Free radicals are molecules having an unsatisfied bond. They are formed by thermal or chemical decomposition so they are actively seeking further electrons to complete their structure. They may add in randomly, alternately, or in blocks of one monomer unit or the other.

Graft copolymers are formed by attaching a side chain to an already polymerized main chain (Fig. 10.1*c*).

**Cohesive Forces.** How do the long molecular chains manage to stay together? There are several forces at work. One of these is molecular attraction. You may have noticed how two very flat metal surfaces seem to stick together. There is no adhesive or magnetic force at work but the situation is simply that the atoms of one block are attracted to those of another in intimate contact much as they are within the same piece of metal. The fact that these attractive forces vary with different materials results in special properties sought in plastics. The strong forces that hold the polymer chains together are called primary bonds whereas the forces between chains are *secondary* and are only 5 to 10 percent as strong.

The pull necessary to separate the building blocks from their neighbors is sometimes called the *cohesive energy density* (CED). If the CED is small, the chains are relatively flexible and have properties associated with elastomers. Stiffer chains make up the plastics group, and chains that resist stretching, and are strong, make up the fibers group.

**Branched and Network Polymers.** Instead of being just long chains, some polymers are more like a tree. Such a structure makes the polymer much bulkier and changes its properties.

Some polymers form networks of crosslinked structures. They are chemically tied together so that in effect a giant molecule is formed (Fig. 10.1*d*). These giant networks become very stable and cannot be made to flow or melt and are called the *thermosets*. Most linear polymers can be made to soften and take new shapes and are called the *thermoplastics*.

## PROPERTIES OF PLASTICS

**Mechanical.** Upon a person's first exposure to the wide variety of plastics, they appear very much alike just as a large group of people upon first introduction. However, as we get to know them better we find there are major differences. Three types of stress-strain curves (Fig. 10.2) typify the various plastic materials. Shown as Type A are polyethylene, acetal, and conditioned nylon (2.5 percent moisture content). These materials and others like them yield gradually. Dry, as molded nylon, yields abruptly (Type B), and acrylic materials usually fracture before yielding (Type C).

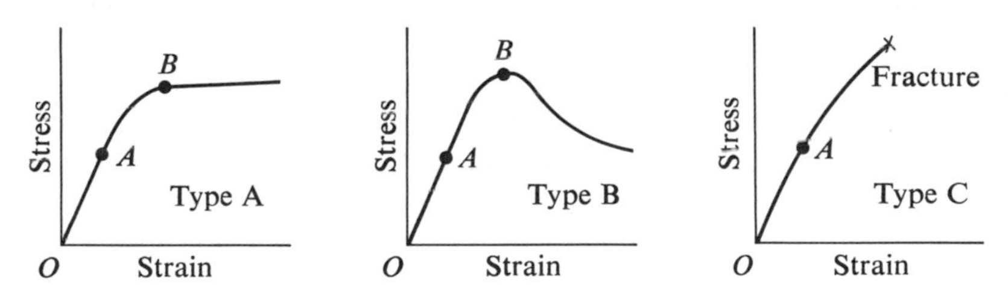


**Fig. 10.2.** Typical stress-strain curves for plastics that yield gradually (Type A) and abruptly (Type B) and that fracture at low strain before yielding (Type C). Courtesy *Machine Design*.

Two distinct regions of the curve are important in design work. The region *OA* is where elastic design principles apply. The area around *B* is the yield point and is important when deformation is of prime importance.

Thermoplastics have no elastic limit, as such. They exhibit, instead, various degrees of deviation with increasing strain.

The ultimate tensile strength of metals and plastics is compared in Fig. 10.3. Of course the plastics are much more temperature sensitive.

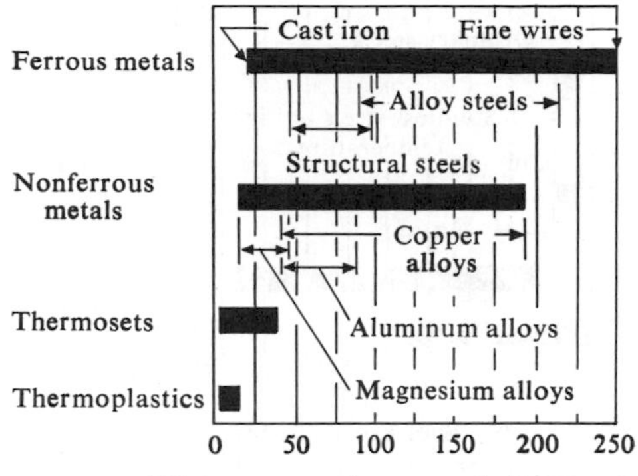


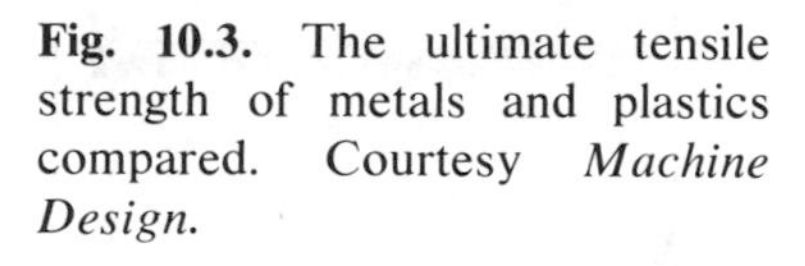
**Fig. 10.3.** The ultimate tensile strength of metals and plastics compared. Courtesy *Machine Design.*

**Thermal.** When a plastic is heated, as in any solid, the molecular action increases. The chains twist and oscillate. As heating continues, the chains become longer and the motion becomes essentially free between the strong bonding points, causing the plastic to become rubbery. More heat causes the strands to slip loose from one another and to slide, forming a viscous liquid. Such polymers are thermoplastic. Further heating causes degradation as side groups are broken off and the backbone gets broken into shorter lengths. Thermosetting plastics chains break or decompose before they slide on each other.

The thermal coefficient of expansion of thermoplastics is about five to ten times higher than that for most metals. The thermosetting plastics range from very little to about half that of the thermoplastics (Fig. 10.4*a*).

The thermal conductivity of plastics and metals is graphically compared in Fig. 10.4*b*), and the useful temperature range is shown in Fig. 10.4*c*.

**Modulus of Elasticity and Cold Flow.** Besides being less stiff, plastics (especially the thermoplastics) are more susceptible to cold flow, or *creep,* under long-term loading than are metals. Fortunately most plastics exhibit cold flow only for a finite period of time and show negligible change beyond this period. Since plastics deflect much more than metals under the same loading (Fig. 10.4*d*), it is important they do not share loads in parallel.

**Optical.** An example of outstanding optical properties is methylmethacrylate. It is crystal clear and superior to all but the finest glass.

**Electrical.** Most polymers resist the flow of dc current. Any current that does pass is probably due to impurities that could not be removed in the manufacturing process. When high voltages exceed the dielectric strength of the material, there is a catastrophic decrease in resistance, and the polymer breaks down physically.

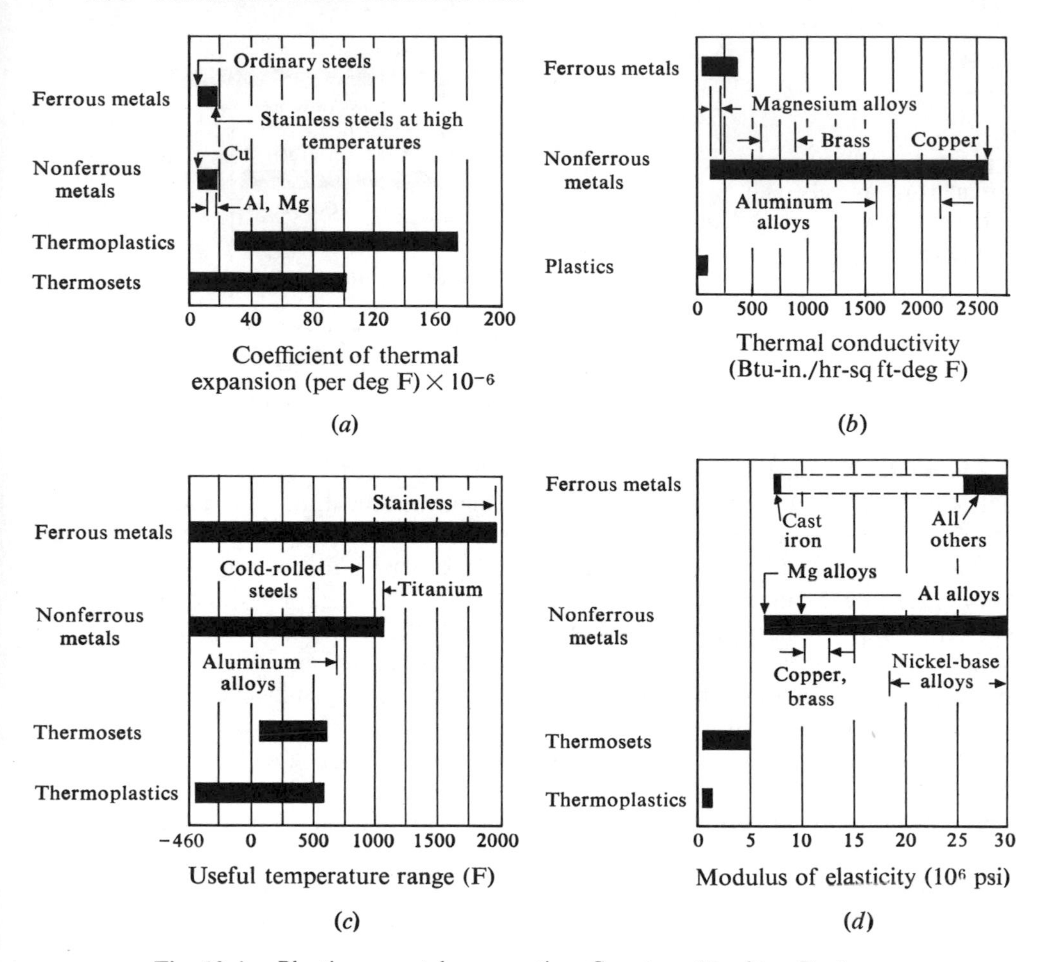


**Fig. 10.4.** Plastic vs metals properties. Courtesy *Machine Design.*

**Chemical.** Most plastics are extremely resistant to deterioration by most chemicals. However, being slightly permeable, active foreign molecules may enter the system and break up the long chains that are indispensable to the polymer structure.

These chemical agents are especially potent for any part of the plastic that is under high stress or has been highly stressed during forming.

## ADDITIVES

**Colorants.** Either natural, organic, or synthetic chemical dyes are used to color plastic materials. Dyes are soluble in many common solvents. They are transparent and have excellent brightness but poor resistance to light and heat. The colorant for plastics may be added at different times by the resin compounder, by the resin manufacturer, or by the molder of the finished part as an integral part of molding or in a surface coating.

**Stabilizers.** To offset the effect of light and heat on polymers, stabilizers are used. One type has the ability to absorb ultraviolet light, which is a high-energy deteriorating agent. Others are transparent to the radiation and as a result are stable even though the radiation level is high.

**Fillers.** Some plastics have limited use in themselves but, combined with fillers, they become very serviceable. Phenolic and amino resins are almost always combined with such substances as wood flour, macerated paper or fabric, powdered mica, pure short-fiber cellulose, or asbestos. These materials improve dimensional stability, impact resistance, abrasion resistance, and strength.

**Plasticizers.** Plasticizers are added to increase flow and therefore processability and to reduce brittleness of the product.

## PLASTIC TYPES

Plastics do not come under any easy-to-learn code such as the SAE steels. They do, however, come in families. Each family is composed of many individual types. To become familiar with these families, it is usually helpful to know the similarities and differences between the monomers. For example, shown in Fig. 10.5 is the ethylene monomolecule and its related monomers. It is also helpful to learn the generic name and the manufacturer's trade name. For example, the generic name of phenol-formaldehyde describes a whole family of synthetic materials under such trade names as Bakelite, Durez, Catalin, etc. These are all given in the general properties charts in the Appendix, which divide the plastics into their two main types, *thermoplastics* and *thermosetting.*

**Thermoplastic Resins.** Thermoplastics, as explained previously, undergo no permanent chemical change during moderate heating. They soften or melt when warmed and can be molded into a shape that is retained upon cooling. If reheated, the parts can be remolded into other shapes. Common examples are: acrylic (Plexiglas), cellulose acetate, fluorocarbon (Teflon), polyethylene, polystyrene, and polyvinyl chloride.

**Thermosetting Resins.** As mentioned previously, when considerable crosslinking takes place so that there is a transition from a reasonably fluid state to a "gel" state, the polymer has passed a critical point known as the thermosetting point. This process is not reversible. Common examples of plastics of this type are: alkyd, amino, epoxy, polyester, silicone, and urethane.

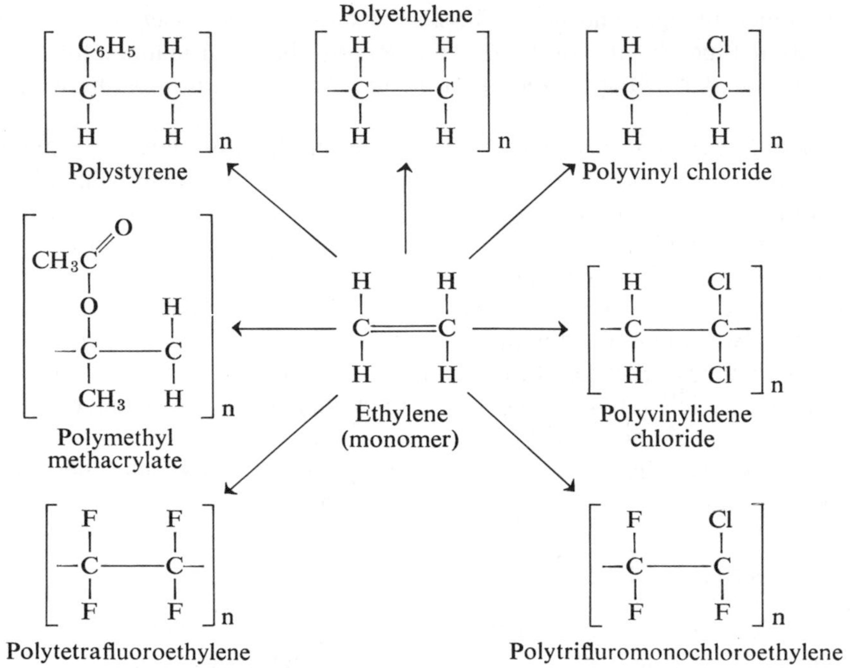

Polyethylene

Polystyrene

Polyvinyl chloride

Ethylene (monomer)

Polymethyl methacrylate

Polyvinylidene chloride

Polytetrafluoroethylene

Polytrifluromonochloroethylene

**Fig. 10.5.** The relationship of various polymers to ethylene.

## REINFORCEMENT MATERIALS

Fiberglas (a trademark of Owens-Corning Fiberglas Corporation) materials are added to the various plastics to give a high strength-to-weight ratio. A strength-to-weight comparison is made between the various Fiberglas-reinforced plastics and various metals (Fig. 10.6).

Other properties that are improved are stiffness (about 50 percent) and impact strength. The thermal coefficient of linear expansion is reduced much below that of unreinforced plastics.

Fiberglas, used as the reinforcing material, is supplied in several basic forms. These allow for flexibility in cost, strength, and choice of manufacturing process. The basic forms, shown in Figs. 10.7 and 10.8, are continuous-strand, fabric, woven rovings, chopped strands, reinforcing mats, and surfacing mats.

Resins used with the glass reinforcement are polyesters, phenolics, silicones, melamines, acrylics, and polyesters modified with acrylics and epoxies. Of these, polyester resins are used about 85 percent of the time because they do a good job and are the most economical. Some thermoplastic materials are also used, including nylon, polystyrene, polycarbonate, and fluorocarbon, or Teflon.

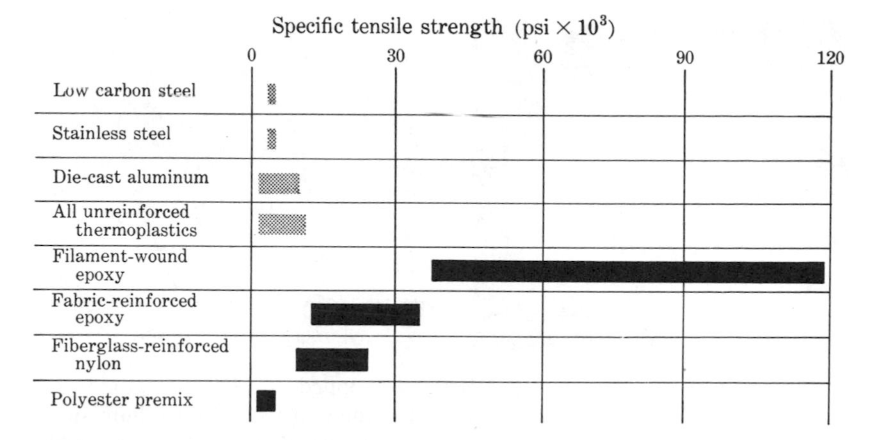


**Fig. 10.6.** Strength-to-weight comparison between Fiberglas-reinforced plastics and some common metals. Courtesy Owens-Corning Fiberglas Corporation.

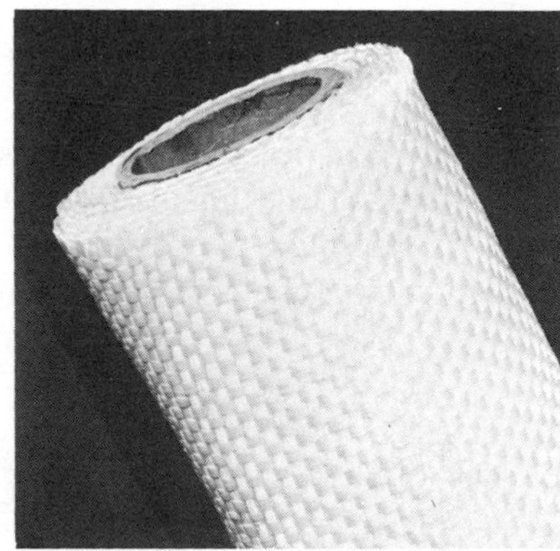

**Fig. 10.7.** Basic forms of Fiberglas used in reinforcing plastics. Courtesy Owens-Corning Fiberglas Corporation.

## MOLDING METHODS

The methods of manufacturing, as applied to plastic powders (both reinforced and unreinforced), include injection molding, jet molding, compression molding, transfer molding, extrusion molding, cast molding, and slush molding. All these processes are based on the fact that when the plastic is heated it will soften to a viscous liquid that can be forced into a mold of the desired shape, where it solidifies.

**Injection Molding.** Injection molding is the method most commonly used for the production of thermoplastic parts.

The process consists of feeding plastic pellets into the hopper above

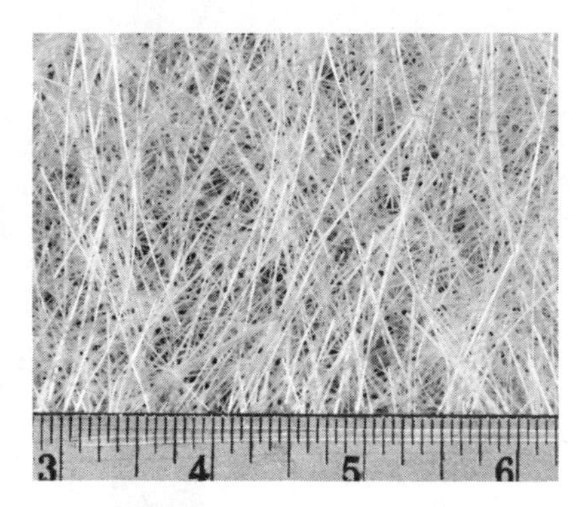


**Fig. 10.8.** (*a*) Continuous-strand Fiberglas chopped into short lengths to facilitate molding complex structures; (*b*) reinforcing mat used for medium-strength structures with uniform cross section. Courtesy Owens-Corning Fiberglas Corporation.

the heating cylinder of the machine. With successive strokes of the injection ram, the material is compacted and forced forward through the thin annular space between a heated "torpedo" in the center of the cylinder and the cylinder walls. There the material is softened. It passes through the cylinder nozzle and is forced into the cold mold (Fig. 10.9).

The fluid plastic completely fills the mold cavity, which is shaped to the contours of the finished product. When the plastic has cooled and solidified, the mold is opened and the finished piece, an exact image of the mold cavity, is ejected. Ejection is aided by knockout or ejector pins.

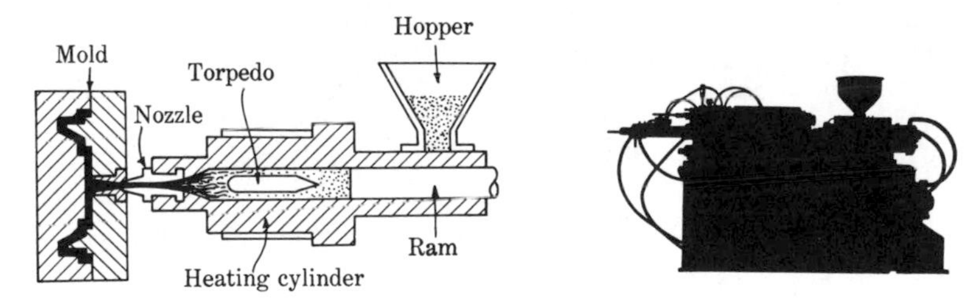


**Fig. 10.9.** Injection molding process and a 4 1/2-oz-capacity Fellows machine.

The temperature to which the material is raised in the heating cylinder is between 350° and 525°F. Every type of material has a characteristic molding temperature, depending on the plasticizer content.

The mold is kept cool so that the injected plastic material will cool quickly and can be removed without distortion. Cooling is done by circulating water through the mold frame.

To speed up molding cycles and facilitate production, a preplasticating chamber is included on many machines. It consists of an auxiliary heating cylinder which melts the plastic granules and transfers the melt to the injection cylinder, where it is further heated and held ready for injection.

MACHINE SIZE. Injection-machine size is usually designated by the maximum amount of material the heating cylinder will deliver in one stroke of the injection ram. This amount is expressed in ounces and may range from a fraction of an ounce, on small hand-operated machines, up to 500 oz. The most popular size is from 8- to 16-oz capacity.

ADVANTAGES AND DISADVANTAGES. Injection molding provides the highest production rate of producing plastic parts. The time required per shot will vary with the material and the size of the mold, but 300 to 400 shots per hr are not uncommon. Metal inserts such as bearings, screws, clamps, etc., can be put into the mold and cast integrally with the plastic. The initial mold cost is usually high, because of the accuracy and high degree of surface finish required.

MATERIAL HANDLING. Various methods are employed to keep the larger-volume machines supplied with material. These methods consist of mechanical conveyors such as screws, paddles, and chains; both low- and high-pressure pneumatic systems; and the newest method, which utilizes the vacuum principle. The vacuum method has the advantage of efficiently handling even dusty materials.

**Jet Molding.** Injection molding is generally limited to forming thermoplastic materials, but some thermosetting materials and rubber compounds are *jet-molded* on the same machines. This, however, requires a change. The torpedo spreader in the heater is taken out so that the main heating takes place at the nozzle passage to the sprue.

**Compression Molding.** Compression molding is the process most widely used for the forming of thermosetting plastic parts. It consists of placing the right amount of plastic compound in an open heated mold. The upper part of the die is brought down to compress the material into the required shape and density. When the mold is closed, the material undergoes a chemical change, or polymerization, that hardens it (Fig. 10.10).

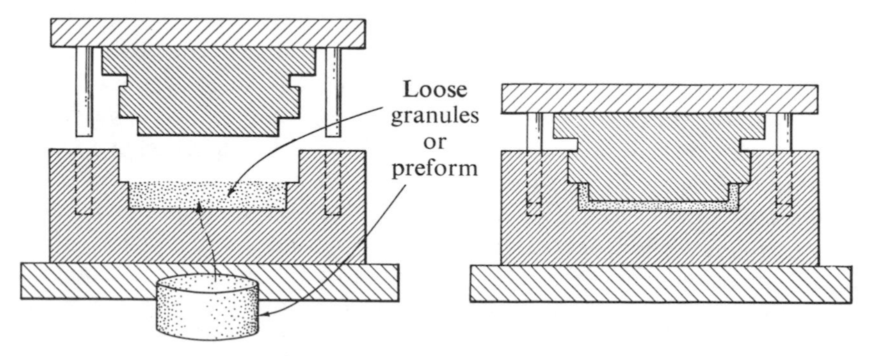


**Fig. 10.10.** Compression molding.

Compression molds may be of the positive or the flash type. In the latter, some of the material is allowed to escape over a land. As the mold closes completely, the material on the land becomes very thin. This is the *flash,* or *fin* (Fig. 10.11). Since the fin solidifies first, it prevents any more material from escaping.

Molding pressures vary from 2,000 to 10,000 psi. A basic figure is 2,000 to 3,000 psi on the mold land. The curing time in the mold may vary from 30 sec on some hard materials to 135 sec on soft materials.

**Transfer Molding.** Transfer molding is a variation of compression molding, in which heat and pressure are applied to the molding materials outside of the mold until they become fluid. The fluid material is then forced through a series of channels from the external chamber to the mold cavity, where final cure takes place.

When complex parts are molded, it may be necessary for the mold to open in several directions in order to remove the part, cores, and insert supports. In ordinary compression molding, these parts must be quite substantial, or they will be damaged before the resin reaches its plastic state. In transfer molding, this problem is largely eliminated, and more intricate and fragile pieces can be produced.

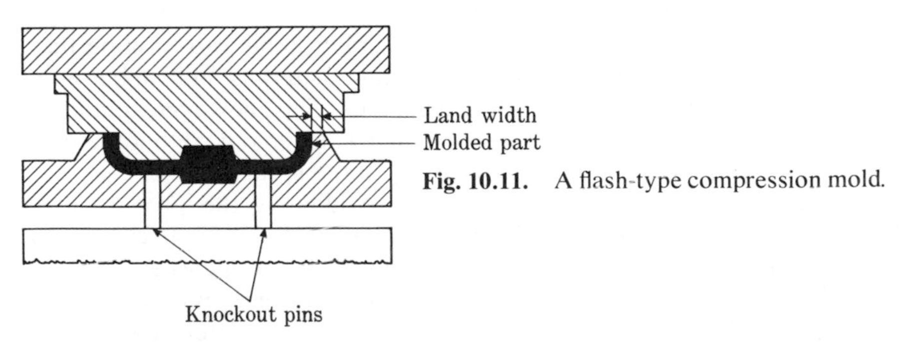


**Fig. 10.11.** A flash-type compression mold.

**Extrusion Molding.** The extrusion of plastics consists of feeding plastic material, in powder or granular form, from a hopper to a heated cylinder. A rotating screw carries the heated plastic forward and forces it through a heated die orifice of the required shape (Fig. 10.12). As it leaves the die, it can be cooled by running it through water or by air blasts, or it can be allowed to cool at room temperature. The cooling period must be carefully controlled, because either too fast or too slow cooling can cause distortion.

Applications. Extrusions are applicable to the continuous production of tubes, rods, sheets, films, pipe, rope, and a wide variety of profiles. Extrusions can be used to coat wire by having it pass through the center of the die, with the plastic extruded around it. Extrusions are thermoplastics and are not recommended for structural members bearing heavy loads.

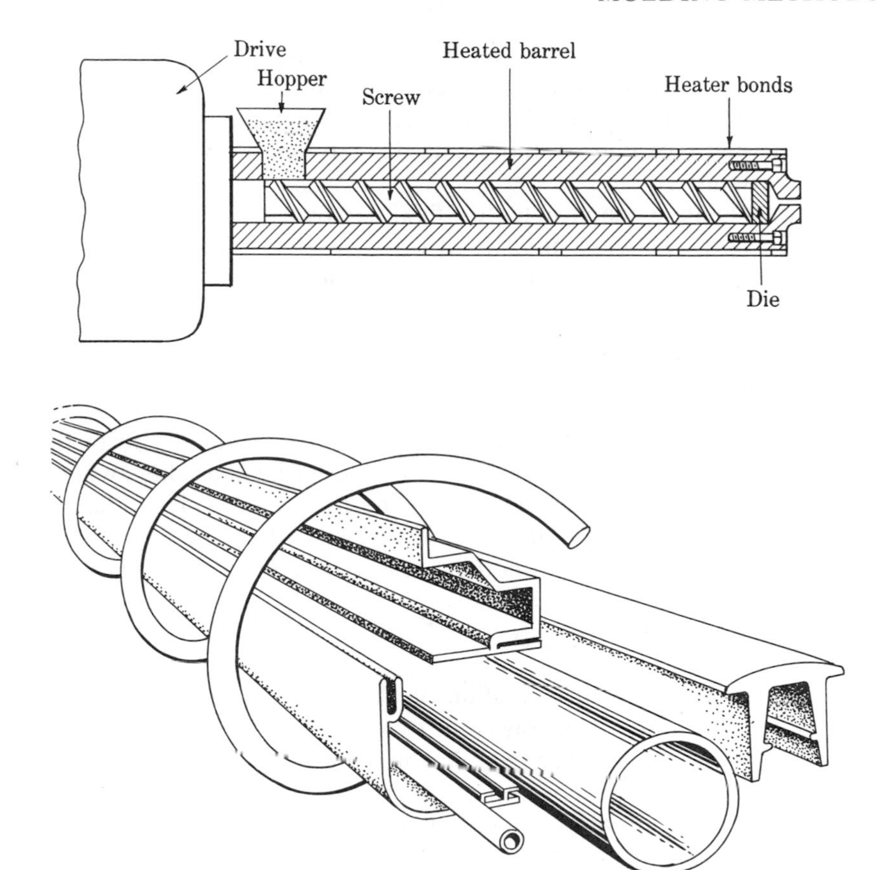


**Fig. 10.12.** Schematic arrangement of a machine for extrusion molding. Also shown are plastic extrusions.

**Casting Molding.** All the methods described thus far use heat and pressure, but *casting* of plastics involves heat without pressure. The plastic, either thermoplastic or thermosetting in granular or powdered form, is liquefied either by heating or with solvents. The liquid is then poured in the mold and cured at room temperature by the addition of catalysts.

Cast acrylic sheets are the highest grade of plastic sheets available. They are superior to other plastic sheets prepared by molding or extrusion because of the optical quality of the surface, freedom from internal stress, and excellent outdoor aging characteristics. Plastic film (sheeting 0.010 in. and under) can be produced by casting it on a moving belt or by precipitation in a chemical bath.

**Slush Molding.** Slush molding is somewhat similar to cast molding in that no pressure is used. In this process a thermoplastic-resin slurry or "slush" is poured into a preheated mold. The heat causes the slurry to set in a viscous layer of the desired wall thickness, after which the excess slurry is poured out. The resin left in the mold is cured by additional heat and then chilled. In its final flexible form it is simply stripped out of the mold.

Slush molding is applicable for flexible toys, insulated overshoes, artificial flowers, squeeze bulbs, and other products where difficulty is experienced in removing the part from the mold.

## CALENDERING

An important method of making film and sheet is known as calendering. In this process, the plastics compound, usually in the form of a warm doughy mass, is passed between a series of heated rollers where it is thoroughly worked. It emerges from the rolls squeezed into flat film or sheet. The thickness of the sheet is determined by the gap between the last pair of rolls—the gaging rolls. It is cooled on a chill roll.

## ROTATIONAL CASTING

Rotational casting is an evolution of slush molding that is relatively new. The mold, which may consist of two or more pieces, is partly filled with plastic material and then closed. The mold, while being heated in an oven, is rotated simultaneously in two or more planes. The rotation of the mold allows material to be deposited and jelled on all its inner faces. Gravity, not centrifugal force, is responsible for coating the mold wall. When the desired wall thickness has been deposited on the mold, it is removed from the oven, cooled, and stripped out. Items made by rotational casting can be completely enclosed, whereas slush molding requires an opening through which the excess material is poured out.

## FABRICATING METHODS

Plastic films, sheets, extrusions, and molded shapes often need other fabricating processes before they are ready for market. Fabricating includes sheet-forming methods, blow molding, laminating and reinforcing (both low-pressure and high-pressure methods), machining, and welding.

**Sheet-Forming Methods.** One of the most important fabricating operations—and one that has virtually grown into an entire industry of its own—is that of heat-forming film or sheeting. The basic principle is this: The heat-softened film or sheet is forced against a cold mold, permanently taking on the contours of the mold. A wide range of forming methods has been developed. In addition to mechanical pressure, air and vacuum are used (Fig. 10.13).

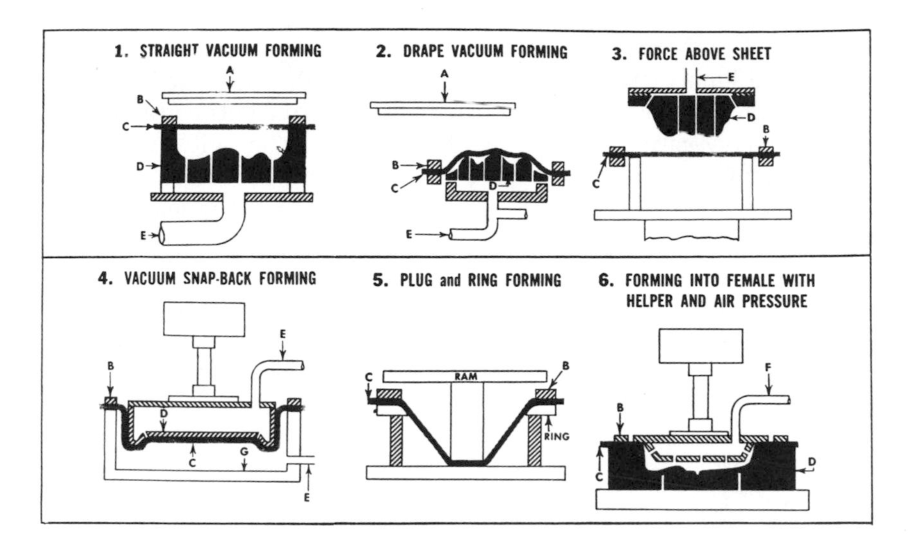


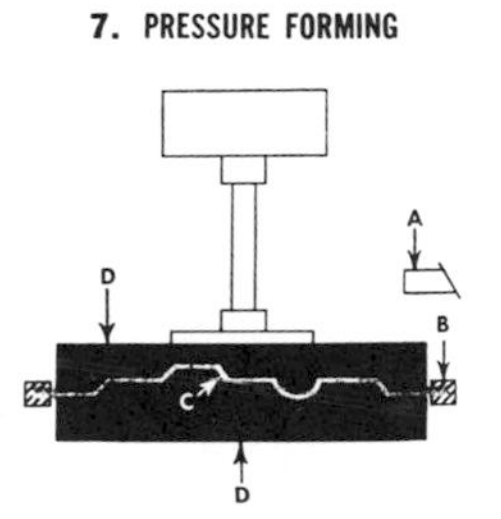


IN THESE SIMPLIFIED schematics of seven basic methods of sheet forming, letters indicate the following: A = Heater; B = Clamp; C = Plastic Sheet; D = Mold; E = Vacuum Line; and F = Air-Pressure Line.

THE methods sketched here are as follows:

(1) STRAIGHT VACUUM FORMING. Clamped in a stationary frame, the heated sheet is vacuum-drawn into the mold.

(2) DRAPE VACUUM FORMING. Moveable frame or clamp drapes the sheet, softened by heat, over male mold before vacuum is applied.

(3) FORCE ABOVE SHEET. Mold descends onto heated sheet, partially forming it; then the vacuum is applied.

(4) VACUUM SNAP-BACK FORMING. Vacuum is applied, drawing preheated sheet into cavity G. Male plug moves down until it reaches a predetermined position. Vacuum is then applied through male plug.

(5) PLUG AND RING FORMING. Heated sheet is placed over a ring and clamped down. Mold mounted on ram is forced into it.

(6) FORMING INTO FEMALE WITH HELPER AND AIR PRESSURE. As press closes, cored plug pushes heated sheet into cavity. Air pressure, introduced through plug, pushes sheet into female mold. Holes in mold let air escape.

(7) PRESSURE FORMING. After heating, framed sheet is formed between matched male and female dies.

**Fig. 10.13.** Basic methods for forming thermoplastic sheets. Courtesy *Modern Plastics Encyclopedia.*

New developments in sheet forming. New developments point the way to higher production through completely automated machines. For example, new double-heater machines heat from both top and bottom to form heavy-gage sheets up to 0.250 in. thick in half the time previously required. The heaters are independently operated so that the bottom heater can be withdrawn automatically at the moment the material begins to sag, while the top heater continues heating to the exact temperature needed.

**Blow Molding.** Blow molding is a simple process of placing a soft, extruded, thermoplastic closed-end tube in a cavity and applying air pressure to inflate it (Fig. 10.14). After allowing sufficient time for the plastic to cool, the mold is opened and the part is removed.

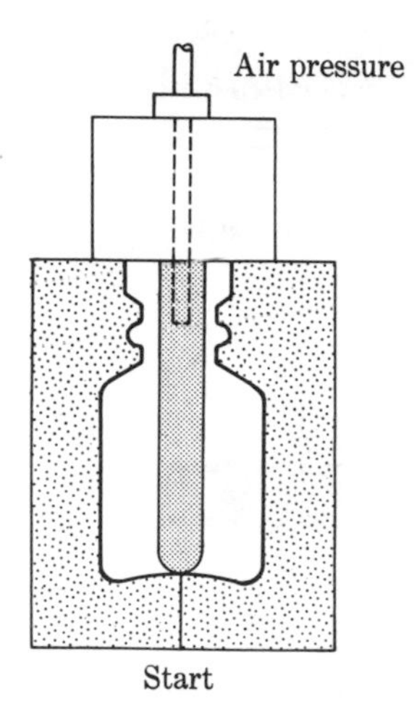


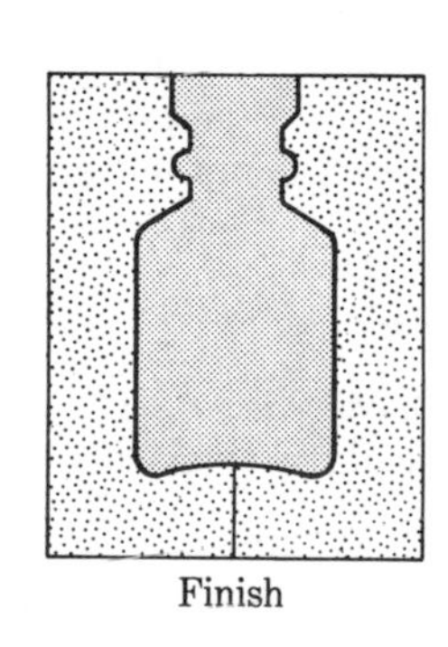


**Fig. 10.14.** Blow molding used to form thermoplastic bottles.

## REINFORCED PLASTIC MOLDING METHODS

Reinforced plastics can be molded in either open or closed molds (Fig. 10.15).

**Open Molds.** Open molds are single-cavity molds, either male or female, with the following characteristics:

(1) The initial investment is low. (Cost mounts up when molds have to match. Since little pressure is exerted, the mold can be made of plaster, wood, or reinforced plastics.)

(2) The labor cost per part is relatively high.

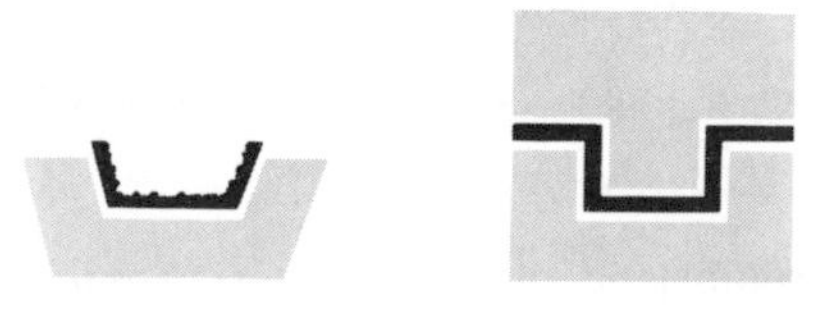

**Fig. 10.15.** The principles of the open mold and the closed mold are schematically shown.

(3) Molds can be made quickly, and design changes are relatively simple.

(4) The production rate is low.

Parts can be made in open molds by a number of different methods, including contact, pressure-bag, vacuum-bag, and autoclave molding (Fig. 10.16).

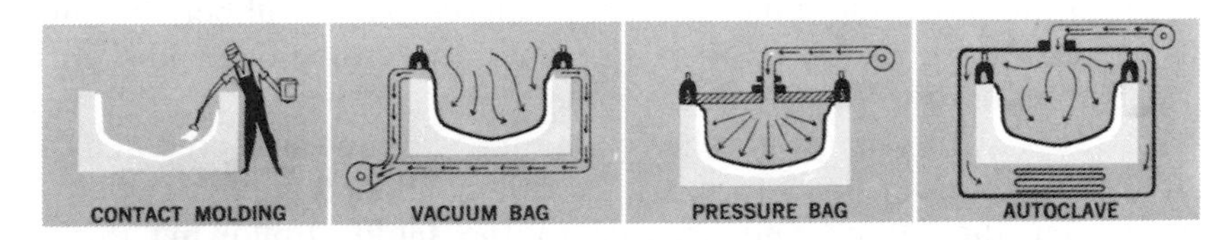


**Fig. 10.16.** Open mold methods used in fabricating reinforced plastic parts. Courtesy Owens-Corning Fiberglas Corporation.

CONTACT MOLDING. Layers of reinforcing material, in the form of glass fabric, mat, chopped strands, etc., are coated with catalyzed resin and placed in layers on the prepared form. The number of layers depends on the strength required in the finished product. The reinforcing material acts like steel in concrete, adding to the mechanical strength, impact resistance, and dimensional stability.

To speed up output of boats and other products and to make possible on-the-site fabricating, the reinforced-plastics industry developed the *spray-up molding* method. It utilizes the same types of molds used in hand lay-up, but the reinforcing material and plastic are applied with spray guns. Several types of guns are used. Some are made to spray both resin and reinforcement simultaneously; others alternate streams of resin and reinforcement (Fig. 10.17).

The resins used are epoxies or polyesters. The guns that deposit

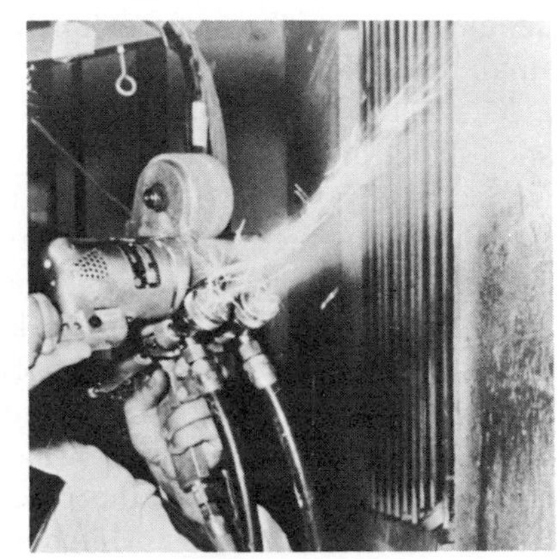

**Fig. 10.17.** A spray gun used to apply both fiber glass and resin. Courtesy Spray-bilt.

reinforcement use continuous-filament glass fibers that are chopped into short lengths as they are applied.

The major advantages of contact molding are the short time it takes to get into production and the relatively low-cost molds. Another factor is that large-size moldings, such as boats, tanks, trucking bodies, and chemical equipment, can be formed.

PRESSURE-BAG FORMING. A lay-up of resin-impregnated material is placed on the mold and covered with a tailored bag of rubber sheeting. Air or steam pressure up to 50 psi is used between the bag and a pressure plate secured to the top of the mold.

VACUUM-BAG FORMING. The mold and lay-up are prepared as described in pressure-bag forming. A cellophane or polyvinyl film is placed over the lay-up and secured to the mold. Following this, a vacuum is drawn through the ports provided, and the resulting atmospheric pressure exerted on the film forces the layers to conform to the mold contours.

AUTOCLAVE FORMING. Autoclave forming is a modification of the pressure-bag method, the main difference being that, after the lay-up has been made and covered with a plastic film, the entire assembly is placed in a steam autoclave at 50 to 100 psi. The additional pressure permits higher glass content and improved air removal.

OPEN-MOLD SUMMARY. The advantages and disadvantages of the various open-mold methods are given in Table 10.1.

**Closed, or Matched-Die, Molding.** MANUFACTURE. Where fully automatic production at rapid cycling rates is desired, matched-metal dies give the best performance. The tooling usually used is solid steel for small products, and cast Meehanite or semisteel for larger parts. The dies are made by machining on duplicating mills or other equipment until they are within close tolerance. After machining, they are filed, polished, and flame hardened. The surface is again polished and buffed to a high luster. A large percentage of the die-manufacturing time is allotted to developing a high surface finish. The molded product will show any small defect left on the die surface.

Matched dies can also be made from reinforced plastics for medium- and short-run production. A wood or plastic pattern is first sealed, waxed, and coated with parting material. A box is built around it, and the surface of the pattern is covered with layers of resin-saturated glass fabric. After the lay-up has had time to set, a mass casting of epoxy resin and filler can be poured in to make up bulk for the remainder of the die. After the die has had time to cure, the pattern is removed, and the process is repeated to make the matching punch. This time, the die serves as the pattern. Sheet wax or felt is placed in the die before the punch is poured to take the place of the material that will be molded.

MOLDING METHODS. Matched-die molding can be done with either reinforced or unreinforced materials. The processes used are quite

**Table 10.1**
**Advantages and Disadvantages of Various Open-mold Fabricating Methods***

| | Contact Molding | Vacuum Bag | Pressure Bag | Autoclave |
|---|---|---|---|---|
| Advantages | 1. Simplest process.<br>2. Low cost molds.<br>3. No size restrictions.<br>4. Max. design flexibility.<br>5. Design changes readily made.<br>6. Min. equipment needed.<br>7. Gel-coats possible. | 1. Higher glass loading.<br>2. Better unfinished side.<br>3. Less air and voids.<br>4. Better adhesion in sandwich constructions possible.<br>5. Retains advantages of contact molding. | 1. Cylindrical shapes can be made.<br>2. Higher glass loading.<br>3. Dense, void-free moldings.<br>4. Undercuts possible.<br>5. Cores and inserts used.<br>6. Retains advantages of contact molding. | 1. Undercuts possible.<br>2. 65 percent glass loading.<br>3. Dense, void-free moldings.<br>4 Cores and inserts used.<br>5. Retains advantages of contact molding. |
| Limitations | 1. Labor per unit is high.<br>2. One finished surface.<br>3. Quality depends on operator. | 1. More labor.<br>2. Surface next to bag not as good as surface next to mold.<br>3. Quality depends on operator. | 1. Only female molds.<br>2. More labor.<br>3. Surface next to bag not as good as surface next to mold.<br>4. Quality depends on operator. | 1. Extra labor to load autoclave.<br>2. Autoclave is expensive.<br>3. Size of autoclave limits size of parts that can be made.<br>4. Quality depends on operator. |
| Output† | 1 boat hull (25 feet).<br>4 tanks (2 × 4 × 2 feet). | Slightly slower than contact molding. | Slightly slower than contact molding. | Slightly slower than contact molding. |
| Mold‡ | Wood, plaster, sheet metal, reinforced plastics. | Wood, plaster, sheet metal, reinforced plastics, castings. | Reinforced plastics, sometimes heated castings. | Sheet metal, castings. |

*Courtesy Owens-Corning Fiberglas Corporation.
†Based on one mold for eight hours.
‡Usual mold materials.

similar, but the reinforced type, described here, is more comprehensive. These processes are preform molding, mat and fabric molding, and premix molding.

PREFORM MOLDING. In reinforced preform molding, the mat or fabric is shaped to conform to the mold in which it will be placed. This is done in three different ways, as shown in Fig. 10.18.

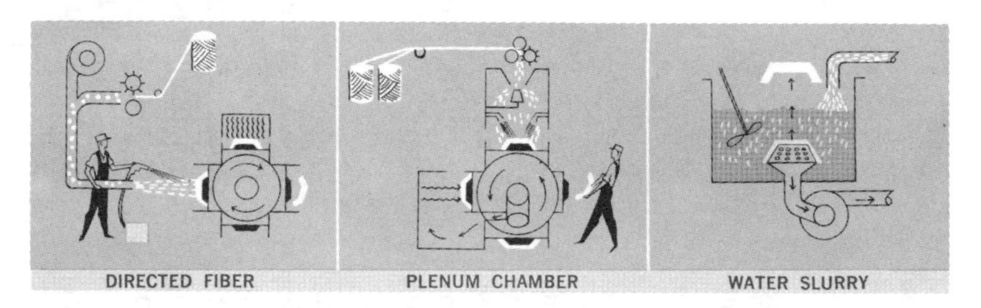


**Fig. 10.18.** When complex cross sections are molded, fabric preforms are made by these methods. Courtesy Owens-Corning Fiberglas Corporation.

The preform, made by one of the described processes, is combined with a resin mix either just prior to or just after placing it in the matched-die mold cavity (Fig. 10.19). The heated metal molds form and cure the part at 100 to 300 psi. Molding temperatures usually range from 225° to 300°F. The cure cycle varies from slightly under 1 min to 5 min, depending on thickness, size, and shape.

MAT AND FABRIC MOLDING. This process is similar to preform molding, except that the material is placed in the mold in chopped-strand mat or fabrics that have been tailored to the die. The resin may be poured on it, as described previously, or the material may be purchased with the resin already in it. This is known as *pre-preg* material. The synthetic resins are partly polymerized. Pre-pregs are made in a variety of materials, from epoxy paper to asbestos phenolic. Pre-preg fabrics

**Fig. 10.19.** The resin mix may be added just prior to or just after the mat fabric or preform is placed in the mold. Courtesy Owens-Corning Fiberglas Corp.

facilitate molding many high-strength containers and structural components (Fig. 10.20).

ADVANTAGES. Matched-die molding is a comparatively high-speed process. The parts have high strength, uniform appearance, and good surface finish. The automobile industry is now one of the large-volume users of matched-die plastic products.

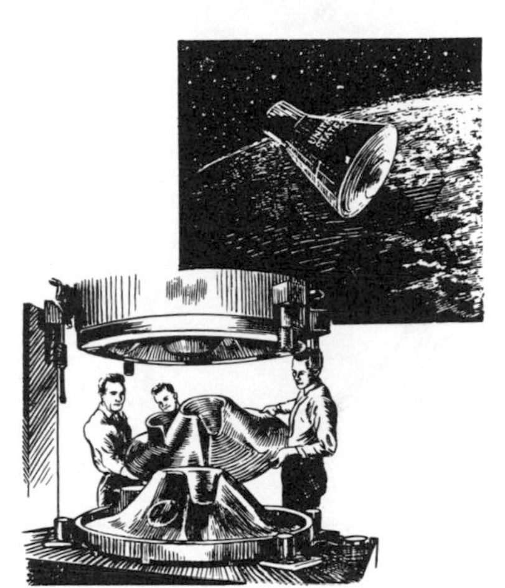

**Fig. 10.20.** Pre-preg plastic materials are being used to form missile containers, access doors for reentry vehicles, structural components for aircraft, etc.

## FILAMENT WINDING

One of the newer methods of building light-weight, high-strength, reinforced plastic structures is by filament winding. This consists of impregnating strands of reinforcing material, usually glass fibers, with epoxy resin. These strands are then wound directly on the basic structure or on a mandrel, which is later removed.

Originally "wet winding," or impregnating the strand just before it was applied, was the only method used. Now, pre-preg material is available; this not only simplifies the process but also makes possible the use of a wider variety of resins.

The filament can be applied in any one of three standard patterns —circumferential, helical, and longitudinal (Fig. 10.21). These methods may be used alone or in combination to give the most effective ratio of loop to longitudinal stress.

The outstanding property of filament-wound structures is their high strength-to-weight ratio—even better than high-strength steel or titanium.

Aircraft-wing specimens are now being wound from glass filaments for jet trainers. It is expected that this will improve efficiency and reliability of the section and to lower costs. The wing specimens include spars, ribs, and attachments.

**Helical pattern is generated on winding machine** that rotates the part and traverses the filament feed back and forth

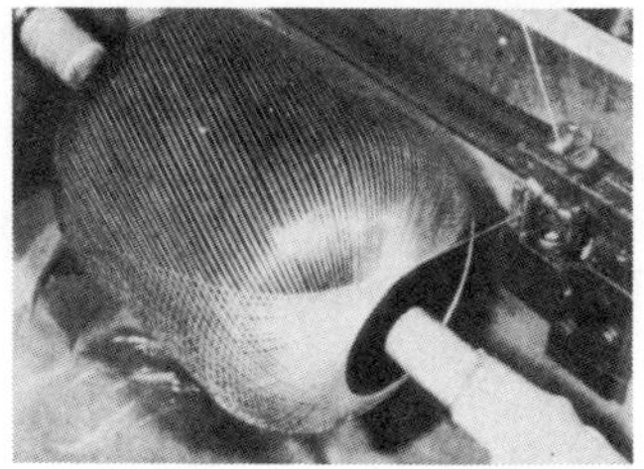

**Longitudinal winding will be overlaid** with circumferential winding, as filament strength is high along axis, low in shear

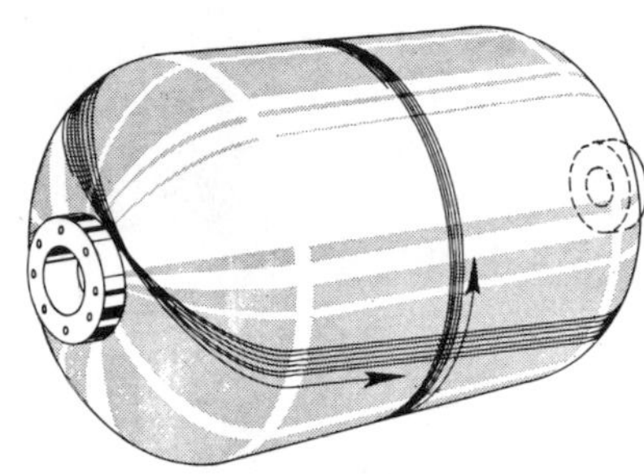

-Two types of filament windings on a pressure vessel.

**Fig. 10.21.** Three types of filament winding: helical, longitudinal, and circumferential.

## MACHINING

Virtually all plastics can be machined with both conventional woodworking or metalworking tools.

The major consideration is that plastics have a greater heat sensitivity and greater thermal expansion than metals. If heat is allowed to build up, the materials will become gummy; also stresses will hamper the performance of the completed part.

Coolants, such as a 10 percent solution of water-soluble oil or compressed air, should be used whenever possible.

Turning tools should have a zero to $-5$ deg top rake to provide a scraping rather than a cutting action. Clearances should be much greater than for metal cutting to prevent any chance of rubbing and heating the plastic. In general the surface speed should be high, ranging from 300 to 800 sfpm for polycarbonate to 1500 to 3000 sfpm for acrylics and styrenes. Light feeds and depth of cut are recommended.

Acrylic and polystyrene plastics can be machined to tolerances of $\pm 0.001$ in. However, to maintain close tolerances, it is necessary to anneal at the time of machining. This is especially true when considerable material is to be removed and there is likelihood of stress buildup. Annealing can be accomplished by heating the plastic to about 10°F below the practical heat-distortion temperature. Because of the poor heat conductivity, the time required for the anneal may be as long as 24 hr.

Laminated plastics are very abrasive. High-speed tools can be used, but carbides, in most instances, prove to be the most economical. A good exhaust system is recommended as a health precaution when machining laminates.

## WELDING

Most thermoplastic materials can be welded by processes similar to those used in welding metals. These include hot-gas welding, heated-tool welding, and friction welding.

**Hot-Gas Welding.** Instead of the oxyacetylene torch, an electrically heated gun is used. Compressed air passes the heating element and strikes the joint area at about 400°F. The gun is held from 1/8 to 1/2 in. away from the filler rod and joint (Fig. 10.22). Nitrogen gas is used in place of compressed air on oxygen-sensitive plastics such as polyethylene. Filler material, when needed, can be a strip of the parent plastic.

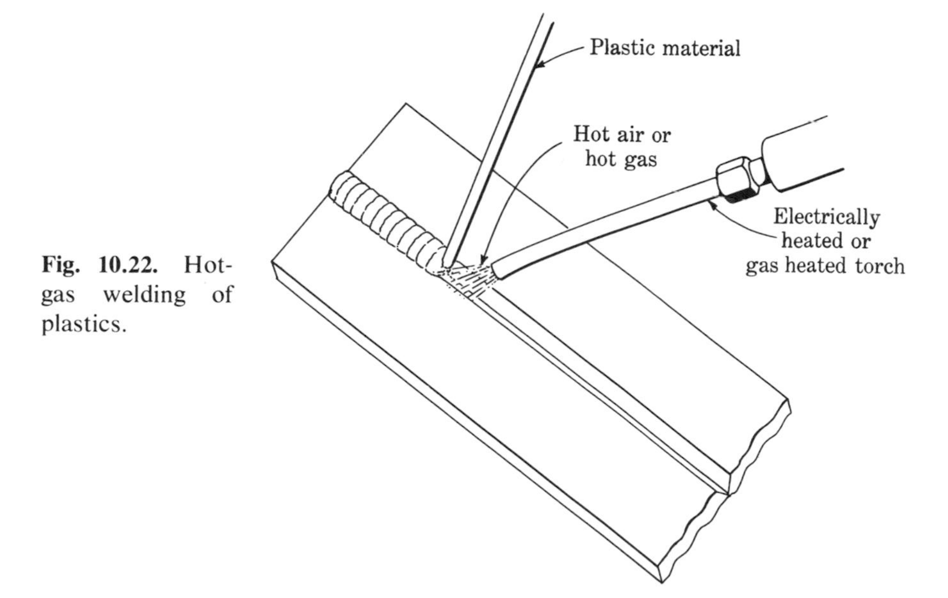


**Fig. 10.22.** Hot-gas welding of plastics.

**Heated-Tool Welding.** Another welding technique employs a heated plate. The plastic films or sheets are brought in contact with the heated surface until they are sufficiently soft, at which time they are removed from the heat and quickly joined together. They are firmly held together until the melted plastic cools and makes a firm joint.

Gas-heated tubes, electrical strip heaters, and soldering coppers are convenient sources of heat. The time required for fusion to take place varies, but an interval of 4 to 10 sec at 3 to 12 lb pressure generally gives good results. The time interval from the time the two pieces are removed from the heat and until they are joined is the most critical factor and should be kept at 1 sec.

**Friction Welding.** Two plastic parts may be fusion-welded together by frictional heat. The easiest way to accomplish this is to have one part spinning in the lathe and bring the stationary part up against it. All that is required is a pressure ranging from 10 to 200 psi for a few seconds. The whole operation, including chucking, seldom requires more than 30 sec.

The parts to be welded require no special cleaning, since the surface film and dirt are squeezed out in the course of the spinning operation.

## DESIGN CONSIDERATIONS

In designing a part to be made out of some type of plastic, one must first consider all aspects of its proposed use and whether it will serve effectively and economically. A preliminary design can then be made, based on shape limitations, dimensional tolerances, critical dimensions, methods of bonding, etc. At this time the designer makes his final selection of material and process, keeping in mind the advantages and limitations of each. Some principles of material and process selection are as follows:

(1) The process of manufacture is based on the material used and on the design.

(2) Chemical, electrical, and thermal performance depend on the type of resin used.

(3) In reinforced plastics, strength results from the arrangement and the amount of glass used.

## RELATIONSHIP OF STRENGTH TO REINFORCEMENT

There is a straight-line relationship between the amount of glass used and the strength of the finished product. For example, a part containing 80 percent glass and 20 percent resin is almost four times stronger than a part containing the opposite proportions.

The arrangement of the reinforcement is just as important as is the amount. There are three standard arrangements: parallel strands, or one-directional; right angle, or two-directional; and random, or in all directions.

One-directional placement of fibers develops maximum strength, but in *one direction*. Examples of uses are rocket-motor housings, golf clubs, and fishing rods.

The two-directional arrangement has less strength than the parallel arrangement, but it is in *two directions*. Some applications are for wing tips, boats, and swimming pools.

The random pattern has the least strength, but it is in *all directions*. This pattern is used in safety helmets, chairs, luggage, and machine-tool housings.

The amount of glass that can be loaded into an item is dependent on the arrangement: when strands are placed parallel, the loading possibility ranges from 45 to 90 percent; when they are at right angles, the loading drops to from 55 to 75 percent; a random arrangement permits only 15 to 50 percent loading.

A summary of the most common thermosetting and thermoplastic materials, their trade names, cost, general properties and applications is given in the Appendix.

## ELASTOMERS

**Thermoplastic Elastomers.** The ASTM definition of an elastomer is "a material which at room temperature can be stretched repeatedly to at least twice its original length and upon immediate release of the stress will return with force to its approximate original length." The thermoplastic elastomers pass this test, whereas the currently available flexible plastics do not. This resiliency makes them an excellent upholstery material.

In addition to excellent elastic recovery, thermoplastic elastomers exhibit the tensile properties typical of rubber vulcanizates.

The relationship of the thermoplastic elastomers to other plastics and rubbers is shown in Table 10.2, where they are classified as to the two main types and also as to their rigidity.

**Table 10.2**
**Classification of Polymers**

| | Thermosetting | Thermoplastic |
|---|---|---|
| Rigid | Epoxy<br>Phenol-formaldehyde<br>Urea-formaldehyde<br>Hard rubber | Polystyrene<br>Polyvinyl chloride<br>Polypropylene |
| Flexible | Highly loaded and/or highly vulcanized rubbers | Polyethylene<br>Ethylene-vinyl<br>Acetate copolymer<br>Plasticized PVC |
| Rubbery | Vulcanized rubbers (NR, SBR, IR, etc.) | Thermoplastic elastomers |

The reason that elastomers exhibit a high degree of elasticity can be explained by their molecular structure. Under load, the chain molecules straighten out. Normally they are in a helical coil arrangement. The process of straightening does not strain the individual bonds.

**Thermosetting Elastomers.** The most widely known elastomer is natural rubber (nr), also known as latex. This material can be cross linked through sulphur atoms in a process known as *vulcanization*. When excessive cross linking occurs, uncoiling of the molecular chain is impossible and a rigid structure is formed. Cross linking takes place naturally due to the presence of oxygen. The process is accelerated by heat or strong sunlight.

Other elastomers are: butyl, used for tires and tubes; ethylene-propylene rubber (EPR), used in steam hose and seals; fluoroelastomer, used in high temperature seals and valve seats; neoprene, used in radiator hose and tank coatings; and polybutadiene, used in truck tires and mechanical goods.

## NEW DEVELOPMENTS

**High-Temperature Plastics.** Scientists have been searching for polymers with higher temperature characteristics. Some polymers can be used for short periods of time up to about 1000°F, but most break down entirely at 450°F. Now new ways of connecting the molecular chain and new elements may raise the temperature to 2000°F.

Under investigation are ladder polymers which are in the early stages of development. Chemists are building on the technology of silicone polymers by incorporating stabilizing elements. The most promising of these agents are boron-hydrogen-carbon compounds called carboranes. The carboranes are said to have opened up a whole new field of polymer chemistry with potentials perhaps as great as those of organic chemistry.

**Whisker-Reinforced Plastics.** The aerospace industry research has created a wide variety of fibers with potentially high strength and stiffness, good thermal resistance, and low specific gravity. The best combination of these properties are found in the single-crystal fibers or "whiskers."

Very recently silicon carbide (SiC) whiskers have become available in large quantities, permitting extensive evaluation. One process that appears promising is that of orienting the (50 percent by volume) whisker composite to give increased directional strength. It has been found that by adding 30 to 50 percent by volume SiC whiskers, the modulus went 100 percent higher than that of the parallel fiber-glass composites.

Highly oriented whisker yarn is now being made which is incorporated into the plastic composites thus solving the more difficult problem of maintaining proper orientation and avoiding whisker damage. It should now be possible to prepare structural composites and determine advanced engineering properties, such as energy absorption, long-time fatigue, dimensional stability, and directional strength and stiffness. There is every reason to expect higher performance composites when processes for whisker and composite manufacture are refined.

## ADHESIVES

The search for faster, more efficient bonding methods has focused manufacturers' attention on adhesives.

Adhesives new to industry are chemical compounds that have replaced some of the older protein glues made from animal hides, soybeans, etc. Present-day industrial adhesives, of which there are many, are based primarily on various epoxy and polyester resins and natural and synthetic rubbers.

## ADHESIVE BONDING

Adhesive bonding, instead of the more conventional methods of assembly such as riveting, bolting, soldering, and welding, is being used in the production of railway cars, boats, refrigerators, storage tanks, microwave reflectors, and countless other items of commercial and industrial use. Many of the joints made in fabricating aircraft wing-and-tail assemblies are of the adhesive-bond type. Increased use is also evident in the fabrication of aircraft internal structures. Cost, in many instances, is cut one third, and time, 75 percent.

Almost any structure within a reasonable temperature service range can be assembled through adhesive-bonding techniques. Most of these bonded structures will show a favorable weight-to-strength ratio and a decrease in cost compared to conventional fastening and joining methods.

In the aircraft industry, enthusiasm for the technique grew with the discovery that "gluing" provided the needed smooth surfaces for supersonic planes, permitted thinner skins with consequent weight reduction, made more complex designs possible, and avoided local stress concentrations around conventional fasteners.

Adhesive bonding increases damping effect and resistance of the joint to fatigue and reduces corrosion. It is also possible to combine the liquid sealing of joints with the actual joining operation, both so important in wing-tank construction.

Adhesive-bonded sandwich-type construction is used in radar reflectors and space communications, because of the close dimensional tolerances that can be maintained. Reflector "horns" up to 180 ft long have employed adhesive construction. The strength of some adhesives under cryogenic (very-low-temperature) conditions has been very good.

The automotive industry has been slower to accept adhesive bonding. Aside from its well-known application of attaching brake linings to brake shoes, other efforts have been largely experimental. More recent applications have been binding stamped X-members to the underside of front hoods and rear decks. The plastic resin is applied to the stamping on the press line, gaining some strength then, but final cure takes place in the heat of the paint-drying booth.

**Advantages.** The advantages gained from this method of joining have made it attractive to the metalworking world. Some of these are as follows:

(1) There is no difficulty in bonding dissimilar materials such as ceramic-to-steel, alnico to zinc, copper to steel, or magnesium to cast iron. These are just a few of the combinations that are now possible with an epoxy adhesive. The brazing of an alnico magnet to a zinc die casting would involve considerable skill with a welding torch.

(2) Materials can be attached to very thin metal parts, far too thin for any other type of fastening, for example, a motor stack laminate, where a sprayable epoxy is used (Fig. 10.23).

**Fig. 10.23.** Thin motor laminates are successfully bonded together with epoxy-resin adhesives. Excellent dielectric strength is furnished, and there is no metallic corrosion between the dissimilar metals. Courtesy Rubber and Asbestos Corp.

(3) In most cases there is no need for high heats in joining.

(4) Adhesives automatically provide heat or electrical insulating layers between two joined surfaces. In the motor laminates of Fig. 10.23, adhesives furnish excellent dielectric strength and low power factors. Such dissimilar metals as beryllium, copper, and stainless steel can be joined without metallic corrosion.

(5) The bonding load is distributed uniformly over a large area. As in a lap joint, the load is distributed over the entire area of the contacting surfaces.

(6) Fewer specialized personnel are needed.

(7) In many cases, weight can be reduced and service life increased, as in aluminum honeycomb assemblies.

(8) Substantially less afterfinishing is required, as compared with welding, soldering, or brazing methods.

(9) Adhesives often permit extensive design simplifications. Laminated wood or metal parts, bent to shape before bonding, require much less machining time than does a part cut from solid stock.

(10) The lighter weight of the completed assembly means greater economy in packing and shipping.

(11) They do not change the contour of parts they join. One aircraft now built uses 900 lb of adhesives.

**Disadvantages.**

(1) Adhesives are not as stable as metal-joining methods.

(2) The joints are difficult to inspect and/or test nondestructively.

(3) At the present time, extensive tests for durability and permanence of adhesive joints have not been conducted. Some standard procedures have been specified by the military and by ASTM.

(4) Some rather elaborate jigs and fixtures may be needed to supply heat and pressure.

(5) Adhesives may be susceptible to high humidity and extremes of temperature.

(6) They may have less strength than some other joining methods.

(7) Strong reliable joints are produced only on clean surfaces.

## PRINCIPLES OF ADHESIVES

Adhesives, whether applied as a liquid or as a film, are dependent for strength on the following factors: chemical forces between the components, cohesive strength of the components and adhesive, and the effective area of the bond.

As an example, a postage stamp is moistened to make tacky the dextrin film on the back. When it is applied to an envelope, the water dries out through the fibers in the paper, and a strong bond results owing to both fibrous entanglement and chemical affinity. If the stamp is applied to glass, there will be no fibrous interlocking; instead, there will be a chemical affinity with the oxygenated materials present. Without this chemical reaction or fibrous interlocking, there will be no effective bond. To illustrate: A crude-rubber cement containing sulfur, accelerators, and carbon black is applied alike to polished steel plate and polished brass. After it is heated for vulcanization to 300°F for 20 min and 300 psi are applied, there is no bond with the steel, but with the brass there is a bond strength of up to 60 psi. This is due to a chemical reaction that took place, forming a strong compatible film of complex sulfides.

Present theory holds that adhesion is primarily due to a chemical affinity of the adhesive for the substrate, and that mechanical action is only incidental. A schematic representation of bonding is shown in Fig. 10.24.

During World War II, German investigators were vulcanizing Buna copolymers by the addition of diisocyanates. It was noted that the two materials reacted but, what was more important, these vulcanized materials adhered strongly to the metal parts of the vulcanizing press. It was soon found that Buna-sulfur mixtures could be attached to iron,

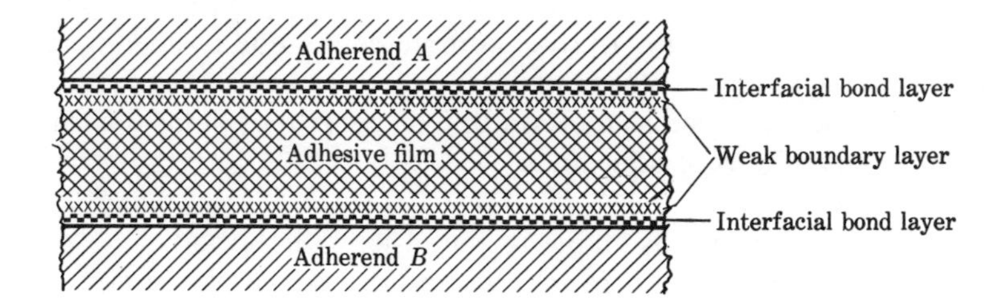


**Fig. 10.24.** Most adhesion is due to molecular attraction between the adhesive and the adherend. Some mechanical bonding is possible in porous materials.

light metals, procelain, etc., with heat-resistant bonds. Adhesive strengths up to 1200 psi were obtained. Rupture occurred in the rubber, not in the bond. The success of diisocyanates during the war led to their use in a broad field of adhesives.

Solutions of diisocyanate in O-dichlorobenzine have been found to produce excellent adhesion between various elastomers and metals. The bonds have excellent resistance to elevated temperatures, flexing and impact, and are not affected by oils and solvents.

New adhesives based on fully aromatic (closed-ring structure of six carbon atoms, Fig. 10.25) have great potential for making structural bonds to withstand elevated temperatures for relatively long periods of time. One of these is Imidite-850, which is based on a polybenzimidazole resin. Tensile shear strengths of 2000 psi at 700°F were obtained with curing pressures as low as 10 psi on lap joints of 17–7 PH steel.

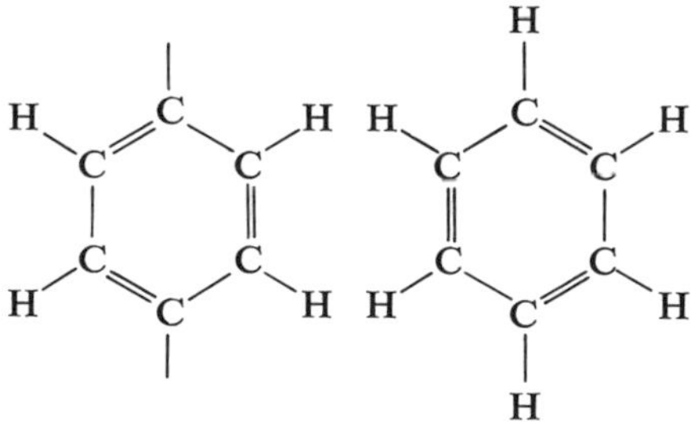

**Fig. 10.25.** Benzine is an aromatic compound consisting of a closed-ring structure with six carbon atoms. Each carbon atom is bonded to only three other atoms, and the bond angle is 120 degrees.

Pressure is generally used with adhesive application to: (1) assure adequate contact between all members of the system; (2) hold the parts together while the solvent is drying or while a cure is taking place that gives the adhesive its strength; (3) offset the destructive effects of gases and steams evolved during the cure, as in most curing of rubbers, phenolics, elastomers, vinyl-phenolics, and amine-aldehyde; and (4) cause the adhesive to flow over the joint area uniformly and expel entrapped air.

## CLASSIFICATION OF ADHESIVES

Adhesives can be divided into two broad groups—structural and nonstructural. In the first group are those used because of their high load-carrying characteristics. In the second group are those, also known as glues or cements, used for low-load applications.

**Structural Adhesives.** Structural adhesives can be further classified as thermoplastic and thermosetting just as plastics are classified. The important difference is that the thermoplastics are reversible; that is, they may be resoftened repeatedly by heat. Of course, any adhesive that is heated to too high a temperature will decompose and lose its bond strength. The too-high temperature is governed by the chemical structure of the adhesive.

THERMOPLASTIC ADHESIVES. The thermoplastic adhesives most commonly used are the polyamides, vinyls, and nonvulcanizing neoprene rubber. In the structural field, the vinyls have proved very versatile. Polyvinyl acetate can be used to form strong bonds with glass, metals, and porous materials.

THERMOSETTING ADHESIVES. Thermosetting plastics can be formulated as strong, waterproof, mold-resistant, and heat-resistant adhesives. In this class, the phenol-formaldehyde resins have wide usage, forming one of the best bonding materials for waterproof plywood.

Resorcinol-formaldehyde resins are similar to phenolic resins but have the advantage of curing at room temperature.

One of the newer thermosetting adhesives that has been widely acclaimed is epoxy resin. It develops good strength, between 2,000 and 4,000 psi at room temperature, and cures without volatile by-products and with little shrinkage.

Silicone rubbers have also been used as bonding agents for nonstructural applications. They have proved successful in extremely low temperatures.

Other thermosetting adhesives are melamine-formaldehyde, polyurethanes, polyesters, phenolic-rubber, phenolic-vinyl and Buna and neoprene rubbers.

Thermosetting adhesives with reinforcing fabrics are best for minimizing *creep* in structural applications. By creep is meant the dimensional change encountered in a material under load, following the initial instantaneous elastic or rapid deformation. Creep at room temperature is sometimes called *cold flow*. Thermosetting adhesives are generally preferred wherever elevated temperatures are encountered.

Structural adhesives are also made from combinations of rubbers and synthetic resins. An example of this is a nitrile-rubber-phenolic-resin combination, which can develop a shear strength of 2,000 to 3,500 psi at room temperature. These adhesives combine the specific adhesion and strength of the phenolic resins with the flexibility and resilience of the rubber. Some structural adhesives used in the lap-joint bonding of aluminum develop 3,000 to 6,000 psi tensile strength at room temperature.

## LEVEL OF STRENGTH REQUIRED

The strength developed in adhesive joining is dependent not only on the inherent strength but on other factors, such as the type of joint used, type of loading, temperature, adherend material, etc. Therefore, it is difficult to give definite figures without citing the environment. The shear-strength ranges given in Table 10.3 are intended as a comparative guide only.

**Table 10.3**
**Comparative Shear Strengths for Adhesives**

| Type of adhesive | Average shear strength, Values at room temperature (psi) |
|---|---|
| Reclaimed rubber-type adhesives | 50–300 |
| Oil-soluble elastomers | 30–200 |
| Oil-resistant elastomers | 500–3,000 |
| Modified epoxies (heat cured) | 3,000–5,000 |
| Phenolic-elastomer films | 3,000–4,200 |
| Vinyl phenolic | 2,000–5,000 |

Structural adhesives, developed to produce high strength, are most often composed of synthetic resins or combinations of synthetic resins and elastomers. Common synthetic resins used are epoxy, urea, phenol, and resorcinol. The thermosetting-resin adhesives are generally hard and somewhat rigid when completely cured. The elastomer-resin combinations have high strength but retain a considerable amount of flexibility even after curing. Almost any adhesive can be made more flexible by formulation. Epoxy resins, for example, are made quite flexible by modification with polysulfide rubbers.

## PERMANENCE REQUIRED

In considering the permanence of an adhesive, environmental factors, such as the chemicals or solvents it will have to withstand, must be taken into account. Many resins are damaged by water, oil, certain acids, and alkalies. Weathering effects and temperature extremes are very important factors in deciding which adhesive is to be used.

Many adhesives can be formulated to match closely the coefficient of expansion of the adherends. This helps to minimize the stresses due to temperature changes.

## SURFACE PREPARATION

Thc first step in preparing the surfaces to be bonded is to see that they are smooth and well fitted. Next, they should be cleaned. Thorough preparation is of the utmost importance. Even a thumbprint on an otherwise clean surface will impair adhesion.

Some metals are cleaned and given an acid etch to provide greater cohesive strength. Because of the likelihood of contamination of the metal surface during storage, it is desirable to use the etched metal within hours after treatment. If storage is necessary, the metal should be kept tightly wrapped or in airtight containers to minimize contamination. The etched surface must never be touched with bare hands; handlers should wear clean cotton gloves.

## ADHESIVE JOINT DESIGN

The most widely used adhesive joint is the lap joint. Factors affecting it are as follows:

(1) The strength is proportional to its width.

(2) The strength is not proportional to its overlap. The strength increases at a slower rate than the overlap. A joint with a 2-in. overlap is less than twice as strong as one with a 1-in. A 5-in. overlap is not appreciably stronger than a 3-in. overlap.

(3) The strength of the lap joint is a function of the stiffness of the members. Stiffer members produce stronger joints.

(4) Joint strength is a function of the type of adhesive and the bonding process.

(5) Strength is not directly related to adhesive layer thickness. It varies with the type of adhesive.

(6) The average tension-shear values of the joint are influenced by the geometry of the joint as well as the elastic contents of the adhesive and adherend. Figure 10.26 shows a simple lap joint loaded in tension with the stress distribution (*a*) and similarly a tapered lap joint (*b*). Tapering the ends of the overlap reduces the shear stress concentration and increases the efficiency of the joint. If bending occurs, maximum tear stresses appear at the same locations.

## NEW AND FUTURE DEVELOPMENTS

Many commercial jet transport planes are now using adhesive-bonded structures. Military planes, rockets, and satellites also use them. The bonding of parts in the electronics field is advancing rapidly. The feasibility of bonding aluminum engines has been demonstrated, and automobile manufacturers have experimentally built up car and truck bodies with no welding, using adhesive bonding exclusively.

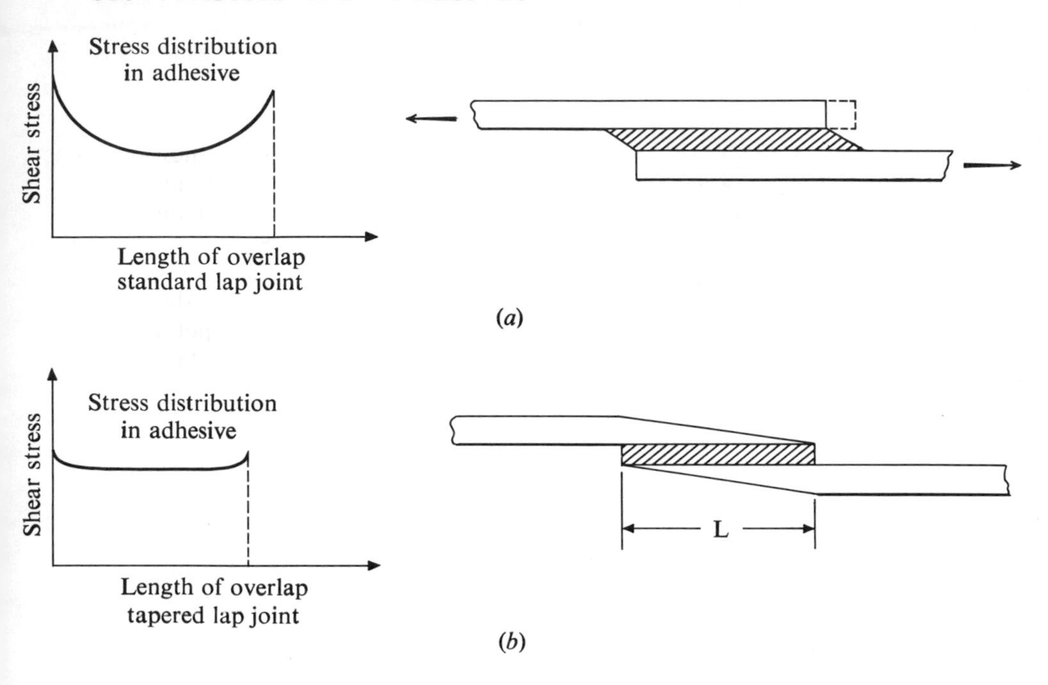


**Fig. 10.26.** A comparison of stress distribution for a standard (*a*) and tapered (*b*) adhesive lap joint.

With the advent of newer adhesives which offer good peel strength, fatigue resistance, and simplicity of tooling beyond anything available in the older materials, this method of joining will be increasingly used.

**Epoxy Pellets.** A newer method of applying adhesives is in the form of epoxy pellets (Fig. 10.27). The pellets are formulated from powdered epoxies that have been partially reacted and arrested. These powders are then pressed into a wide variety of shapes to suit the job. The size of the pellet will be such that, when heat is applied, the liquid formed will be just enough to fill a particular cavity.

**Fig. 10.27.** Cast-epoxy pellets can be obtained in a variety of sizes and shapes. Courtesy Joseph Waldmann and Sons, Epoxy Products Division.

A typical application of an epoxy pellet is shown in Fig. 10.28, where two aluminum tubes are joined. The epoxy pellet has been preshaped to fit the flared end of the tube. Brazing could probably have been used, but there might have been danger of melting the thin wall.

Pellets will cure at 300°F or as low as 185°F. Of course, the higher the heat, the faster the cure. A pellet that cures at 300°F requires only 3 hr, whereas at 212°F, 24 hr are required. The heaters most commonly used are induction, dielectric, oven, and infrared types.

The simplicity of handling makes the pellets much more adaptable to large runs and even to automated ones.

**Fig. 10.28.** An epoxy pellet is used to make a strong bond between two pieces of fitted aluminum tubing. The pellet, shaped to fit the large end, melts under heat and gives just enough liquid to fill the cavity. It then cures to form a bond. Courtesy Waldmann and Sons, Epoxy Products Division.

## QUESTIONS

**10.1.** Explain the basic structure of a polymer.

**10.2.** What is the significance of molecular orientation in polymers?

**10.3.** How are monomers able to add into an existing polymer chain?

**10.4.** Where do primary and secondary bonds exist in polymer chains?

**10.5.** Name an example of each of the plastic types A, B, and C according to their stress-strain curves.

**10.6.** Compare aluminum and magnesium with thermosetting plastics as to ultimate tensile strength.

**10.7.** What happens in the molecular structure to make some plastics thermosetting and others thermoplastic?

**10.8.** Compare the thermal coefficient of expansion of thermoplastics with most metals.

**10.9.** What is the maximum useful temperature range of plastics?

**10.10.** Compare the strength-to-weight ratio of reinforced Fiberglas and steel.

**10.11.** What type of compression mold is shown in Fig. 10.10?

**10.12.** What is one advantage of transfer molding?

**10.13.** Compare slush and rotational molding.

**10.14.** Why may vacuum snap-back forming be preferred for some items over straight vacuum forming?

**10.15.** If you were asked to build the following quantities of Fiberglas boats, which method would you choose and why? Boat A, 10; Boat B, 2; Boat C, 30; Boat D, 5,000. All boats are 16 ft long.

**10.16.** What are major considerations in machining plastics?

**10.17.** What development may bring a new family of high temperature plastics to the market?

**10.18.** What are the principles involved in adhesives being able to bond to a perfectly smooth surface?

**10.19.** What is one of the main steps in preparing an adhesive joint of any type?

## PROBLEMS

**10.1.** What would be the average molding time for a reinforced polyester chair seat? If the chair area were equivalent to 1/10 the area of the tank mentioned in Table 10.1, how would the molding time compare?

**10.2.** If the contact mold used to make the chair in Prob. 10.1 costs \$500.00 to construct and the matched die molds cost \$8,000.00, how many chairs will have to be made to warrant the matched dies? Labor and overhead can be figured at \$10.00 per hr. The cycle time for matched die molding is 4 min. A break-even formula that can be used is:

$$E_b = \frac{\text{extra cost of tooling}}{\text{cost/pc without tooling} - \text{cost/pc with extra tooling}}.$$

**10.3.** How many sheets of laminated Fiberglas paneling 4 ft $\times$ 8 ft could be produced per hour if the machine's rated output is 40 to 50 ft$^2$/min?

**10.4.** The injection molding cycle varies from 0.25 min. to 1.5 min. What would be the average molding time to produce 500 plates in a 3-cavity die?

**10.5.** Design an adhesive lap joint that will withstand a shear stress of 6,000 psi using a vinyl phenolic adhesive. Make a three-dimensional drawing.

**10.6.** What would the maximum shear strength be of a joint having a lap area of 2 sq in. if a vinyl phenolic adhesive were used and tested at room temperature? How would this compare with epoxy resin?

## REFERENCES

Burlant, W. J., and A. S. Hoffman, *Block and Graft Polymers,* Reinhold Publishing Corp., New York, 1960.

Dombrow, B. A., *Polyurethanes,* Reinhold Publishing Corp., New York, 1965.

Elliott, S. B., "ABC's of Plastics," *Machine Design,* April 28, 1966.

Fenner, O., "Plastics for Process Equipment," *Chemical Engineering,* Nov. 9, 1964.

"Joining of Materials for Aerospace Systems," Society of Aerospace Materials and Process Engineers, 9th National Symposium, Dayton, Ohio, 1965.

Lewis, T. J., and P. E. Secker, *Science of Materials,* Reinhold Publishing Corp., New York, 1965.

Luftglass, M. A., W. R. Hendricks, G. Holden, and J. T. Bailey, "Thermoplastic Elastomers," *Machine Design,* April 14, 1966.

*Machine Design,* Plastics Reference Issue, June 1966.

*Modern Plastics Encyclopedia,* Plastics Catalogue Corporation, 1966.

Rondean, H. F., "Synthetic Materials, Plastics, Elastomers, Adhesives," *Machine Design,* July 21, 1966.

Wohrer, L. C., F. J. Frechette, and J. Economy, "Whisker-Reinforced Plastics," *SPE Paper Regional Technical Conference,* Cleveland, September 1966.

# 11

# POWDER METALLURGY

THE MANUFACTURE of parts from powdered metals can be briefly described as a process of placing metal powder in a mold and compressing it. After the compressed powder is removed from the mold, it is heated or sintered to give it strength.

Powder metallurgy is not a discovery of modern times. It had its more notable beginnings in ancient Egypt around 3000 B.C. The Incas of Peru also used it. However, it was a "lost art" until a brief rebirth in 1829, when an Englishman cold-pressed and sintered some platinum powder to produce the first ductile platinum. It was used again in 1916 to produce the first commercial tungsten wire.

During World War II, when new methods for mass production of machine parts were being investigated both here and abroad, the process came into its own. Germany was short of copper. German scientists found that shell driving bands, usually made of copper, could just as well be made of porous iron impregnated with paraffin wax. The good forming qualities of this material soon led to other applications. The past two decades have seen major advances in the use of powder-metal parts in consumer, industrial, and military applications.

## PROCESS DESCRIPTION

The manufacture of parts by the powder-metallurgy process usually involves a series of steps, as follows: the manufacture of metal powders, blending, briquetting, presintering, and sintering, and a number of secondary operations including sizing, coining, machining, impregnation, infiltration, plating, and heat treatment. These processes are shown schematically in Fig. 11.1.

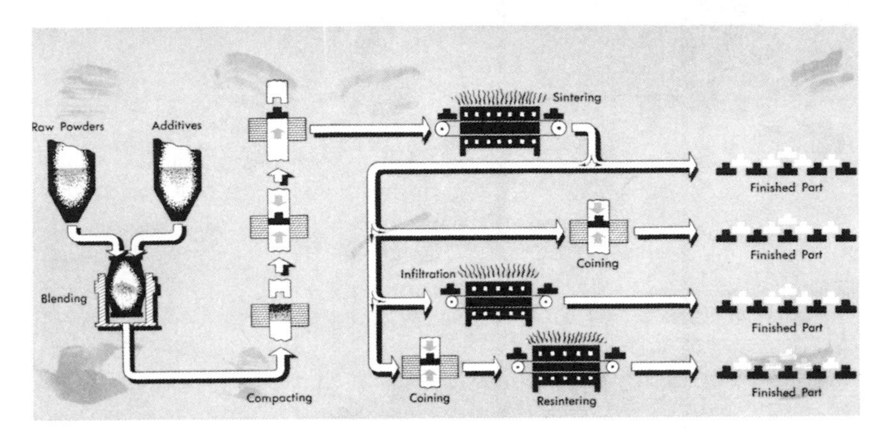


**Fig. 11.1.** Schematic diagram of the powder-metallurgy process. Courtesy *Machine Design.*

## MANUFACTURE OF METAL POWDERS

A wide range of metal powders having a nearly infinite range of properties has been developed in recent years. The powders most commonly used are copper-base and iron-base materials. Other powders not used extensively, but supplying a constantly growing need, are stainless steel, titanium, nickel, beryllium, chromium, and other refractory and exotic metals. Powder metallurgy is the only way of making some items that are commonplace today as, for example, tungsten carbide cutting tools. This is the only way of supplying the high demand for certain high-purity metals.

The particle size of powders falls into a range of 1 to 100$\mu$ ($1\mu = 10^{-6}$ meter), with the range of 10 to 20$\mu$ being predominant. There are various methods of manufacturing powders of this size, but those most commonly used are atomization, reduction, electrolysis, crushing, and milling (Fig. 11.2).

**Atomization.** In the atomization process, molten metal is forced through a nozzle into a stream of air or water. Upon contact with the stream, the molten metal is solidified into particles of a wide range of sizes. The size is controlled by varying the nozzle size, metal flow rate, and the temperature and pressure of the stream. This process is used mostly for low-melting-point metals because of the corrosive action of the metal on the nozzle at high temperatures.

**Reduction.** Sponge-iron powder is a pure, porous iron obtained when iron oxide or iron ore is reduced at temperatures below the melting point

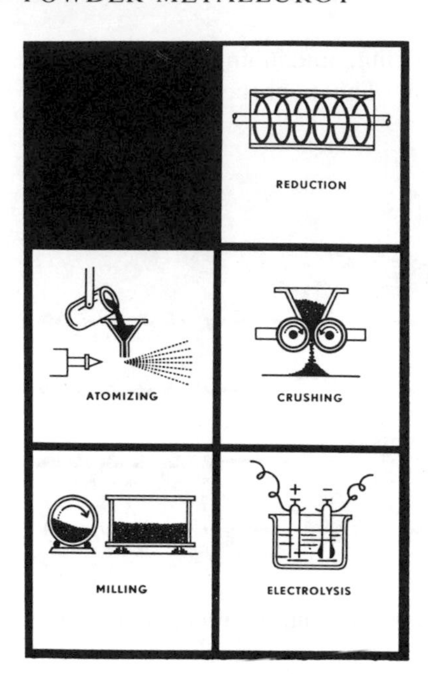


**Fig. 11.2.** Various methods of manufacturing metal powders. Courtesy Hoeganaes Sponge Iron Corp.

of the iron in an atmosphere-controlled furnace. The sponge iron is made into powder by crushing and grinding. The process is shown pictorially in Fig. 11.3.

**Electrolysis.** The electrolytic process is similar to electroplating. In this process the metal plates are placed in a tank of electrolyte. The plates act as anodes, while other metal plates are placed into the electrolyte to act as cathodes. High amperage produces a powdery deposit on the cathodes. After a period of time, the cathode plates are removed from the tank and rinsed to remove the electrolyte. After a drying period, the deposit is scraped off and pulverized to produce powder of the desired grain size. An annealing process follows pulverization to remove work-hardening effects.

## MIXING OR BLENDING OF POWDERS

The blend of the powders determines the many different properties that can be obtained. During the blending operation, lubricants are added to reduce friction during the pressing operations. This both reduces die wear and lowers the pressure required for pressing. Through blending, uniform distribution of particle size is obtained, and the different powders are thoroughly mixed. The mixing may be done either wet or dry. Wet mixing has the advantage of reducing dust and the danger of explosion which is present with some finely divided powders.

**Fig. 11.3.** The ingredients for making iron powders and some finished parts. Courtesy Hoeganaes Sponge Iron Corp.

## BRIQUETTING

Briquetting is the art of converting loose powder into a "green" compact of accurately defined size and shape. Owing to interparticle friction, pressure applied from one direction will not be distributed uniformly throughout the part. Metal powders do not *flow* much under pressure, so density variation is kept to a minimum by multiple-action briquetting presses which apply pressure from both the top and bottom of the die (Fig. 11.4). The briquette is considered fairly fragile, but it can be handled.

For high-volume parts, tungsten carbide is used for the die material. Although the cost is higher, it will outwear other materials by a ratio of roughly 10:1. Some of these dies can produce a million parts before they

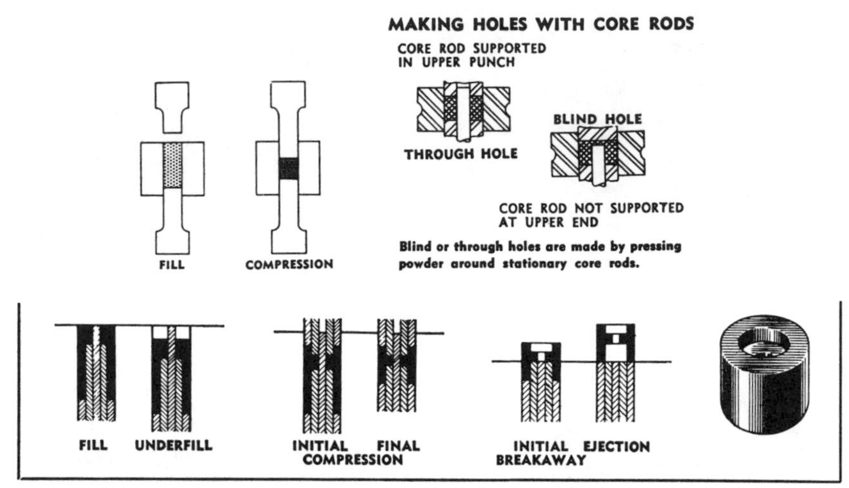


**Fig. 11.4.** Briquetting dies for powdered-metal parts use pressure from both the top and the bottom to achieve more uniform density. Courtesy *Metalworking.*

become worn beyond tolerance limits. High-carbon, high-chrome vanadium steels are second-choice die materials. High-carbon tool steels can be used for low-volume and light-duty applications. The punches have to be made with some sacrifice of hardness for toughness.

The dies must be highly polished to aid in briquetting and ejection. Clearances between punches, dies, and core rods must be held to a minimum to maintain alignment and concentricities, but they must be large enough to permit free movement under all operating conditions. Figure 11.5 shows how a core rod is used to make a thin-walled bushing.

The powder is compressed to approximately one third of its original volume. Slight pressure on the punches results only in denser particle packing. As pressure is increased (up to 200,000 psi maximum), plastic deformation of the particles takes place. Thin sections bend and break.

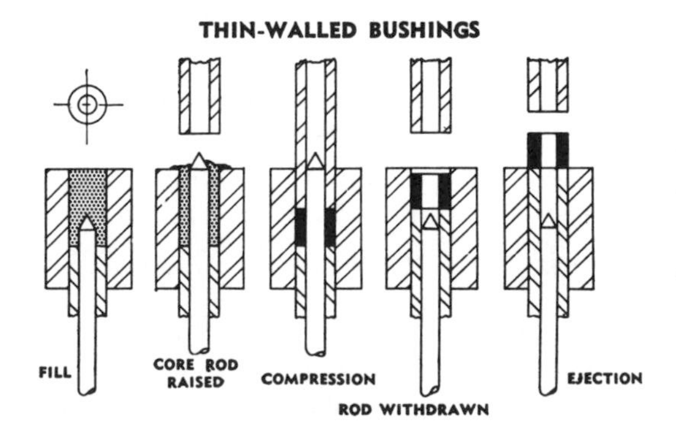


**Fig. 11.5.** Core rods and a five-step sequence make thin-walled bushings. Courtesy *Metalworking.*

Cold welding and interlocking of adjacent grains takes place, and voids are filled.

The presses used for briquetting may be either mechanical or hydraulic, or a combination of the two.

Mechanical presses usually operate in the lower pressure ranges and rely on fast action for high-production rates. These presses operate on a given stroke length; therefore, the parts produced are of uniform volume, but it is difficult to control density.

Hydraulic presses are slower but can develop high pressures. Since stroke is adjustable, parts of uniform density can be produced. Capacities of these presses range from 150 to 5,000 tons.

## PRESINTERING

Presintering is done at a temperature below the final sintering temperature. Its purpose is to increase the strength of the briquette for handling or to remove lubricants and binders added to the powders during the blending operation. Some materials, such as tungsten carbide, become very hard after sintering, but they are relatively easy to machine at this stage. If no machining is required, presintering can be eliminated.

## SINTERING

Sintering is performed to achieve all possible final strength and hardness needed in the finished product. The three most important variables governing the sintering process are temperature, time, and sintering atmosphere. Other factors that have an influence are the green density, and the size composition, and desired properties of the compacts.

By keeping the part at the right temperature for a proper period of time, bonding of the particles is accomplished by the exchange of atoms between the individual particles.

Sintering is accomplished in high-volume, continuous furnaces (Fig. 11.6). Because of the large exposed areas of the briquettes, the sintering furnaces contain controlled atmosphere for protection against oxidation and other chemical reactions.

## SECONDARY OPERATIONS

Many powder-metal parts may be used in the "as sintered" condition. However, when the desired surface finish, tolerance, or metal structures cannot be obtained by briquetting, additional finishing operations must follow. These operations include sizing, coining, machining, impregnation, infiltration, plating, and heat treatment.

**Fig. 11.6.** A controlled-atmosphere, continuous roll furnace used to obtain high production of powder-metal parts. Courtesy Hoeganaes Sponge Iron Corp.

**Sizing.** When a part must be dimensionally correct, sintering is followed by a sizing operation. In this, the sintered part is placed in a die, which is designed to meet the required tolerances, and is re-pressed.

**Coining.** The coining operation is very similar to sizing, except that this time the sintered part is re-pressed in the die to reduce the void space and impart the required density. After coining, the part is usually resintered for stress relief. Oftentimes, the sizing and coining operations are combined in the same die.

**Machining.** The principal object of powder metallurgy is to be able to press metal powders directly into dimensionally accurate finished shapes. However, certain features, such as threads, reentrant angles, grooves, side holes, and undercuts, are usually not practical for powder-metallurgy fabrication. These features are generally machined on the semifinished sintered blanks (Fig. 11.7).

Boring, turning, drilling, tapping, and all the conventional machining operations are performed easily on sintered metal parts. Although high-speed tools prove satisfactory for short runs, tungsten carbide cutting tools are recommended because they retain a sharp cutting edge for a much longer time. When finishing, very sharp tools, fine feeds, and high speeds are essential if the open-pore structure necessary for filters and the self-lubricating qualities of bearings are to be preserved.

Ordinary coolants are not recommended for the machining operation, since they cannot be removed easily from the porous structure and may cause internal corrosion of the part. Volatile coolants like carbon tetrachloride can be used because they vaporize readily and leave no residue. Coolant oil of the same type used to impregnate a self-lubricating bearing can also be used.

Reaming is not recommended unless the tool is very sharp. The cut should be limited to 0.005 in. in diameter. A better tool for obtaining desired hole diameters is the button-type burnishing tool shown in Fig. 11.8. These tools are recommended for holding bore tolerances of 0.0005 in. or less. Normally the burnishing tool is 0.0003 in. larger than the finished size required, to compensate for springback of the metal.

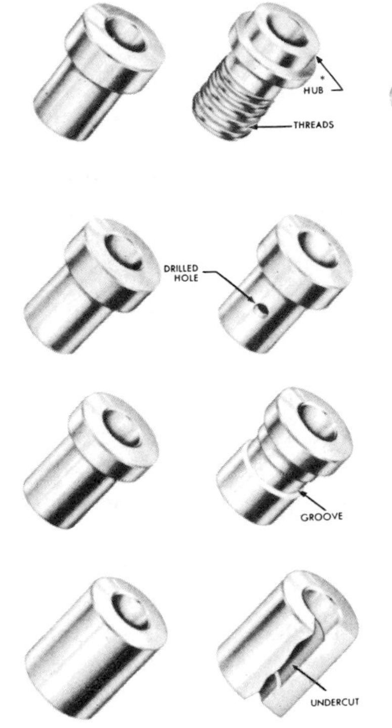


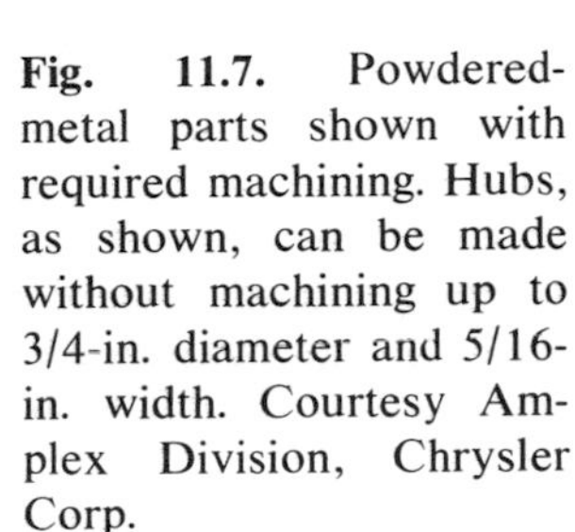

**Fig. 11.7.** Powdered-metal parts shown with required machining. Hubs, as shown, can be made without machining up to 3/4-in. diameter and 5/16-in. width. Courtesy Amplex Division, Chrysler Corp.

For example, if the bore diameter is to be 1.001 in., the burnishing-tool diameter should be 1.0013 in.

**Impregnation.** When self-lubricating properties are desired, the sintered parts are impregnated with oil, grease, wax, or other lubricating materials. In this process, the parts are placed in tanks of lubricants heated to approximately 200°F. The porous structure is completely impregnated in 10 to 20 min. The lubricant is retained in the part by capillary action, until external pressure or heat of friction draws it to the surface.

**Infiltration.** An infiltrated part is made by first pressing and sintering iron powder to about 77 percent of theoretical density. A replica or

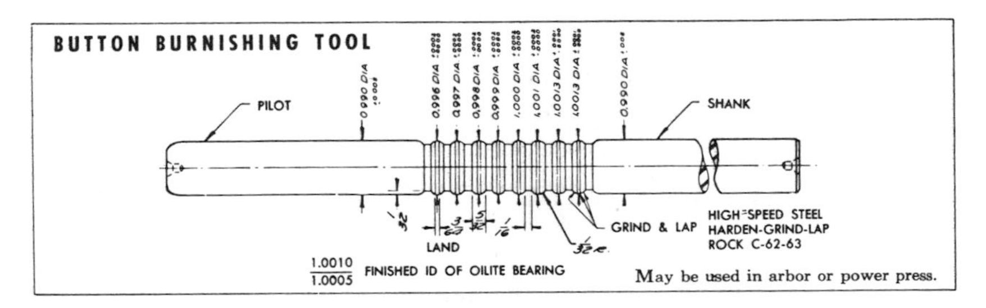


**Fig. 11.8.** A burnishing tool used to finish bearing diameters to close tolerances. Courtesy Amplex Division, Chrysler Corporation.

infiltration blank of copper or brass is then placed on the part, which is sent through the furnace a second time. The infiltrant melts and soaks through the porous part, producing close to 100 percent density.

Infiltration provides increased strength, hardness, and density not obtainable by straight sintering.

**Plating.** Plating of any part usually has two objectives—pleasing appearance and protection from corrosion. The procedures for plating powdered-metal parts are quite different from those used for wrought- or cast-metal parts. Peening, tumbling, or other methods of mechanical smoothing of the part prior to plating provide a more even surface for smoother plating.

These methods, however, do not take care of inner porosity, and any electrolyte which has been entrapped in the porous structure will cause internal corrosion, leading to failure. Therefore, before the part can be plated, porosity must be eliminated. Impregnation with molten metal helps, but, since the pores are not filled completely, galvanic corrosion is likely between the dissimilar metals. Also, the difference in coefficients of expansion and contraction will eventually lift the plating material. Impregnating the part with a plastic resin will overcome most of these difficulties. The resins have low coefficients of expansion, do not react galvanically with the metal part, are low in cost, and have good filling properties. After the porosity has been eliminated, regular plating procedures can be used.

**Heat Treatment.** Powdered-metal parts are heat treated to improve grain structure, strength, and hardness. The conventional methods of heat treatment used for wrought metals can also be used for powdered metals but care must be taken in several steps of the process. Porosity decreases the heat conductivity; therefore, longer heating and shorter cooling periods are required. Heat treatment must be carried on in a controlled atmosphere to prevent oxidation of the internal structure.

Carbon, in the form of graphite, added to iron-powder mixtures to produce medium- or high-carbon steel is an inexpensive way of improving the physical properties of a part.

The tensile and yield strengths of iron-graphite compacts are related directly to density. The higher-density materials have higher yield strengths, since a greater volume of material is stressed. To develop high tensile strength, high-density materials must be allowed sufficient time to attain complete diffusion of the carbon.

## PROPERTIES OF METAL POWDERS AND FINISHED PARTS

**Metal Combinations.** Powder metallurgy makes it possible to unite materials that cannot be alloyed in the usual sense or would not yield the desired characteristics if they were joined mechanically. Examples of this are lead dispersed in copper to form bearing surfaces, copper and carbon combinations for commutator brushes, steel and copper combined to make a self-brazing alloy, and a combination of ceramics and metals to produce refractory cermets.

**Particle Size and Distribution.** Particle size and distribution are important factors in the control of porosity, density, and compressibility. They also control the amount of shrinkage when sintering. If all coarse or all fine particles are used, the overall porosity is greater than with a mixture of different grain sizes. A range of grain sizes makes for a close-fitting pattern. The density of a part is reduced when all large particles are used, owing to the large void space. When all small particles are used, the density is good, but higher briquetting pressures are required because of the large amount of contact area between the particles and the small deformation. Proper particle size and distribution are determined by a series of standard screens. The amount held back by each screen allows the percentage of each size of particle, on a weight basis, to be estimated.

**Density and Flow Rate.** The apparent density of the material is determined from the variation in the particle size and is given by the volume-to-weight ratio of a loosely filled mixture. To manufacture a part of a certain density, a metal powder of a corresponding apparent density is used. An increase in density of the product can be obtained, but the power required will also increase.

Through controllable density, one area of a part can be made hard and dense while another portion of the same piece is soft and porous. Porosity can be varied from practically nil to so much that the part can be used as a filter (Fig. 11.9).

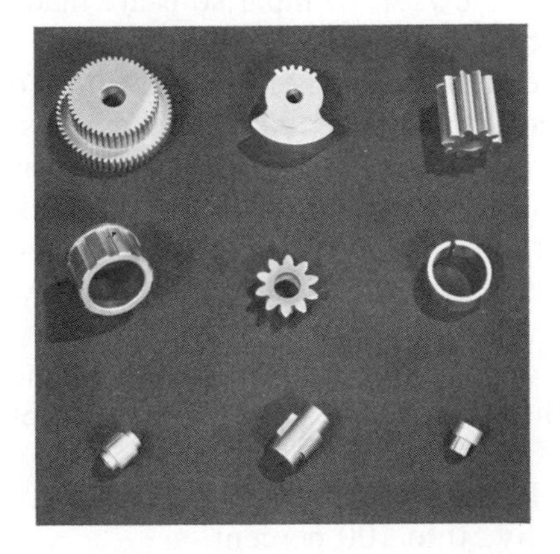

**Fig. 11.9.** A wide variation in particle size and density is obtainable in powdered-metal parts. Courtesy Amplex Division, Chrysler Corporation.

The flow rate of the powder governs the amount of time required to fill a cavity and is important for equal filling of narrow cavities. The flow rate also helps determine the possible production rate.

**Purity.** Impurities in the powder have various effects on the briquetting and sintering operations. Foreign substances cause wear on the die parts and thus reduce the useful life of the die. Oxides and gaseous impurities can be removed from the part during sintering by the use of a reducing atmosphere.

**Green Strength.** Green strength refers to the mechanical strength of a compacted part prior to sintering. This factor is very important for convenient handling during mass production.

## DESIGN CONSIDERATIONS

When powdered metals are considered for a new or existing part, a number of factors must be evaluated. These include quantity, tolerances, physical characteristics, shapes, and size.

**Quantity.** The quantity must be sufficient to justify the necessary investment in tools and dies. A die may cost from several hundred dollars for a small simple part to several thousand dollars for a larger, more complicated one. Die life varies greatly; it may range from 100,000 parts for a tool-steel die to over a million parts for carbide-lined dies with high-carbon, high-chrome punches.

**Tolerances.** Pressed and sintered ferrous parts can be controlled to about ±0.002 in. per in. on diameters and other dimensions formed by the die. Sizing can be used to cut this tolerance in half.

**Physical Characteristics.** Physical properties of powdered metals vary with density, composition, and processing. A partial list of nonferrous powdered materials is shown in Table 11.1. Comparisons are made with the same composition in wrought material where applicable.

Table 11.2 shows a partial list of ferrous materials in the as-sintered state. The iron-carbon group is the most widely used alloy in the industry. The addition of graphite to the iron doubles the strength and wear resistance. Various amounts of copper enhance the strength and hardness of steel parts but reduce ductility and toughness. However, when copper is infiltrated into the part, the result is the toughest of the PM materials.

The austenitic stainless steels are ductile, but more important is the yield strength, which is higher than that of the same material in wrought form.

Table 11.3 contains a partial list of heat-treatable materials. Generally heat treatment can improve strengths by 50 to 100 percent.

The proper control of furnace atmosphere and sintering cycle is important to gain the full strengthening effect of the copper. If carbon

**Table 11.1**
**Properties of Some As-Sintered, Nonferrous Powdered Materials**

and some comparisons with wrought and cast materials (courtesy of *Machine Design*).

| PMPA designation | Nominal composition | Density | Tensile strength 1000 psi | Yield strength 1000 psi | Elongation % | Transverse fiber strength 1000 psi | Hardness |
|---|---|---|---|---|---|---|---|
| BT-0010-N | Bronze | 5.8-6.2 | 8 | 7 | 1 | 30 | Rb-11 |
| BT-0010-R | 90-Cu- | 6.4-6.8 | 14 | 11 | 1 | 36 | Rf-30 |
| BT-0010-S | 10Sn | 6.8-7.2 | 20 | 20 | 2-3 | 42 | Rf-43 |
| BT-0010-W | | 8.0 | 45 | 30 | 11-15 | 90 | Rf-80 |
| Wrought | | 8.8 | 85 | 65 | 68 | – | – |
| Cast | | 8.80 | 46 | 26 | 15-25 | – | – |
| BZ-0218-T | Brass | 7.2 min | 20 | 15 | 10 | 31 | Rf-37 |
| BZ-0218-U | 90 Cu + | 7.7 min | 23 | 18 | 12 | 65 | Rf-42 |
| | 20Zn + | | | | | | |
| BZ-0218-W | 1½ Pb | 8.0 min | 37 | 28 | 21 | 80 | Rf-50 |
| Wrought | | 8.8 | 38 | 12 | 53 | – | Rf-61 |
| Cast | | 8.8 | 34 | 17 | 25 | – | Rf-61 |
| | Copper | 8.0 | 23 | 15 | 8 | – | Rf-50 |
| | | 8.5 | 30 | 15 | 29 | – | Rf-70 |
| Wrought | | 8.92 | 32 | 10 | 45 | – | BHN-42 |

**Table 11.2**
**Properties of Some As-Sintered Powdered Materials**

| PMPA designation | Nominal composition | Density | Tensile strength 1000 psi | Yield strength 1000 psi | Elongation % | Transverse fiber strength 1000 psi | Hardness |
|---|---|---|---|---|---|---|---|
| F-0010-P | Fe99-1C | 6.1-6.5 | 35 | 27 | 1.0 | 89 | B-50 |
| F-0010-S | | 7.0 | 60 | – | 3.0 | 120 | – |
| F-0010-T | | 7.3 | 68 | – | 3.0 | 140 | – |
| SAE-1080 | | 7.8 | 90 | 54 | 24.0 | – | C-15 |
| FC-1000-N | 90Fe-10 Cu | 5.8-6.2 | 30 | 25 | 0.5 | 75 | – |
| FC-0710-N | 92Fe-7Cu-1C | 5.8-6.2 | 50 | 40 | 0.5 | 115 | B-70 |
| FC-0710-S | | 6.8 | 83 | 63 | 1.0 | 131 | B-73 |
| FX-2000-T | 80Fe-20Cu | 7.1 min | 70 | 70 | 1.0 | 140 | B-75 |
| FX-2010-T | 79Fe-20Cu-1C | 7.1 min | 110 | 90 | 1.0 | 190 | B-95 |
| SS-303L-P | 18Cr-8Ni SS | 6.0 | 35 | 32 | 2.0 | – | B-52 |
| SS-303L-R | | 6.6 | 52 | 47 | 7.0 | – | B-55 |
| SAE-303 | | 7.9 | 90 | 40 | 50.0 | – | B-72-80 |
| SS-316L-P | 18Cr-12Ni-2 Mo | 6.16 | 38.5 | 35 | 2.0 | 95 | B-55 |
| SS-316L-R | | 6.65 | 58 | 51 | 8.0 | 135 | B-65 |
| SAE-316-L | | 7.9 | 78 | 32 | 50.0 | – | B-74-90 |

**Table 11.3**
**Properties of Some Heat-Treated Powdered Materials**

| PMPA designation | Nominal composition | Density | Tensile strength 1000 psi | Yield strength 1000 psi | Elongation % | Transverse fiber strength 1000 psi | Hardness |
|---|---|---|---|---|---|---|---|
| F-0010-P | Fe99-1C | 6.1-6.5 | 47 | – | 0.5 | – | B-90 |
| F-0010-S | | 7.0 | 65 | – | 0.5 | 120 | B-100 |
| F-0010-T | | 7.3 | 127 | – | 2.5 | 235 | B-105 |
| SAE-1080 | | 7.8 | 185 | 140 | 12.0 | – | C-53 |
| FC-1000-N | 90Fe-10Cu | 5.8-6.2 | 54 | – | 1.0 | 103 | C-30 |
| FC-0710-N | 92Fe-7Cu-1C | 5.8-6.2 | 85 | – | 1.5 | 180 | C-30 |
| FC-0710-S | | 6.8 | 110 | – | 1.5 | 209.5 | C-40 |
| FX-2000-T | 80Fe-20Cu | 7.1 min | 128 | – | 0.5 | 209.5 | C-35 |
| FX-2010-T | 79Fe-20Cu-1C | 7.1 min | 152 | – | 1.0 | – | C-40 |
| SS-410-N | 12.5Cr-0.15C | 5.9 | 85 | – | – | – | C-15 |
| SS-410-P | | 6.4 | 110 | – | – | – | C-29 |
| SAE-410 | | 7.9 | 200 | 150 | 20.0 | – | C-45 |
| | 1.5Ni-0.5Mo-0.6C | 6.8 | 90 | 80 | 0.5 | 150 | C-25 |
| | | 7.2 | 140 | 120 | 0.5 | 207 | C-35 |
| AISI-4660 | | 7.9 | 275 | 250 | 12.0 | – | C-57 |
| | 7.0Ni-2.0Cu-1C | 6.8 | 135 | – | 1.5 | 262 | C-42 |
| | | 7.2 | 157 | – | 2.0 | 285 | C-44 |

control is not exact, the full strengthening effect is lost and a lower strength results. Carbon analysis of test bars should show no measurable uncombined carbon or graphite or indications of carburization or decarburization if maximum strength is to be obtained.

Other than the normal properties given for wrought materials, transverse fiber strength is shown. A knowledge of this property is widely used for the more brittle materials. The test used for it is the "transverse" test, since the specimen is suspended from its two ends and loaded in a transverse direction to failure. The load deflection curve can also be plotted up to the time the material reaches its proportional limit. The maximum outer fiber stress for a rectangular cross section can be calculated from the equation

$$S_m = \frac{3PL}{2wd^2}$$

where $S_m$ = maximum fiber stress; $P$ = load in lb; $L$ = length of span in in.; $w$ = width of specimen; $d$ = depth of specimen.

**Shapes.** Metal powders do not flow easily, even under high pressures. Best results are obtained when narrow and deep passages are avoided. Holes cannot be placed at right angles to the direction of pressing. Inside fillets are preferred to sharp corners, both for ease in fabrication and for strength in the finished part. Sharp points on the punches and dies should be avoided, as they wear quickly.

Substantial economies are frequently realized by revising an original design to adapt it for manufacture by powder metallurgy.

**Size.** The size of the part that can be made is limited by the size of the available presses. Practical size limitations, at the present time, are from about 1/8 to 25 sq in. in area and from 1/32 to 6 in. in height. By rule of thumb, the ratio of height to effective diameter is never greater than 4:1 for parts less than 1 in. in diameter; as the diameter increases, this ratio decreases considerably.

## APPLICATIONS OF POWDERED METALLURGY

The applications of powdered metals can be divided into two broad categories—those items that are dependent on the unique properties of the materials, and those that are used for structural or machine parts.

**Parts Dependent on Properties of Materials.** BEARINGS. Porous oil-impregnated metal is an ideal bearing material. It is especially recommended where positive lubrication is left to the human element. The impregnated oils or greases frequently provide lifetime lubrication. The lubricant is metered to the bearing surface by capillary action when heat or pressure is applied (Fig. 11.10).

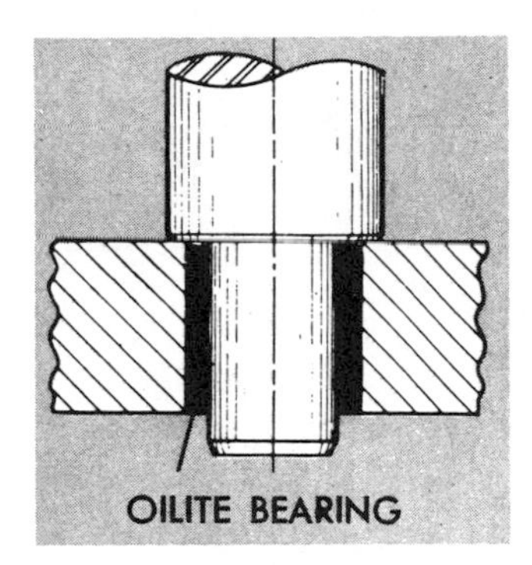


**Fig. 11.10.** Self-lubricating, oil-impregnated, powdered-metal bearings and bearing stock. Drawing shows bearing pressed in place. Courtesy Amplex Division, Chrysler Corporation.

METAL FILTERS. Powdered-metal filters provide uniform depth filtration of particles as small as $5\mu$. These filters may be cleaned by reversing the flow or by back-flushing with solvent. Other uses include filtering, diffusing, and controlling the flow of gases and liquids; separating liquids having different surface tensions; removing moisture from airstreams; acting as sound deadeners; and serving as wicks for lubricating airstreams (Fig. 11.11).

WELDING ELECTRODES, FLAME CUTTING, AND SCARFING. Large quantities of powdered iron are used in the manufacture of electrodes for arc welding. Powdered iron is also used in flame cutting and flame scarfing.

**Structural or Machine Parts.** Perhaps the outstanding reasons for using powdered metals for machine parts are the facts that there is no scrap, and very little, if any, machining is required.

Machine parts are produced in a broad range of ferrous and nonferrous alloys. The physical properties range up to the equivalent of low-carbon steel. Factors in favor of powdered-metal parts are: (1) elimination of most machining operations, (2) low cost per part in large quantities, (3) quiet operation, (4) low wear because of self-lubrication, and (5) vibration dampening. A group of miscellaneous machine parts is shown in Fig. 11.12.

## ECONOMICS

As stated previously, to be profitable, powdered-metal parts must be made in fairly large quantities. However, if tooling is simple, these may be as low as 500 or 1,000 pieces. In runs of 10,000 parts, powdered-

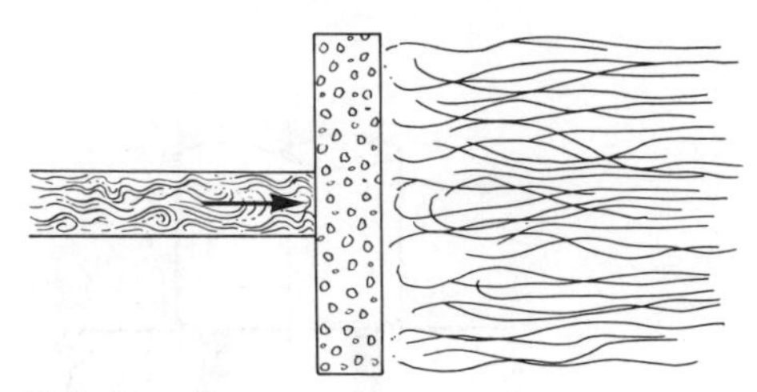

Diffusing air

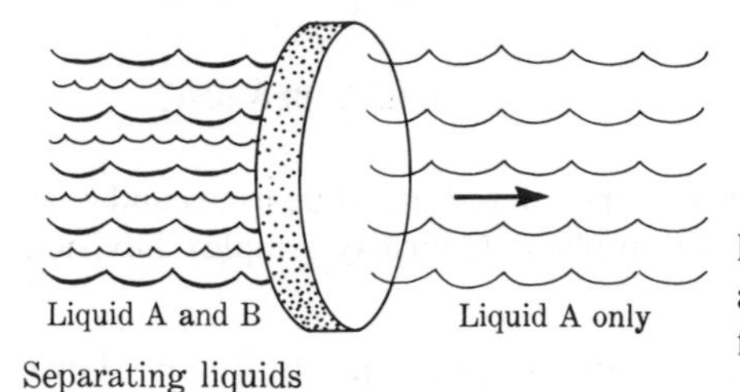


Separating liquids

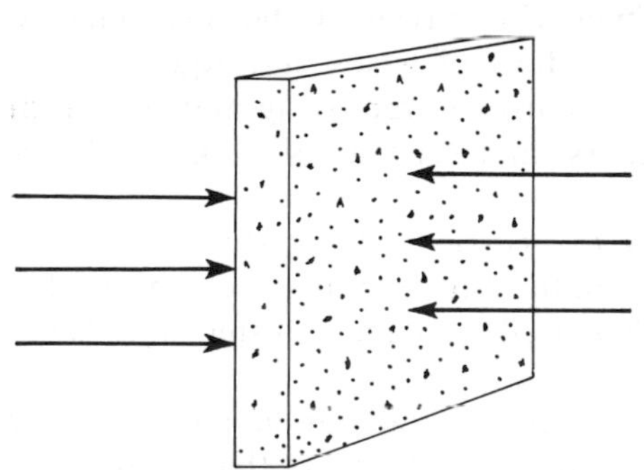


Filters

**Fig. 11.11.** Powdered metals are used as air diffusers, liquid separators, and filters.

metal applications are likely to have the advantage over conventional processes.

Labor and machine-time costs are low. Semiskilled labor is all that is required, and production rates of 500 to 50,000 pieces per hour can be obtained. Another major economic consideration is the elimination of scrap loss owing to the machining.

## FUTURE DEVELOPMENT

Sintered metal parts are being made in progressively larger components. At one time the process was confined to ounces; now some parts are being measured in pounds.

**Fig. 11.12.** Miscellaneous machine parts made by the powder-metallurgy process. Courtesy Graphite Co., division of the Wickes Corporation.

At the present time the automobile industry is the largest user, averaging about 8 to 10 lb per car (Fig. 11.13). In the next few years this average is expected to be 30 to 35 lb per car. Similarly, others such as the appliance, business machines, and industrial equipment industries are expected to expand their use of sintered metal parts.

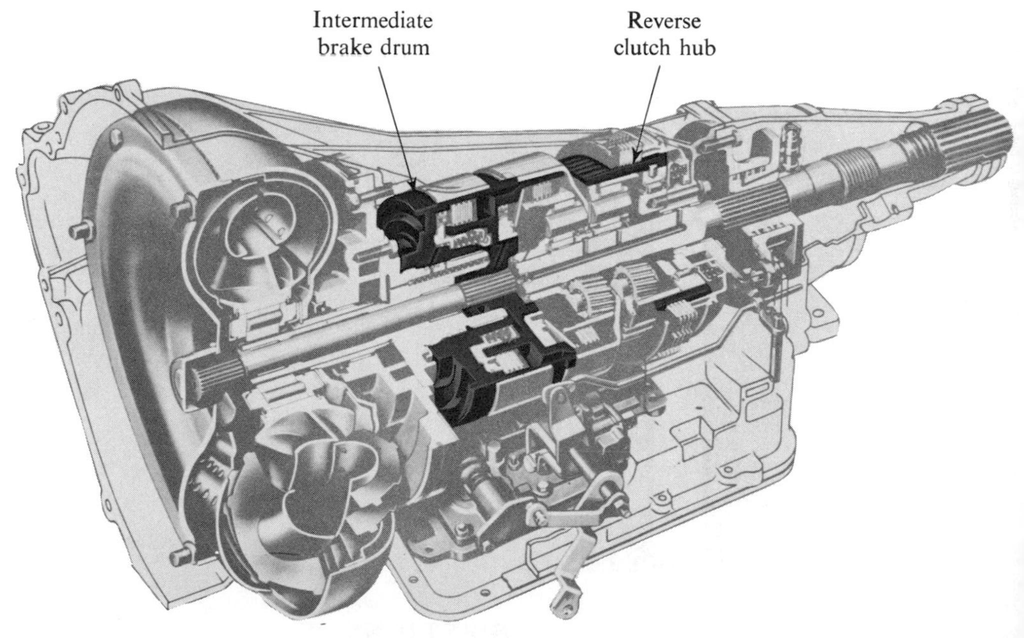


**Fig. 11.13.** Automotive manufacturers are now one of the largest users of powdered metals, with individual parts weighing up to 6 lb being used in the automatic transmission. Courtesy Ford Motor Company.

Work is being done on continuous rolling of particle matter into strip, tubing, and rods.

A newer material is being studied, a dispersion-hardened combination of ceramics and metal. Basically the process consists of dispersing ultrafine insoluble particles throughout the matrix. After working the mixture, the high-temperature strength improves noticeably. Apparently the dispersed phase simultaneously performs "locking" action to hinder deformation and increases the recrystallization temperature.

Larger parts will be made as larger presses become available. A new press recently announced delivers 550,000 ft lb of kinetic energy, or the equivalent of a 7,000 ton forging press.

**Isostatic Pressing.** Another new approach has been in the opposite direction, away from large presses, but incorporating isostatic methods. Briefly, it consists of loading the powdered metal into a rubber mold which is then immersed in a hydraulic fluid and subjected to pressures from 10,000 to 100,000 psi (Fig. 11.14). There are other variations of the process, one of which is shown in Fig. 11.15, consisting of a core and rubber bag charged with powder and placed in a tank. Hydraulic fluid is then introduced into the tank to create a pressure build-up on the bag.

Advantages claimed for this method are uniform strength and shrinkage, close tolerances, inexpensive equipment, low space requirements, and the ability to form otherwise impossible parts.

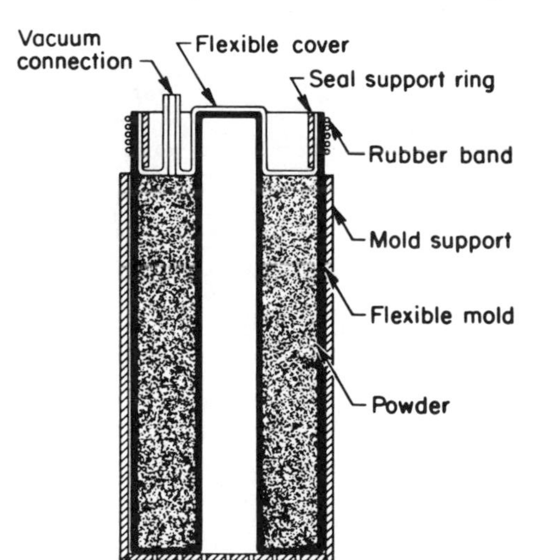


**Fig. 11.14.** An example of true isostatic compacting. Greater accuracies can be obtained by providing a core to support the center section.

## QUESTIONS

**11.1.** What is meant by powder metallurgy?

**11.2.** What metals are most commonly made in powder form?

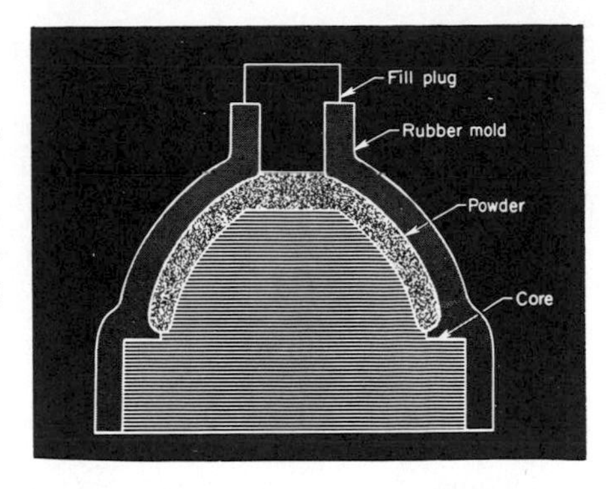


**Fig. 11.15.** A cup mold prepared for isostatic compacting by immersion in a hydraulic pressure chamber. This is known as the wet-bag method. The previous figure shows the dry-bag method.

**11.3.** Describe two methods used in making powdered metals.

**11.4.** Why is pressure applied from both the top and the bottom to produce briquettes?

**11.5.** What is the purpose of presintering?

**11.6.** What is accomplished during the sintering process?

**11.7.** Explain briefly what is accomplished by sizing, coining, impregnation, and infiltration of powdered-metal parts.

**11.8.** Why must the cutting tools used in machining powdered-metal bearings be very sharp?

**11.9.** What preparation is necessary before powdered-metal parts can be plated?

**11.10.** What special precautions are necessary when heat treating powdered-metal parts?

**11.11.** What is the significance of particle size in compacting the powders?

**11.12.** What tolerance can be expected on powdered-metal parts?

**11.13.** Why is a 25-sq-in. cross section about the largest powdered-metal part produced?

**11.14.** Why do powdered-metal structures make good bearings?

**11.15.** Why are powdered metals used for machine parts?

**11.16.** When is it economically feasible to use powdered-metal manufacture?

**11.17.** What are some new advances in the manufacture of powdered-metal parts?

**11.18.** What are some restrictions as to design of powdered-metal parts?

**11.19.** What are the advantageous of isostatic pressing?

## PROBLEMS

**11.1.** What volume of powder would be required to make 100 bearings 1 in. long with a 1-in. outside diameter and 5/8-in. inside diameter if the metal compresses to 77 percent of its theoretical density?

**11.2.** What size press would be required to make the bearing mentioned in Prob. 11.1 if a compacting pressure of 75,000 psi is required?

**11.3.** How high could the rotor for a gasoline pump shown in Fig. 11.12 be if it is 2 in. in diameter?

**11.4.** What would be the approximate diameter of the largest gear that could be made in powdered metals if the inside diameter was 2 in. and the thickness 1 in.?

**11.5.** Compare the following properties of FC – 1000 – N, from Table 11.3, with handbook values of SAE 1020 cold drawn steel as to T.S., elongation, and hardness.

**11.6.** Compare the transverse fiber strength of a specimen of F – 0010 – S 1/2-in. square suspended between two points 2, 3, and 4 in. apart, loaded to 50 lb.

## REFERENCES

*Powder Metallurgy Design Guidebook,* Powder Metallurgy Parts Manufacturers' Association, New York, 1962.

Feir, M., "P/M Parts—Today's Properties," *ASME Paper No. 66 – MD – 6,* May 1966.

Jackson, H. C., "Advances in Powder Metallurgy," *The Tool and Manufacturing Engineer,* November 1965.

Kobrin, C. L., "New Status, New Uses for Versatile Metal Powders," *Iron Age,* May 1963.

Sherman, J. V., "Powder Metallurgy Has Made Great Strides in Recent Years," *Barron's National Business and Financial Weekly,* February 7, 1966.

# 12

# COLD FORMING AND STAMPING

COLDWORKING OF METAL is possible through the formation and motion of slip planes and dislocations as explained in Chap. 2. Also mentioned was the piling up of dislocations, making the metal progressively harder to form. Work hardening is associated with certain beneficial results such as increases in tensile and yield strengths. An outstanding example of this is high strength wire often termed "piano wire." This wire is drawn from steel of approximately eutectoid composition (0.80 carbon) which is in the form of fine pearlite. As it is drawn through successive areas of reduction, the strength increases so that it may ultimately attain a tensile strength of 700,000 psi. The material still has sufficient ductility to provide a 20 percent reduction in area.

This strengthening mechanism is of considerable interest and has recently been studied with the aid of electron microscopy. Shown in Fig. 12.1*a* is the fine pearlite structure as taken from a 0.024-in.-dia wire. Since this is only a very small section, the randomness of the lamellar structure does not show and must be judged by Fig. 12.1*b*. After drawing the wire down to 0.017 in. dia, considerable alignment of the lamellar structure takes place parallel to the drawing axis. The electron microscope reveals that the original cementite lamellae are fragmented into long ribbons and numerous short segments (Fig. 12.2). Higher magnifications indicate the structure consists of tangled dislocation grain or cell walls and cementite fragments. These boundaries block the movement of slip dislocations.

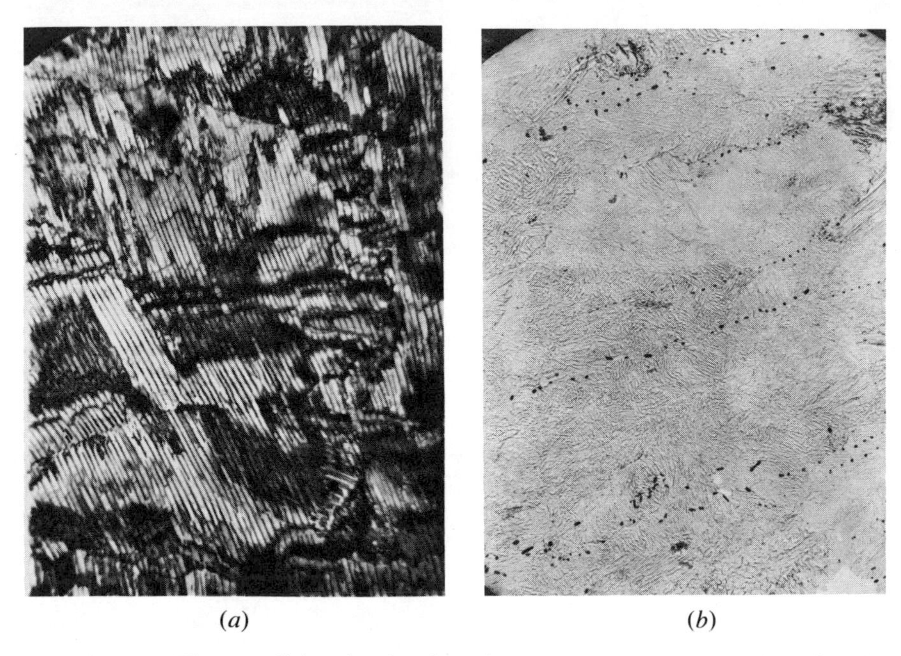

(*a*) (*b*)

**Fig. 12.1.** (*a*) Fine pearlitic wire developed by a patented process of heating the previously austenetized 0.024-in.-dia wire to 496°C, original magnification 24,000×. (*b*) Pearlitic wire prior to drawing, original magnification 6,400×. Courtesy United States Steel Corporation.

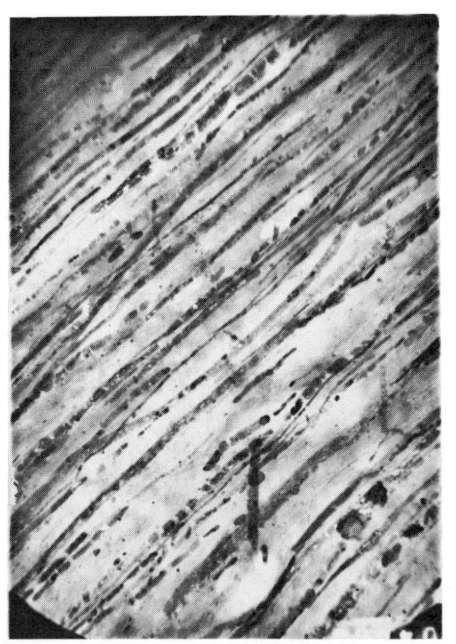

**Fig. 12.2.** The pearlitic structure after drawing shows the cementite lamellae are in long ribbons with numerous fragmented short segments, original magnification 40,000×. Courtesy United States Steel Corporation.

The same operation was studied using coarse pearlite and bainite (Fig. 12.3*a*, *b*). In each type of structure there was a linear relationship between the size of the elongated deformed grains and its developed strength. The fact that bainite showed the same results as pearlite indicates that other ferrous structures can be subjected to large plastic strains and are not limited to the lamellar type. The foregoing discussion serves to point out how recent research now bears out some earlier theory of how coldworking affects the basic structure of metals.

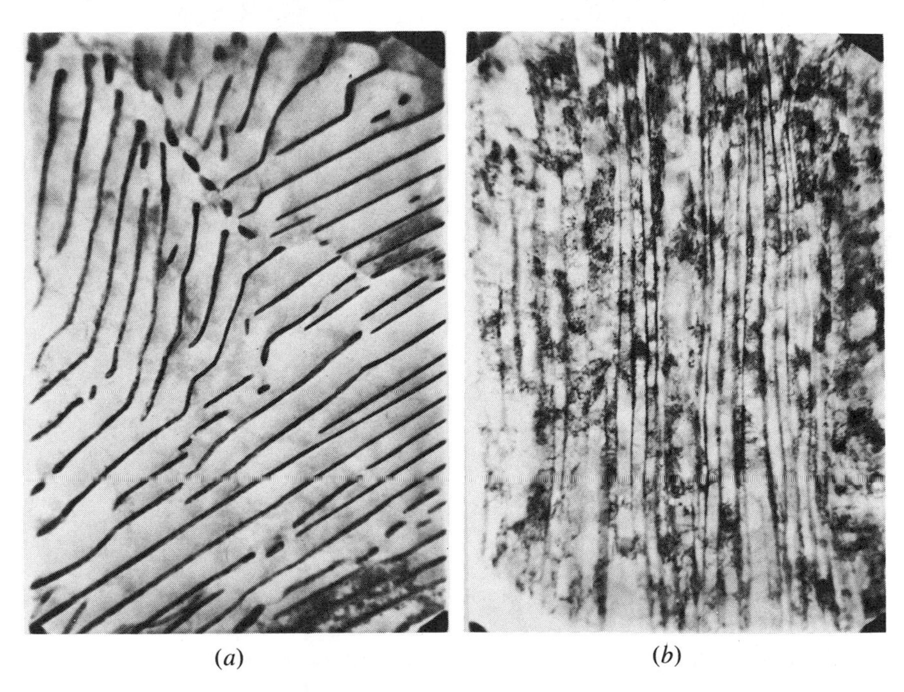

(*a*) (*b*)

**Fig. 12.3.** (*a*) Coarse pearlite before drawing, original magnification 32,000×, (*b*) after drawing, original magnification 32,000×. Courtesy United States Steel Corporation.

## PHYSICAL PROPERTIES AND DESIGN PRINCIPLES

In cold forming, it is important to know how the properties of the material will be affected during fabrication. Many materials can now be purchased with varying degrees of coldworking. For example, aluminum, stainless steel, and brass may be obtained in sheet form designated as fully annealed, 1/4H, 1/2H, 3/4H, and fully hardened by coldworking. The designer and fabricator must be careful to select the material so that after completion it will have the required ridgidity without excess weight. Knowing which gage and condition of metal to select can produce significant savings.

Many design and fabrication techniques are available that add strength and rigidity to simple parts fabricated in sheet metal. For example, strength can be incorporated into the structure by means of flanges, ribs, corrugations, beads, etc. (Fig. 12.4).

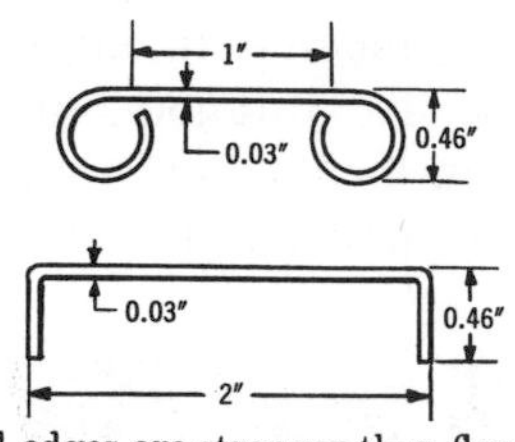


Curled edges are stronger than flanged and present a smooth, burr-free edge. Production, however, may require one more operation since the curl is usually started as a flange.

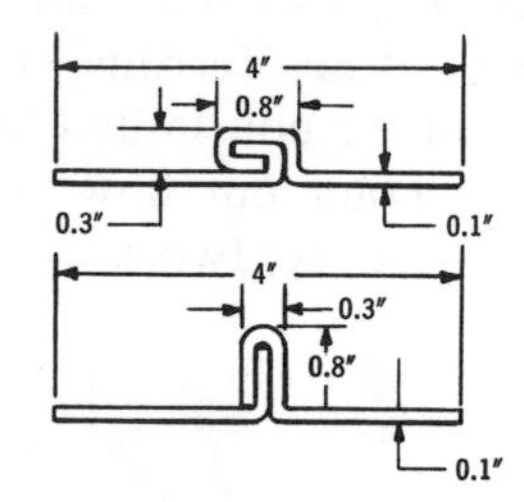


Although locking ability is sacrificed, vertical standing seams are 3 times as strong as flattened seams.

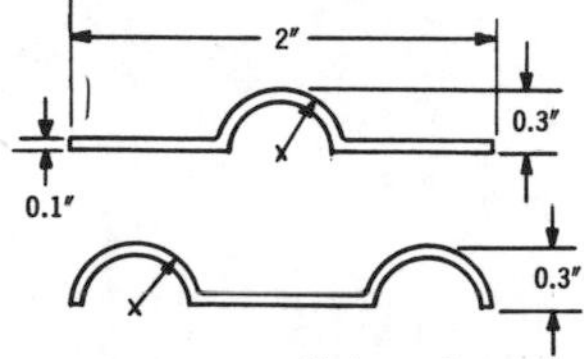


Ribs are even more efficient than flanges. Dual-rib design yields 56.5 percent more strength for 10.8 percent more material.

Corrugated sheets are common examples of ribs, used as a continuous form.

**Fig. 12.4.** Methods of increasing strength and rigidity on sheet metal parts. Courtesy *Machine Design.*

A comparison of the physical properties of hot-rolled and cold-drawn steels is given in Table 12.1. You will notice, in comparing the materials listed, that there is considerable increase in tensile strength, yield strength, hardness, and machinability of the cold-drawn as compared to the hot-rolled material. The most marked change is in yield strength, which increases from 43,000 to 66,000 in the case of the low-carbon steel and from 62,000 to 90,000 for the alloy steel.

**Table 12.1**
**Comparison of Physical Characteristics of Hot-Rolled and Cold-Drawn Steel**

| AISI number | Condition of steel | Tensile strength (psi) | Yield point (psi) | Brinell hardness |
|---|---|---|---|---|
| C1020 | Hot rolled | 65,000 | 43,000 | 143 |
| C1020 | Cold drawn | 78,000 | 66,000 | 156 |
| 4140 | Hot rolled | 89,000 | 62,000 | 187 |
| 4140 | Cold drawn | 102,000 | 90,000 | 223 |

Many similar metalworking operations can be done either hot or cold. Hot working is done at a temperature high enough to produce recrystallization. If the metal is worked at the time recrystallization takes place, little or no strain hardening results. Cold forming, on the other hand, is done below the recrystallization temperature of the metal, and considerable strain hardening results. Recrystallization occurs at widely differing temperatures; tin, for example, recrystallizes at 25°F, aluminum alloys at about 600°F, and low-carbon steel at about 1000°F.

## BENDING, FORMING, AND DRAWING

Metal is formed by both bending and drawing. To determine feasibility limits for each, certain procedures and formulae have been worked out.

**Bending.** When a material is bent, the outer fibers, beyond the neutral axis, are placed in tension and inside the neutral axis, in compression (Fig. 12.5).

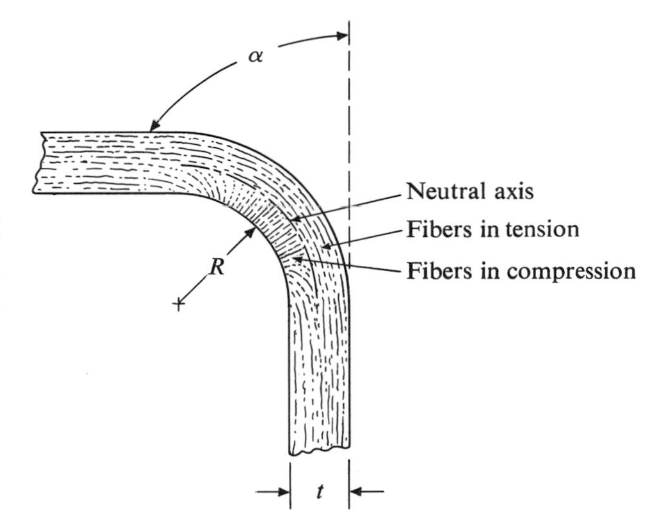


**Fig. 12.5.** Bending places the outer fibers in tension and the inner ones in compression.

**Forming and Drawing.** When a blank of metal is drawn into the cavities of a die, the metal is made to flow on a plane parallel to the die face, with the result that the thickness and surface area do not change appreciably from what was on the original blank.

In drawing, there is considerable metal flow, and both thickening and thinning of the blank take place. As the blank is first pulled over the draw radius, the excess metal, due to shrinking, is forced into the surrounding portion of the blank. Consequently, as each annular segment of material is drawn over the die radius, it thickens, work hardens, and becomes more resistant to forming. Thus the formed metal is subject to increasing stresses and strains. This causes thinning. A balance between the thickening process of going over the die radius and subsequent thinning due to strain should come in about the middle of a deep-drawn cup wall.

**Deep Drawing Test.** There are a number of tests that have been designed to test the suitability of a metal for deep drawing and stretch forming. The Erichsen test is one, as shown in Fig. 12.6. Although this is one of the earliest developed, it is, with modifications, still used today.

Blanks of metal are placed in the small press and seated with a holder so that a certain amount of inward flow is allowed under the action of the punch.

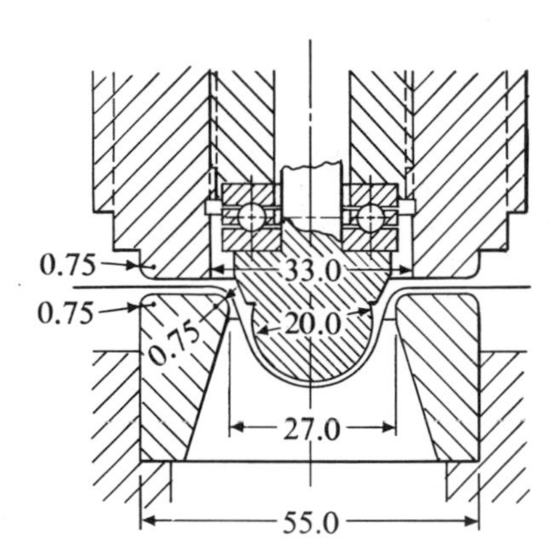


**Fig. 12.6.** The Erichsen Test for examining the deep-drawing qualities of sheet metal.

The test has proven popular as it is simple, has low cost, is easily evaluated, is quickly made, and uses very little material. Experienced operators are able to judge average grain size of the sample by the roughness of the formed dome. The roughness or surface texture of the formed part can also be estimated. The type of fracture is useful in judging the directional properties of the materials.

## DIE FORMING OF METALS

Metal forming generally refers to the shaping of sheet or plate material. This can be done between two dies (closed-die method) or with a single (open) die and some external means of applying pressure to the work material. This latter method is of lower initial cost, but production is slower.

**Matched Dies.** Dies usually represent a considerable investment. This is particularly true if the dies are matched. The term *matched dies* refers to dies in which the punch matches the die to very close tolerances. These dies are used for accurately forming parts in high production, such as household appliances and auto-body parts. They are usually cut out of steel, cast iron, or semisteel on duplicating mills or die-sinking machines. Many hours of filing and polishing are needed to bring them to the desired finish and tolerance.

Shorter-run matching dies can be made by casting both components. In this process, whether the dies are made out of epoxy resin or low-melting-point zinc alloys, the method is much the same.

The pattern is madc of wood or plaster. It is given several protective coatings and is covered with a parting agent. A box is built around it, and the die is cast over it. The pattern is removed, and the punch is then cast into the die. Provision must be made for the material thickness between the punch and die. This is taken care of by layers of sheet wax or, in some cases, felt cloth (Fig. 12.7).

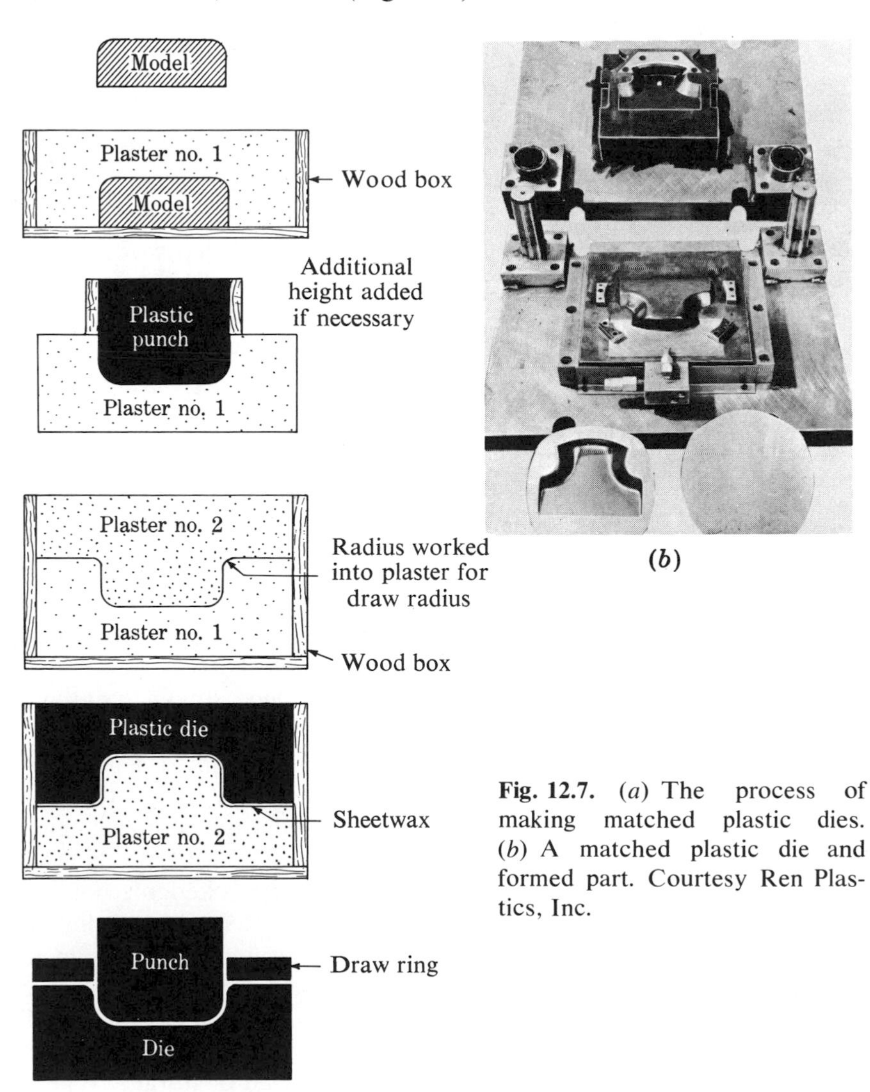


**Fig. 12.7.** (*a*) The process of making matched plastic dies. (*b*) A matched plastic die and formed part. Courtesy Ren Plastics, Inc.

If the materials are resilient, as in some epoxy-resin mixtures, no allowance is made for metal thickness between the punch and die. The punch is cast right on the die, with only parting material in between.

Another material that is easily cast to shape to make forming dies is *Kirksite*. It is a low-melting-point zinc alloy which can be cast and mounted in the press the same day. Little or no machining is required. Kirksite dies can be used for forming several thousand nonferrous metal parts. After use, they are easily melted down and recast for other dies.

**Drawing Dies.** Drawing of sheet metal parts consists of pulling a sheet metal blank over a draw ring radius into a die cavity. The process is shown in Fig. 12.11 as a compound die.

Deep draws usually require several redraws before the diameter-to-height proportion can be greatly reduced. A simple drawn shell, produced from 3003–0 aluminum, is shown in Fig. 12.8*a* with the necessary redraws. Formulas have been worked out for various materials (these may be found in handbooks), and thus the amount of reduction per draw can be quite accurately predicted. Material that work hardens readily must be annealed between draws. The blank diameter required for making some standard drawn containers are as shown in Fig. 12.8*b*.

Pressure required for drawing may be calculated as follows:

$$Pd = \left(\frac{D_1}{D_2} - 0.7\right) \pi D^2 t S_t$$

where $D_1$ = blank diameter; $D_2$ = cup diameter; 0.7 = constant; $t$ = thickness of material; and $S_t$ = tensile strength of material.

Clearances required for drawing are: *1st draw,* die dia = punch die + 2.2$t$; *2nd draw,* die dia = punch die + 2.3$t$; *any following draw,* die dia = punch die + 2.4$t$.

The relationship of the drawn shell diameter to the blank diameter is termed the *drawing ratio* and is expressed as $R_d = D_2/D_1$.

IRONING. Sometimes it is desirable to produce drawn parts with considerable variation in the thickness of metal. For example, it may be necessary to have a relatively heavy end on a shell casing or a heavy top to allow for machining, or it may be necessary simply to thin the metal and strain harden it. The drawing process used to accomplish this is known as *ironing*.

**Die Sets.** In general practice, when dies are spoken of, both the punch and the die are meant, as shown in Fig. 12.9. Not all dies are mounted in die sets as shown; they may be mounted directly in the press. This requires careful alignment of the punch and die each time they are taken out or put in. Since this can be very time-consuming, die sets are used. The punch and die need only be aligned in the set once. After the run is completed, the entire unit is placed in storage. When it is to be used again, the setup time required will be only a few minutes.

The function of the stripper shown in Fig. 12.9 is to keep the metal

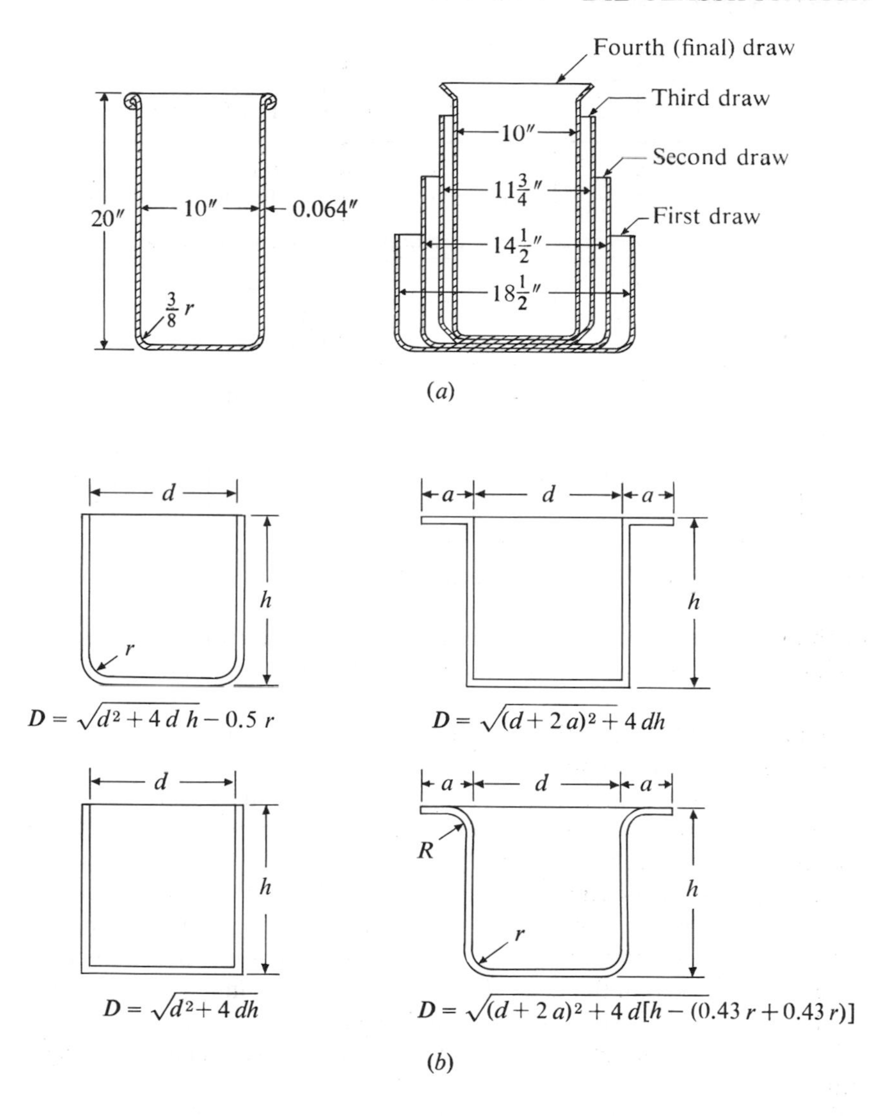


**Fig. 12.8.** (*a*) The number of redraws to produce a shell out of 3003-0 aluminum. (*b*) Blank diameter formulas for common drawn shells.

from sticking to the punch as it is withdrawn from the die. Spring-action strippers are often used around the punch rather than the solid type that is shown.

## DIE CLASSIFICATION

Dies are classified both as to operation and as to their construction.

**Classification as to Operation.** All die work can be classified into four basic types of operations: cutting, bending, drawing, and squeezing.

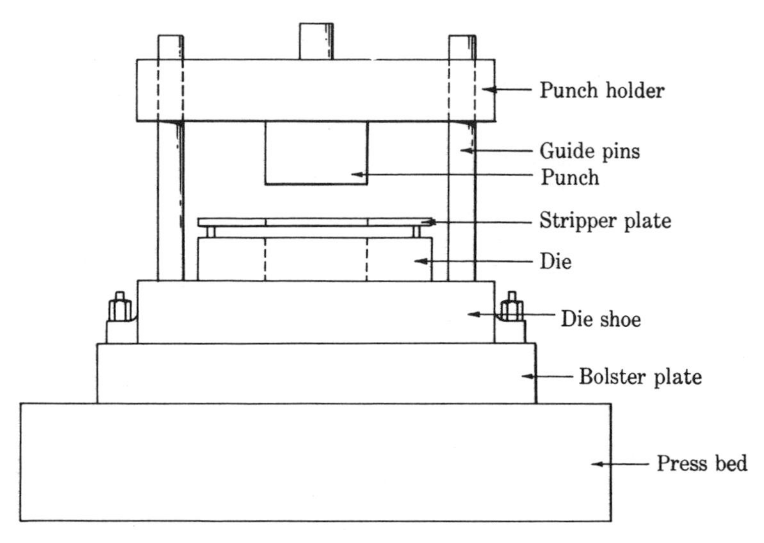


**Fig. 12.9.** Standard die set with punch and die.

Each of these operations has several subdivisions as follows:

Cutting—blanking, piercing, notching, shearing, trimming, and shaving.

Bending—folding, seaming, curling, and angle bending.

Drawing—forming, embossing, and ironing.

Squeezing—coining, sizing, swaging, extruding, and upsetting.

The four main types of die operations and some of the variations are shown schematically in Fig. 12.10.

**Classification According to Construction.** The die construction may be such that various operations are performed as the material moves in progressive steps through the die or that more than one operation can be done at one station in the die. Dies are classified as simple, compound or combination, progressive, transfer, and special.

SIMPLE DIES. Simple dies perform one operation, usually forming or cutting.

COMPOUND OR COMBINATION DIES. These dies perform more than one operation in one location; that is, a part can be blanked and formed in one stroke of the press, as shown in Fig. 12.11.

PROGRESSIVE DIES. The combination dies have the disadvantage of crowding too many die elements into a limited area, which makes construction and maintenance costs higher. A progressive die is arranged with two or more stations so that the operations can be spread out over more area to reduce maintenance costs and increase operating speeds. Examples of progressive die work are shown in Fig. 12.12.

TRANSFER DIES. Transfer dies are used where the complexity, shape, or size of the part does not permit it to be fed from station to station.

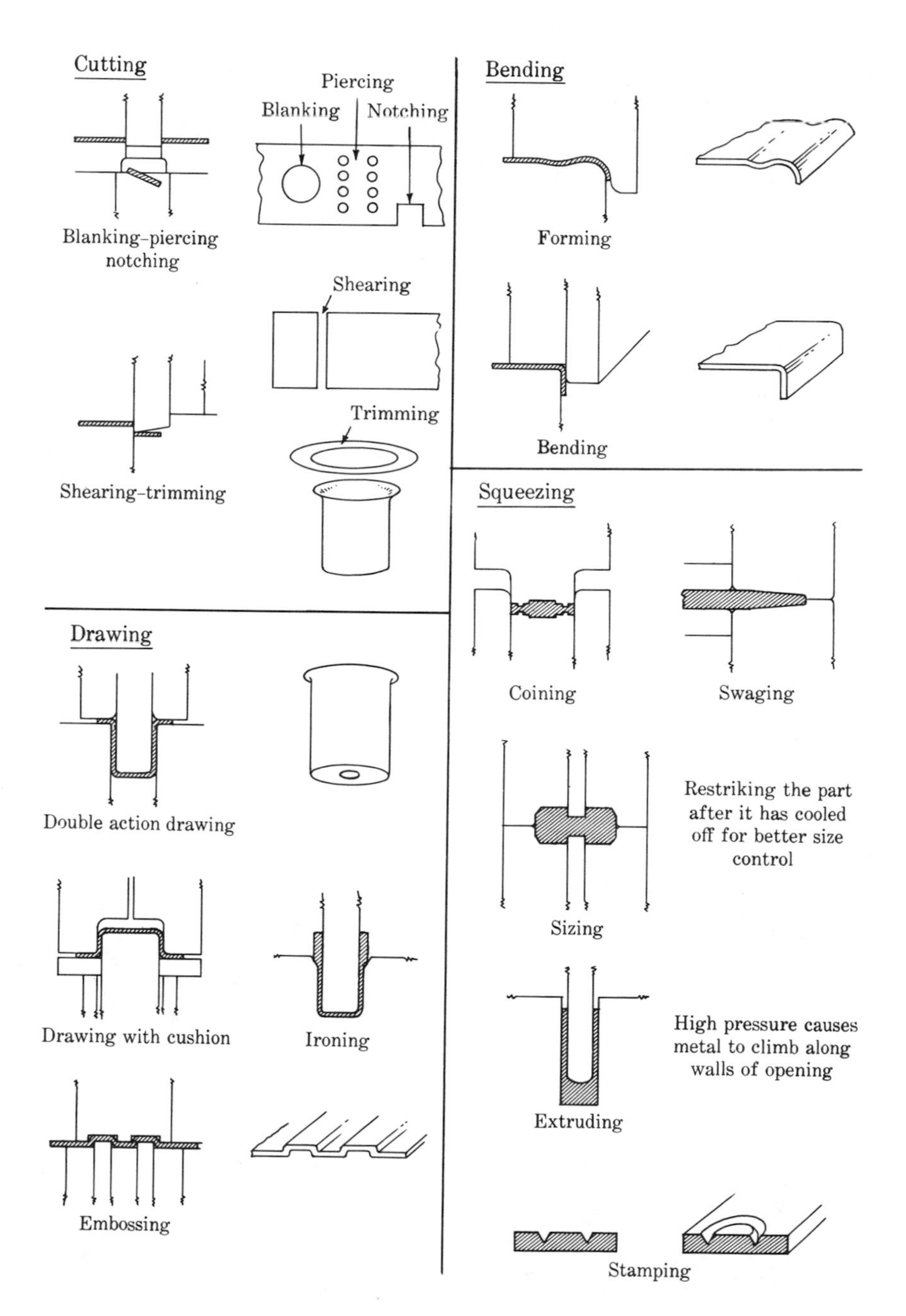


**Fig. 12.10.** Four main pressworking operations.

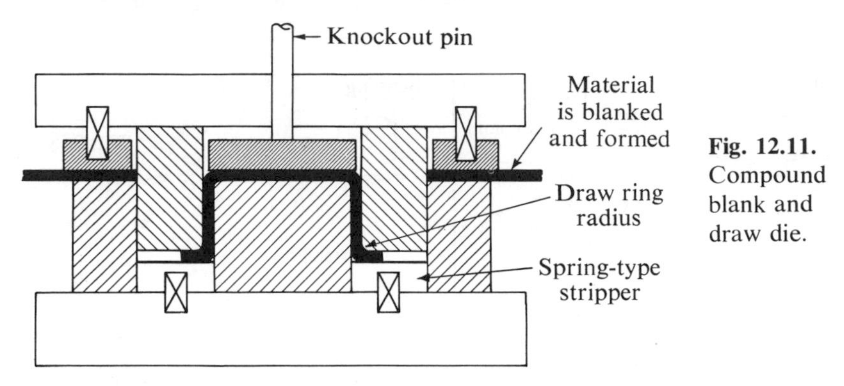


**Fig. 12.11.** Compound blank and draw die.

Transfer dies incorporate a mechanism for moving the part from one die to the next (Fig. 12.13). The transfer mechanism is powered by the stroke of the press ram. Also shown in Fig. 12.13 is a "press pacer" used to move parts from one press to another. Feed fingers pick the part out of the die and move it to the next press.

## PRESSES

**Types and Uses.** Press classification is a broad subject, since much of it has resulted from common usage. There are many ways by which a

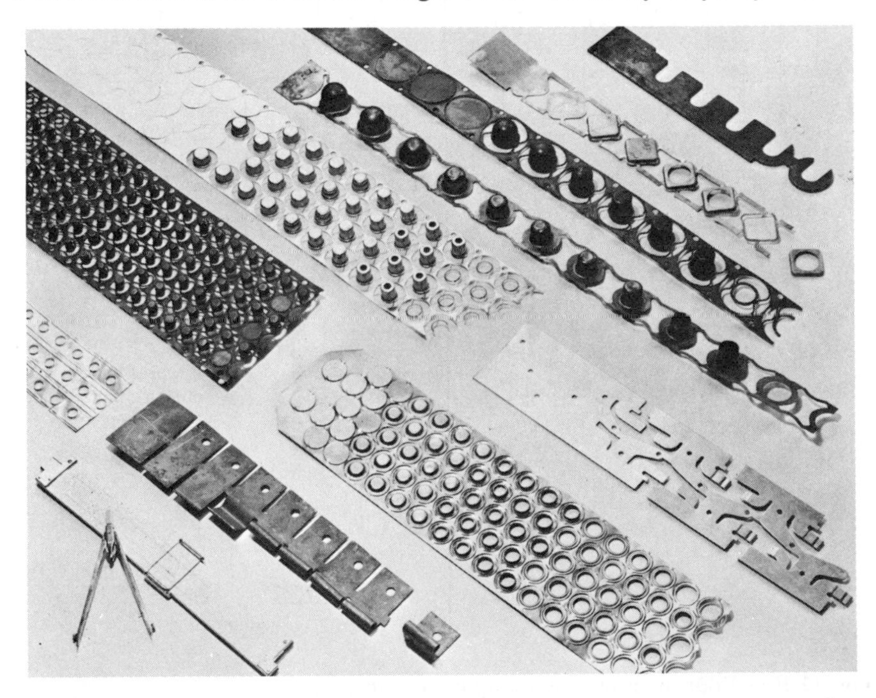

**Fig. 12.12.** Examples of progressive die forming-and-cutting operations. Courtesy The Brandes Press, Inc.

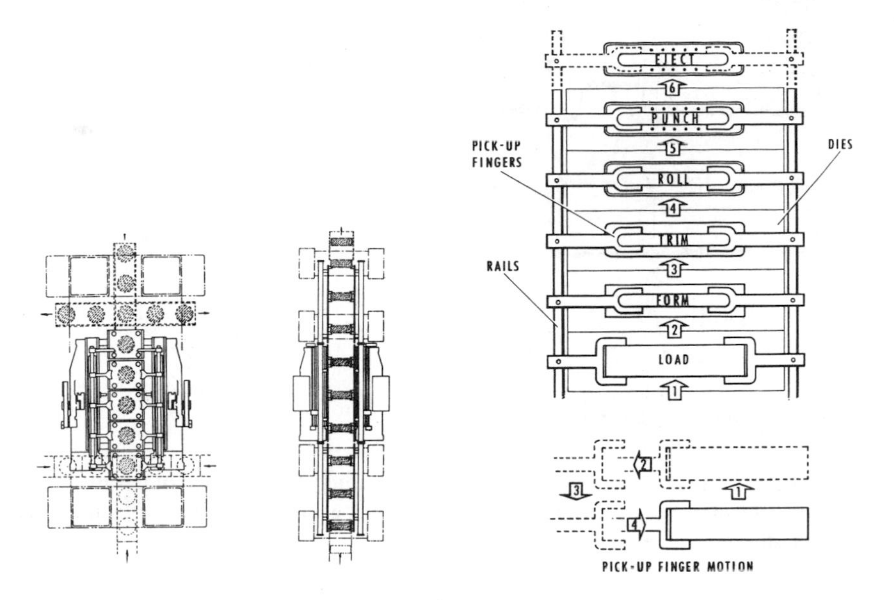


**Fig. 12.13.** Transfer dies are shown in the press at left. A press-to-press transfer mechanism is shown in the center, with a more detailed view at the right.

press can be classified. Some of these are: the drive mechanism, or method of moving the ram; the number of drives; the frame type; the frame position; the action, or number of rams; the size, including tonnage, bolster area, and shut height; and the suspension, or number of connections for the ram.

DRIVES. Press drives can be broadly classified as mechanical or hydraulic. There are several types of mechanical drives, such as crankshaft, eccentric, cam, toggle, rack and pinion, screw, and knuckle. The most widely used of these are the crank, eccentric shafts, and toggle mechanism, as shown in Fig. 12.14.

The crank method of converting rotary motion to straight thrust provides variable mechanical advantage between the drive and the ram. Near the bottom of the stroke, a toggle effect is obtained, giving a high mechanical advantage. Presses are rated by the power they develop at or near the bottom of the stroke.

NUMBER OF DRIVES. The number of drives refers to the number of places from which the crankshaft, cam, or whatever driving mechanism is used, is being driven. For example, a small press may be driven from one flywheel, whereas large presses may have as many as four drives and four flywheels.

FRAME TYPE. The most easily distinguished characteristic of a press is its frame. Common types of frames are C-frame and straight-side.

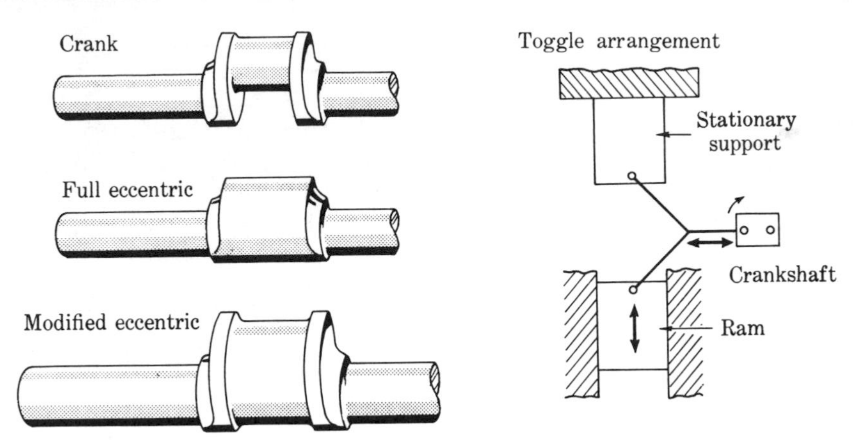


**Fig. 12.14.** The crank and eccentric are among the most commonly used methods of imparting motion to the ram. The eccentric and toggle are used where a slower squeezing action is desired.

*C-frame presses.* C-frame presses get their name from the C-shaped upright frame (Fig. 12.15). They include open-back inclinable (OBI), gap-frame, adjustable-bed, and horning presses.

The OBI press is one of industry's most versatile presses. It is usually a low-tonnage press, ranging from a 1-ton bench model to a 250-ton floor press. The fact that it has an open back and can be inclined backward up

**Fig. 12.15.** C-frame presses. Courtesy Johnson Machine and Press Corp.

to 45 deg facilitates the removal of both the manufactured part and the scrap.

The gap-frame press is essentially the same as the OBI but is not inclinable. The frame is of one solid piece, and it may or may not have an open back.

Adjustable-bed presses provide for raising and lowering the bed to suit the required die space. Some of these presses allow the bed to be swung out from under the slide so that a horn can be used in place of the bed (Fig. 12.16).

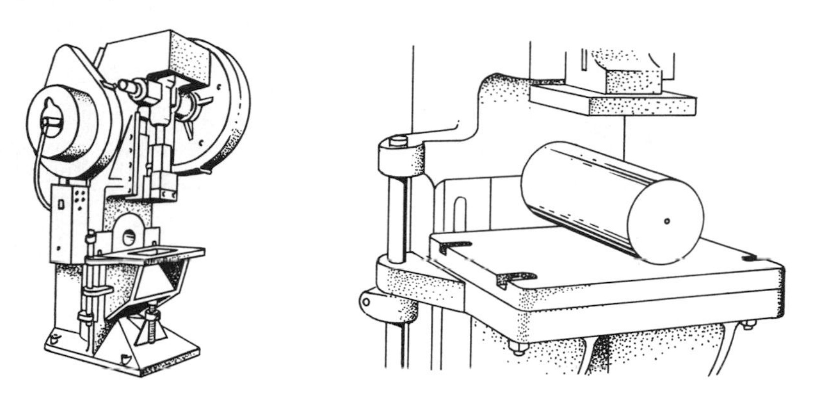

**Fig. 12.16.** Adjustable-bed and horning-press combination.

Horning presses are useful for secondary operations such as punching holes in formed parts. If the holes are made prior to forming, distortion results. Oftentimes, a die on the ram is all that is needed.

*Straight-side presses.* Straight-side presses are made for larger die areas and higher tonnages than the C-frame presses can accommodate. The straight-side frame gives a strong, rigid construction. Two common types of straight-side presses are shown in Fig. 12.17.

FRAME POSITION. The three common positions of the press frame are vertical, inclined, and horizontal. Vertical is the most common. The OBI is an example of the inclined. Some presses are made with a fixed angle.

The horizontal press is usually small and of the high-speed type. Ejection of piece parts and scrap presents little difficulty owing to the free fall of both items. Although horizontal presses can be counted on for high production, there is also more wear on the sliding parts than on a vertical press. A versatile horizontal press is the multislide press shown in Fig. 12.18.

These machines are made so that several slides can be used to complete the forming operation around a stake. Strip material is fed into the press, progressively stamped, notched, bent, formed around a stake, cut

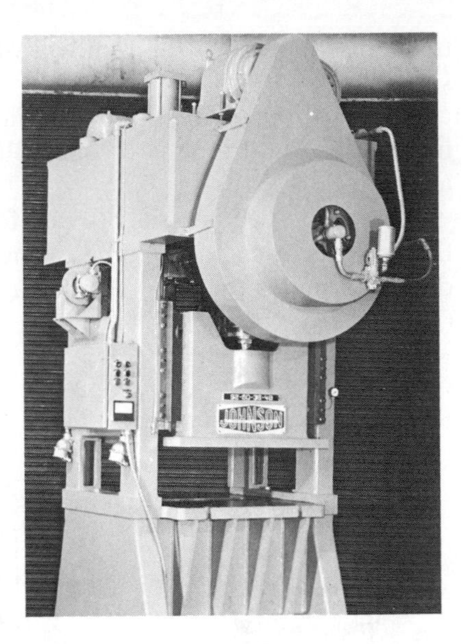

**Fig. 12.17.** Two common types of straight-side frame presses. Courtesy Johnson Machine and Press Corp.

off, and ejected. Examples of the type of work done on strip material are shown in Fig. 12.18. Some rather complicated parts can be made at rates of 125 per min. These machines are also used for wire-forming operations.

ACTION. Press action refers to the number of rams or slides on the press. There may be one, two, or three as follows:

*Single action.* A single-action press is the conventional type, having one ram located in the top, or crown, of the press.

*Double action.* A double-action press has a ram within a ram. The outer ram comes down first, to seat the material and keep the right amount of pressure on it. This is espccially important, in deep-drawing or forming operations.

*Triple action.* A triple-action press has the arrangement of the double action plus an additional ram that is located in the bed of the press. This third ram can be made to move upward after the other two have completed their action.

**Press Identification.** With so many factors used to designate press size and type, an industry standard has been worked out. A Joint Industry Council (JIC) standard press designation is as shown in Table 12.2.

Other designations in place of the OBI shown are: S, single action; T, triple action; D, double action. A number following the letter indicates the points of suspension used. For example, T2 would be the code for a triple-action press with two points of suspension from the drive to the ram.

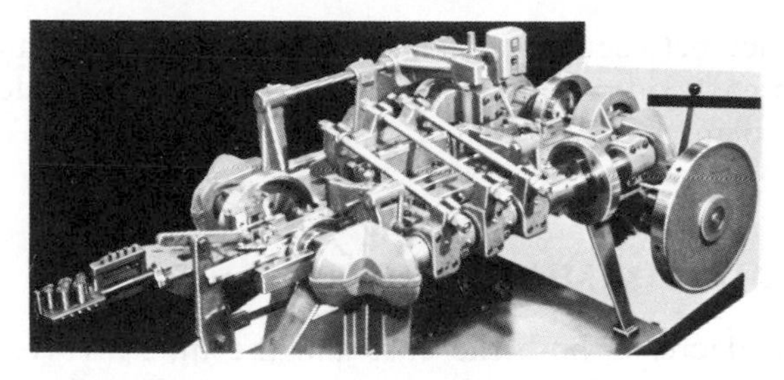

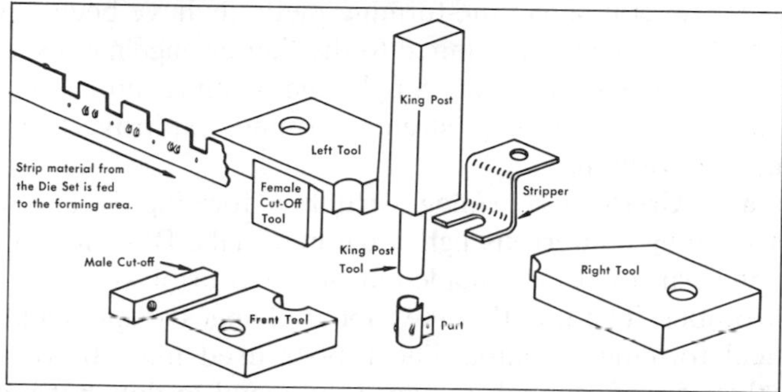

**Fig. 12.18.** A four-slide automatic combination press can be used for forming either wire or ribbon stock. Strip material is fed in and formed by the four tools around the king post. Courtesy A. H. Nilson Machine Company.

**Table 12.2**
**JIC Standard Press Designation**

*OBI – 22 – 12 3/4 × 17 1/2*

Bed size left to right and front to back
Tonnage capacity
Type of press

**Press Size.** The press size in tons required for blanking is based on the formula

$$F_b = (ltS_s/2{,}000) \times F_s$$

where $F_b$ = blanking force; $l$ = length of sheared edge; $t$ = material thickness; $S_s$ = ultimate shear strength of the material; and $F_s$ = safety factor (usually 20 to 50 percent).

The clearance between a punch and die is about 7 percent of the metal thickness (on each side of the punch). This clearance is added to the die diameter for blanking. The punch diameter must remain equal to the desired blank.

## SINGLE DIE-FORMING METHODS

In order to decrease both the cost and the time involved in making matched dies, several single die-forming methods have been developed. These methods are generally limited to the lighter-gage metals, except in the case of explosive forming. Single die-forming processes include rubber forming, Hydroforming, stretch forming, explosive forming, and electromagnetic forming.

**Rubber and Urethane Forming.** Rubber forming is a convenient method of forming nonferrous light-gage materials. Basically, it consists of having one half of the die made out of metal or plastic and the other half out of rubber. The fact that the rubber changes shape readily makes it a practical forming medium. The rubber used must be of the right hardness, that is, soft enough to cause the metal to flow but not so soft that it fills up the cavities before maximum pressure is applied.

GUERIN PROCESS. The Guerin process, shown in Fig. 12.19, is used to form sheet metal into relatively shallow parts. The male dies are made of metal, and several layers of rubber, held in a retainer, serve as the female die. The male dies are usually grouped on sliding platens so that they can be easily moved in and out under the press ram. Materials formed by the Guerin process are mostly aluminum and magnesium alloys; however, stainless steel as thick as 0.1875 in. has been used successfully. Holes can be punched in annealed aluminum and stainless up to 0.050 in. thick.

The male dies for rubber forming can be made of steel, Kirksite,

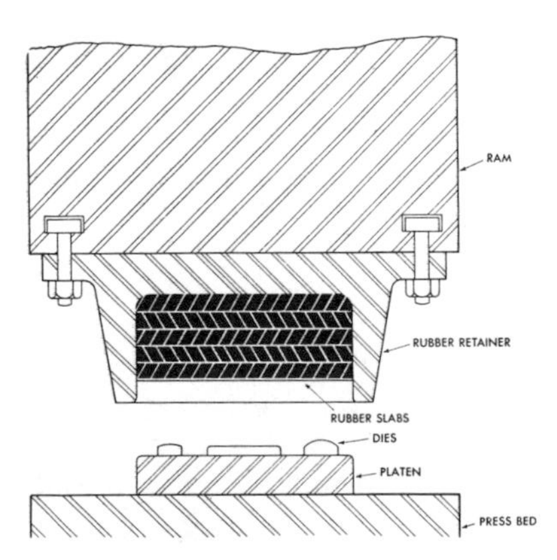


**Fig. 12.19.** Punch and die arrangement for the Guerin process.

Masonite, plastic, or aluminum. It is quite often desirable to "face" plastic or Masonite dies with thin sheet steel.

Sharp corners should not be used in the die design, since the rubber will not flow into them. Where sharp corners are necessary, they can be taken care of by metal die inserts. Locating pins are used to keep the metal from sliding around before it is formed.

URETHANE FORMING. Urethanes are resilient elastomers that are tough enough to withstand repeated deflections. Urethane forming is based on somewhat the same idea as the Guerin process. Most of the work has been done using urethane as the female die on press brakes. The urethane can be cast to almost any shape. Air space (Fig. 12.20)

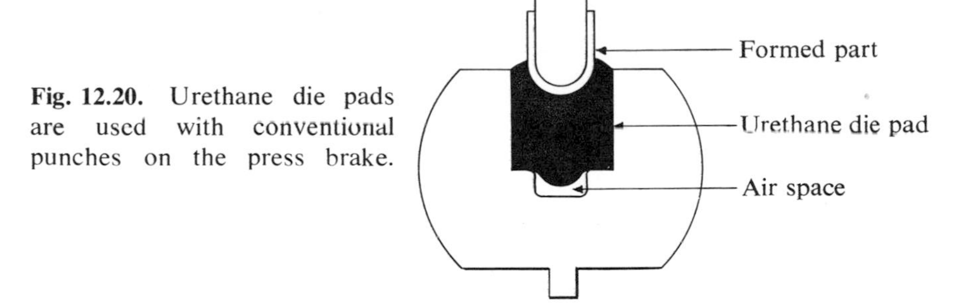


**Fig. 12.20.** Urethane die pads are used with conventional punches on the press brake.

beneath the pad helps to concentrate forming pressures in the right area and direction. Without air space, the urethane would bulge over the top of the retainer, which would cause substantial loss of pressure and might cause the urethane to exceed its deflection limits. Penetration should be limited to one third of the total thickness of the pad plus the air space under it.

Deflector bars are sometimes used under the pad to concentrate forming pressures (Fig. 12.21).

New polyether urethanes have higher hardness, which permits

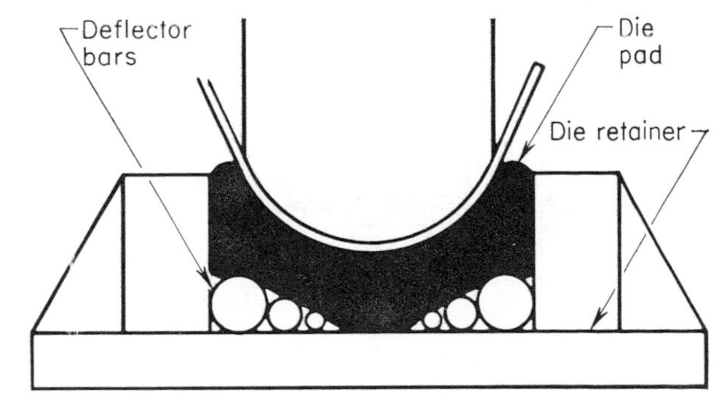


**Fig. 12.21.** Deflector bars are used under the urethane die pad to concentrate the forming pressure. The bars are easily changed when other part configurations are needed.

heavier-gage materials to be formed. Some of these dies have been used to produce over 100,000 pieces and are still in good condition.

WHEELON PROCESS. The Wheelon process is a refinement of the Guerin process (Fig. 12.22). The specially designed press is made to operate at 5,000 to 10,000 psi, which is sufficient for sheet metal forming of shallow parts.

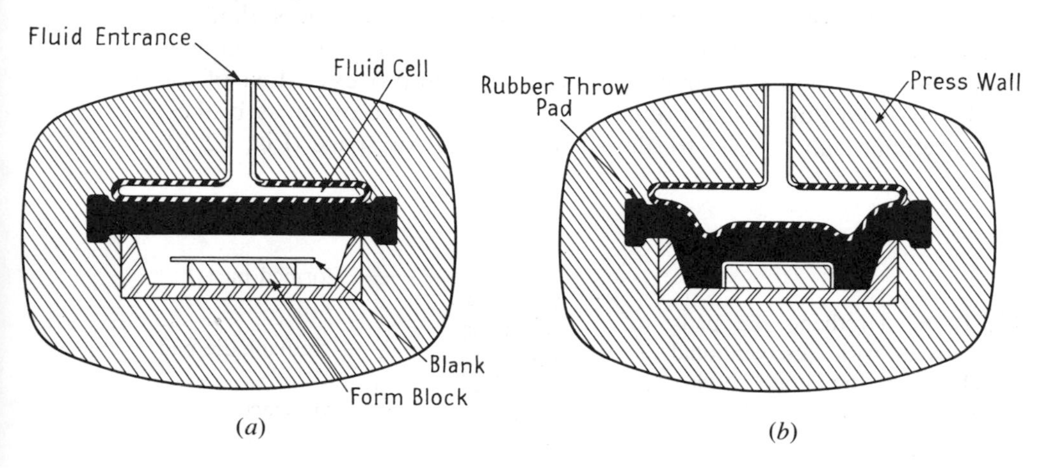


**Fig. 12.22.** The Wheelon process uses a fluid cell, a rubber pad, and form blocks for shallow draws of sheet metal parts. (*a*) Before operation; (*b*) at completion of operation.

**Bulging.** The process of bulging, or making a cylinder with enlarged diameter, can be readily accomplished on nonferrous metals by the use of a rubber punch and an appropriate die (Fig. 12.23) or with hydraulic pressure.

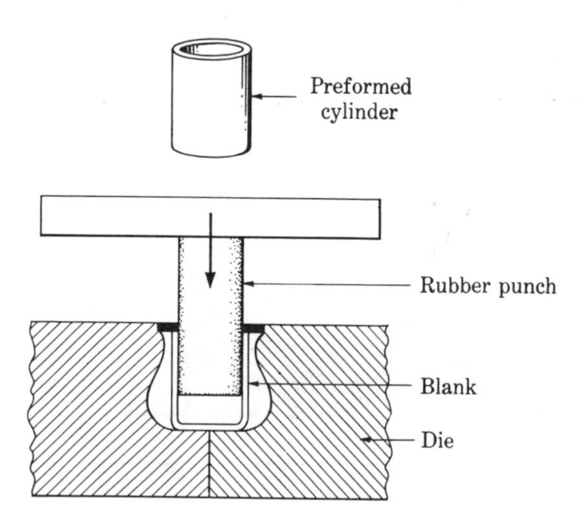


**Fig. 12.23.** Bulging. The rubber punch is brought down, causing metal to flow out against the walls of the die. The die halves are then separated, and the bulged part is removed.

**Hydroforming.** Hydroforming, as the name implies, is forming with fluid pressure. In this case, the pressure is furnished by hydraulic fluid on one side of a rubber diaphragm, the part to be formed being located on the other side (Fig. 12.24). The diaphragm is sealed to withstand hydraulic pressures up to 15,000 psi when backed up by the workpiece.

In the Hydroforming process, the blank is placed on the blank holder. Enough pressure is pumped in to hold it in place. The punch is then moved up against the metal blank. As the punch moves up into the forming-dome cavity, pressure continues to increase, causing the metal to flow evenly around the punch. After forming, the hydraulic pressure is released and the punch is retracted, leaving the formed part on top of the blank holder.

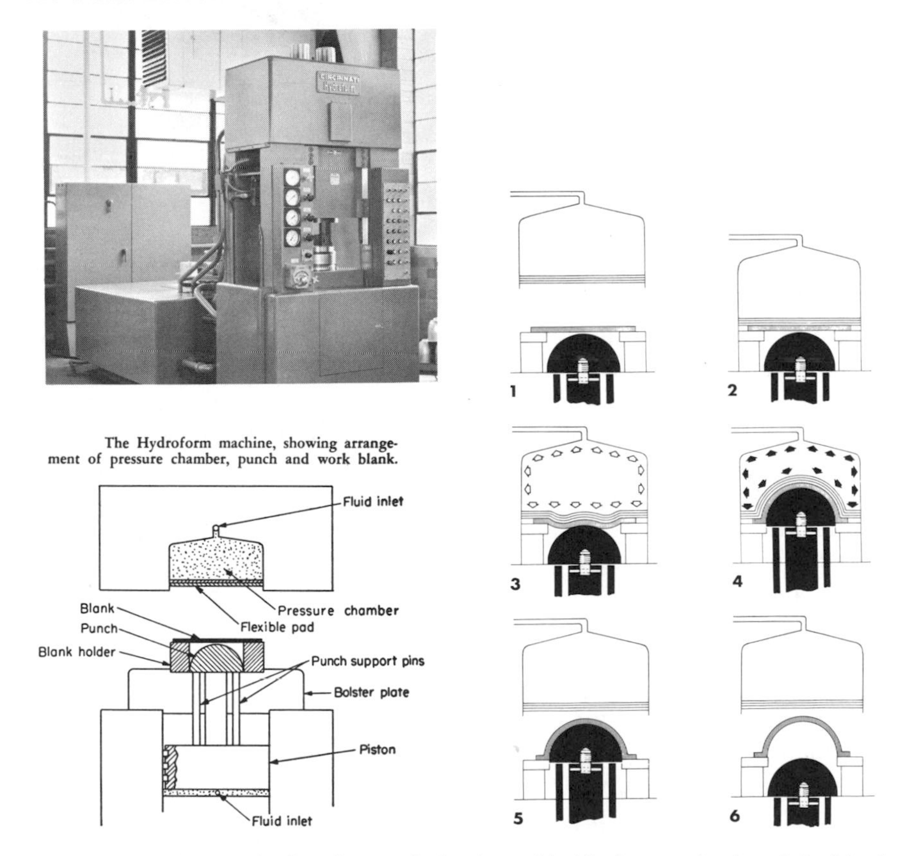


The Hydroform machine, showing arrangement of pressure chamber, punch and work blank.

**Fig. 12.24.** The Hydroforming cycle begins with (1) dome raised and the blank in place. (2) The dome is lowered, clamping the blank to the holder, and (3) starting pressure is applied to force the blank against the punch. (4) The punch moves upward, increasing the pressure and forming the part. (5) After the forming pressure is released, the dome rises and (6) the work is stripped from the punch. Courtesy the Cincinnati Milling Machine Co.

The advantage of Hydroforming is that there is very little, if any, thinning of the metal that is being formed. The reason for this is that the metal is tightly gripped between the punch and the diaphragm, allowing the metal to be eased over the form without slipping. Hydroforming can usually perform, in one or two operations, what would take four or five draws by ordinary pressworking methods, as shown in Fig. 12.25.

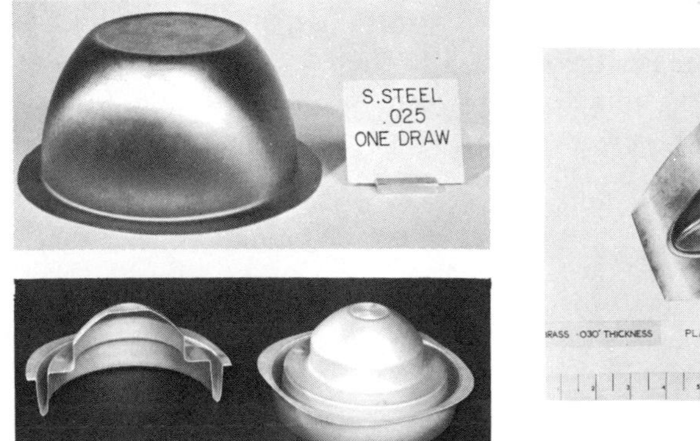


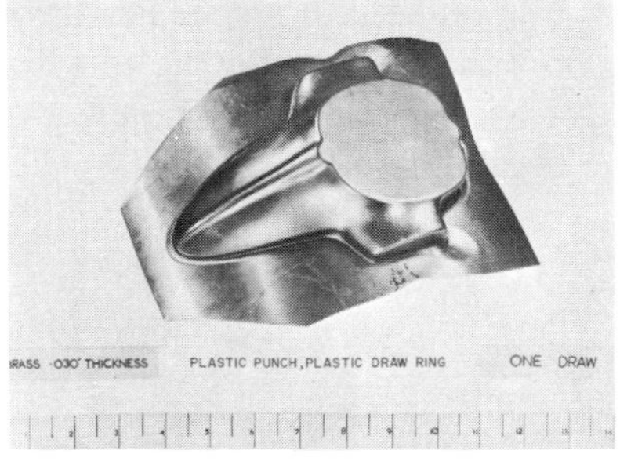


**Fig. 12.25.** Parts formed by one or two draws with the Hydroforming process. Courtesy The Cincinnati Milling Machine Co.

**Stretch Forming.** Stretch forming of metal consists of placing the sheet material in tensile load over a form block. The material is stressed beyond its elastic limit, causing it to take a permanent set. In stretch forming, all of the fibers are stretched, but those on the inside radius of the bend are stretched less than those on the outside (Fig. 12.26). Since the metal is placed under one type of load in stretch forming, there is no *springback* (attempt to regain some of its former position). However, allowance must be made for dimensional changes that occur in the metal that is being used; that is, during stretching, the length of the part increases and the width decreases.

There are several methods of stretch forming, but only contour-forming types will be discussed here.

CONTOUR FORMING. Contour stretch forming is used to produce compound curves in sheet stock (Fig. 12.27). The ends of the stock are gripped in hydraulically operated serrated jaws. Stretching is accom-

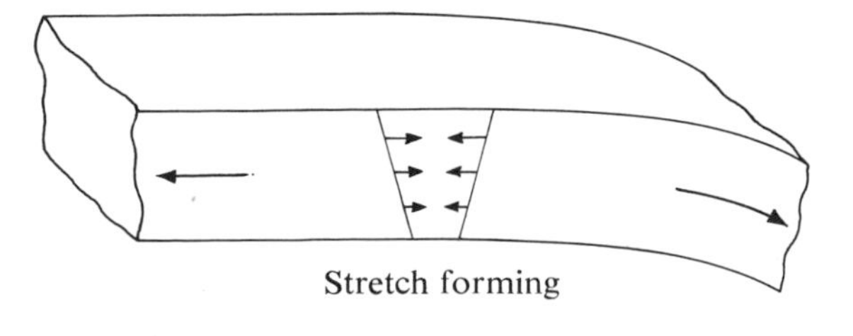


**Fig. 12.26.** Theory of stretch forming.

**Fig. 12.27.** Stretch forming a compound curve in sheet stock. Courtesy Sheridan-Gray Inc.

plished by one or more hydraulic cylinders moving up under the die. Forming by this method is restricted to parts that do not have sharp edges, which tend to restrict the stretch in local areas that would lead to fractures.

The forming block is usually highly polished and lubricated. Sometimes, a blanket of thin rubber or fiber glass is used to avoid the cost of a highly polished surface or the need of lubricants.

Contour stretch forming is very useful in making preliminary models of aircraft and automotive parts. It is also used on a production basis for forming truck and trailer bodies.

**Explosive and Electromagnetic Forming.** Both explosive and electromagnetic forming are single-die forming operations. Since they are relatively new processes they are described in Chap. 20, where new manufacturing processes are discussed.

## METAL SPINNING

Spinning is a method of forming symmetrical shapes such as spheres, cones, parabolas, etc., in sheet metal by means of a rotating form or chuck. The forming is done by application of pressure on the metal with a roller or spinning tool that causes it to conform to the shape of the rotating wood or metal chuck. The spinning process can be divided into two main methods – conventional spinning and displacement spinning (often referred to as power roll forming).

**Conventional Spinning.** The spinning method originated to fill a need for the small-quantity production of cylindrical shapes, but it now can produce parts in lots of several thousand. This advancement has resulted from the development of semiautomatic and automatic spinning equipment and from the realization by methods engineers that spinning, combined with other operations such as roll forming, can produce parts that are difficult or impossible to make by any other method. Spinning may be used after drawing as a final operation in perfecting contours, trueing diameters, and removing tapers from cylindrical draws. Drawn articles may also be spun to form odd shapes, necks, and flanges.

Hand spinning requires skill and experience. The metal must be made to flow at the proper rate or wrinkles and tears will result. A blank somewhat larger than the diameter of the finished shell is placed between a live center and the chuck. The tools used consist mainly of steel rods with various-shaped ends and long wooden handles (Fig. 12.28).

Other tools are used to perform various operations such as trimming the top edge or rolling a bead to give strength and rigidity to the upper edge.

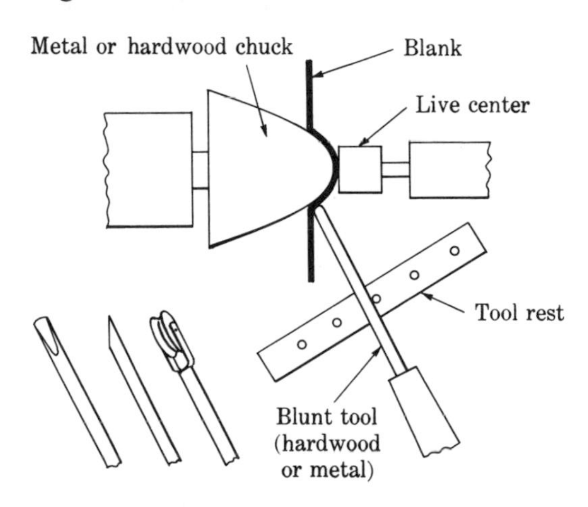


**Fig. 12.28.** Forming the metal to the shape of the chuck on a metal-spinning lathe.

The size of parts that can be spun is limited only by the equipment available. Gap lathes are used for medium-sized jobs, but large jobs (16 ft in diameter or more) require the construction of special equipment (Fig. 12.29).

**Displacement Spinning or Power Roll Forming.** Power roll forming is a cold-rolling process in which material in a blank is caused to flow over a rotating mandrel. Forming is accomplished by one or more hardened-steel rollers which travel parallel to and at a preset distance from the surface of the mandrel during the forming operation (Fig. 12.30). As the mandrel and workpiece are revolved, the rollers, which may be either hydraulically or mechanically actuated, displace the blank material in a spiral manner. The contours of the formed piece are identical with those of the mandrel, and the wall thickness can be closely controlled. When precision parts are desired, 0.015 or 0.020 in. can be left for machining.

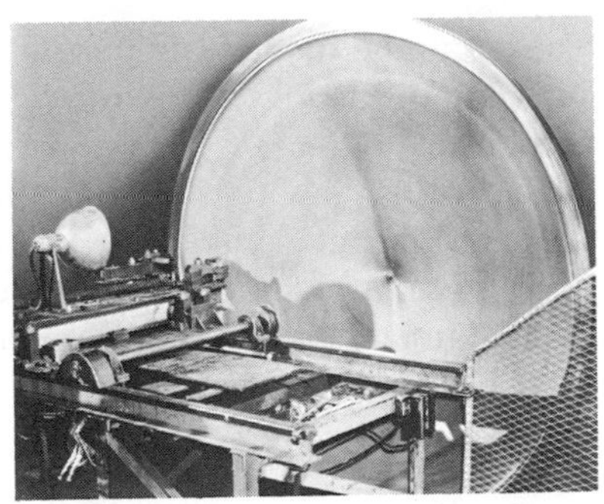

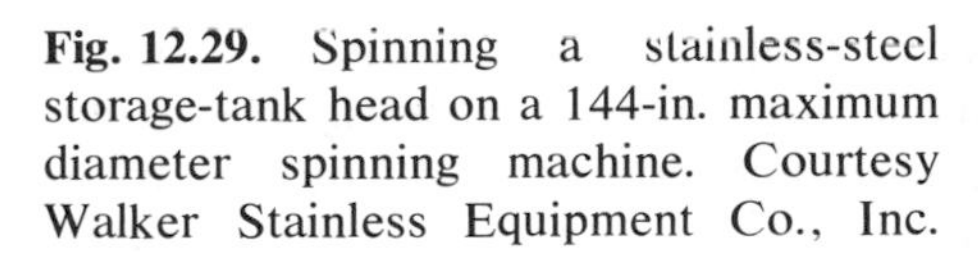

**Fig. 12.29.** Spinning a stainless-steel storage-tank head on a 144-in. maximum diameter spinning machine. Courtesy Walker Stainless Equipment Co., Inc.

The control over workpiece contours and wall thickness gives the process significant advantages over drawing and conventional spinning. Springback, encountered in both pressworking and conventional spinning, is eliminated. Rolling the metal increases its strength and hardness. There is also a beneficial effect on the granular structure of the metal. Tests have shown that flow turning can increase the tensile strength of the finished product by as much as 100 percent over that of the original metal.

In roll forming, no calculations as to blank diameter are needed, since the starting blank is the same diameter as the finished part. The workpiece material is drawn from the thickness of the blank rather than from the diameter. Cone-shaped parts are usually produced from flat blanks. Some parts, such as cylinders, require preformed blanks so that a sufficient volume of metal can be maintained for wall thickness. Preformed

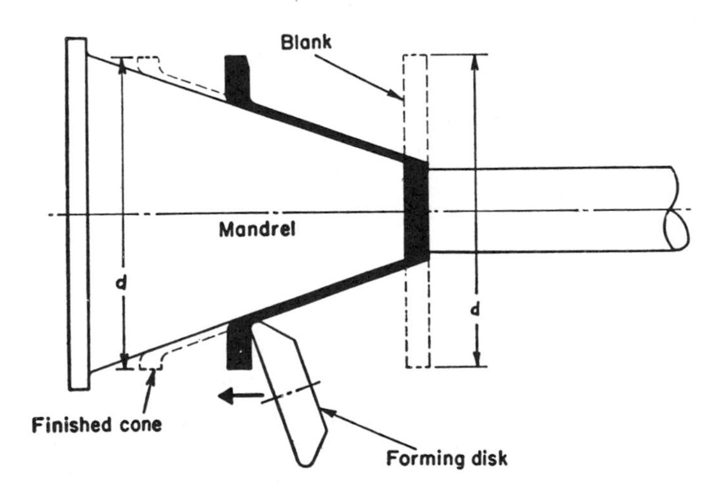


**Fig. 12.30.** Schematic diagram showing the principle of the flow-turning process. Note that the diameter of the blank is the same as that of the finished part.

blanks can be made from partially machined forgings, partially machined centrifugal castings, drawn or machined cups, and welded cylindrical sections.

Generally, flow turning is limited to symmetrical cylindrical or conical shapes. Special setups can be made for elliptical shapes. The metal thickness that can be worked is usually not more than 1/2 in. The minimum diameter of the mandrel is also 1/2 in.

## ROLL FORMING

There are two main types of roll forming: one uses continuous-strip material for high-production work; the other uses sheet and plate stock.

**Continuous Roll Forming.** Continuous roll forming utilizes a series of rolls to gradually change the shape of the metal. As the metal passes between the rolls in a fast-moving continuous strip, the cross-sectional shape is changed to the desired form. The forms are almost limitless in variation (Fig. 12.31).

**Fig. 12.31.** A wide variety of shapes can be produced in a continuous strip by roll forming. Courtesy The Yoder Co.

The intricacy of the shape, the size of the section, the thickness, and the type of material will determine the number of rolls required. A simple angle or channel with straight web and flanges can usually be formed with 3 or 4 pairs of rolls, whereas the more complicated shapes require up to 12 or more roll passes (Fig. 12.32). In addition, some straightening rolls and some idle-station rolls may be needed.

**Sheet and Plate Bending Rolls.** Plain and beaded cylinders, cones, ovals, etc., are fabricated on bending rolls. Bending-roll machines range in size from those that are able to roll heavy steel plate 1 in. thick to

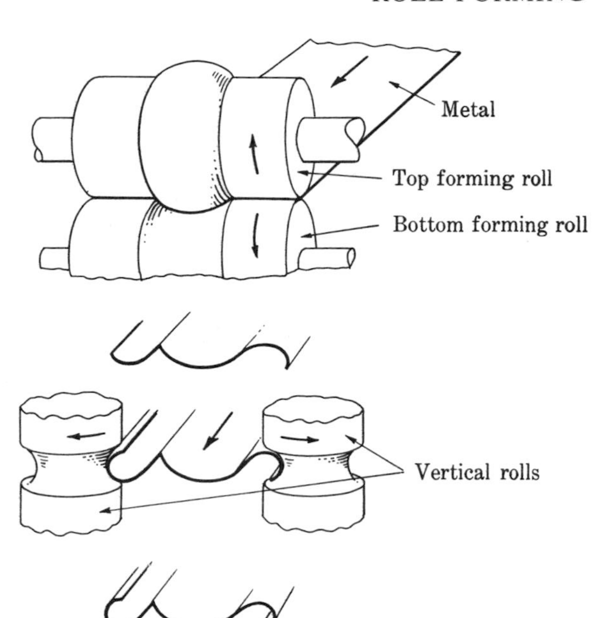


**Fig. 12.32.** Progressive rolls needed to form a curtain rail.

small bench models used for light-gage sheet metal work. In either case, the bending action takes place between rotating horizontal bending rolls. The most commonly used type is shown in Fig. 12.33. By adjusting the lower front roll up or down, various thicknesses of metal can be accommodated.

Bending rolls are used to fabricate a wide variety of parts, including production quantities of tanks and pipe. In heavy-gage pipe and tank manufacture, no attempt is made to close the circle completely. After the part is removed from the rolls, it is placed in a fixture to close the gap prior to welding.

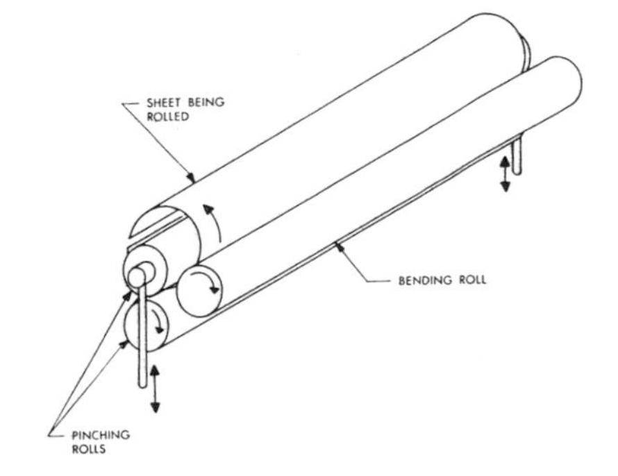


**Fig. 12.33.** Roll forming of sheet metal.

## OTHER FORMING AND DRAWING METHODS

**Rotary Swaging.** Rotary swaging is the process used to reduce the cross-sectional area of rods and tubes (Fig. 12.34).

The operations shown in Fig. 12.34 can be closely controlled, maintaining tolerances within ±0.0005 on small diameters and within ±0.005 on large diameters.

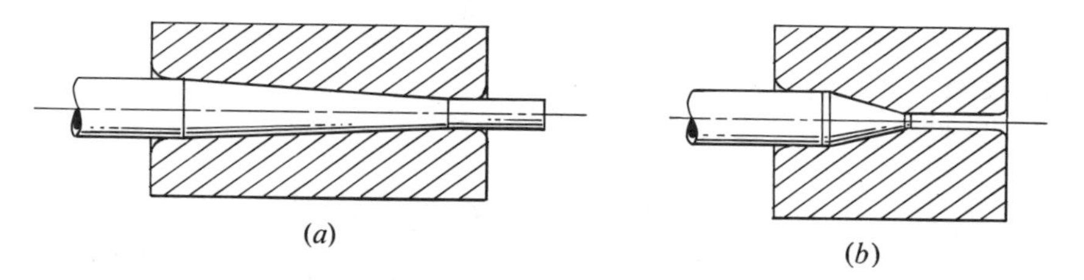


**Fig. 12.34.** Examples and uses of swaging. (*a*) Long tapered tubular legs for furniture, made by swaging in multiple, overlapping passes, are light and sturdy. (*b*) Ball point pens and automatic pencils are tapered and pointed by simple swaging procedures at high production rates.

Swaging is often spoken of as a cold-forging operation because the metal forming takes place under the hammering blows of die sections. The swaging machine consists mainly of a hollow spindle which carries the die sections and rollers (Fig. 12.35).

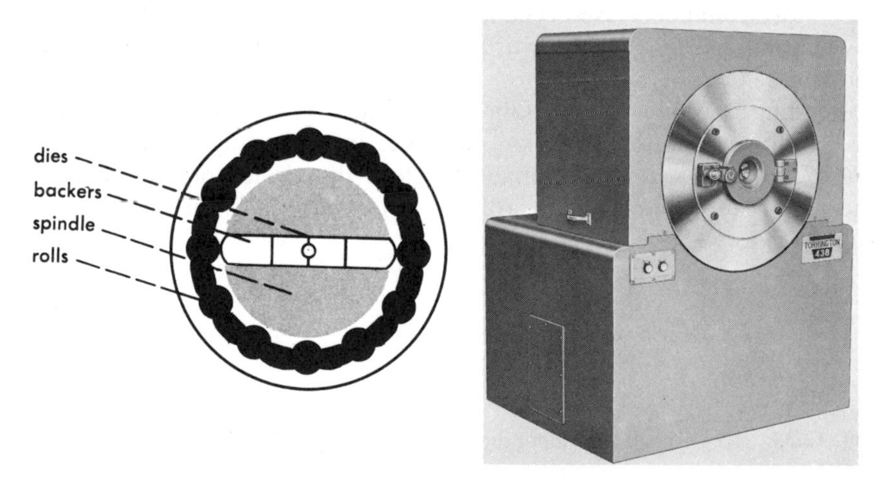


**Fig. 12.35.** The rotary swaging machine reduces the end of stock by a series of rollers that rotate over backer, forcing the dies together. Courtesy The Torrington Co.

As the machine is started, the outer rolls are moved outward by centrifugal force, but, as the hammers contact the opposite rolls, they are forced together, moving the dies together simultaneously. Approximately 1,200 to 6,000 blows per minute are struck by the dies. Under normal conditions the metallurgical qualities of the metal are improved. However, this is dependent on the condition of the metal before starting and on the amount of reduction desired, since too much reduction without proper annealing will result in fractures. The finish also is generally improved.

**Cold Drawing.** Rods, tubes, and extrusions are often given a cold-finishing operation to reduce the size, increase the strength, improve the finish, and provide better accuracy.

Bars or tubes that have been hot rolled are prepared for drawing by first removing the oxide scale in a pickling acid. After the scale is removed, the material is washed in lime to remove the acid. The lime, plus soap or oil, acts as a good drawing lubricant. The process is shown schematically in Fig. 12.36.

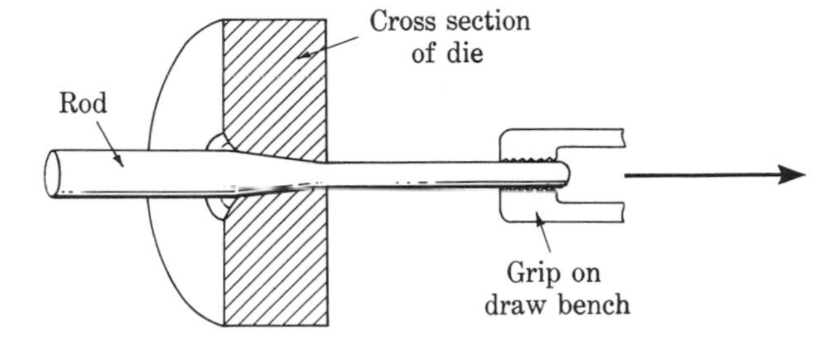


**Fig. 12.36.** Methods of cold drawing and rolling. Courtesy *Steelways,* published by American Iron and Steel Institute.

**Drawing and Extrusion Forces.** Many factors must be taken into consideration when calculating drawing and extrusion pressures such as the die design, speed of extrusion, lubricant, temperature of the metal and its corresponding yield stress, the volume of the material compared to its percent reduction, the friction at the die surfaces, the nonhomogeneous flow of the material, and strain hardening.

The ratio of the large diameter to the small diameter, or reduction of area (Ar), is used in determining the strain imposed upon the rod and can be calculated as follows:

$$A_r = \frac{D_b^2 - D_a^2}{D_b^2} \times 100 \quad \text{(see Fig. 12.37)}$$

$$\epsilon = \ln\left(\frac{1}{1 - A_r}\right).$$

For example, a rod 0.8 in.$^2$ in area is drawn to 0.6 in.$^2$ in area:

$$A_r = \frac{.8 - .6}{.8} \times 100 = 25\%$$

$$\epsilon = \ln\left(\frac{1}{1 - .25}\right) = .29.$$

Thus a 25 percent reduction corresponds to a true strain of 0.29 in./in. This amount of strain can be interpreted in terms of pounds per square inch by means of a true stress strain curve plotted for the material. If the same material were used as in Problem 2.1, this amount of strain would correspond to approximately 40,000 psi.

Because drawing is very similar to extruding, one formula, as given by Hoffman and Sachs [ref 6], can be used to find the principal stress. In the case of drawing, the principal stress is tensile ($\sigma_\times$). In the case of extrusion, the principal stress is equal to the front pressure ($\sigma_{\times a}$), Fig. 12.37.

$$\text{Drawing stress} = \frac{\sigma_{\times a}}{\sigma_0} = \frac{1 + B}{B}\left[1 - \left(\frac{D_a^{\,2}}{D_b^{\,2}}\right)^B\right] + \frac{\sigma_{\times b}}{\sigma_0}\left(\frac{D_a^{\,2}}{D_b^{\,2}}\right)^B$$

where $B = \mu/\tan\alpha$ and $\sigma_0$ = yield stress in tension as a common example, $\alpha = 6°$ and $\mu = 0.1$, which gives a maximum $A_r$ of 1/2.

If the backpull ($\sigma_{\times b}$) is considered to be 0, the equation can be reduced to that shown for extrusion. Backpull can, however, be applied to increase die life and to speed up the drawing operation.

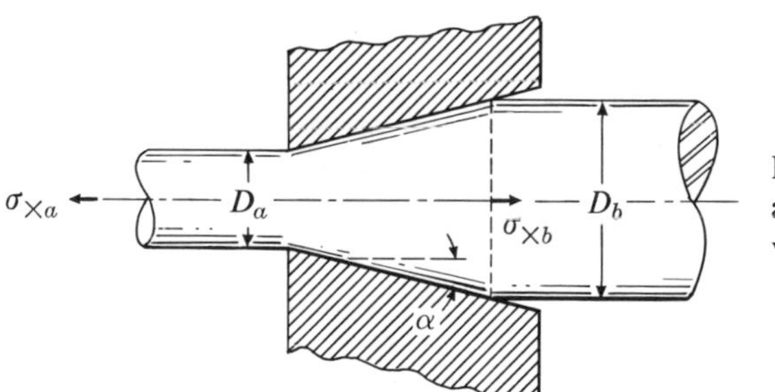


**Fig. 12.37.** The extrusion and drawing processes are very similar in principle.

In extruding, the process is reversed and the front pressure $\sigma_{\times a}$ is considered to be 0, and the formula used is:

$$\text{Extrusion pressure} = \frac{\sigma_{\times b}}{\sigma_0} = \frac{1 + B}{B}\left[1 - \left(\frac{D_b^{\,2}}{D_a^{\,2}}\right)^B\right].$$

If strain hardening is small, the average of the flow stresses at the beginning and the end is used in place of the (initial) yield stress $\sigma_0$.

Not all the factors mentioned are taken into consideration in this formula. However, it does serve to get an order of magnitude of the forces that will be involved, and after that empirical work will have to be done.

**Cold Heading.** Cold heading is a comparatively old process that has been used on nails and rivets for many years. However, it is only in more recent years that the process has been looked upon as a versatile competitive process.

The greatest factor in the recent advancement of cold heading has been the development of high-impact carbide header tools that are able to maintain very close tolerances. Impact forces range from 3 tons on the small header to 100 tons on the very large headers. Cold heading is seldom done on wire diameters larger than 1 in.

COLD-HEADING PROCESS. Cold heading is basically cold forging or cold upsetting. It is done on two types of cold-heading machines—the open die and the solid die (Fig. 12.38). The wire is fed through the open

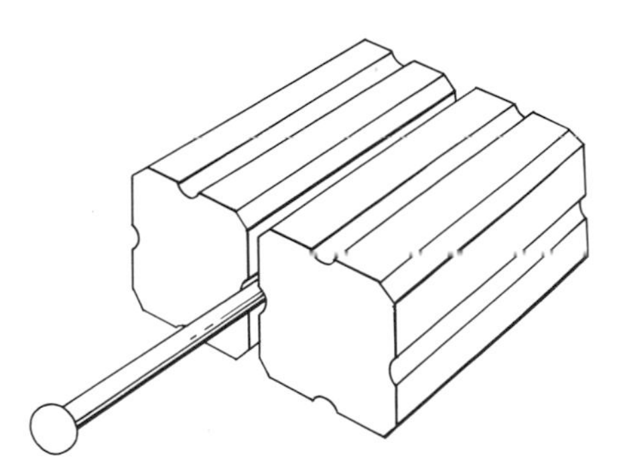


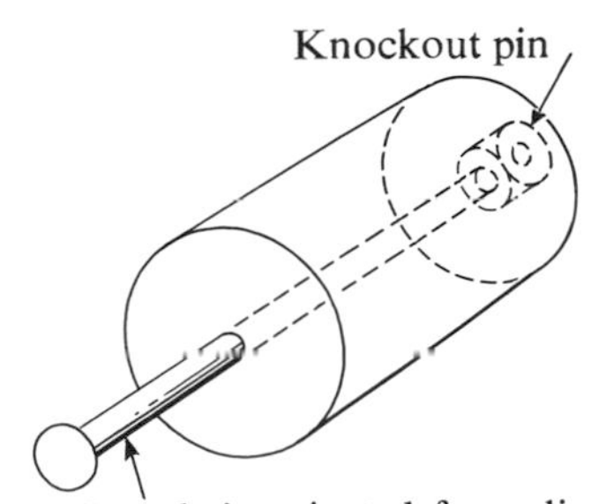


**Fig. 12.38.** Split and solid dies used to hold the wire for cold heading.

dies and against a stop which measures the correct amount for cold heading. The dies close on the wire and move laterally to shear the wire. At the same time, the heading punch moves forward to form the head. As the heading punch moves off, the dies open, and the incoming wire ejects the finished part (Fig. 12.39). The whole operation can be performed at the rate of 400 parts per min on small machines. Speeds are considerably reduced on the very large cold headers, ranging down to 50 blanks per min.

ADVANTAGES AND LIMITATIONS. As with other chipless machining operations, there is a gain in tensile strength and in fatigue and shock

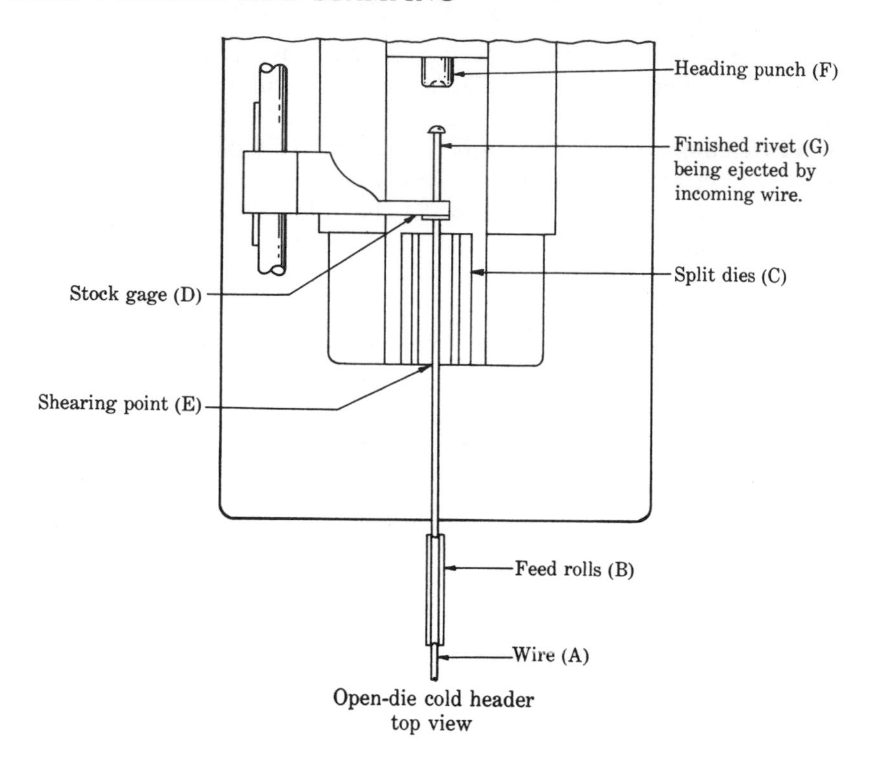


**Fig. 12.39.** Cold-heading-machine operation.

resistance because the material flow lines follow the contour of the upset section. Better properties result also because fillets are required for all inside corners.

Cold-headed parts can take a wide variety of shapes (Fig. 12.40) and

**Fig. 12.40.** A variety of cold-headed parts. Courtesy Industrial Fasteners Institute.

need not be symmetrical, as they must be in most regularly machined parts.

When volume is large, production costs are relatively low because of the high speed and very small scrap factor. The material used must be ductile and highly resistant to cracking.

The maximum amount of material that can be upset in a single blow is 2 1/4 times the diameter of the wire. If two blows are used, 4 1/2 times the wire diameter may be upset. Extending the wire too far out of the die will cause folds or laps.

A comparison of the amount of material needed to make a part on a lathe and by cold heading is shown in Fig. 12.41. Some design limitations are also indicated by the rounded corners and fillets.

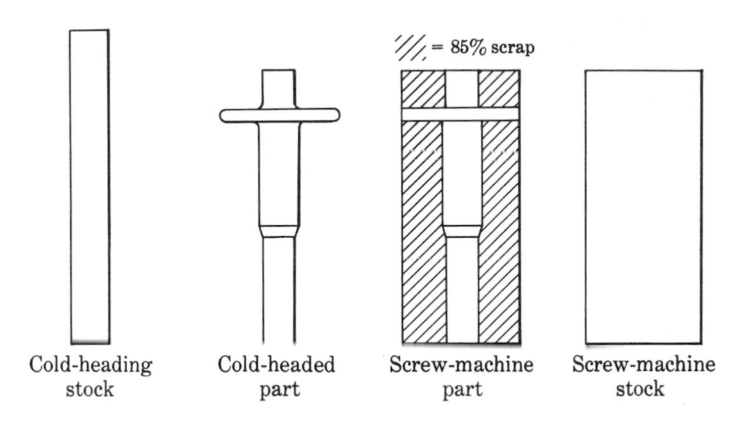


**Fig. 12.41.** Comparison between lathe and cold heading as to material savings and design limitations. Courtesy Townsend Co.

Cold-headed parts can sometimes replace several pieces of an assembly, as shown in Fig. 12.42. The part was redesigned to incorporate the bolt, washer, and spacer in one cold-formed part, which reduced the cost over 50 percent.

**Impact Extrusion – The Process.** Impact extrusion is a combination of both forging and extruding. The process consists of hitting a slug, held in a die, with a punch. The metal flows into the die and through an orifice or around the punch. The metal plastically deforms when the yield point is exceeded but extrusion does not start until the pressure becomes seven to fifteen times the initial yield strength of the alloy.

Although the blank may be loaded cold, the severity of the operation may cause the temperature to rise 500°F. The formed blank has a forged base and extruded sidewalls. The process can be classified as reverse, forward, or combination impact extrusion. The terms *forward* and *reverse* refer to the direction of metal flow. When the metal flows in the

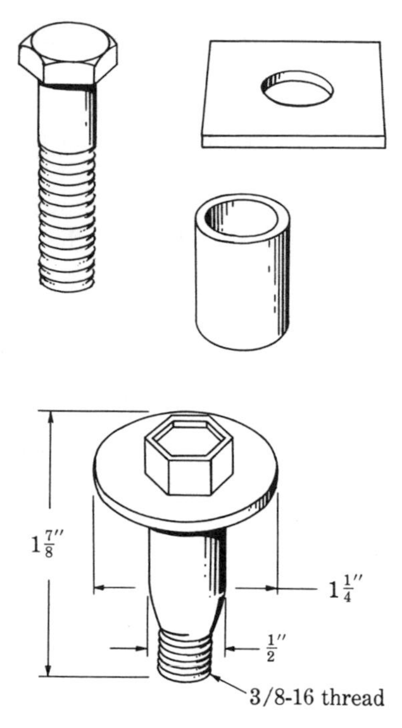


**Fig. 12.42.** A three-piece assembly redesigned for a one-piece cold-headed part. Courtesy Townsend Co.

same direction as the movement of the punch, it is termed forward; when it moves opposite to that of the punch, it is known as reverse extrusion. The three methods are shown in Fig. 12.43.

ADVANTAGES AND LIMITATIONS. Metals particularly adapted to impact extrusion are the softer, more ductile ones, such as aluminum, copper, and brass. However, low-carbon steels 1010 and 1012 have been used successfully. Higher alloys such as 4130 are being worked experimentally. The harder metals present tooling difficulties, since the process is the most severe of all forming operations.

Lubrication of the workpiece, especially in the case of steel, tends to eliminate galling and welding between the die and the work. One of the best methods of providing lubrication is by surface coating the blank with zinc phosphate and then adding a reactive soap-type lubricant to the treated surface.

Generally, steel impacts are limited to 2 1/2 times the punch diameter on reverse extrusions. Approximate limits for aluminum extrusions are 14 in. in diameter and 60 in. in length. These limitations are due largely to present demand and the equipment available. Mechanical presses are used where they can provide sufficient capacity. For loads over 2,000 tons, hydraulic presses are used because of longer stroke and economic advantages.

Tolerances will vary with materials and design, but production runs calling for 0.002 to 0.005-in. tolerance are regularly made.

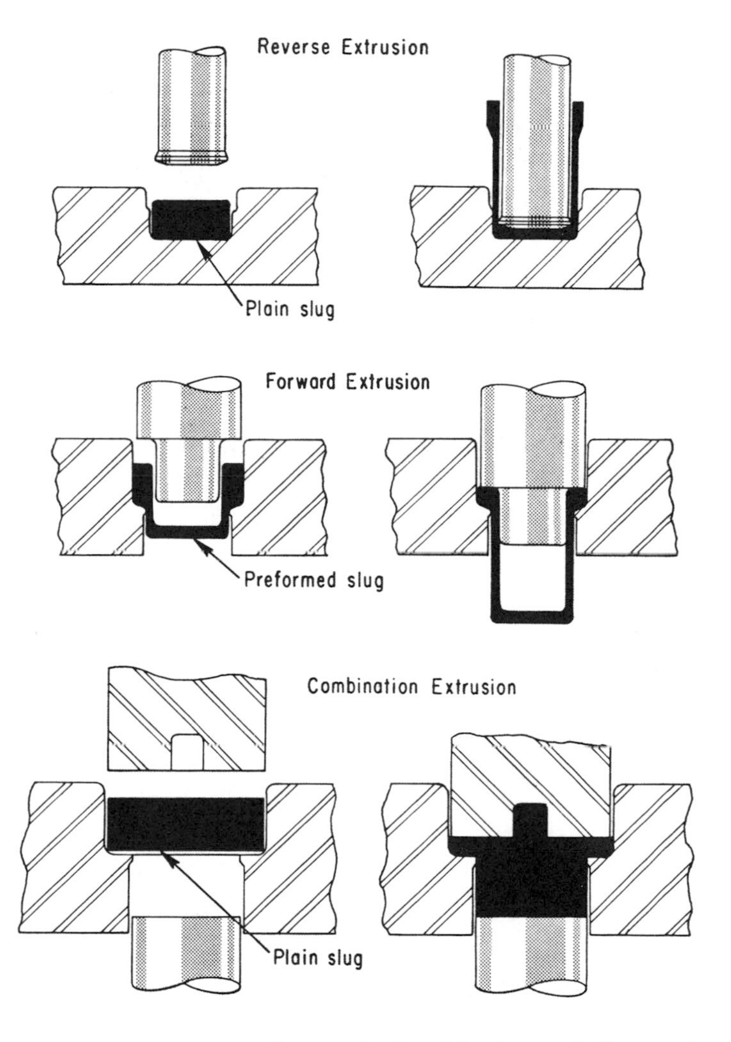


**Fig. 12.43.** Impact extrusion methods. Aluminum shells can be produced under 2 in. in dia at the rate of 3,600 pieces/hr; 2 to 4 in. in dia at 1,200 pc/hr; 4 to 6 in. in dia at 800 pc/hr; and 7 to 8 in. in dia at 350 pc/hr. The rate for parts over 10 in. in dia is 90 pc/hr.

APPLICATIONS. Impact extrusion competes with other press-working operations and with metal machining. Common products are aerosol cans, cocktail shakers, coffee makers, fire extinguishers, flashlight cases, lipstick cases, pen and pencil shells, and vacuum bottles. Secondary operations, such as beading, thread rolling, dimpling, and machining, are sometimes needed to make the completed item.

## QUESTIONS

**12.1.** Why is metal able to reach such high strengths when it has undergone several drawing operations?

**12.2** What is the lowest temperature at which a metal may be worked without producing any appreciable strains in it?

**12.3.** What is the relationship between minimum radius of bend and tensile strength?

**12.4** Explain why a sheet metal blank may become both thicker and thinner as it is drawn into the cavity.

**12.5.** How would you go about determining the drawability of a metal other than by the Erichsen test mentioned?

**12.6.** (*a*) Why are open or single die forming methods used for short runs? (*b*) What die making methods are used?

**12.7.** What is meant by "ironing" in a drawing operation?

**12.8.** Explain the function of a die set.

**12.9.** What is the basic difference between coining and embossing?

**12.10.** When is a progressive die used to best advantage?

**12.11.** Where do the initials OBI come from in describing a type of press?

**12.12.** For what type of work is the four-slide press well adapted?

**12.13.** What type of press *action* is best for deep drawing of sheet metal parts?

**12.14.** What are two outstanding advantages of the Guerin process?

**12.15.** How does the Wheelon process differ from the Guerin process?

**12.16.** What is the outstanding advantage of Hydroforming?

**12.17.** Explain the basic difference between bending metal and stretch forming it.

**12.18.** (*a*) Where does spinning fit into the metal forming picture for low, medium, or high production? (*b*) Why?

**12.19.** For what type of work is swaging used?

**12.20.** On what type of work is cold heading competitive with machining?

## PROBLEMS

**12.1.** The size press needed for a particular blanking operation can be calculated by the following formula: Blanking force = $(LtS_s)/2000$, where $L$ = length of cutting edge, $t$ = thickness of material in in., and $S_s$ = ultimate shear strength in psi of the material being blanked. A safety factor is added to the answer to allow for die wear and changes in the material. Thus, the calculated blanking force is usually doubled. (*a*) What size press is required for making 6-in.-dia blanks in 1/16-in.-thick 1100–0 aluminum with a shear strength of 9,500 psi. Include a 30% safety factor.

**12.2.** When a part is to be formed, the amount of material to allow for bending is:

$$A_b = (A/360)2\pi(\mathrm{IR} + Kt)$$

where $A$ = angle of bend; IR = inside radius; and K = constant = 0.33 when IR is less than $2t$ and 0.50 when IR is more than $2t$.

The required strip length before bending can then be calculated by:

$$L_s = \Sigma l + \Sigma A_b - (2t + 2R)n$$

where $L_s$ = strip length; $\Sigma l$ = the sum of the individual leg lengths; $\Sigma A_b$ = sum of the bend allowances; $t$ – thickness of the material; and $n$ = number of bends. (See Fig. P12.2.)

**Fig. P12.2**

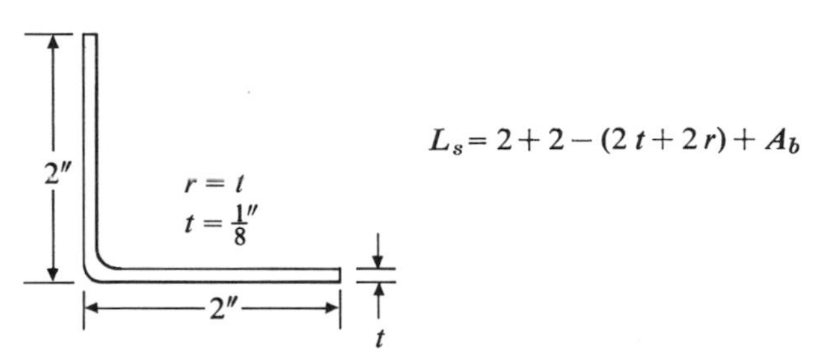


(*a*) How wide should the strip material be before forming it into a 2-in. angle iron? The material is 1/8 in. thick. Angle iron size is given by the outside dimension. Bend radius = $t$.

(*b*) How wide would strip material have to be before forming 3-in.-wide channel iron 1 1/2 in. high and 1/8 in. thick? Bend radius = $t$.

**12.3.** A 20-in.-dia aluminum blank is to be drawn into a shell. A 50 percent reduction is made in the first draw; the two succeeding draws each make an additional 15 percent reduction. What is the size of the finished shell?

**12.4.** A cover 8 in. in diameter and 2 in. deep, outside dimensions, is to be formed from mild steel having an ultimate shear strength of 60,000 psi. The inside radius of bend = $2t$. The material is 1/8 in. thick. (*a*) What is the size of the blank? (*b*) What is the size of the press required for the blanking operation?

**12.5.** Long, solid, low carbon steel rods (0.2%C), 0.5 in. dia, are to be cold drawn through a tapered die. The total die angle, $\alpha$, is 16° and the die throat dia is 0.375 in. (*a*) Show diagramatically the external forces that act on the rod and die surfaces while drawing. (*b*) Assuming a suitable value of the coefficient of friction between the die and rod surfaces, calculate the drawing force required for the operation. (Yield strength of the steel rod = 24 tons/in.$^2$)

**12.6.** (*a*) Cylindrical 2 1/2-in. dia flat-bottom steel cups are to be produced. The deep drawn cups, 2 1/2 in. high, are to be formed from blanks cut out of 0.050-in.-thick aluminum strip. Calculate the following: (*a*) Initial blank diameter. Allow 1/4 in. for trimming at the top after forming. The bottom corner radius can be neglected. (*b*) Select the most suitable one of the following available strip widths: 4, 6, 8, 10, 12, and 16 in. Tell why this strip width is chosen. (*c*) Calculate the pressure required to punch the blanks and (*d*) to perform the drawing operation, given that: ultimate tensile stress = 13,000 lb/in.$^2$; shear stress = 8,000 lb/in.$^2$; max drawing ratio = 2.2 for the first draw and = 1.8 for any following draw. Assume that the outside surface area remains constant for all the blanks and formed cups.

**12.7.** What is the pressure needed to extrude aluminum in a direct extrusion process, given that the coefficient of friction between the billet and the container walls = 0.12; $\bar{\sigma} = 32{,}000\ \bar{\epsilon}^{0.28}$; billet dia. = 8 in.; height = 8 in., and the extruded bar dia is to be 0.75 in.

**12.8.** In the production of steel strip of 1/8-in. thickness, a continuous hot rolling mill composed of several stands is to be used. Initial material thickness is 1/2 in. The allowable rolling reduction per stand lies between 20 and 22 percent. Find the number of stands necessary for the purpose. If the linear roll speed of the first stand is 350 ft/min, calculate the speed of the strip at the exit of the different stands. (Take the linear roll speed $\cong$ the speed of the strip leaving the rolls.)

## REFERENCES

1. Achler, H. S., "Low Cost Forming With Urethane Die Components," *The Tool and Manufacturing Engineer*, June 1964.

2. *Computations for Metal Working in Presses*, E. W. Bliss Company, Hastings, Michigan.

3. Datsko, J., *Material Properties and Manufacturing Processes*, John Wiley & Sons, Inc., New York, 1966.

4. Embury, J. D., and R. M. Fisher, "The Structure and Properties of Drawn Pearlite," *ACTA Metallurgica*, vol 14, February 1966.

5. Hall, F. J., "Aluminum Impact Extrusions," *Machine Design*, August 8, 1966.

6. Hoffman, O., and G. Sachs, *Introduction To The Theory of Plasticity for Engineers,* McGraw-Hill Book Company, New York, 1953.

7. Pearson, C. E., and R. N. Parkins, *The Extrusion of Metals,* Chapman and Hall, London, 1960.

8. Shawki, G. S. A., "Assessing Deep-Drawing Qualities of Steel," *Stamping Guide,* Metal Stamping Association, Cleveland, Ohio, 1966.

# 13

# CUTTING-TOOL PRINCIPLES AND MACHINABILITY

ON FIRST OBSERVATION it seems somewhat miraculous to see one metal able to cut another metal and do it consistently and precisely. The removal of metal at present production rates has indeed been a miracle of combined science and engineering. We no longer stand in amazement when metal is being cut, but expect it to be done at hundreds and even thousands of feet per minute.

A tool that cuts with one point or edge is referred to as a *single-point tool* as contrasted with drills, reamers, and milling cutters which have more than one cutting edge. The cutting action of a single-point tool is shown in Fig. 13.1. The metal is severely compressed in front of the cutting edge. The plastic deformation results in work hardening, temperature rise, and plastic shear. Most of the heat is confined to the shear zone; however, there is considerable friction as the metal rubs along the surface of the tool, which also accounts for the temperature rise. If the material being cut is soft there is a tendency for it to build up in front of the tip making a temporary weld to the cutting tool. The welds are intermittently sloughed off on the back side of the chip.

## CUTTING ACTION

As the metal moves through the shear zone and up along the face of the tool, three distinct types of cutting action take place (Fig. 13.2) depending on the type of material being cut. The chips produced

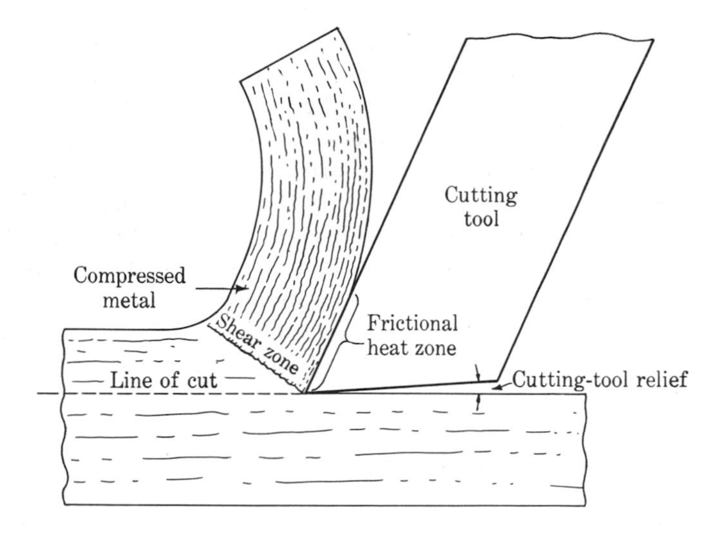


**Fig. 13.1.** The cutting action of a single-point tool.

are *continuous, continuous with a built-up edge* (BUE), and *segmented.*

**Continuous Chips.** Continuous chips are those that do not fracture as the metal moves up across the face of the tool. The back side of the chip that has passed the cutting tool is smooth. Unless some obstruction is placed in its path, the chip will probably not curl but will form a snarl around the cutting-tool point and the work.

**Continuous Chips with a Built-Up Edge.** Soft, ductile materials can be strained in shear to a high degree without rupture. The internal shear

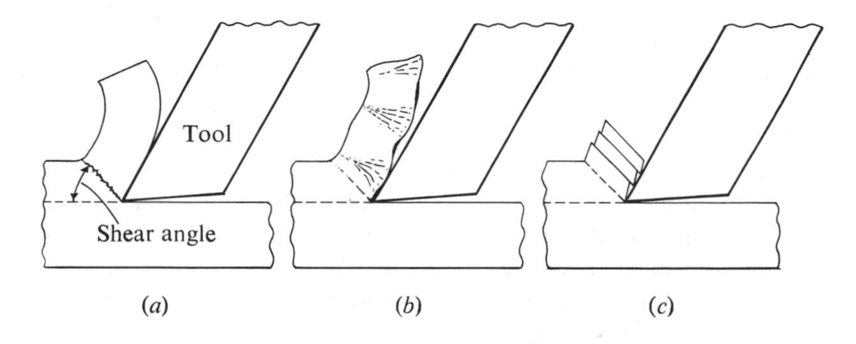


**Fig. 13.2.** Cutting tool action on various materials. (*a*) Medium hard materials with low coefficient of friction (BHN 170–240); continuous-type chip. (*b*) Soft ductile materials with high coefficient of friction (BHN 107–163); continuous-type chip with built-up edge. (*c*) Brittle material; chip is segmented.

action is followed by a secondary flow along the base of the chip, at the tool face. Owing to the high heat and pressure being generated, small particles of metal become welded to the cutting tip. This built-up edge changes the cutting action considerably, increasing the friction and therefore the heat generated, thus contributing to a shorter tool life. The built-up edge remains only momentarily. It is then sloughed off on the back side of the chip, leaving an inconsistent finish on the metal surface. Proper tool geometry along with the right cutting speed and cutting fluid can reduce built-up edge conditions to a minimum.

**Segmented Chips.** Brittle materials, such as gray cast iron, are removed by a combination of shear and fracture, causing the chips to come off in segments.

## CUTTING-TOOL GEOMETRY

The basic angles needed on a single-point tool may be best understood by removing the unwanted surfaces from an oblong tool blank of square cross section (Fig. 13.3).

**Relief Angles.** Relief on both the side and end of the tool is necessary to keep it from rubbing. These relief angles are usually kept to a minimum to provide good support when machining hard metals. They may be increased to produce a cleaner cut on some soft materials. Secondary clearances are used below the cutting edge to reduce the amount of regrinding needed for touching up the cutting edge.

**End and Side Cutting-Edge Angles.** The end of the tool is cut back approximately 15 deg to relieve the amount of surface contact. The side-cutting edge angle serves two purposes: it protects the point from taking the initial shock of the cut, and it serves to thin out the chip. For a given depth of cut, the material will be spread over a greater cutting surface (Fig. 13.4).

**Back-Rake Angles.** Rake angles cause the top surface of the tool to slope. Sloping the top surface back (positive back rake) is necessary in machining soft, ductile, draggy materials if excessive built-up edge is to be avoided. Although too much back rake helps prevent any built-up edge, it also reduces the strength of the tool and its ability to conduct the heat away from the cutting edge. Negative back rake (the top surface slopes toward the cutting edge) is used for heavy intermittent cutting. It is also used in very high-speed machining to lend more area to the included angle of the cutting tip.

**Side-Rake Angles.** Positive side rake, as shown in Fig. 13.5, is used to direct the flow of chips and make the cutting edge smaller. The higher the rake angle, the less cutting force is required; however, the cutting edge becomes fragile. Negative side rake is sometimes used to increase the cutting-edge strength, although it also increases the force required

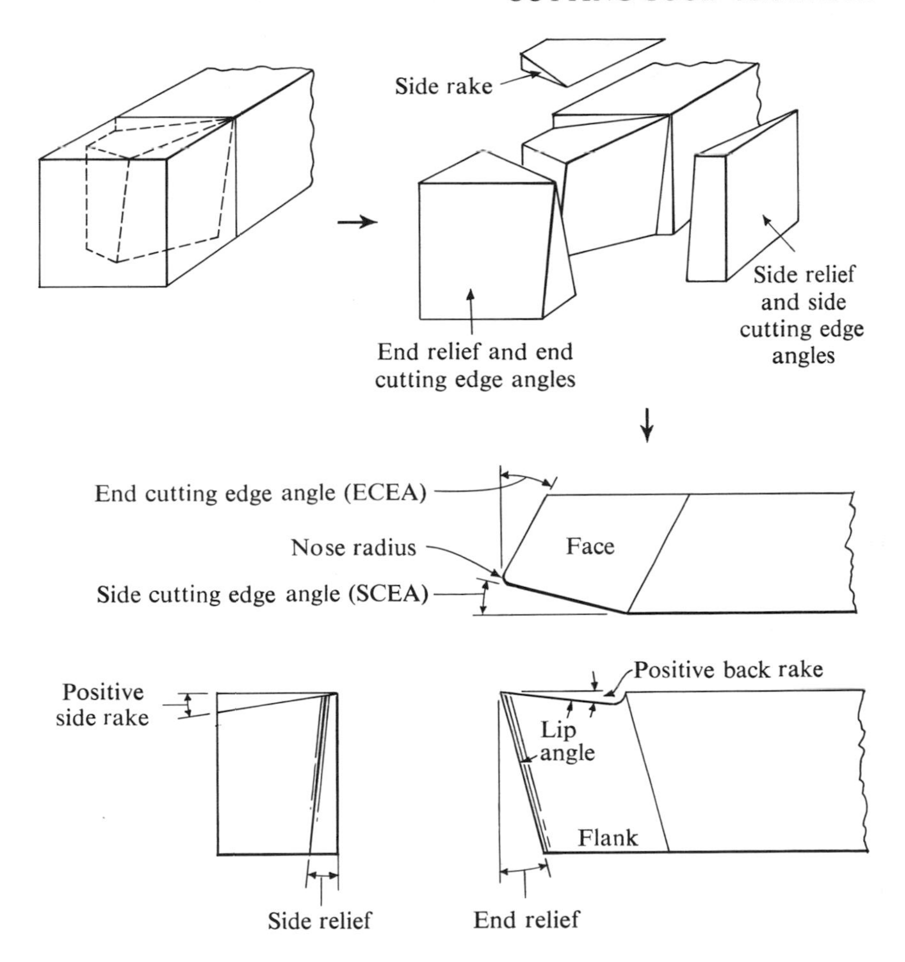


**Fig. 13.3.** The development of essential tool geometry.

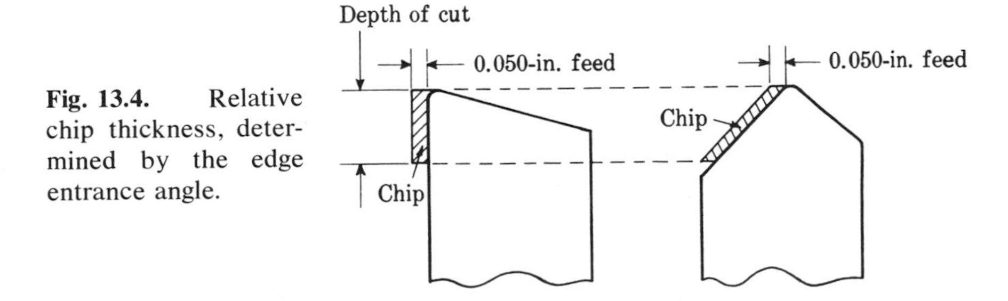


**Fig. 13.4.** Relative chip thickness, determined by the edge entrance angle.

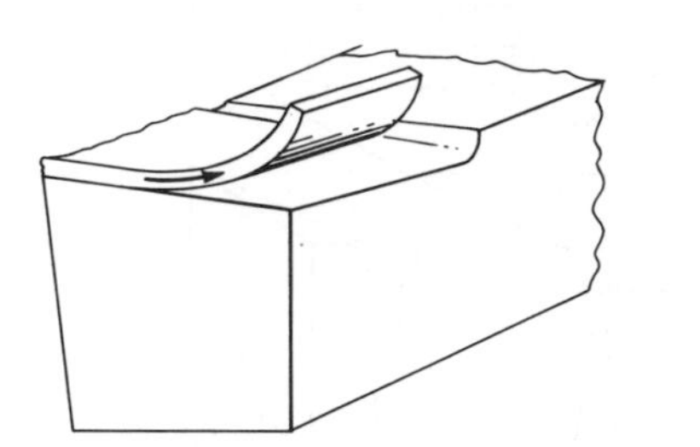

Positive side rake

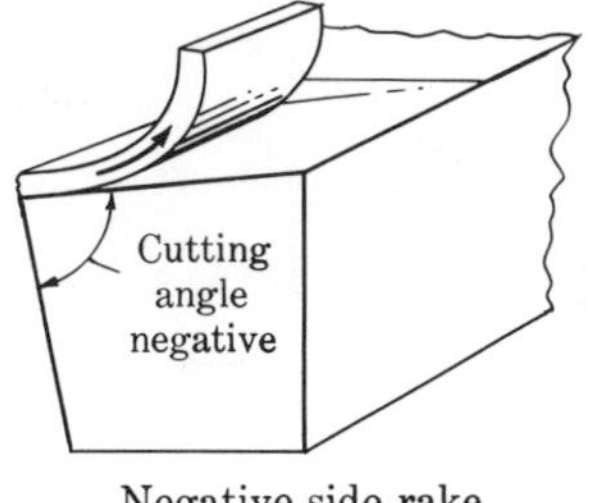


Negative side rake

**Fig. 13.5.** Side rake.

for the cut. The added strength needed for making heavy intermittent cuts that severely shock the tool, as in milling or planing, is obtained with negative side rake. The material undergoes greater compression and work hardening.

**Nose Radius.** The nose radius is the rounded tip of the cutting end. The amount of the radius varies with the general depth of the cut for which the tool is to be used. A large radius provides a strong tool, more contact area, and a smoother cut. Pressures on the cutting tool and the stock being machined are proportionate to the contact area. For example, small-diameter parts require a small nose radius with light cuts to minimize tool pressures and obtain a true surface.

The tool angles of a carbide tool used to cut low-carbon steel are shown in Fig. 13.6.

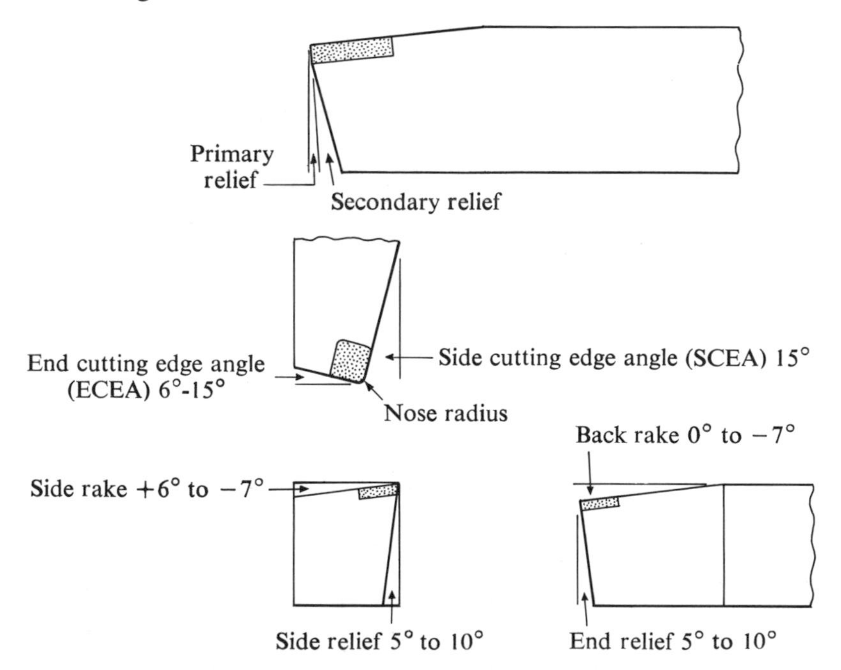


**Fig. 13.6.** Cutting-tool geometry for a carbide tool used to cut mild steel.

**Tool Signatures.** The Cutting Tool Manufacturers Association has established a standard system of tool signatures in which various angles that make up a single-point tool appear in a definite order. The order used is back rake, side rake, end relief, end clearance, side relief, side clearance, end cutting edge, side cutting edge, and nose radius. A given tool may be designated as 10, 12, 8(10), 8(10), 6, 6, or 3/64. When clearance below the cutting edge is used on relief angles, it is placed in parentheses—the value (10) in this signature. References to clearance angles are often omitted in actual practice.

**Chip Breakers.** As the metal is separated from the parent material, it should be broken into comparatively small pieces for ease of handling and to prevent it from becoming a work hazard. A chip breaker forms an obstruction to the metal, as it flows out across the face of the tool, causing it to curl and break. The fact that the metal is already work hardened helps the chip breaker to perform effectively. Various types of chip breakers are made: some consist of a step ground into the leading edge of the tool; others are made by clamping a piece of carbide on top of the cutting tool; and still others have a groove formed just in back of the cutting edge (Fig. 13.7). The last mentioned type is now the most popular for carbide insert tools. The insert is held in place by a cam-locking arrangement operating from the center pin.

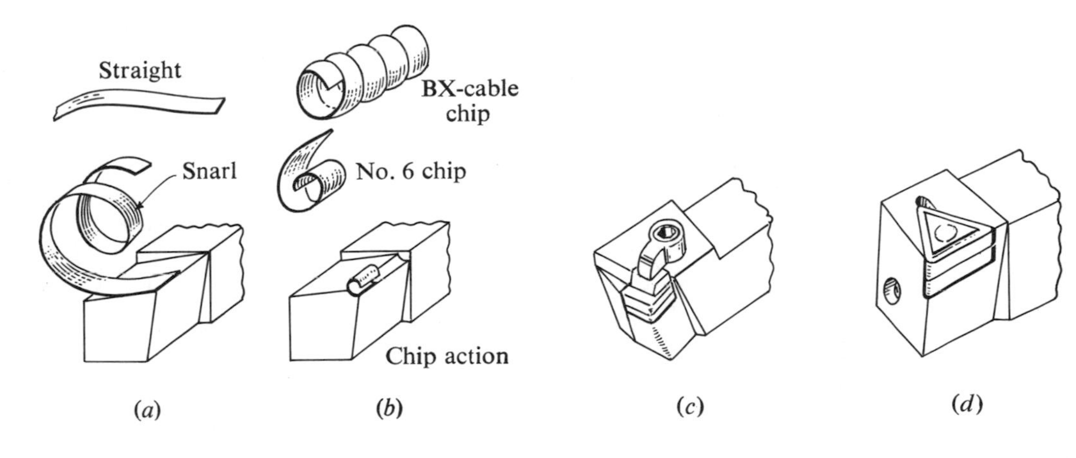


**Fig. 13.7.** Common chip breakers. (*a*) No chip breaker; (*b*) step type; (*c*) clamp-on; and (*d*) formed-groove.

The efficiency of the chip breaker is primarily a function of its width as compared to the feed being used. A light cut with a fine feed allows the chip to travel over a flat surface before it is made to curl. Under these conditions, the chip would only be lightly curled or might even cause a snarl.

**Chip Forms.** The most desirable type of chip is termed a *number 6* chip, since it resembles that numeral. A long continuous chip, rather tightly curled, is termed a *BX-cable chip.* These are not too objectionable from the standpoint of safety, but they are harder to remove from the machine. A chip forming no curl at all will usually become entangled with the work or cutting tool and is referred to as a *snarl* (Fig. 13.7).

## FUNDAMENTAL NATURE OF CUTTING-TOOL MATERIALS

Today's most used cutting-tool materials (high-speed steel, cast alloys, and cemented carbides) have basically the same type of microstructure. That is, they consist of hard, brittle, refractory carbide particles in a lower-melting-point metallic matrix. The major difference lies in the volume of the carbide phase. For example, high-speed steels will have about 10 to 15 percent of this hard compound held in a ferrous matrix (Fig. 13.8).

(*a*) (*b*)

**Fig. 13.8.** Photomicrographs of (*a*) conventional high-speed steel and (*b*) super high-speed steel showing the difference in size and distribution of carbide particles, original magnification 1000×. Courtesy *The Tool and Manufacturing Engineer.*

The carbide phase contributes mainly to the wear resistance of a tool. The matrix phases of the various cutting-tool materials have similar melting points, approximately 2,350°F to 2,550°F, and in general they are quite hard, strong, and relatively tough. Thus the overall properties of the two-phase tool materials are governed by the relative volume of the individual phases.

**High-Speed Steel.** High-speed steels that have the lowest carbide volume are the toughest and the least wear resistant. The most common type, until more recent years, was 18 - 4 - 1. It is composed of 18 percent tungsten, 4 percent chromium, and 1 percent vanadium. It is now being replaced by lower alloy compositions. Another type, also known by its main alloying element, is molybdenum HSS. The molybdenum, or M-type, high-speed steels are generally better for impact resistance, and the tungsten, T-types, are better for heat resistance. High-speed tools find their best application in drills, reamers, counterbores, milling cutters and single-point tools since they perform reasonably well, are less expensive than carbide tools, and can be quite easily ground. Hardness is in the range of $R_c$ 65 to 67. The cutting speed range is from 40 to 90 sfpm for most steels.

**Cast Alloys.** The cast nonferrous alloy tool materials get their name from the fact they are cast directly to shape, and they contain no iron. They consist of cobalt, chromium, tungsten, and carbon. The carbide phase makes up 25 to 30 percent, by volume, of the matrix. After casting, the tool material is heat treated to break up the network of carbides and produce high hardness values throughout.

The cast alloys are capable of taking a high polish, which helps prevent the metal from sticking to the face of the tool and forming a built-up edge. The built-up edge, which is a characteristic result of cutting soft, ductile materials, impairs the cutting efficiency of the tool. These tools are also extremely corrosion resistant, which makes them desirable in machining chemical-bearing rubbers and plastics. Cast-alloy tools are better known by the trade names of *Stellite, Blackalloy, Crobalt,* and *Tantung.* Cast alloys are ordinarily run between 100 and 180 sfpm for most steels.

**Cemented Carbides.** Cemented carbides have the highest ratio of carbides to matrix, about 80 percent. Carbides are made by mixing pure tungsten powder under high heat with pure carbon (lampblack) in the ratio of 94 percent and 6 percent by weight. The new compound, tungsten carbide, is then mixed with cobalt until the mass is entirely homogeneous.

The powdered mixture is then compacted at high pressure and sintered in a furnace at 2,500°F. After cooling, the carbide tools are ground and in some cases lapped or honed before being used as a cutting tool.

Straight tungsten carbide grades are the strongest and most wear resistant, but they are not very satisfactory for machining steels. The hot steel chip welds itself to the carbide and pulls out microscopic particles of the cutting tool. This forms a crater that causes rapid tool failure. To prevent this, tantalum carbide and titanium carbide are added to the basic composition. These materials greatly reduce cratering action; however, some abrasion resistance and strength is sacrificed.

A straight titanium grade of carbide is also made, primarily for fine finishes on steel or ultra-high-speed machining. It has excellent crater resistance and freedom from the built-up edge. Carbides are generally used at 200 to 1,500 sfpm.

Most carbide tools are of the insert type, brazed in place or clamped on as shown in Fig. 13.7. Since carbide tools are more difficult to grind, the clamped-on inserts are not resharpened. After all cutting edges have been used, they are discarded; hence the name "throw-away" inserts. They are made in square, diamond-shaped, triangular, and round sections. Inserts on negative rake holders can be turned over to double the amount of usable cutting edges. Triangular, 60-deg inserts do not have the same strength as the square, 90-deg inserts but are used when the workpiece geometry calls for it.

Round inserts give good finish and are ideal for forming corner radii. But, because of the greater surface contact, cutting pressures can build up to cause deflection of the workpiece and chatter.

**Sintered Oxides.** Sintered oxides (often referred to as ceramics) are a relatively new cutting-tool material. They are made from sintered aluminum oxide and various boron nitride powders that are mixed together, compacted, and sintered at 3,000°F.

These tools have very low heat conductivity and extremely high compressive strength, but they are quite brittle. Generally, where shock and vibration are not factors, ceramic tools can outperform carbides. A rule of thumb indicates a cutting speed two to three times that of carbides. Because of their extreme hardness, ceramic tools have proved successful where they are exposed to abrasive wear, such as is encountered when machining cast iron, plastic materials, and heat-treated steels. Some fully hardened tool steels can be machined in production. Even steels as hard as $R_c$ 65 have been machined. Ceramics are used in the cutting speed range of 400 to 3,000 sfpm.

Although tool costs are a consideration, it must be emphasized that performance is far more important. For example, a 3/8 in. carbide square insert 1/8 in. thick costs about $1.00 whereas a ceramic insert of the same size and shape costs approximately $2.00. However, on many jobs, the ceramic tool will outperform the carbide tool by a ratio of 5:1.

**Tool Surface Treatment.** Surface treatments such as salt-bath nitriding and steam oxidizing are often used on high-speed steels to give them better performance characteristics. The familiar blue oxidized surface is produced by steam in a sealed retort. The oxide layer about 0.0002 in. thick is quite soft and wears away as the tool is used. As it wears away, it burnishes the tool and reduces the tendency for galling and chip welding.

Cutting tools can be honed, lapped, and superfinished to produce an extremely smooth surface which reduces chip flow friction.

**Sound Wave Treatment.** One of the newest treatments for tools is that of sound waves to reduce residual brittleness usually encountered in conventional heat treating methods. The process allows the molecular structure to rearrange itself and yet maintain maximum hardness for use. Two-year tests have shown increased tool life ranging from 30 to 400 percent.

## TOOL LIFE

After the selection of the right tool geometry and material, the next consideration should be performance or, as it is termed, *tool life*. Obtaining proper tool life poses a complex problem, involving such factors as the machinability of the material being cut, relative speeds and feeds, and the economics of how long the tool must last to get maximum performance out of both the tool and the machine. Common ways in which tools fail are top or crater wear, flank or edge wear, and chipping or spalling (Fig. 13.9).

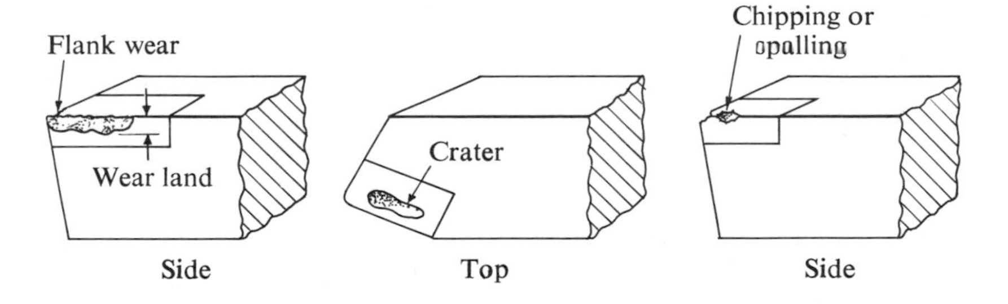


**Fig. 13.9.** Common types of tool failures.

**Crater Wear.** A cavity that develops on the face of the tool as it is being used is termed a *crater*. Microscopic particles of the tool-face material are heated by the ductile chip material as it is rubbing over them. As the temperature continues to increase, a point is reached in which diffusion takes place. First, atoms of the matrix material are transferred across the diffusion zone, weakening the carbide. As other chips continue to pass over the area, particles of carbide are swept away, forming a crater. The crater is formed behind the cutting edge since this is the area of highest temperature. Once the tool chip interface temperature is high enough for diffusion to take place, about 1,050°F for HSS tools and 1,200°F to 1,500°F for carbides, crater wear increases rapidly. Studies by Meyer and Wu (Ref. 8) show that when cutting at speeds sufficient to cause accelerated crater wear, the most dependent variables are feed and nose radius; feed varies directly, and nose radius inversely (Fig. 13.10).

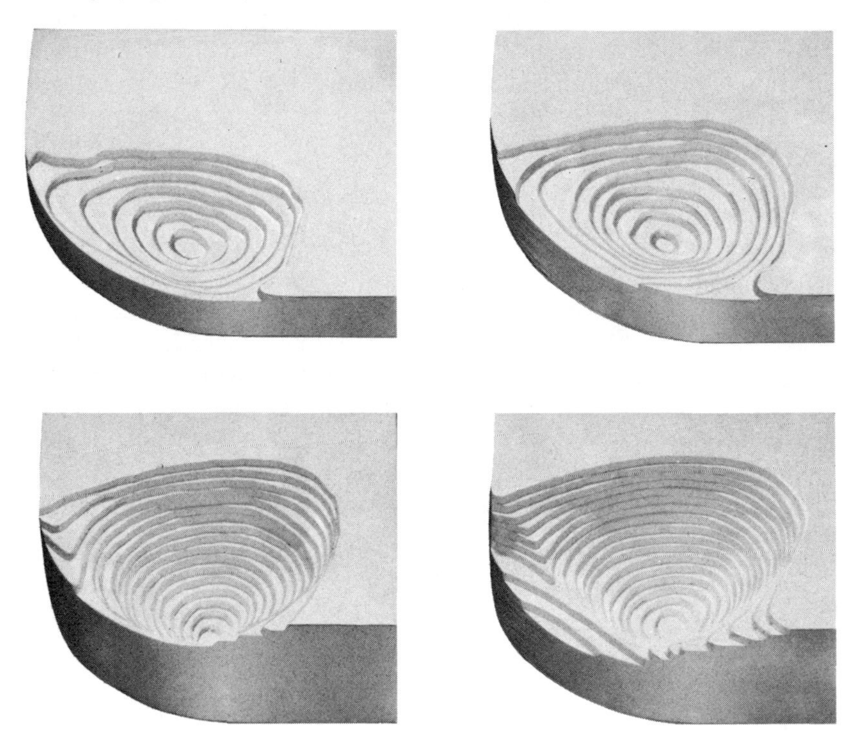

**Fig. 13.10.** Models showing the topography of progressive crater wear. The cut was made with a 370 grade carbide tool at 900 fpm and 0.011 ipr feed. The material cut was SAE 1048 HR steel.

Diffusion does not occur at the same temperature for all materials being cut. The temperature for steels is about 1,050°F, but for stainless steels it is well above 1,200°F. At higher temperatures, the tool material is more susceptible to being plowed out without regard for diffusion.

Moderate crater wear can be taken as an indication that the metal is being cut at a near optimum rate, provided the tool is of the proper grade for the work.

**Flank Wear.** Flank wear is always present regardless of tool materials or cutting conditions. It is due to the friction between the cutting edge and the material being machined. The wear is not confined to the flank but continues on around the nose radius. However, it is usually more pronounced on the flank, particularly when abrasive materials are cut and when fine feeds are used. The wear land is usually not uniform, hence an arithmetical mean value is used to represent it. For carbide tools a flank wear of 0.030 in. has been designated as a standard for tool life. A typical flank wear curve is shown in Fig. 13.11*a*. In region *A*, the wear is rapid, the sharp edge is soon removed; *B* represents the normal wear period; and region *C* again shows rapid wear. The end of the *B* range can also be represented as being the 0.030-in. wear land. Flank wear is accelerated at higher cutting speeds. Also shown is the relationship of both crater and flank wear (Fig. 13.11*b*).

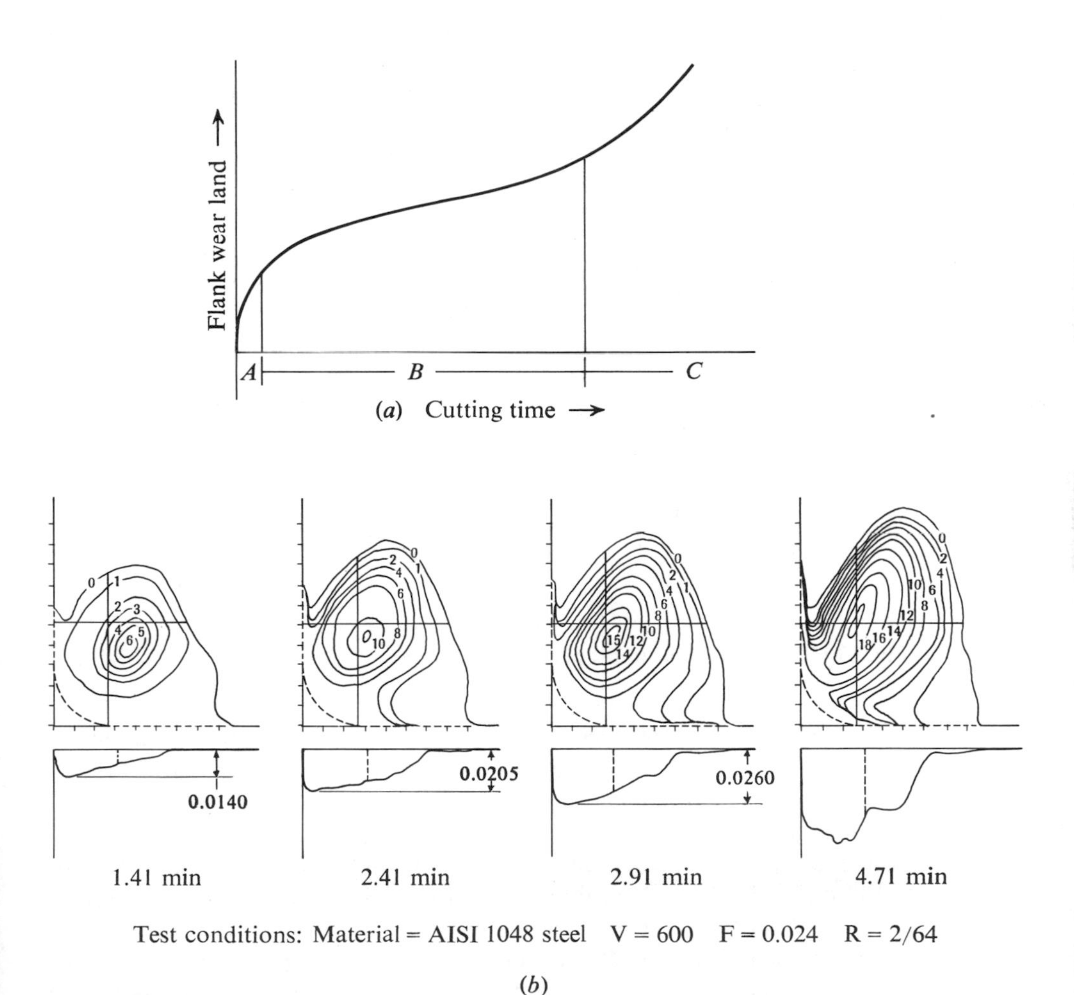


**Fig. 13.11.** (*a*) The development of the flank wear curve. (*b*) Crater wear contours are plotted showing the relationship to nose radius. The simultaneous flank wear is also shown.

**Chipping or Spalling.** Breakage of tools may occur long before the service life limits are reached. Many factors are involved such as heat, pressure, cutting speed, and coolant.

Assuming the tool has the proper geometry for the work and that it is set at the right height (usually center or slightly above), some conditions that cause breakage are: the tool size is too small, overhang on the tool is too great (Fig. 13.12), vibration in the setup or machine, carbide not of the right grade, speed either too high or too low, and improper use of coolant. A wide chip passing over the top of the tool will prevent the coolant from reaching the cutting edge. The edge becomes very hot. As the chip breaks off, a splash of coolant hits the carbide tool, quenching it. Constantly repeated occurrences such as this will cause heat checks to develop which eventually become deep so that particles of the tool break from the tip. Under these conditions, it is better not to use coolant.

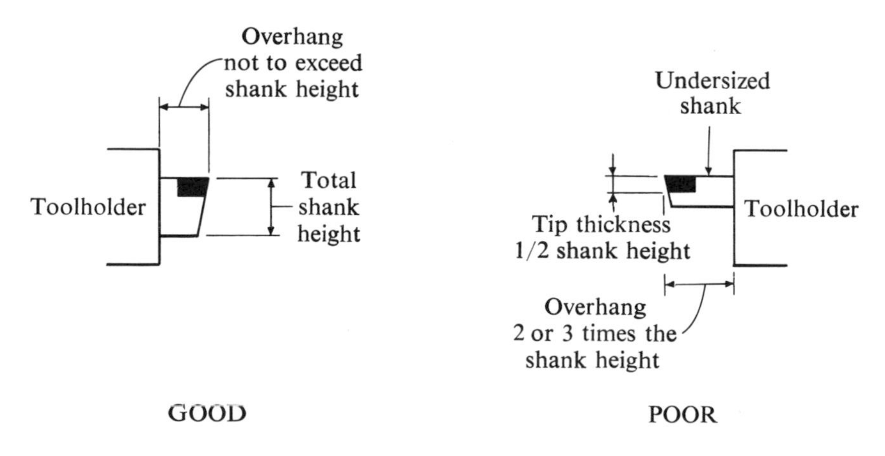


**Fig. 13.12.** Proper and improper tool tip support.

**Tool Life and Microstructure.** The microstructure of the metal that is being cut has a pronounced effect on tool life. Two pieces of steel may have the same Brinell hardness but differing microstructures. The tool reaction will be different, and the surface finish will also be different. Of course the process of making microstructure specimens for each piece of metal that is cut is not feasible, so the Brinell hardness number (BHN) is often used as the next best indication of how the material will machine.

**Cubic Inch Removal Rate.** Tool life is also expressed in terms of cubic inch removal rate. This is the preferred method since it takes into consideration both the velocity and the amount of material that pass the cutting tool. However, a true picture cannot be had without relating the material to a machinability factor, usually given as a constant.

**Tool Life Formula.** The following tool life formula, which ties tool life factors together, was developed by F. W. Taylor, who made extensive tests in the area of metal cutting:

$$VT^n = C$$

where $V$ = velocity of the work material in sfpm; $T$ = tool life in minutes (0.030) wear land; $n$ = the slope of the curve on log-log coordinate paper and can be found as follows:

$$n = \tan \theta = \frac{\log V_1 - \log V_2}{\log T_2 - \log T_1};$$

and $C$ = a constant representing the velocity of the work in surface feet per minute (sfpm) for one minute of tool life.

Significant changes in tool geometry, depth of cut, and feed will change the value of the constant $C$ and cause slight changes in the exponent $n$. Each tool material has a different tool life curve as shown in Fig. 13.13.

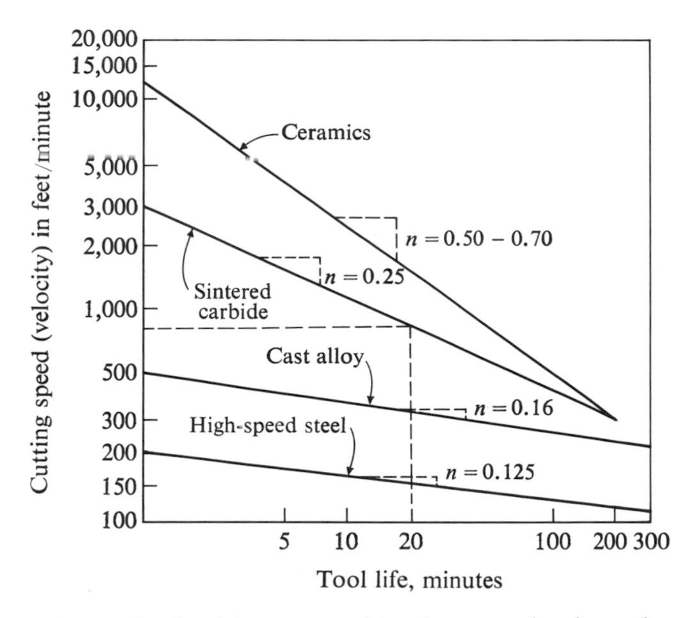


**Fig. 13.13.** Tool life curves for different cutting tool materials.

The points plotted are obtained by measuring the wear land on the tool. For example, at approximately 825 fpm, a tool life of twenty minutes was plotted as shown in Fig. 13.13. At this time a 0.030 in. wear land could be measured on the tool. (It should be noted that although flank wear is taken as a standard, crater wear may be a better indication of tool life for machining steel, but it is more difficult to measure accurately. For accuracy, tool life curves should be made under actual operating conditions.) The values for $n$ shown in Fig. 13.13 may be accepted as standard.

## TOOL FORCES AND HORSEPOWER

**Tool Forces.** The forces acting on the tool in oblique cutting, as done on a lathe, are shown in Fig. 13.14. Each of these forces may be measured with a strain gage type of tool dynamometer shown in Fig. 13.15. A study of many machining variables (such as tool geometry, materials, speed, and feeds) may be done with a tool dynamometer and suitable recording equipment or with an oscilloscope.

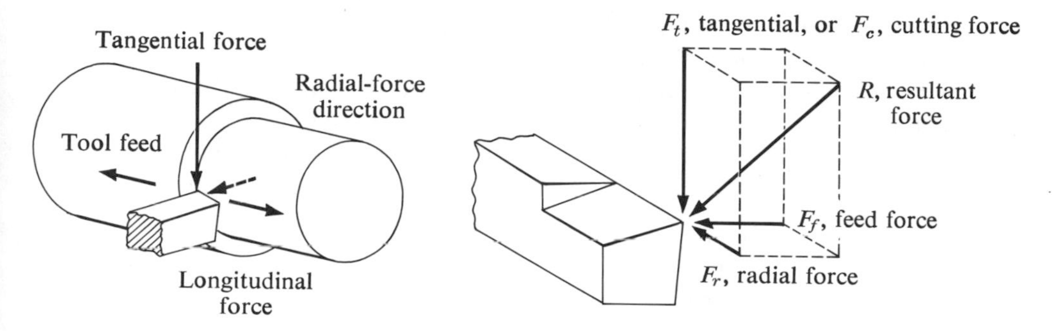


**Fig. 13.14.** Forces that act on the cutting tool during a turning operation.

**Horsepower.** Horsepower may be defined as: $hp_c = F_tV/33{,}000$ where $hp_c$ = horsepower at the cutting tool; $V$ = cutting speed in feet per minute; and $F_t$ = tangential force component also referred to as $F_c$ or cutting force.

**Fig. 13.15.** A three-dimensional strain gage type dynamometer used to measure tangential, longitudinal, and radial cutting forces on a lathe.

This basic formula does not take into consideration radial and longitudinal forces. However, they are not of sufficient magnitude so that they would alter the results appreciably. Another formula, probably more useful, does not require the measurement of the tangential or cutting force and is given for the horsepower required at the motor as

$$\mathrm{hp_m} = 12 \times V_c \times f \times d \times P_u$$

where $P_u$ = unit power required per cubic inch per minute corrected for dull cutter and 80 percent spindle drive efficiency; $d$ = depth of cut; $f$ = feed rate in inches per revolution; and $V_c$ = cutting speed in feet per minute.

A study of the unit horsepower values as given in Table 13.1 will show the relationship of the BHN to the power required for making a cut.

The horsepower for milling and drilling can also be determined with unit power values given in Table 13.1. The formula used for milling is given as:

$$\mathrm{hp_m} = f_m \times d \times W \times P_u$$

where $f_m$ = feed in inches per minute; $d$ = depth of cut in inches; and $W$ = width of cut in inches.

For drilling,

$$\mathrm{hp}_m = 1/4(D_d^2 \times \pi) \times f \times P_u \times \mathrm{rpm}$$

where $D_d$ = drill diameter in inches and $f$ = feed per revolution in inches.

## MACHINABILITY

Machinability, as stated in Chap. 3, is an involved term with many ramifications but, briefly defined, it is the ease with which metal can be cut. There are definite means by which it can be measured as well as predicted.

**Measurement of Machinability.** Methods of measuring the machinability of a given material are based on: cutting ratio, shear angle, horsepower, surface finish, tool temperature, and tool life.

**Cutting Ratio.** Cutting ratio is based on the thickness of the chip in relation to the depth of cut in orthogonal cutting, shown in Fig. 13.16, or the ratio of feed to the chip thickness in turning on a lathe. The ratio is expressed as

$$r = (t_1/t_2)$$

where $r$ = cutting ratio; $t_1$ = depth of cut (or feed); and $t_2$ = chip thickness. The nearer $r$ approaches one, the better the machinability.

**Shear Angle.** The shear angle, $\phi$, can be determined from the cutting ratio by the formula

**Table 13.1**
**Unit Horsepower Values**
used in calculating motor horsepower for turning, drilling, and milling, corrected for dull cutting tools and 80 percent spindle drive efficiency

| Material | Hardness BHN | Unit power, hp/in.³/min | | |
|---|---|---|---|---|
| | | Turning $P_t$ HSS & CARBIDE TOOLS Feed 0.008–0.010 ipr | Drilling $P_d$ HSS DRILLS Feed 0.002–0.010 ipr | Milling $P_m$ HSS & CARBIDE TOOLS Feed 0.004–0.008 ipt |
| STEELS *wrought & cast* | 85–200 | 1.2 | 1.0 | 1.2 |
| Plain carbon | 35–40 $R_c$ | 1.6 | 1.4 | 1.5 |
| Alloy steels | 40–50 $R_c$ | 2.0 | 1.7 | 2.0 |
| Tool steels | 50–55 $R_c$ | 2.2 | 2.0 | 2.2 |
| CAST IRONS | 110–190 | 1.0 | 1.0 | 1.0 |
| Gray, ductile, & malleable | 190–320 | 1.8 | 1.6 | 2.0 |
| STAINLESS STEELS *wrought & cast* | 135–275 | 1.5 | 1.3 | 1.4 |
| Ferritic, austenitic, & martensitic | 30–45 $R_c$ | 1.5 | 1.4 | 1.6 |

| | | | | |
|---|---|---|---|---|
| PRECIPITATION HARDEN- ING STAINLESS STEELS | 170–450 | 1.5 | 1.4 | 1.7 |
| TITANIUM | 250–375 | 1.0 | 1.0 | 1.2 |
| HIGH TEMPERATURE ALLOYS Nickel & cobalt base | 200–360 | 2.0 | 2.0 | 2.5 |
| REFRACTORY ALLOYS | | | | |
| Tungsten | 321 | 3.5 | 3.3 | 3.6 |
| Molybdenum | 229 | 2.4 | 2.0 | 2.5 |
| Columbium | 217 | 2.2 | 1.7 | 2.3 |
| Tantalum | 210 | 2.7 | 2.1 | 2.7 |
| NICKEL ALLOYS | 80–360 | 2.2 | 2.2 | 2.6 |
| ALUMINUM ALLOYS | 30–150 500 kg | 0.3 | 0.2 | 0.4 |
| MAGNESIUM ALLOYS | 40–90 500 kg | 0.2 | 0.2 | 0.2 |
| COPPER | 80 $R_b$ | 1.2 | 1.1 | 1.2 |
| COPPER ALLOYS | 20–80 $R_b$ | 0.8 | 0.6 | 0.8 |
| | 80–100 $R_b$ | 1.2 | 1.0 | 1.2 |

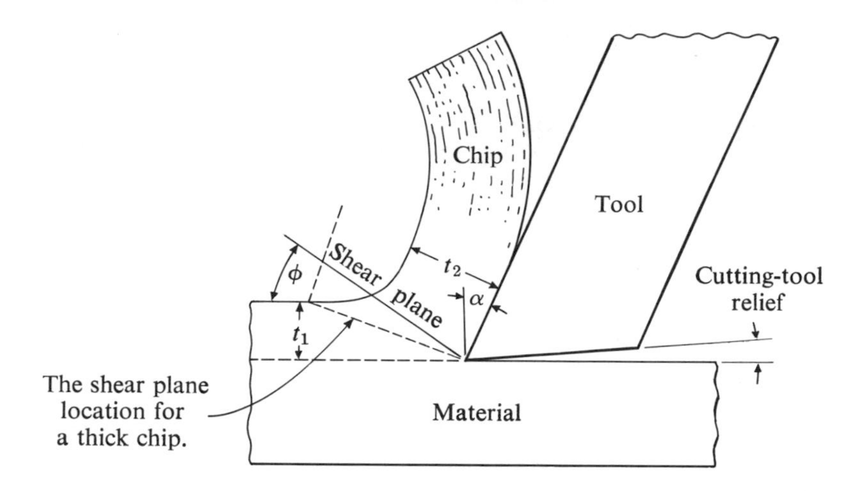


**Fig. 13.16.** Schematic view of tool and shear plane in orthogonal cutting. The dotted line shows how long the shear plane becomes when the chip is thick.

$$\text{Tan}\ \phi = \frac{r \cos \alpha}{1 - r \sin \alpha}$$

The closer the shear angle approaches 45 deg, the better the machinability.

When oblique cutting is done, as on lathe work, compound angles on the tool have to be considered, and other formulas involving true rake angles must be used. These can be found in the references listed at the end of this chapter. However, an approximation can be obtained by using the side rake angle if the back rake is neutral.

Coolants are used in machining primarily to cool the tool and workpiece but they also reduce friction on top of the tool face and affect the shear angle. Cutting oils are the best lubricants, but they do not carry heat away fast enough for the high speeds used with carbide tools. Water-base soluble oils and the newer type synthetic coolant mixtures are used. Shown in Fig. 13.17 is the effect on shear angle of machining with a good coolant and with no coolant.

**Horsepower, Surface Finish, and Tool Life.** Each of these variables may be taken as a measure of machinability by comparing the results obtained for each one on a given material with that obtained under the same conditions for SAE B1112 steel (rated as 100 percent machinable).

**Prediction of Machinability.** Oftentimes it is desirable to predict how a material will perform when subjected to various cutting actions. Two methods used to predict machinability are microstructure examination and Brinell hardness tests.

Microstructures can be altered by heat treatment so that machinability

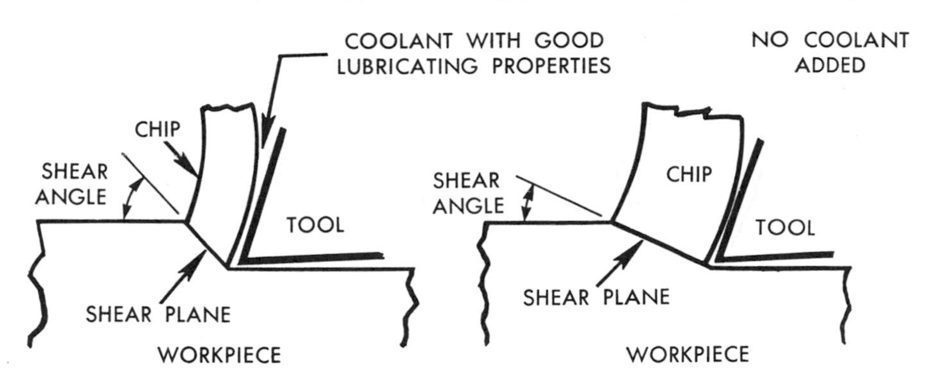

**Fig. 13.17.** An effective coolant properly applied can reduce the friction between the chip and the tool face, changing the shear angle. Other beneficial results are a decrease in the horsepower required, a cooler, more accurate operation, and better surface finish.

is improved in some cases as much as 100 to 200 percent. However, the physical properties required in the finished part should be the governing criteria rather than machinability.

## METAL-CUTTING ECONOMICS

All the foregoing information is of little use unless it is intelligently applied to obtain the *lowest possible unit cost and the highest possible production rate consistent with quality for any given operation.* We know that, at high cutting speeds, we can expect increased tool cost owing to shorter tool life. We can, at the same time, expect the machine cost per piece to go down. Throwaway carbide tips are preferable to brazed-type tooling from the standpoint of downtime. These factors are shown graphically in Fig. 13.18.

As shown in Fig. 13.18, machine cost goes down rapidly with increase in cutting speed. Although the increased speed causes tool cost to rise, this factor is relatively insignificant when compared to the costs of labor and equipment.

The total cost per piece is based on the machining cost, time, labor, and overhead, plus tool-changing cost, plus regrinding and depreciation cost, plus nonproductive cost. (Nonproductive cost is the total of the labor involved in loading, unloading, placing the tool in position for the cut, etc.) The total cost, as plotted, is a summation of all the costs and represents the surface foot speed that can best be used to produce a part at minimum cost and in minimum time.

So much emphasis has been placed on tool life that sometimes other factors, such as the amount of money invested in the machine and its

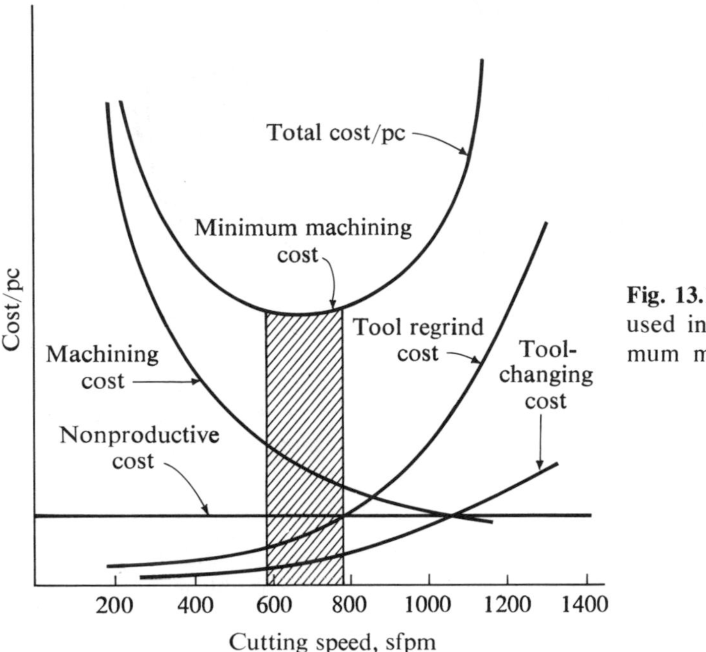


**Fig. 13.18.** Factors used in plotting minimum machining cost.

ever-increasing obsolescence, fade into the background. Sometimes the answer has been to work machines that require a large capital outlay continuously in order to make them pay off during their expected life. This, in turn, often requires premium labor rates. An alternative answer may be to make the same machines work harder.

The optimum cutting speed (Fig. 13.18) is in the range of about 580 to 770 sfpm. The exact speed selected will depend on whether *low cost* is emphasized or *high production.* The cutting speed for minimum cost per piece is expressed as follows:

$$V_m = \frac{C}{\{[(1/n) - 1]\ [(K_2/K_1) + \mathrm{TCT}]\}^n}$$

where $K_2$ = tool cost/sharp cutting edge in dollars, sharpening included; $K_1$ = machine labor and overhead rate in dollars/min; TCT = tool changing time in min; $n$ = slope of tool life curve on log-log paper; $C$ = the cutting speed that equals 1 min of tool life.

Comparing this formula with $VT^n = C$, the tool-life formula, it can be seen that the value under the line in the braces is equal to tool life for minimum cost per piece ($T_m$).

In Fig. 13.13, the values of $n$ for various tool materials were given. The values of $1/n - 1$ may be taken as follows: HSS 7, cast alloy 5, carbide 3, and ceramics 1. With this information, the formula may be simplified. Thus, for an HSS cutting tool, the tool life for minimum cost per piece would be:

$$T_m = 7[(K_2/K_1) + \text{TCT}].$$

The tool life for maximum production is given as follows:

$$T_{max} = [(1/n) - 1]\text{TCT}.$$

Again, simplified for HSS tools:

$$T_{max} = 7 \times \text{TCT}.$$

It is important that care be taken in using the proper units. For example, if TCT is in minutes, then $K_1$ must also be converted to dollars/min rather than the more usual dollars/hr. Also, tool costs/cutting edge must be in dollars rather than cents.

In actual practice a tool life chart should be made from experiments conducted under the required cutting conditions. The tool life vs speed relationship can be obtained by making several tests at high, medium, and low speeds and recording the resultant data on log-log graph paper. Cutting speed is plotted against time. If this data is not available, it may be possible to obtain a value of $V_{60}$—the cutting speed for a tool life of 60 minutes—from a handbook (see section 18-32 of *The Tool Engineer's Handbook*, second edition). This value can then be corrected to give the required tool life with the following formula:

$$V_t = V_{60}(60/T)^n$$

For example, if the velocity for a tool life of 60 min was given for a carbide tool as 550 sfpm, for 6.4 min it would be

$$V_t = 550\left(\frac{60}{6.4}\right)^{0.3} = 1{,}078 \text{ fpm}.$$

## NEW CONCEPTS IN MACHINING

In recent years great emphasis has been placed on *optimum* machining. However, to successfully achieve this goal presents many difficult problems. The latest and most promising approach has been adaptive control.

**Adaptive Control.** The concept of *adaptive control* means, in metal cutting, that the tool can react for the optimum advantage to unpredictable changes in cutting conditions. An operator usually selects speeds on the conservative side to avoid risking damage to the part or machine. Adaptive control, on the other hand, responds to the actual cutting conditions. When the tool is sharp and the workpiece not too difficult to machine, the speed will be high, but as the tool becomes dull or hits hard spots the automatic control will cut the speed back. In this way production can be maintained at an optimum rate.

Initial speeds and feeds do not have to be carefully chosen. Sensors, usually based on spindle torque, tool vibration, and tool-tip temperature, will feed the required information back to the performance measurement computer. The computer calculates the metal removal rate, tool wear rate, and surface finish. This information in turn is given to an optimizing controller (Fig. 13.19). Based on the information received, the controller

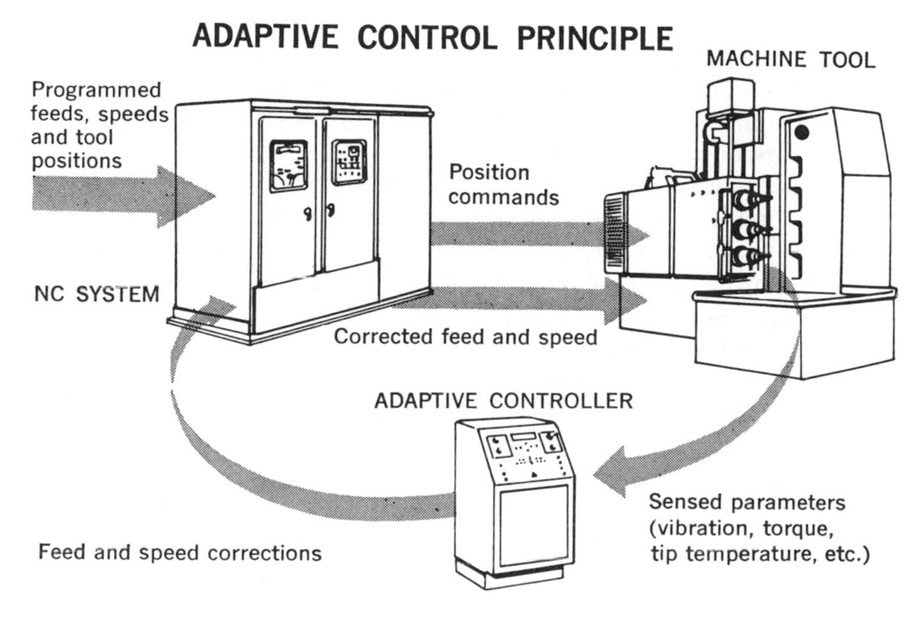


**Fig. 13.19.** Adaptive control adds a lobe to NC system's brain. It continually modifies feed and speed commands as cutting conditions vary to maintain top machine performance. Courtesy *Metalworking*.

will modify the initial feed and speed to optimize the overall machining performance. Shown in Fig. 13.20 is a schematic of a system used for adaptive control based on tool-tip temperature.

In order to assure product quality, certain boundary limits must be set up within which the controller must operate.

At the present time the system has been demonstrated and its feasibility acknowledged, but there are still many problems to be worked out before it becomes a full-fledged production item.

Work is now being done in "training" general purpose computers to recognize the patterns that correspond to good and poor results. This is done by building up a network of data for the computer memory.

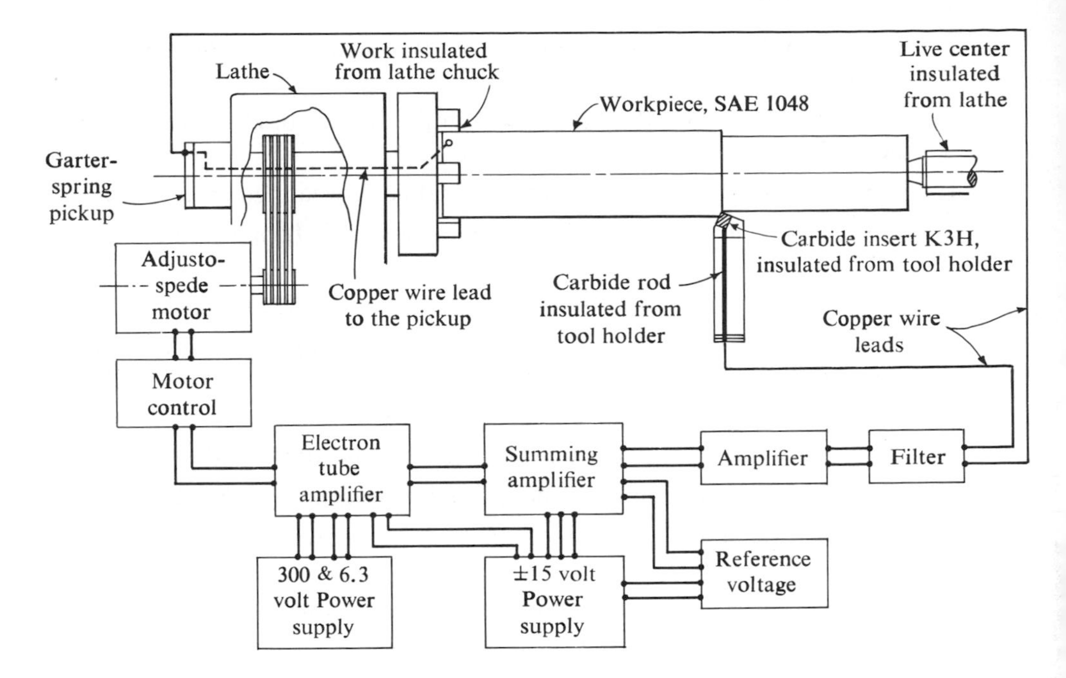


**Fig. 13.20.** Schematic of physical system incorporating tube control feedback and garter-spring pickup.

## QUESTIONS

**13.1.** What is the main cause of high temperatures in metal cutting tools?

**13.2.** (*a*) What causes BUE? (*b*) How can it be reduced?

**13.3.** What two tool angles have the most effect on the formation and thickness of the chip?

**13.4.** Why is the newest development in chip breakers considerably better than the previous ones?

**13.5.** If an adjustable chip breaker were used and the chip, before adjustment, turned out to be a snarl, how could the proper correction be made?

**13.6.** What is the basic difference in structure between the most used cutting tool materials today?

**13.7.** Give a general classification as to use for 18–4–1 HSS steels as compared to M types.

**13.8.** Why are straight tungsten carbide tools not well suited for cutting steel and other high strength ductile materials?

**13.9.** What are two advantages of "throw-away carbide and ceramic inserts"?

**13.10.** How does the blue oxide on the surface of HSS aid its cutting action?

**13.11.** What is the main cause of: (*a*) flank wear, (*b*) crater wear?

**13.12.** What are some factors you can use in judging whether a tool is set right or not?

**13.13.** How can cu in. removal rate be related to tool life?

**13.14.** (*a*) How can the individual forces acting on a cutting tool be measured? (*b*) Which one should be the largest on a lathe tool?

**13.15.** What is meant by unit horsepower?

**13.16.** What is the relationship between the BHN and horsepower?

**13.17.** What are some methods used to measure machinability?

**13.18.** What are some effects that coolants have, both good and bad, on cutting operations?

**13.19.** How can machinability of a metal be predicted?

**13.20.** What may be the result of overemphasizing tool life?

**13.21.** What is meant by "training" a computer for adaptive control in metal cutting?

## PROBLEMS

**13.1.** Would a lathe with a 3-hp motor be large enough to make a cut under the following conditions: material, stainless steel, 40 $R_c$; cutting speed, 200 sfpm; depth of cut, 0.050 in.; feed, 0.020 in./rev; diameter, 3-in. bar?

**13.2.** (*a*) What horsepower milling machine would be required to make the following cut: material, magnesium alloy; cutting speed, 450 fpm; cutter, 3/4-in.-wide end mill; number of teeth in the cutter, 4; chip load/tooth, 0.002 in.? The material to have a slot cut in it is 6 in. long. The cut is 0.250 in. deep.

(*b*) How much more or less power would be required to make the same cut in an aluminum alloy?

**13.3.** If the milling machine given in Problem 13.2*a* has a maximum rpm of 1,500, what would be the $hp_m$ requirement?

**13.4.** (*a*) There is no back rake on the tool given in Problem 13.1, but a 10 deg side rake. The measured chip thickness turned out to be 0.030 in. What is the cutting ratio? (*b*) What is the shear angle? (*c*) How would you rate its machinability?

**13.5.** (*a*) Find the tool life for minimum cost/pc to machine the following part using an HSS tool: a 6-in.-dia bar is to be machined for

30 in.; feed, 0.020 in./rev; depth of cut, 1/4 in.; machine overhead rate, $6.00/hr; toolroom overhead rate, $3.50/hr; machine labor rate, $3.00/hr; tool grinder's rate, $2.50/hr; HSS tool, $3.50; cutting edges, 7; TCT, 2 min; tool grinding time/edge, 1.5 min. (Use tool-life curves in text to convert min. to spm.)

(*b*) Find the cutting speed for minimum cost/pc.

**13.6.** (*a*) Use the information given in Problem 13.5 to obtain the tool life for maximum production using a carbide tool. (*b*) What is the rpm used at maximum production? (*c*) How long would it take to machine 20 inches of the bar, given in Problem 13.5, down to a 5 inch diameter if the tool life for maximum production were used?

## REFERENCES

1. Wennberg, J. L., "How To Get The Most Out of Metal Cutting," *The Tool Engineer*, March 1961.

2. Black, P. H., *Theory of Metal Cutting*, McGraw-Hill Book Company, Inc., New York, 1961.

3. Boston, O. W., *Metal Processing*, John Wiley & Sons, Inc., New York, 1951.

4. Centner, R. M., "What's Ahead in Adaptive Control," *Metalworking*, November 1966.

5. Cook, N. H., *Manufacturing Analysis*, Addison-Wesley Publishing Company, Inc., Reading, Massachusetts, 1966.

6. *Fundamentals of Tool Design*, ASTME, Prentice-Hall, Inc., Englewood Cliffs, New Jersey, 1962.

7. Merchant, M. E., "Mechanics of the Metal Cutting Process," *Journal of Applied Physics,* vol 16, no 6, June 1945.

8. Meyer, R., "A Quantitative Analysis of the Development of Carbide Tool Crater Wear," *Ph.D. Theses*, University of Wisconsin, under the supervision of S. M. Wu, Mechanical Engineering Dept., December 1966.

9. Taylor, F. W., "On the Art of Cutting Metals," *ASME*, 28, 1907.

10. Trigger, J. K., "How Heat Affects Tool Wear," *American Machinist,* July 18, 1966.

11. Troup, G. B., "A Metallurgical Guide To Machinability," *American Machinist,* October 14, 1963.

12. Troup, G. B., "Machinability Tests: A Critique," *American Machinist,* April 11, 1966.

13. Zimmerly, R., "Automatic Feedback Control for Maintaining Constant Cutting Tool Temperature," *ASTME Research Study* conducted at University of Wisconsin, August 1966.

# 14

# TURNING AND RELATED OPERATIONS

THE LATHE FROM ITS EARLY BEGINNINGS has been one of the most important metal cutting machines.

Some lathes have only one cutting tool and are generally used for limited or low production. Others have several cutting tools but only one work-holding station, and these are usually considered machines of medium production. Those that have several work-holding stations and many tools are called high-production lathes.

## LOW-PRODUCTION LATHES

Low-production lathes are those that can easily be set up for a few parts. The various types are engine lathe, toolroom lathe, bench lathe, speed lathe, and special-purpose lathes.

**Engine Lathes.** The most common metalworking lathe is called an engine lathe. This term dates back to the time when the early lathes were powered by steam engines. The engine lathe (Fig. 14.1) is capable of performing a wide variety of operations.

A careful study of the operations shown in Fig. 14.2 will help you understand the capabilities of this machine. The setups shown in this figure are for limited quantities.

**Basic Structure of the Engine Lathe.** The five main parts of the modern engine lathe are the headstock, tailstock, bed, carriage, and quick-change gearbox, as shown in Fig. 14.3.

**Fig. 14.1.** Engine lathe making a cut on stock held in a chuck and supported by the tailstock center. Courtesy the Monarch Machine Tool Co.

HEADSTOCK. The headstock contains the driving mechanism—either pulleys or gears—to turn the work. Provision is made so that 8 or 10 different speeds may be obtained on the smaller sized lathes, and several times this number and variable speeds on larger ones. The headstock furnishes a means of support for the work, either with a center fitted into the spindle or by a chuck.

TAILSTOCK. The tailstock is the most common means of supporting the outer end of the stock. It may be positioned upon the bed at any point and locked securely in place. It contains the dead center, which is held in place by a taper. Adjusting screws in the base of the tailstock are used to shift it laterally on the bed. This provides a means of maintaining accurate alignment with the headstock center. It also makes it possible to introduce a small amount of taper into the work (Fig. 14.4).

BED. The bed makes up the basic structure of the lathe, on which all other parts are mounted. It is usually made of aged gray cast iron in the smaller sized lathes and may be of welded steel construction in the larger sizes. At the top of the bed are the ways, which act as guides for accurate movement of the carriage and tailstock.

CARRIAGE. The carriage, as the name implies, is used to carry the cutting tool along the bed longitudinally. A cross slide is mounted on top of the carriage. It moves the tool laterally across the bed. Another slide is mounted above the cross slide. This is referred to as the *compound rest.* The compound rest can be swiveled in a horizontal plane at any desired angle with respect to the work, thus furnishing another method of cutting tapers.

QUICK-CHANGE GEARBOX. The quick-change gearbox allows the operator to change the longitudinal or cross feed quickly. Feed refers to the amount the tool is made to advance per revolution of the work. The

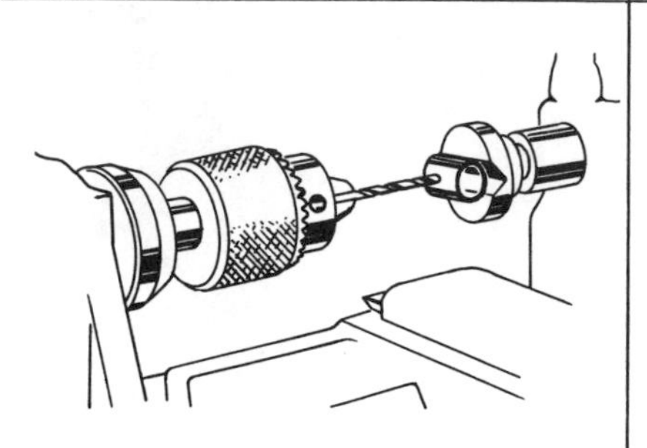

Drilling an oil hole in a bushing with crotch center in tailstock.

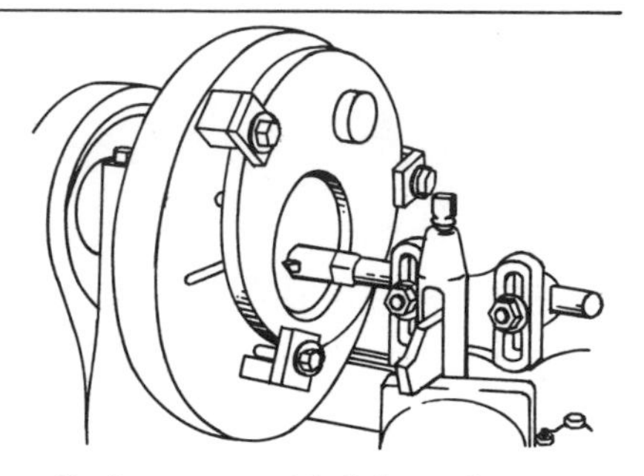

Boring an eccentric hole on the faceplate of the lathe.

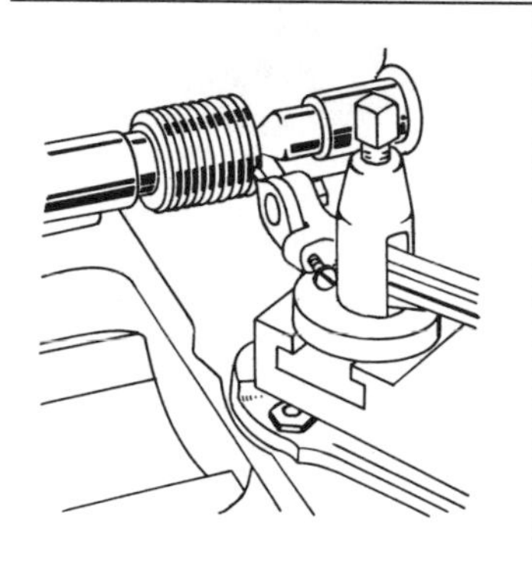

Cutting a screw thread with compound rest set at 29 deg.

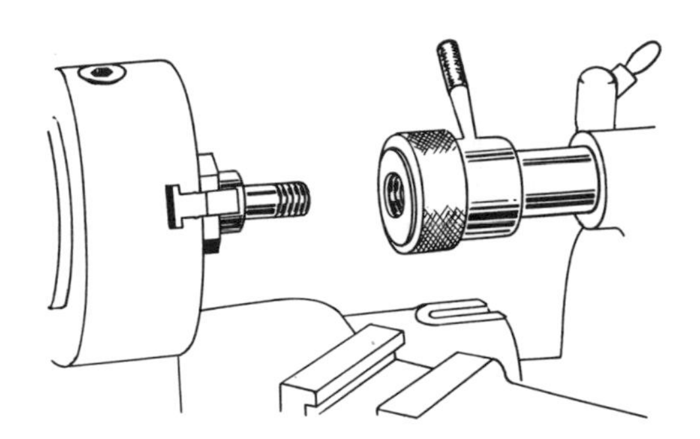

Die mounted in tailstock of lathe for threading studs.

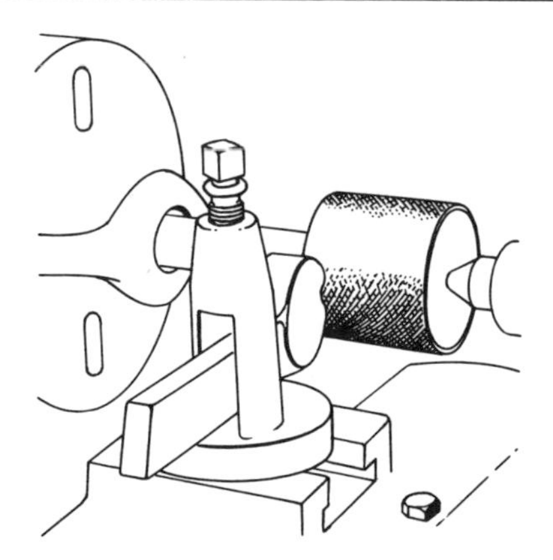

Knurling a steel piece in the lathe.

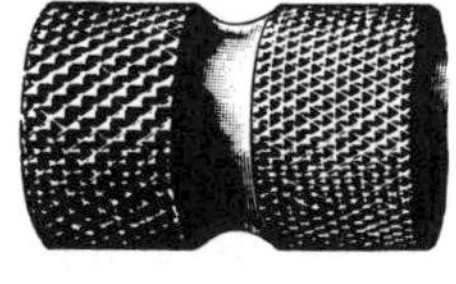

Coarse Medium

Sample of knurling.

**Fig. 14.2** A variety of operations commonly performed on the engine lathe.

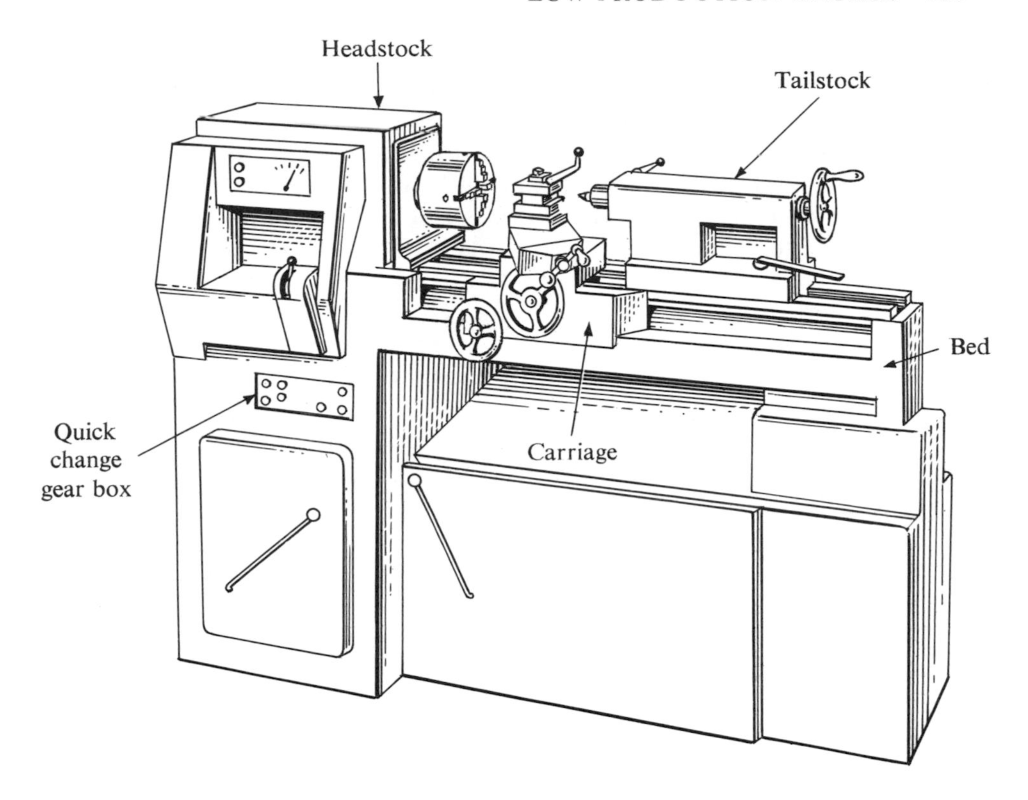


**Fig. 14.3.** The five main parts of an engine lathe.

quick-change gearbox also provides settings for cutting all the commonly used threads.

**Lathe Size and Accuracy.** The *size* of a modern engine lathe is given by the maximum diameter of swing, in inches, and the maximum length of bar that can be turned between centers, in inches. *Swing* refers to twice the distance from the lathe center point to the top of the ways. An engine-lathe size specification may read: swing 12 3/4 in. over bed, 7 5/8 in. over cross slide; length 47 in.

As for *accuracy,* lathe dials are graduated in increments of 0.001 in. Some are made to read directly; that is, for each 0.001-in. depth of cut, a

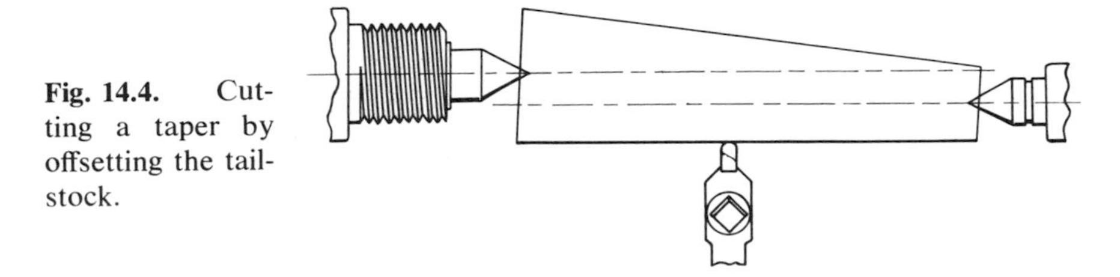

**Fig. 14.4.** Cutting a taper by offsetting the tailstock.

corresponding 0.001 in. will be removed from the stock. Other lathes are made so that the tool moves in the amount registered on the dial. Since the material is removed from both sides of the stock, the amount taken off will be twice the dial reading. Engine-lathe work can, with good equipment, be maintained within 0.001- or 0.002-in. accuracy.

**Work-Holding Methods.** Work is held in lathes by several devices: lathe centers, chucks, faceplates, and mandrels.

LATHE CENTERS. The work to be turned is first cut to length, and then center holes are drilled into each end (Fig. 14.5*b*). A dog is securely fastened to one end of the work, which is then placed between the lathe centers. The tail of the dog fits in a slot of the drive plate for positive rotation.

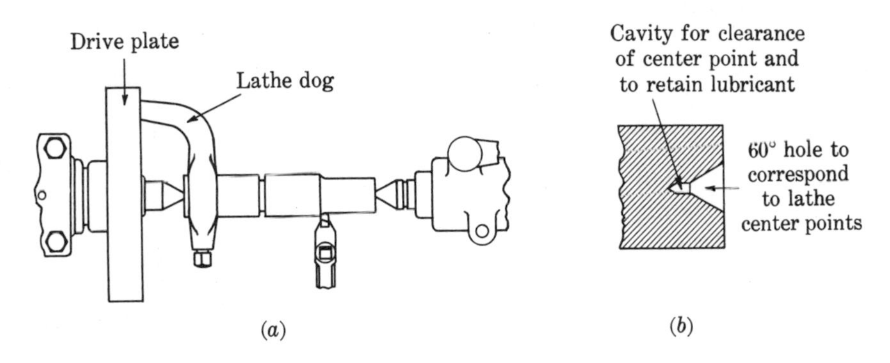


**Fig. 14.5.** (*a*) Stock mounted between centers on a lathe. (*b*) Center holes drilled into each end of the stock.

CHUCKS. Another common method of mounting work in a lathe is by means of a chuck (Fig. 14.6). The universal chuck shown in the figure is self-centering; all three jaws move toward the center on a scroll arrangement. This permits the operator to load and unload work quickly and easily.

A more versatile chuck is the four-jaw independent type (Fig. 14.7). Loading and unloading the work is slower in this chuck, since the jaws must be adjusted separately; however, it has more holding power and can be used for both on- and off-center work. It can also be used to advantage on odd-shaped pieces.

The collet chuck (Fig. 14.8) is the most accurate of all chucks. The collets have split tapers that are made to close by means of a draw-in handwheel which fits through the spindle of the headstock. The work mounted in the collet should not be more than a few thousandths of an inch larger or smaller than the size stamped on the collet.

A newer, more versatile collet chuck is the rubber-flex type (Fig. 14.9). The parts to be chucked may vary as much as 1/8 in. in diameter

**Fig. 14.6.** Round work held in a universal chuck. Courtesy South Bend Lathe, Inc.

**Fig. 14.7.** Centering work with a dial test indicator using an independent four-jaw chuck. Courtesy South Bend Lathe, Inc.

for each collet, except in the smallest size, 1/16 in. The standard range of sizes for collets of this type is 1/16 to 1 3/8 in.

FACEPLATES. The faceplate shown in Fig. 14.10 is made with bolt slots so that flat work may be clamped or bolted to it. More elaborate setups may call for angle plates, as shown, or for fixtures attached to the faceplate.

MANDRELS. It is often desirable to machine a surface concentric and true with the inside diameter. This can be accomplished with the aid of a mandrel. Several types of mandrels are shown in Fig. 14.11.

The solid mandrel (*a*) has an 0.008-in.-per-ft taper. The part to be machined is placed on the small end of the mandrel and forced on with an arbor press until it is firmly in place.

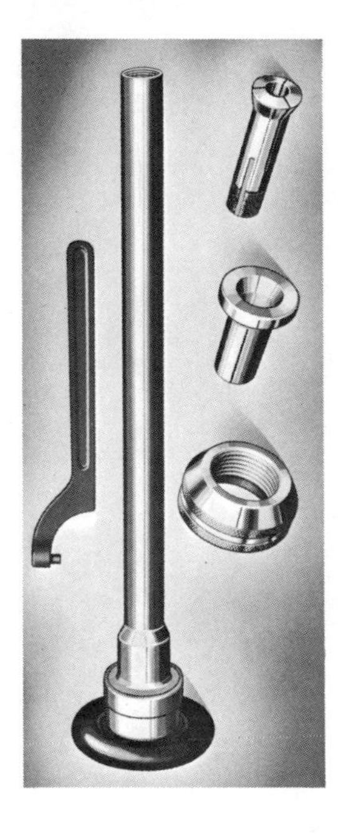

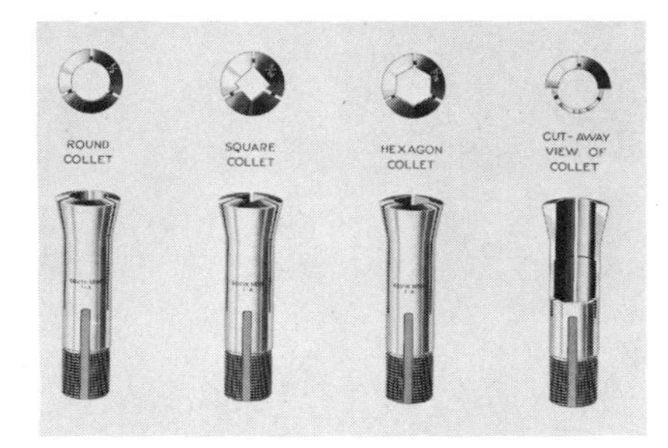


**Fig. 14.8.** Various types of collet chucks and their uses. Courtesy South Bend Lathe, Inc.

**Toolroom Lathe.** There is very little difference in appearance between an engine lathe and a toolroom lathe. However, a toolroom lathe is built with greater precision, has a greater range of speeds and feeds, and has more attachments. It is more expensive than the engine lathe and is designed for precision toolroom work.

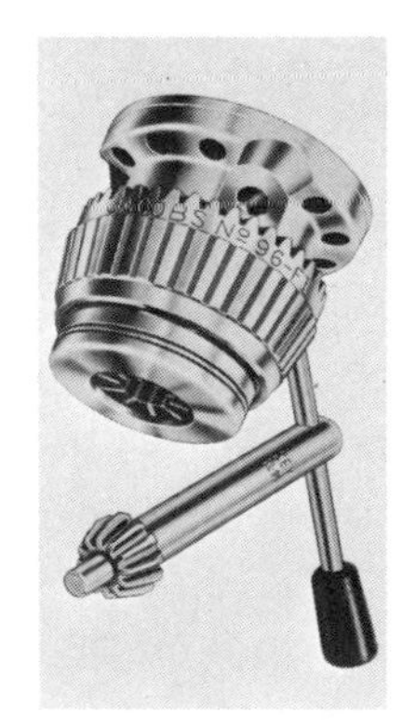

**Fig. 14.9.** Rubber-flex collets. Courtesy The Jacobs Manufacturing Co.

**Fig. 14.10.** Boring a bracket with an angle plate attached to the faceplate.

**Bench Lathe.** The bench lathe is a small version of the engine lathe. It is usually mounted on a bench; hence its name. The bed length seldom exceeds 6 ft.

**Speed Lathe.** Speed lathes are built to satisfy light machining operations that require a comparatively high speed. High surface foot speeds are required for buffing, polishing, spinning, and wood turning, and these operations can be done on a lathe of this type.

**Duplicating Lathe.** The duplicating lathe is very much like an ordinary engine lathe except that it is equipped with a tracer attachment (Fig. 14.12). A template, which may be cut out on a metal-cutting band saw and then filed for accuracy, is attached to a stand so that a stylus can trace its contour. The stylus, in turn, actuates a pilot valve that is used to control the hydraulic cylinder attached to the cutting tool.

Tracer attachments may be set up for between-center work or facing operations. An application of tracer turning is shown in Fig. 14.13. Time savings that might be realized by using this method instead of the conventional engine lathe are also shown. Note that the accuracy is indicated to be within 0.0002 in. on some dimensions.

## ENGINE LATHE EVALUATION

Lathes should be evaluated on their ability to provide long-term geometric accuracy, reliability, and ease of maintenance.

The key to long term reliability lies in the lathe's proportions that are built to cope with the variety of forces created by cutting (Fig. 14.14). Specifications will give a great deal of information as will a visual inspection. The proper gear spacing will prevent stresses from reinforcing each other. Other points to check are quality and type of spindle bearings,

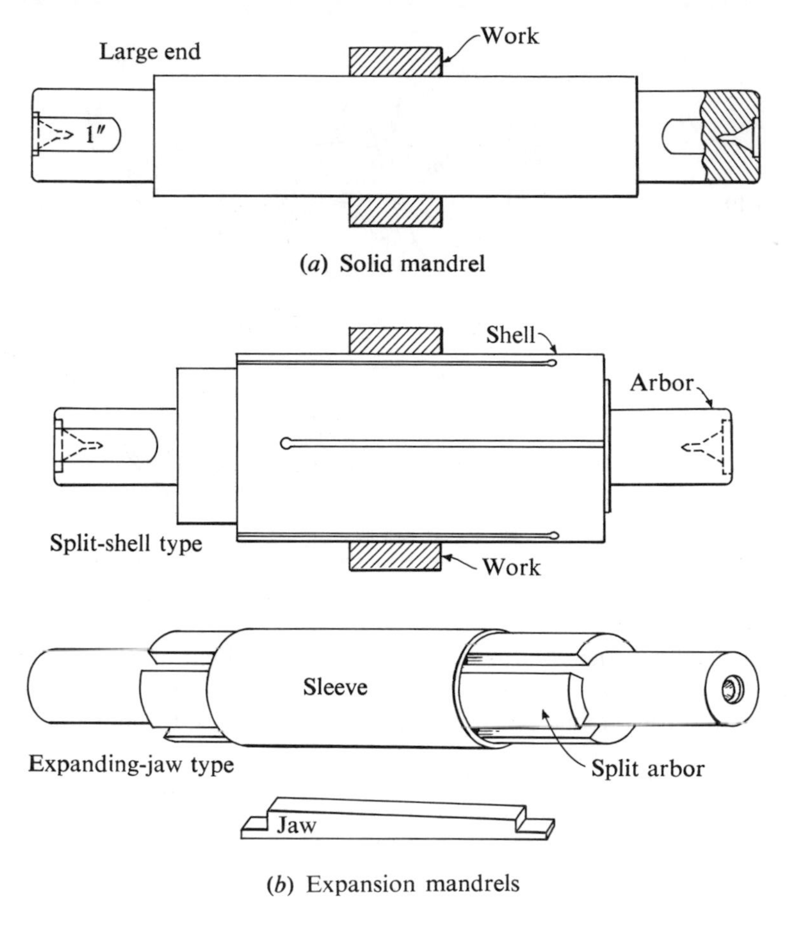


**Fig. 14.11.** Solid and expansion mandrels used to hold parts for machining so the outside surface is concentric with the bore.

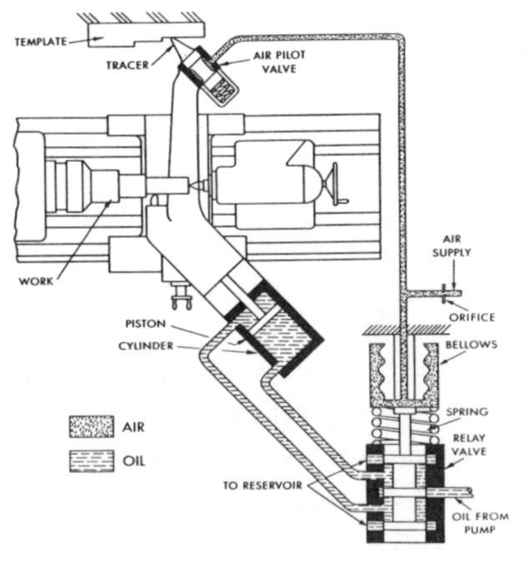


**Fig. 14.12.** An air-oil circuit applied to lathe work. Courtesy The Monarch Machine Tool Co.

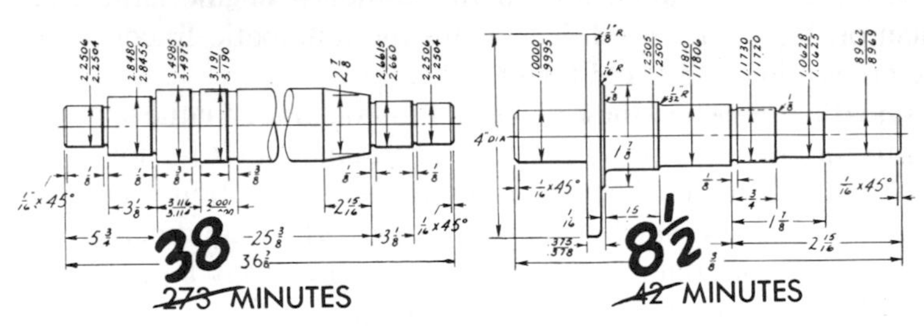


**Fig. 14.13.** A lathe tracer arrangement with typical parts, and the time required compared to the same operation by a standard engine lathe. Courtesy Sidney Machine Tool Co.

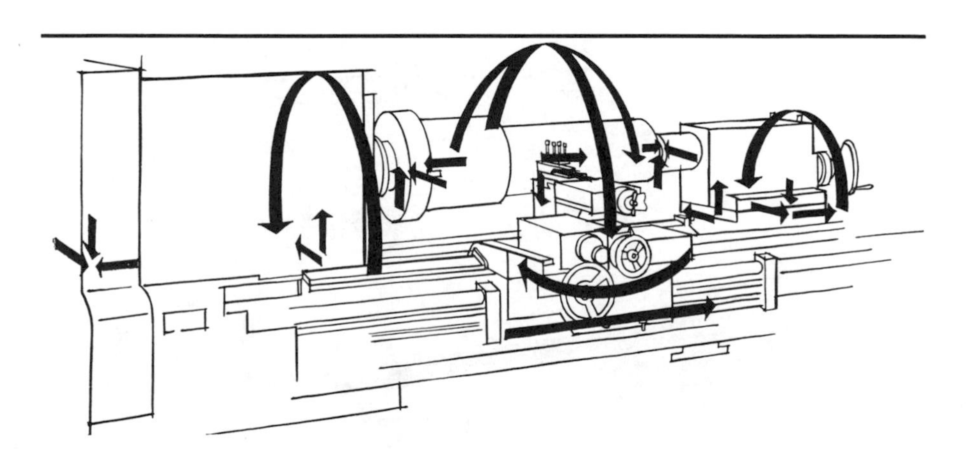

**Fig. 14.14.** A lathe must be designed and constructed to withstand a variety of cutting forces like those shown by arrows.

available horsepower, speeds, feeds, types, and location of controls.

American Standards Association accuracies call for a bed level to 0.0005 ipf in both longitudinal and transverse directions.

Although massiveness and weight are important, they do not solve all problems of rigidity. Weight must be distributed where it will absorb and resist distorting forces.

## MEDIUM-PRODUCTION LATHES

When the volume of production of certain machine parts is such that it does not warrant the use of fully automatic lathes but is greater than can be handled economically on the standard engine lathes, medium-production lathes are used. These are the automatic tracer lathe, automatic cross-slide lathe, and turret lathe.

**Automatic Tracer Lathes.** The automatic tracer lathe may be compared to an engine lathe with a tracer attachment. It is, however, much larger, and is capable of considerably higher production. The one shown in Fig. 14.15 is the single-spindle type, which is equipped with both a vertical and a horizontal bed. Several tool slides may be engaged at one time with one or more tools in a block (Fig. 14.16).

**Automatic Cross-Slide Lathe.** The automatic cross-slide lathe was built to provide multiple tooling on both the front and the rear sides of the stock (Fig. 14.17).

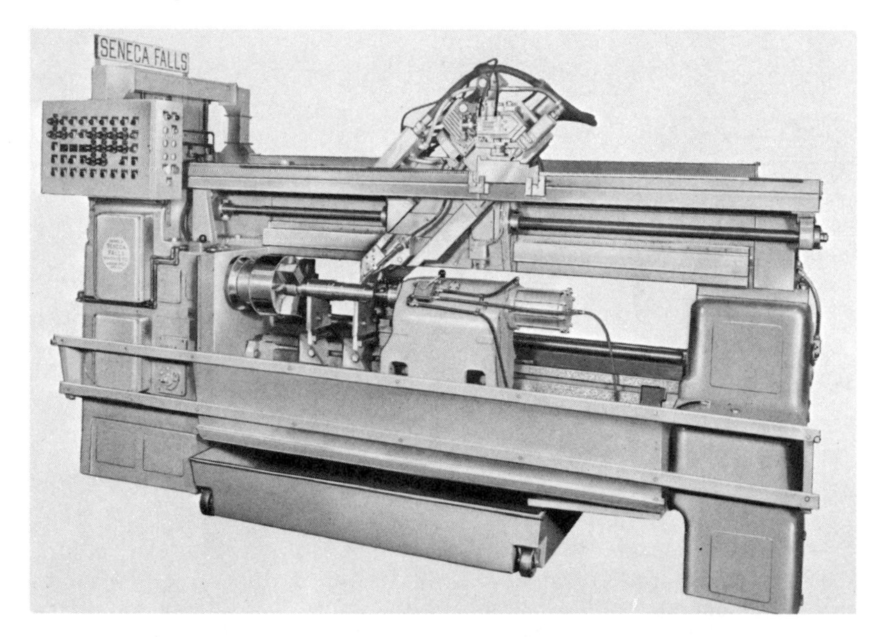


**Fig. 14.15.** Single-spindle tracer lathe. Courtesy Seneca Falls Machine Co.

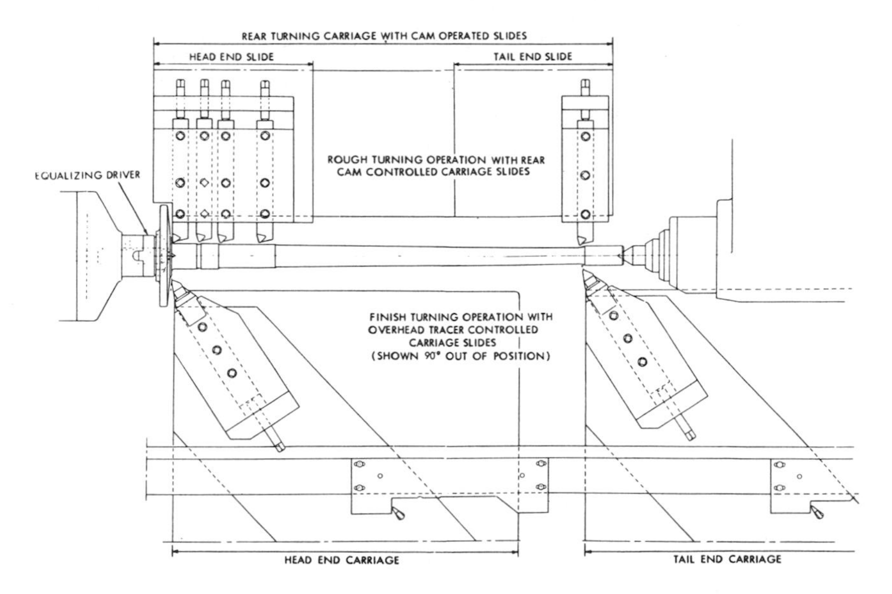


**Fig. 14.16.** Tool slides at rear of lathe are cam operated; those on overhead carriage slides are template controlled. Courtesy Seneca Falls Machine Co.

Turning multiple diameters on long stock is difficult unless the stock is properly supported. Cutting tools located on both sides of the stock at one time will balance thrust and minimize the deflection caused by tool forces. In addition to the supporting effect of opposing tools, auxiliary

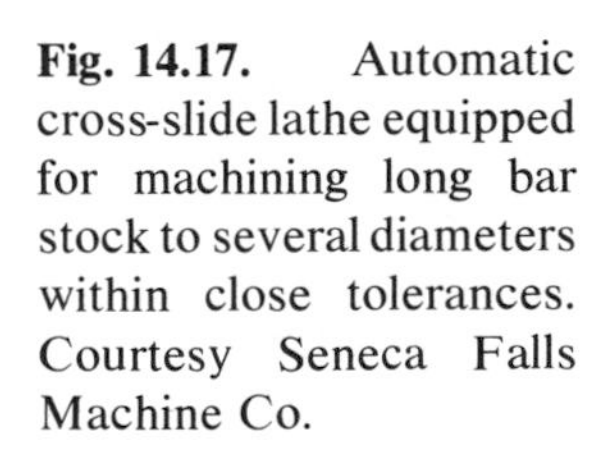
**Fig. 14.17.** Automatic cross-slide lathe equipped for machining long bar stock to several diameters within close tolerances. Courtesy Seneca Falls Machine Co.

rollers are used (shown supporting the stock). The tooling layout (Fig. 14.18) shows the arrangement of tools for a common production part.

Notice that the amount of travel for each tool slide is also shown. The various tool slides can be made to cut simultaneously or in sequence.

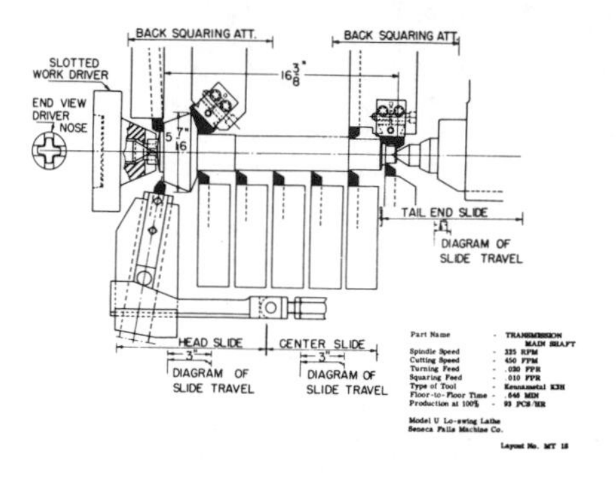


**Fig. 14.18.** Tooling layout for automatic cross-slide lathe. Courtesy Seneca Falls Machine Co.

This type of lathe is not limited to between-center work but may be used for facing and boring operations, as shown by the tooling layout in Fig. 14.19. Some parts require more than one layout and setup, since it is not possible to reach all surfaces in one operation.

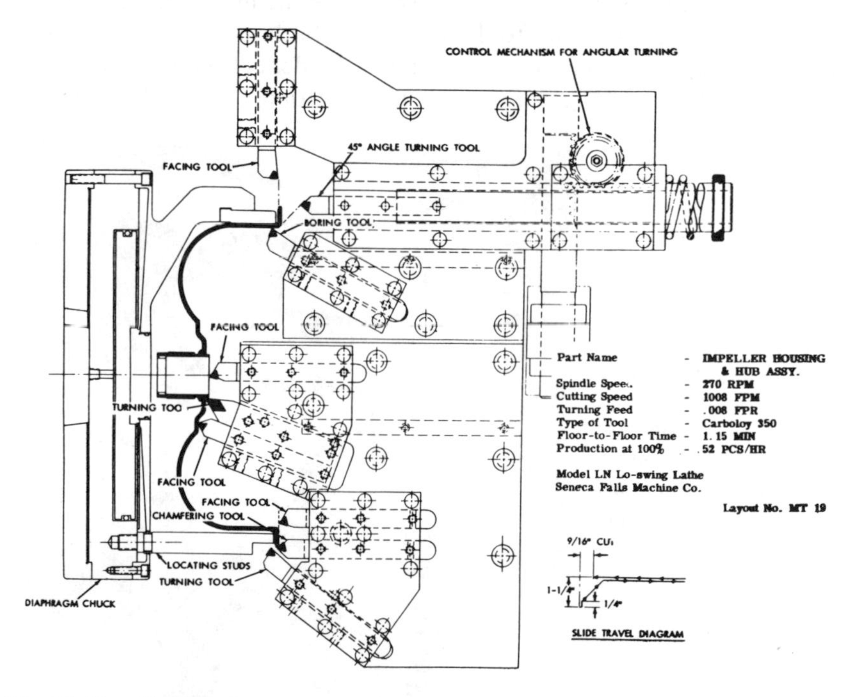


**Fig. 14.19.** Tooling layout for facing and boring with an automatic lathe. Courtesy Seneca Falls Machine Co.

**Turret Lathes.** Turret lathes are an outgrowth of a demand for a variety of repetitive turning operations on one machine. Although the engine lathe is versatile, it requires time to change the tool for each different operation. Also, some repetitive accuracy is lost. The turret lathe derives its name from the fact that the tailstock has been converted into an indexing turret with five or six stations. On the front side of the cross slide is mounted a square turret, which is indexed or turned by hand. On the back of the cross slide is a single toolholder. Once the tools have been properly placed on these tooling stations, they need not be removed, other than for sharpening, for the entire run of parts. This makes for a high degree of repetitive accuracy and for parts that can be used in interchangeable manufacture.

**Types of Turret Lathes.** The turret lathe is manufactured in a wide variety of types which may be broadly classified as horizontal, including ram type (bar or chucking) and saddle type (bar or chucking); and vertical, with single- or multistation models.

The economic lot size to be run on a particular type of machine varies with work shape, changeover time, and machining time. The hand operated turret lathe is generally considered economical for lot sizes of 25 to 50 pieces. Above this quantity, the automatic type is usually more economical. Sometimes the automatic type can be economical for lot sizes of less than 25.

The *break-even cost* in comparing two machines is the point at which the total and unit costs are the same for both machines. One machine is more economical for small quantity production and the other for larger quantity. This may be plotted as shown in Fig. 14.20. The break-even cost may be calculated with the formula

$$E_b = \frac{E_c}{A_c - B_c}$$

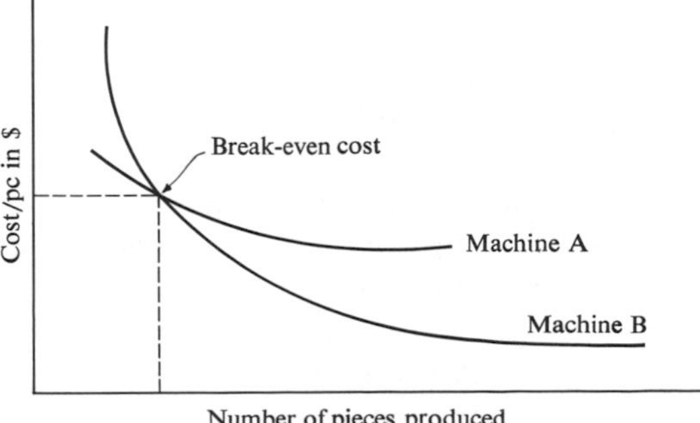


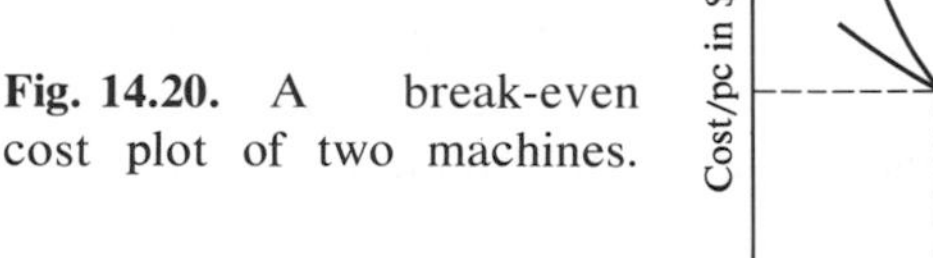
**Fig. 14.20.** A break-even cost plot of two machines.

where $A_c$ = cost of producing the part on Machine A; $B_c$ = cost of producing the part on Machine B; and $E_c$ = additional cost of tooling and setup time to produce the part on Machine A.

HORIZONTAL TURRET LATHES. *Ram type.* Turret lathes have been developed to handle, efficiently, several types of work. The ram-type bar machine (Fig. 14.21) is made to handle bar stock which is fed through the collet-type chuck either by feed fingers actuated by a hand lever or by hydraulic or airdraulic arrangements.

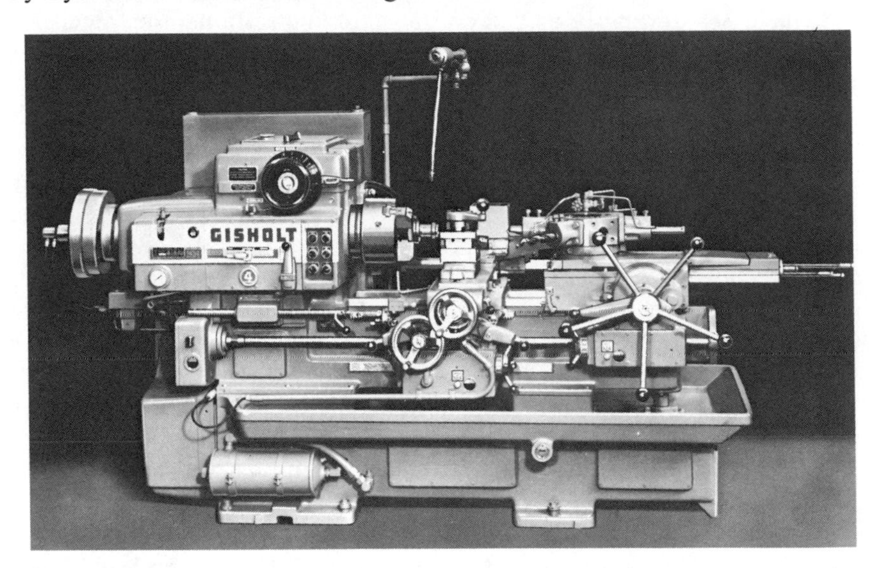


**Fig. 14.21.** Ram-type turret lathe. Courtesy Gisholt Machine Co.

The principal difference between a bar and a chucking machine is in the way the stock is held, whether this is by collet chuck for bar stock or by the larger 2-, 3-, or 4-jaw chucks for individual parts. Some machines are made so that they may be quickly adapted to either type of work.

Parts turned out by the ram machine, both bar and chucking, are shown in Fig. 14.22.

The distinguishing feature between a ram-type and a saddle-type machine is in the way the hexagon turret is mounted. The ram-type machine is so called because the hexagon turret is mounted on a slide or ram that fits in the saddle bolted to the bed of the machine. This arrangement makes for ease of operation, since only the ram and the turret move back and forth, and the heavier saddle can remain in one position for the entire run of parts.

*Saddle type.* In the saddle-type turret lathe, the hexagon turret is mounted directly on the saddle of the machine (Fig. 14.23). This arrangement, although slower and more cumbersome to operate, provides

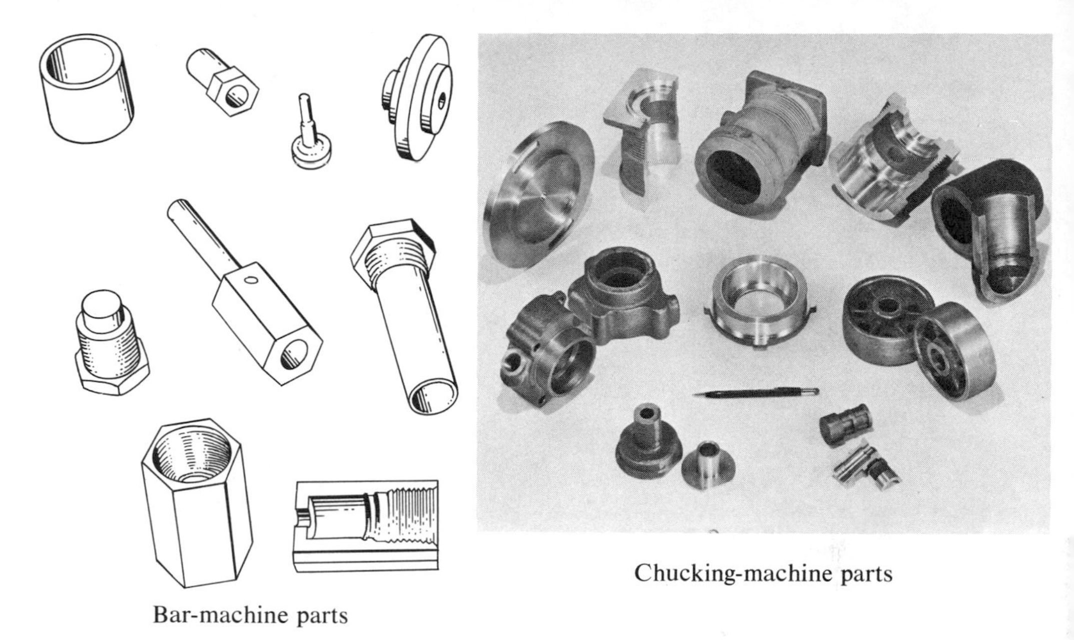


**Fig. 14.22.** Ram-type turret-lathe production parts. Courtesy The Warner and Swasey Co.

for a heavy-duty setup and is able to handle larger parts than the ram-type machine.

*Tooling.* Special tooling has been developed for the turret lathe to make for higher production and ease of handling. Some of these tools are now associated with other machines, but they remain essentially turret lathe tools. Figure 14.24 shows turret lathe tools, with an explanation of their use and typical applications. Study the tooling carefully to gain an understanding of the type of work that can be done without difficulty on the turret lathe.

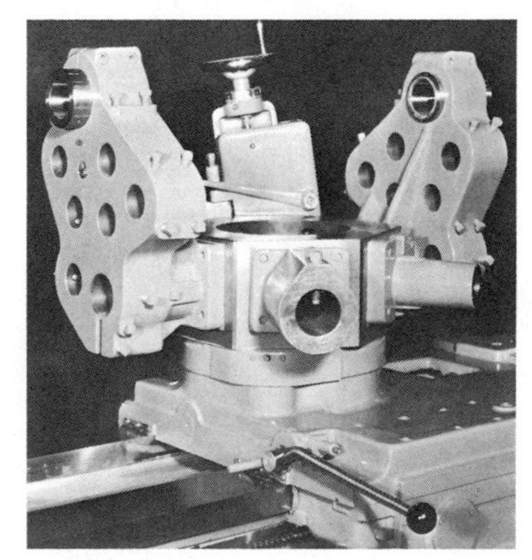

**Fig. 14.23.** The hexagon turret of the saddle-type machine is mounted directly on the saddle. Courtesy Gisholt Machine Co.

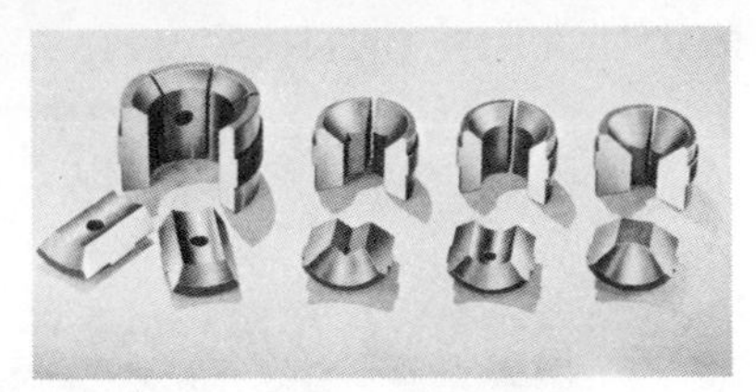

(*a*) Collet pads

Round, square, and hexagonal collets are used to hold the bar stock.

(*b*) Combination stockstop and starting drill

The combination stockstop and starting drill is used to let the stock extend a designated distance from the chuck. The drill is then brought out to make a start for a drilled hole.

(*c*) Chamfering tool

The chamfering tool is used to bevel the end of the bar stock. This operation is always used before the box tool and the self-opening die head, (*d*), (*e*), (*g*).

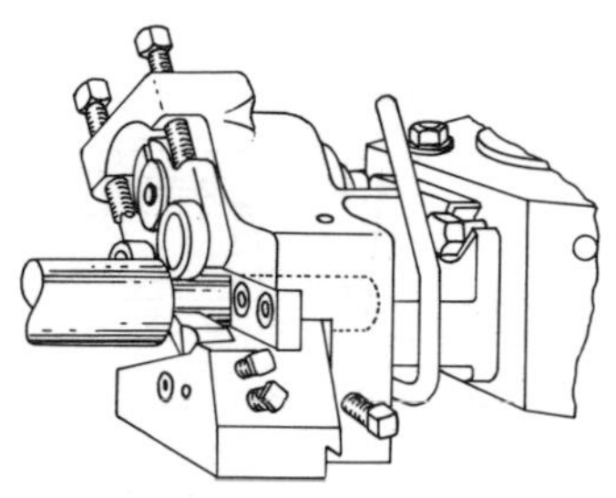

(*d*) Single-roller turner or box tool

The box tool furnishes support for the bar stock as it is being cut, maintaining accuracy of the machined surface.

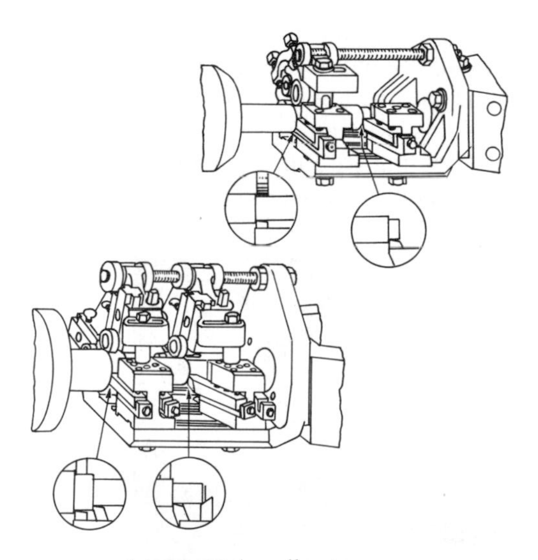

(*e*) Multiple-roller turner

The multiple-roller turner has positions for several tools to turn several diameters at one time.

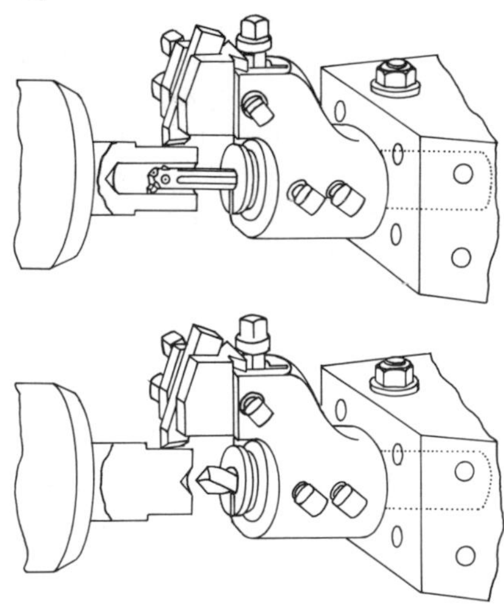

(*f*) Adjustable knee tool

The adjustable knee tool can be used for both internal and external machining. The vertical external tool can be adjusted to turn different diameters.

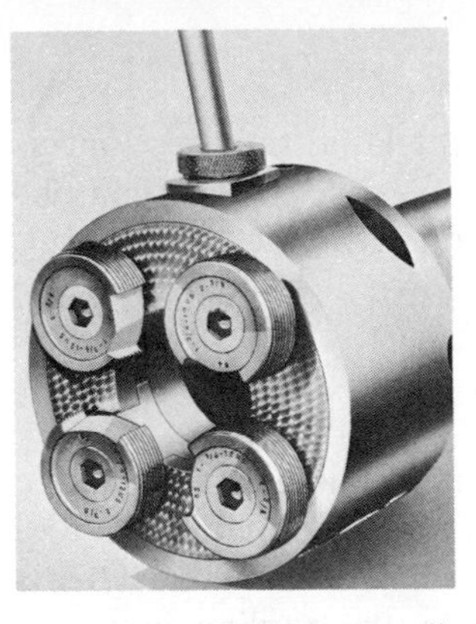

(*g*) Self-opening die head

The self-opening die head is used for making threads. The finished thread can be produced in one or two passes.

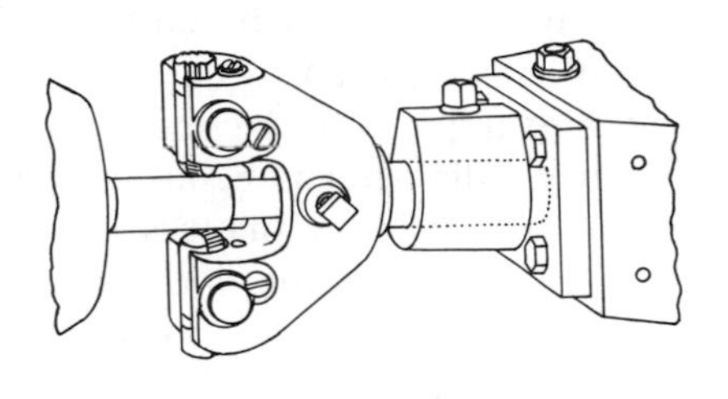

(*h*) Knurling tool

The knurling tool is used to make a diamond-shaped or a serrated surface.

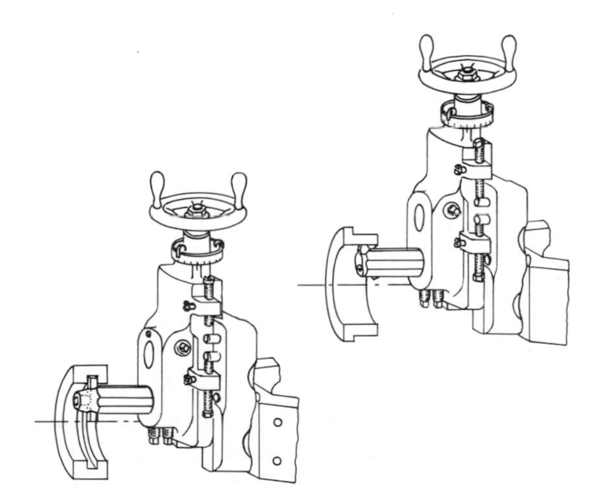

(*i*) The adjustable vertical slide

The adjustable vertical slide allows the tool mounted on the hexagon turret to be raised and lowered. Adjustable stops are also provided for ease of tool resetting. Boring, recessing, and external turning may be done with this type of tooling.

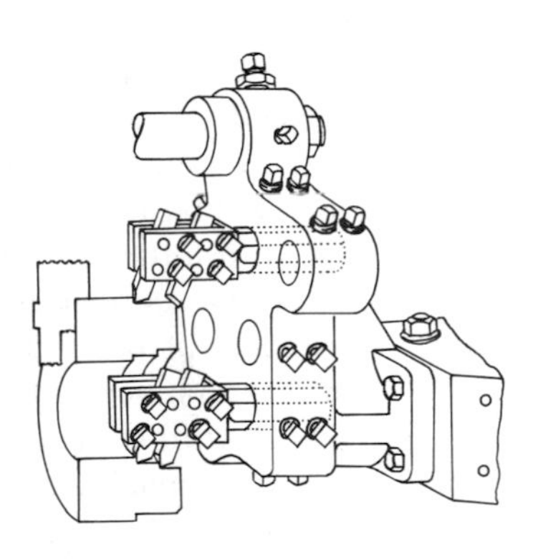

(*j*) The multiple turning head

The multiple turning head provides mounting holes for several tools to cut at one time from one tooling station or one face of the hexagon turret. The bar at the top is an overarm support which slides into a bearing on the headstock to give rigidity to the setup.

**Fig. 14.24.** Turret lathe tools: (*a*) to (*h*), tooling used in machining bar stock on a turret lathe; (*i*) and (*j*), turret lathe tooling commonly used for chucking work on the saddle-type machine.

*Tooling principles.* Efficient use of the turret lathe requires that as many tools should be cutting at one time as is feasible. More than one tool may be used to cut from one face or station of the hexagon turret; for example, on the multiple turning head shown in Fig. 14.24*j*, an outside diameter is being turned, the end of the part is being chamfered, and two inside diameters are being bored. In turret lathe terminology, this is known as *multiple tooling*.

Tools may be made to cut from both turrets at one time; that is, a hole may be bored by a tool on the hexagon turret at the same time that a tool on the square turret is being used to cut the outside diameter. This is known as a *combined cut*. Careful planning in the tooling layout is necessary to take advantage of these principles. The lot size will also decide how much time can be spent on making the setup and if any special tooling is warranted.

VERTICAL TURRET LATHES. *Single-station vertical lathes.* Large castings, forgings, and other parts are difficult to mount and hold in chucks or between centers for machining on the horizontal turret lathe. The vertical turret lathe (Fig. 14.25) was developed to overcome some of the difficulties encountered in mounting and holding large workpieces. The tabletop is a combination three-jaw chuck and faceplate. The size of this lathe is given by the diameter of the table, in inches, which usually ranges from 30 to 46 in. The main structure of the machine, in addition to the table, consists of a column which supports a crossrail and a side rail. On the horizontal crossrail are mounted one or two vertical slides, one with a turret toolholder, the other with a square toolholder. The side rail supports a square turret used for holding tools to cut the outside diameter of parts mounted on the table. As with the horizontal turret lathe, combined cuts and multiple tooling are possible.

*Multistation vertical lathes.* The six- or eight-station vertical lathe or chucking machine is designed for high production. The six or eight independently driven spindles are mounted on a carrier around a stationary column. This is somewhat similar to the multispindle tracer lathe described previously, but it is now in a vertical position. As the machining operations are completed, the tools retract and the spindle carrier indexes one station. Thus one part is completed with each indexing cycle. These machines have the advantage of requiring only the minimum amount of floor space (Fig. 14.26).

## HIGH-PRODUCTION LATHES

The borderline between medium- and high-production lathes is not well defined, but the distinction is made here to help the reader consider the machines in terms of relative quantity production. Generally, high-production machines are those that run continuously, with very little operator attention. The amount of production depends on the difficulty

**Fig. 14.25.** The vertical turret lathe with a part being machined. Courtesy Giddings and Lewis Machine Tool Co.

of the job, the degree of automation, and the number of tooling stations available. Lathes placed in this category are: automatic turret lathes (including electric, hydraulic, and tape-controlled lathes, and single-spindle, automatic-chucking-type lathes), and single- and multispindle automatic screw machines.

**Electric, Hydraulic, and Tape-Controlled Automatic Turret Lathes.** Many turret lathes are now equipped with automatic indexing devices for the hexagon turret which minimize the machine and work

**Fig. 14.26.** Multistation vertical turret lathe. Courtesy The Bullard Co.

handling responsibilities of the operator. Dial-in programming systems are available as shown on the machine in Fig. 14.27. Some programming systems are more sophisticated than others, but in the main they eliminate cam and gear changing and permit preselection of an operation by means of panel switches. Factors that can be preset include spindle speeds, cross-slide feed, feed start points, bar advance feed, collet open and close, turret index, and other motion sequences.

**Fig. 14.27.** A single spindle automatic turret lathe with selector switches for all machine functions. Courtesy Pratt & Whitney.

Both electric and hydraulic drives have been used to make these machines automatic. Punched cards and tape are also used to provide a wider variety of operations, including contour turning.

**Single-Spindle Automatic Bar and Chucking Lathes.** Automatic turret lathes have long been referred to as single-spindle automatic bar or chucking machines. These machines are of two distinct types, depending on the position of the turret. They may be automatic ram or saddle machines, with the turret conventionally mounted as just described, or they may have the turret vertically mounted (Fig. 14.28). Both the front and back cross slides can be used simultaneously, and they can be combined with cutting from any one of the turret faces. Speeds, feeds, and cutting distances are set by adjustable trips (Fig. 14.29) on the selector drum.

On this machine, the larger the diameter of the work, the greater the rigidity, since the tools need not be extended so far from the overhead mounting. Also, chips, coolant, and dirt tend to fall free of the machined surfaces.

Single-spindle automatics of this type may be obtained with 8 1/2- and 10 1/2-in. swing over the cross slides; these are designated for size as No. 1 and No. 2, respectively. Bar capacities range from 1/2 up to 2 3/8 in. dia and will turn at spindle speeds of 10,000 rpm on 1/2-in. dia capacity model and 2,000 rpm on the 2-in. dia model. These speeds place this type of machine in the high production class.

**Fig. 14.28.** Single-spindle automatic chucking machine. Courtesy The Warner and Swasey Co.

**Fig. 14.29.** Setting the stops of a selector-drum switch of the single-spindle automatic. Courtesy The Warner and Swasey Co.

**Single-Spindle Automatic Screw Machines.** Screw machines of the single-spindle type may be thought of as small cam-operated turret lathes. The distinguishing features of these machines are the position of the turret and the tool mountings. The automatic screw machine (Fig. 14.30) has the turret mounted on a horizontal pin so that it may rotate on a vertical axis.

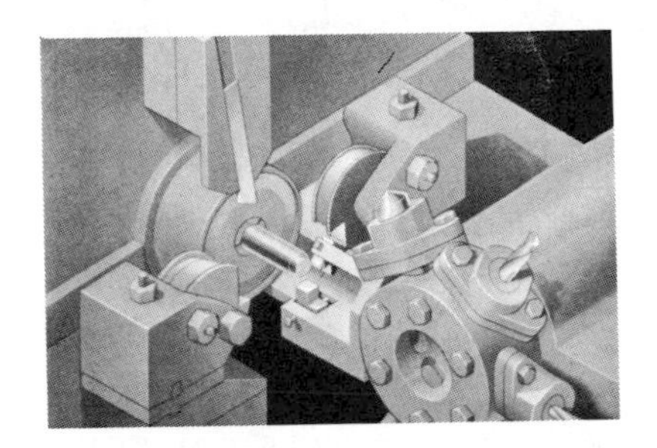

**Fig. 14.30.** Single-spindle automatic screw machine, showing turret and front, rear, and vertical cross slides. Courtesy Aluminum Company of America.

The front, top, and rear slides are fed at predetermined rates by means of disk cams. The three superimposed cams needed to perform the four operations of rough turn, finish turn, form-the-head, and cutoff are shown in Fig. 14.31. You will note that the cam perimeter is divided into 100 parts. Each part is assigned a certain number of seconds, depending on the speed of the camshaft.

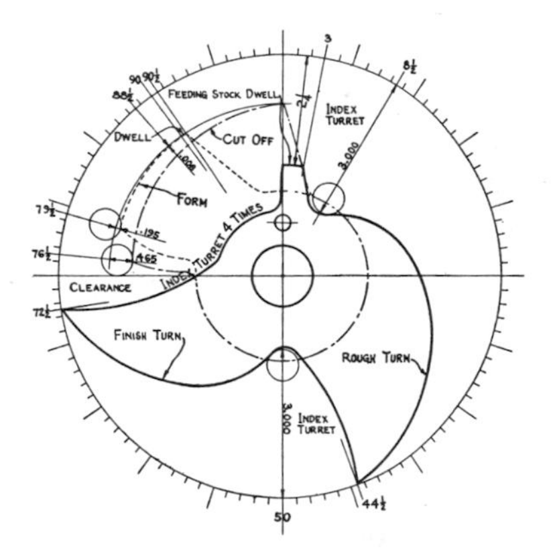


**Fig. 14.31.** Various cams required to make a part automatically on a Brown and Sharpe automatic screw machine.

Although single-spindle screw machines are generally associated with small parts, the bar capacity through the headstock may range from 5/16 in. on a No. 00 Brown and Sharpe machine to 2 in. on a No. 6 machine. The tolerance that can be maintained ranges from 0.0005 in. on small diameters to 0.002 in. on the larger diameters. This machine provides for rapid production of a wide variety of identical parts from rod, bar, or tube stock (Fig. 14.32).

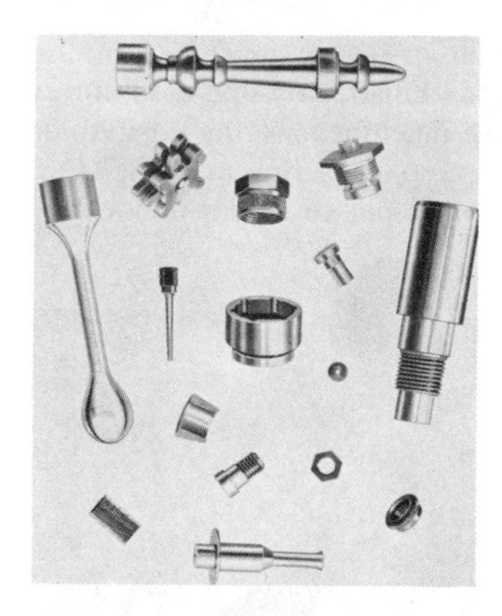

**Fig. 14.32.** Types of work performed on an automatic screw machine. Courtesy Brown & Sharpe Mfg. Co.

AUTOMATIC SCREW MACHINE ATTACHMENTS. Many attachments are available for automatic screw machines that can extend its usefulness and eliminate second machining operations. Some of them are for screw slotting, burring, nut tapping, rear-end threading, and cross drilling. Most of these attachments operate automatically at the same time that another piece is being produced in the machine spindle, thus reducing overall production time.

SWISS-TYPE SCREW MACHINE. The Swiss-type screw machine has five radially mounted tools (Fig. 14.33) that are cam controlled. The single-point tools mounted at each of these five stations operate close to a carbide-lined guide bushing that supports the work. By coordinating the forward movement of the stock with the single-point cam-controlled tool, the desired shape may be generated. Accuracy can be maintained, on small diameters, to 0.0005 in. The part shown in the figure indicates the variety of operations that can be done on this type of machine.

Attachments may be mounted in front of the spindle for internal operations. American-made machines of this kind range in size from a 3/32- to 1/2-in. maximum bar capacity; Swiss makes run as large as 1 1/4-in.

**Multispindle Automatic Screw Machines.** The principle of the tooling arrangement of the single-spindle automatic has been broadened to include four, six, eight, and nine spindles, with corresponding cross slides (Fig. 14.34), to make a high-production lathe. This lathe may be of either the bar or the chucking type. In the bar-type machine, the stock moves up by cam arrangement; in the chucking machine, the stock may be either hand or magazine loaded.

The tooling needed for nine operations performed automatically in 3 sec on a carburetor adjusting screw is shown in Fig. 14.35. An example

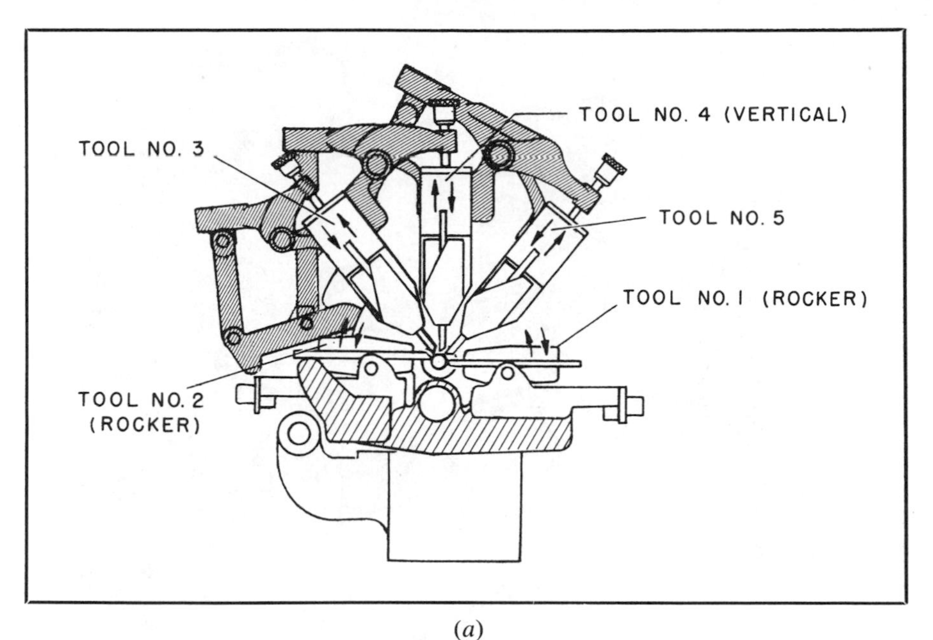


(*a*)

(*b*)

**Fig. 14.33.** The Swiss automatic screw machine. (*a*) shows the relative tool locations, and the arrows indicate the movement. Also shown (*b*) are typical parts produced.

**Fig. 14.34.** Multispindle automatic bar machine. Courtesy The National Acme Co.

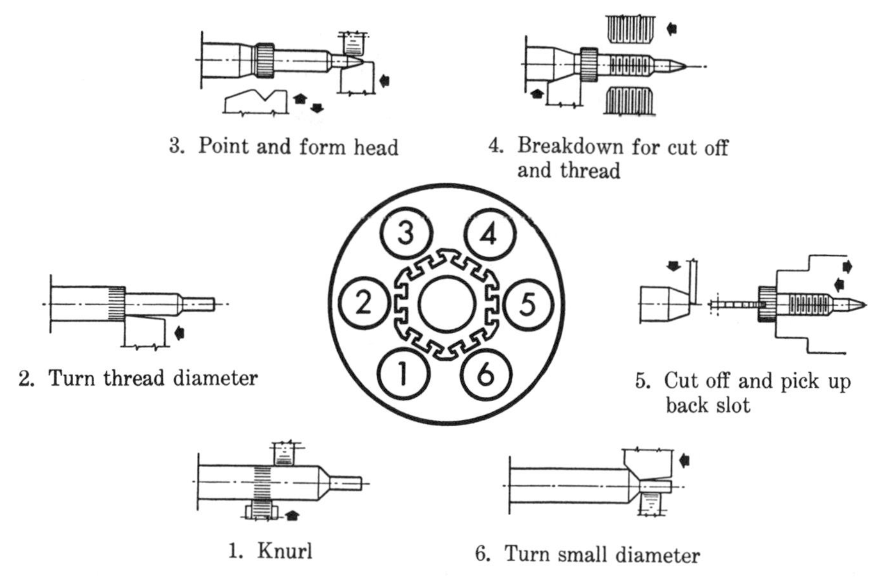


**Fig. 14.35.** Tooling for six-spindle automatic lathe used to complete a part in nine operations. Courtesy The National Acme Co.

of automatic magazine-load chucking work is shown as a tooling layout in Fig. 14.36.

Typical bar production parts (Fig. 14.37) give some idea of the wide variety of work performed on these versatile machines. Note that many of the parts have milling, cross drilling, and slotting operations. This covers only a few of the attachments that may be used to avoid second-step machining operations. Shown in Table 14.1 are some operations that can be performed on automatic screw machines and the tolerances that can be expected.

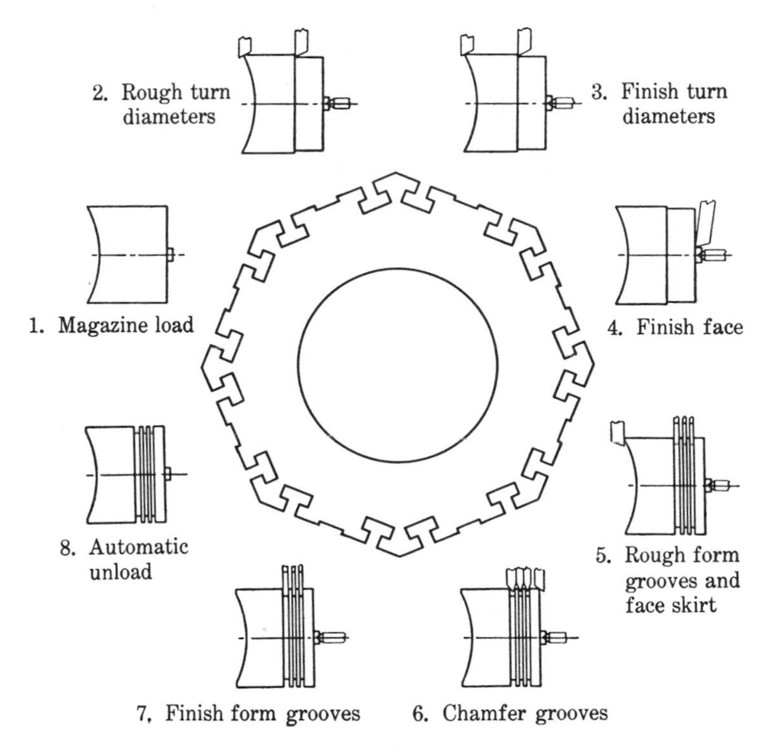


**Fig. 14.36.** An example of tooling used on a multispindle automatic chucking-type machine. Castings are conveyed to the magazine, which loads the machine. After machining (16 operations in 8 sec) is completed, the piston is automatically unloaded, and the castings are carried on to assembly. Courtesy The National Acme Co.

## MACHINE TIME

The time required to make a cut on a lathe is based on the speed in revolutions per minute and the feed used. Use the following procedure:

(1) Find the cutting speed of the material for the type of tool used.

**Table 14.1**
**Automatic Screw Machine Operations**
C = common category; S = special category. Courtesy, *Machine Design*

| Operation and Category | Special Fixtures | Commercial Tolerance in. | Commercial Finish $\mu$in. rms | Remarks and Limitations |
|---|---|---|---|---|
| INTERNAL MACHINING OPERATIONS | | | | |
| Drilling C | No | ±0.003 | 63 to 125 | Maximum practical depth, eight diameters. Over five diameters in brass and over four in steel requires pullout of drill. Same general limitations apply to multidiameter drilling. |
| Reaming C | No | ±0.001 | 32 to 63 | Maximum practical depth, 8 dia. |
| Cross Drilling S<br>Cross Reaming S | Yes<br>Yes | ±0.003<br>±0.001 | 63 to 125<br>32 to 63 | Requires special, separately driven attachment, usually warranted only by long runs. |
| Broaching S | Yes | ±0.006 | 32 to 63 | Not a common operation. Requires special tools. Most adaptable on single-spindle machines. |
| Counterboring C | No | ±0.004 | 63 to 125 | Performed on all machines. Operation is special for depth over 4 dia. |
| Recessing C | No | ±0.010 | 63 to 125 | Standard on all machines. |

| | | | | |
|---|---|---|---|---|
| Tapping C | Sometimes | Class 2 | —— | Both RH and LH tapping standard on all machines. Blind holes require 1/32 in. or more untapped length at bottom. |
| Boring C | No | ±0.001 | 63 to 125 | A standard operation. Hole depth over four diameters requires special attention. |
| Chamfering C | No | As required | 63 to 125 | Standard operation on all machines. No limitations, except back-end chamfering usually requires secondary, transfer operation. |
| Back-End Drilling C | Yes | ±0.003 | 63 to 125 | A common operation, but requires special attachment. Limited to small holes on light parts. |
| EXTERNAL MACHINING OPERATIONS | | | | |
| Turning | | | | |
| Rough C | No | ±0.005 | 63 to 125 | Standard on all machines. Only limitation is machine capacity. |
| Finish C | No | ±0.005 to 0.002 | 16 to 63 | |
| Form turning | | | | |
| Rough C | No | ±0.005 | 63 to 125 | Requires special cutting tool. Length limitation approximately four times smallest diameter. |
| Finish C | No | ±0.001 to 0.002 | 16 to 63 | |
| Roller Shaving C | No | ±0.001 | 16 to 63 | Requires special cutting tool. Length limitation approximately four times smallest diameter. |

**Fig. 14.37.** Typical parts for automatic multispindle bar machine. Courtesy The National Acme Co.

(2) Use this formula for finding the speed:

$$\text{rpm} = \frac{S \times 12}{D \times \pi} \qquad \text{or, simplified,} \qquad \text{rpm} = \frac{S \times 4}{D}$$

where $S$ = surface feet per minute, or cutting speed; $D$ = diameter of the material being turned; and $\pi = 3.1416$.

(3) Select a feed rate. This is based on the type of cut made and on experience. For average work, it may range from 0.001 in. for the feed of a cutoff tool to 0.020 in. for a roughing cut.

(4) After the speed is known and the feed rate is established, calculate the time for machining.

$$\text{Time in minutes} = \frac{\text{length of cut}}{\text{feed in thousandths} \times \text{rpm}}.$$

## QUESTIONS

**14.1.** Name and tell the function of the five main parts of the lathe.

**14.2.** Compare the uses of each of the chucks used on an engine lathe.

**14.3.** What is the purpose of a mandrel?

**14.4.** To what accuracy can you expect to work to on an engine lathe?

**14.5.** What is the big advantage of a tracer lathe?

**14.6.** What are some of the main things you would look for in evaluating an engine lathe?

**14.7.** For what type of work is the automatic cross-slide lathe particularly well adapted?

**14.8.** What are the main differences between an engine lathe and a turret lathe?

**14.9.** What is meant by the "break-even point" when considering two different machines for the same job?

**14.10.** What are the basic differences between a ram-type and a saddle-type turret lathe?

**14.11.** What is meant by the terms *multiple tooling* and *combined cuts?*

**14.12.** What are some tools that are distinctly turret lathe tools?

**14.13.** What is a practical drilling depth for an automatic lathe?

**14.14.** Compare the uses of the Swiss-type screw machine with the single spindle automatic (Brown and Sharpe) screw machine.

**14.15.** What finish and accuracy can be expected on external screw machine operations?

**14.16.** What are some unusual or special operations that can be performed on the automatic screw machine?

**14.17.** What is meant by a Class-2 tapping tolerance as shown in Table 14.1?

## PROBLEMS

**14.1.** Find the rpm that would be used in turning an 8-in.-dia piece of 1020 steel 10 in. long with (*a*) a high-speed tool, (*b*) a carbide tool. Assume a cutting speed of 100 fpm for the HSS tool and 300 fpm for carbide.

**14.2.** Find the time required to make one cut over the stock described in (*a*) and (*b*) of Problem 14.1 if the feed used is 0.005 in. per revolution.

**14.3.** What would be the time required to machine 100 shafts of 4130 steel 3 in. in diameter and 8 in. long. A carbide tool is used. The cutting speed is 300 fpm and the feed is 0.010 in. Allow 1 min for center drilling each piece and 2 min for handling.

**14.4.** What would the cost of the previous job be if the material cost is 15¢ per lb? The time required for sawing each piece to length is 15 sec. Allow 1/8 in. of length for each saw cut. The labor and overhead charge is $9 per hr. Steel weighs 0.2816 lb per cu in. Lathe setup time is 10 min.

**14.5.** It takes 20 min to turn out a bolt on an engine lathe. The same bolt can be turned out on a turret lathe in 3 min. The setup time for the turret lathe is 45 min. The engine lathe has to have all tools changed each time a bolt is made, so setup time is included in the time for the part. Labor and overhead rates on the engine lathe = $9.00/hr and $12.00/hr on the turret lathe. The additional tooling cost on the turret lathe is $50.00. How many pieces would have to be made before it would pay to set up on the turret lathe?

**14.6.** Compare the production obtainable between a brazed-type carbide tool (A) and a throw-away insert-type tool (B). Tools A and B are of comparable cost. After 1 hr and 100 pieces, both tools have a 0.030 wear land. It takes 15 min to regrind tool A and 3 min to reset it in the machine. It requires only 30 sec to reindex tool B. One part is produced every 2 min. (*a*) How many more pieces can be turned out using tool B than using tool A in an 8-hr day? (*b*) If labor and overhead are charged at the rate of $12 per hr, what would be the savings with tool B in fulfilling a contract for 1,000 pieces?

**14.7.** What size motor would be required to make the following cuts on a lathe? (*a*) Material, hot-rolled AISI 1020 steel; depth of cut, 0.125 in.; feed, 0.0625 in./rev; cutting speed, 300 fpm.

(*b*) Same as in (*a*) but change the material to austenitic stainless steel of 35 $R_c$ hardness and the cutting speed to 200 fpm.

## REFERENCES

Design Guide, "Screw Machine Parts," *Machine Design,* September 27, 1962.

Selection Guide, "Engine Lathes," *Metalworking,* July 1963.

Selection Guide, "Hand Operated Horizontal Turret Lathes," *Metalworking,* February 1963.

# 15

# HOLE MAKING AND FINISHING

THE MOST COMMON METHOD of originating a hole in metal and many other materials is by drilling. For precise hole location and size, drilling is often followed by boring and reaming operations. Less common, but becoming more competitive to many drilling applications, is hole punching, discussed later in this chapter.

## DRILLING

The present standard drill (Fig. 15.1) has progressed through a long history of design development. There are now many variations, some of which will be discussed in this chapter.

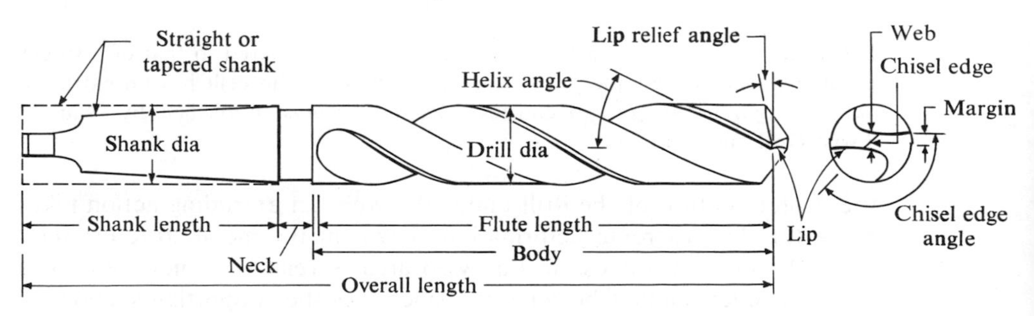


Fig. 15.1. A standard general-purpose drill is the most versatile and useful.

**Theory of Drill Action.** The cutting action of a drill can be compared to that of two tools. The main cutting action is along the lip of the drill, which is similar to that of a conventional single-point tool that is cutting obliquely (Fig. 15.2*a*). The formation of the chip is shown at (*b*) and, in

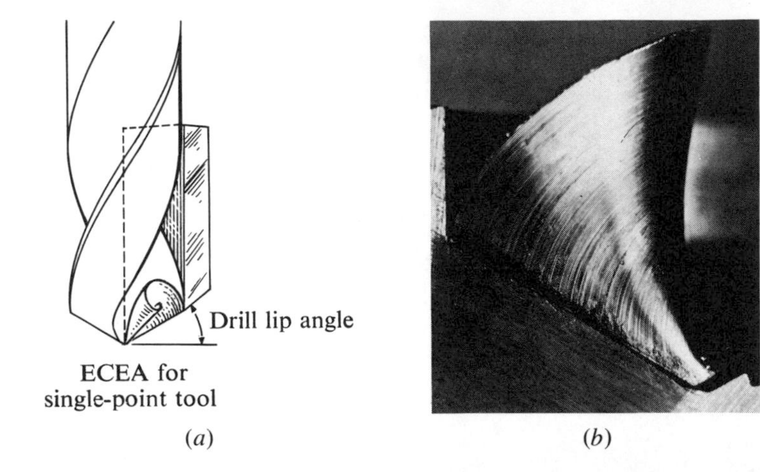


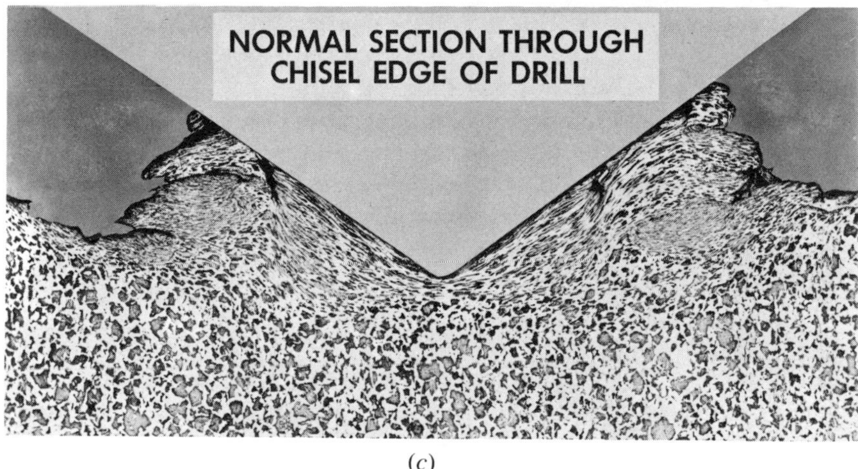


**Fig. 15.2.** (*a*) The cutting lip action of a drill is compared to that of a single-point tool cutting obliquely. (*b*) The web action of the drill has an extruding action. Figures (*b*) and (*c*) courtesy *Metal Cuttings*, published by National Twist Drill and Tool Co.

the center section of the drill under the web, an extruding action takes place (*c*). The extreme deformation shown in the metal structure indicates the cutting process in this web area is relatively inefficient. The web, however, cannot be removed since it is the supporting section for the two main cutting edges. A study of the drill point contact area and wear show that relatively small portions are involved, the leading edge of the lip and the web (Fig. 15.3). The shape of the inactive areas is not critical other than that they should not interfere with the cutting action but form good support for it.

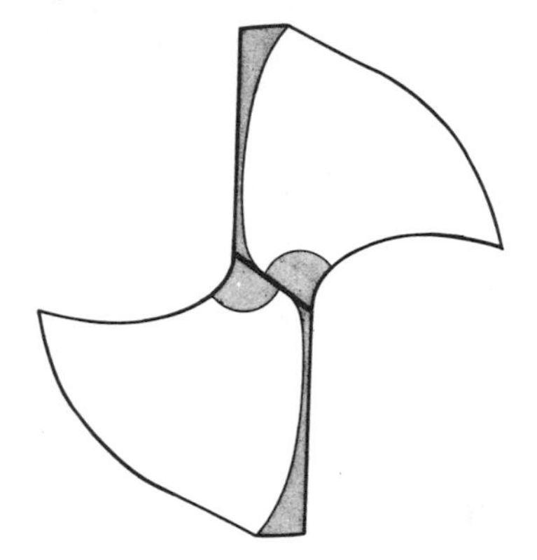

**Fig. 15.3.** Drawing of the end of a worn drill with shading to emphasize wear and contact regions. Note that these regions involve a relatively small portion of the total drill point surfaces.

Since the web of a drill does not have efficient cutting action and because of the fact that it becomes thicker toward the shank, periodic thinning is necessary (Fig. 15.4). This is an operation that can be performed free hand by tool grinding personnel or, if many drills have to be ground, with the aid of a fixture to provide accurate centering.

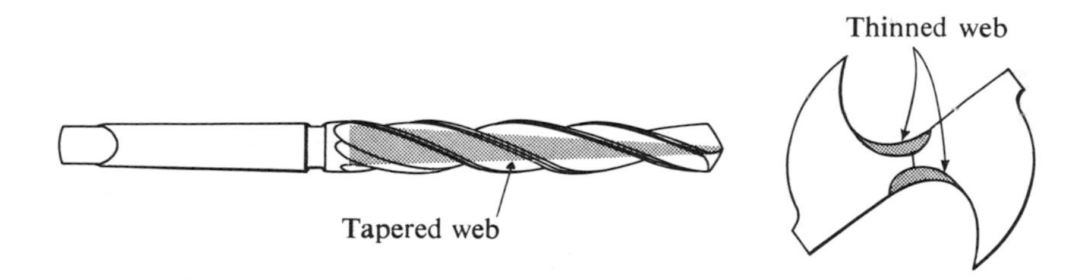


**Fig. 15.4.** Periodic thinning of the web is done to reduce excessive thrust forces especially on heavy web drills and as the drill becomes shorter.

To reduce drill thrust and accompanying machine deflection, split point (or crankshaft) drills have been developed (Fig. 15.5) to deal with the web problems. These drills have conventional cutting action almost to the center of the hole.

**Hole-Locating Methods.** In a simple small quantity production, a layout is first made on the part to be drilled. At the intersection of two layout lines, a prick-punch mark is made, which, if well centered, is enlarged with a center punch. The drill is carefully positioned over the

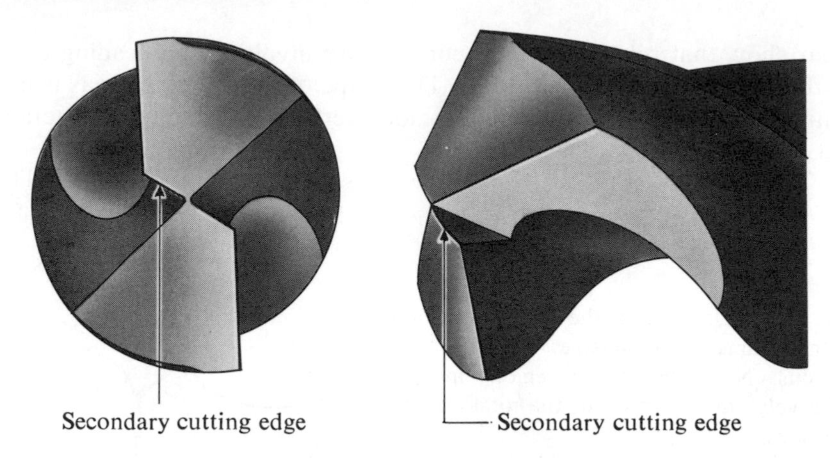


**Fig. 15.5.** The split (crankshaft) drill point. The secondary cutting edges permit conventional cutting (rather than extrusion).

center-punch mark, and the hole is drilled. Although this method is slow, for small quantity production, the tool cost is low.

Large quantity production requires quick and accurate hole location. One method of accomplishing this is by means of a drill jig (Fig. 15.6). The part to be machined is placed in the drill jig against a number of locating pins. Clamping holds the part tightly against these pins to maintain constant location of the holes. Notice that the work does not rest against a flat surface. A well-designed drill jig makes provision for chip clearance, chip removal, and quick and easy loading and unloading of the piece parts. The hardened and ground drill bushing is accurately located in the jig and acts as a guide for the drill. Drill bushings are

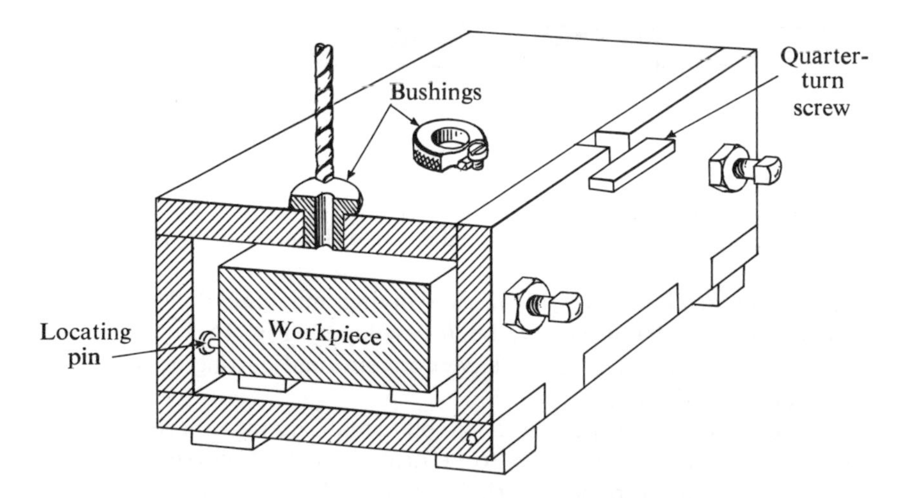


**Fig. 15.6.** The drill jig is used for accurate hole location.

pressed in for a semipermanent arrangement, or they may be of the slip-renewable type held in place by a screw, as shown in Fig. 15.6. A slight turn of the bushing allows the slot on the bushing head to clear the screw and be removed. This type of bushing is used if another operation, such as tapping or reaming, follows drilling.

A newer method of automatic location of the part for each hole to be drilled makes use of punched or magnetic tape. The table of the drill press is made to move along $x-y$ coordinates according to a previously programmed tape (Fig. 15.7). The drill is automatically fed down and retracted after each table movement.

**Fig. 15.7.** On this tape-controlled drill, the table is automatically moved on the x – y coordinates for accurate hole location.

**Variations in Standard Drills.** A change in the helix angle of the flutes makes a difference in the ease with which the chips are removed from a hole and in the rake of the cutting edge or lip of the drill. For materials of low tensile strength that tend to pack and do not climb out easily, such as some aluminum, magnesium, copper, and thermoplastics, a high helix angle is recommended. The low helix is used for materials that clear the hole readily. It minimizes the danger of "hogging" into the work and breaking the drill. (Fig. 15.8).

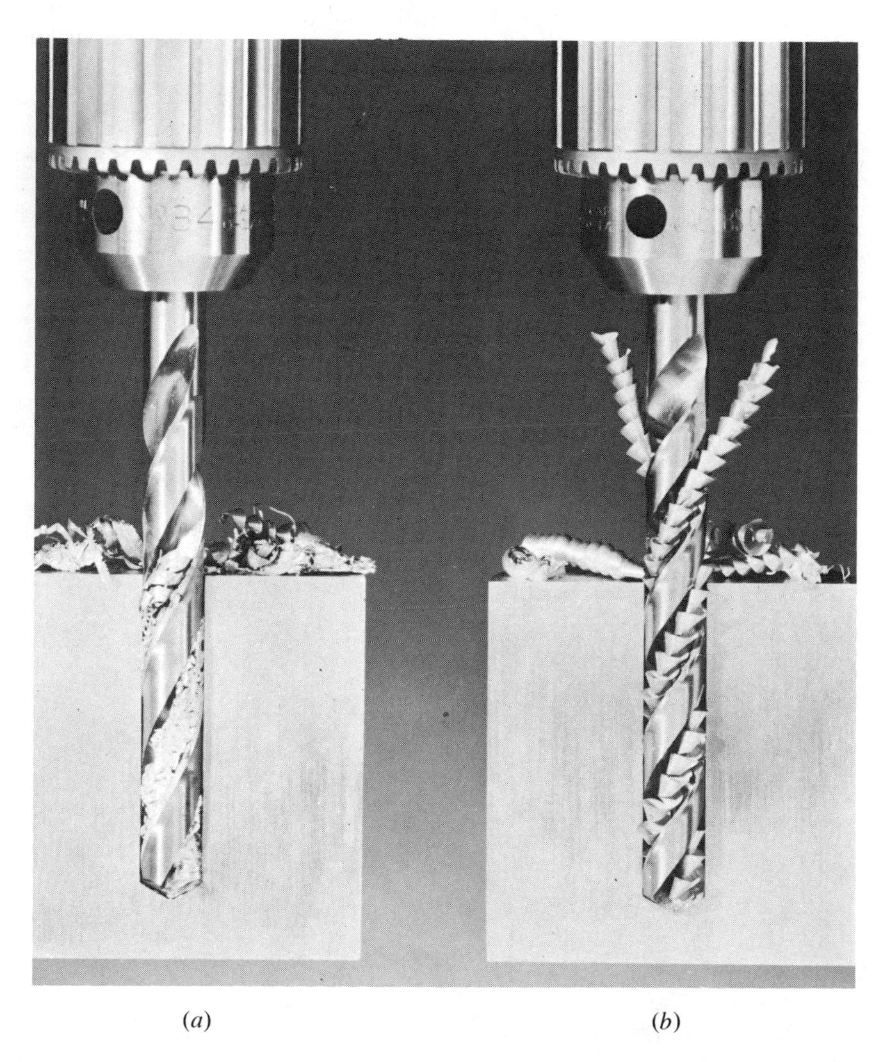

(*a*) (*b*)

**Fig. 15.8.** Difference in chip-removal action between a low-helix angle drill (*a*) used for material that does not climb out easily, and a regular drill (*b*). Courtesy Greenfield Tap & Die.

## SPECIAL DRILLS

**Gun Drills.** Gun drills (Fig. 15.9) differ from conventional drills in that they are usually made with a single flute (Fig. 15.9*a*) or a tubular shape. The hole in the single-flute type provides a passageway for the pressurized cutting fluid, which serves to keep the cutting edge cool as well as flush out the chips. Single-flute gun drills form a wedge-shaped chip that tends to curl and break easily. Because of this chip-generating action, single-flute gun drills are effective in drilling holes in low-carbon steels or other tough, stringy materials where two-flute designs tend to clog.

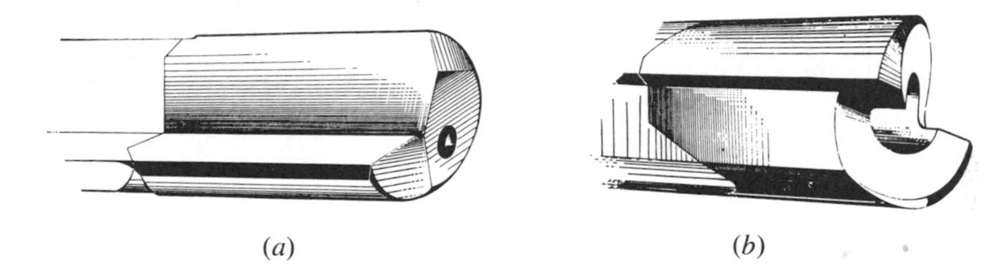


**Fig. 15.9.** (*a*) Single-flute gun drill and (*b*) two-flute pin-cutting gun drill.

The two-flute pin-cutting type (Fig. 15.9*b*) has an off-center coolant through-hole in the molded carbide tip which generates a pin (Fig. 15.10). The two cutting edges provide a balanced cutting action which allows extremely high feed rates and excellent size control.

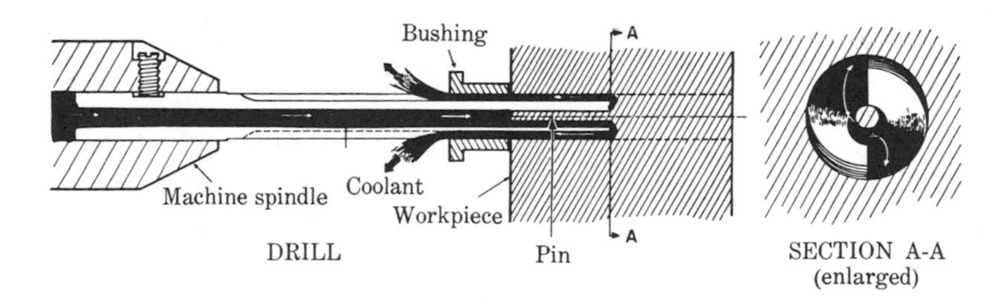


**Fig. 15.10.** Coolant flow and pin generation of a two-flute pin-cutting gun drill. Courtesy Star Cutter Company.

Gun drills are used to produce true holes to depths and tolerances not obtainable with ordinary drills. Depths from 6 dia to 130 in. are possible in a single uninterrupted operation. Even though the drill is made to advance rapidly—from 15 to 60 in. per min—good tolerance and finish are maintained. This is due, in part, to the high (60 to 1,500 psi) cutting-fluid pressure. The efficient metal-cutting operation is due to the thorough cooling of the drill point and to the continuous scavenging of the chips by the high-pressure cutting fluid. A general rule for the maximum depth that may be attained by deep-hole drills is 100 times the drill diameter. Some parts that require deep-hole drilling include gun barrels, camshafts, crankshafts, machine-tool spindles, and connecting rods. Deep-hole drills may develop a runout of as much as 0.0005 in. per in. of depth. The general hole tolerance, however, is 0.002 in. for most materials. Fairly good finishes are obtained, usually in the range of 65 rms.

**Trepanning Drills.** Trepanning is a method of producing holes, ordinarily more than 2 in. in diameter, by cutting a narrow annular ring of material from the work so as to leave a center core or plug. The drill is similar to the two-lip gun drill mentioned previously. A trepanning drill, however, is made for larger holes and has a separate head keyed or threaded to the end of a tube (Fig. 15.11).

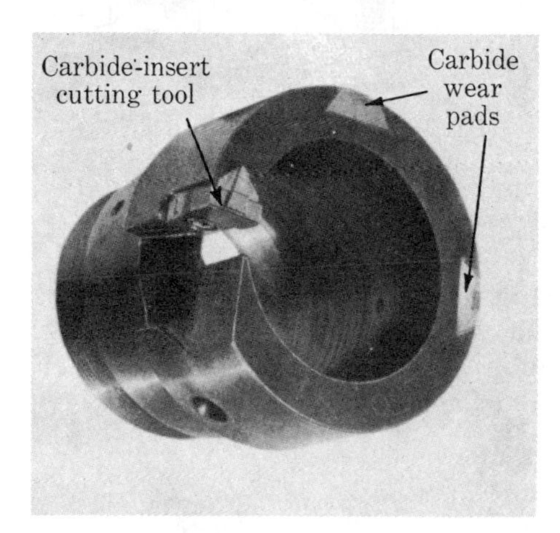


**Fig. 15.11.** A trepanning head used with inserted wear pads. It is threaded on the end for inserting into a drive tube. Courtesy *The Tool and Manufacturing Engineer.*

On large holes, trepanning offers considerable advantage. A 6-in.-dia hole 10 in. deep requires the removal of 282.6 cu in. of metal as chips with a conventional drill. A trepanning drill, using a 3/4-in.-wide cutter, calls for removing only 126.6 cu in. of metal as chips. Normally, the tolerance may be held to ±0.003 in. on the hole diameter.

## DRILL MATERIALS AND COST

**Materials.** Comparatively few metal-cutting drills are now made from low-capacity carbon steel. Production demands have made high-speed steel, carbide-tipped, and solid-carbide drills imperative.

The high speed steels M1, M2, M8, M10, and T1 are most commonly used. These steels combine toughness, edge strength, and wear resistance with good red hardness. When these prove inadequate, as in drilling stainless steel, special high speed steels containing more vanadium and carbon may be used.

Carbide-tipped drills give excellent results in drilling cast iron, highly abrasive materials, and certain nonferrous alloys.

Carbide-tipped and solid-carbide drills do not give good performance unless the setup is extremely rigid, the spindle is precisely aligned, and there is a minimum of tool overhang.

**Cost Comparison.** An example of the cost comparison for three different 3/16-in.-dia drills: HSS, 48¢; carbide-tipped, $4.10; solid-carbide, $11.70. Although initial cost of the three drill materials shows a wide disparity, the higher cost carbide drill may be the right one to lower hole costs on many applications.

**Drilling Tolerances.** In most material the drilled hole will be oversize by 1/3 to 1 percent of the drill diameter. Shown in Fig. 15.12 are summary curves of the normal oversize encountered in drilling cast iron and steel without guide bushings.

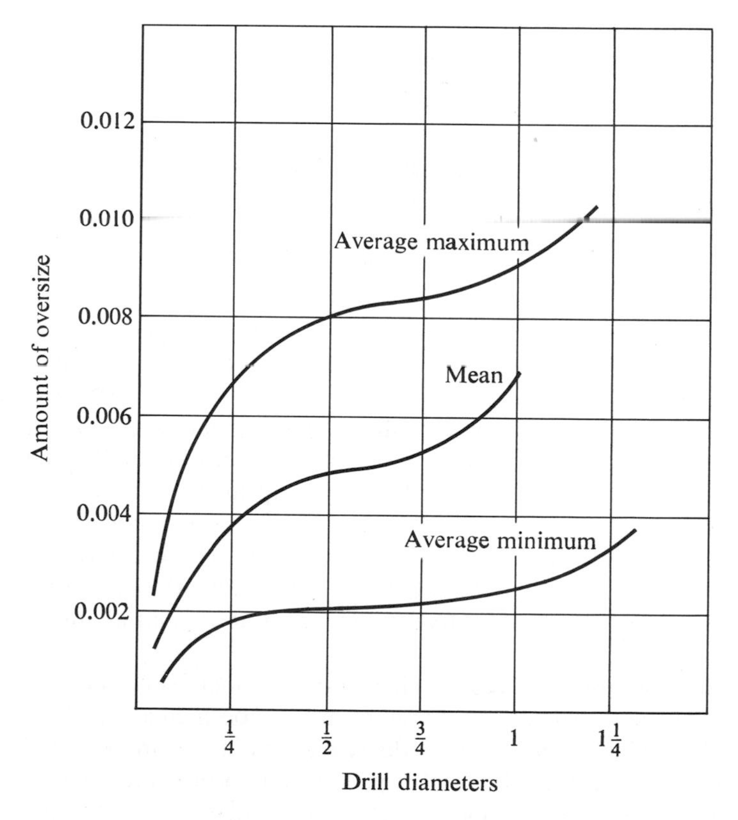


**Fig. 15.12.** Summary of normal oversize encountered in drilling cast iron and steel without the use of guide bushings.

Where accurate size and location are required, holes are commonly drilled and then followed by one or more operations such as reaming, boring, and honing.

## REAMERS

**Types.** Reamers are used to size a hole and give it a good finish. The hand, machine, and shell reamers are shown in Fig. 15.13. Hand reamers are recognized by their square shank, whereas machine (or chucking) reamers have a tapered shank to fit into the machine spindle. The shell

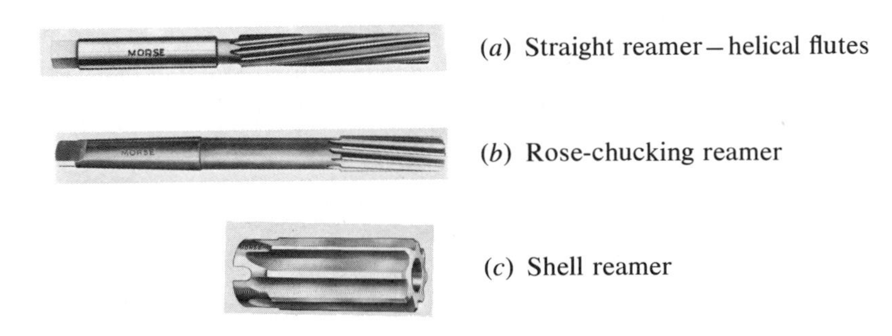


**Fig. 15.13.** Hand, machine and shell reamers. Courtesy Morse Twist Drill & Machine Co.

(or hollow) reamers are used with arbors that have driving lugs. In using shell reamers, the size can be quickly changed, and the cost of the cutting tool is less since only the shell needs to be replaced when dull or damaged. Other common types of reamers are tapered, tapered-pin, and adjustable.

**Material Removal.** To work efficiently, a reamer must have all teeth cutting. This depends on proper stock allowance. For semifinish reaming most metals, 0.015 in. is allowed on holes up to 0.500-in. dia. For holes over 0.500 in. in dia, 0.030-in. allowance is recommended. For finish reaming, the allowance is 0.005 and 0.015 in., respectively. The tolerance expected is normally $\pm 0.0005$ in. Although the size tolerance is good, reaming does not improve the location or angular alignment of the hole.

**Reaming Speeds and Feeds.** High-speed reamers should be run at two thirds to three fourths of the speed of drills. Too high a speed causes the material to cling to the cutting edges, resulting in premature dulling and rough hole diameters. The feed is generally two or three times that used for a drill of corresponding diameter. As with other machining operations, feed is related to the finish desired. Too fine a feed, however, will cause the tool to idle in the cut and produce undue wear.

## COUNTERBORING, COUNTERSINKING, AND SPOT-FACING

Closely associated with the operations of drilling and reaming are those of counterboring, countersinking, and spot-facing. Counterboring is done so that a bolt head will be flush with the surface. Countersinking provides a tapered recess for flathead screws and bolts. Spot-facing is used to smooth the surface around a hole so that the washer and bolt head may have a level surface.

## DRILLING EQUIPMENT

Drilling equipment ranges from the portable hand drill to the large multispindle machines able to drive hundreds of drills at one time. The main types are sensitive, upright, radial, gang, turret, multispindle, special drilling and tapping machines, self-contained drilling heads, and deep-hole drilling equipment.

**Sensitive Drill Press.** The most common type of drill press (Fig. 15.14) is known as the sensitive, or hand-feed, type. The operator is sensitive to the rate at which the drill is cutting and can regulate it

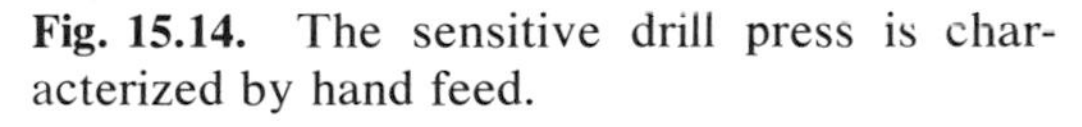

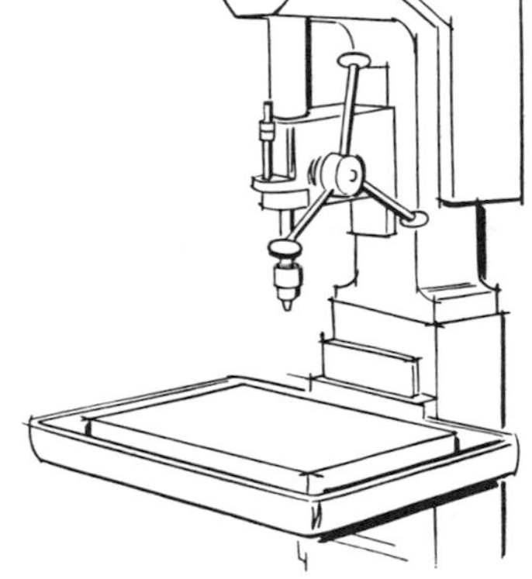

**Fig. 15.14.** The sensitive drill press is characterized by hand feed.

according to the cutting conditions. The careful operator will watch for excess drill deflection and correct it before the drill breaks or inaccuracies result. Machines of this type have a speed range from 300 to 30,000 rpm. The size is designated by twice the distance from the centerline of the spindle to the column. Thus, a 16-in. drill press would be able to drill to the center of an 8-in. plate.

The work is usually held on the drill press table by means of a vise. If the work is large and the holes to be made are also large, it is necessary to bolt the work to the table, making use of the T-slots and holes provided. Drill jigs and fixtures are used for accurate hole placement when the work is of sufficient quantity to warrant the cost of such tooling.

**Upright Drill Press.** The upright drill press is quite similar to the sensitive drill press in appearance. It is equipped with power feed and a speed-change gearbox. The feeds may range from 0.004 to 0.025 in. per revolution. This machine is of heavier construction and is better suited to a wide range of jobs than is the sensitive type. A universal table may be used in place of the standard T-slot table, providing accurate lateral and longitudinal movement.

**Radial Drill Press.** Work that is large and requires several holes is difficult to position for each operation. The radial drill press (Fig. 15.15) is equipped with a radial arm that can be swung through an arc of 180

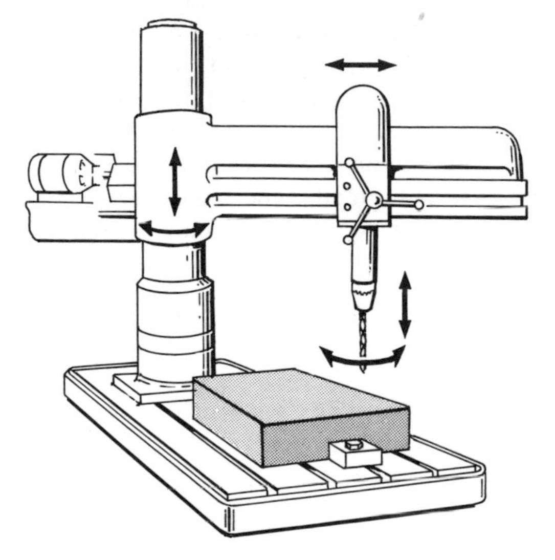

**Fig. 15.15.** Components and movements of a radial drill press.

deg or more. On the radial arm, which is power driven for vertical movement, is an independently driven drilling head equipped with power feed. The drilling head may be moved along the arm by hand or power on a gear and rack arrangement. To drill a hole, the following procedure is used: The arm is raised or lowered as needed, the drill head is positioned and locked on the arm, the column is locked, the spindle speed and feed are adjusted, and the depth is set. The drill will then feed down and retract when the proper depth has been reached. The arm and column may then be unlocked and the drilling head moved to a new position without disturbing the work.

Universal radial drills allow the radial arm to be rotated on a horizontal axis, providing for angular-hole drilling.

The size of a radial drill is designated by the radius, in feet, of the largest plate in which a center hole can be drilled, and the diameter of the supporting column, in inches. Sizes range from 3 to 12 ft for arm capacity and 4 to 34 in. column diameters.

**Gang Drill Press.** The gang drill press is made up of a number of upright drill presses placed side by side, with a common table and base (Fig. 15.16). The parts being drilled, reamed, counterbored, tapped, etc., can be easily transferred from one spindle to the next with no lost time for tool changing.

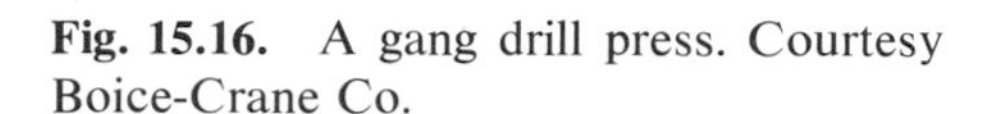

**Fig. 15.16.** A gang drill press. Courtesy Boice-Crane Co.

**Turret-Type Drill Presses.** When a number of drill sizes or other tools are needed to complete a part, the turret-type drill is very useful. Any one of the six, eight, or ten tools can be quickly indexed into place and used (Fig. 15.17). The turret drill should be used in combination with a rapid or automatic positioning table.

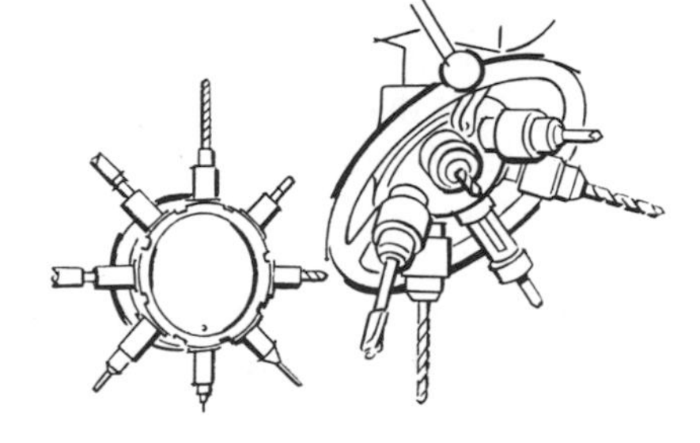

**Fig. 15.17.** Turret type drilling machines have six, eight, or ten tools ready to use.

**Multispindle Drilling Heads.** Multispindle drilling heads can be attached to a single-spindle machine to provide a means of drilling several holes at one time. These heads are of two main types—adjustable, for intermediate production, and fixed, for long high-production runs.

The adjustable head (Fig. 15.18*a*) is driven by a gear from the spindle through a universal-joint linkage to each drill. This arrangement allows maximum variation in drill placement with minimum center distances ranging from 5 to 12 in. As many as 15 drills can be held at one time in a head of this type. Previously drilled and bored templates are used to position the drills for accurate center distances.

Oscillator heads (Fig. 15.18*b*) permit drills to be located anywhere within the head area with drills as close as the sum of the two drill diameters. Drills can vary in size from 1/64 in. to 1 1/2 in. in diameter.

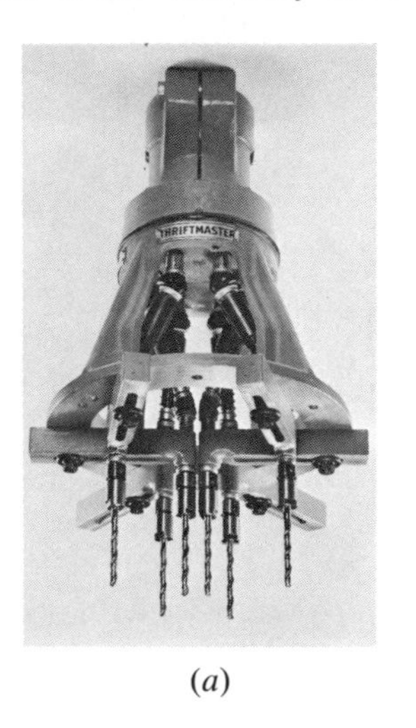


(*a*)

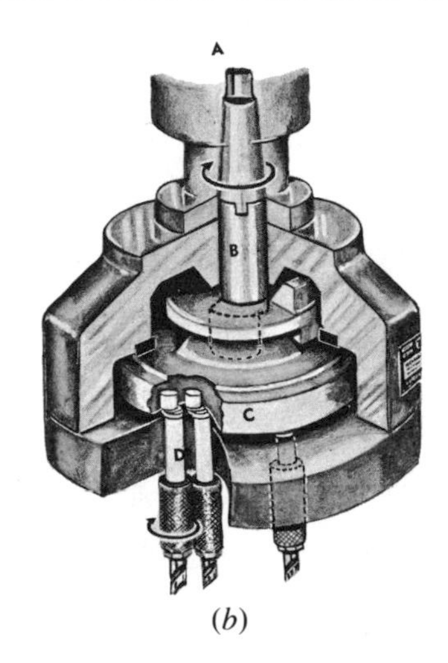


(*b*)

**Fig. 15.18.** Two types of adjustable multiple-spindle drilling heads. Spindle (*a*) has adjustable arms driven by universal joints from gears in the head. Spindle (*b*) transmits power through the drill sleeve (A) which turns the drive crank (B) and oscillator (C). The oscillator turns the individual drill spindles (D) in the same direction and at the same speed as the drive crank.

Fixed-head drills are engineered for a specific job. When the production run for that pattern is over, the heads are torn down and rebuilt to suit the new requirements. Each job requires a jig-bored bushing plate and gearing or wobble plates to suit.

Whether a fixed- or an adjustable-type head is used is determined by the length of the run or the cutting load. Adjustable heads are, theoretically, subject to more wear and may require more maintenance. One approach to the problem of selecting the proper type of production drilling equipment is to temporarily substitute an adjustable head for individual or gang drilling. If it proves satisfactory, the more expensive fixed-type head can be made up.

**Special Drilling and Tapping Machines.** The next step beyond the multispindle head for higher production is the multispindle automatic indexing machine built to suit the customer. Automatic indexing of a piece part is usually done with a rotating table that indexes in either the vertical or the horizontal position. Individual motor-driven spindles equipped with automatic feed are positioned to work on parts as they are indexed.

**Self-Contained Drilling Units.** Self-contained units consist of a power source, a feeding arrangement, and a chuck. These units come in many sizes from small air powered to large electric drive motor type. The feed may be by cam, screw, or by hydraulic or air pressure. Units can be put together in any combination, often using standard type stands and slides. A schematic arrangement of self-contained multiple-spindle units is shown in Fig. 15.19.

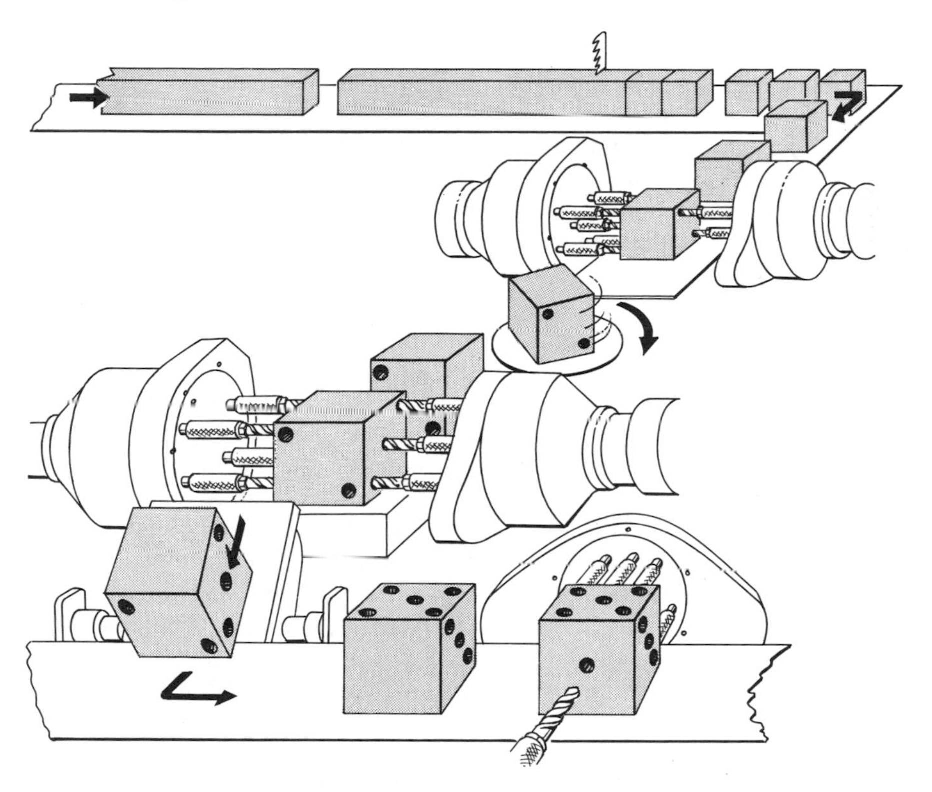

**Fig. 15.19.** Schematic representation of how independently powered drilling heads can be placed along inline and rotary indexing stations to make for complete automation. Courtesy Zatar Inc.

## DRILLING SPEEDS AND FEEDS

**Speeds.** The sfpm for drilling should be less, about 60 to 80 percent of that required for turning. There are several reasons for this reduced speed. (*a*) The lip of the drill is a relatively small mass of material and cannot dissipate much heat. (*b*) The drill point is in constant contact with the material being cut in a very confined area. It is difficult to do an effective job of cooling the tip unless cooling is done through the drill itself as in gun drills and oil hole drills that provide a passage for this purpose. (*c*) The tool regrinding time is longer than for single-point tools.

**Feeds.** Feeds are dependent on what the drill will take without deflection. A general guide is 0.001 to 0.002 in./rev for drills for 1/8 in. to 1/4 in.; 0.004 to 0.007 in. for drills 1/4 in. to 1/2 in.; 0.007 to 0.015 in. for drills 1/2 in. to 1 in.; and 0.015 to 0.025 in. for drills larger than 1 in.

Carbide drills are run at approximately the same speed as HSS drills but at higher feeds.

## DRILLING ECONOMICS

**Drill jigs.** As mentioned at the beginning of the chapter, the hand layout method is slow but suitable when the production volume is low. How much production is necessary before money can be invested in a drill jig? The answer can be calculated with the aid of the following formula:

$$N = \frac{C(I + T + D + M) + S}{s(1 + p)}$$

where $N$ = number of parts or pieces required to pay for the jig; $C$ = first cost of jig; $I$ = annual interest rate, percent, on investment; $T$ = annual taxes, percent; $D$ = annual depreciation, percent; $M$ = annual maintenance allowance, percent; $S$ = setup cost, yearly; $s$ = savings in labor cost/unit; and $p$ = percentage of overhead applied on labor saved. *Note: I, T, D,* and *M* are sometimes added together for a total of 70 percent. If a jig must pay for itself in a single run then $D$ = 100 percent, and the sum of $I$, $T$, $D$, and $M$ = 120 percent. $1$ = constant.

**Drilling Time.** Drilling time can be calculated by first finding the rpm.

$$\text{rpm} = (S \times 4)/D$$

where $S$ = sfpm; 4 = approximation for 12 in./$\pi$; $D$ = drill dia in inches.

$$T = \frac{L + A + O}{\text{rpm} \times f}$$

where $T$ = time in minutes to drill the hole; $L$ = length of the hole; $A$ = approach; $O$ is the overtravel; and $f$ = feed in in./rev.

*Note:* The approach is usually considered as the amount the drill travels before it is cutting at full diameter. The overtravel is the amount of travel necessary to clear the point of the drill or for the body of the drill to go completely through the hole. An approximation for both together is about 1/2 the dia. of the drill.

**Power Requirements.** One might expect that the torque required for drilling would be proportional to the amount of metal removed per revolution, which would be $(\pi/4)fd^2$, where $f$ = feed, ipr, and $d$ = drill dia. It has been found that the drill torque is not proportional to the volume removed/rev; but $(\pi/4)f^{0.8}d^{1.8}$. This means that doubling the feed/rev will increase the torque by a factor of 1 3/4 instead of by 2. Similarly, doubling the drill diameter while maintaining a given feed will increase the torque by a factor of 3 1/2 instead of by 4. This is related to the fact that metals in very small sections are appreciably stronger than in large sections.

The horsepower requirements for drilling and reaming can be quite critical at times, especially when multiplespindle drilling heads are used. In order to insure efficient cutting action, the horsepower can be calculated through the use of a formula developed by Shaw and Oxford.

$$M = 23{,}200f^{0.8}d^{1.8}\left[\frac{1 - (c/d)^2}{1 + (c/d)^{0.2}} + 3.2(c/d)^{1.8}\right]$$

where $M$ = torque, in./lb; $d$ = drill dia, in.; $c$ = chisel edge length, in. (approximately 1.15 times the web thickness for normal sharpening); and $f$ = feed/rev, in.

For drill of normal proportions the ratio $c/d$ can be set equal to 0.18 and the equation simplified to

$$M = 25{,}200f^{0.18}d^{1.8}.$$

Thrust forces are important when determining drill, jig, and machine rigidity but are only 2 percent or less than the total power requirement and so can be disregarded.

Thus the horsepower required per drill is computed from the torque as follows:

$$\text{hp} = MN/63{,}025$$

where hp = horsepower; $N$ = rpm; $M$ = tool torque in in./lb. At least 25 percent should be added to the hp formula to allow for tools becoming dull.

If torque measurements are not available, the horsepower can be calculated from the cubic inch removal rate as follows:

$$\text{hp}_{\text{m}} = (\pi D_d^2/4) \times f \times \text{rpm} \times P_u$$

where $D_d$ = drill dia. in inches; $f$ = feed rate in inches; $P_u$ = unit power (see Table 13.1).

## BORING

James Watt and others tried to produce more efficient steam engines but were unsuccessful until John Wilkinson improved a boring machine to such an extent that it could make large bores (48 to 72 in. in diameter) with a degree of accuracy described as "true within the thickness of an old shilling." Up until 1774, there had not been much demand for accurately bored holes. Early cannons were made of bronze with carefully cored cylinders which were gradually cleaned out. Accuracy was not considered the greatest factor, since the cannonballs were used in the "as-cast" condition. After seeing James Watt's early attempts to build a steam engine, an English civil engineer by the name of John Smeaton reported to the Society of Engineers, "Neither the tools nor the workmen exist that can build so complex a machine with sufficient precision."

Fortunately, Wilkinson, Watt, Smeaton, and the others continued to design and build machines of increasing accuracy. Today's boring machines are able to work with an accuracy of a few ten thousandths of an inch on production work.

Boring is used to enlarge and locate accurately a previously drilled or cored hole. Drills tend to wander or drift, making hole placement inaccurate. Also, if the lips of the drill are not equal, the hole will be oversized. When accurately positioned holes are needed, they are first drilled and then bored. Boring a previously drilled hole has the advantage of accurate location, since the tool does not follow the hole but bores on its own center or axis (Fig. 15.20).

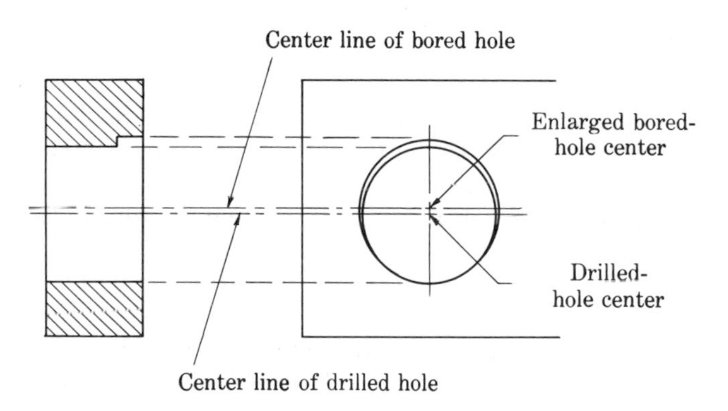


**Fig. 15.20.** A boring tool is used to correct the location of a drilled hole.

Boring tools are single-point, high-speed, carbide or ceramic tools held in a supported or nonsupported bar. The single-point tool shown in Fig. 15.21 is mounted in a nonsupported bar and is of the adjustable type. The vernier adjustment allows the tool to be set to within 0.0001 in.

## BORING MACHINES

The main types of boring machines are the horizontal and vertical mills and the precision and jig boring machines. In addition to these, boring operations are often performed on a lathe and milling machine.

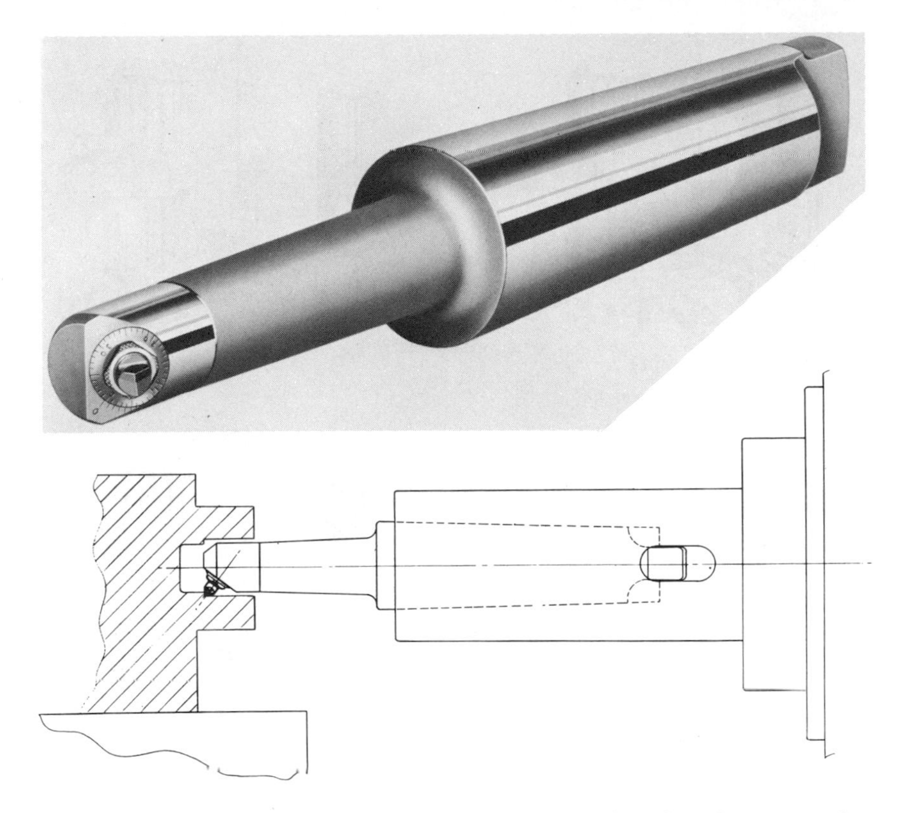

**Fig. 15.21.** Boring bar with vernier adjustment on tool setting. Courtesy Microbore.

**Horizontal Boring Mill.** The horizontal boring mill, shown in Fig. 15.22*a*, is a versatile machine used for milling, drilling, and boring. It is especially well adapted for in-line boring, as shown in the figure. The boring bar can be supported either in a fixture or, on long work, in the support column at the end of the machine. The boring-bar extension is removed when closeup operations of milling and drilling are performed, as shown in Fig. 15.22*b* and *c*. Mills of this type are used extensively in machining diesel-engine bearing seats, cylinder bores, and block surfaces. The size of the horizontal boring mill is designated by the diameter of the spindle, in inches, which ranges from 3 to 11 in.

**Vertical Boring Mill.** The vertical boring mill (Fig. 15.23) is designed to handle large outside turning and inside boring applications.

Large parts are rather difficult to mount in a lathe chuck for machining. With the vertical boring mill, the task is simplified since the table becomes the chuck, and the weight of the part is not critical. The parts to be machined are put on the faceplate table surface and centered according to the bore or outside diameter. T-slot bolts and clamps are

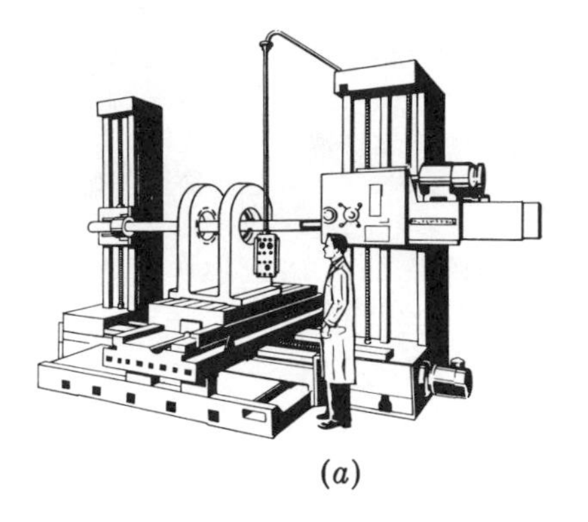

(*a*)

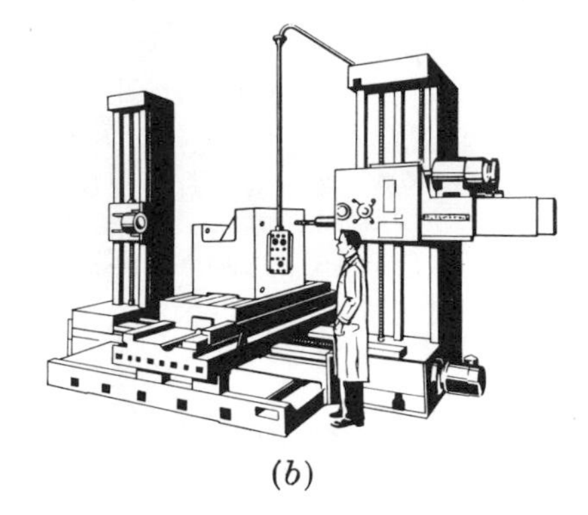

(*b*)

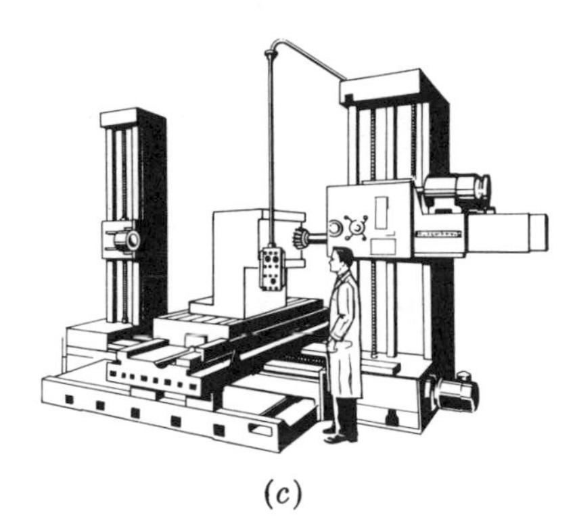

(*c*)

**Fig. 15.22.** Basic operations performed on the horizontal boring mill. (*a*) Boring mill used for incline boring with a supported spindle, (*b*) used for drilling, and (*c*) used for milling. Courtesy Giddings & Lewis Machine Tool Co.

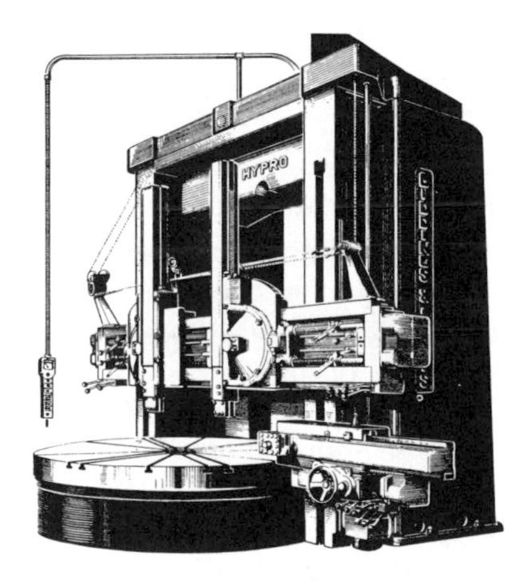


**Fig. 15.23.** Vertical boring and turning mill. Courtesy Giddings & Lewis Machine Tool Co.

then used to hold the part in place for machining. Cutting is done from either of the two vertical heads or from the horizontal cross slide. Angular cuts up to 30 deg can be made on one of the vertical slides. These machines may be equipped with tracer controls operating from a template to speed production of the contour cuts. The size of the vertical boring mill is designated by the diameter of the table, which ranges from 54 in. to 40 ft.

**Precision Boring Machines.** Precision boring machines have comparatively small high-speed spindles (Fig. 15.24). These machines vary in design, depending upon the way the spindles are mounted and the number of them. Basically, all the machines are equipped with a number of fixed spindles and a reciprocating table.

**Fig. 15.24.** Precision boring machine with part ready to be bored. Courtesy The Heald Machine Co.

**Jig Boring Machines.** Because of its accuracy and versatility, the jig boring machine (Fig. 15.25) is generally considered a toolroom rather than a production-line piece of equipment. Whereas, the average milling machine is capable of an 0.002-in. accuracy, the jig boring machine works to accuracies of 0.0001 in. or less.

As the name implies, this machine is often used for making drill jigs, as shown in various steps in Fig. 15.26. Drill jigs and templates are usually made at one tenth the tolerance allowed on the parts they will be

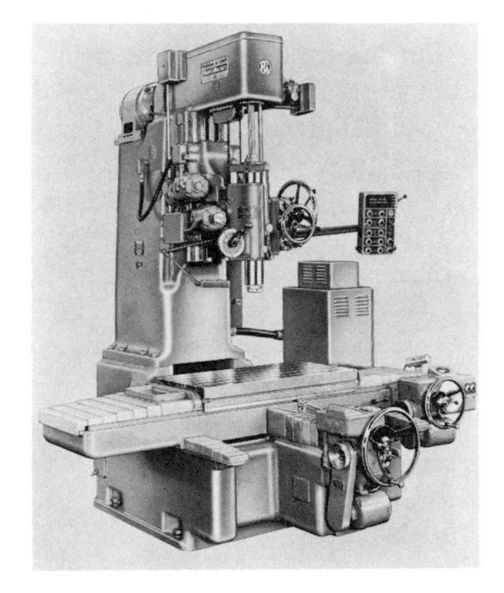

**Fig. 15.25.** A size-2 jig boring machine, table size 22 by 44 in. Courtesy Pratt & Whitney Co., Inc.

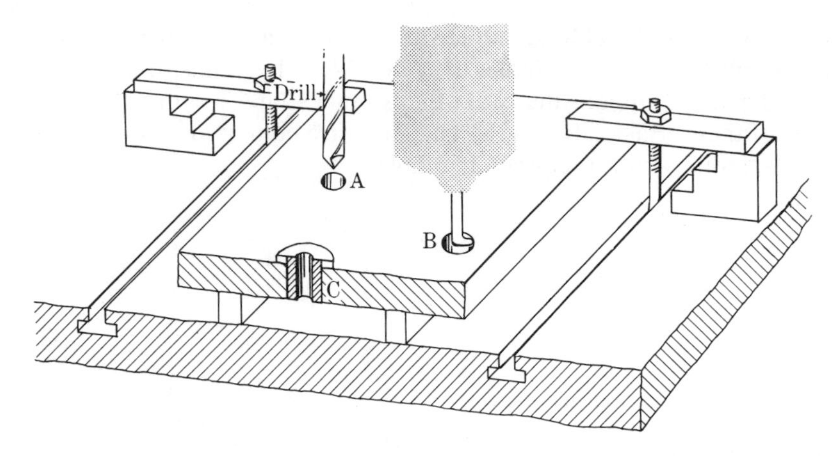


**Fig. 15.26.** Drill jig, in sequence of construction and use. A drilled hole, B drilled and bored hole, C drilled hole with hardened bushing pressed in place.

used to produce. This, plus the fact the jigs and templates will control the accuracy of thousands of parts, accounts for the painstaking care with which they are machined. Jig boring machines are used whenever close tolerances are required in drilling and boring operations.

Several methods are used to obtain accurate table settings. One system employs precision-length rods supplemented with vernier micrometers and dial indicators. Another system uses an electromagnetic head with a magnetic center, together with a master bar. Still another system makes use of a photoelectric optical centering device which guarantees repeat table movement within 0.00015 in.

## HOLE PUNCHING

Punching holes in other than sheet metal has not in the past been competitive to drilling since the equipment needed had neither the capacity nor the versatility required. Now, however, with heavy-duty portable hydraulic punches, these conditions have been rectified. Shown in Fig. 15.27 are two hydraulic punch models of this type capable of

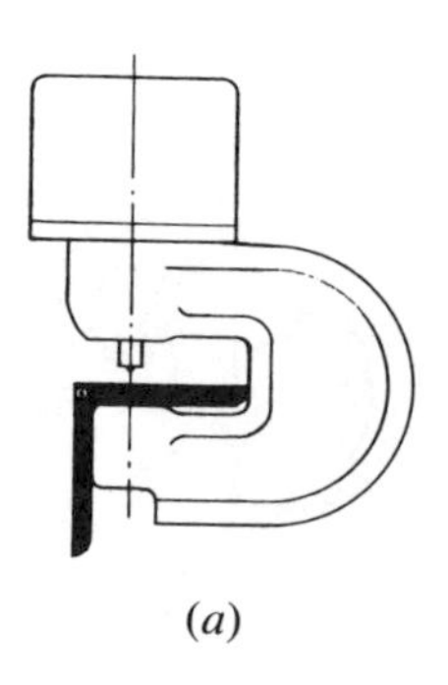

(*a*)

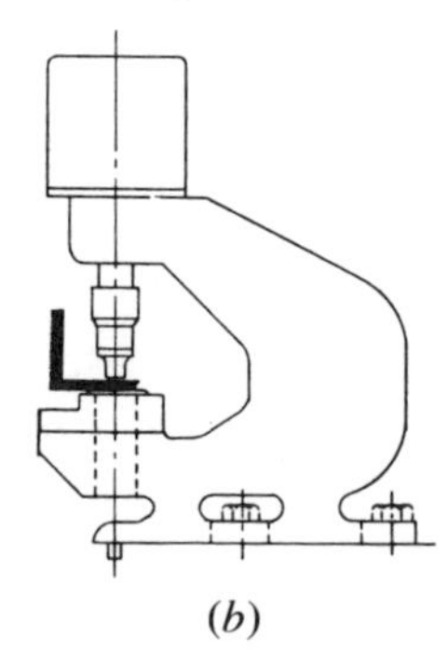

(*b*)

**Fig. 15.27.** Two types of hydraulic punching units. Some are designed to be used as portable punching units (*a*), while others are bolted to rails (*b*) for gang punching units.

handling stock from 12-gage up through 1 1/16 in. thick and up to 1 1/16 in. dia holes. Models range in size from 6 to 96 ton capacity.

In addition to being very portable, punches of this type can easily produce a wide variety of hole configurations (Fig. 15.28). Also the units

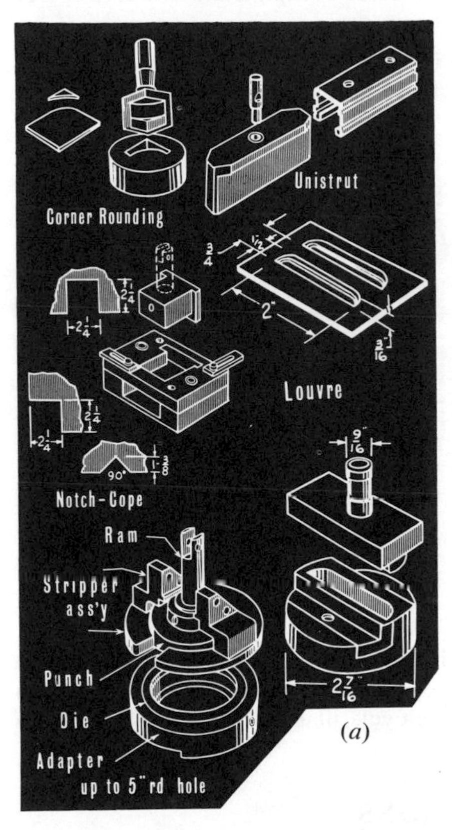


(*a*)

**Fig. 15.28.** A wide variety of punching operations can be performed by changing the tooling as shown (*a*) on either the portable punches or the automatic table type (*b*). Courtesy W. A. Whitney Mfg. Co.

(*b*)

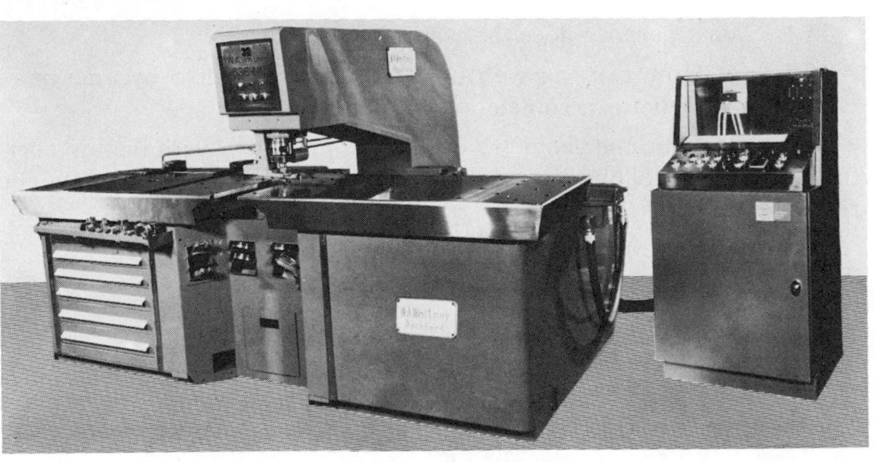

can be ganged together in an endless variety of arrangements to provide production facilities (Fig. 15.29).

Still another advantage of punching over drilling, where applicable, is the time savings. For example, a 1-in.-dia hole can be punched in 1-in.-thick steel plate in approximately 2 sec compared to about 46.8 sec for drilling.

Similar hydraulic units are built for cutting off and notching angle iron, cutting off bar stock, and many other operations.

**Fig. 15.29.** A variety of individual punch units can be mounted on common base with a single power unit. All units operate at the same time to complete a part in a single cycle. With all units operating the cycle time is 3.5 seconds. Courtesy W. A. Whitney Mfg. Co.

## QUESTIONS

**15.1.** What purpose does the web of a drill serve?

**15.2.** (*a*) How can the web of a drill be changed to become more effective? (*b*) What is a commercial example of this?

**15.3.** Compare the geometry of a single-point tool with that of a drill for each of the following angles: (*a*) end cutting edge, ECEA, (*b*) end relief, (*c*) back rake.

**15.4.** What are the advantages of a split-point drill?

**15.5.** What is the importance of the helix angle on drills?

**15.6.** How does a gun drill differ from an ordinary drill?

**15.7.** What is the advantage of trepanning?

**15.8.** Why is trepanning not applicable to most drilling situations?

**15.9.** What average tolerance can be expected when using a 3/4 in. drill without a guide bushing?

**15.10.** How much material removal can normally be left for finish reaming holes over 1/2 in. in dia?

**15.11.** What tolerance can be expected from a reaming operation?

**15.12.** Why are the smaller size drill presses termed sensitive?

**15.13.** For what type of drilling operations and work would you use the radial drill press?

**15.14.** What are some methods of increasing drilling production?

**15.15.** What should the limit of feed be for a 1/4-in. dia drill?

**15.16.** Why are thrust forces neglected in calculating drill power requirements?

**15.17.** Why is it often necessary to bore a hole after drilling it?

**15.18.** What are the basic applications of the horizontal boring machine?

**15.19.** Why is a jig boring machine used for accurate low production work in place of a precision boring machine?

**15.20.** What are some of the advantages of hole punching?

## PROBLEMS

**15.1.** Find the break-even point for using a carbide-tipped 3/16-in. dia drill as compared to an HSS drill with cost as given in the chapter and the cutting conditions as in Table P15.1. The material is AISI 1020 steel, 1 in. thick; labor and overhead, $12/hr; approach and overtravel 2 drill dia. *Note:* The shorter break-even formula as given in Chap. 14 may be used.

**Table P15.1**

| | sfpm | feed | regrind time |
|---|---|---|---|
| HSS | 60 | 0.003 | 5 min/100 holes |
| Carbide tip | 80 | 0.006 | 6 min/200 holes |

**15.2.** A drill jig that costs $400 to make saves 10¢/pc. Interest, taxes, depreciation, and maintenance add up to 70 percent of the cost. The setup cost is mainly getting it out of storage and checking it, which costs $10/yr. Of the overhead, 90% is applied on the labor saved. How many pieces will have to be made before the jig can be considered paid for?

**15.3.** Compare the punching and drilling time in seconds for 10 3/4-

in. dia holes in 1-in.-thick steel plate. The cutting speed for drilling is 80, and the feed 0.006 in./rev. The approach and overtravel for drilling can be taken as one half the drill dia. The time for punching each hole is 2 sec.

**15.4.** A single-spindle 3-hp drill press is to have a 6-spindle multiplespindle head attached to it. Will it have sufficient hp to turn 4 1/2-in.-dia drills and 2 3/4-in.-dia drills? The cutting speed used for the mild steel plate is 80 fpm, and the feed is 0.005 in./rev. The larger drill will govern the speed used. Table 13.1 may be used to obtain unit power numbers.

## REFERENCES

American Society of Tool and Manufacturing Engineers, *Tool Engineer's Handbook,* McGraw-Hill Book Company, Inc., New York, New York, 1959.

"How to Select Twist Drills," *Metalworking Specification Guide,* June 1962.

"How Twist Drills Work," *Metal Cuttings, National Twist Drill,* January 1954.

"Power Requirements for Gun Drilling," *Metal Cuttings, National Twist Drill,* July 1958.

Selection Guide, "Drilling Machines," *Metalworking,* October 1963.

# 16

# STRAIGHT AND CONTOUR CUTTING

MOST MACHINES are normally associated with the main type of work they perform. For example, the lathe is normally associated with cylindrical work although it can be used for a wide variety of operations. In this chapter, the machines normally associated with both straight and contour cutting will be discussed. Some machines are associated with both contour and flat surfaces.

## STRAIGHT OR PLANE SURFACE MACHINING

Plane surfaces are normally associated with shapers and planers.

**Shapers.** Shapers are made to cut by means of reciprocating single-point tools. The tool is held in a tool post on the end of a ram. Typical surfaces that can be cut with a shaper are shown in Fig. 16.1.

A shaper is usually not considered a production machine; however, it is widely used in machine shops and toolrooms, since it is easy to set up and operate. The cutters are low in cost and are easily sharpened. Because the amount of metal removed at one time is relatively small in area, little pressure is imposed upon the work, and elaborate holding fixtures are not needed. Although many shaper operations can be more rapidly performed by other machining processes, usually more costly tooling and setups are involved. These alternate methods are economical where the number of parts to be machined is large enough to justify a greater initial investment.

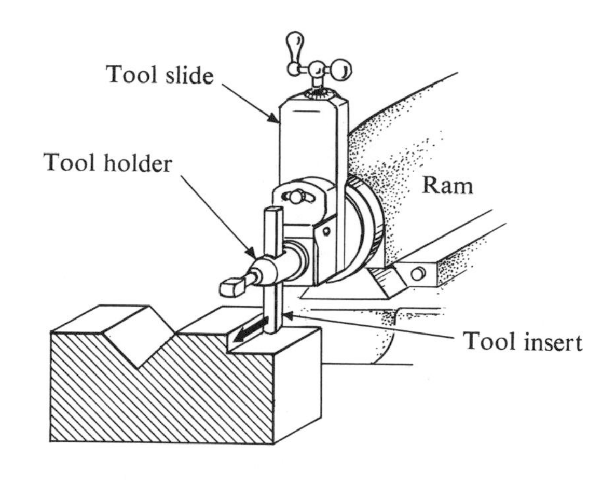


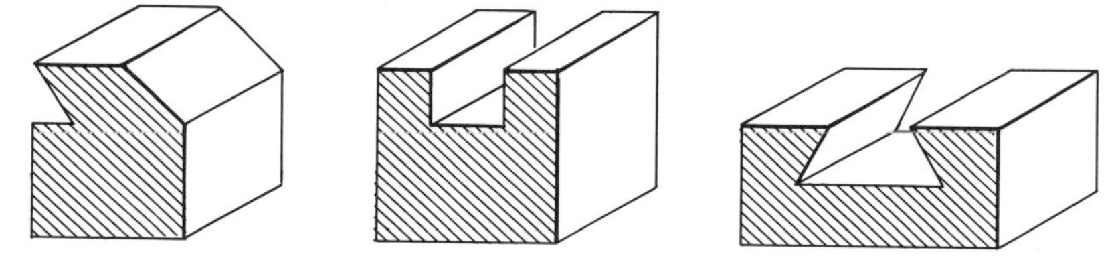

**Fig. 16.1.** Types of external and internal surfaces best produced by a shaper.

**Planers.** Planers are used for much the same type of surface cutting as shapers are, but on a larger scale. On the planer the tool remains stationary, and the work reciprocates back and forth. At the end of each cutting stroke, the tools are fed a small distance across the work so as to give a fresh bite in the work. The work is fastened to the table which, in turn, rides over the bed of the machine.

Three or four tooling stations are available, depending on the type of planer. Two toolheads are mounted on the overhead crossrail and one on each of the columns. When more than one surface is to be cut, three or four cutting stations can be engaged at one time (Fig. 16.2).

**Planer Action.** The single-point tool used in the planer imposes relatively little pressure on the work, and only a small amount of heat is generated. This is important when producing large, accurate, distortion-free surfaces. Modern planers are able to move carbide cutting tools in semisteel at the rate of 400 fpm. Large heavy castings usually receive their first cuts on a planer. A roughing cut used to true the casting surface will be approximately 3/4 in. deep, with a feed of 1/16 to 1/8 in. per cut. Two or three finishing cuts are taken to obtain accuracy on large surfaces. As little as a 0.001- or 0.002-in. depth may be left for the last finishing cut. Although planers are most often associated with the ma-

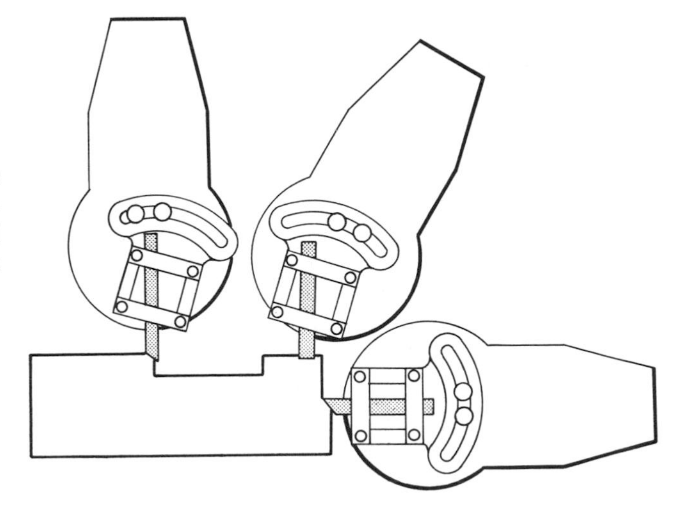

**Fig. 16.2.** Three or four tooling stations may be used at one time on a planer.

chining of large parts, such as machine tables, beds, rams, and columns, they may be used economically on many similar parts. The practice of setting up duplicate pieces in tandem is known as *string planing.*

**Planer Classification.** Planers are classified according to their main structures—*openside, double-housing, pit,* and *edge* or *plate* type. The double-housing type shown in Fig. 16.3 is of extremely solid construc-

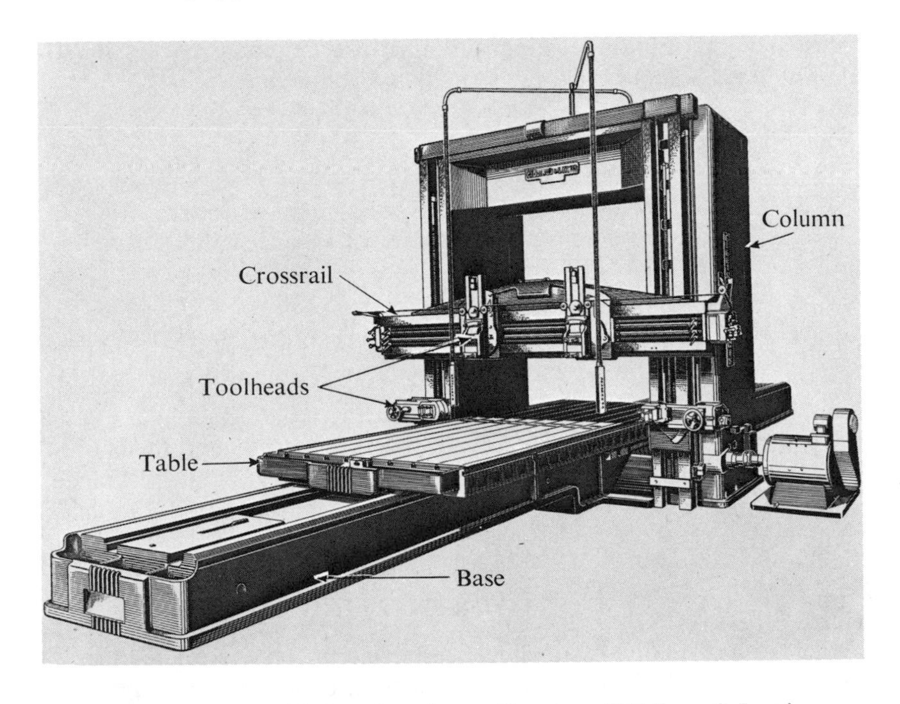


**Fig. 16.3.** Double-housing planer. Courtesy Giddings & Lewis.

tion. The support furnished the tool by the two columns and heavy crossrail allows heavy cuts to be taken even far above the table.

The openside planer is made with only one supporting column. Its main advantage is that work may be extended out over the bed. It is also somewhat easier to get at for setup work.

The pit-type planer is more massive in construction than the double-housing planer. It differs also in that the tool is moved over the work by means of a gantry-type superstructure which rides over the ways on either side of the table. The clapper boxes, which are of the double-block type, allow planing in both directions. Since the pit-type planer has its bed recessed in the floor, loading and unloading are facilitated.

Another more specialized planer is the *plate planer,* used for squaring or beveling the edges of heavy plate stock.

**Work-Clamping Methods.** The planer table is made with evenly spaced T-slots and holes. Use of the T-slot is the most common method of holding work against the surface of a machine. Shown in Fig. 16.4 are various T-slot clamping methods. Blocking materials are used to support the heel of the clamp at the proper height. These may be any scrap materials or, more conveniently, step blocks, parallels, and jack screws. Flat work is held on the table by means of toe dogs, chisel points, and poppets.

Other methods of holding work to the planer table include electromagnetism and hydraulic and vacuum clamping. Owing to large inertial forces and intermittent cutting action, these methods usually must be aided by mechanical blocking methods at each end.

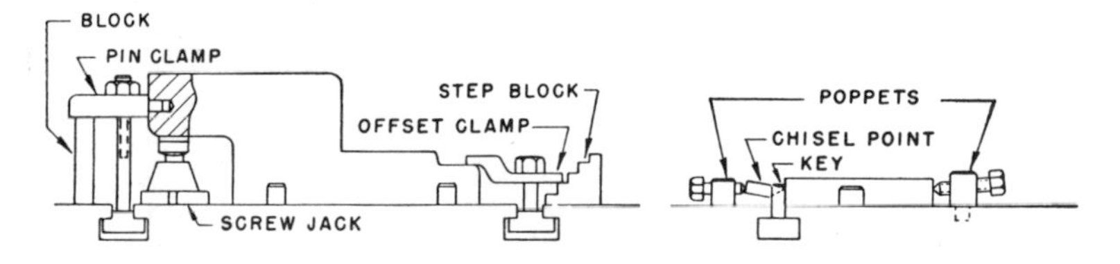


**Fig. 16.4.** Various methods used in clamping work to the planer table.

**Speeds and Feeds.** It is difficult to make definite speed and feed recommendations for planers, since cutting conditions vary widely. Some of the considerations involved in setting speeds and feeds are the size of the workpiece, clamping facilities, ability of the workpiece to withstand the pressure of the cut, and the type and number of tools used at one time.

## STRAIGHT AND CONTOUR CUTTING MACHINES

Machines normally associated with contour type surfaces are also used for plane surface machining. These are: milling, broaching, and sawing machines.

## MILLING-MACHINE CLASSIFICATION

Many machine-tool builders are engaged in the manufacture of milling machines. Each machine has a special purpose, requirement, or condition. As a result, we have a wide variety of milling-machine types. These can be classified, depending on their use, in three general groups—low-production, high-production, and special types. In the first category are the column-and-knee machines. Fixed-bed machines comprise the second group. Special types include duplicating or profiling machines, pantograph mills, rotary-table mills, and others especially adapted for certain jobs.

**Column-and-Knee Milling Machines.** The column-and-knee mills are so named because of their two main structural elements (Fig. 16.5).

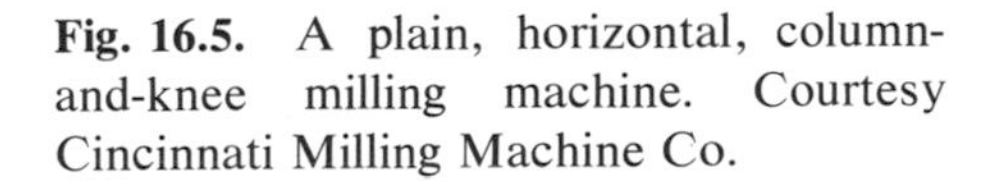

**Fig. 16.5.** A plain, horizontal, column-and-knee milling machine. Courtesy Cincinnati Milling Machine Co.

The column-and-knee mill is classified as plain or universal, depending on whether or not the table can be swiveled in a horizontal plane. The table on the universal machine can be swiveled up to 45 deg to the right or left, making possible angular and helical milling.

**Horizontal and Vertical Mills.** The table of a column-and-knee milling machine has a wide range of feeds that can be applied in all three directions—vertically, longitudinally, and laterally. The table can also be quickly positioned by means of a rapid traverse mechanism, and some incorporate automatic table movements. Graduated dials allow the operator to set the table to within 0.001 in. accuracy. The rigid overarm and bearing support used on horizontal machines also increase accuracy.

The vertical milling machine is so called because the spindle extends in a vertical position (Fig. 16.6). Some vertical spindles are adjustable only vertically; others can swivel through 360 deg across the face of the machine.

A variation is a combination vertical and horizontal machine (Fig. 16.7). Some machines of this type have an independently driven head mounted on the overarm ram. This head is of the universal type and can be swiveled for cuts at any angle between the cutter and a horizontal plane.

**Fig. 16.6.** A column-and-knee milling machine of the vertical type. Courtesy Cincinnati Milling Machine Co.

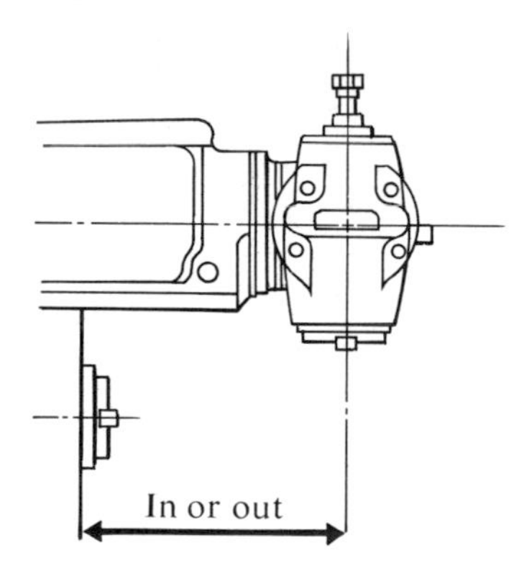


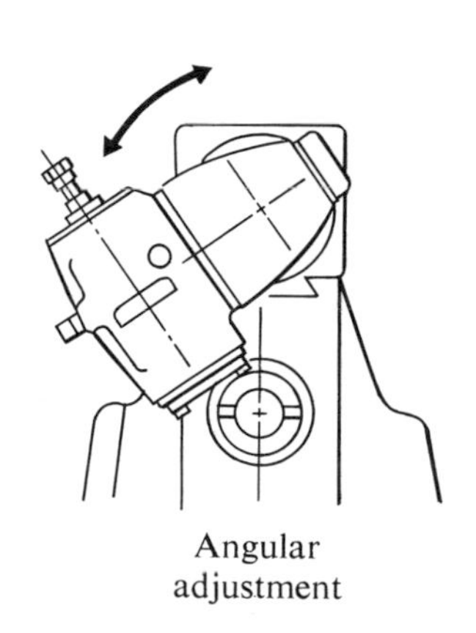

**Fig. 16.7.** A combination vertical and horizontal column-and-knee milling machine. The Universal ram-mounted head adds range and versatility to the machine. Courtesy Cincinnati Milling Machine Co.

To gain a better understanding of the use of both the vertical and the horizontal milling machines, study the types of work done by each as shown in Fig. 16.8. You may see that there is some overlapping in the type of work performed; however, each kind has its own advantages.

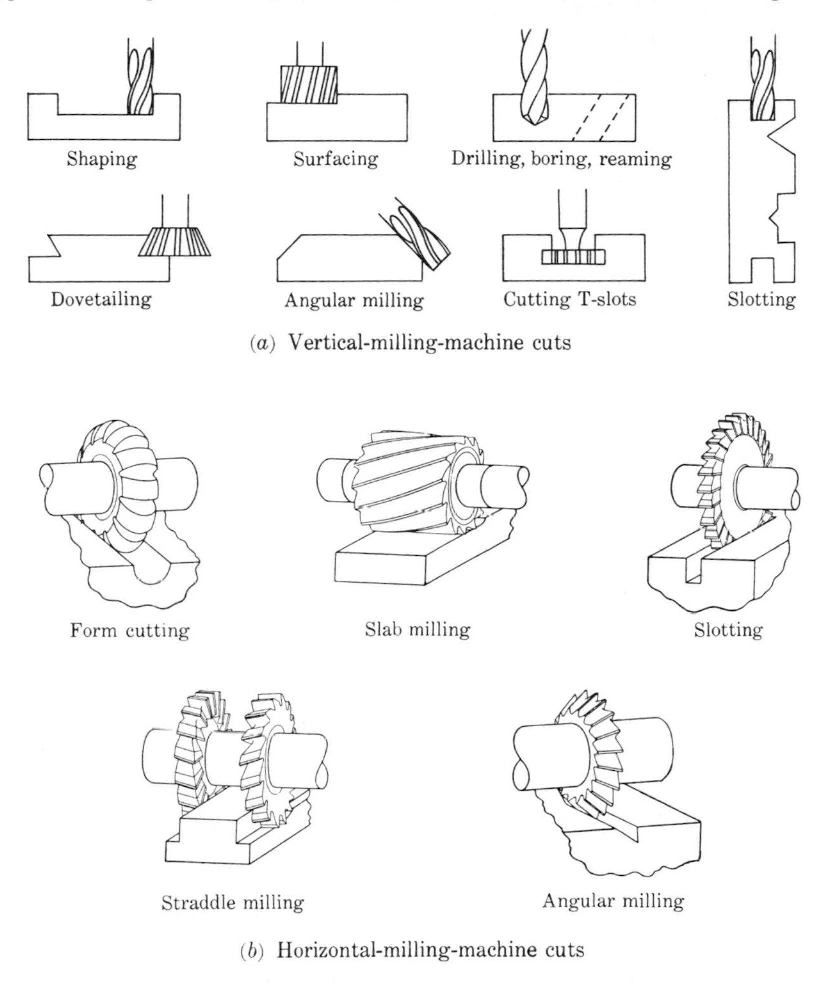


**Fig. 16.8.** Applications of vertical and horizontal milling machines.

**Fixed-Bed Milling Machines.** Fixed-bed milling machines are made, as the name implies, with a stationary bed. All vertical movements for depth of cut and lateral cutter positioning are made on the spindle. This arrangement makes for more rigidity and accuracy in the heavy-production cutting required of these machines (Fig. 16.9).

The distinctive feature of fixed-bed machines is the automatic cycle. The work may be set to approach the cutter at rapid traverse, make the

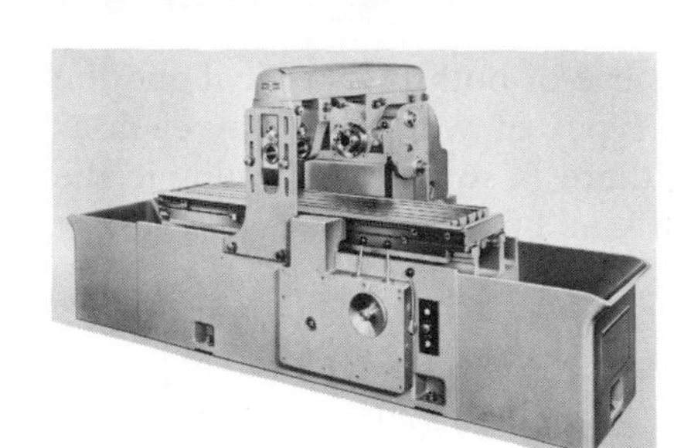

**Fig. 16.9.** The fixed-bed type of milling machine is made for production milling with such features as rapid traverse approach, feed, and automatic rapid traverse return stroke. Courtesy Cincinnati Milling Machine Co.

cut at a predetermined feed, and then automatically return to the starting position. Thus, once the machine is set up, the operator is required only to clamp the part in the milling fixture and start the automatic cycle. If continuous operation is desired, a workpiece is placed in a fixture on one end of the table while cutting is taking place on the workpiece at the other end. This is known as *continuous reciprocal milling*.

**Planer Mill.** A planer mill is very similar in structure to the double-housing planer discussed previously. On the planer mill, the single-point tools are replaced by rotary milling heads (Fig. 16.10).

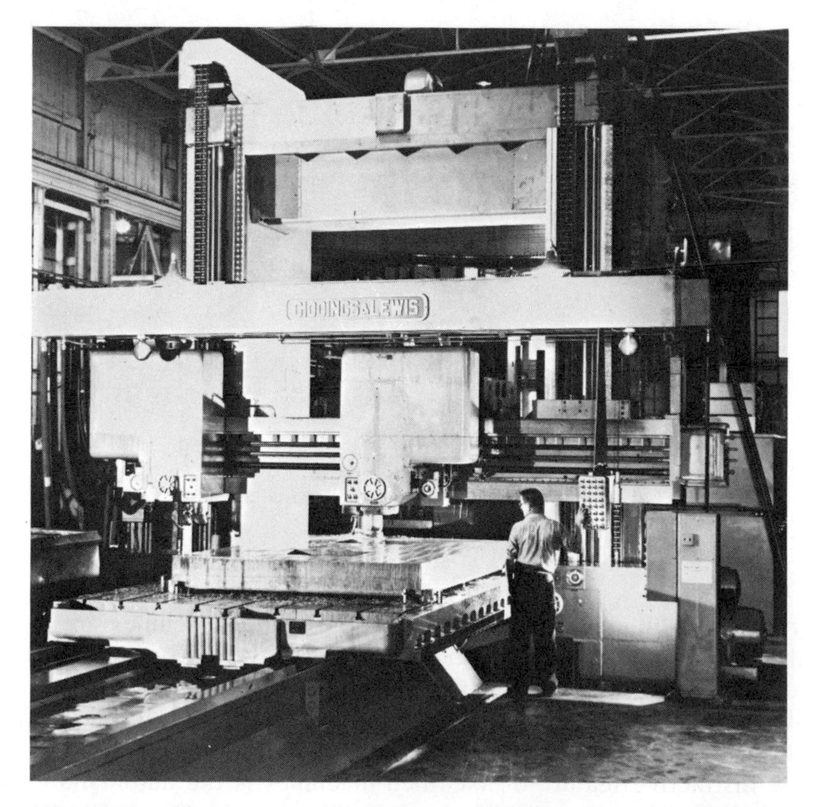


**Fig. 16.10.** Planer-type mill. Courtesy Giddings & Lewis Machine Tool Co.

This mill is usually equipped with two cutting heads on the overhead crossrail and one on each of the columns. The overhead units can be positioned for angular cuts. When several cutting heads are employed at one time, the metal-removal rate is high. This is necessary on a machine of this type in order to justify the large initial investment, the cost of operation, and the special tooling.

**Special Types of Milling Machines.** Many special types of milling machines are made to accomplish specific kinds of work more easily than present standard machines. In this category are the duplicating mills, die sinkers, profiling machines, and pantographs. Most of these machines are vertical mills that have been adapted to reproduce accurately by means of a tracer, the forms and contours from a master pattern.

**Duplicating Mills.** In the hand-type duplicating mill, the operator must guide the tool back and forth across the pattern (Fig. 16.11). Automatic machines are made to traverse the pattern either electrically or

**Fig. 16.11.** Mechanical-type duplicating mill with twin spindles profiling 15 blades on two impeller rings from a single-blade master. Courtesy George Gorton Machine Co.

hydraulically. Once the cycle is started, the tracer will automatically follow the master pattern until it is completed. The pattern may be of easily formed material such as wood or plaster. Large electrically controlled machines of this type, used extensively in making sheet metal dies for aircraft and automotive bodies (Fig. 16.12), are better known by their trade name of Kellering machines.

**Rotary-Table Mills.** To facilitate the loading and unloading of a large number of parts while the machine is cutting, a rotary-type table is used for holding the work. Figure 16.13 shows a view of a rotary-table machine. These individual fixtures are being used to hold the work. Each fixture locates and holds one block, which is clamped automatically as it approaches the cutters. After the cut is completed, the part is automatically unclamped. The table may be made to travel continuously or on a cycle of alternate feed, dwell, and rapid traverse.

(*a*)

(*b*)

**Fig. 16.12.** (*a*) The Keller mill being used to finish a propeller forging, using a finished propeller as a master; (*b*) a steel mold for a precision aluminum casting being cut using a wood model. Courtesy Pratt & Whitney Co., Inc.

**Fig. 16.13.** Rotary-table mill. Courtesy Ingersoll Milling Machine Co.

## THE MILLING PROCESS

**Milling-Cutter Rotation.** Milling cutters may be mounted on the arbor so that the cutting action will bring the cutting forces down into the work (called *climb milling*), or the forces may be directed up as in *up*, or *conventional, milling* (Fig. 16.14). The advantage of climb milling is that the downward action of the cutter helps hold the material in place. Thus the fixture can be less rugged and of simpler design. This is especially helpful in milling thin materials that are hard to clamp. Higher

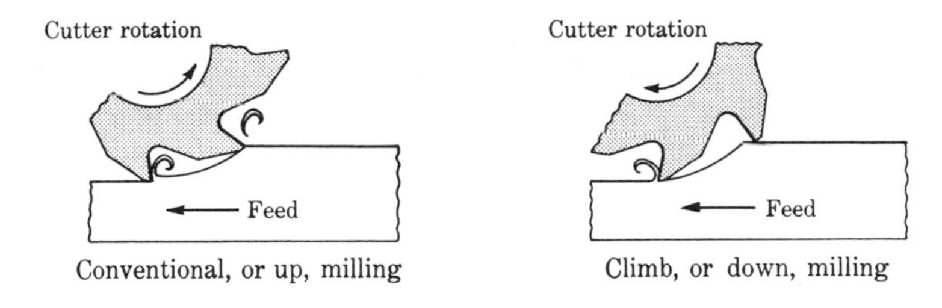


**Fig. 16.14.** Cutter relationship to work in conventional and climb milling.

feed rates may be used, with a corresponding increase in production, increased cutter life, and lower horsepower per cubic inch of metal removed. However, as the name implies, there is a tendency for the cutter to climb over the work. This imposes undue strain on both cutter and arbor. The climbing action may be eliminated if the work can be made to feed into the cutter at a steady pace. On production machines, this is taken care of by a hydraulic backlash eliminator.

Up milling tends to pull the work up away from the fixture. However, where no provision has been made to eliminate the lead-screw backlash, this method is used exclusively.

## MILLING METHODS

The methods used to mill a part or a number of parts may vary widely. Wherever possible, ways of increasing production are given prime consideration. The practice of placing more than one cutter on the arbor at one time or more than one workpiece on the milling-machine table may be the deciding factor as to whether or not a job will prove profitable.

**Straddle Milling.** When two cutters are mounted on an arbor and are spaced so as to cut on each side of the material (Fig. 16.15), the term applied is straddle milling. The cutters are spaced very accurately and are held rigidly. Therefore, the parts produced are uniform.

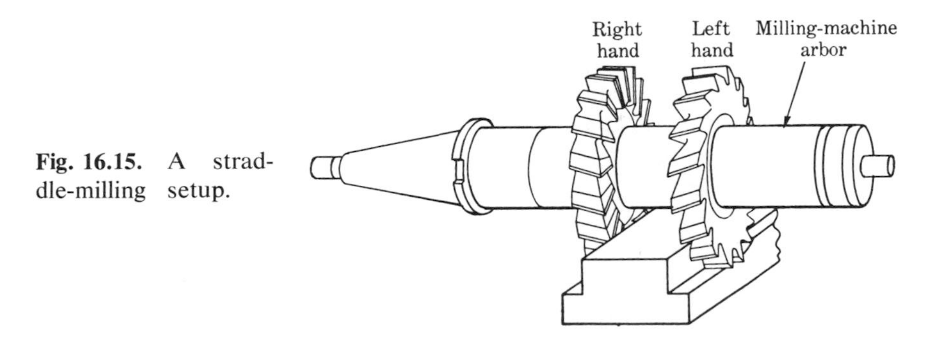


**Fig. 16.15.** A straddle-milling setup.

**Gang Milling.** Many cutters may be placed on the same arbor to make an entire surface or contour change in one pass (Fig. 16.16). A combination of gang and straddle milling cutter arrangement is also shown.

**Fig. 16.16.** Gang-milling setups used to machine a combination of surfaces simultaneously. Courtesy Ingersoll Milling Machine Co.

## PLANNING FOR PRODUCTION MILLING

After a preliminary study of the part prints, it is necessary to specify the most suitable type of milling-machine fixture, tools, cutters, speeds, and feeds to produce the part at the lowest unit cost. The following outline may aid in accomplishing this.

(1) Determine the type of machine to be used. Keep in mind, when selecting a machine, that the column-and-knee machines are easily set up and operated. They can be changed quickly and are therefore ideal for short-run jobs. The bed-type machines require more setup time, since feeds and speeds are set by change gears; therefore, longer runs are desirable. Specialized machines are considered only for parts not conveniently handled on a regular machine and where the quantity is large enough to warrant the extra cost.

(2) Determine the type of milling to be done—whether straddle, contour, etc. Also, decide on the method of mounting the work—whether several parts can be placed on the table for abreast milling, string milling, or reciprocal milling, or if indexing or other fixtures are to be used.

(3) Check the Brinell hardness of the part in several places to obtain a basis for cutting speed.

(4) Choose the type of cutters needed, including the style, size, and material. Milling-cutter dimensions are given by outside diameter, width,

and inside diameter. The cutter diameter is based on the depth of cut plus the clearance between the arbor collars and the work. Too large a cutter adds to the time of the job and to the horsepower requirements. Heavy cutting requires rigid cutter mounting. A 2-in.-dia arbor is 16 times as rigid as a 1-in. arbor. Face mills mounted on stub arbors (Fig. 16.17) have minimum overhang and can be counted on for heavy rough-

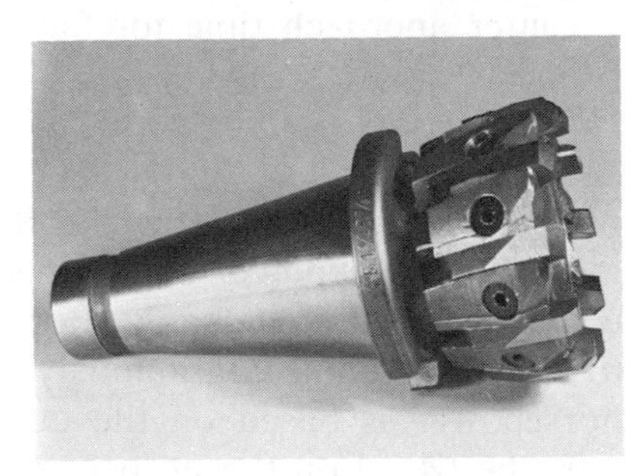

**Fig. 16.17.** A face mill mounted on a stub arbor. Courtesy Kearney & Trecker Corp.

ing cuts as well as for accuracy in the finished cut. The cutting-tool material, whether high-speed steel, carbide, or ceramic, will be governed by the rigidity of the machine and the setup. Carbides and ceramics demand rigid setup of both the cutter and the work, or chipping of the tool will occur.

(5) Determine proper speeds and feeds. The speed at which the cutter rotates is based on the formula

$$\text{rpm} = \frac{S \times 12}{D\pi}$$

where $S$ = surface feet per minute; $D$ = cutter diameter; and $\pi = 3.1416$. The Brinell hardness and the cutting-tool material may be used to determine $S$, as mentioned previously.

The feed, in inches per minute of table travel, will be based on the machine horsepower, the finish desired, and the chip clearance of the cutter. The feed rate will be calculated for the formula

$$F_r = f_t N_t R$$

where $F_r$ = feed rate of milling-machine table, inches per minute; $N_t$ = number of teeth in cutter; $R$ = rpm of cutter; and $f_t$ = feed per tooth in inches.

(6) The time required to make a cut with a plain milling cutter can now be obtained by the formula

$$T_m = \frac{L + A + O}{f}$$

where $T_m$ = time, in minutes; $L$ = length of material being cut, in inches; $A$ = approach—this includes the distance before the cutter actually touches the work and until it is cutting at full depth; $O$ = overtravel, in inches; and $f$ = feed in inches per minute. Thus

$$A = \sqrt{d(C_d - d) + Cl}$$

where $A$ = approach; $d$ = depth of cut; $C_d$ = cutter diameter; Cl = clearance (distance cutter is from work).

Cutter approach time for face milling cuts may be calculated by the formula

$$A = (D/2) - \sqrt{1 - (W/D)^2}$$

where $D$ = cutter diameter and $W$ = width of cut.

(7) The horsepower of the machine should be considered, since it is economical to take advantage of the amount available, but it is not wise to run the machine continuously over its rated capacity. Milling-machine horsepower is based on the cubic-inch removal rate, which in turn, must be based on the proper cutting speed for the material and cutter. The cubic-inch removal rate is determined by the formula cim = depth × width × feed in inches per minute, and

$$hp_c = (cim)/K$$

where $hp_c$ = horsepower at cutter and $K$ = a factor based on the material and the feed rate used.

Tables giving these values may be found in the *Tool Engineer's Handbook*. A few are given in Table 16.1 for some common materials.

**Table 16.1**

| Material | K |
|---|---|
| Alloy steel: 300–400 BHN | 0.5 |
| Aluminum | 2.5–4.0 |
| Cast iron, soft | 1.5 |
| Stainless steel, austenitic free-machining | 0.83 |

An alternate method that can be used is to multiply the cubic inch removal rate (cim) by the unit horsepower constant found in Table 13.1. The results will be higher since the constant has been corrected to allow for dull cutters and 80 percent spindle drive efficiency.

When all the above factors have been considered, a reasonable estimate as to the type of equipment needed for a given production rate can be made. For the overall time of a given item, other factors should be included, such as start, rapid traverse time, rapid return indexing, etc. Much of these data may be obtained from books on standard times.

## BROACHING

The broach is used equally well on straight or irregular surfaces, either externally or internally, and it can perform many of the operations that are done more laboriously on milling, drilling, boring, shaping, planing, or keyway-cutting machines.

**Internal Broaching Tools.** Internal broaching tools are designed to enlarge and cut various contours in holes already made by drilling, punching, casting, forging, etc. The tool shown in Fig. 16.18 has the

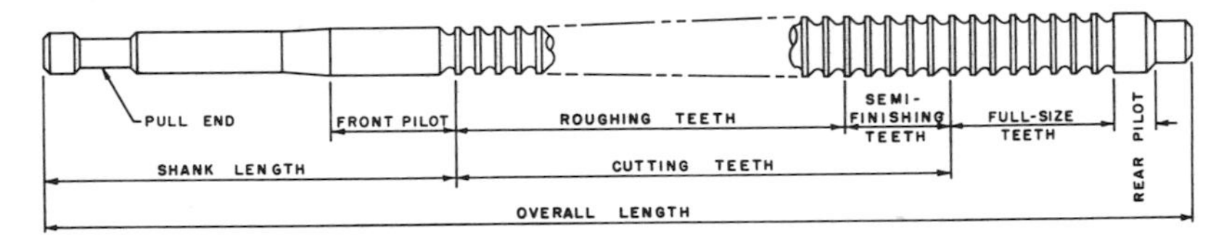

**Fig. 16.18.** Internal broach with sample workpieces. Courtesy Sundstrand Machine Tool Co.

essential elements found in most broaches. The first teeth are designed to do the heaviest cutting and are called the *roughing teeth.* The next portion has the *semifinishing teeth,* followed by the *finishing teeth* with progressively lighter cuts. Some broaches are made so that the last few teeth do no cutting at all but have rounded edges for a burnishing action.

The amount of stock that each tooth can remove varies with the type of operation and the material. A general average is 0.002 or 0.004 in. per tooth for high-speed-steel broaches. The space between the teeth must be sufficient to provide ample chip room, since the chip is carried by each tooth until it clears the stock. On ductile materials, it is necessary to have staggered grooves in the cutting teeth to act as chip breakers.

The internal forms shown in Fig. 16.19 represent only a few of the

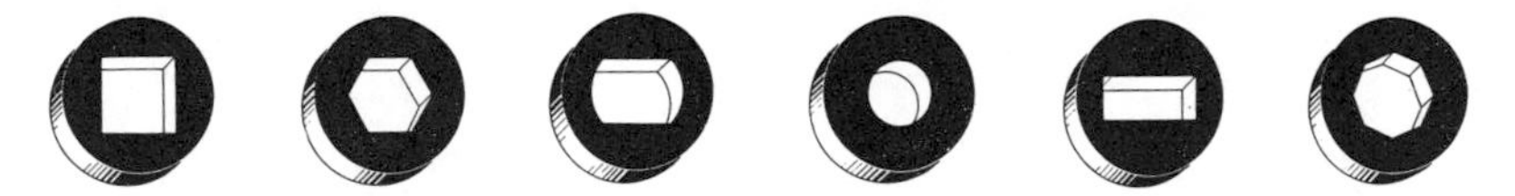

**Fig. 16.19.** A variety of forms produced by internal broaches. Courtesy Colonial Broach Co.

shapes that have become standard. Special broach shapes are built by manufacturers for odd and difficult contours.

Spline and gear broaching. The involute type of spline, shown in the center sample of Fig. 16.20, is an improvement over the regular straight-sided spline in that it is much stronger and tends to centralize under a turning load. When there is an endwise movement under load, the spline acts as a gear, transferring the load uniformly from tooth to tooth.

**Fig. 16.20.** Internal gears and splines cut with internal broaches, some to extremely close tolerances. Courtesy Colonial Broach Co.

The broaching of internal gears is similar to that of internal splines. Usually, however, involute form, tooth spacing, runout, and finish must be held to closer limits. Broaching of internal gears may be applied to either straight or helical gears. If the helix angle is less than 15 deg, the broach will turn as it cuts its own path. Angles greater than 15 deg require a spiral-lead-type drive head to turn the broach as it is pulled through the work.

**External Broaching Tools.** External surface broaching competes with milling, shaping, planing, and similar operations. It offers a combination of a high degree of accuracy and excellent surface finishes, combined with high output rates and low downtime.

Automobile manufacturers have replaced many milling operations with surface broaching because of the combined speed and accuracy. An example of this kind of work is shown in Fig. 16.21. A maximum of 3/16 in. of stock is removed in a single pass. The broach consists of carbide-tipped inserts made up of five sections in combinations of half-round and facing broaches. As many as 22,000 cast-iron engine blocks can be run

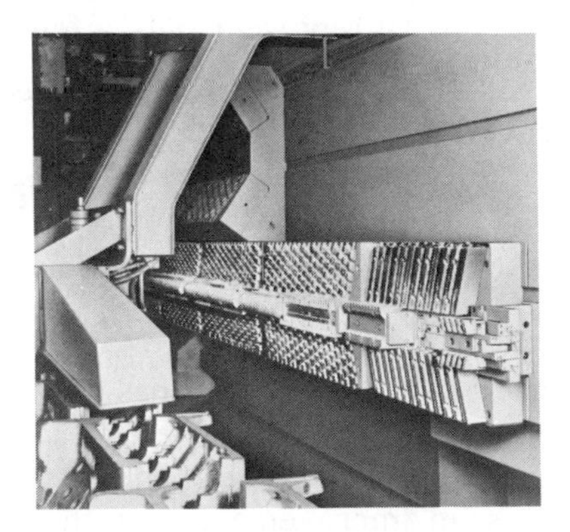

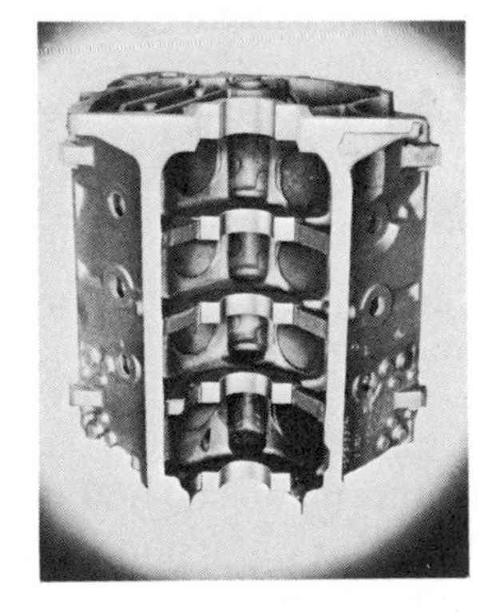

**Fig. 16.21.** A large carbide-tipped sectional broach used to machine several surfaces of a V-8 engine block in one pass. Courtesy Ex-cell-o Corporation.

before the tool needs resharpening. Some broaches of this type are made on the same principle as the single-point lathe tool with throwaway inserts. As the carbide tips become dull, they can be unclamped and indexed 90 deg; they will then be ready to cut again. By incorporating negative rake into the toolholder, six or eight cutting edges can be utilized before the insert is discarded.

Broaching speeds have increased tremendously with this kind of tooling. Formerly, 20 to 40 fpm was considered average; now, speeds in excess of 200 fpm are used on cast iron. A further timesaving is made on large surface broaches by mounting them in pairs. With this arrangement, cutting can be done on both the forward and the return strokes. The top broach on a horizontal machine may be cutting on the forward stroke and the lower broach on the return stroke.

Some idea of the wide variety of surface broaching operations can be gained by referring to Fig. 16.22.

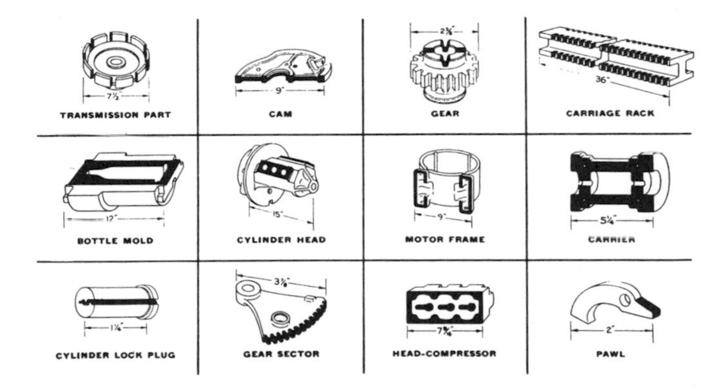


**Fig. 16.22.** A variety of surfaces prepared by broaching. Courtesy *Cincinnati Milling Report*.

## BROACHING MACHINES

The main types of broaching machines are vertical ones, including pullup, pulldown, and ram; and horizontal ones, both plain and the continuous-chain-type surface broach. These operate primarily as a means of either pushing or pulling the tools through or over the work.

**Vertical Broaches.** The various vertical machines are similar in appearance. In fact, one broach builder makes a convertible machine that can perform all four of the basic operations—push down, surface, pull down, and pull up (Fig. 16.23).

**Horizontal Broaches.** PLAIN MACHINES. Horizontal broaching machines (Fig. 16.24) are used primarily for broaching keyways, splines, slots, round holes, and other internal shapes or contours. They have the

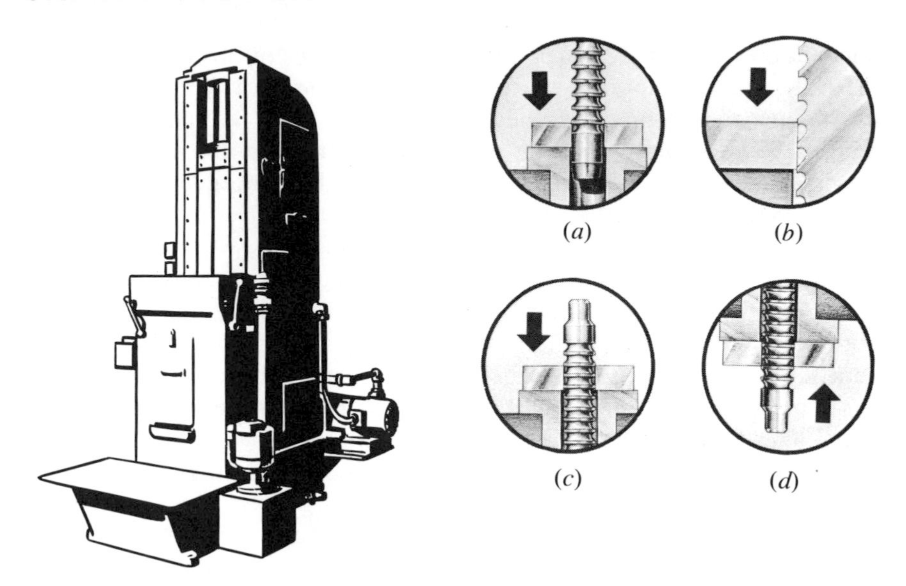


**Fig. 16.23.** Vertical broaching machine convertible to four basic operations. (*a*) Push down, (*b*) surface, (*c*) pull down, (*d*) pull up. Courtesy Sundstrand Machine Tool Co.

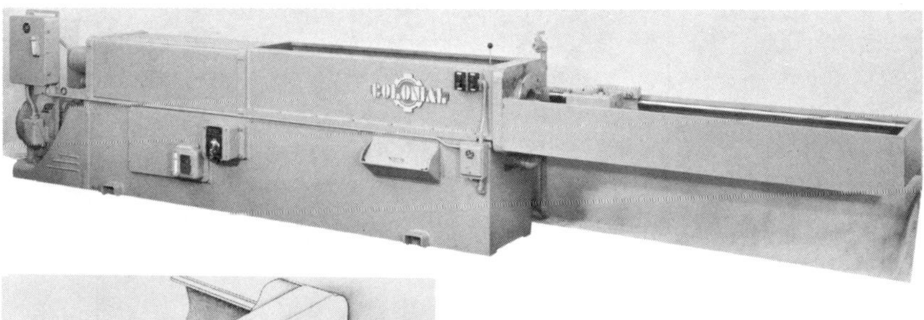

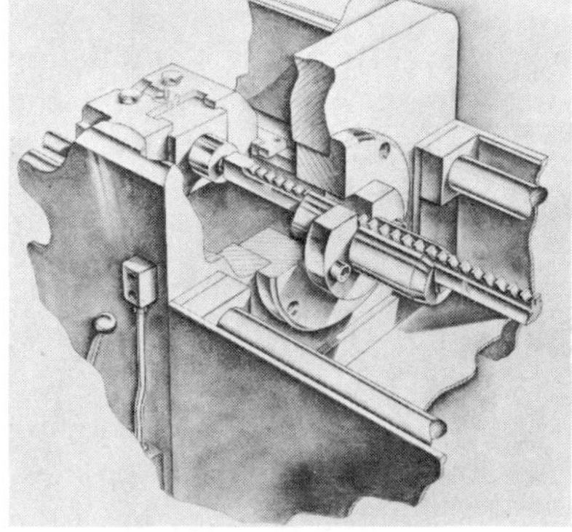

**Fig. 16.24.** Horizontal broaching machine and operation. Courtesy Colonial Broach Co.

disadvantage of taking more floor space than do the vertical machines. However, long broaches and heavy workpieces are easily handled on the horizontal machine.

CONTINUOUS-CHAIN-TYPE SURFACE BROACH. The continuous-chain-type surface broach (Fig. 16.25) is used where extremely high

**Fig. 16.25.** Continuous-chain surface-broaching machine.

production is desired. This machine consists mainly of a base, driving unit, and several work-holding fixtures mounted on an endless chain. The work is pulled through the tunnel where it passes under the broach. The operator has only to place the work in the fixture as it passes the loading station. Work is automatically clamped, machined, and unloaded.

**Broaching Machine Sizes.** The size of a broaching machine is expressed by the total ram travel, in inches, and the maximum pressure exerted on the ram, in tons.

**Finish and Accuracy.** A 30-$\mu$in. finish can be consistently held when broaching steel of uniform microstructure. Better finishes are available, but costs are higher. Irregular and intricate shapes can be broached to tolerances of $\pm$ 0.001 in. from a location established on the part. Tolerances of $\pm$ 0.0005 in. can be held between surfaces cut simultaneously.

**Broach Length and Machining Time.** The length of a broach will determine the cutting cycle. Also, the length of the broach must be suited to the machine on which it is to be used. By way of example, if the average amount of stock removed per tooth is 0.003 in., the pitch (or tooth spacing) is 1/2 in., and the material to be removed is 1/8 in., the *effective length* of the broach can be calculated. By effective length is meant the area containing the broach teeth.

$$L_e = \frac{C_d}{C_t} \times p \quad \text{or} \quad L_e = \frac{0.125}{0.003} \times 0.500 = 20.8 \text{ in.}$$

where $L_e$ = effective length of broach; $C_d$ = depth of cut to be made; $C_t$ = cut per tooth (average); and $p$ = pitch.

The length of the broach will change with the cut per tooth and the pitch. On internal operations, the amount of metal to be removed from the diameter will be divided in half.

The pitch of the broach varies with the length of the cut to be made, for example, a cut only 3/32 in. long would require a 3/64-in. pitch, whereas a cut 1 in. long should have a pitch of 3/8 in.

**Economic Considerations.** Broaching tools are usually more costly than milling cutters. However, this initial outlay can be offset by other factors. The machine time for broaching is generally less, and the piece part is often completed in one pass. Further time savings per piece are realized when more than one part is broached at a time, by stacking. Stack broaching generally is best adapted to internal operations.

The relatively low speed and small cut per tooth give broaches a long life. Contour broaches can be resharpened by grinding on the face of the teeth. Generally, special fixtures are needed for sharpening the broach, making it a rather expensive operation. Surface broaches are often made in sections so that, as the teeth are reground and become shorter, they can be moved forward. The last section will be all that has to be replaced. The cost of broaching tools runs high because of the care needed in forging and grinding. Generally, a high quantity of production is necessary to justify the cost of the tool.

## SAWING

Many machining operations start with the sawing of the stock to length or size. Improvements in the speed, accuracy, and the variety of operations that can be performed on sawing machines give this process an important place in the field of manufacturing.

**Basic Types of Sawing Equipment.** Sawing equipment may be classified by the motion used for the cutting action. There are reciprocating saws, represented by the power hacksaw; band saws, including cutoff and contour (traditional types and other applications); and circular saws, such as the cold saw and those with friction disks or abrasive disks (Fig. 16.26).

**Reciprocating Saws.** The reciprocating motion in sawing is familiar to all who have used the hand hacksaw. Although it is hardly a production process, its usefulness is appreciated in many toolroom and maintenance situations.

**Power Hacksawing.** The power hacksaw is probably the simplest of the metal-cutting machines. It consists of a vise for clamping the work and a means of reciprocating the saw frame. Hydraulic pressure is used to regulate the downward feed force on the blade. A hydraulic or mechanical arrangement is also incorporated for lifting the blade on the return stroke. This prevents the teeth from being dragged backward over the work, which would cause them to become prematurely dull.

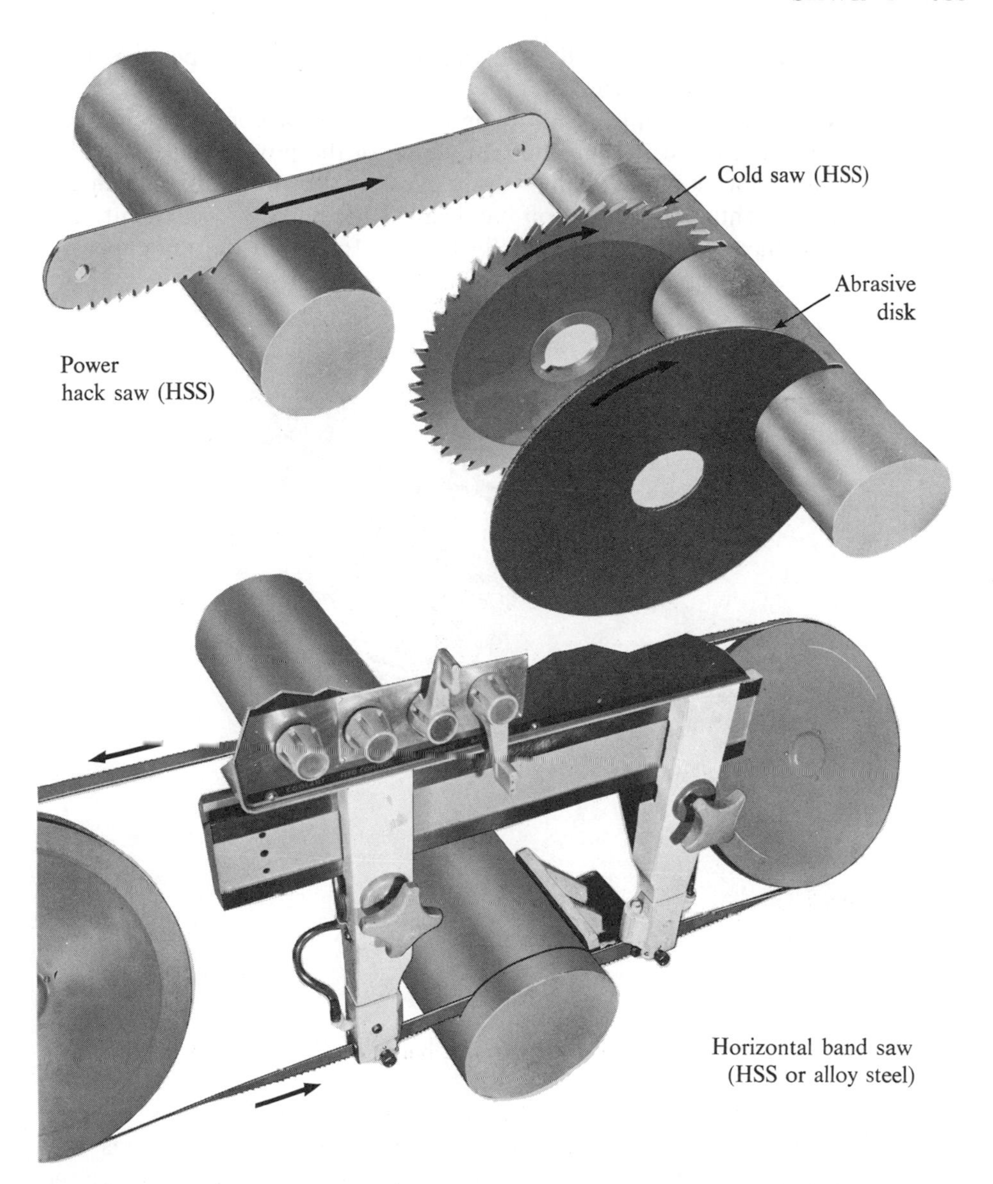


**Fig. 16.26.** Four types of cut-off machine action and cutters. Courtesy DoAll Company.

The stock to be cut is held between the clamping jaws. Several pieces of bar stock can be clamped together and cut at the same time. Both square and angular cuts can be made. Some of the larger heavy-duty hacksaws have hydraulic feeding and clamping devices which automatically move the bar forward to the correct length, and clamp it. After the cut, the stock is automatically unclamped, and the cycle repeats as long as the stock lasts (Fig. 16.27).

**Band Saws.** CUTOFF TYPE. Band saws have a continuous cutting action in contrast to the intermittent action of the power hacksaw. This makes for shorter cutting time—up to 10 sq in. per min on some steels. Another advantage of the cutoff band saw is the narrow *kerf,* or cut, that it makes. Saws of this type usually have a 16-in.-dia cutting capacity.

**Fig. 16.27.** Power saw, showing hydraulic bar feed. Courtesy DoAll Company.

CONTOUR TYPE. The metal-cutting contour band saw is an outgrowth of the woodworking band saw. One of the first models was produced in 1935. It provided the advantage of being able to remove large pieces of unwanted metal without reducing them to chips, and thus an immediate market for this machine (Fig. 16.28) was created.

Many developments followed the early models. First among these improvements was a variable-speed control. A wide choice of speeds resulted in greater accuracy, faster cutting, and increased tool life. Other refinements were added gradually, such as the flash butt resistance welder used to join the saw bands right on the machine for internal cutting. A grinder mounted on the column of the machine is used to

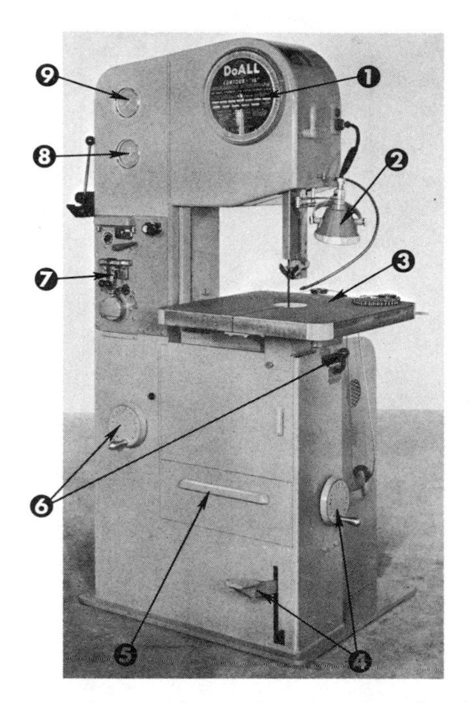


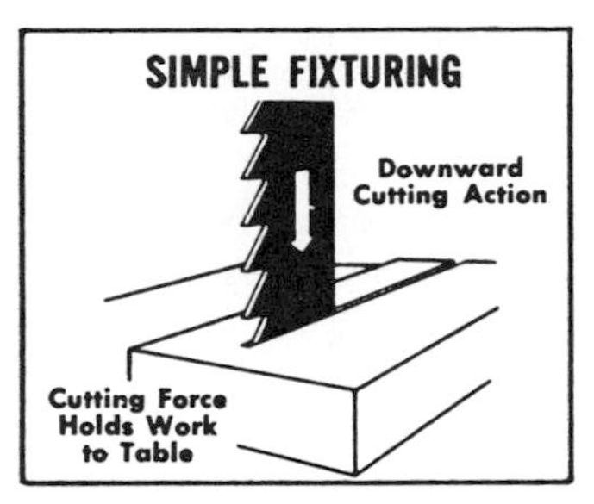


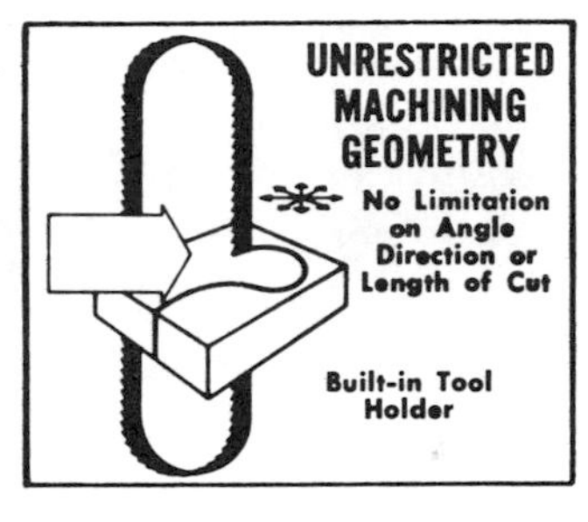


**Fig. 16.28.** A standard metal-cutting band saw is normally used for straight, angular, and contour cutting but may also be used for filing and polishing. (1) Job selector dial; (2) reflector flood lamp; (3) tilt table—45 deg right, 10 deg left; (4) work feed controls; (5) chip drawer; (6) speed control and indicator, (7) saw blade welder; (8) band tension indicator; (9) blade sfs indicator. Courtesy DoAll Company.

smooth the weld joints, permitting them to pass through the saw guides.

Numerous attachments added to the versatility of the machine. Among these are the circle-cutting attachment, ripping fence, and magnifying lens. Power feeds and coolant systems have also been added. On some of the larger machines, the work can be bolted directly to the power-fed table. This reduces operator fatigue and increases accuracy.

Other operations performed on the regular contour band saw are filing and polishing.

*Friction contour-sawing machines.* Special contour-cutting band saws are those that are capable of doing operations beyond the scope of the regular machine. For example, friction sawing is a comparatively new term when applied to contour-sawing machines. The process consists of a fast-moving blade that produces enough friction to heat the material to the softening point. The heat of friction is confined to a small area just ahead, and a little to each side, of the blade. When the material becomes soft and loses its strength, it is removed by the saw teeth.

The right blade for friction sawing is more critical than in ordinary sawing. Too few teeth for the material thickness will tend to remove more material than is softened, which will ruin the blade. Too many teeth, on the other hand, will tend to clog and will not remove the material at all. Specially made blades are required for most friction-sawing operations. Saw velocities range from 6,000 fpm on 1/4-in.-thick carbon steel to 13,500 fpm on 1-in.-thick armor plate. The latter can be cut at the rate of 5 in. per min.

The chief advantage of friction sawing is that it can cut many times faster than conventional methods on hard materials (Fig. 16.29).

**Fig. 16.29.** This fast-action photo stops the saw blade traveling 9,000 fpm in a high-speed milling cutter. Heat penetration is at the immediate saw point. Courtesy DoAll Company.

**Circular Saws.** Circular saws cut by means of a revolving disk. The disk may have rather large teeth, as in the case of cold saws, or almost no teeth, as in the friction disk.

COLD SAWING. Cold sawing would imply that no heat is generated in the cutting operation. Since this is not true, the term is used only to distinguish it from friction-disk cutting, which heats the metal to nearly the melting temperature at the point of metal removal. Also the blades are usually large, which allows cooling time for the tooth after leaving the cut until reentering.

FRICTION SAWS. Circular friction saws operate on the same principle as the friction band saw. However, since the saw friction heat is high and natural cooling is insufficient, an auxiliary coolant must be forced against the blade to keep it from becoming red hot. The saws may be as large as 6 ft in diameter, operating at speeds up to 25,000 sfpm.

ABRASIVE-DISK CUTTING. Abrasive-disk cutting, frequently used in cutoff operations, is not a true sawing technique, but, since it serves the same purpose, it is frequently classified with the cutoff sawing operations.

Thin resinoid or rubber-bonded wheels rotating at 12,000 to 14,000 sfpm are used. The cutting action is fast and accurate. For example, ordinary 2-in.-dia pipe can be cut in 8 sec. For most work, the cut surface need not be refinished (Fig. 16.30).

Machines for abrasive-disk cutting are fairly simple and usually feature a powerful drive motor with a belt-driven wheel head. Most of the machines use a swing frame and are fed into the work manually.

**Fig. 16.30.** The abrasive cut-off wheel can be used to cut both soft and hardened materials. In most cases no further machining needs to be done on the cut ends. Courtesy Wallace Supplies Mfg. Co.

To speed cutting action and handle diameters from 2 to 12 in., an oscillating wheel is used (Fig. 16.31). The wheel is power fed and made to travel back and forth across the material. Cutting capacities are high. Wheels range from 16 to 34 in. in diameter, with corresponding motors of 7 1/2 to 30 hp.

**Fig. 16.31.** A swing-frame-type cutoff machine that is designed to move in and out as well as up and down. Units of this type are available that can travel up to 12 ft for cutting off plate and other special applications. Courtesy Wallace Supplies Mfg. Co.

## QUESTIONS

**16.1.** What are some advantages and disadvantages of the shaper?

**16.2.** Compare the use of a planer with that of a shaper.

**16.3.** What advantage has a double-housing type planer over an openside type?

**16.4.** Why is the standard or universal column-and-knee milling machine considered to be extremely versatile?

**16.5.** Sketch four distinct surface types and show how they can be done on both the vertical and horizontal mill by showing the cutter in place.

**16.6.** What is the main advantage of a fixed-bed milling machine?

**16.7.** What types of mills are used to sink die cavities?

**16.8.** Why must one know the machine characteristics before down milling can be used?

**16.9.** What are some methods used to increase production on a standard column-and-knee mill or a fixed-bed mill?

**16.10.** What is the advantage of using a face mill over a slab mill for a roughing cut on the same type of work?

**16.11.** Why must some distance other than the length of the piece be used to find the time required for milling?

**16.12.** How is it possible to broach an internal gear with a helix angle of more than 15 deg?

**16.13.** Why has broaching become competitive to milling?

**16.14.** (*a*) Which type of broaching machine do you feel would be the most suitable for making a 1/4-in.-dia hole into a square hole if the material is one inch thick? (*b*) Why?

**16.15.** What are two advantages of the continuous broaching machine?

**16.16.** Why can fixturing for parts for internal broaching be comparatively simple compared to those needed for surface broaching?

**16.17.** Explain the principle of friction sawing.

**16.18.** What are some advantages of the abrasive-disk cutting machine?

**16.19.** What tolerance would you expect to hold on a production run of internal splines on a cast-iron gear with a 1-in. I.D. and 1 in. thick?

**16.20.** What is the major disadvantage of broaching?

**16.21.** What are the advantages of a horizontal-type band saw over a power hacksaw?

## PROBLEMS

**16.1.** A large, medium-hard, cast-iron casting is machined on a planer. It is 20 ft long and 48 in. wide. Two roughing and two finishing cuts are required. Find the machine time needed if carbide tools are used at 250 fpm. Both of the cross-rail tools are used at the same time in tandem, set at different depths with the cutting edges 8 in. apart. The roughing feed is 0.060 in. and the finish feed is 0.020 in. The return stroke is made at two times the speed of the cutting stroke.

**16.2.** The horsepower for planing can be estimated using the cubic-inch removal rate per min as the basis. Thus the depth of the cut in in. × feed in in. × (speed in fpm × 12) × a constant = the approximate horsepower required. The constant for free cutting steel is 0.6 and for medium cast iron is 0.3. Find the approximate horsepower required for a roughing cut on the casting described in Prob. 16.1. The depth of cut for roughing is 1/2 in. for each tool.

**16.3.** A hole in a 1-in.-thick casting is 7/8 in. in diameter and is to have a 3/16-in.-deep, 1/8-in.-wide keyway broached in it. If the average amount of material removed by each tooth is 0.003 in. and the pitch of the broach teeth is 1/2 in., how long will the cutting portion of the broach have to be? How long will it take to broach 1,000 parts if the broaching speed is 30 fpm and a total of 1 min is allowed for the return stroke, handling each part, and placing the broach in position for the cut. Allow 2 in. for approach and overtravel at each stroke.

**16.4.** A cylinder head 8 in. wide and 20 in. long with 1/4 in. of material to be removed can be surface broached or milled. Milling is done with a 10-in. dia face mill in one pass. The carbide mill has 25 teeth with a chip load per tooth of 0.009. Use an overtravel of 1 in. The cutting speed is 300 fpm. No return of table is required since milling can be done both directions. The carbide broach has an average chip load of 0.005 in. per tooth with a pitch of 5/8 in. and moves over the surface at 150 fpm. One foot of travel is required before the broach is in full cut. Which method will be faster? If labor and overhead at $10 per hr are the same for each machine, how long will it take to pay for the broach, which costs $300 more than the milling cutter?

**16.5.** What would be the horsepower required for the milling operation as given in Problem 16.4 if the material being milled is soft gray cast iron?

## REFERENCES

The American Society of Tool and Manufacturing Engineers, *Tool Engineers Handbook,* McGraw-Hill Book Company, Inc., New York, 1959.

*Broaching Practice,* National Broach and Machine Co., Detroit, Michigan, 1953.

*Manual of Broaching,* Detroit Broach Co., Detroit, Michigan, 1948.

*Sawology,* a Nicholson Handbook, Nicholson File Co., Providence, Rhode Island, 1960.

*The DoAll Friction Sawing Manual,* The DoAll Company, Des Plaines, Illinois, 1953.

*Treatise on Milling and Milling Machines,* Cincinnati Milling Machine Company, Cincinnati, Ohio, 1951.

Warner, J. H., "Broaching Saves Time in Small Part Production," *The Tool and Manufacturing Engineer,* July 1958.

Wharen, H. S., "Throw-away Inserts or Inserted Blades Selection Affects Milling Costs," *Metalworking,* November 1962.

17

# GRINDING AND RELATED ABRASIVE-FINISHING PROCESSES

IT WAS WITH THE ABRASIVE WHEEL that precision was first brought to the metalworking industry. Tolerances on machine parts have become continually smaller, until we are no longer satisfied with one thousandth or one ten-thousandth, but, in many cases, hundred thousandths and millionths of an inch are required. These exacting tolerances are not achieved by grinding alone, but with the related processes of lapping, honing, and superfinishing.

Abrasives are also used to improve product appearance. These processes are barrel finishing, vibration finishing, polishing, and buffing.

## ABRASIVES

Abrasives are of two main types—natural and synthetic.

**Natural Abrasives.** Abrasive stones found in nature are emery, sandstone, corundum, and diamonds.

Both Turkish and American emery are natural mixtures of aluminum oxide and magnetite ($Fe_3O_4$). Due to the presence of some softer accessory minerals, both types have a milder cutting action than synthetic abrasives have. Emery is often preferred for use as an abrasive on coated cloth and paper as well as in many buffing compositions.

The large sandstone wheels used to grind hand tools have now almost passed out of existence. Emery and corundum, however, are still being used. Though these two products are found in nature and have different characteristics, they both contain aluminum oxide. The main difference is that corundum crystals are much larger, which makes it more suitable for fast, rough cutting (or *snagging*) on soft-steel and annealed malleable castings.

**Synthetic Abrasives.** SILICON CARBIDE. The search for a means of making precious stones by artificial methods led Edward Acheson to some experiments. In 1891, by means of a very crude carbon-arc furnace, he developed the first silicon carbide crystals. Although the industry is now large, the same basic method of mixing pure white quartz sand, petroleum, coke, salt, and sawdust is still used. Huge electric furnaces are used to drive off the impurities and fuse the silicon and carbide into a mass of interlocking iridescent crystals which vary in color from light green to black, depending upon the chemical composition.

ALUMINUM OXIDE. At the time silicon carbide was being discovered, work was being done on aluminum oxide. Calcined bauxite was mixed with coke and iron filings and subjected to a high temperature. This resulted in a glassy mass which, when cooled and crushed, became the most widely used abrasive in industry today—aluminum oxide. Today, precise selection of additives and the use of advanced technical controls has made it possible to have the physical characteristics specifically tailored to the abrasive need.

DIAMONDS. Diamonds can be classified as both natural and artificial. The natural stones unsuitable for gems are crushed down into a series of sizes known as *bort*.

The General Electric Company was the first to manufacture diamonds on a commercial scale. In 1955 they announced a successful commercial diamond that could sustain pressures up to 470,000 psi for long periods of time and extremely high temperatures.

Shortly after GE's successful manufacture of diamonds, the Carborundum Company fabricated the first man-made diamond wheel.

**Abrasive Properties.** Certain desirable properties inherent in abrasives are hardness, toughness, and fracture qualities.

Hardness allows the abrasive to enter the material, toughness keeps it from fracturing with the impact of the cut, and fracture allows part of the grain to leave the wheel so as to present a new cutting surface.

If the abrasive and the material holding it are too tough, the grains become dull and glazed. Under these conditions the work heats up, and very little material is removed. For this reason, silicon carbide, the harder abrasive, is considered the best for hard and brittle materials. Aluminum oxide is best for the tough, high-strength materials.

The natural diamond does not fracture to present new cutting faces, but becomes dull and tends to glaze after continued use. There are, however, applications where the diamond is superior to all other abrasives. An example of this is the finish grinding done on carbide tools.

There is some controversy over the comparative qualities of natural and artificial diamonds. Artificial diamonds are now being made in greater quantities, and the price is approaching that of natural diamonds. Man-made diamonds have greater *friability,* or breakdown, for cooler cutting action.

## GRINDING

**Grinding Wheels—Cutting Action.** The cutting action of a grinding wheel is dependent on the abrasive type, grain size, bonding materials, wheel grades, and wheel structure.

ABRASIVE TYPE. Industrial use of abrasives is largely limited to emery, aluminum oxide, silicon carbide, and diamonds.

GRAIN SIZE. After fusing the necessary ingredients in a furnace, the manufactured abrasive is taken out, cooled, and crushed. The grains are then cleaned and screened into many different size groups. Grit sizes are standard throughout industry. They are based on the number of openings per linear inch of screen (Fig. 17.1). A 60 grit size, for example, has approximately 60 grains per linear inch.

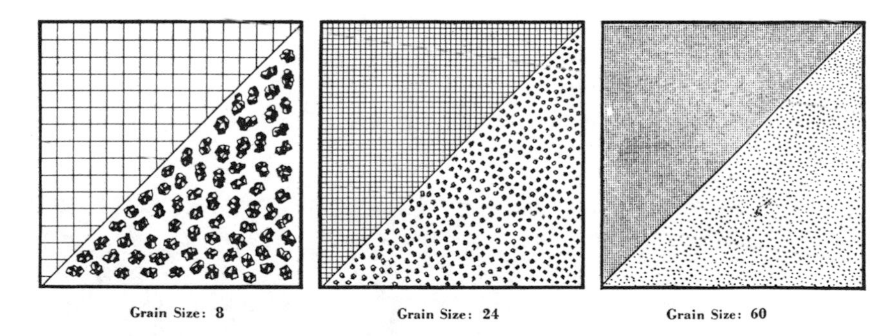


**Fig. 17.1.** Typical screens through which have been sifted grain sizes 8, 24, and 60. Screens shown represent 2-in. squares. Courtesy The Carborundum Co.

BONDING MATERIALS. The abrasive grains are held together in the grinding wheel by what is known as the bonding material. This material is of six main types. The four most extensively used—vitrified, resinoid, rubber, and shellac—will be discussed here.

VITRIFIED BOND. Although vitrified is actually a process (firing at about 2,300°F) rather than a material, it has become the name of the principle type of bond. The material is a blend of ceramic materials such as clay, feldspar, silica, talc, etc. When the wheel mix, comprising abrasive grains and bond, is fired, it becomes a fused glassy structure with predetermined strength (or weakness) according to closely controlled patterns for porosity or air spaces between grains and bond posts. Because of this porosity and strength, a wheel with this type of bond has a very favorable metal removal ratio, or grinding ratio ($R_g$). The $R_g = M_v/W_w$ where $M_v$ = volume of metal removed and $W_w$ = wheel wear (volume).

Vitrified bonded wheels are made to operate at 5,000 to 6,000 sfpm.

Resinoid bond. The resinoid type bond is next to the vitrified in order of usage. It is composed of a blend of synthetic resins of the phenol formaldehyde type such as Bakelite and Resinox. The bond is extremely strong and is capable of standing much higher operating speeds than the vitrified type, ranging up to 12,500 sfpm. It is used on high-speed roughing operations, cutoff wheels, and cup wheels for portable grinders.

Rubber bond. Either natural or synthetic rubber is used to make a more specialized bonding material. When used as wet cutoff wheels, rubber-bonded abrasives can make nearly burr-free cuts. Rubber-bonded wheels are also used for regulating wheels on centerless grinders and specialized applications where a high finish is required.

Shellac bond. Shellac, a natural resin, is now being replaced by resinoid; however, shellac is still used in some specialized finishing jobs.

**Wheel Grades.** The grade of a grinding wheel refers to its strength in holding the abrasive grains in the wheel. This is largely dependent on the amount of bonding material used (Fig. 17.2). As the amount of bonding material is increased, the linking structure between grains becomes larger, which makes the wheel act harder.

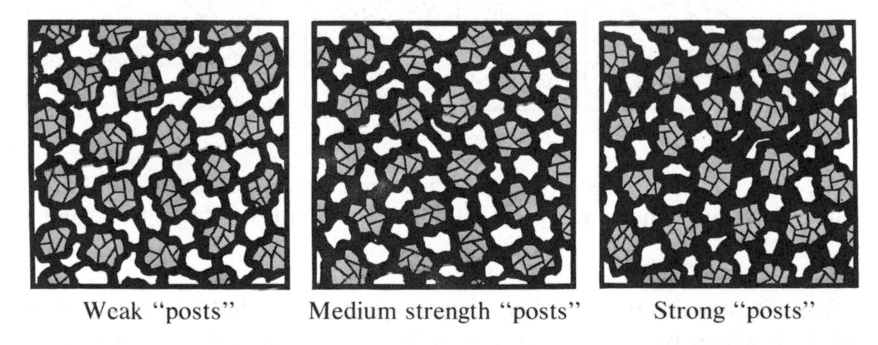


**Fig. 17.2.** The grade of the grinding wheel is based on the strength of the bond posts. Courtesy The Carborundum Co.

**Wheel Structure.** Grinding should not be thought of as a rubbing action. The grinding wheel is more like a milling cutter with hundreds of teeth. You will notice that the wheel must have voids to allow space for the chips and for proper cutting action (Fig. 17.3). This space must not be too small or the chip will stay in the wheel, causing it to "load up." A

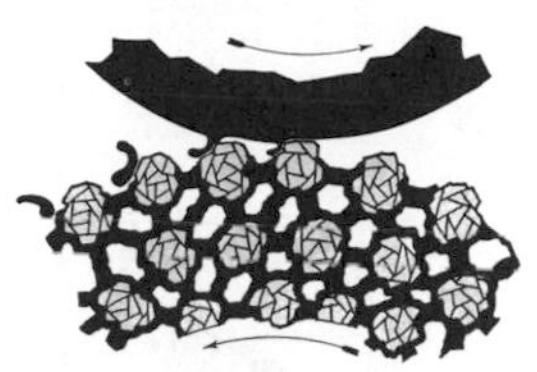

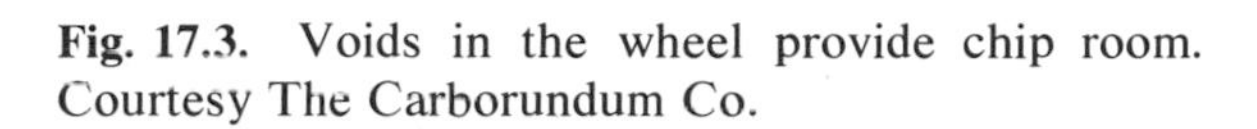

**Fig. 17.3.** Voids in the wheel provide chip room. Courtesy The Carborundum Co.

loaded cutting wheel heats up and is not efficient in cutting action. Too large a space is also inefficient, as there will be too few cutting edges.

**Grinding Wheel Selection.** From the foregoing discussion, you can see that there are many factors which must be taken into consideration in choosing a wheel for a given job.

Grinding-wheel manufacturers developed codes for putting on the particular wheel all the information for its identification. Each company had its own system, and this was very confusing. However, in 1944, a standard code of seven main parts, as shown in Fig. 17.4, was adopted by all the grinding-wheel manufacturers.

Among the factors that help decide what type of wheel to use are the material to be ground, the type of operation, and the machine condition.

MATERIAL TO BE GROUND. If the material to be ground is hard, the bond or grade of the wheel must be relatively soft to ensure that the abrasive grains will be able to break away before they become dull and glazed. If the material is soft, the grade of the wheel can be relatively hard, since the abrasive will stay sharp for a longer time. High-tensile-strength material should be ground with aluminum oxide abrasive. For most low-tensile-strength materials, silicon carbide should be used.

TYPE OF OPERATION. The type of operation will determine the wheel contact area. A straight, or plain, wheel, used to grind the outside diameter of a cylinder, has only a small contact area. Under these conditions, the wheel breaks down much faster than when it has a larger contact area, as in grinding flat or internal surfaces. Since the relative pressure on the wheel is high for a small contact area, the structure must be of a harder grade than is needed for grinding the same material with a larger surface contact.

MACHINE CONDITION. The type and condition of the machine will determine the size of wheel that can be used, the work speed, the feed, etc.

JOB SPECIFICATIONS. Since so many varying conditions exist in grinding, manufacturers often urge their customers to send in the exact job specifications. As with many other processes where variables are numerous, grinding-wheel manufacturers have turned to punched cards for help. The five main grinding-wheel elements—abrasive type, grain size, grade, structure, and bond—make up the main code. With these five main elements, over 150 variables can be matched according to the customer's job specifications. One change in any of these variables

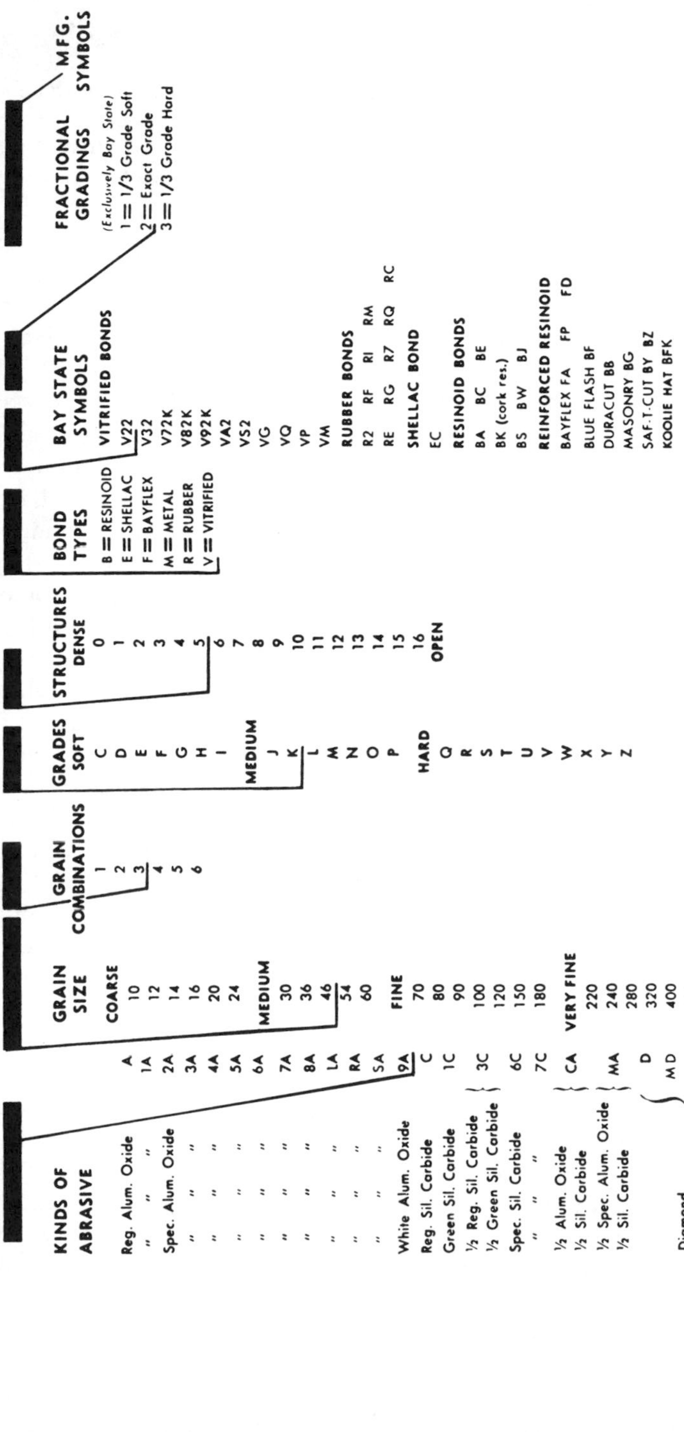

# 9A-463 K5-V22 AE

**KINDS OF ABRASIVE**

|  |  |
|---|---|
| Reg. Alum. Oxide | A |
| " " " | 1A |
| Spec. Alum. Oxide | 2A |
| " " " | 3A |
| " " " | 4A |
| " " " | 5A |
| " " " | 6A |
| " " " | 7A |
| " " " | 8A |
| " " " | LA |
| " " " | RA |
| " " " | SA |
| White Alum. Oxide | 9A |
| Reg. Sil. Carbide | C |
| Green Sil. Carbide | 1C |
| ½ Reg. Sil. Carbide / ½ Green Sil. Carbide | 3C |
| Spec. Sil. Carbide | 6C |
| " " " | 7C |
| ½ Alum. Oxide / ½ Sil. Carbide | CA |
| ½ Spec. Alum. Oxide / ½ Sil. Carbide | MA |
| Diamond | D, MD, BD, SND |

**GRAIN SIZE**

COARSE: 10, 12, 14, 16, 20, 24
MEDIUM: 30, 36, 46, 54, 60
FINE: 70, 80, 90, 100, 120, 150, 180
VERY FINE: 220, 240, 280, 320, 400, 500, 600

**GRAIN COMBINATIONS**

1, 2, 3, 4, 5, 6

**GRADES**

SOFT: C, D, E, F, G, H, I
MEDIUM: J, K, L, M, N, O, P
HARD: Q, R, S, T, U, V, W, X, Y, Z

**STRUCTURES**

DENSE: 0, 1, 2, 3, 4, 5, 6, 7, 8, 9, 10, 11, 12, 13, 14, 15, 16 OPEN

**BOND TYPES**

B = RESINOID
E = SHELLAC
F = BAYFLEX
M = METAL
R = RUBBER
V = VITRIFIED

**BAY STATE SYMBOLS**

VITRIFIED BONDS: V22, V32, V72K, V82K, V92K, VA2, VS2, VG, VQ, VP, VM
RUBBER BONDS: R2 RF RI RM, RE RG R7 RQ RC
SHELLAC BOND: EC
RESINOID BONDS: BA BC BE, BK (cork res.), BS BW BJ
REINFORCED RESINOID: BAYFLEX FA FP FD, BLUE FLASH BF, DURACUT BB, MASONRY BG, SAF-T-CUT BY BZ, KOOLIE HAT BFK

**FRACTIONAL GRADINGS**

(Exclusively Bay State)
1 = 1/3 Grade Soft
2 = Exact Grade
3 = 1/3 Grade Hard

**MFG. SYMBOLS**

**Fig. 17.4.** The standard marking system for grinding wheels. Courtesy Bay State Abrasive Products Co.

will make a totally different wheel. A computer is used to relate the basic information to the variables, thus producing a grinding-wheel formula. This formula is then used in weighing out and mixing the ingredients that make up the wheel. Examples of grinding-wheel manufacturers' recommendations are shown in Table 17.1.

**Table 17.1**

| | |
|---|---|
| General purpose surface-grinding aluminum | 37C24J8 |
| General purpose surface-grinding steel | 32A60G12 |
| General purpose surface-grinding brass | 37C24L5B |

**Wheel Care and Use.** SAFETY. Grinding wheels run at comparatively high speeds, the recommended speed for most operations being 5,000 to 6,000 sfpm. Grinding-machine manufacturers follow safety codes in providing proper guards. Wheel manufacturers' specifications as to testing, balancing, mounting, and operating must be followed closely to ensure safe practice.

BALANCING. Wheels that are out of balance not only produce poor work but may put undue strains on the machine. Large wheels should be placed on a balancing stand (Fig. 17.5) and balanced by moving weights around a recessed flange. Some manufacturers now provide for dynamic balancing right in the machine.

DRESSING AND TRUING THE GRINDING WHEEL. *Dressing.* After use, the wheel becomes dull; that is, the individual abrasive grains become rounded over. The wheel may also become loaded. It is then necessary to cut a portion of the face off the grinding wheel.

**Fig. 17.5.** The weights in the flange are shifted until the wheel is in balance, as shown. Courtesy The Carborundum Co.

Abrasive sticks and wheels are used for dressing purposes. For precision grinding, the wheels are dressed with a diamond as shown in Fig. 17.6.

A good supply of coolant should be used when dressing with a diamond, as overheating can cause the diamond to fracture or drop out of its setting. Cluster or multipoint diamonds are used on larger wheels.

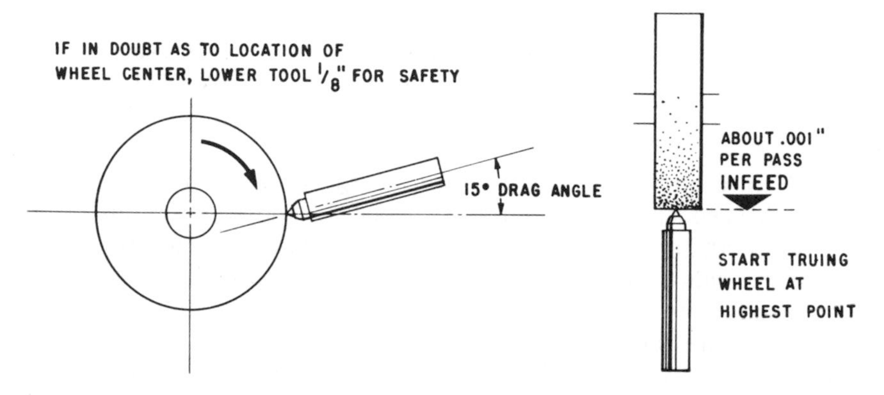


**Fig. 17.6.** Diamond position, and recommended infeed per pass. Courtesy The Desmond-Stephan Mfg. Co.

*Truing.* Truing is the process of changing the shape of the grinding wheel. This is done to make the wheel true and concentric with the bore or to change the face contour for form grinding. Truing and dressing may be done with the same tool, but they are not done for the same purpose.

**Diamond Wheel Use.** Diamond wheels are made with three different types of bonds: resinoid, vitrified, and metallic. Each has particular applications, with some uses overlapping. Vitrified wheels cut well but tend to be brittle. Resinoid wheels can take a lot of abuse and are good for heavier metal removal, as in cutting chip breakers in carbide tools. The metallic-bonded diamond wheels will outlast, by 10 times, the resinoid-bonded wheels, but they are more apt to cause heat checks.

Diamond and other wheels are made in a variety of standard shapes (Fig. 17.7). In order to conserve diamonds, wheels larger than 1 in. in

**Fig. 17.7.** Standard diamond grinding-wheel types. Courtesy The Carborundum Co.

diameter are produced with a bonded diamond layer at the cutting surface.

All diamond wheels operate at greater efficiency when used wet. Either flood- or mist-coolant applications are effective. Plain water is an effective coolant. However, if rusting is a problem, water-soluble oils or mineral-seal oils are used. They also provide a better finish.

Diamond wheels can be cleaned with a lump of pumice or a stick of fine soft silicon carbide. A diamond wheel should not be dressed with a diamond tool.

Only hard materials should be ground on diamond wheels. Soft materials tend to load the wheel quickly. For this reason, carbide tools are often ground with three wheels: (1) aluminum oxide for the secondary clearance on the tool shank, (2) silicon carbide wheels for rough-grinding the carbide, and (3) diamond wheels for the finish grind of the carbide tip.

Ceramic cutting tools are similar to carbide cutting tools in grinding characteristics. Either silicon carbide or diamonds are used as the abrasive. The moderately hard ceramic materials, such as porcelain, can be satisfactorily ground with silicon carbide, but ceramics of high aluminum oxide content should be ground with metal-bonded diamond wheels.

**Grinding Operations and Equipment.** The wide variety of grinding operations can be classified as to the main types of surfaces to be ground. These include cylindrical, surface, and internal (Fig. 17.8).

CYLINDRICAL GRINDING. Cylindrical grinding is usually characterized by having the work mounted between centers (Fig. 17.8*a*) and traversing back and forth parallel to the wheel axis (Fig. 17.9). Work produced is true and concentric with the centerholes.

CENTERLESS GRINDING (Fig. 17.10*a*) does not require mounting of the work, which makes such machines capable of higher production. The work rests on a support between the grinding wheel and the regulating wheel. Feed is accomplished by setting the regulating wheel at an angle (Fig. 17.10*b*).

PLUNGE-CUT GRINDING (Fig. 17.11) is a variation of cylindrical grinding in which no table traverse movement is used. The wheel is fed straight into the work. The wheel face may be straight or formed by *crushing*.

A hardened-steel or carbide roller of the same profile as the desired workpiece is fed into the abrasive wheel until the wheel takes on the reverse form of the roll.

Either of the wheels may be rotated at approximately 300 fpm, with an infeed of a few thousandths per revolution. Forming the wheel takes only a few minutes, and redressing it only a few seconds.

The best wheels for crush forming are the vitrified bonded ones, although resinoid wheels are also used for this purpose.

(*a*) Cylindrical grinding

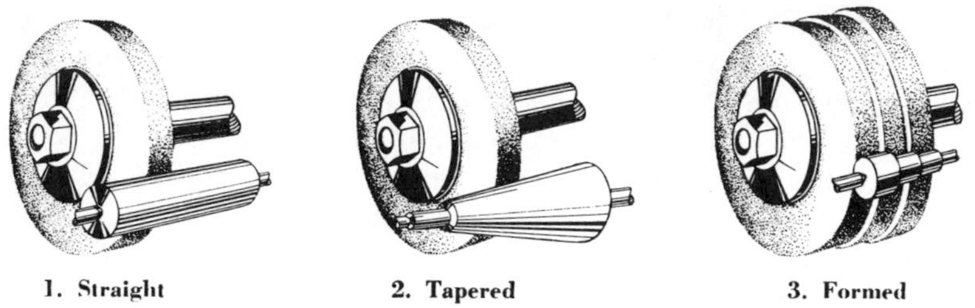

1. Straight

2. Tapered

3. Formed

(*b*) Surface grinding

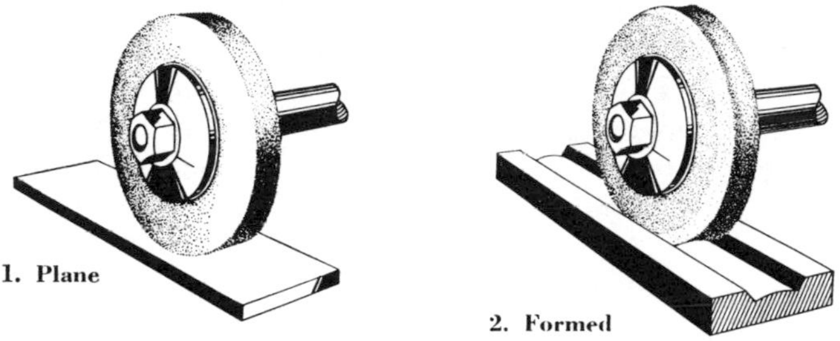

1. Plane

2. Formed

(*c*) Internal grinding

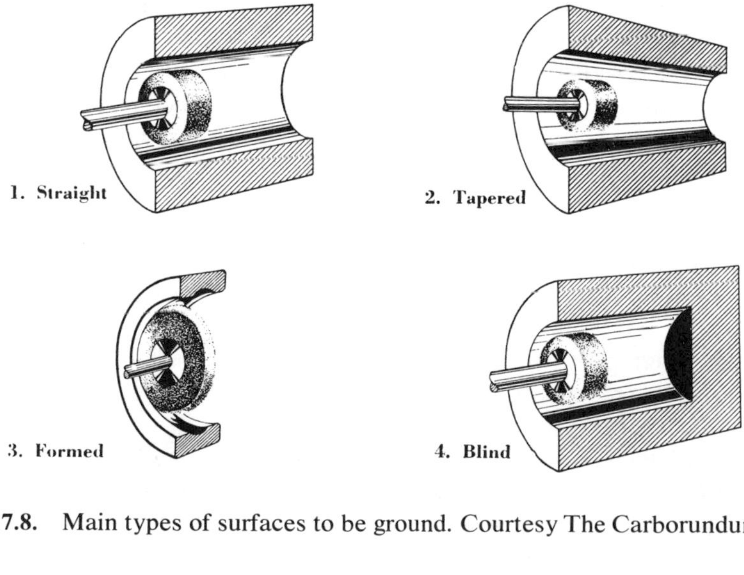

1. Straight

2. Tapered

3. Formed

4. Blind

**Fig. 17.8.** Main types of surfaces to be ground. Courtesy The Carborundum Co.

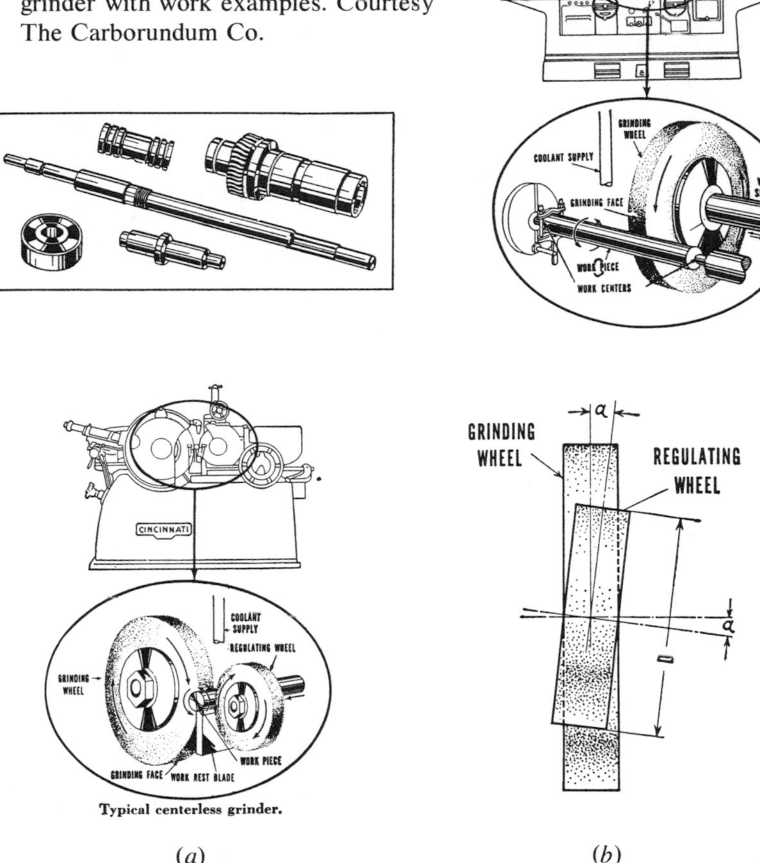


**Fig. 17.9.** Center-type cylindrical grinder with work examples. Courtesy The Carborundum Co.

**Fig. 17.10.** (*a*) Centerless grinding as applied to straight cylindrical work in "through-feed" grinding. The regulating wheel (*b*) is set at an angle to control the rate at which the stock is fed past the surface of the grinding wheel.

Crush forming of the grinding wheel has many advantages. It is not only faster than diamond dressing, but it is possible to make forms that are impossible by any other method. When a wheel has been dressed by crushing, it has been found that the wheel cuts better since many more sharp cutting points are left than by conventional dressing. Close tolerances can be maintained, and grooves as narrow as 0.020 in. can be cut. Surface finishes as low as 8 $\mu$in. are obtainable.

**Surface Grinding.** All grinding could be construed to be surface grinding. However, the term refers specifically to flat or plain surfaces.

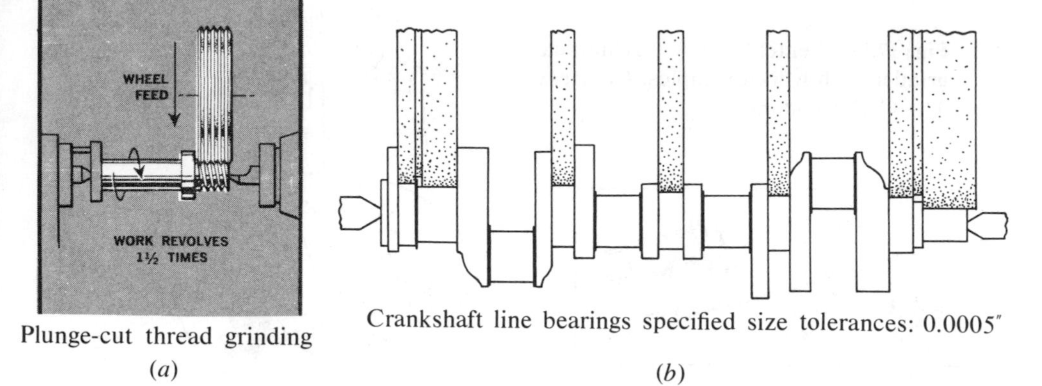


Plunge-cut thread grinding
(*a*)

Crankshaft line bearings specified size tolerances: 0.0005″
(*b*)

**Fig. 17.11.** (*a*) Plunge-cut thread grinding with a wheel face that has been crush formed. (*b*) Plunge-cut crankshaft grinding with both straight and formed wheels.

The two main types of machines are classified, by the position of the spindle, as horizontal or vertical (Fig. 17.12).

The horizontal-type surface grinder uses a plain wheel. Special shapes of wheels, such as the cup and cylinder type, are used on thc vertical surface grinder. The tables shown in Fig. 17.12 are of the reciprocating

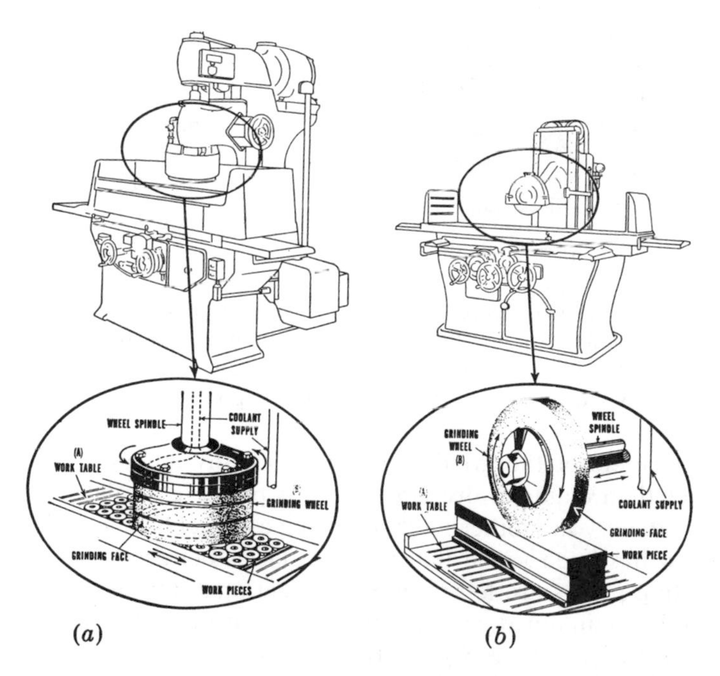


(*a*) (*b*)

**Fig. 17.12.** (*a*) Vertical and (*b*) horizontal surface grinders. Courtesy The Carborundum Co.

kind, but both types of grinders are also built with rotary work tables. Ferrous materials are usually held on the table by electromagnetic action. Nonferrous materials can be held by a vacuum chuck, a vise, or a fixture.

**Tool and Cutter Grinding.** Tool and cutter grinding refers to sharpening all types of metal-cutting tools. Single-point tools used in lathe work are often sharpened by hand on a bench- or pedestal-type grinder. This is called *offhand* grinding. Carbide and ceramic throwaway inserts have eliminated a large amount of this type of grinding.

Cutter grinding refers more specifically to the sharpening of milling cutters, hobbing cutters, drills, reamers, etc. It is considered economical to grind a metal-cutting tool *before* it gets very dull. Sharpening at the proper time will give longer tool life, better finish, and closer tolerances. The horsepower required will also be less.

Cutters are ground on special tool and cutter grinders or universal grinders. The wide variety of attachments available for these machines makes it possible to hold the work either between centers, in a chuck, in a vise, or directly in the headstock spindle.

**Snag Grinding.** Snag grinding is done where a considerable amount of metal is removed without regard to the accuracy of the finished surface. Examples of snag grinding are trimming the surface left by sprues and risers on castings, removing the excess metal on a weld, or grinding the parting line left on a casting.

Three types of machines are used for snag grinding; these are stand or bench grinders, portable grinders, and swing-frame grinders (Fig. 17.13).

Castings and rough workpieces that can be handled are brought to the stand grinder. The wheels may be as large as 36 in. in diameter on these solidly built machines.

Castings that are too large for the operator to hold up to the wheel are ground with a swing-frame grinder. This machine is moved around with a jib crane suspended from columns, or by mobile units.

Portable grinders are self-contained or flexible units used for miscellaneous grinding operations such as removing excess weld material, defective spots needing repair, parting lines, and fins. They are easily moved about and are used to the best advantage in removing comparatively small amounts of stock from widely separated areas.

## ABRASIVE MACHINING

*Abrasive machining* is a comparatively new term (1961) intended to denote that grinding is competitive with metal cutting by the more conventional means.

Grinding has not been associated with higher rates of metal removal due to the possibilities of burning the work and of loading or stalling the

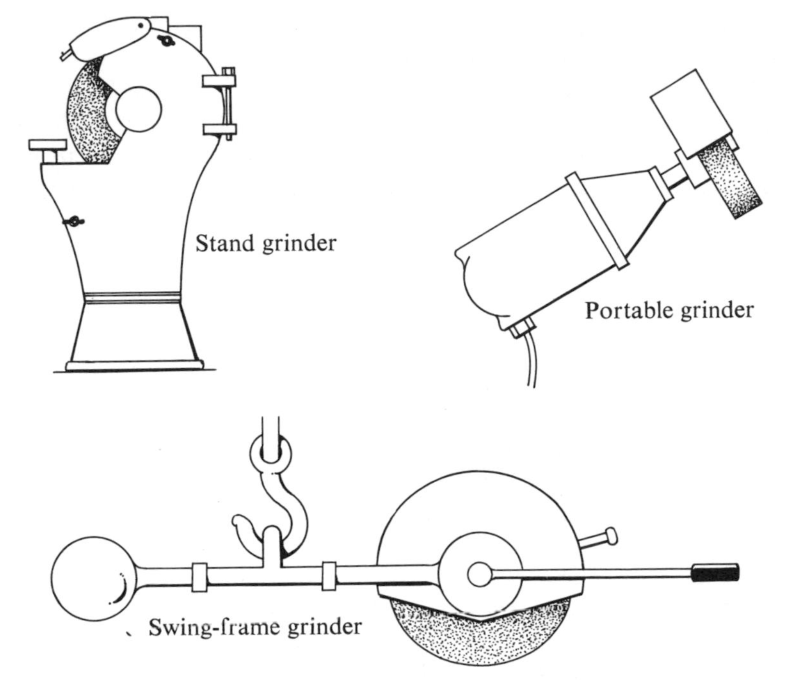


**Fig. 17.13.** Types of machines used for snag grinding.

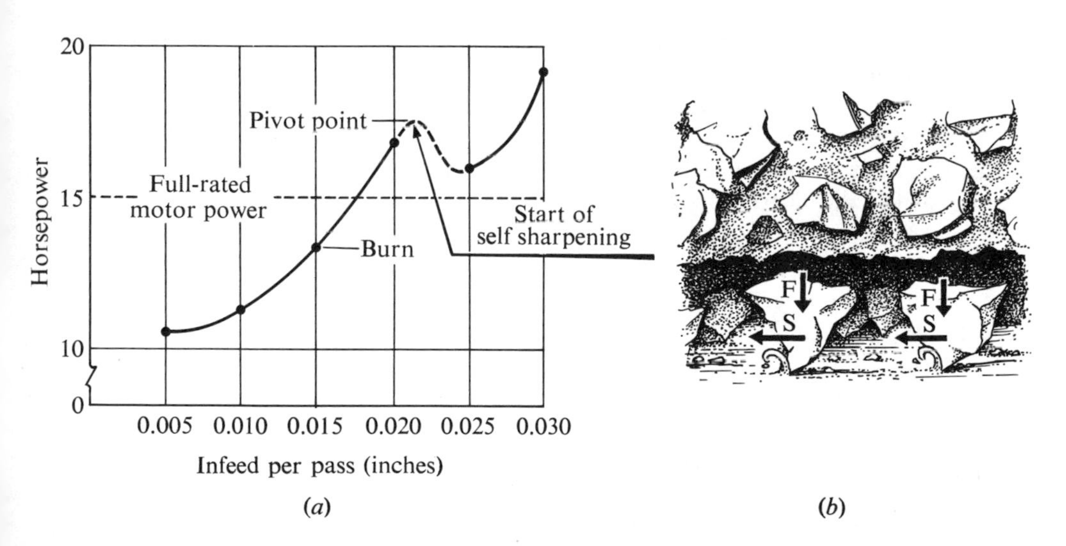


**Fig. 17.14.** (*a*) As the infeed of the wheel increases, a point of burn is reached; however, beyond this point the wheel goes into more effective self or reactive *dressing*, and a "pivot point" is reached at which cutting is cooler and the energy required is less. *(b)* The abrasive grain cutting action sketch shows the force (F) acting on the grain and its shear plane (S).

wheel. Now, however, wheels are made to break down at just the right rate to keep the cutting edges sharp. This has become known as *reactive dressing.* Shown in Fig. 17.14, as the infeed rate is increased, the power goes up. At 0.015 in. it burned the work, at 0.020 in. the burn disappeared, and between 0.020 and 0.025 in. the power required dropped off. In this last area, reactive dressing takes place.

**Economic Consideration.** Shown in Fig. 17.15 and Table 17.2 is a comparison of a part machined on a vertical mill and by abrasive ma-

**Table 17.2**

| Job data | Present method | Proposed method |
|---|---|---|
| Machine | vertical mill | Blanchard no. 18 |
| Tool | 7″ face mill | segments, no. 18A |
| Horsepower | 30 hp | 35 hp |
| How loaded | 1 pc @ time | 6 pc @ time |
| Speeds | cutter: 250 sfpm | wheel: 720 rpm |
| Feeds | table: 25″/min | table: 18 rpm<br>wheel: 0.048″/min |
| Stock removal per piece | 6.75 in.$^3$ | 6.75 in.$^3$ |
| Machining time per piece | 1.76 min | 1.62 min |
| Rate of stock removal | 3.84 in.$^3$/min | 4.16 in.$^3$/min |
| Floor-to-floor time per piece | 4.76 min | 3.70 min |
| Total cost per piece | 55.3¢ | 32.3¢ |
| Setting-up costs | $2.07 | $1.00 |

Remarks: Abrasive machining reduced costs by 42% and increased production by 29%.

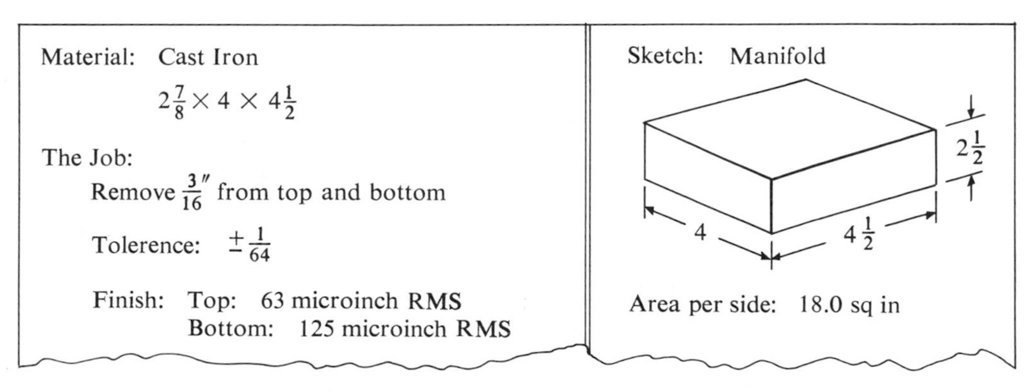


**Fig. 17.15.** A comparison of vertical milling and abrasive machining costs for the manifold part shown. See data in Table 17.2.

chining. There are a number of factors that make abrasive machining competitive such as: (*a*) rough and finish cuts can be done in one operation; (*b*) fixtures can be largely eliminated; (*c*) oftentimes cutting tools and their maintenance are carried as overhead costs whereas the abrasive wheel is charged as a direct expense to the job; fixtures can be eliminated or made much simpler; and the handling time is greatly reduced.

Another consideration which does not show up in direct comparison of machining methods is that less stock needs to be allowed for abrasives than for cutters. Thus there is also an initial savings.

High feed rates decrease the time cost, but increase the wheel cost (Fig. 17.16). The sum of the two shows the total unit cost excluding the portion independent of the feed rate, such as material handling.

An important indication of wheel wear efficiency is the "*G* ratio" or grinding wheel wear to volume of metal removed, which was mentioned previously. Some typical grinding ratios for gray iron castings are shown in Fig. 17.17. As much as 45 cu in./min have been removed economically by abrasive machining in gray iron castings.

## LAPPING, HONING, AND SUPERFINISHING

Grinding is considered an accurate method of producing surface finishes. However, it is not always possible or economical to achieve the

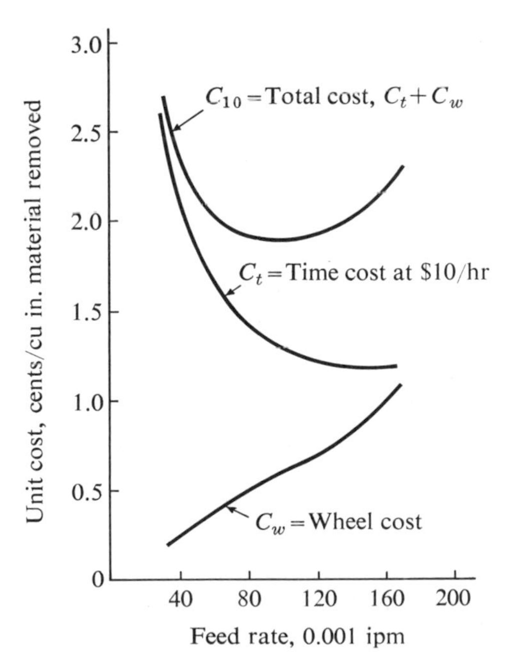


**Fig. 17.16.** Abrasive machining cost as a function of feed rate. Increasing the feed rate increases wheel cost but decreases time cost for unit material removed. The summation of the two cost curves gives the total unit cost and shows at what feed rates the most economical operation would be.

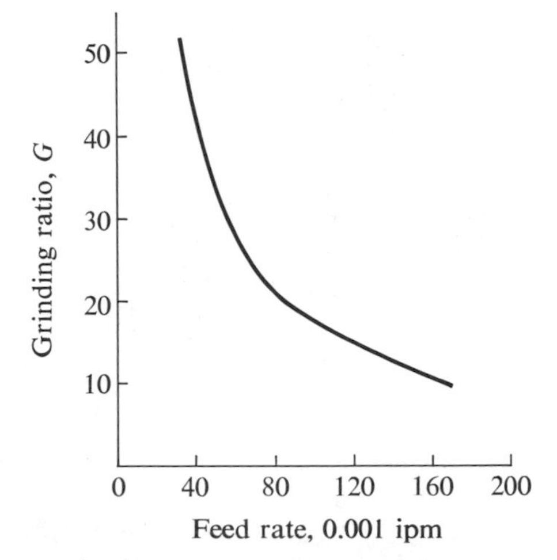


**Fig. 17.17.** Some $G$ values for various feed rates obtained in abrasive machining gray iron castings.

degree of accuracy and finish desired. The more refined methods of lapping, honing, and superfinishing are used to obtain the ultimate in finishes.

**Lapping.** Lapping is an abrading process done with either abrasive compounds or solid bonded abrasives. Lapping is performed primarily to increase accuracy, improve the surface finish, and match mating surfaces. Gage blocks are lapped to ± 0.000002 in. per in. of length, and they are also parallel within this dimension. Surface finishes obtained are within 1 or 2 rms. The reason lapping is more accurate than grinding and some other finishing operations is that very little heat and pressure, which induce strains in the finished part, are involved.

THE LAPPING PROCESS. Lapping consists of using a loose abrasive to wear a surface to the desired contour and finish. A soft porous metal such as gray cast iron holds the abrasive temporarily and rubs it against the surface to be cut. For example, gear teeth are lapped by running them with a *master* gear. Lapping compound consists of fine-grain abrasive mixed with a liquid such as cutting oil to act as a carrier. The type of abrasive and carrier are dependent on the materials to be lapped. Soft materials can be lapped with aluminum oxide or emery, and hard materials with diamond or silicon carbide grit. Valves are often lapped to valve seats by rotating the two surfaces together with lapping compound in between.

*Machine lapping.* In machine lapping, the lapping plate becomes a rotating table (Fig. 17.18). The parts to be lapped are confined to cages that impart a rotary and gyratory motion at the same time, covering the entire surface of the lapping table. Parallelism is maintained by having a stationary lapping plate on top of the workpieces. Cylindrical piece-part lapping is done by a cage arrangement as shown in Fig. 17.19.

A newer type of lapping machine uses vibratory rather than rotary

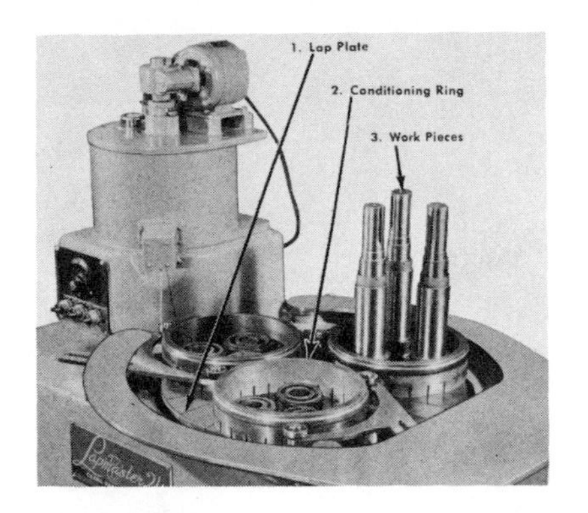


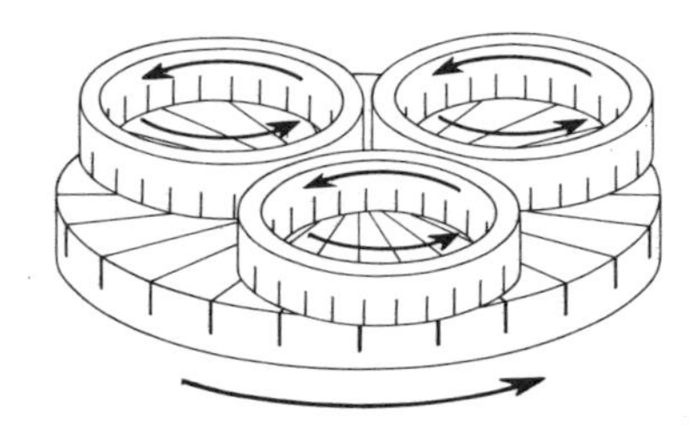

Examples of work pieces with lapped surfaces

**Fig. 17.18.** Work to be lapped is placed within the conditioning rings on two of the rotating lap plates. The conditioning rings are held in place but are free to rotate. The work tends to wear the lap plate, but the rotating action of the conditioning rings causes the lap plate to wear evenly, maintaining a flat surface. A wide variety of shapes and surfaces can be lapped. Standard machines handle parts from 1/8 in. to 32 in. Steel, tool steel, bronze, cast iron, stainless steel, aluminum, magnesium, brass, quartz, ceramics, plastics, and glass can be lapped by the same lap plate. Courtesy Crane Packing Co.

motion. In this case the lapping plate, or pan as it is called, is covered with an abrasive cloth that is cemented down. Light oil is used as a lubricant. Parts are held down on the abrasive paper by means of one or more pressure plates. There are no rotating parts. An electromagnet is used to furnish 3,600 vibrations per minute. A combination of drive components causes the parts to flow around the bottom of the pan.

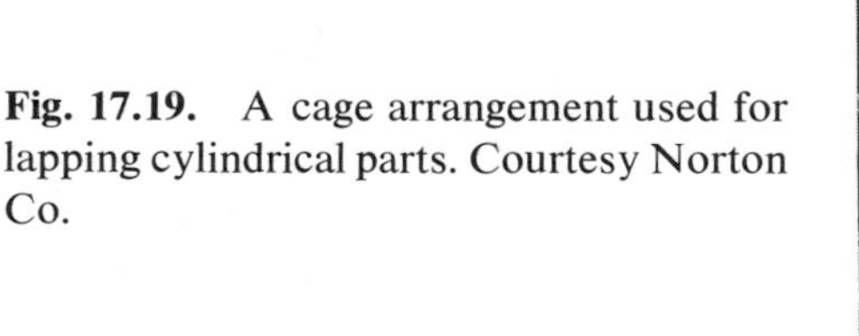

**Fig. 17.19.** A cage arrangement used for lapping cylindrical parts. Courtesy Norton Co.

*Bonded-abrasive lapping.* Lapping compounds tend to leave a dark, dull surface that is hard to remove. Solid bonded abrasives are used on machines, similar to the vertical type shown, to remove the lapping compound and improve the finish.

**Honing** can be defined as *controlled low velocity abrasive machining.* It is applied to many different types of surfaces but most often to ID cylindrical type. The three main functions of honing are to:

(1) correct workpiece geometry, that is, shape (Fig. 17.20*a*);

(2) produce accurate size (Fig. 17.20*b*); and

(3) produce the desired surface character—(*a*) pattern, or lay, (*b*) structure, (*c*) roughness (Fig. 17.21).

The control mentioned in the definition is the most important aspect of honing. It refers to the self-sharpening aspect of the abrasive stones used. If they are not breaking down at the right rate, they will either get dull and not cut or they will break away too soon and cause needless expense.

In cylindrical honing, the stones are inserted in the tool and equally spaced around its periphery (Fig. 17.21*a*). A cone-shaped wedge is used to force the stones out radially.

A combination of motions gives the stones a figure-eight travel path (Fig. 17.21*b*). This motion causes the forces acting on the stones to be continually changing their direction.

To assure equal pressure on all stones and all areas of the cylinder, either the stones, the tool, or the workpiece is permitted to float. The axes of the tool and the workpiece align themselves automatically.

Accurate size control is provided by a number of methods, both contact and noncontact types. An example of the noncontact type is feeding the coolant fluid through the honing tool. As the bore becomes larger, the pressure builds up to a predetermined size and then terminates the cycle.

TOLERANCE AND FINISH. Honing, unlike grinding, is done at low speeds. Very little heat is generated. Therefore there is no submicro-

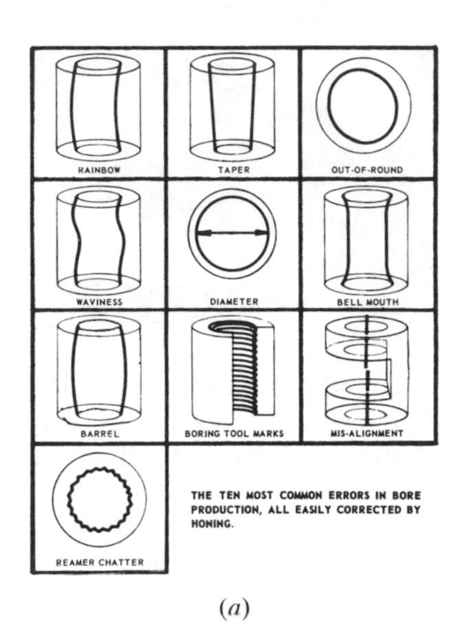


(*a*)

Work exampie

*Part:* Comparator column
*Use:* Gage
*Material:* Cast iron
*Hole diameter:* 0.813″ ± 0.0002″, tandem
*Length:* 1″ each dia
*Finish:* Smooth base metal
*Stock removal:* 0.0015″ to 0.002″
*Gage:* Ames Comparator
*Prev. operation:* Grind
*Production Rate:* 50 per hr

(*b*)

**Fig. 17.20.** (*a*) Bore geometry that can be corrected by honing. (*b*) An example of accuracy obtained and production rate. Courtesy Sunen Products Co.

scopic damage to the workpiece surface. Tolerances are easily maintained within 0.0001 in.

The surface finish produced can be very smooth (1 μin.) or rough (100 μin.). A crosshatch pattern of varying angles is imparted to the work surface to enhance the lubrication-holding qualities (Fig. 17.22).

In repair work, cylinder walls are honed not only to correct the geometry but to replace the glazed surface with a crosshatch pattern.

STOCK REMOVAL. The rate of stock removal varies with the material and the size of the work. A hone that would be used for a 3/4-in.-diameter hole 2 1/2 in. long could be expected to remove metal at the following rates: hardened tool steel, 0.0015 in. per min; annealed steel SAE 1020, 0.004 in. per min; 3004 ST aluminum, 0.010 in. per min.

**Superfinishing.** Superfinishing is somewhat similar to honing but is applied primarily to external surfaces. The process was first conceived in 1936 in answer to the age-old problem of obtaining maximum wear on mating parts. Extensive studies indicated that the more closely the bore and shaft diameters approximate each other, the greater will be the load-carrying capacity of the bearing. The oil film between the mating members is in danger of being punctured by any sharp points left by the

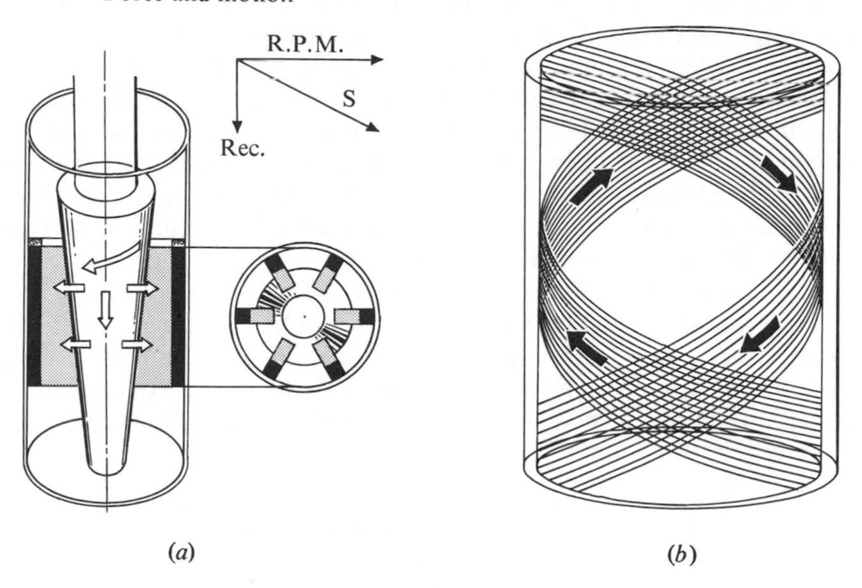


**Fig. 17.21.** (*a*) The hone stones are equally spaced around the periphery. A cone-shaped wedge in the center is used to force the stones out radially. (*b*) A combination of motions gives the stones a figure-eight travel path which develops the crosshatching pattern.

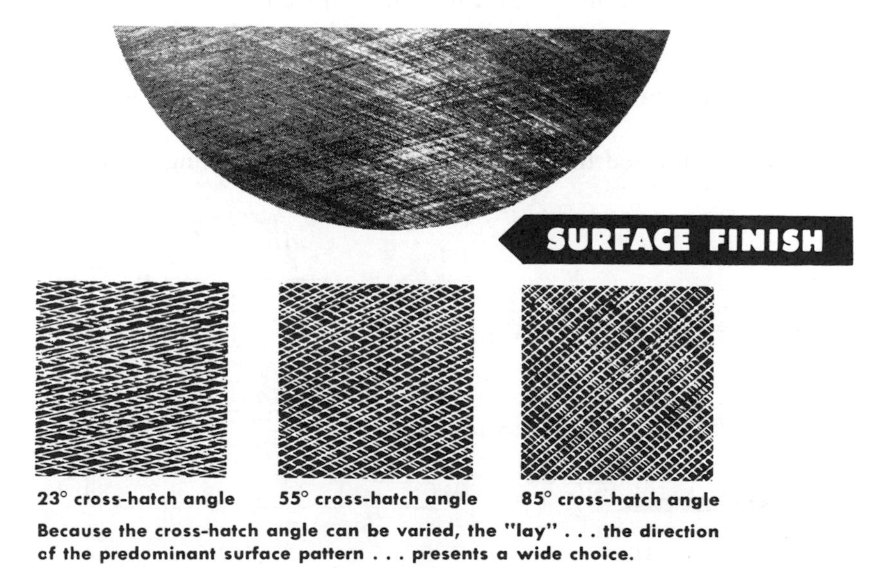


**Fig. 17.22.** Crosshatch pattern obtained by honing. Courtesy Micromatic Hone Corp.

finishing process (Fig. 17.23). As these sharp points penetrate the oil film, the concentration of weight and friction cause the two metals momentarily to weld and tear apart, causing rapid wear conditions. Wear will help break down the high points and correct some of the out-of-roundness. The difficulty lies in the fact that the foreign particles score the smooth surfaces. For this reason, hardened rotating shafts should have the smoothest surface possible. Owing to the difficulty in maintaining an oil film on reciprocating parts, a crosshatch pattern is recommended. Scratches at a substantial angle to the motion of the two bearing surfaces will break up minute weldments from one part to the other. It is also easier for the oil contained in the valleys to move across the angular ridges than if the ridges were in the lengthwise direction. The depth of the crosshatch pattern will depend on the severity of conditions that tend to produce welding, but the depth should seldom exceed 10 to 12 $\mu$in.

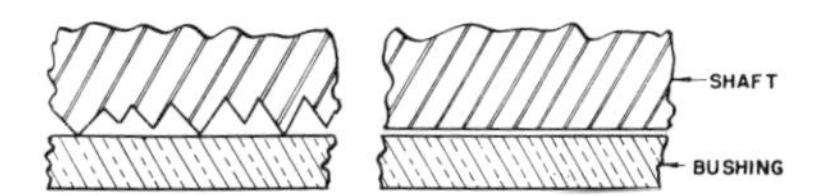


**Fig. 17.23.** A smooth surface does not have high load-concentration points that puncture the oil film. Courtesy Gisholt Machine Co.

EQUIPMENT. The superfinishing equipment for cylindrical surfaces consists of an abrasive stone mounted on the end of a spring-loaded quill (Fig. 17.24). The assembly has an oscillating mechanism by which the stone can move back and forth a maximum distance of 3/16 in. at 425 cycles per min. The assembly can be mounted on the cross slide of a lathe or on a machine especially designed for the purpose. If the part to be superfinished is not longer than the stone, traverse motion will not be necessary.

The stone is formed to the radius of the work by putting emery cloth

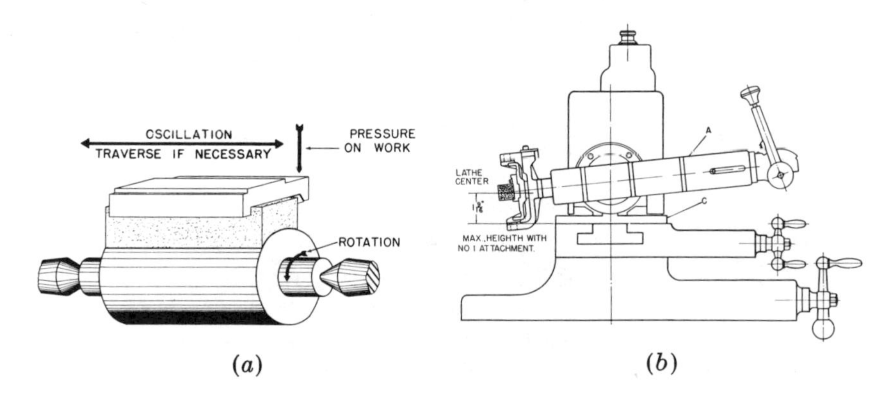


**Fig. 17.24.** (*a*) Basic equipment used for cylindrical superfinishing. (*b*) Superfinishing attachment for a lathe. Courtesy Gisholt Machine Co.

over the workpiece, face out. As the stone contacts the emery cloth, it is soon worn in to the approximate radius needed. The rest is taken care of during the operation.

After the stone is dressed, it is brought in contact with the work at about 20 or 30 psi pressure. The part is rotated at a speed equal to 50 or 60 fpm. The oscillation motion is started, and coolant is allowed to flood the area. If traverse is not needed, the part can be finished in less than a minute.

Flat surfaces are superfinished with a machine containing both an upper and a lower spindle (Fig. 17.25). The upper spindle is a spring-loaded quill on which a stone is mounted. The lower spindle carries the work. Both spindles are parallel so that, when they rotate, the result is a very flat smooth surface.

TOLERANCE AND FINISH. The superfinishing process is not designed to remove material or correct part geometry. The amount of material removed from the part may range from 0.0001 to 0.0004 in. on the diameter. The surface finish produced ranges from less than 1 to 80 $\mu$in., with the average around 3. When a surface smoother than 3 $\mu$in. is desired, the time needed to produce it goes up rapidly. As an example, a part that can be brought to a 3-$\mu$in. finish in 1 min may require 3 min more to bring it to a 2-$\mu$in. finish or better. Figure 17.26 shows a range of finishes that can be produced by this process.

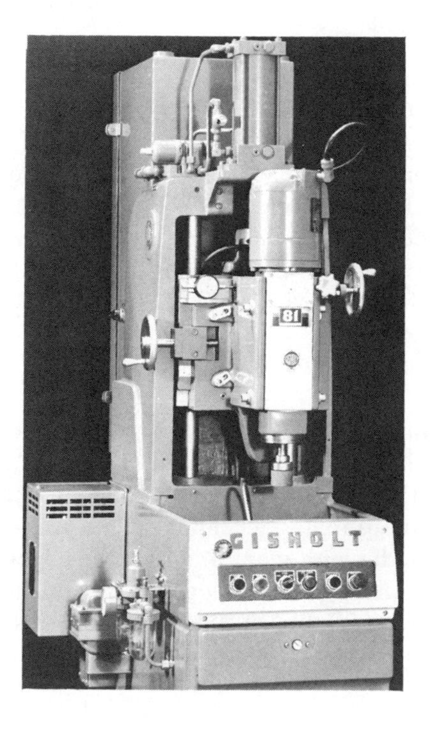

**Fig. 17.25.** The type of superfinishing machine used for flat surfaces. Courtesy Gisholt Machine Co.

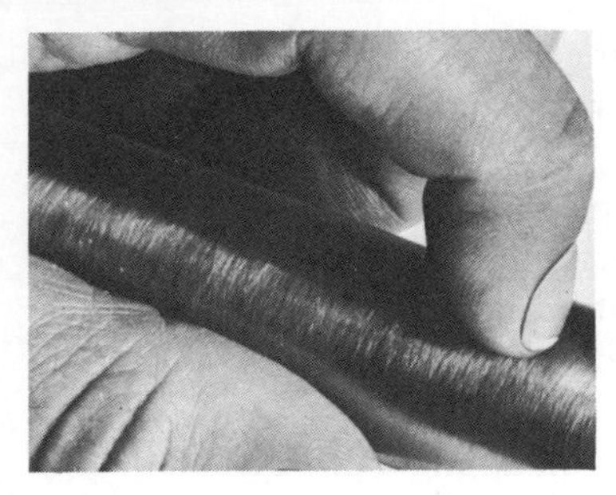

**Fig. 17.26.** Superfinishing can be used to produce a preselected surface finish from 1 to 80 $\mu$in. Courtesy Gisholt Machine Co.

## BARREL FINISHING

Basically, barrel finishing consists of putting a number of workpieces in a six- or eight-sided barrel together with an abrasive medium. The barrel is rotated so that superfluous stock can be removed by the abrading action of the medium used.

One important aspect of barrel finishing is that it can replace hand filing, wire brushing, polishing, and buffing operations. Parts may be finished by the dozen or by the thousand, with practically no increase in labor cost. Another important factor is uniformity; barrel finishing makes the parts completely uniform, whereas with hand finishing this is impossible.

**Barrel-Finishing Media.** Figure 17.27 shows a number of different media used in barrel finishing.

**Natural and Man-Made Stones.** Natural and artificial stones are also called *nuggets* or *chips*. The natural stones consist of granite, limestone, gravel, and sand. Aluminum oxide is the principal artificial stone. It has considerable cutting ability, as, with the grinding wheel, it can be made constantly self-sharpening. Heavier media with more and sharper edges will cut faster than lighter media with more rounded edges. The configuration of the part, if it includes areas which present lodging problems, will dictate the size and shape of the media.

**Fig. 17.27.** Various types and grades of synthetic aluminum tumbling media. Courtesy Pangborn Corporation.

**Compounds.** In addition to the abrasives in the media, various types of compounds are used, such as alkaline and acid descaling, derusting, burnishing, and polishing solutions. The compounds are usually used as a separate treatment; however, they may sometimes be combined into one operation.

**Applications.** Barrel finishing is limited only by the size of the equipment available. Small parts are shown in Fig. 17.28 with data to give the reader an idea of some of the most important factors involved.

## VIBRATION FINISHING

A newer concept of finishing that depends on vibrated rather than tumbled abrasive is shown in Fig. 17.29, with finished aluminum castings. It consists basically of a "tub" or container to hold the media and parts. When the tub is vibrated at a suitable amplitude and frequency, the entire parts-media mass rolls over continuously. The vibration causes a peening action to take place as well as the overall rotation.

**Advantages of Vibratory Finishing.** The biggest advantage has been the reduction of cycle time, anywhere from 50 to 90 percent. Other advantages of the method are:

(1) it can handle parts that tend to lock or ball up;
(2) it works well on recesses and internal areas;
(3) it handles larger parts without damage;

*Part:* Machine gear
*Operation:* Remove burrs caused in hobbing operation
*Medium:* 3/8 in.-diameter pellets. 2 hr
*Compound: Carbofast* No. 4
*Former method:* Remove burrs with hand file. 10 min required per gear
*Cost savings:* 24¢ each (approx.)

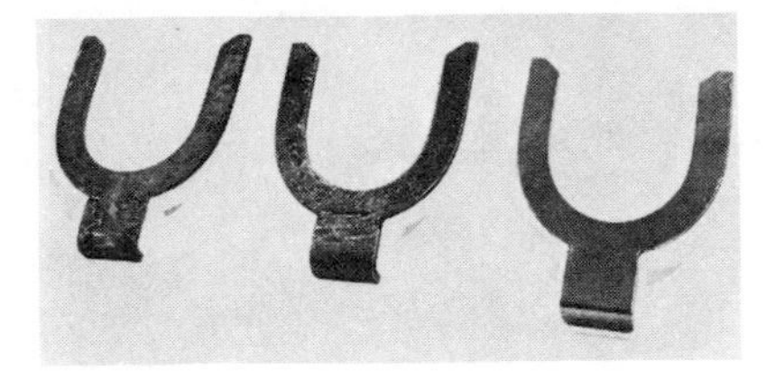

*Part:* Hydraulic-press shims
*Operation:* Removal of heat-treat scale and surface roughness. Original at left
*Medium and compounds:* 1 hr in No. 3 *Nuggets* and *Carboscaler* No. 61. Step 2 – 1 hr combination neutralizing and polishing
*Former method:* Clean with hand file and coated abrasive. Time – 2 min each
*Cost savings:* 1.3¢ per piece

**Fig. 17.28.** Examples of barrel finishing with time cost data. Courtesy The Carborundum Co.

**Fig. 17.29.** Vibratory abrasive-finishing equipment and examples of finished parts. Courtesy Pangborn Corporation.

(4) larger amounts of metal are removed compared to the amounts removed in the corners, thus decreasing the amount of radius formed; (5) it can be used for handling semiautomatic as well as fully automatic parts.

## POLISHING

Coated abrasive cloths are used to remove material and improve the finish of a wide variety of products. The abrasive applied to the cloth or paper backing can be either natural or manufactured. Natural abrasives used are flint, emery, and garnet. Garnet is usually thought of as a precious stone; however, common garnets, as used in polishing, are found in many kinds of silicate rocks. Manufactured abrasives are aluminum oxide and silicon carbide. The mesh size ranges from 12 to 400.

Many kinds of machines have been built to bring the coated abrasive in contact with the workpiece. They may be broadly classified in two groups—the endless-belt machines and the coated abrasive wheels.

**Endless-Belt Machines.** The five main types of belt machines—platen, contact wheel, formed wheel, centerless, and flexible belt—are shown in Fig. 17.30.

**Platen Grinder.** The platen machine provides a support or platen for part of the belt surface. It is designed for light cuts, with stock removal seldom exceeding 1/32 to 1/16 in. on ferrous metals and twice this on nonferrous metals. Platen grinders permit semiautomatic operations, with pressure, depth of cut, and rate of feed all constant.

**Contact Wheel.** There are many variations of the contact-wheel or wheel-and-idler machine as it is sometimes called. An adaptation of this machine is the swing grinder shown schematically in Fig. 17.30. Swing-frame grinders are used to advantage when the workpiece is difficult to move. They may also be used above or beside conveyor lines. Stand-mounted arrangements provide for greater movement of the part around the contact wheel. A typical machine of this kind uses a 1 1/2-hp motor for driving and a 2 1/2- by 72-in. belt over a 6- by 2 1/2-in. contact wheel.

**Formed Wheel.** The formed wheel acts as a special platen. By forming the fiber-backing wheel, it is possible to direct the abrasive belt into the irregular surface, as shown.

**Centerless Machine.** Several arrangements are made for centerless grinding with the abrasive belt. One setup uses a resilient contact wheel with a conventional regulating wheel. Through-feeds of 35 fpm are available, with tolerances of 0.001 in. on small diameters. The maximum diameter reduction on steel is about 0.005 in. per pass at 2 fpm.

**Flexible-Belt Machines.** Flexible-belt machines are made in several ways, but they usually consist of a wheel-and-idler arrangement. These machines are often used for polishing difficult inside and outside contours. Large abrasive-belt machines are now available which are capable of handling stock up to 80 in. wide and 10 ft long. They are used to remove surface defects and to produce fine finishes, often to close tolerances.

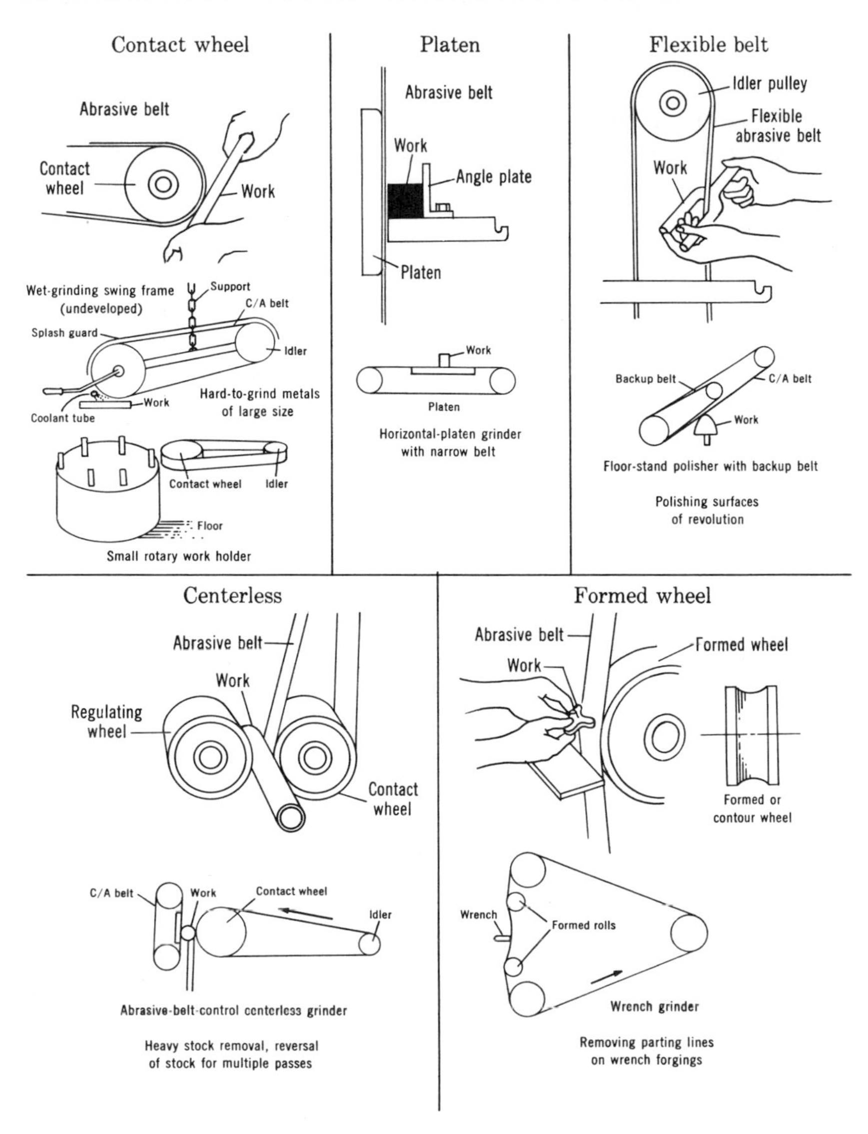


**Fig. 17.30.** The five main types of endless-belt polishing machines. Variations of each kind are also shown.

## BUFFING

Buffing wheels are usually made from muslin or canvas. They are designated by the way in which they are sewed—concentric, radial, radial arc, parallel, and square. Some loose wheels have only a row of stitching around the hub (Fig. 17.31).

Buffing is used to give a much higher, lustrous, reflective finish than

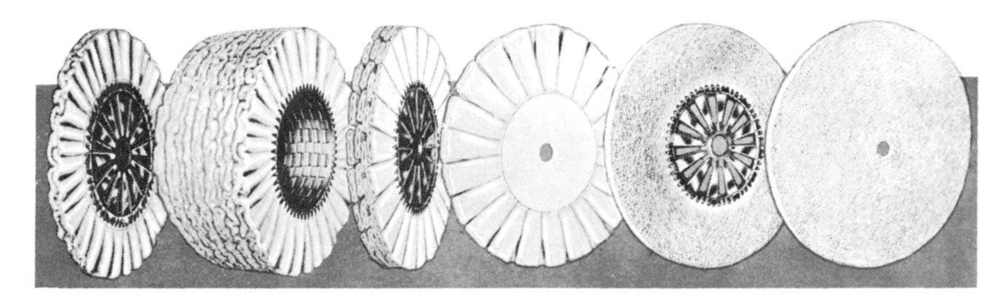

**Fig. 17.31.** A variety of buffing wheels, ranging from the hard-stitched wheel to the soft-pleated wheel. Courtesy American Buff Company.

can be obtained by polishing. For a mirrorlike finish, the surface must be free of defects and deep scratches. Abrasives embedded in wax are applied directly to the rotating wheel. The cutting action, whereby a small amount of metal is removed, is accomplished with an aluminum oxide abrasive. A "coloring" operation, that brings out the best color and luster of the metal, is done by using a white compound of levigated alumina type of abrasive.

Cutting is done at approximately 10,000 sfpm. Speeds higher than this are hazardous for handwork. Surface speeds of 9,000 fpm should not be exceeded for color buffing.

## QUESTIONS

**17.1.** Why does aluminum oxide abrasive work better than silicon carbide in grinding steel?

**17.2.** What is meant by *friability* of a cutting abrasive?

**17.3.** Why is it necessary to have a certain range of porosity in a wheel for grinding a given type of metal?

**17.4.** Why would you expect to be able to use a resinoid-bonded wheel for rough grinding operations?

**17.5.** What is the difference between truing and dressing a wheel?

**17.6.** Describe how you would go about selecting a wheel to grind a solid carbide cylinder $R_c$ 68, 4 in. in dia and 8 in. long. The outside dia is to be finish ground.

**17.7.** (*a*) What type of abrasive is recommended for ceramic tools? (*b*) Why?

**17.8.** What is the main difference between cylindrical grinding and centerless grinding?

**17.9.** Why is "plunge cut" grinding well adapted to crankshaft grinding?

**17.10.** What is meant by the term "abrasive machining"?

**17.11.** What is meant by the "pivot point" in abrasive machining?

**17.12.** What is meant by the "*G* ratio" in abrasive machining?

**17.13.** Other than accuracy, why is lapping one of the best finishing methods for metal working?

**17.14.** What is meant by "control" in the honing process?

**17.15.** How is the pressure equalized on all stones for the honing process?

**17.16.** What is the advantage of a crosshatch pattern in honing?

**17.17.** Compare the tolerances and finishes obtained in lapping, honing, and superfinishing.

**17.18.** What are some of the advantages of vibratory finishing over barrel finishing?

**17.19.** What is the difference in purpose between polishing and buffing?

## PROBLEMS

**17.1.** Check the example of abrasive machining given in Fig. 17.15 and Table 17.2 with the following added information: (1) It is found that 6 pieces can be milled per setup on the mill, making the total travel 24 in. plus 1 in. for approach and 1 in. for overtravel. (2) Assume that the cutter has 12 teeth and that the chip load is 0.015 in.

Compare the milling information given in Table 17.2 with your computation as to: (*a*) the table feed; (*b*) stock removal rate; (*c*) machining time/part; and (*d*) the practicality of doing it this way on the mill.

**17.2.** (*a*) What is the cost of metal removal per cubic inch by milling as shown in Fig. 17.15, not counting setup time? (*b*) What is it for grinding?

**17.3.** What is the cost of metal removal based on in.$^3$/min for (*a*) milling, (*b*) grinding? (*c*) What would it be for milling as proposed in Problem 17.1?

**17.4.** What would be the approximate production rate/hr for honing a 3/4-in.-dia hole 2 1/2 in. long in hardened tool steel if the amount of metal to be removed were 0.003 in.? Assume that the additional thousandth takes 10 percent more time.

**17.5.** If a pin 4 in. long and 1 in. in diameter can be superfinished to about 3 $\mu$in. in 1 min, how long will it take to bring it to 2 $\mu$in. or better?

**17.6.** What would the maximum production rate be in pieces per hour for steel pins 3 in. long and 1 in. in diameter on a centerless abrasive belt machine?

## REFERENCES

Blake, K. R., M. P. Ellis, and R. W. Militzer, "Low Velocity Abrasive Machining For Precision Production," Paper at ASAM National Technical Sessions, 1964.

Haggett, J. E., and R. L. Smith, "How To Find Profit in Abrasive Machining," *Steel,* September 16, 1963.

Haggett, J. E., and S. I. Berman, "The Economies of Abrasive Machining," *Grinding and Finishing,* October 1964.

Patterson, M. M., "How Does Abrasive Machining Save?" *Grinding and Finishing,* June 1963.

# 18

# METROLOGY AND QUALITY CONTROL

MEASUREMENT HAS PLAYED an important role in man's scientific advancement. Early attempts at standardization of length measurements were based on the human body. The width of a finger, for example, was termed a digit, and a cubit was the length of the forearm from the end of the elbow to the tip of the longest finger.

From these simple beginnings, man progressed gradually toward standardization of basic measurements. The most significant advancements were made in the metric system. It not only provides an easily usable scale of linear measurements but of volume and weight measurements as well and all units are in multiples of 10. Some comparisons of the English system of measurement with the metric system are shown in Fig. 18.1.

The metric system is based on the earth's measurements, the meter determined astronomically as being one ten-millionth part of the distance from the north pole to the equator, measured on a line running along the earth's surface through the city of Paris. At the Convention of the Meter in Paris in 1875, several platinum-iridium bars of modified X cross section were made. Fine lines were etched on these bars to represent the standard meter length when measured at 0°C. Most nations have accepted this as the national standard of measurement. A new standard for the inch has now been issued, making it 41,929.399 wavelengths of krypton light. Although these standards were necessary, they were not available to the average manufacturer. The task of bringing a workable standard of length to industry still remained.

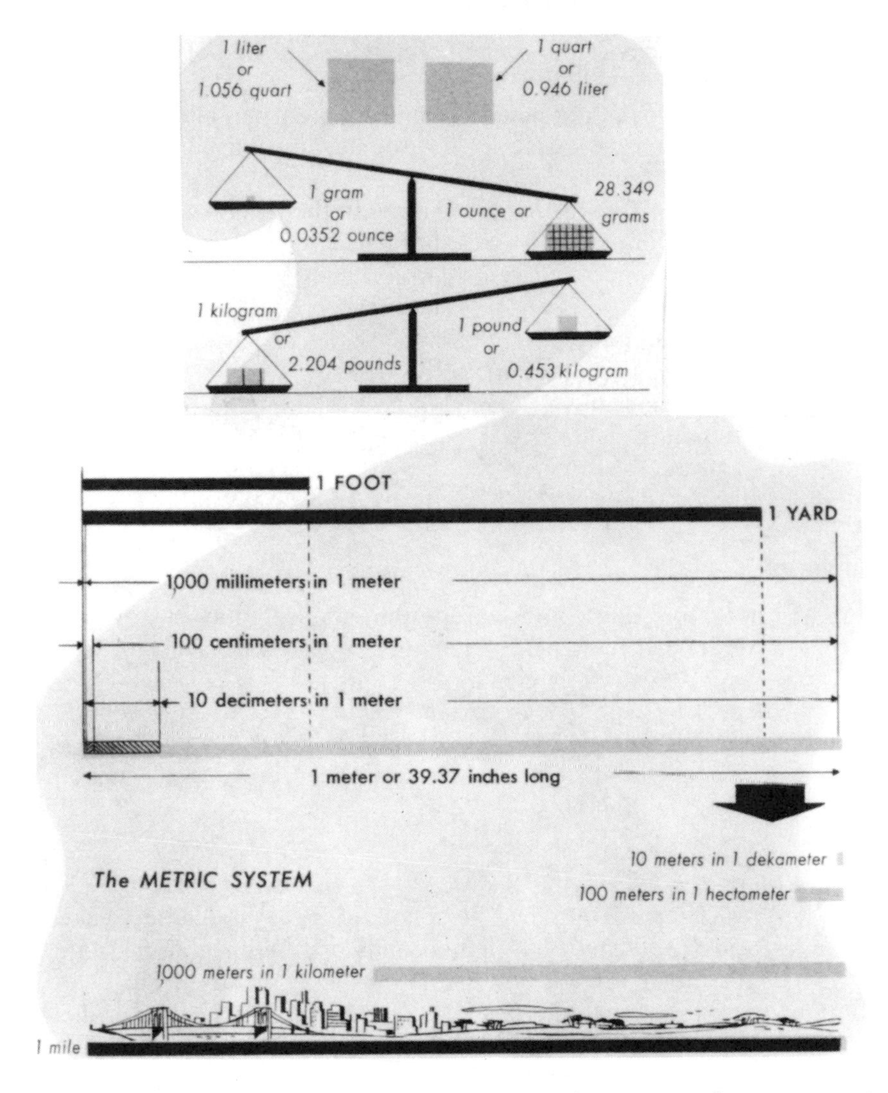


**Fig. 18.1.** A comparison of the English and metric systems of measurement. Courtesy General Motors.

## THE DEVELOPMENT OF GAGE BLOCKS

Toward the close of the 19th century, Carl Johansson of Sweden decided to make measurement standards that would be available to industry. These were to duplicate the government standards as nearly as possible. To accomplish this, he painstakingly made a set of steel blocks that were accurate to within a few millionths of an inch. This was an unheard of tolerance, and how it was achieved remained a secret with Johansson for many years. Since these blocks were so accurate, they could be used to check the manufacturers' gages and other measuring instruments.

Modern gage blocks are made of a medium carbon-alloy steel, hardened to a Rockwell C65, stabilized, ground, and lapped to exact size. Stabilization is a carefully controlled heat treatment performed to relieve internal stresses so that the blocks will retain their exact size over a long period of time. More recently, case-hardened stainless-steel gage blocks have been placed on the market. These have the added advantage of corrosion resistance (an ever present problem with steel gage blocks) and much greater stability. Solid carbide blocks are made for use where wear is likely to be an important factor.

**Gage Block Classification.** Gage blocks are classified by the National Bureau of Standards according to accuracy as shown in Table 18.1.

**Table 18.1**

| Letter | Meaning |
|---|---|
| AA | Laboratory gage blocks are within plus or minus two millionths of an inch per inch of length; they are also flat and parallel within this tolerance |
| A | Inspection gage blocks are similarly accurate to +6 millionths or −2 millionths of an inch |
| B | Working gage blocks are similarly accurate to +10 millionths or −6 millionths |

**Gage Block Use.** Gage blocks are sold in sets of from 5 to 103 pieces (Fig. 18.2). The predetermined selection of sizes in the set makes it possible to arrive at any one of thousands of reference dimensions by using combinations of blocks. The blocks may be used singly or in combinations to check micrometers, vernier calipers, vernier height gages, and fixed gages and to set comparators, snap gages, and other adjustable gages (Fig. 18.3), to mention a few of their uses. Thus we see that gage blocks are the connecting link between the international standard of measurement and the manufacturers' measuring and gaging equipment.

Gage blocks maintain accuracy even though several have to be used in combination to give a specified dimension. Plus and minus dimensions of blocks tend to balance out. The air can be eliminated from between the blocks by what is known as a *wringing-in process*. The procedure used is to clean the blocks thoroughly and then slide them into full contact while pressing them firmly together (Fig. 18.4).

Precision gaging is done in an air-conditioned room kept at 68°F. This was the temperature specified at the Convention of the Meter in Paris,

**Fig. 18.2.** Gage blocks may be either square or rectangular. There are usually 88 blocks in a set. Courtesy Dearborn Gage Company.

and it has since been adopted internationally. By using 68°F as a standard, materials other than steel can be checked with gage blocks, even though they may have a different coefficient of expansion.

Oftentimes, it is desirable to have an accurate reference in a shop which is not kept at 68°F. Because the coefficients of expansion of steel parts are so nearly alike, the gage blocks will correspond to the material in the measuring instruments so that accurate measurements and setups will result.

**Gage Block Calibration.** We have seen how gage blocks are the usable measurement standard for industry. The question arises as to just how the gage blocks are measured by light waves to a high degree of accuracy. Professor A. A. Michelson, of the United States, developed the optical arrangement, shown in Fig. 18.5, for studying the various phenomena in the interference of light. A monochromatic light emanating from a source, as shown, is passed through a condenser lens so that the light rays are parallel. A sodium light having a wavelength of 0.00002 in. is usually used. The light waves fall at 45 deg on a semimirror *A*, where they are partly reflected and partly transmitted. The transmitted rays are reflected normally from the full mirror *D*, and, after a second reflection at the mirror *A*, pass into the telescope and eyepiece. The reflected portion of the original rays traverses an inclined transparent compensating quartz plate *B*, which is exactly the same thickness as mirror *A*, and is reflected normally at the fully silvered mirror *C*. The

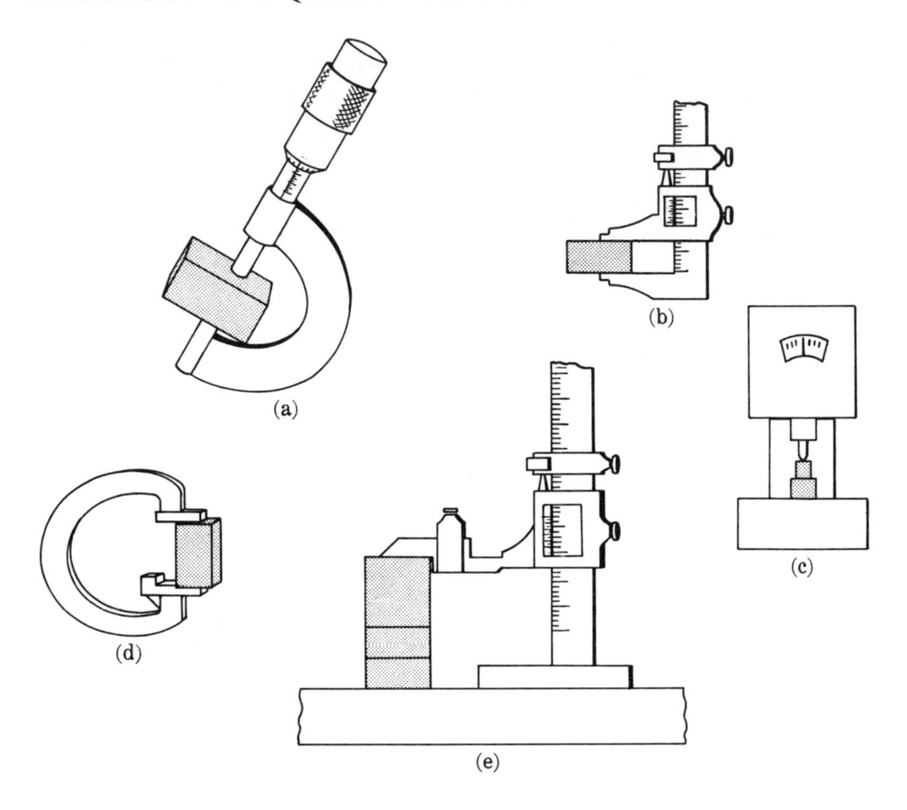


**Fig. 18.3.** Gage blocks are used as a standard for comparison in many ways. Gages shown are (*a*) micrometer, (*b*) vernier calipers, (*c*) comparator, (*d*) snap gage, and (*e*) vernier height gage.

rays returning from mirror *C*, after passing through *B* and *A*, enter the telescope and thence pass to the eye.

The two sets of rays, one reflected from *D* and the other from *C*, when united, produce interference effects, or fringe lines.

Fringe lines may be understood more easily by studying the optical flats shown in Fig. 18.6. The monochromatic light goes through a quartz flat *A* to the surface of the work and is reflected back through the quartz flat. If the wave from the bottom of the flat and the wave reflected from the work surface are exactly out of phase, they cancel each other, so that they appear as dark bands. Such interference will occur at every point where the distance from the flat to the work is in odd multiples of one-half the wavelength of the light used.

The picture at the right of Fig. 18.6 shows the top view of a gage block under the optical flat.

In the interferometer, these fringe lines are picked up in the telescope

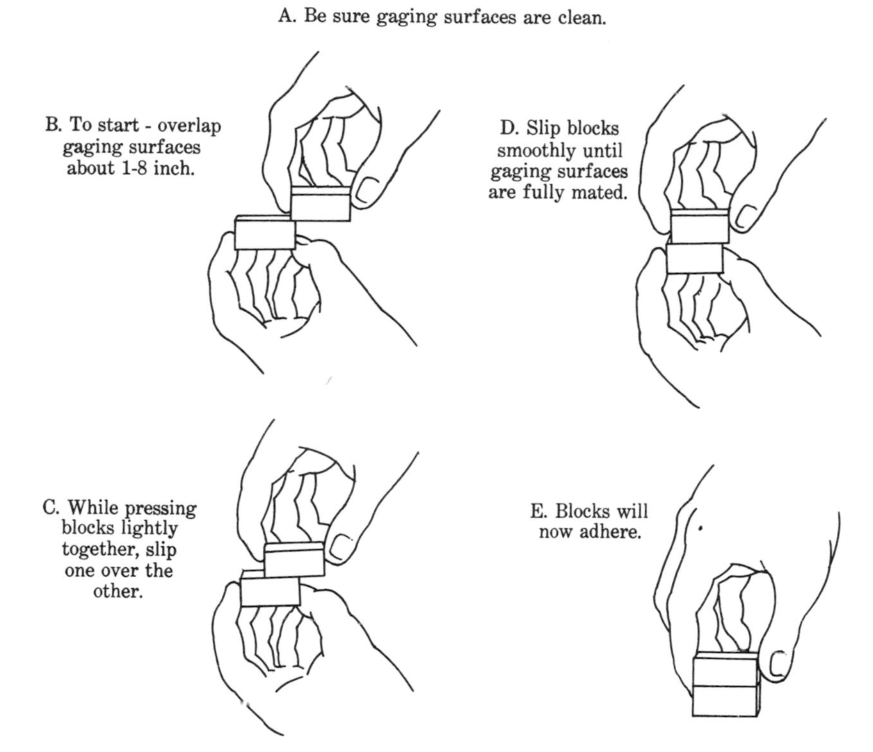


**Fig. 18.4.** The procedure used in wringing gage blocks together. Courtesy DoAll Company.

by wringing a gage block to mirror *D* (Fig. 18.5). As one of the mirrors is slowly displaced to cover the surface of the gage block, the observer counts the lines as they pass.

This procedure will give the length to within one-half wavelength of light, but still further accuracy can be obtained. The spectrum colors of red, yellow, green, blue-green, blue, and violet from the helium light in the interferometer have definite relations with one another at various distances from the source. Constellations of fractional divisions or pattern arrangements of the spectrum recur at uniform intervals. Since the recurrence of constellations is absolutely known, lengths may be determined to a very high degree of accuracy.

## MEASUREMENT

Measuring tools and instruments are of two broad types – direct and indirect. Direct measuring instruments obtain the measurement without the aid of other equipment. Indirect measuring tools transfer the given

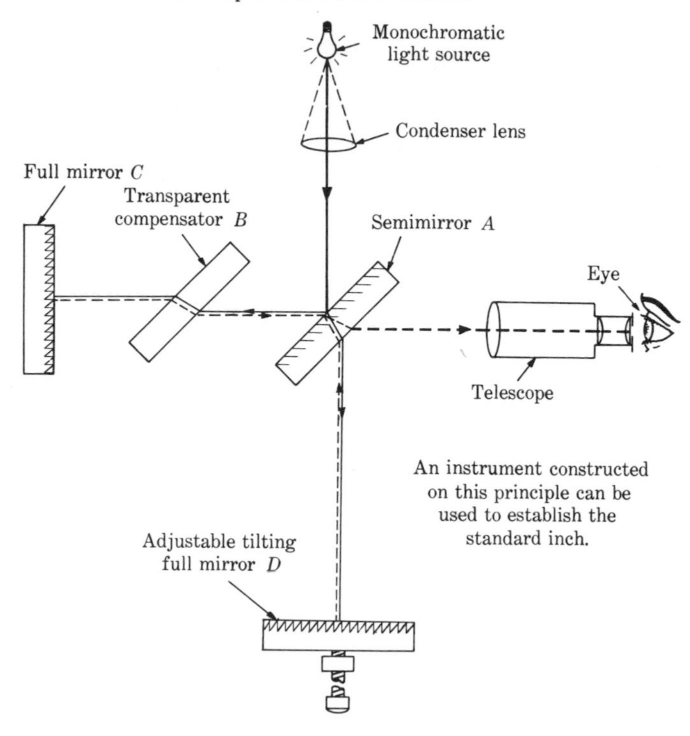


**Fig. 18.5.** Schematic diagram showing the principle of the interferometer. Courtesy DoAll Company.

size to a measuring instrument for the actual size reading. Measurement may be further divided into four main areas – linear, angular, roundness, and surface. Each of the four areas will be discussed in the order given. with direct measurement first in each case.

## LINEAR MEASUREMENT

**Direct Measurement.** The following hand-operated measuring tools are used for linear measurement.

STEEL RULE. The simplest measuring device is the steel rule. It is divided into fractions of an inch, usually by 64ths but sometimes in 100ths of an inch. Rules are made with various attachments so that they can be used as depth gages and hook rules.

MICROMETER CALIPERS. As mentioned, the steel rule is used successfully for measurements to within 1/100th of an inch. In most machine-tool work this is far from satisfactory, as the work must be held to a few thousandths or even to 1/10,000 of an inch. In 1638 an astrono-

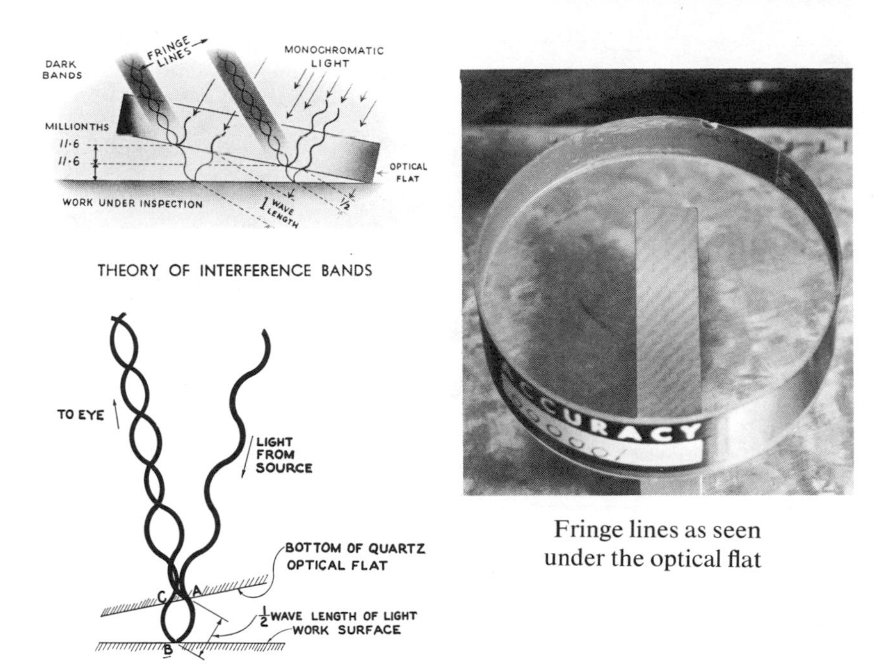


Fringe lines as seen under the optical flat

**Fig. 18.6.** Monochromatic light passing through an optical flat is partly reflected by the lower surface of the flat, while the remainder passes through. As the two reflected light rays travel different distances, they create a series of alternating bright and dark bands. Courtesy DoAll Company.

mer by the name of William Gascoigne was interested in measuring the size of the sun and moon. To do this he chose the screw-thread principle (Fig. 18.7), which became the forerunner of the present-day micrometer caliper. The micrometer screw has 40 threads to the inch. This means that each revolution of the screw moves the spindle 1/40 of an inch toward the anvil or away from it. Therefore one revolution will move the spindle 1/40 of an inch or 0.025 in. Rather than move one revolution, the distance may be changed to 1/25; this amount may be accomplished by having the sleeve graduated into 25 parts (Fig. 18.8).

The 0.001 reading may be further refined by adding a vernier scale. The vernier in this case consists of 10 divisions marked on the barrel which equal, in overall dimension, 9 divisions on the thimble (Fig. 18.9). Notice that the fourth-place number is obtained by observing which of the vernier lines coincides with a line on the thimble. This number is added to the thousandths reading in the example:

$$0.369 + 0.0005 = 0.3695.$$

*Indicating micrometer comparator.* The fact that the micrometer is made to measure to 0.001 or 0.0001 in. does not mean that it will guar-

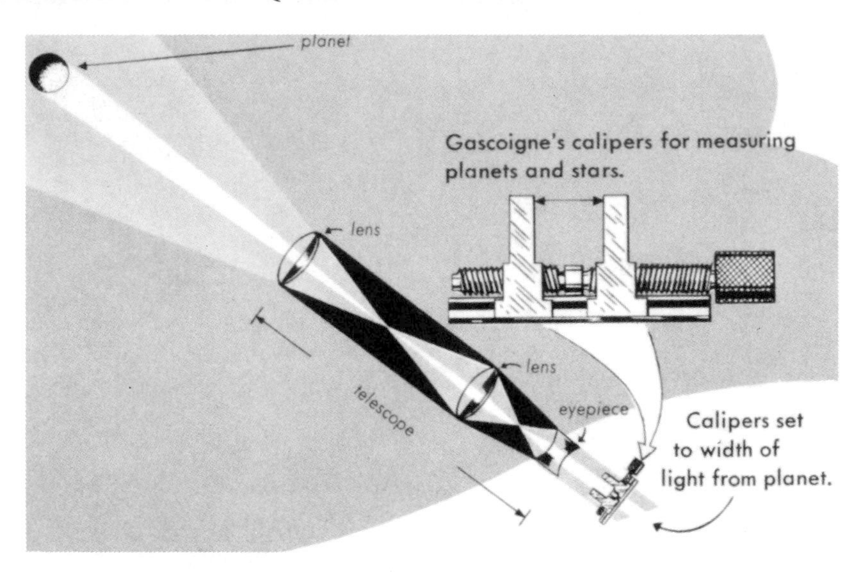


**Fig. 18.7.** Gascoigne's calipers for measuring planets and stars. Courtesy General Motors.

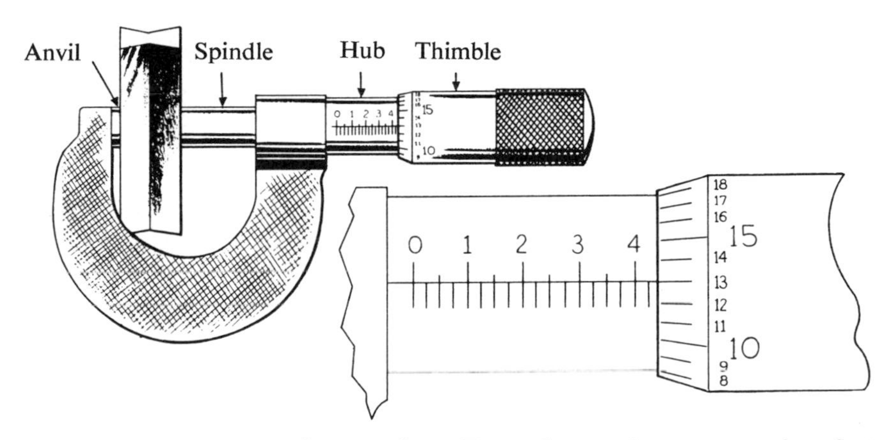


**Fig. 18.8.** Micrometer-caliper readings. To read any micrometer setting: first, read scribe marks along hub—this one reads 425 thousandths (0.425) inch. Second, read scribe marks around thimble—this one reads 13 thousandths (0.013) inch. Add reading of thimble to reading of hub—0.013 + 0.425 = 0.438. This micrometer is set at 438 thousandths (0.438) inch.

antee this accuracy to every user. There is a difference in the amount of pressure used in closing the micrometer on the workpiece. To overcome this objection, an indicating micrometer (Fig. 18.10) was developed.

*Super micrometers.* A still more advanced step, using the micrometer principle, is the super micrometer, which has a variable pressure

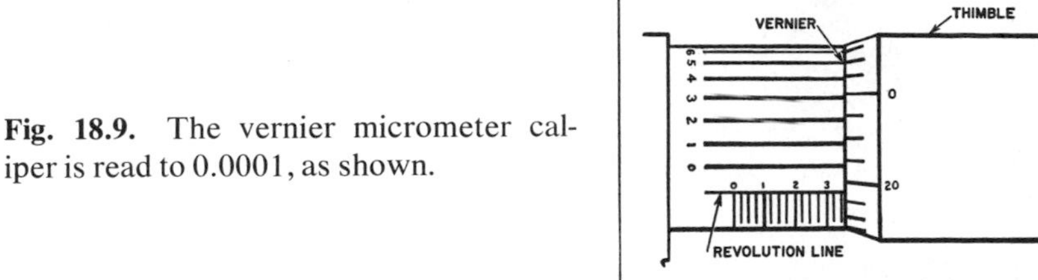


**Fig. 18.9.** The vernier micrometer caliper is read to 0.0001, as shown.

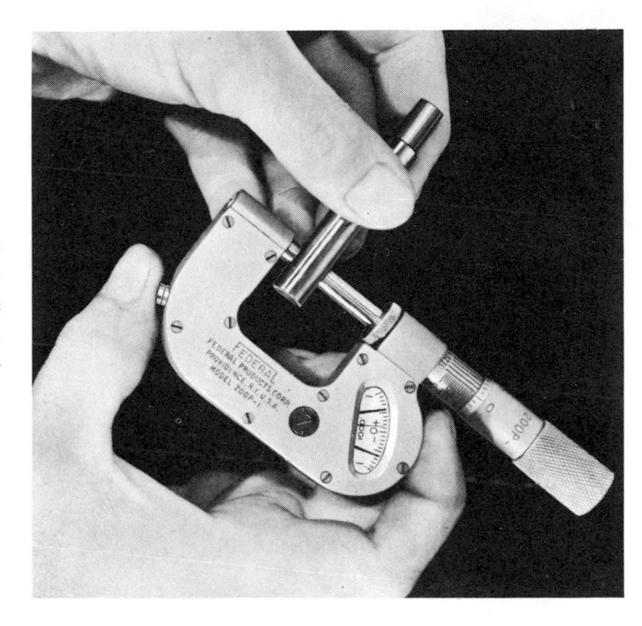


**Fig. 18.10.** The indicating micrometer is used to a high degree of uniform pressure. Courtesy Federal Products Corporation.

tailstock or anvil. The pressure may be set to anywhere from 1 to 2 1/2 lbs. pressure by spring loading or an electrolimit arrangement. When the desired pressure is reached on the part being measured, a pointer on the tailstock or on a milliampmeter dial will be in the center of its scale range. The actual size reading is then read to one ten-thousandth on the large micrometer dial.

Many other micrometers are made such as inside, depth, and special anvil types. Special anvil types allow accurate measurements to be made on threads, inside curvatures, hole edge distance, etc. A depth micrometer is shown in Fig. 18.11.

VERNIER CALIPERS. Another instrument that is capable of measuring to 0.001 in. accuracy is the vernier caliper. The inset picture of the vernier scale (Fig. 18.12) is used as an example of how the total vernier reading is obtained. Notice that the zero on the vernier is past the 2-in. mark on the scale, thus we have:

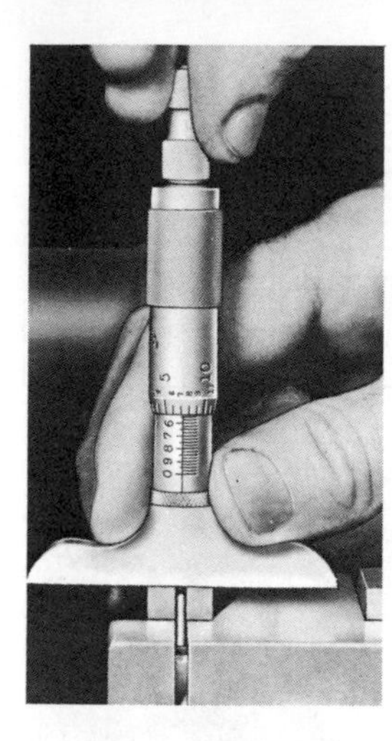

**Fig. 18.11.** Depth micrometer used to measure the depth of the milled slot. Courtesy The L. S. Starrett Co.

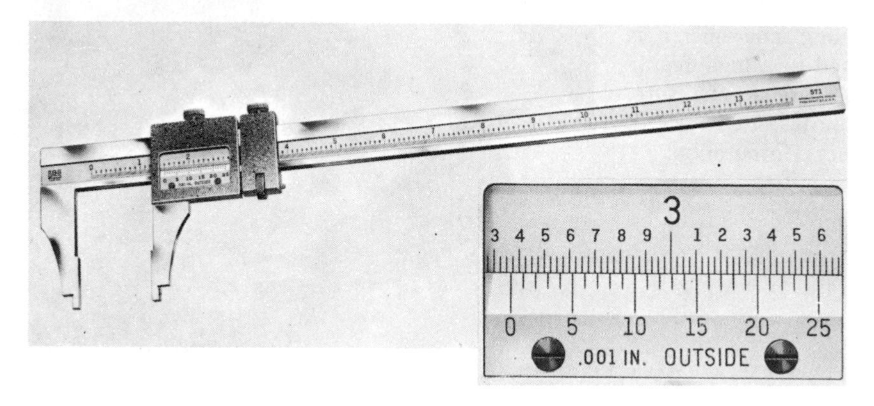


**Fig. 18.12.** The vernier caliper, with enlarged scale inset. Courtesy Brown & Sharpe Mfg. Co.

| | |
|---|---|
| 2.000 | inches |
| 0.300 | hundred-thousandths on scale |
| 0.050 | two twenty-five thousandths lines |
| 0.018 | the line on vernier that coincides with a line on scale |
| 2.368 | total |

Both inside and outside diameters may be measured with a vernier caliper.

*Vernier height gage.* The vernier height gage uses the same principle as the vernier caliper. It is, however, mounted on a base and is used to measure and scribe accurately layout lines of a given height from a plane (Fig. 18.13).

**Indirect Measurement.** Indirect measuring tools are used to transfer or compare distances and sizes.

SMALL-HOLE GAGES. It is often difficult to check small-hole diameters accurately with either an inside micrometer or inside calipers and micrometer. The best combination for this purpose is the small-hole gage (Fig. 18.14). The ball contacts are made to spread apart to the diameter of the hole by means of adjusting the knurled handle. A micrometer is then used to measure the distance over the split curved surfaces. As shown in the figure, the small-hole gage is also used for gaging slots and recesses.

Somewhat similar to the small-hole gages but made for holes from 1/2 to 6 in. are *telescoping gages.*

Other common tools widely used for making indirect measurements are *inside, outside,* and *hermaphrodite calipers.*

DIAL INDICATORS. Dial indicators are used to compare a part with a master setting. The master setting may be a combination of gage blocks (Fig. 18.15). By means of rack-and-pinion arrangement, the movement of the contact pointer is greatly amplified so that dimensional variations are easily read on a dial. The value of each graduation on the face of the dial may vary from 0.00005 to 0.005 in. In addition to comparing parts with a master setting, dial indicators are

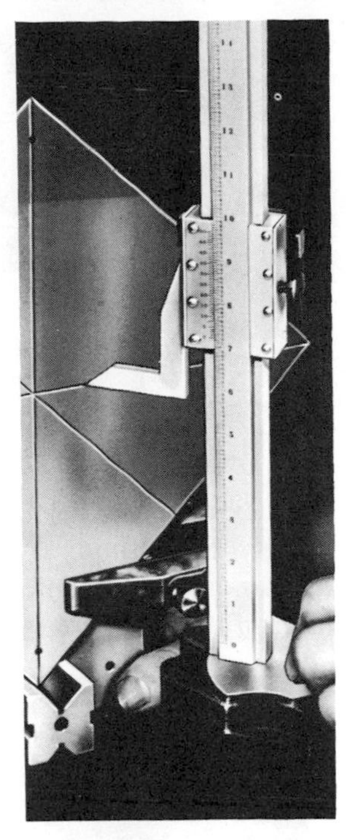

**Fig. 18.13.** Vernier height gage used in making layout lines on a steel plate.

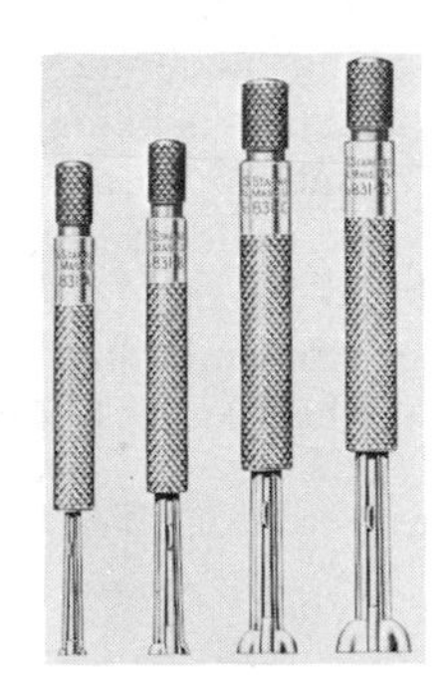

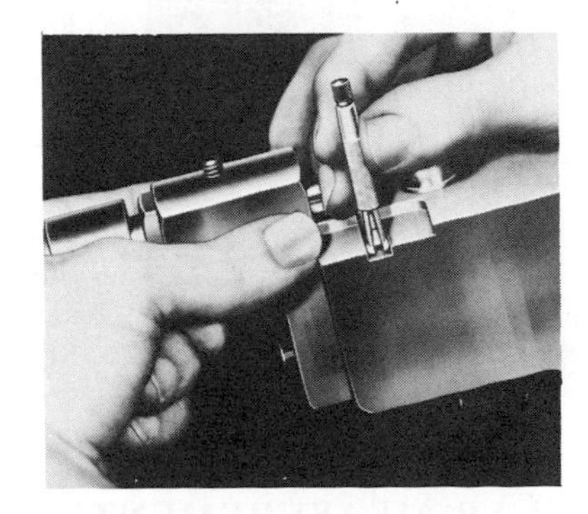

**Fig. 18.14.** Small-hole gages are used to check small holes, slots, and recesses in the range of 0.125 to 0.500 in. Courtesy The L. S. Starrett Co.

**Fig. 18.15.** A dial indicator used to check the height of a ground ring against a gage block. Courtesy The L. S. Starrett Co.

used extensively in making machine-tool setups (Fig. 18.16). Notice the variety of methods used to hold the indicator. Dial indicators are used as standard mechanical-type comparators and in other gages discussed later in this chapter.

(*a*)

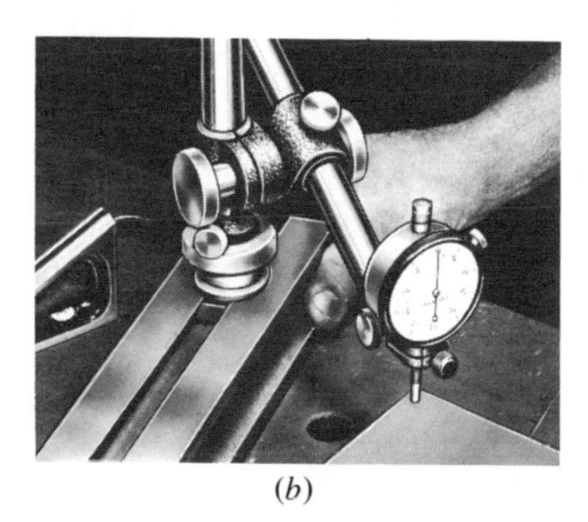

(*b*)

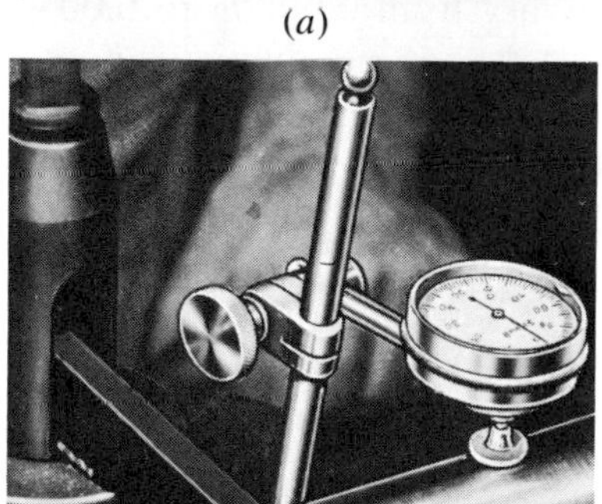

(*c*)

**Fig. 18.16.** The dial indicator is used in making accurate machine setups. At (*a*) a part is being checked for center location in a three-jaw chuck. (*b*) The indicator is used to check the depth of a cut. (*c*) Checking the runout of work mounted between centers on a lathe. Courtesy The L. S. Starrett Co.

## ANGULAR MEASUREMENT

**Direct Measurement.** Angular measurements can be made in a variety of ways, ranging from the use of a simple protractor to that of an angle-calibrating interferometer.

PROTRACTORS. Angular measurements are made with an ordinary bevel protractor (Fig. 18.17).

A more precise measurement can be obtained with a vernier bevel protractor which can be accurately read to within 5 min, or 1/12 deg.

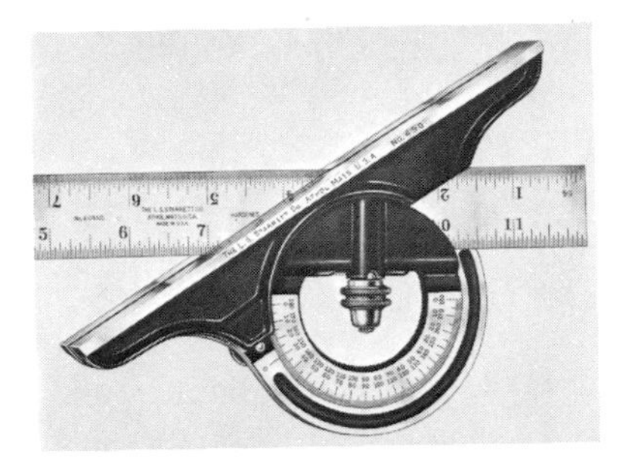

**Fig. 18.17.** A protractor head and 12-in. rule. Courtesy The L. S. Starrett Co.

Reading the inset vernier of Fig. 18.18, we notice that the zero on the vernier scale lies between 50 and 51 on the protractor dial, or 50 whole deg. The line from the vernier scale that coincides with the protractor scale is 20. Therefore the reading is 50 deg and 20 min (50°20′).

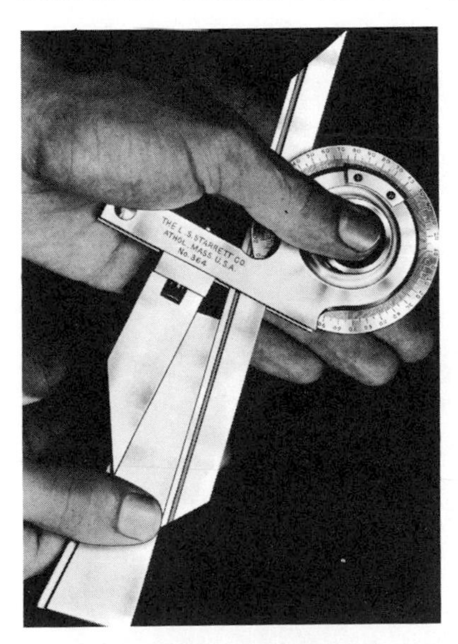

**Fig. 18.18.** A universal bevel protractor being used to measure an acute 10-deg angle. The enlarged dial shows the reading of the vernier scale at 50 deg 20 min. Courtesy The L. S. Starrett Co.

**Indirect Measurement.** SINE BARS. A sine bar is a very accurate straightedge to which two plugs have been attached (Fig. 18.19). These two plugs are very accurate in diameter, and the center distance is within 0.00002 in., or less, per inch of length. The length between centers is given in multiples of 5 in. from 5 to 20 in.

The sine bar makes use of the known relationships between the sides of a right triangle and its angles. Measurements are made by using the principle of the sine of a given angle or the ratio of the opposite side of the right triangle to the hypotenuse. To check the accuracy of the 45-deg angle of a squaring head, using a 10-in. sine bar (Fig. 18.19), the following procedure would be used:

(1) Find the sine of 45 deg from the tables (0.70711).

(2) Use the procedure given earlier for selecting and wringing in the proper gage blocks to obtain the height,

$$\begin{array}{r} 0.121 \\ 0.950 \\ 2.000 \\ \underline{4.000} \\ 7.071 \end{array}$$

(3) Use a dial indicator, as shown, to check to see if the 45 deg is now parallel to the top of the surface plate. Any change in the dial reading would indicate an error in the angle. The amount would be according to the change necessary in the gage block height.

A variety of sine-bar fixtures have been made to make the setting up and checking of angles easier.

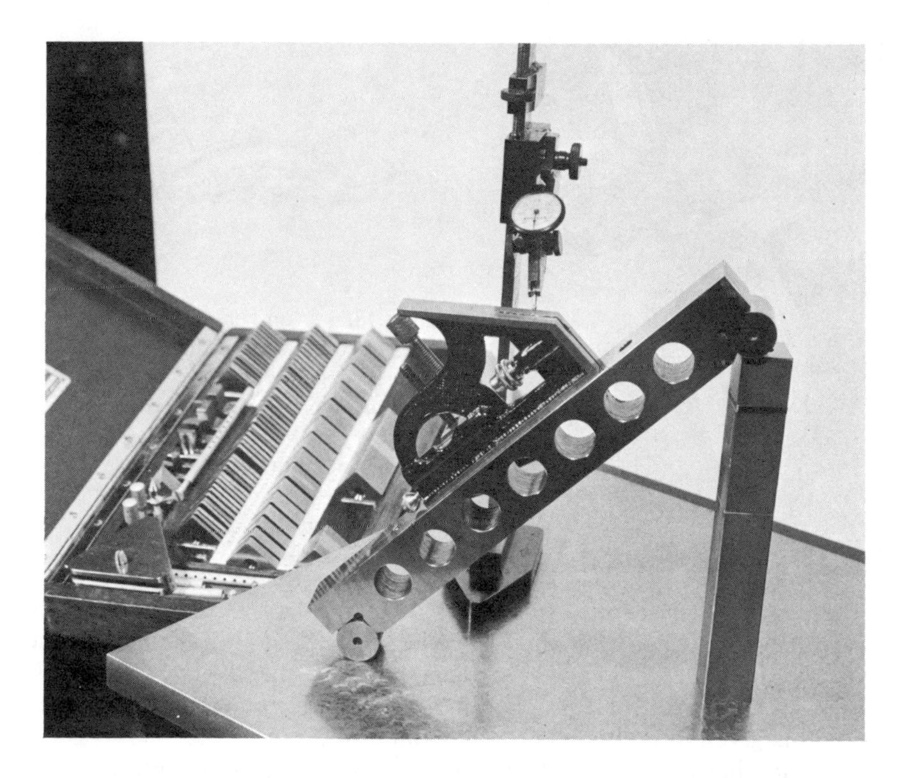

**Fig. 18.19.** A sine bar used to check the 45-deg angle of a squaring head.

ANGULAR GAGE BLOCKS. With a set of angle gages consisting of 10 pieces, it is possible to make any angle between 0 and 180 deg by 5-min intervals. These blocks may be set up in combination and used as a standard to check a given part, with the aid of a dial indicator.

## ROUNDNESS MEASUREMENT

Roundness may be measured indirectly. For example, the Talyrond (trademark of Taylor, Taylor and Hobson) instrument (Fig. 18.20)

**Fig. 18.20.** The Talyrond used for precision roundness measurement. Courtesy Engis Equipment Co.

is used for measuring the roundness of parts such as balls, roller races, pistons, and cylinders. Both external and internal diameters can be checked. Diameters range from 1/16 to 14 in., with the maximum height of the specimen being 10 in.

An electric displacement indicator, in operation, is carried on an optically guided precision spindle of extreme accuracy, which is rotated around the outside or inside of the part to be examined; the part itself remains stationary on the table. The signal from the indicator is amplified and then applied to a polar-coordinate recorder.

In Fig. 18.21 are various graphs showing the degree of amplification that can be obtained on one surface. The one on the left shows only major errors; the one in the center shows all the errors; the one on the right shows an average between the two.

## SURFACE MEASUREMENT

The control of surface finish involves much of the metalworking industry. A few years ago it was not possible for many manufacturers to measure the finishes produced with any degree of accuracy. The only

**Fig. 18.21.** Talyrond trace charts can be used to record either major or minor errors, depending upon the amplification used, and can also record the average of these two records. Courtesy Engis Equipment Co.

means of control was to specify certain speeds and feeds to be used.

Today the designer can ask for a given finish with confidence that it will be delivered as specified. Underfinishing, like overfinishing, can prove to be very expensive. Underfinishing will not produce the required service life or accuracy needed. Overfinishing has been found to increase costs on an average of one percent per microinch in the upper ranges.

The term "surface texture" is replacing "surface finish" because it describes the total condition of the surface better. Surface texture is defined as *random deviations from the nominal surface that make up the pattern of surface.* It includes roughness, waviness, lay, and flaws.

**Lay.** Lay is the predominant direction of the surface pattern. This is normally determined from the method by which the material was produced.

**Flaws.** Irregularities that occur at one place, or at relatively infrequent intervals on the surface, are called flaws, for example, a scratch, ridge, hole, peak, crack, or check.

**Surface Texture Characteristics.** The American Standards Association has defined surface texture characteristics in ASA B10.1 – 1962. These characteristics, shown in Fig. 18.22, may be described as follows.

The term *roughness* refers to the finely spaced irregularities, the height, width, and direction of which establish the predominant pattern. These irregularities are produced by the cutting tool and the machine feed. The height is measured in microinches, using an arithmetical average. This average, as shown in Fig. 18.23, is the average vertical distance from a datum line of every point of the profile occurring in the length of surface sampled. The datum line or "mean" is not a constant. Its "elevation" varies with each specific length of profile being measured.

To check a part for average roughness, one must also know the "roughness width cutoff." This value establishes the maximum spacing

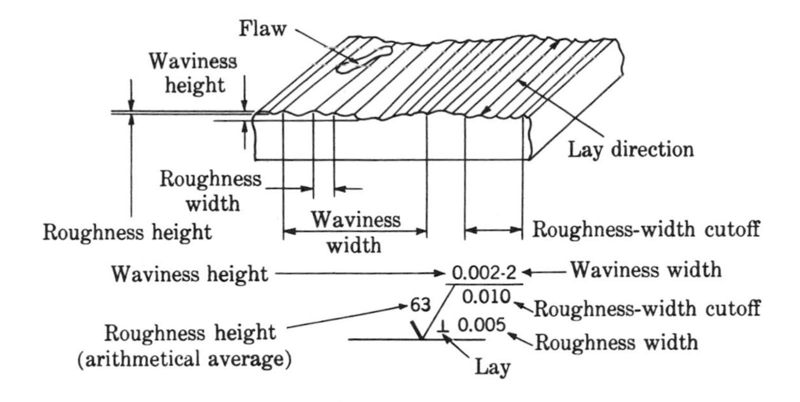


Symbols indicating direction of lay

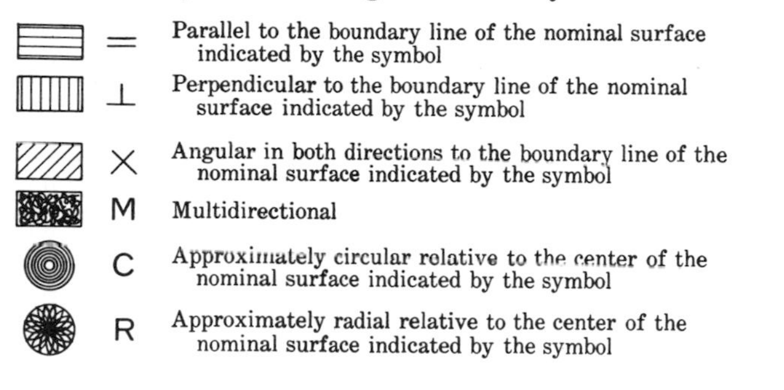


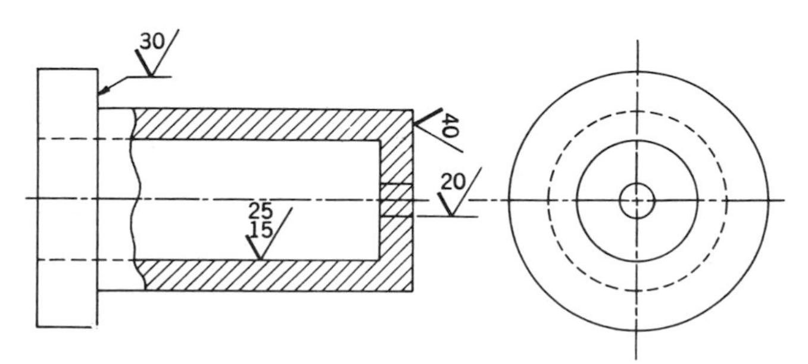


How surface roughness is specified on working drawings. Figures in V-shaped symbols show roughness in microinches

**Fig. 18.22.** Surface texture characteristics and symbols as defined by the American Standards Association, through whose courtesy this is used.

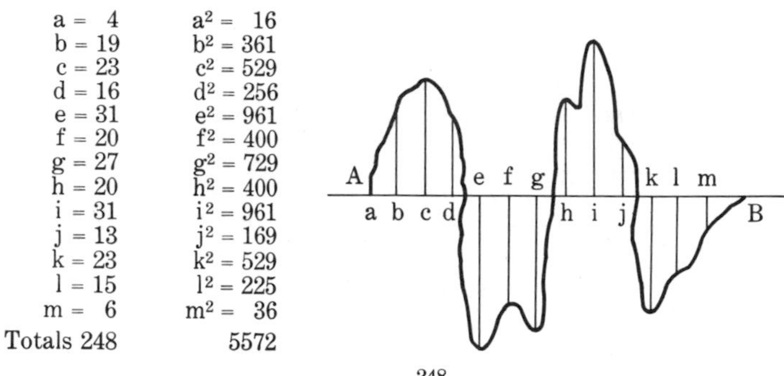

| | |
|---|---|
| a = 4 | $a^2$ = 16 |
| b = 19 | $b^2$ = 361 |
| c = 23 | $c^2$ = 529 |
| d = 16 | $d^2$ = 256 |
| e = 31 | $e^2$ = 961 |
| f = 20 | $f^2$ = 400 |
| g = 27 | $g^2$ = 729 |
| h = 20 | $h^2$ = 400 |
| i = 31 | $i^2$ = 961 |
| j = 13 | $j^2$ = 169 |
| k = 23 | $k^2$ = 529 |
| l = 15 | $l^2$ = 225 |
| m = 6 | $m^2$ = 36 |
| Totals 248 | 5572 |



$$\text{Arithmetical average} = \frac{248}{13} = 19.1\ \mu\text{ in.}$$

$$\text{Root-mean-square average} = \sqrt{\frac{5572}{13}} = 20.7\ \mu\text{ in.}$$

**Fig. 18.23.** Arithmetical average roughness and rms average roughness can be calculated from profile charts of the surface. Calculations are simple but time-consuming. Direct-reading stylus instruments are preferred for mass inspection of ferrous parts. Courtesy *American Machinist.*

between irregularities that will be included in the measurement. There is no rule of thumb for selecting the cutoff value other than that it should always be larger than the roughness width. If no cutoff value is specified, the standard states that the 0.030 in. cutoff shall be used.

**Trace-Type Average Surface Measurement.** A number of instruments have been made that operate on the principle of a fine (0.0005-in.-radius tip) stylus that is moved over the surface to be checked. The vertical movements of this tracer point are transmitted to a coil inside the tracer body. The coil moves in a field of a permanent magnet, and this produces a small fluctuating voltage that is related to the height of the surface irregularities (Fig. 18.24).

The trace may be moved either manually or mechanically over the surface of the work. Mechanical movement gives a more consistent and dependable roughness measurement. Manual operations, however, make for convenience and timesaving setups. The surfaces measured may range from 1 $\mu$in. or less up to 1,000 $\mu$in. of roughness. Figure 18.25 shows the profile of a machined surface with both roughness and waviness.

Other trace-type instruments are the Surfindicator, with a range of 0 to 1,000 $\mu$in., and the Talysurf, with a range of 0 to 200 $\mu$in.

These instruments give an average surface indication. Flaws, waviness, or some scratches too minute to measure will not show up on the record.

**Profile Recorders.** Profile recorders are also stylus-type instruments but are made to provide a highly magnified profile as seen by the moving

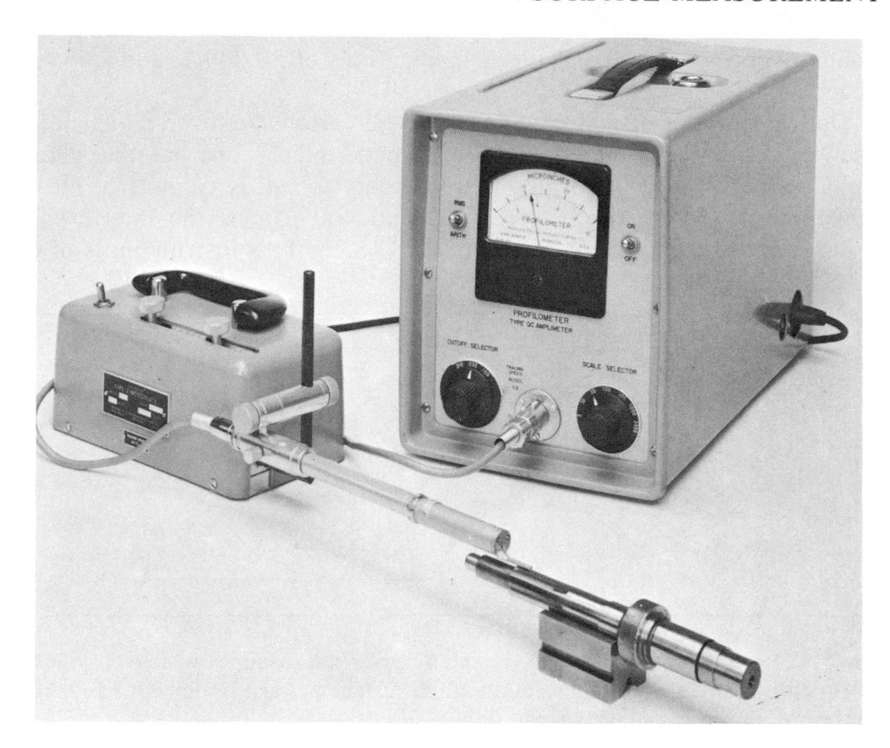


**Fig. 18.24.** The profilometer being used manually to check the roughness of a cylindrical surface. The meter on the panel shows roughness directly in micro-inches as the tracer is moved along the work. Courtesy Micrometrical Mfg. Co.

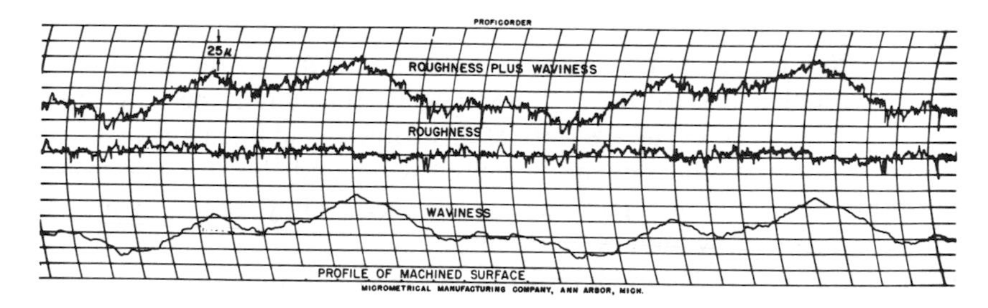


**Fig. 18.25.** The recorded profile of a machined surface, showing both roughness and waviness. Courtesy Micrometrical Mfg. Co.

stylus. The trace shows the height, spacing, and shape of individual peaks and valleys as well as overall patterns of roughness and waviness. It also indicates the presence of flaws and scratches and, if the stylus is fine enough, actual scratch depth.

Considerable skill is needed to interpret the profile trace, so these recorders are better suited to inspection and laboratory use than to routine shop work. Most have magnifications of 50,000:1, and some go much higher.

**Optical Instruments.** Optical instruments provide a three-dimensional view of a surface texture. The interference microscope has the widest application for overall surface texture analysis and is capable of measuring roughness depths between 1.2 and 400 μin. is the interference microscope shown schematically in Fig. 18.26. This instrument is often

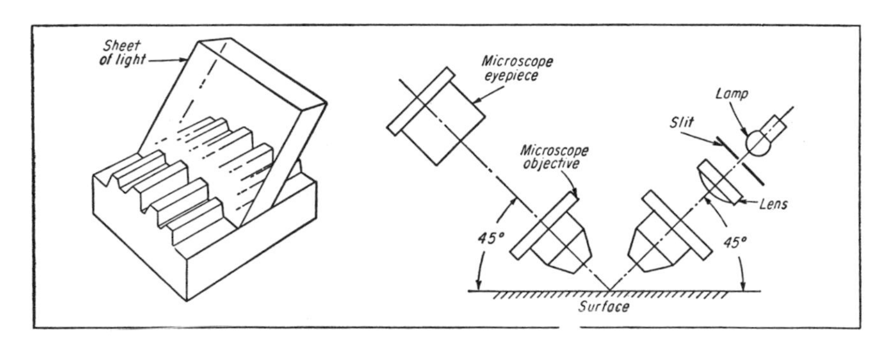


**Fig. 18.26.** Schematic view of the interference microscope, which provides a sheet of light upon which the section of the surface pattern is viewed. The height and width of the irregularities can be measured.

referred to as the 45-deg microscope or the light-section microscope. The light is permitted to shine through a slit, falling on the work in the form of a fine band. This band of light traces out the profile of the surface, that is, its peaks and valleys. A reticle in the microscope can be shifted within the field of view to measure the height (or the width) of the surface irregularities. Both roughness and waviness may be determined.

Two advantages of using optical instruments in measuring surface roughness are that they do not mar the work and they can be used to measure more than one parameter at a time. They can also measure surface roughness, waviness, and flaws.

**Machined-Surface Characteristics.** There are so many factors, within one machining area, affecting surface finish that it is difficult to make a conclusive statement of what to expect. On a lathe, for example, the feed, depth of cut, condition of the tool, and other factors beyond the operator's control will influence the surface finish. The American Standards Association recognized the need of a guide, and established a surface-finish range for each type of machine operation, as shown in Fig. 18.27.

**Surface Finish Costs.** A machined surface is often a basic requirement of manufactured parts in order to insure proper alignment, tolerances, wear, etc. Usually a primary cutting operation if performed with

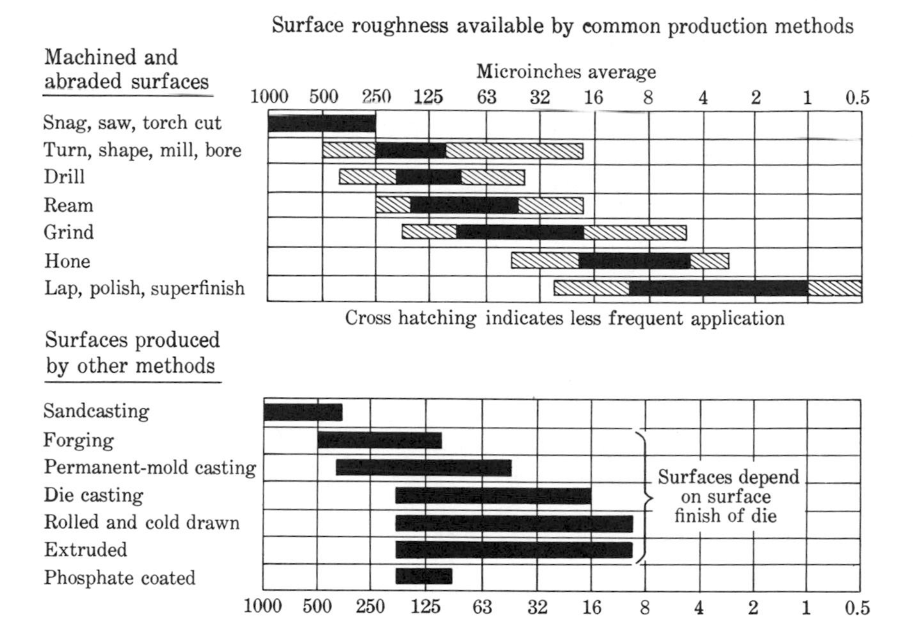


**Fig. 18.27.** The general range of surface finishes that can be obtained by various machining, casting, and other methods. Courtesy American Standards Association.

care on a mill or lathe can produce a surface finish of 63 microinches (AA classification as in Table 18.1) at reasonable cost. When finer finish values are selected, costs rise rapidly (Fig. 18.28).

**Optical Tooling.** When large, accurate, machine-tool surfaces or jig-and-fixture points and alignment have to be checked, optical methods are used; that is, a line of sight is established and used somewhat as a surveyor uses a theodolite. There are other ways of checking large flat surfaces or points of alignment, such as with stretched wires or large block levels, but they are more cumbersome.

The basic instruments needed in optical tooling are a theodolite, a microalignment telescope, and an autocollimator. The theodolite is used for measuring horizontal and vertical angles. The microalignment telescope is an internally focusing telescope with an erecting eyepiece that is used in conjunction with a target. Built-in micrometers enable the operator to observe both horizontal and vertical displacement of the target and make readings as close as 0.0002 in. The autocollimator combines a collimator (an instrument for projecting parallel light rays) and a telescope.

The principle of the *autocollimator* can best be explained with the aid of the diagram in Fig. 18.29. Light projected from the front lens, in the form of parallel rays, is reflected from a front-surface mirror or other reflector. After reentering the objective lens, the light forms an image of

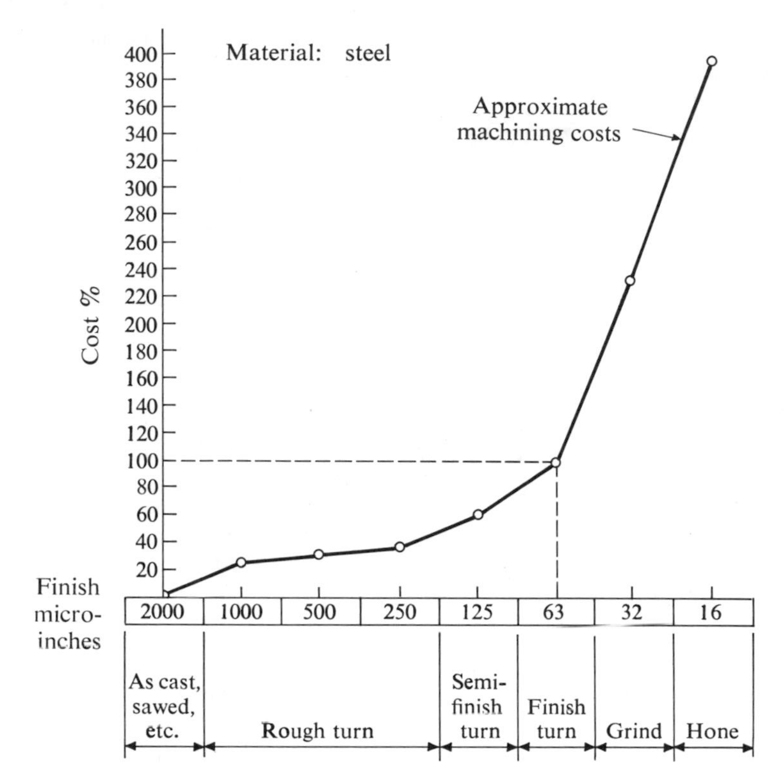


**Fig. 18.28.** Machining costs vary with surface finish requirements. A rapid rise is noted after a 63 AA finish.

target wires in the same plane as the wires in the graticule. The microscope eyepiece enables the observer to measure any angular displacement caused by misalignment at the point where the mirror is stationed.

The autocollimator can also be used to measure the flatness of surfaces or the accuracy with which locating points lie within a straight line. High accuracy is possible because the autocollimator can measure angles as small as 1 sec of arc.

An example of the use of the autocollimator is shown in Fig. 18.30. The autocollimator is mounted on the surface plate. A line across the plate is calibrated by moving a reflector carriage across the table in lengths equal to the carriage base. A straightedge clamped to the surface plate is used in guiding the reflector carriage. Tilts of the carriage indicate a variation in the surface. Several tracks are made across the surface plate to determine its overall accuracy. A corner mirror is used when the line of sighting is changed. The results are correlated, and the deviation from the mean true plane is established.

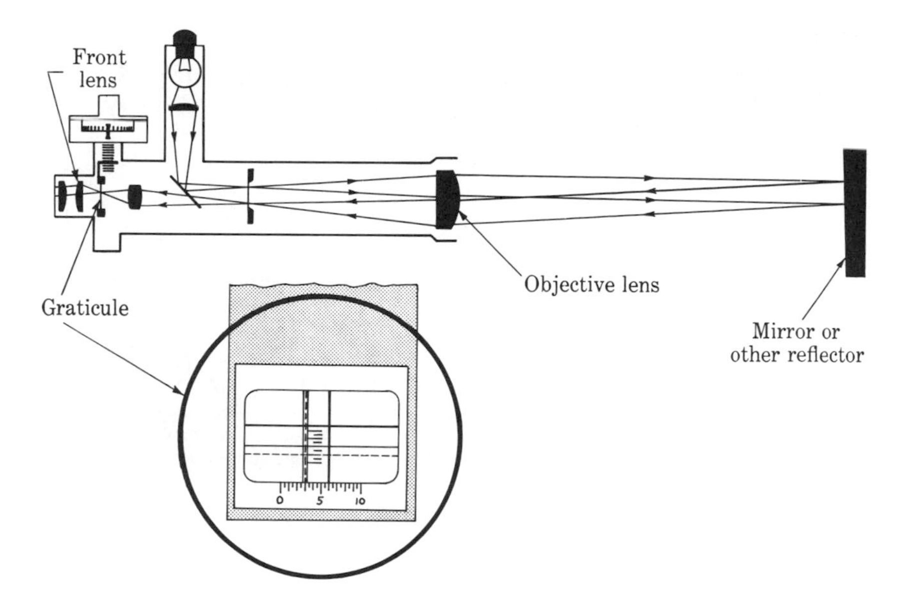


**Fig. 18.29.** The principle of the autocollimator. Courtesy Engis Equipment Co.

An example of the use of the *microalignment telescope* is shown in Fig. 18.31. Two alignment telescopes are mounted on the tooling fixtures with targets located in the templates. The targets may be of the cross-hair type for alignment only, or they may have scales that allow measurement of displacement. Mirrored targets make it possible to determine the tilt.

The telescopes are removed from the tool upon its completion, but the telescope mounting and the targets remain an integral part of the tool. These mountings allow replacement of the telescopes for rechecking the tooling as needed.

**Coordinate Measuring Machines.** Considerable progress has been made in developing coordinate measuring machines in recent years due to the trend to dimension-from-datum elements. Since most engineering drawings are dimensioned from datum elements, such as points, lines, and planes, the dimensions must be directly related. To relate all measurement to a surface plate makes the measurement unnecessarily burdensome. The coordinate measuring machines perform the dual function of both staging and measuring the workpiece. They may be classified as two- or three-axes machines.

TWO- AND THREE-AXES COORDINATE MEASURING MACHINES. The two-axes machine measures along two mutually perpendicular axes parallel to the staging plane (Fig. 18.32). The movable head may be aligned with any selected reference point on the part, and, as it is moved

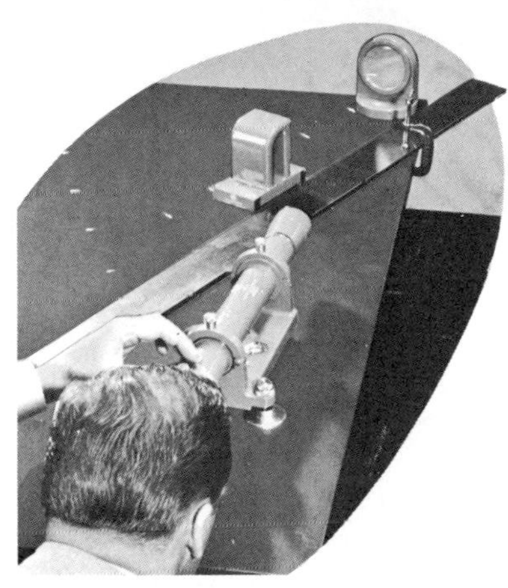

**Fig. 18.30.** The autocollimator is used to calibrate a surface plate. Courtesy Engis Equipment Co.

to a second point, a digital readout will permit the operator to get the dimension quickly and accurately to within ±0.0001 in. with repeatability to 0.000025 in. at speeds up to 6,000 ipm.

The three-axes machine is capable of measuring in *x*, *y*, and *z* dimensions to an accuracy of 0.0005 in. repeatable to ±0.0002 in. at speeds up to 6,000 ipm over all three axes. Automatic position printout is available for permanent records.

## GAGING

Most of the tools and instruments discussed thus far are used for fine accurate work on an individual basis. Even though more and more accuracy is being called for and some of the tools mentioned are moving into the work areas, it remains for specially built gages to handle the bulk of production inspection.

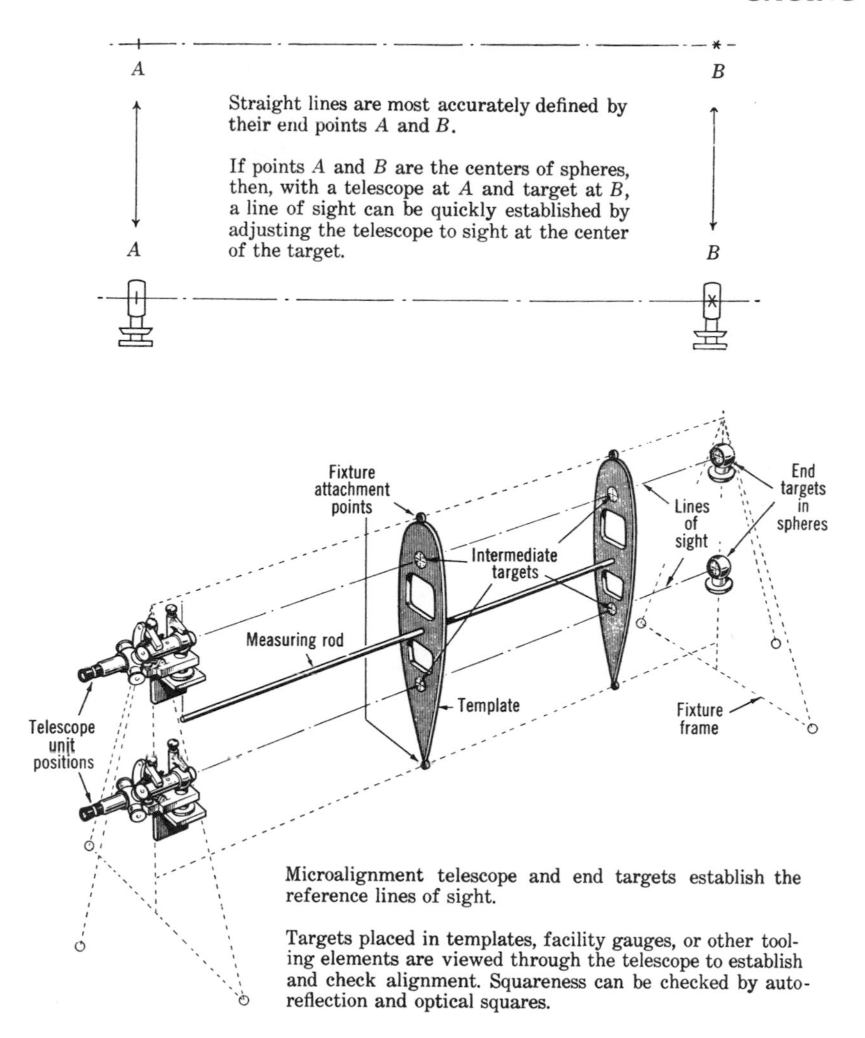


**Fig. 18.31.** Optical tooling used for alignment of aircraft-wing templates.

A gage is a device used by a machine operator or inspector to determine whether the manufactured part is within the prescribed dimensions. In mass production, gages provide the best means of securing interchangeability. Gaging is done as much to *prevent* unsatisfactory parts from being made as to sort out the correct from the incorrect.

Gages are built on a system of *fits* which, in turn, have prescribed limits, tolerances, and allowances. These and related terms are identified as follows:

*Tolerance*—the amount of size variation which is permitted on a part without interfering with its functional operation.

**Fig. 18.32.** An example of a two- or three-axis coordinate measuring machine with digital readout. Courtesy Farrand Controls, Inc.

*Limits*—the extreme dimensions of the tolerance.

*Clearance*—the difference in size between mating parts when the critical dimension of the male part is smaller than the corresponding internal dimension of the female part (Fig. 18.33*a*).

*Interference*—the difference in size between mating parts when the critical dimension of the male part is greater than the corresponding internal dimension of the female part (Fig. 18.33*b*).

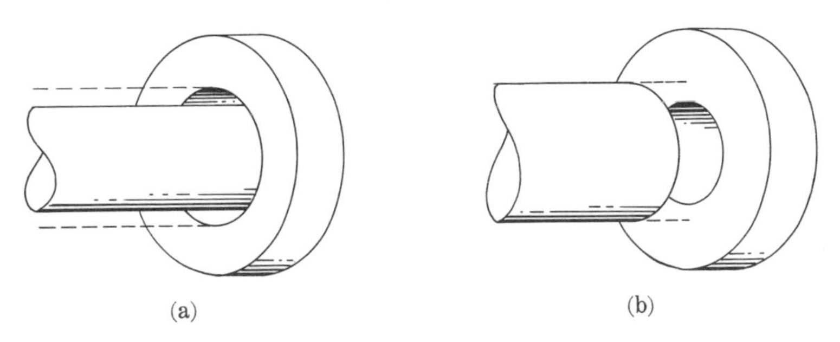


**Fig. 18.33.** (*a*) Clearance fit and (*b*) interference fit.

*Allowance*—the prescribed difference (minimum clearance or maximum interference) between mating parts to attain a specific class of fit.

*Basic size*—the size from which the limits of size are derived by the application of allowances and tolerances.

*Unilateral tolerance*—tolerance in which the variation is permitted in one direction only.

*Bilateral tolerance*—tolerance in which the variation from the design size is permitted in both directions.

**Classification of Fits.** Fits are of three general types—running, locational, and force.

RUNNING OR SLIDING FITS. Running or sliding fits are broken down into seven gradations, starting with the designation RC1 and running through RC7, each with an increasing clearance. The first classification in this series is intended for accurate location of parts without perceptible play; the last in the series is for a free-running fit where accuracy is not important.

LOCATIONAL FITS. The locational fits are designated by letters, as follows: LC for locational clearance, LT for locational-transition fits, and LN for locational-interference fits. This group of fits, intended to determine the location of mating parts, varies from the clearance type with free assembly to the interference type used to provide rigid or accurate location.

FORCE FITS. Force fits are designated by the letters FN, followed by a number indicating the degree of force needed to assemble the parts. FN1 indicates a light drive fit, whereas FN5 is a force fit used for parts able to take a high degree of stress or for shrink fits when heavy pressing forces are impractical.

The design engineers, along with the various draftsmen, are responsible for selecting the limits or tolerances according to the functioning of the equipment. There are no set rules for establishing limits and tolerances for all phases of mechanical design, so careful judgment must be exercised in specifying the proper amounts. The greater the permissible limits or tolerances, the less costly the part is to produce; there will be less scrap and lower labor costs, and less expensive tools will be needed.

**Gage Types.** The designer may specify the proper tolerances and allowances, but it remains for the production department to see that they are maintained. To ensure production standards, gages used to inspect the work are made to very rigid tolerances, usually about 10 percent of the workpiece tolerance (Fig. 18.34).

Gages may be divided into mechanical and comparator types.

**Mechanical Gages.** Mechanical gages are often termed *fixed-limit gages*. These gage by attributes, meaning that the part is either passed or is not passed by a single gage setting; there is no variation. These gages

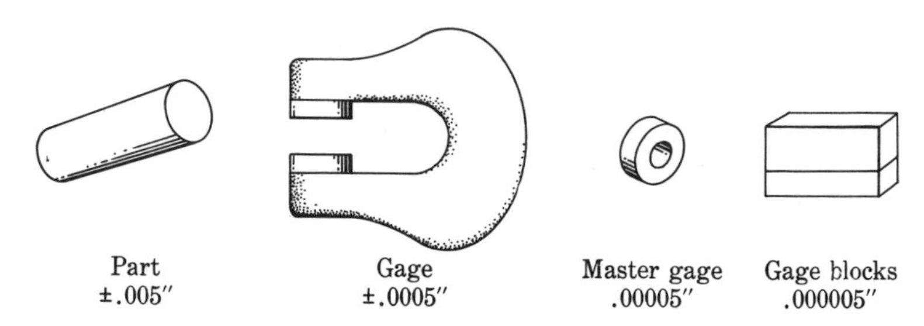


**Fig. 18.34.** A gage must be ten times more accurate than the dimension being checked for reliable quality control. Courtesy DoAll Company.

may be subdivided into groups, according to the purpose for which they are used.

PLUG GAGES. Plug gages are used to check internal diameters. A go – not-go plug gage is a double-ended gage used to control minimum looseness or maximum tightness of mating parts. Plug gages may also be tapered or threaded, as shown in Fig. 18.35.

RING GAGES. Ring gages are used to check external diameters. They may be used to check straight, tapered, or threaded external diameters.

SNAP GAGES. A snap gage is used to check the outside diameter of a part, somewhat like the action of a caliper. It may be made adjustable within small limits. With two pairs of anvils, it also becomes a go – not-go snap gage, as shown in Fig. 18.36.

**Comparator Gages.** Comparators are instrument-type gages that make use of a standard, such as a template or gage blocks, for a reference. They differ from the fixed-limit or attribute gages in that they are able to show the amount of variation from the standard. For this reason they are also called *variable* gages.

The four main types of comparators, based on the amplification mechanism, are mechanical, electrical, air, and optical.

Many special gages have been made using the dial indicator to measure directly in thousandths or ten-thousandths of an inch. One such gage, the adjustable snap gage, is shown in Fig. 18.37.

ELECTRICAL COMPARATORS. The electronic comparator and height gage (Fig. 18.38) use an amplifier to get high magnifications (from 1,000 to 100,000 times). Repeat readings can be made within a millionth of an inch.

The Multichek (Fig. 18.39) employs signal lights to show whether any given dimension shown on the profile drawing is under, over, or within tolerance. Tolerances may be checked up to 0.000050 in. A red master light will automatically go on if any one of the dimensions is out of tolerance. A quick check of the small lights on the panel will indicate exactly which dimension or dimensions are off.

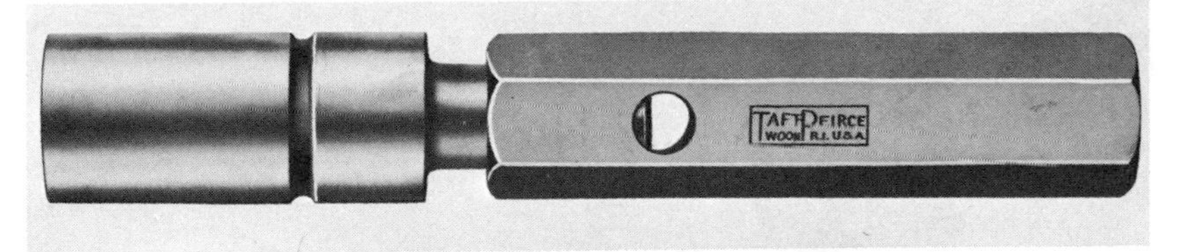


A go – not-go progressive plug gage

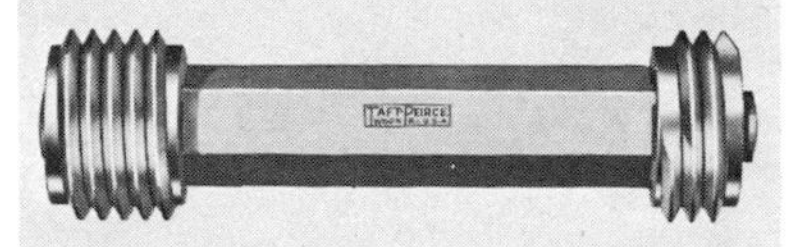

A go – not-go double-ended-thread plug gage

A not-go ring gage

A go ring gage

**Fig. 18.35.** Various plug and ring gages. Courtesy The Taft-Peirce Manufacturing Co.

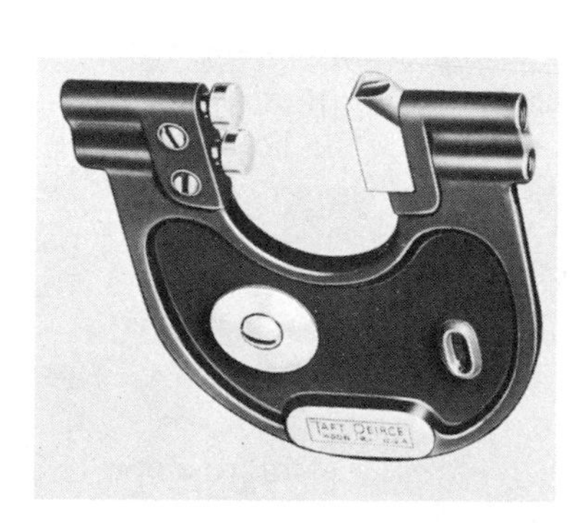

**Fig. 18.36.** Go – not-go snap gages of the adjustable type, used to check plain and threaded outside diameters. Courtesy The Taft-Peirce Manufacturing Co.

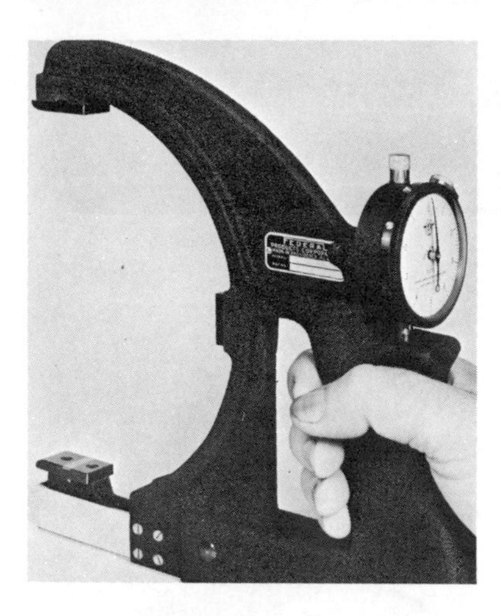

**Fig. 18.37.** A dial indicator snap gage.

AIR COMPARATORS. Air comparators are gaging devices that utilize the effect of minute dimensional changes on metered air. The air gage consists principally of an air plug with small orifices, a gage to register the rate of the escaping air, and pressure regulators.

The gaging plug is first placed in a master ring gage, at which time the pointer is set to zero on the dial. The difference in size of the workpiece and the master will then show up as an amplified reading on the dial indicating the clearance between the plug and the work. The air gage is useful in showing out-of-round, taper, irregularity, and concentricity of a bored hole (Fig. 18.40). Magnifications ranging from 1,250 to 1 and 20,000 to 1 may be obtained, with a full-scale range of 0.006 to 0.0003 in., respectively. Measurement is in 0.0001- and 0.000005-in. increments. Air-gaging principles are also used on snap gages, depth gages, and in multiple-diameter checking.

OPTICAL COMPARATORS. With the aid of lenses and mirrors, an enlarged "picture image" of an object may be projected on a screen and measured, as shown both schematically and pictorially in Fig. 18.41.

A master chart of the magnified image, usually containing tolerance limits, is superimposed over the image on the viewing screen. By visually comparing the chart lines, an operator can quickly determine if the part meets prescribed tolerance standards. Direct measurements may also be made with accuracy in the order of 0.0001 in. The optical comparator is particularly useful in checking contours and a number of dimensions at one time. A tracing attachment makes it possible to check optically hidden surfaces, as shown in Fig. 18.42.

THE TOOLMAKER'S MICROSCOPE. The toolmaker's microscope (Fig. 18.43) may be thought of as both a direct and a comparative measuring instrument. It is principally a microscope that is equipped with a protractor head and a stage or table that can be moved in increments of

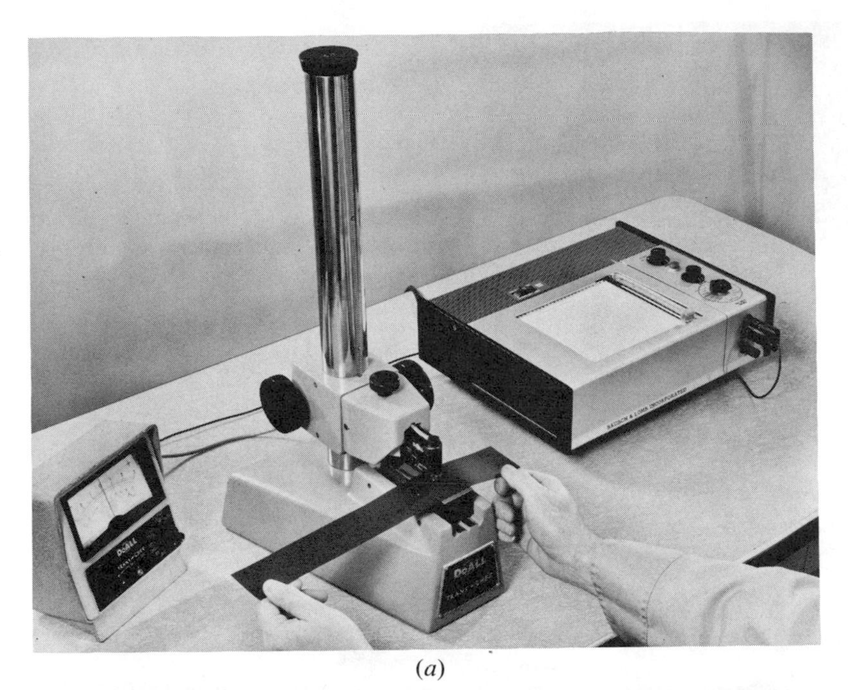


(*a*)

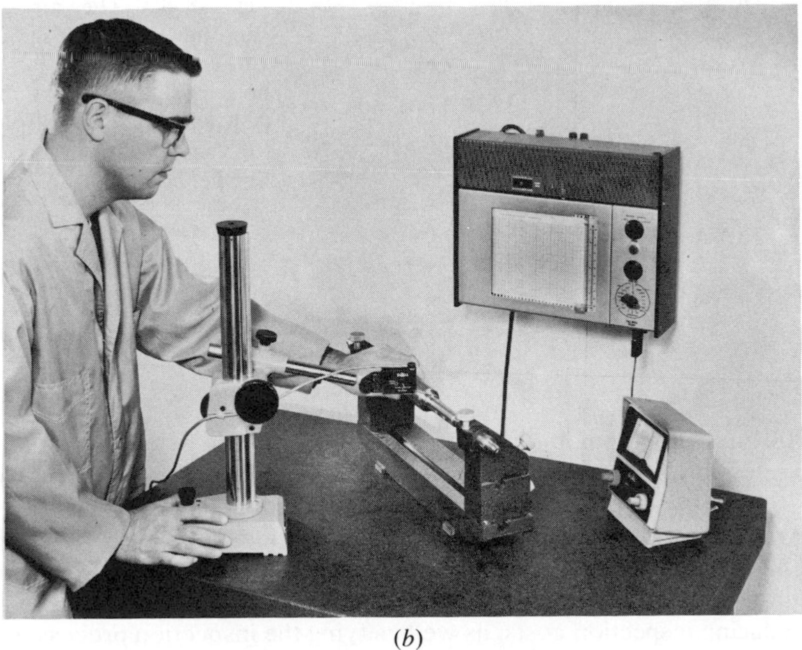

(*b*)

**Fig. 18.38.** The electrical comparator (*a*) and height gage (*b*) can be used for extremely accurate dimensional comparisons with printed readout. Courtesy DoAll Company.

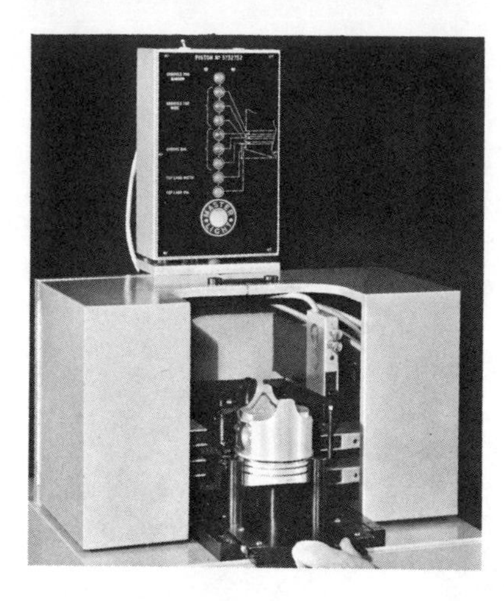

**Fig. 18.39** A Multichek used to check the grooves in a piston. Courtesy The Sheffield Corp.

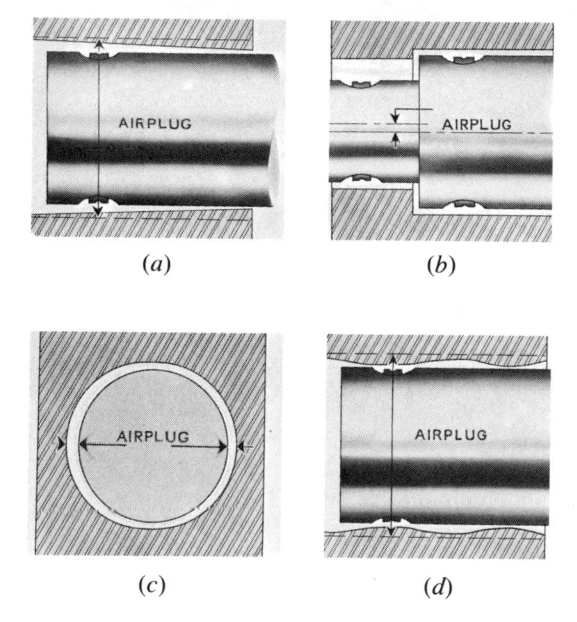


**Fig. 18.40.** The air gage can be used to indicate various conditions of the hole: (*a*) taper, (*b*) concentricity, (*c*) out of round, (*d*) waviness. Courtesy The Sheffield Corp.

0.0001 in., either longitudinally or laterally. A part to be checked may be mounted in a fixture on the table and viewed in magnifications of from 10× to 100×. Typical applications are also shown in Fig. 18.43.

**Automatic Gaging.** Automatic gaging provides a means of checking each part, as it is being made or immediately afterward. Automatic gaging has come into being through a series of progressive steps aimed at reducing inspection costs, as well as tying the inspection process more closely to the operator and machine. It is estimated that, since World War II, 85 percent of measurement has changed from mechanical to air and electrical-electronic gaging.

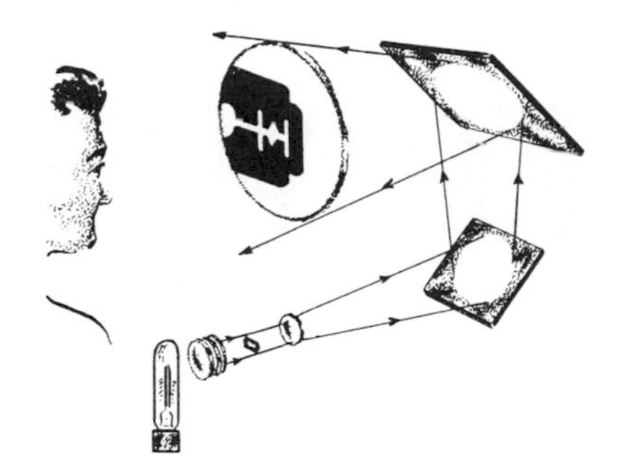

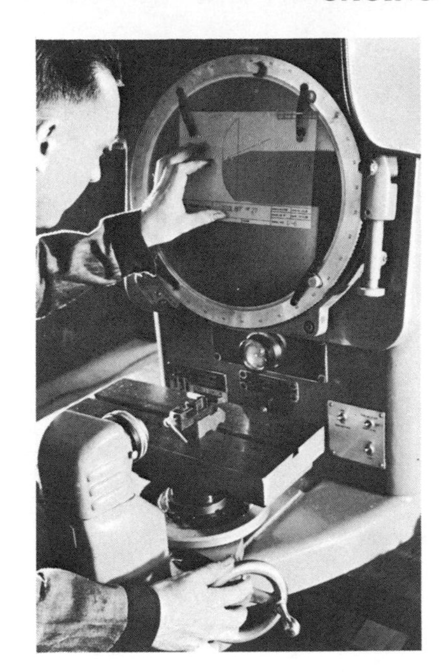

Fig. 18.41. The principle of the optical comparator is shown schematically and as ordinarily used to check contours. Courtesy Jones and Lamson Machine Co.

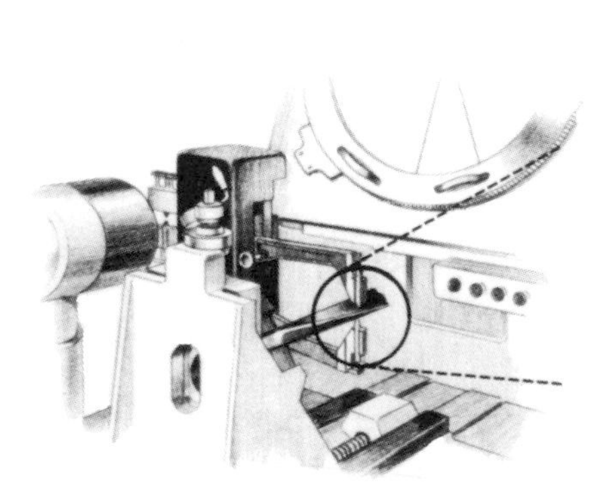

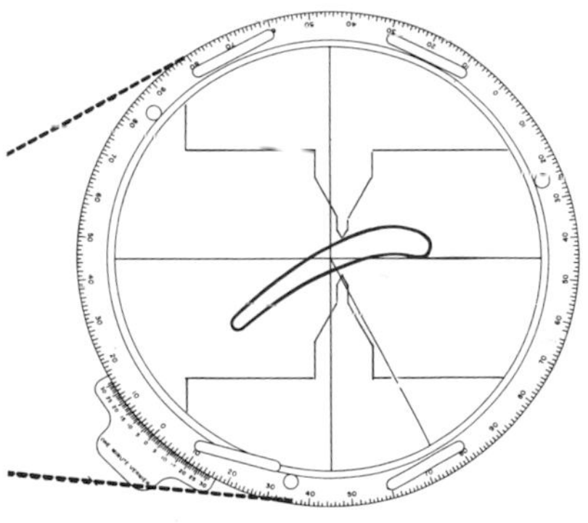

Inspecting optically-hidden surfaces which cannot be projected or reflected (such as the contours of dies, molds, airfoils, etc.) can be done with the Universal Tracing Attachment.

Adjustable carbide styli are fastened to tracing arms which are mounted on a coordinate slide. The styli of one arm are in contact with the object surface, the others lie in the focal plane of the comparator. The movement of the tracing styli over the object surface is duplicated by the projected styli. The magnified shadow of the projected styli appears on the viewing screen, where its path of travel can be compared to a master outline chart.

**Fig. 18.42.** Optically hidden surfaces can be inspected by means of a tracing stylus. The magnified shadow appears as shown on the viewing screen. Courtesy Jones and Lamson Machine Co.

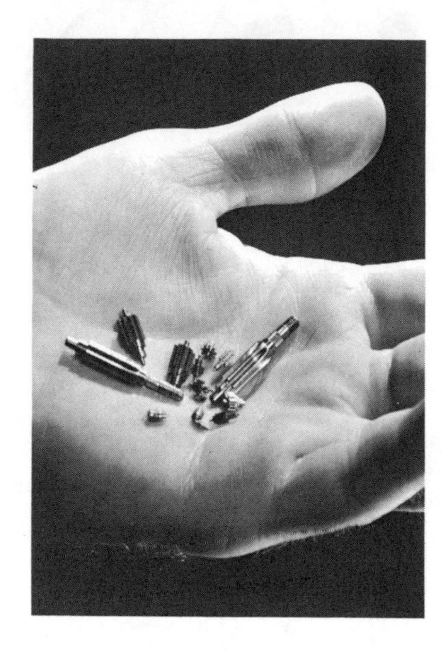

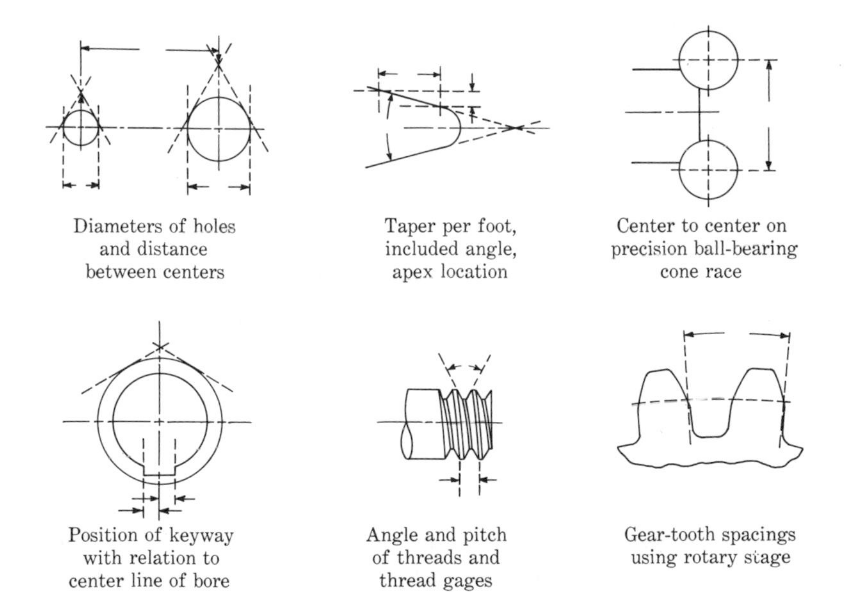


**Fig. 18.43.** A toolmaker's microscope and typical applications for measurement. Courtesy The Gaertner Scientific Corporation.

The first step gave the operator some of the standard inspection equipment, such as mechanical comparators. With this equipment near his machine, he could check quickly and accurately the parts produced.

A second step was to place the gage directly on the machine (Fig. 18.44), which was termed *in-process* control. By this method, the size of the part is shown continuously as the metal is being removed. The cutting tool is directed automatically to retract when the part is finished to size. The control circuit may be made to adjust the tool automatically when the size is toward either limit of tolerance.

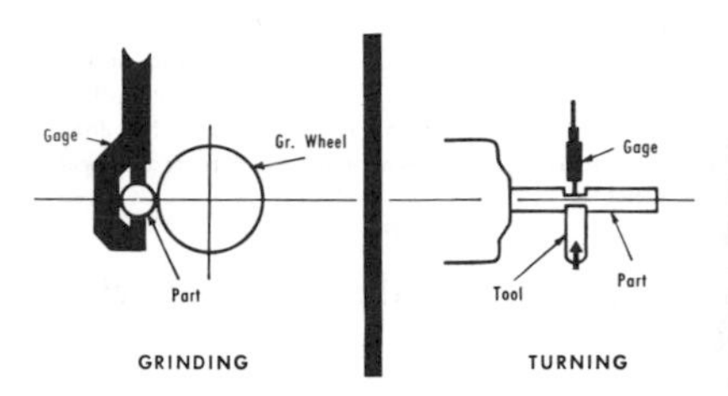


"In-process" control or gaging during machining provides the machine or cutting tool with signals to:

- **Automatically stop or change the tooling from a roughing to a finishing cut.**
- **Automatically retract the cutting tool when the part is to finished size.**
- **Automatically indicate and adjust the tool when the size trend is toward either limit of tolerance.**
- **Automatically stop the machine after the tools become worn and parts no longer can be machined within limits.**

"In-process" control prevents faulty parts from being made by initiating signals for tool correction or replacement before part size is out of control.

**Fig. 18.44.** The gage is placed right on the machine for continuous measurement, or *in-process* control. Courtesy The Sheffield Corp.

A third step for closer product control made use of charts to plot the results of the operator's inspection. By watching the process, with respect to control limits, the operator could follow the trend toward the upper or lower tolerance limit. Before limits were exceeded, the necessary machine adjustments could be made and scrap parts avoided.

**Postprocess Control.** A fourth step was postprocess control. In this method, the parts move out of the chucking or holding device into the gaging station to be inspected before being transferred to the next operation. The postprocess inspection makes use of a *feedback* control that can automatically warn the operator or adjust the tool when the part size is approaching the extreme limits of control (Fig. 18.45). Feedback means that the information from the gage is given to the machine control center, which, in turn, can bring about the necessary tool adjustments.

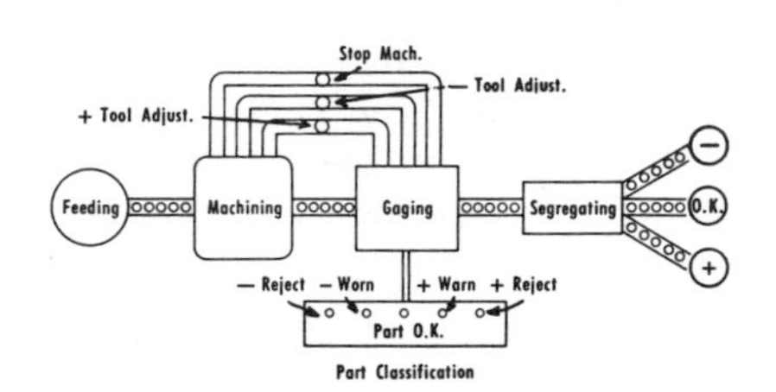


With "Post-process" control or gaging, the part is machined, moved out of the chucking or holding device (leaving the machining station open for another part) and into the gaging station where it is inspected before being transferred to the next operation.

A major advantage of "Post-process" Control over "In-process" Control is that the parts are gaged in a free state—without influence from chuck or machine.

In "Post-process" Control, signals are initiated to:

- **Automatically warn and/or adjust the tool when part size is approaching the extreme limits of control.**
- **Automatically stop the machine in event of tool failure.**
- **Automatically stop the machine if a specified number of consecutive parts is beyond tolerance.**
- **Automatically actuate segregating mechanism so that faulty parts are rejected before reaching the next operation.**

"Post-process" Control means on-the-spot inspection and rejection of faulty parts.

With this type of control, the gaging station is generally located adjacent to the machining operation. Or, it may be remotely located and the parts from several machines channeled to it in sequence, so that the machine that produced the faulty part is indicated and shut down after a specified number of consecutive parts has been rejected.

**Fig. 18.45.** Postprocess gaging. Parts have passed from the immediate vicinity of the machine. Courtesy The Sheffield Corp.

**Other Automatic Gaging Functions.** SALVAGE CORRECTIONS. Automatic-gaging units are used not only to separate parts into their various categories but also to salvage parts that have been rejected. The schematic diagram (Fig. 18.46) shows how distorted connecting rods are checked for parallelism between the crank and pin bores. Two air spindles determine the amount and direction of the "out-of-parallelism" through a differential head. The resulting signal actuates a ram that exerts predetermined increments of force on the part. Other heads may be brought in for further corrective action if required.

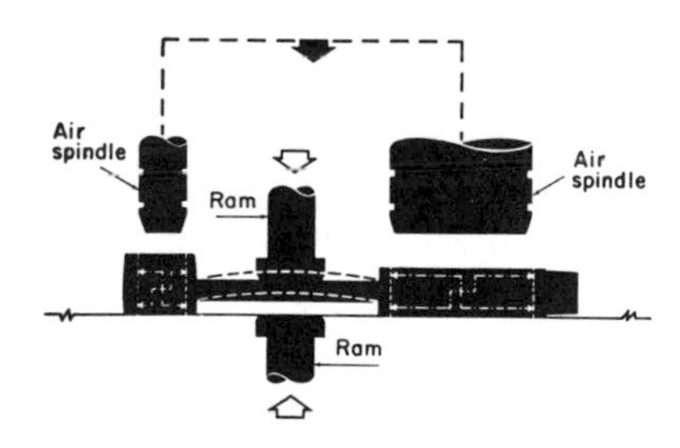


**Fig. 18.46.** Automatic salvaging operation of distorted connecting rods. Courtesy *The Tool and Manufacturing Engineer.*

SELECTIVE FITTING. Although all the parts made may be of such tolerance as to make them interchangeable, selective assembly adds quality to the product. To illustrate, if a pulley of maximum allowable diameter were to be assembled with a shaft representing the minimum allowable diameter, the fit would be loose. The opposite condition would give a tight fit. Selective assembly would narrow this margin. In some assemblies the tolerance limit can mean the difference between a long or a short life.

The automatic gage shown in Fig. 18.47 checks glass medical syringes. The glass plungers are gaged and classified into seven size ranges at the rate of 3,600 per hr. A companion unit is utilized to find matching glass cylinders for selective assembly.

**Future Automatic Gaging.** Pneumatic-electronic gaging systems will continue to be made in the decision-making field. These will accurately control high-speed production lines, or they may be used to control entire plants through a centralized system.

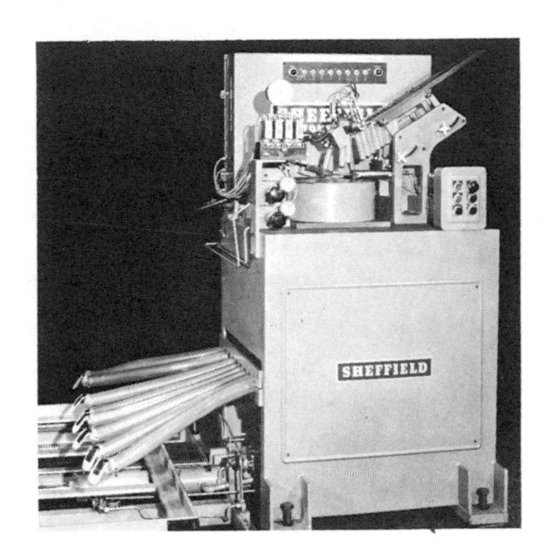


**Fig. 18.47.** This gaging machine automatically classifies glass plungers in seven sizes. They can then be given to an assembly unit for close-tolerance fitting. Courtesy The Sheffield Corp.

## INSPECTION AND QUALITY CONTROL

Closely associated with metrology are inspection and quality control. The terms *inspection* and *quality control* may, at first, seem somewhat synonymous. They are, however, quite different in application. Inspection is concerned with *how well* the physical specifications for a product are being met. Quality control is not concerned with physical aspects, as such, but rather with the *amount* of inspection that will be necessary to produce a good product.

## INSPECTION

Inspection has two broad and equally important tasks. One of these is to detect errors in workmanship and defects in materials so that the item will meet specifications; the other is to help minimize future errors. It is the latter which occasionally leads the inspection function to go beyond workmanship and materials into design specifications. For example, inspection may reveal tolerances that are impractical or materials that are inferior, etc.

**First-Piece Inspection.** Before an inspector can check an item that has been produced, he must have certain specifications and standards for reference. The specifications are usually obtained from the part print. Not all the dimensions listed on the print are checked. Except for the first piece, critical dimensions are selected and are used thereafter as the basis of inspection.

Specifications may be abbreviated on the print by a reference to a standard. An example of a standard reference for a thread may be 3/4-16 NC-3. The inspector knows that this refers to the American National Coarse Thread 16 pitch series, class 3 fit, as found in the National Bureau of Standards *Screw Thread Handbook*. The size in this standard is specified as 0.7500 maximum diameter, 0.7410 minimum diameter, tolerance 0.0090. Many other standards are established by engineering societies and by plant usage. These standards are necessary so that the inspector can be sure that items conform to specifications; they also prevent much duplication of effort.

First-piece inspection refers to the first good part that is turned out after a new setup has been made. The inspector may be called for a first-piece check after a new tool has been installed, after a new shift has started, or at any time when there is a change of operator.

**Sampling Inspection.** A roving inspector is called by the setup man when he has finished the first production part. After that, the inspector may return at regular set intervals to check a sample number of parts. The sample is taken to show what the machine is doing. This method is useful not only to show what is being produced at that time, but also to predict any trouble in the near future.

Since so much depends on the inspector's measuring tools, they too must be checked at set intervals or whenever there is any doubt as to their accuracy. Gage departments or laboratories are established in the plant to collect and check regularly the various measuring instruments and tools. After calibration and repair, they are returned to the job.

**Batch Sampling.** Batch sampling generally concerns parts that are removed from the manufacturing floor and sampled according to statistical methods (to be discussed later in this chapter).

**Final Inspection.** Inspection of parts and assemblies just before they are prepared for packaging is called final inspection. It is vitally important to the reputation and goodwill accorded to the producer. With many products, the inspection will be visual—a check for flaws, defects, and missing or damaged parts. Products such as engines and motors may be given a brief operational test. Some products, such as ammunition, require destructive tests. Destructive testing or inspection points up the need for the proper use of sampling plans as developed in statistical quality control.

**Inspection Requirements.** The inspection department must have a broad knowledge of the products being manufactured and of their use. The principal functions of the inspection department are to be familiar with the product design and specifications; to know the relationship of each component part of an assembly and how it functions in respect to the completed assembly; to set up a sequence of inspection operations, designating the type and the point in the manufacturing process at which each is to take place; and to determine the equipment needed for inspection.

PRODUCT DESIGN. It is advantageous for the inspection department to learn as much about the product as possible. Advanced planning can be done by checking both the prints and the product in the early stages of manufacture.

RELATIONSHIP OF COMPONENTS TO FUNCTION OF THE PRODUCT. It is necessary to understand the relationship of each component to the proper functioning of the completed assembly. The performance of the assembled end product will be governed by the type and amount of inspection carried out on the various components.

INSPECTION SEQUENCE. The manufacturing process is analyzed to establish the best sequence of inspection operations. It must be determined whether one of the in-process types of inspection can be used or whether the product should be checked in the final assembly.

## QUALITY CONTROL

As stated previously, quality control is concerned with *how much* inspection is necessary to achieve the desired product standards agreed upon by engineering and management. By statistical means, quality control is able to set up sampling plans which will assure that the process quality is being controlled and, at the same time, that inspection costs are being kept at an economical level. The brief description given here is intended only to present an overview of some of the tools and methods available. For a more complete discussion of the subject, consult the references at the end of this chapter.

**The Normal Frequency-Distribution Curve.** To understand why sampling plans are not a hit-or-miss proposition but, rather, a carefully worked out mathematical approach, one should know something about a normal frequency distribution. To illustrate, suppose that 50 parts were turned out on a lathe and measured with a micrometer. It is probable that most of the parts would fall close to the size the operator was trying to obtain. A few parts might be larger and a few smaller, as shown in Fig. 18.48.

The normal frequency-distribution curve may be divided into six zones that are mathematically equal in width. Each part is represented by the Greek letter sigma ($\sigma$). By referring to Fig. 18.49, we can see that plus or minus two sigma ($\pm 2\sigma$) from the center will include 95.5 percent of the parts. Likewise, $\pm 3\sigma$ will include 99.75 percent of the parts. The

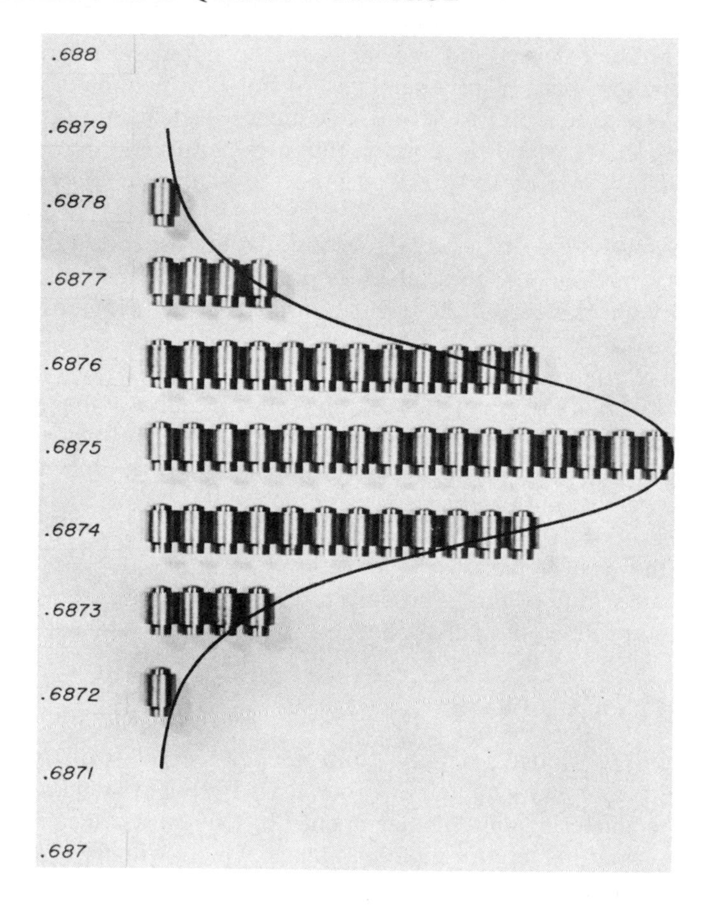


**Fig. 18.48.** Shafts turned out on a lathe, showing normal size variation. Courtesy Federal Products Corp.

$3\sigma$ limit is usually referred to as the *natural tolerance* limit. The peak of the curve shows where the process tends to center itself. The shape and spread of the curve give useful information to the engineering department in helping them decide the limitations of a process.

The distribution-curve spread may vary considerably from the "normal" curve as shown in Fig. 18.50. This variability is termed *measure of dispersion.*

One measure of the variability or spread of the distribution is the *range* which is the difference between the largest and smallest measurements in the group.

Another measure of dispersion involves the average distance (or deviation) of the sizes from the *mean.* The mean refers to the sum of all

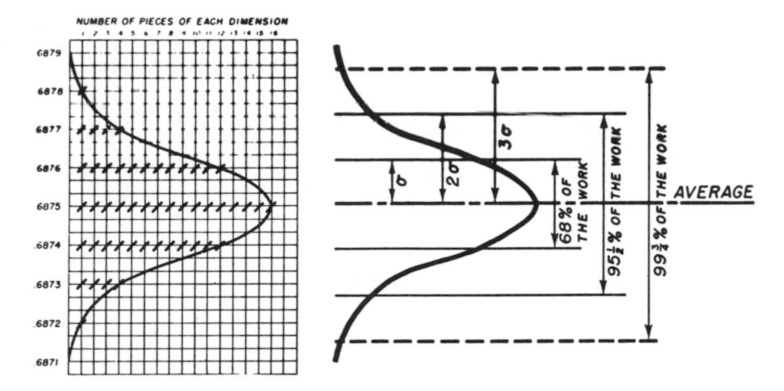


**Fig. 18.49.** The normal curve, showing six mathematical divisions. This breakdown applies to Fig. 18.48. Courtesy Federal Products Corp.

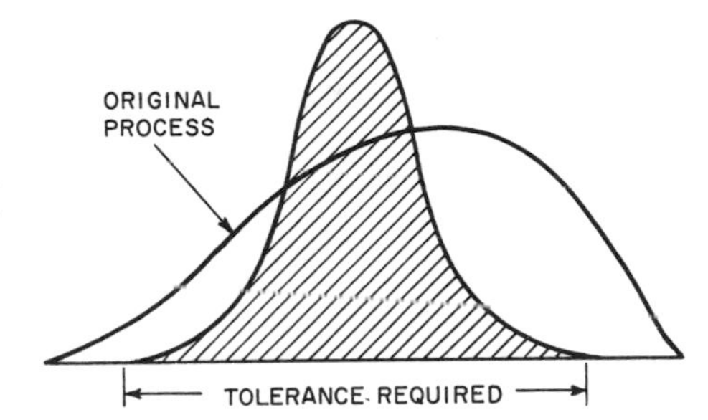


**Fig. 18.50.** The process has been refined to bring it within the acceptable tolerance zone.

the part sizes divided by the number of parts. The two terms are written mathematically as:

$$\text{Mean} = \Sigma X/n \qquad \text{AD} = \Sigma d/n$$

where $\Sigma$ = add or sum up; $X$ = individual sizes; $n$ = number of sizes; and $d$ = deviation or distance from the mean.

In finding the average deviation all sizes are added without regard to sign. Since an algebraic addition is more normal, the *standard deviation* is generally used. A somewhat simplified formula is:

$$\sigma = \sqrt{\Sigma d^2/(n-1)}.$$

The data as represented in Fig. 18.49 can be used as an example for finding the mean. In this case, we have seven sizes with a varying number or frequency in each group. The mean can be computed by multiplying each value by the frequency with which it occurs, adding the products, and dividing by the sum of the frequencies, as shown in Table 18.2.

**Table 18.2**
**Finding the Mean ($\bar{x}$) and Standard Deviation ($\hat{\sigma}$)**

| Size interval | Midpoint | $F$ | $d$ | $Fd$ | $Fd^2$ |
|---|---|---|---|---|---|
| 68715–68725 | 0.6872 | 1 | −3 | −3 | 9 |
| 68725–68735 | 0.6873 | 4 | −2 | −8 | 16 |
| 68735–68745 | 0.6874 | 12 | −1 | −12 | 12 |
| 68745–68755 | 0.6875 | 16 | 0 | 0 | 0 |
| 68755–68765 | 0.6876 | 12 | 1 | 12 | 12 |
| 68765–68775 | 0.6877 | 4 | 2 | 8 | 16 |
| 68775–68785 | 0.6778 | 1 | 3 | 3 | 9 |
| Totals | | $n = 50$ | | $\Sigma Fd = 0$ | $\Sigma Fd^2 = 74$ |

$$\text{Mean} = \overline{X} = \alpha + \delta\overline{d}$$

$$\text{Standard Deviation} = \hat{\sigma} = \left\{\frac{\delta^2}{n-1}\left[\Sigma Fd^2 - \frac{1}{n}(\Sigma Fd)^2\right]\right\}^{1/2}$$

where $\alpha$ = some cell midpoint; $\delta$ = cell length; $d$ = negative or positive integers or zero; and $\overline{d} = (1/n)\Sigma Fd$. Now

$$\alpha = 0.6875 \qquad \delta = 0.0001 \qquad \overline{d} = (1/50)(0) = 0.$$

$$\therefore \overline{X} = 0.6875 + (0.0001)0 = 0.6875.$$

$$\hat{\sigma} = \left\{\frac{(0.0001)^2}{49}\left[74 - \frac{1}{50}(0)^2\right]\right\}^{1/2}$$

$$= \left[\frac{(0.0001)^2(74)}{49}\right]^{1/2}$$

$$= \frac{0.0001}{7}(8.602)$$

$$= 0.0001$$

**Quality and Cost.** The terms quality and cost are closely related. The problem is to get as much as possible of the first without raising the second excessively.

From a quality standpoint, the ideal part is well within the capability of the existing process (Fig. 18.51). Under these conditions, the need for inspection is minimized.

The worst conditions are having a part that is produced by processes having wide variations in accuracy and capability, causing high inspection screening, rework, and scrap costs (Fig. 18.52).

When the repeatable accuracy of a given manufacturing process is uncertain, it is necessary for the quality control department to specify large inspection samples with consequent increased inspection costs.

The specified tolerance spread may be within the capabilities of the

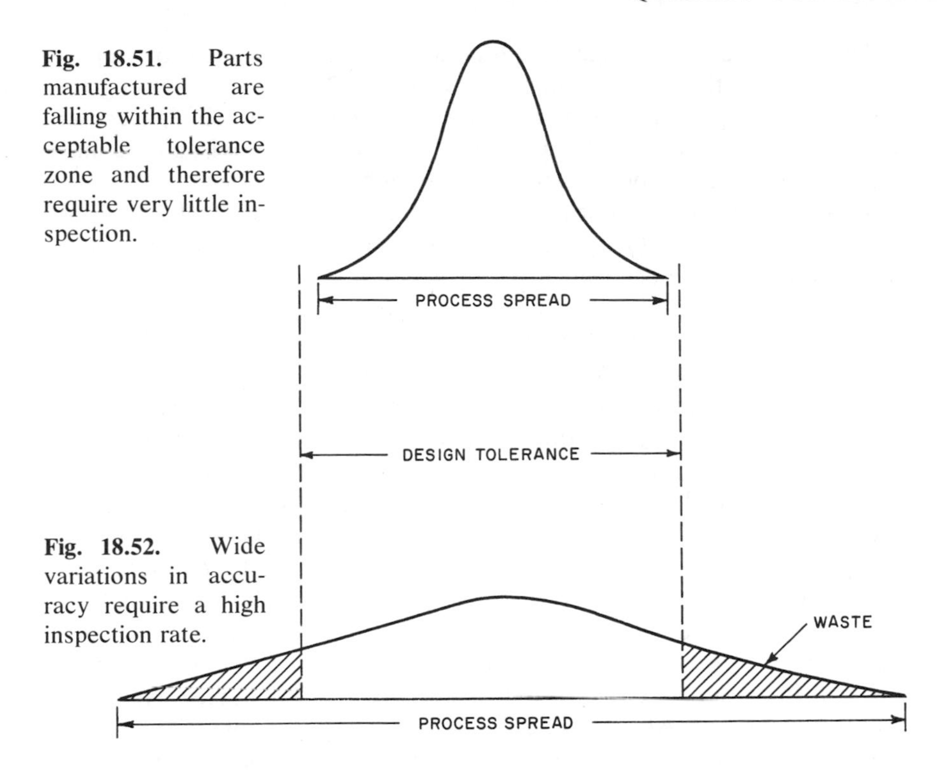


**Fig. 18.51.** Parts manufactured are falling within the acceptable tolerance zone and therefore require very little inspection.

**Fig. 18.52.** Wide variations in accuracy require a high inspection rate.

process but located so close to the upper or lower limit of the range that some of the pieces are not within the acceptable tolerance. The quality yield of a process is dependent, not only on the magnitude of the range, but also on the position of the process relative to the drawing dimension limits.

A method used to help the designer, the manufacturing engineer, and the quality control engineer work towards reducing costs is to classify all features of a part as to their functional importance as follows:

*Class 1:* Features which, if out of specification, could cause malfunction of the product.

*Class 2:* Features which, if out of specification in conjunction with other out-of-specification features, could cause malfunction.

*Class 3:* Features which, if out of specification within reasonable limits, will not affect the function of the product.

These numbers can be penciled alongside the appropriate features on the drawing to help the quality control engineer prepare inspection sheets. Processes may also be classified as to expected quality. For example, parts that are entirely controlled by the set tooling tend to be more consistent than those controlled by an operator. Where alternate methods of manufacture are provided, the least operator influence will produce parts more nearly within the process average range.

Thus we see that design and production engineers need to keep the capabilities of their plant constantly in mind. The production engineer must also know the tolerance capabilities of equipment being purchased. If special equipment must be built, it would be advisable to have tolerance-capability specifications clearly defined.

**Methods of Quality Control.** Quality control is concerned with establishing a certain quality level both for the incoming materials and for the manufactured product. Two broad aspects are involved: establishing a reasonable number of defective parts for a given quantity (acceptance sampling) and watching the manufacturing process and recording the results (control charts). Acceptance sampling may be further subdivided into the various plans that have been formulated, such as single sampling, double sampling, multiple sampling, and sequential sampling.

**Sampling Plans.** BATCH OR LOT SAMPLING. Batch or lot sampling usually refers to large quantities of parts that have left the machining area, have gone through the necessary cleaning, and have been brought to the inspection room. Sampling plans are based on the premise that a number of parts taken from a lot will be representative of this lot. Since 100 percent inspection of all parts is not practical, except on an automatic basis, a certain small percentage of defectives is allowable. This is referred to as the acceptable quality level (AQL). From past experience and trial runs, the AQL can be determined. For example, the AQL for small parts made on an automatic screw machine may be 1.2 to 2.2 percent defective. The average outgoing quality limit (AOQL) for the same lot will be from 1.5 to 2.5 percent defective. This means that the parts leaving the inspection station will average no more than a certain percentage of defects, either because all the lots met the sampling requirements or because any lots that did fail at sampling were reinspected and sorted before they were allowed beyond the inspection station.

A typical problem will serve to illustrate the principles used in various sampling plans. Table 18.3 is for a batch or inspection lot of 800 to 1,300 pieces with an AQL of 1.2 to 2.2. You will note that in the single-sampling plan a sample size of 55 is required and that 3 defective parts are acceptable but 4 will cause the lot to be rejected. A rejected lot is not scrapped; it is screened; that is, each part of the sample is examined, and the defective parts are replaced. You will notice the data for double and sequential sampling in the same table. These plans operate in much the same way, the main difference being in the size of the first sample. In the double-sampling plan the first sample size is 35. The acceptance number is 1 and the rejection number is 5. If more than 1 but fewer than 5 defectives are found, we neither accept nor reject the lot. In this case a second sample of 70 must be taken, making the total number of parts checked 105. There is now no spread between the acceptance and the rejection numbers. If there are 4 or fewer defective,

**Table 18.3**
**Sampling Plans for Various Lot Sizes Ranging from 800 to 1,300**
*Acceptance not permitted until two samples have been inspected.

| | | | Combined Samples | | |
|---|---|---|---|---|---|
| Type of Sampling | Sample | Sample Size | Size | Acceptance Number | Rejection Number |
| Single | First | 55 | 55 | 3 | 4 |
| Double | First | 35 | 35 | 1 | 5 |
| | Second | 70 | 105 | 4 | 5 |
| Sequential | First | 14 | 14 | * | 2 |
| | Second | 14 | 28 | 0 | 3 |
| | Third | 14 | 42 | 1 | 4 |
| | Fourth | 14 | 56 | 3 | 5 |
| | Fifth | 14 | 70 | 3 | 5 |
| | Sixth | 14 | 84 | 3 | 5 |
| | Seventh | 14 | 98 | 4 | 5 |

we accept the lot; if there are 5 or more, it is rejected and subjected to screening.

FACTORS AFFECTING THE CHOICE OF A SAMPLING PLAN. One ever-present consideration is cost. Sampling plans always add expense and man-hours for the user but nothing to the price received for the product. Incoming parts such as small machine screws are standard items that need little if any inspection. On the other hand, precision parts such as crankshafts will need a close check. Single-sampling plans are the easiest to use. Double- and sequential-sampling plans are more involved.

Single sampling makes use of a larger original sample. It has been shown that where a single sample would require 100 pieces to be inspected, double sampling will provide the same protection with 74 pieces, and sequential with 55.

Where parts are of rather homogeneous large lots and a relatively quick, inexpensive check on them is needed, sequential sampling will fill the bill. If the parts need rather close checking, as for the crankshafts mentioned, the larger single-sampling plan will be necessary.

**Control Charts.** While there are many kinds of control charts, they all provide a graphical representation of the process undergoing inspection. Commonly used control charts are shown in Table 18.4. Since the average-and-range charts are usually used together and are very common, they will be discussed briefly.

**Table 18.4**
**Control Charts**

| | |
|---|---|
| *p* charts | show the percent defective in a sample |
| *c* charts | used in plotting the number of defects in one piece |
| $\overline{X}$ charts | used in plotting the variations in the averages of samples |
| *R* charts | show variations in the ranges of samples |
| $\underline{\overline{X}}$-and-$\underline{R}$ charts | average-and-range charts |

X-AND-R CHARTS. Samples of a part that is being manufactured are taken at regular intervals. The size of the sample is usually four or five items. The average of the sample is plotted for the given time interval on the chart as shown in Fig. 18.53. Care must be exercised to get samples of the same size each time and to get a random sample. At the same time, the range (the largest dimension minus the smallest dimension) is plotted as shown in Fig. 18.53.

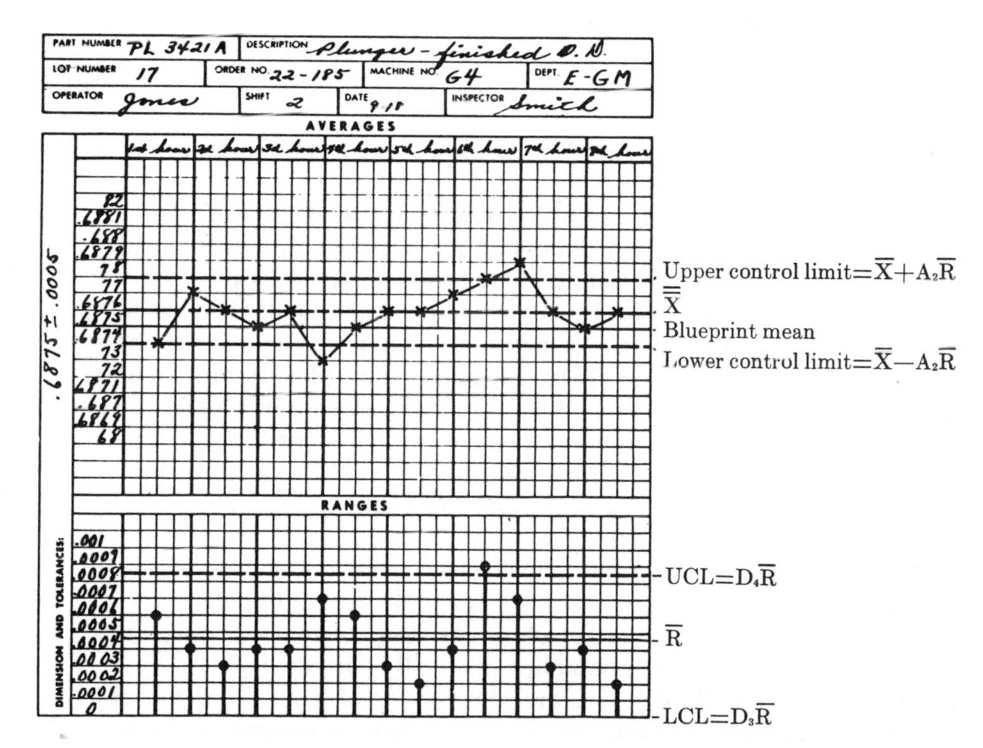


**Fig. 18.53.** An average-and-range chart with sample averages posted each half hour. The range of each sample is also shown. Courtesy Federal Products Corp.

After all the samples have been plotted

the grand average $\overline{\overline{X}}$ of all the samples

is computed and marked on the chart as a red dashed line. The ranges are averaged, and the average range line is drawn on the chart. The upper and lower control limits are obtained by multiplying the average range by a constant based on the sample size as shown in Table 18.5:

$$\overline{x} + A_2 \overline{R}$$

If our sample size were 5 and the average range turned out to be 0.00048 in., the upper control limit would be

$$0.6875 + 0.577 \times 0.00048 = 0.6877$$

This figure becomes the upper control limit (Fig. 18.53). The lower control limit is obtained in the same way except that the resulting number is subtracted from the grand average.

The range-control limits are obtained by multiplying the average range by the constant $D_4$ (Table 18.5) for the upper limit and by the constant $D_3$ for the lower limit.

The blueprint mean is drawn on the chart as the solid zero line. Control limits can be established after a reasonable number of samples have been taken, say 20 to 25. Too few samples will not give a true picture, and too many will add to the cost.

USING THE CHARTS. The information plotted on the control chart forms the basis for action not only by the inspector, operator, and foreman, but also by engineering and management. Inspectors can watch

**Table 18.5**
**Control-Chart Constants for Small Samples**

Courtesy American Society for Testing Materials

| Number of observations in sample, $n$ | Chart for averages: Factors for control limits | | | Factor for central line | Chart for ranges: Factors for control limits | | | |
|---|---|---|---|---|---|---|---|---|
| | $A$ | $A_1$ | $A_2$ | $d_2$ | $D_1$ | $D_2$ | $D_3$ | $D_4$ |
| 2............... | 2.121 | 3.759 | 1.880 | 1.128 | 0 | 3.686 | 0 | 3.268 |
| 3............... | 1.732 | 2.394 | 1.023 | 1.693 | 0 | 4.358 | 0 | 2.574 |
| 4............... | 1.500 | 1.880 | 0.729 | 2.059 | 0 | 4.698 | 0 | 2.282 |
| 5............... | 1.342 | 1.596 | 0.577 | 2.326 | 0 | 4.918 | 0 | 2.114 |

the chart and notice if the process is erratic and needs close attention, or they can shut down a machine before too much scrap occurs.

The engineering department can use the information to specify tolerance limits more favorably. This does not mean that engineering will set its specifications in accordance with the process-control chart. If a process constantly goes beyond tolerance limits, perhaps another method will be used for manufacture, or perhaps a better machine can be substituted. On the other hand, parts may be made that are within the control limits but outside the tolerance limits. This may call for a reexamination of the specifications. A process that does not come near either tolerance limit may be speeded up.

## QUESTIONS

**18.1.** About how many centimeters are there in 1 in.?

**18.2.** What purpose do gage blocks serve in industry?

**18.3.** What is now the standard for the inch?

**18.4.** (*a*) How are gage blocks checked for dimensional and surface accuracy? (*b*) To what accuracy can they be measured?

**18.5.** What is the basis of micrometer accuracy?

**18.6.** What is the principle of a vernier?

**18.7.** How can micrometers be made to give a high degree of repeatability?

**18.8.** (*a*) What is meant by indirect measurement? (*b*) Give some examples of indirect measurements.

**18.9.** What is the principle of the sine bar?

**18.10.** Why is surface texture a better term than surface finish?

**18.11.** Why is it necessary to have a "roughness width cutoff"?

**18.12.** What is meant by "optical tooling"?

**18.13.** Why have coordinate measuring machines replaced many of the previous methods of measurement?

**18.14.** What classification of fit should be used for a gear that is to be permanently assembled to a shaft?

**18.15.** What is meant by the rule of "ten to one" in gaging?

**18.16.** What are other names for fixed gages and comparator gages?

**18.17.** Compare the magnifications that can be obtained with electronic and air gaging equipment.

**18.18.** What is the basic difference between quality control and inspection?

**18.19.** How can the manufacturing engineer, the design engineer, and the quality control engineer work together to best advantage?

**18.20.** What is the advantage of a sequential sampling plan?

**18.21.** What does the average-and-range chart accomplish?

## PROBLEMS

**18.1** (*a*) What is the percent increase in cost to go from a 63 $\mu$ finish to a 32 $\mu$ finish? (*b*) What is the percent cost/$\mu$?

**18.2** (*a*) In what range can a machined finish be improved the most without adding too much cost? (*b*) What percent of the surface finish range does this represent?

**18.3.** If a 10-in. sine bar is used to measure a 30-deg angle, what would be the total height of the gage blocks under the opposite side?

**18.4.** If a closed 1-in. micrometer is opened 3 1/2 turns, what will the reading be?

**18.5.** After inspection of 750 manufactured parts, it is found they fall into a normal distribution. How many parts will be within plus or minus three sigma?

**18.6.** How many parts would have to be inspected out of a lot of 1,000 pieces if a single-sampling plan were used?

**18.7.** (*a*) How many parts would have to be inspected out of a lot of 1,000 pieces if a double-sampling plan were used? (*b*) How many parts would have to be inspected if five defective parts were found in each sample?

**18.8.** (*a*) If a sequential sampling plan were used, what would the sample size be if five defective parts were found in each sample? (*b*) How many parts would be inspected if three samples had to be taken?

**18.9.** In constructing an average-and-range chart, the grand average was computed to be 8.502 in. and the range average was 0.008 in. What would the upper and lower control limits be if the sample size were 4? What would the upper limit be on the range chart?

## REFERENCES

Bowker, A. H., and G. J. Lieberman, *Engineering Statistics*, Prentice-Hall, Inc., Englewood Cliffs, N.J., 1959.

Burr, I. W., *Engineering Statistics and Quality Control*, McGraw-Hill Book Company, Inc., New York, 1953.

*Dimensional Quality Control Primer*, Federal Products Corporation, 11th Printing, Providence, 1959.

Duncan, A. J., *Quality Control and Industrial Statistics,* Richard D. Irwin, Inc., Homewood, Ill., 1959.

Farago, F. T., "An In-depth Review of Measuring," *Metalworking,* November 1966.

Grant, E. L., *Statistical Quality Control,* 2nd ed., McGraw-Hill Book Company, Inc., New York, 1952.

Schrock, E. M., *Quality Control and Statistical Methods,* 2nd ed., Reinhold, New York, 1957.

# 19

# AUTOMATION AND NUMERICAL CONTROL

AUTOMATION IS a relatively new word, coined in 1935 by D. S. Harder, then vice president in charge of manufacturing for the Ford Motor Company. The word did not receive popular acceptance until 1947. Automation is a substitute word for "automatization," which means to make something operate automatically. This definition, when applied to manufacturing, refers not only to individual machines but also to the linking of automatic machines together until a continuous process emerges.

## AUTOMATION

In recent years we have witnessed a very much accelerated movement of science, engineering, and technology. Great breakthroughs (to paraphrase Sir Isaac Newton) usually stand on the shoulders of many other achievements that have come before. A phenomenon such as automation is ultimately propelled into being by dozens of preliminary accomplishments, all of which combine to permit another giant step forward in man's march to control his environment. Automation is in debt to literally hundreds of developments ranging from the refinement of diverse metals, ceramics, and plastics to the academic elaboration of mathematical theories.

One of the major resources behind the present outpouring of technology throughout all industries is the vastly expanded educational base in the industrial nations of the world. Many automated techniques would not yet be realized had there not been a large reservoir of engineers and technicians.

The symbol of the age of automation is the computer, and with good reason. Computers can perform, in minutes and with superhuman accuracy, prodigious calculations that would take a man a lifetime to complete.

Computers now enable us to run massive machinery by remote control or, through mathematical "models," to simulate manufacturing processes under all possible variables. Thus months of time and millions of dollars can be saved before actual operations begin. Engineers sometimes get tired of striving for the optimum; fortunately computers do not.

Whatever the device or process, the essence of automation is either (*a*) the continuous or periodic scanning of results, with corrections following automatically whenever deviations exceed established limits, or (*b*) the sequential accomplishment of a series of operations according to prearranged programs. The latter of these two is most predominant in the metalworking field. However, adaptive control, as explained in welding and metal cutting, denotes the entrance of (*a*) into this field also.

Although many forms of automation are in evidence in metalworking plants, there are as yet no automatic factories. No manufacturer can produce an item of any consequence without knowing that it may be obsolete tomorrow, through a competitor's ingenuity. Therefore the metalworking industry has been content to take automation in smaller steps, employing individual-machine automation and multiple-machine automation. In the former, individual machines may be cam-, electric-, air-, hydraulic-, or tape-controlled. Circular-indexing automation also is included in this category. Multiple-machine automation is described by the kinds of lines that are used to connect the machines.

## INDIVIDUAL-MACHINE AUTOMATION

**Cam-Operated Machines.** Some automatic screw machines, the multispindle automatic lathe, and gear-cutting machines are examples of individual-machine automation. In a true sense of the word, unless machines have automatic stock loading and unloading, they should be classified as semiautomatic. Magazine-type feeders are now used to supply lathes with bar stock, and gear blanks to the gear-cutting machines, etc., to make them fully automatic.

**Electric, Air, and Hydraulic Controls.** The basic machine movements are linear, as in the travel of a drill, or rotary, as in the case of an indexing table. The power used to accomplish these movements can be furnished by electricity, air, or hydraulic fluid. The selection of the power source will depend on what is readily available, the relative cost, amount of power required, space available for actuation, and speed requirements.

Space does not permit an explanation here of each of these systems, but, because of its relative simplicity, compactness, versatility, low initial cost, and wide acceptance, the principles of pneumatic automation will be discussed briefly.

Rotary motion is produced by air motors and linear motion by air cylinders. This motion is controlled by valves which with air lines, filters, and lubricators make up the pneumatic circuit.

The successful application of pneumatic power depends on good circuit design. Despite their complex appearance, all pneumatic circuits are essentially simple. In fact, the simpler the circuit, the better the results. A pictorial sketch of an automatic electric circuit is shown in Fig. 19.1. This simple pneumatic system serves to illustrate how a variety of automatic operations, such as drilling, assembly, packaging, stamping, etc., can be accomplished. A gravity conveyor is used to feed the parts. When toggle switch (1) is placed in the *on* position, current flows through

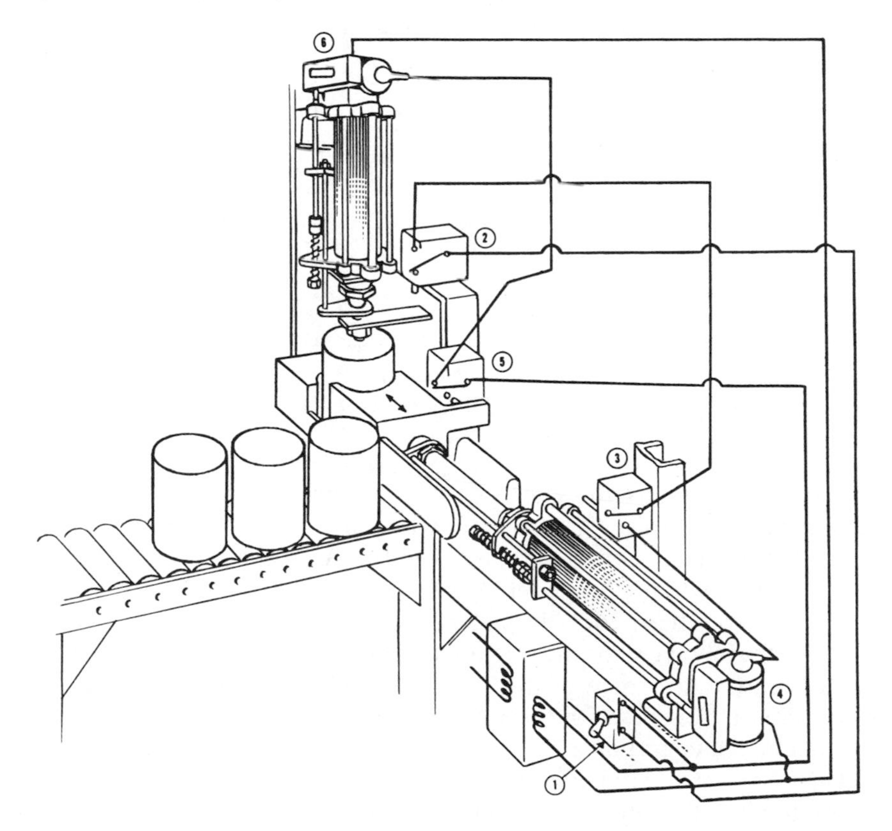


**Fig. 19.1.** A pneumatic electric circuit used to automate a simple operation of stamping, assembly, punching, etc. Courtesy The Bellows Co.

switches (2) and (3), advancing the piston of cylinder (4), which trips switch (5), advancing the punch. The clamping piston returns automatically and does not advance again until cylinder (6) has completed its work and returned to trip switch (2). The operation is continuous until the toggle switch (1) is thrown to the *off* position.

Pneumatic power is becoming an important item in helping manufacturers automate many operations, but, because air is compressible, it is generally impractical to use it for the smooth, even movement of machine feeds with varying loads. Air and hydraulic circuits can be combined for this purpose. In the sample circuit shown in Fig. 19.2,

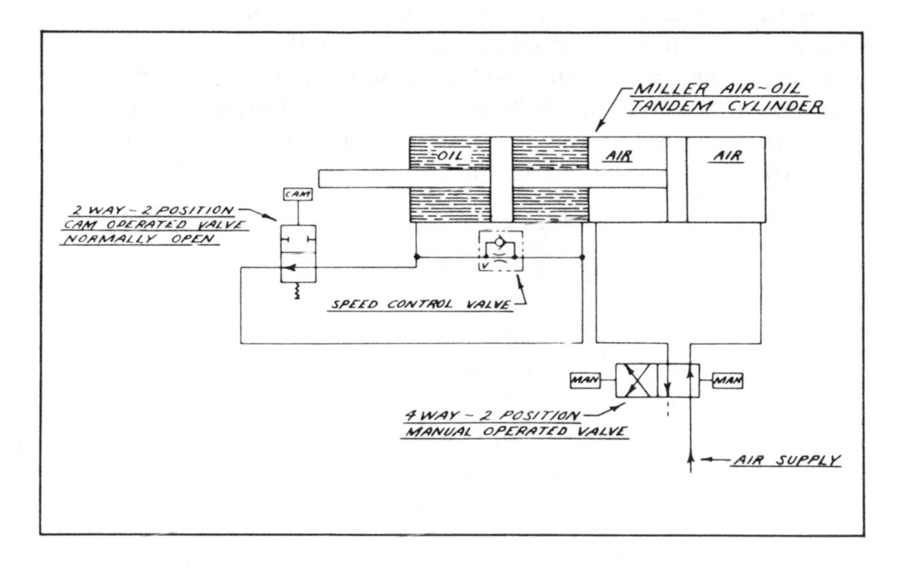


**Fig. 19.2.** A combination air-hydraulic circuit. Courtesy Miller Fluid Power Division.

air alternately enters the ends of the air cylinder through the four-way valve. The piston used to feed the work or tool is attached to the rod that extends through the hydraulic cylinder. A pipeline connects the two ends of the hydraulic cylinder. When the air cylinder cycles back and forth, it drives the hydraulic piston, forcing oil through the connecting pipeline from one end of the hydraulic cylinder to the other. This arrangement incorporates some of the advantages of both systems, namely the lower-cost air circuit with the smoother-acting hydraulic system needed for machining operations.

**Tape Control.** Although numerical control (N/C) will be discussed in more detail in the latter half of this chapter, it is introduced here to show how it fits into individual-machine automation. An individual machine of this type is versatile and is referred to as an *N/C machining center.*

Machining centers usually perform all operations with the part held in a single fixture. Tolerance buildup due to varying fixture locations is eliminated, and piece-part handling is minimized.

The machine shown in Fig. 19.3 is equipped with an indexing table and a tool-selector magazine. The machine has automatic tape-controlled

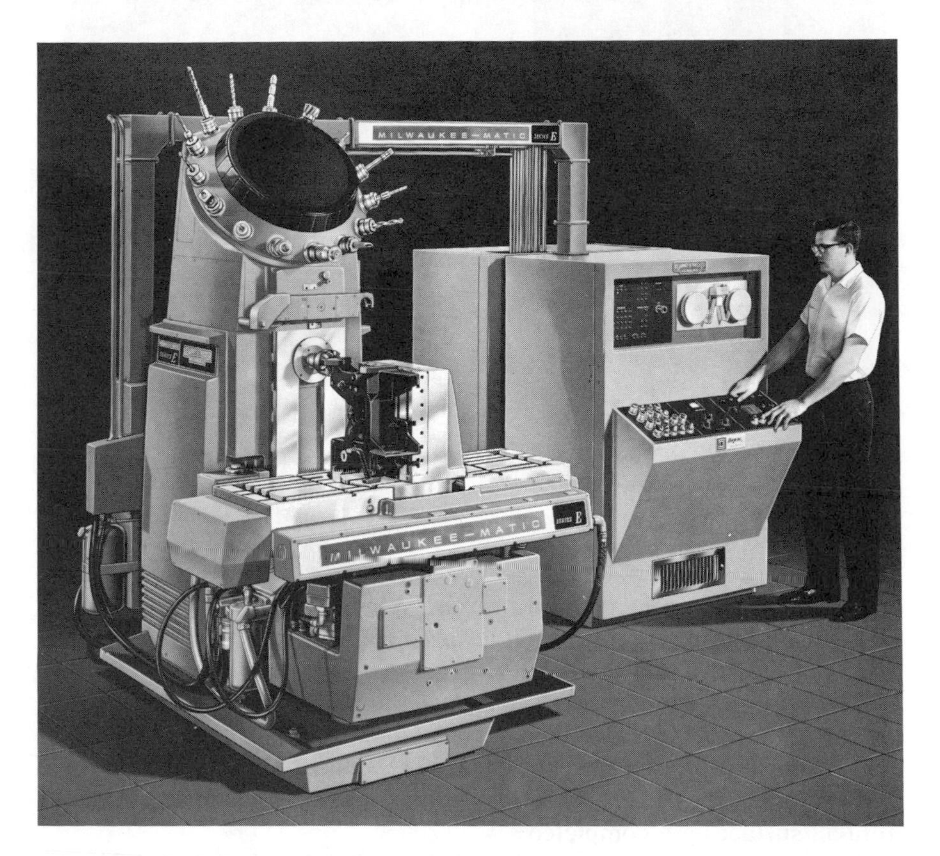


**Fig. 19.3.** An N/C machining center. Courtesy Kearney & Trecker Corp.

movements in three axes, $x$, $y$, and $z$. The table can be automatically rotated in increments of 5 deg. Examples of the type of work done are shown in Fig. 19.4.

Fixturing for the part is designed to provide maximum accessibility consistent with good rigidity.

The operations performed on the switch bracket are shown in Fig. 19.5. Each of the operations requires a number of machine movements which are directed by the tape. Operation 1, for example, calls for rough and finish milling of two surfaces. The steps involved are:

(1) Select the required face mill from the tool magazine.

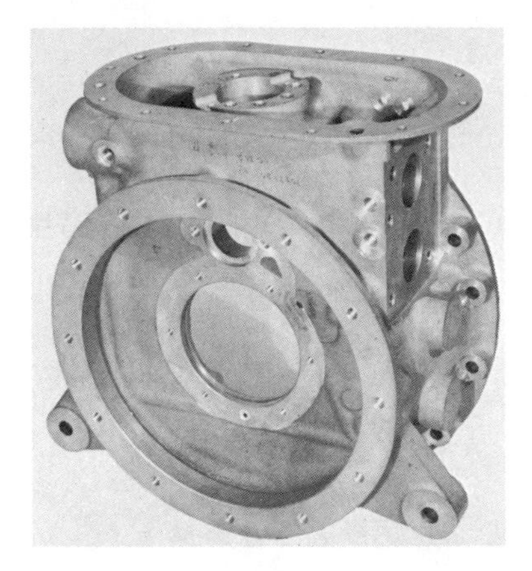

**Fig. 19.4.** Examples of parts that can be handled on an N/C machining center. Courtesy Kearney & Trecker Corp.

(2) Transfer the selected tool into the spindle.

(3) Bring the tool into position at rapid traverse rate. Mill the surface at 15 in. per min.

(4) Turn the coolant on, and preselect the tool for operation (2).

(5) Rapid traverse to the next surface.

(6) Mill the surface at 15 in. per min.

This sequence, or similar ones, is repeated for each of the surfaces until all surfaces are completed.

Figures given for the machining of the switch bracket, comparing the N/C machining center with conventional machining methods, are as in Table 19.1.

**Table 19.1**
**Comparison of Conventional and N/C Machining**

|  |  |
|---|---|
| Number required per year | 90 |
| Conventional tooling cost | $1,890.00 |
| Conventional manufacturing time | 85.38 min |
| Machining center tooling cost | $260.00 |
| Machining center manufacturing time | 19.18 min |

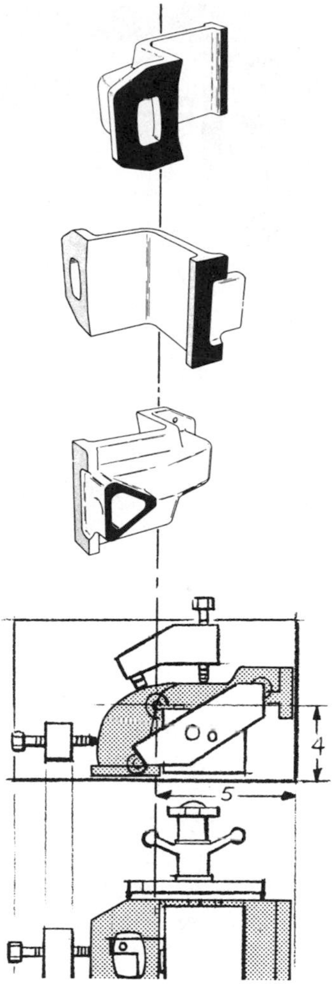

ESTABLISH PART ORIENTATION--Orientating this part horizontally as illustrated exposes all the machining operations to the spindle when the part is properly indexed. When the part must be held more than once for completion, all critical surfaces should be machined on the final holding if possible. The sequence of operations is established at this time.

SKETCH HOLDING FIXTURE--A rough layout is prepared of the fixture or holding device. This layout serves useful purposes:

1. It is used for estimating the MILWAUKEE-MATIC tooling cost.

2. It serves as a check for validity of the machining process--most tool access or interference problems will be apparent on the layout.

3. It establishes the reference dimensions used in estimating the cycle time.

**Fig. 19.5.** Operations, tools, and times needed to complete the switch bracket on the N/C machining center. Courtesy Kearney & Trecker Corp.

**Circular-Indexing Automation.** Circular-indexing automation consists of intermittently moving the work from one station to the next in a circular path, Fig. 19.6.

Among the favorable factors of circular-indexing automation are the following: floor-space requirements are lessened; holding fixtures always return to their starting position; the work is clamped only once; loading and unloading are accomplished during the machine-cycle time; and a finished part is produced with each index.

Less favorable aspects include the need for balancing operation times,

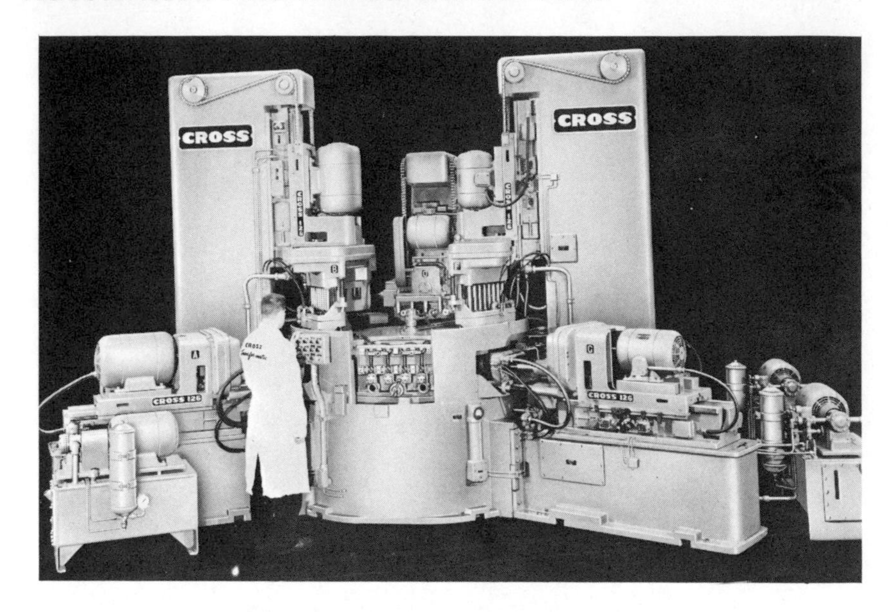


**Fig. 19.6.** Circular-indexing automation. Courtesy The Cross. Co.

the practical size limit of the indexing table, the vibration transmitted between stations, and the difficulty of power clamping. The production rate of the machine is established by the longest operation, plus indexing. It is therefore necessary to plan operations carefully to avoid one of long duration. Tool changes idle the entire machine, cutting down on overall efficiency.

Similar automatic indexing systems have been worked out in which the work is moved in a vertical plane. These are referred to as *trunnion-type machines*. Machine heads are mounted on each side of the vertical table, permitting simultaneous machining of opposite sides of the work.

## MULTIPLE-MACHINE AUTOMATION

The discussion thus far has been concerned with circular- and linear-movement controls applied to individual machines. The next step is the integration of a number of machines performing a wide variety of operations on a given product.

**Integrated Lines.** The first major step in the integration of individual machines into an automatic line was in 1940. At that time, some modifications were made on existing machines, and a means of connecting them by suitable conveyors was worked out. The grouping consisted essentially of drilling, reaming, tapping, and milling, arranged along both sides of an automatic conveyor. The workpiece was attached to a special locating plate and was pushed along a section of roller conveyors until it was engaged by a catch on a reciprocating transfer bar. The bar pulled the work into its approximate position at the first machine station, and hydraulically operated pins were raised to engage with the locating holes for exact positioning. Hydraulic clamps locked the plate, and the first set of tools performed the first machining operation. When the operation was completed, the tools withdrew, the clamps released, and the work was pulled into the next position by the transfer bar. At the same time, a new workpiece was pulled into the first position.

In spite of its tremendous potential, this transfer device had inherent weaknesses. If a tool became worn or broken, there was no way to detect this until inaccurate pieces were discovered by inspection at the end of the line.

Improved transfer-line designs incorporated automatic inspection stations at various places. After drilling operations, a jet of compressed air blew out the chips, and metal fingers probed to make sure that proper depths had been reached and that no broken tool or tightly packed chips remained. Only as all intermediate checks were completed and found satisfactory was the machine able to index and repeat the sequence of operations.

Tool life expectancy can now be predicted, based on past experience and statistical data. This information is incorporated into a special control board (Fig. 19.7) that is used to maintain a cumulative record of each tool. Rather than wait for a tool to wear excessively or to break, a predetermined number of revolutions is set for a given tool. When this amount of cutting has been done, the board signals the machine to stop. Preset tools are then taken from the board and used to replace tools that have run their predetermined number of cycles. Gages are kept near the tool board so that all tools can be accurately set before they are placed in the machine.

The integrated line is still used. It has a great deal of flexibility, since machines can be taken out and substitutions made when the part changes design.

**Unitized Lines.** Several improvements have been made in the integrated line. Special handling equipment, such as chutes, conveyors, and elevators, have been added, as shown in the partial line of Fig. 19.8. Work starts at one end of the main conveyor and flows along to takeoff stations which supply parts to the machine units. The work is processed automatically through these independent machines, and then goes back up to the conveyor line by means of elevators. In the unitized system, one group of machines does the first operation, another group does the second operation, the third group does the third operation, etc.

A number of favorable factors influence the choice of the unitized system. It provides a truly continuous work flow, because the individual units are releasing work to the main conveyor at different times, and

**Fig. 19.7.** Tool board and gaging equipment used to maintain an automatic line. Courtesy Scully, Jones & Co.

the conveyor itself is moving the work along at all times. One can have complete freedom in planning the number of stations required.

Opportunities to reorient the part are practically unlimited, since these can be varied from machine to machine. Consequently, the approach to the workpiece is unrestricted. The machine units can readily be designed for automatic loading and unloading. The most suitable type of loading can be used for each unit in the line.

The use of separate machines makes for easier application of a variety of machine types, including standard models. If suitable units are available, it is possible to fit them into this system. Furthermore, the unitized arrangement gives the user an opportunity to employ specialized machines made by different manufacturers, thus taking full advantage of their particular abilities. Aftergaging and feedback can be fully utilized to increase efficiency and improve quality by making corrections in successive cycles.

Banking of parts between operations is provided automatically and allows one group of machine units to continue functioning even though machines on prior operations are stopped temporarily.

It is entirely possible to operate this system efficiently at production rates below the normal maximum. In this case, a portion of the units performing each operation can be made inoperative, the remaining units being allowed to perform normally. This is the only system that allows such flexibility in adding or removing operations, increasing production, or accommodating part changes.

Considerations affecting the use of unitized automation are not all favorable. There are restrictions on the workpiece sizes and design. As an example, it would be rather inefficient to handle a large casting such as an automotive engine block. In general, small parts create fewer problems, and the ideal parts are those that can easily pass through the simple chutes.

Floor-space requirements present another problem, since a number of independent units will not be so compact as some of the other systems discussed previously. There is also the need for making duplicate equipment to orient the part, for gaging, and for other functions needed owing to part design.

**Transfer Machines.** The specialized in-line transfer machines probably come nearer the average conception of automation than any other system in the metalworking industry. It is in these specially built machines that hundreds of operations are performed on rough-cast engine blocks, and they emerge as a finished component without the aid of human hands. To illustrate some of the steps involved in designing such a line, let us examine the casting shown in Fig. 19.9.

The first step is to list all the machining operations which, in this case, are 8 milling, 30 boring, 40 drilling, 4 recessing, 24 chamfering, 24 tapping, and 32 checking operations. The second step involves the design of work-holding fixtures and pallets (Fig. 19.10) so that as many operations as possible can be done with one clamping. This helps preserve accuracy and prevents marring the machined surfaces.

The actual machine units may then be chosen from basic "building-block" units such as the vertical and horizontal bases, slide units, and boring heads used for drilling, reaming, chamfering, spot facing, and counterboring (Fig. 19.11).

The number of stations necessary is determined. The transfer mechanism to be used will help decide the line layout (Fig. 19.12).

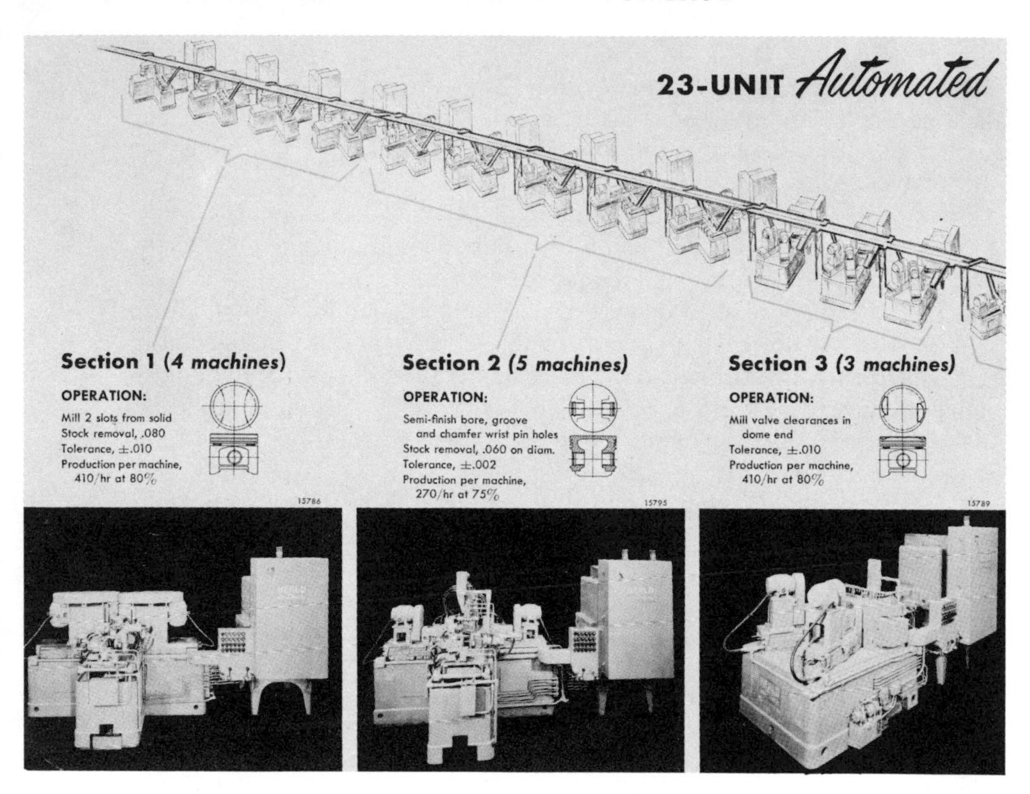


**Fig. 19.8.** The sketch shows a part of a unitized type of transfer line. Note

**Fig. 19.9.** Casting considered for automatic transfer-line machining. Courtesy Greenlee Bros. & Co.

You will note that provision is made for probing, rotating the part, and inspection. Other considerations are cleaning, deburring the casting, chip removal, coolant distribution, and safety interlocks. The manual operation consists of removing the finished parts and placing the rough

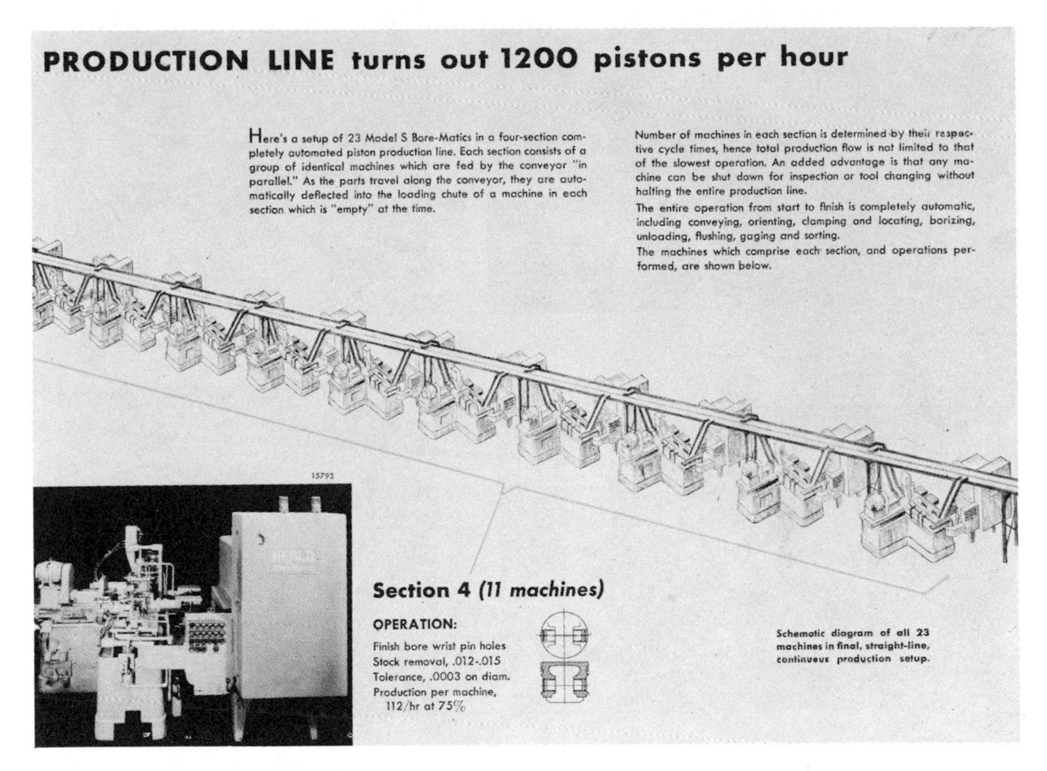


the machines are arranged in groups or units. Courtesy The Heald Machine Co.

**Fig. 19.10.** Casting fixtured and palletized in preparation for placing it on the line. Courtesy Greenlee Bros. & Co.

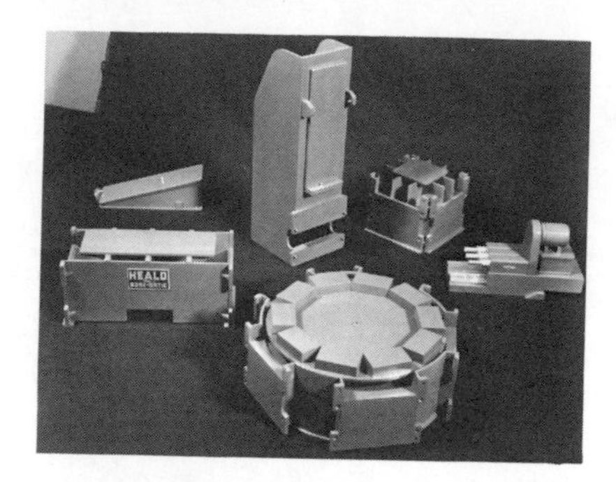


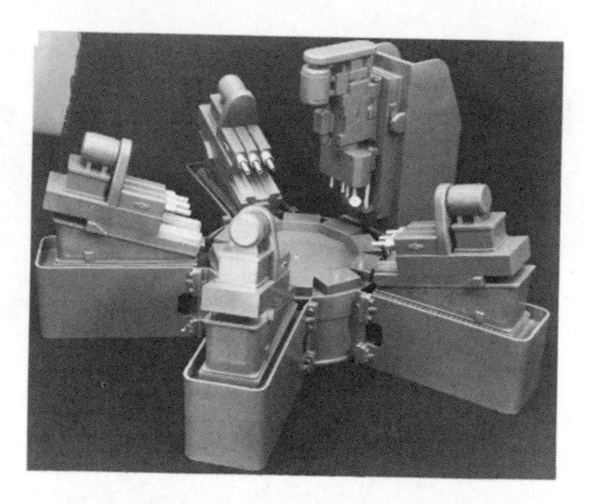

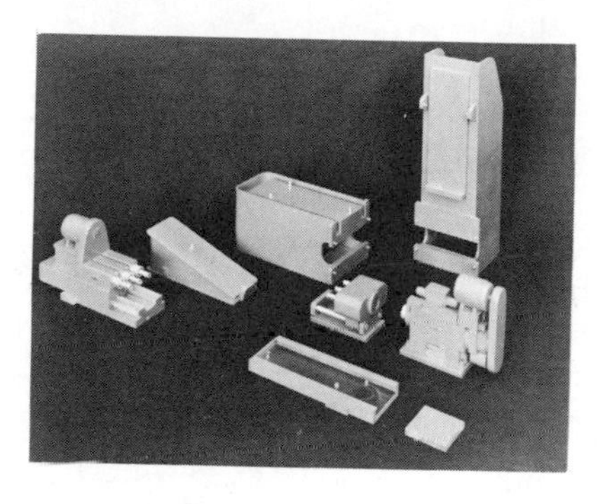

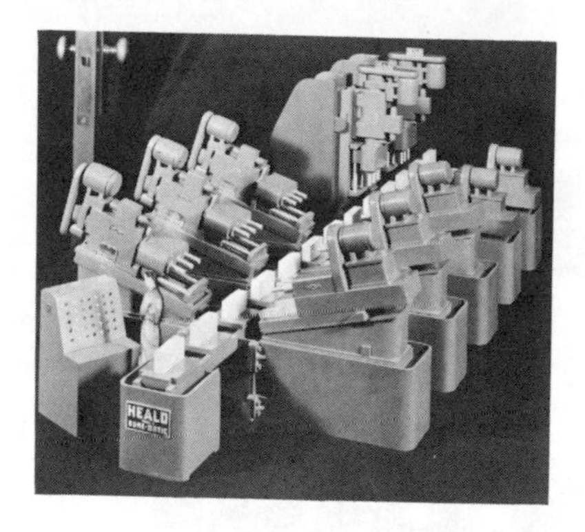


**Fig. 19.11.** Some basic "building block" units used to make up circular automation and transfer lines as shown in the assembled views. Courtesy The Heald Machine Co.

castings on the pallet. Locating the part in the pallet, clamping, and unclamping are automatic.

Transfer-line movements are of three types—continuous, intermittent, and advance-and-return.

CONTINUOUS LINE. The continuous conveyor is the simplest of the three basic types of line movements. The cost is generally lower, and the cycle speed is higher. Some assembly and machining operations, such as milling and broaching, are done on a continuous-type line. The continuous movement of parts limits the kinds of operations that can be performed. Since clamping and locating are done only once, not all surfaces can be reached. Loading and unloading must be done "on the fly" or during movement of the line.

A unitized line may be considered a continuous-line operation, but the parts leave the conveyor for the machining operations via outlet chutes and return by means of elevators. The part banks at each station ensure continuous machine utilization.

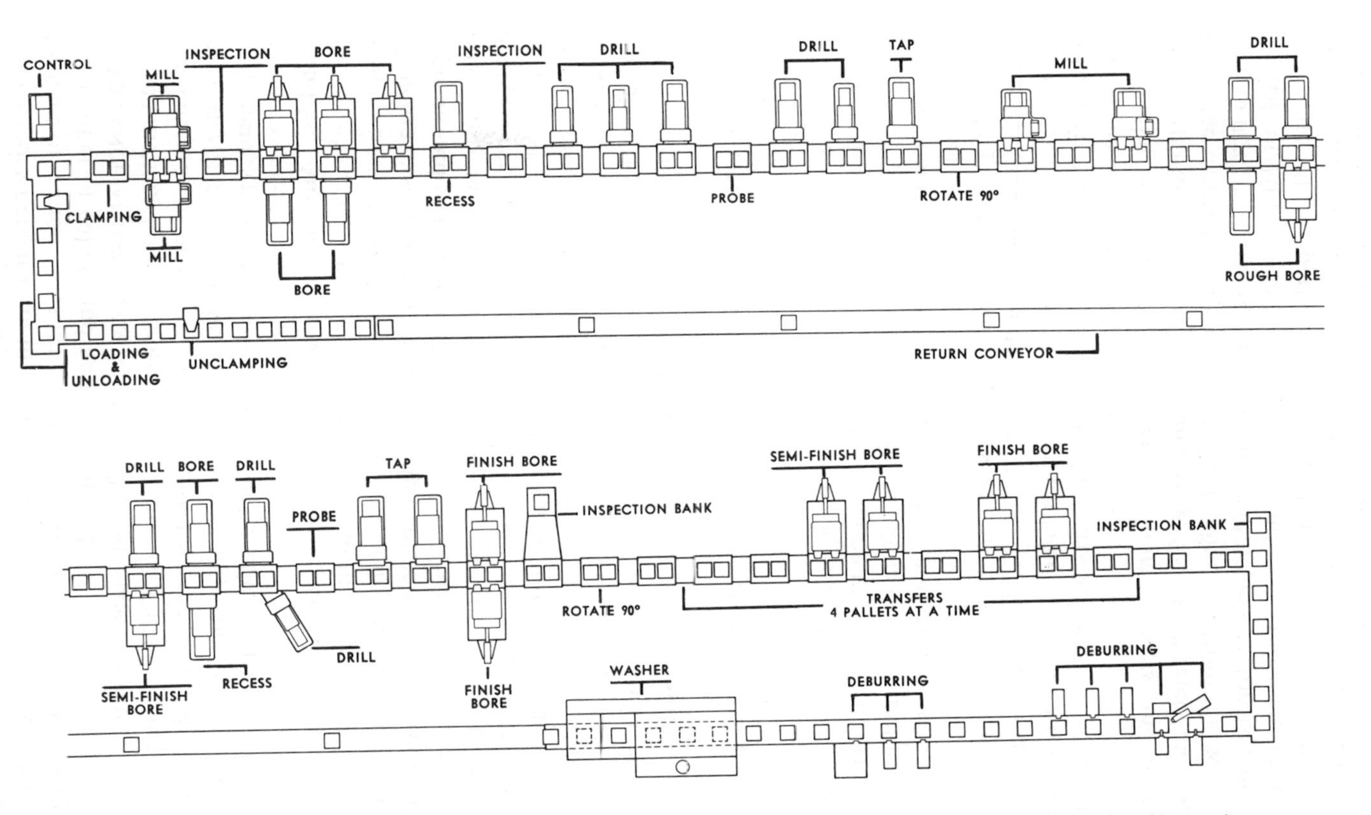


**Fig. 19.12.** Plan view of the completed transfer line used to machine and inspect the casting shown in Fig. 19.9.

INTERMITTENT LINE. The intermittent line is the conventional movement used where a time interval is required at specific stations to perform an operation. This type is usually expensive from the standpoint of the number of pallets and fixtures required.

ADVANCE-AND-RETURN LINES. The advance-and-return movement is used in many large pressworking operations. The part is moved through the sequence of operations by means of transfer bars and fingers which grasp the part during the transfer movement. This type of moving arrangement is generally more economical than the interrupted- or indexing-type movement.

**In-Line Automation Evaluation.** Floor-space requirements and operating efficiency are the two principal limiting factors of in-line automation. Some lines allow considerable flexibility in the number and type of stations used. With the proper equipment, work may be reoriented and repositioned so that all surfaces can be machined and inspected. Continuous chip removal is relatively easy to incorporate, since the chips can be dropped on straight-line chutes and conveyors. The effect of vibration between stations is slight, because machine units are separated. The in-line machine is easy to load and unload by automatic means; it can also be adapted to manual handling at the beginning or end of the line. Heat and distortion are greatly minimized owing to unclamping of the workpiece and the time interval needed to progress from station to station.

## WORKPIECE ORIENTATION AND MOVEMENT FOR AUTOMATION

Many devices have been developed to make piece-part orientation and movement automatic. Best known among these for part orientation is the vibratory hopper. Newer material-handling devices are the robot and the systems that employ electric clutches and brakes, photoscanners, counters, etc.

**Vibratory Hopper.** Bowl feeders are used to orient and feed in a straight line pieces from 1/8 to 11 in. long. The pieces may be either metal or plastic and of almost any conceivable shape. Vibration, induced by electromagnetic action, causes the parts to feed up a spiral ramp around the inside of the bowl, Fig. 19.13.

Orienting devices such as *wipers, dish-outs, pockets,* and *hold-downs* are incorporated into the spiral ramp (Fig. 19.14) to reject parts that are not in the desired position. The track leading out from the bowl can be used to turn the parts to any desired angle necessary for assembly.

Feeds of 10 to 15 fpm are considered normal. However, this does not mean that all the parts are oriented at this rate. A part takes several positions as it travels along the track. Therefore, the desired feed rate is set at the number of correctly oriented parts needed per minute. More than one track can be used at the same time, as shown.

**Fig. 19.13.** Two-track feeder bowl is used to feed incandescent lamp bases to an automatic lamp machine at a rate of 300 parts per minute. Fully oriented in a base-down vertical position as they are jogged forward, the bases travel up both the inclined tracks which are machined in the sloping sides of a cast aluminum alloy bowl. A chute built into each of the tracks relieves pressure buildup by allowing excess bases to slide back into the bowl. Courtesy Automation Devices.

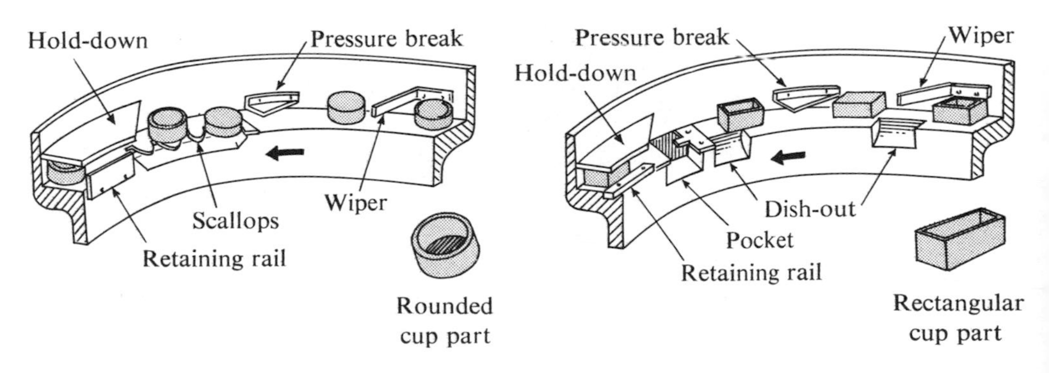


**Fig. 19.14.** Examples of round and rectangular piece-part orientation on the track of a vibratory bowl.

Vibratory-hopper feeders, combined with air cylinders, are widely used for assembly as well as for small machining operations. A bowl feeder is used to load gear blanks into the machine in Fig. 19.15, where they are automatically bored and gaged.

**Robots.** A few automatic positioning machines are beginning to be placed on the market. The Versatron, shown in Fig. 19.16, is used to

**Fig. 19.15.** Parts are dumped into vibratory hopper of machine (*a*). They move down the chute to enter an automatic boring operation. After boring, the hole is gaged for size. If it is not correct, the information is fed back to the tool block which automatically corrects its setting before the next part is started. The correct size parts go to the vibrating loader of machine (*b*) and then to the final machining operation. Courtesy The Heald Machine Co.

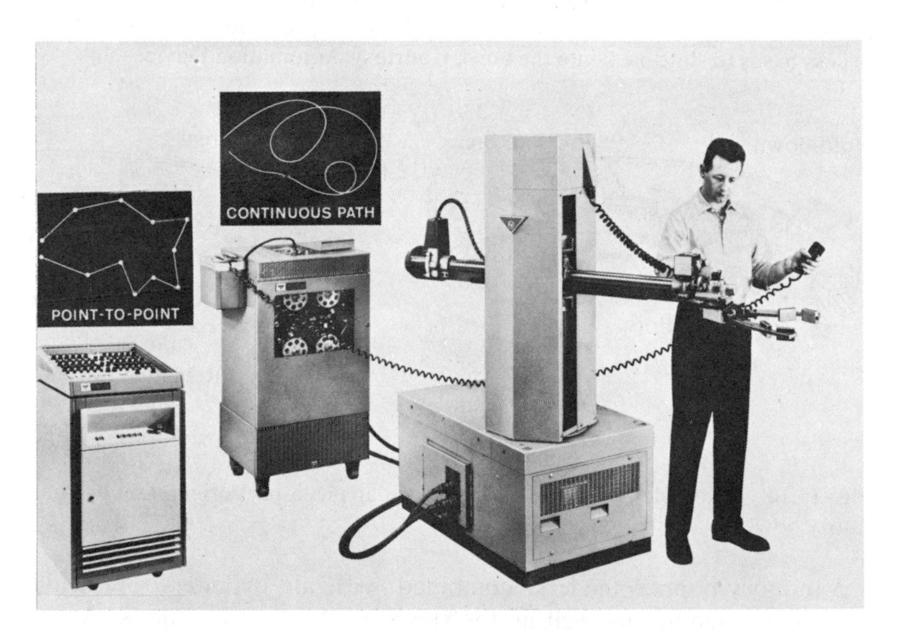


**Fig. 19.16.** The Versatron is an example of automatic equipment used to handle and position parts on both a point-to-point and a continuous-path basis. Courtesy American Machine and Foundry Company.

load and unload and operate a wide variety of production machines. It can also hold and operate conventional hand tools, spray guns, and power tools.

Two types of control are available for the machine shown, point-to-point (*a*), and continuous path (*b*).

POINT-TO-POINT. The point-to-point console provides an easily programmed, less sophisticated machine with 12 discrete transfer positions and up to ten commands in each position. For hazardous environments it can be operated by remote control. The manipulator motions are controlled and memorized by adjusting the console controls.

CONTINUOUS-PATH. The continuous-path control has five degrees of freedom to reach a specified point and two "hand motions," wrist twist and bend. It is controlled by digital command servos. To "train" it to perform a specific job, it is first led through the required motions by a human operator while position signals are stored on a magnetic drum. Thereafter, motions are repeated exactly as they were made the first time through.

A variety of hands and grippers can be installed on the Versatron's arm that have their functions precoded on tape. They grasp, rotate, or swing in whatever sequence required.

**Electric Clutch-brake Control Systems.** Custom automation can be fabricated for assembly and conveying operations by electric clutch-brake systems along with sensors, control modules, and switching units (Fig. 19.17).

The electric clutches and brakes are turned "on" and "off" in response to input signals from photo scanners, magnetic pickups, limit switches, or similar devices.

## FURTHER DEVELOPMENT AND USE OF AUTOMATION

The incentive behind automation has been the desire to make things faster and more cheaply. However, the greatest benefit has been in doing jobs that are beyond the power of human workers. Instrumentation, mechanical gaging, and perception have made possible closer tolerances than human hands can achieve. The blessing of automation is that it eliminates many of the tedious, monotonous jobs formerly done by hand.

Automation does not replace human brainpower. Instead, it imposes an even greater demand on the mental skills. The need for absolutely clear and thorough planning has put a new and tougher discipline on human intelligence. Automation demands engineers and technicians of broad backgrounds to design, to build, and to maintain the future automatic factory.

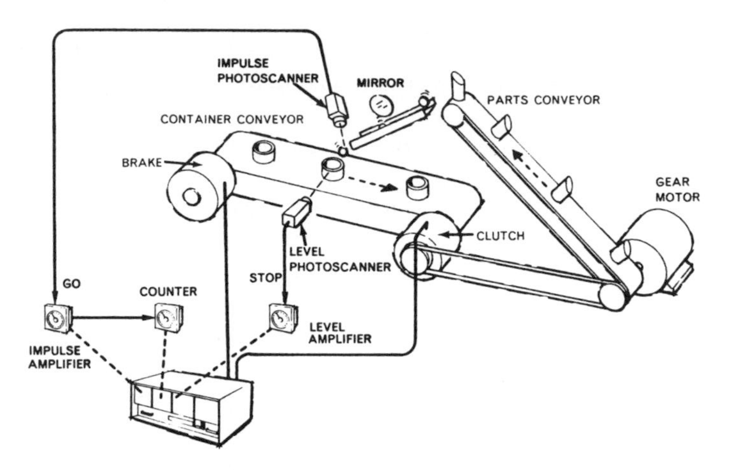


**Fig. 19.17.** The container conveyor is stopped, and the parts conveyor is moving, dropping one part at a time into the container. An impulse photoscanner and light source are positioned to detect the movement of the parts as they are dropped from the parts conveyor. The pulse from the photoscanner is fed to a counter module. After a predetermined number of pulses is counted, output signal directs basic control to switch from *brake on* to *clutch on*. The conveyor starts to bring the next container into position. A level photoscanner detects the next container, sending a signal to basic control. The clutch is then disengaged, and the brake engaged to stop the container in position before the next part is dropped. Courtesy Warner Electric Brake & Clutch Company.

## NUMERICAL CONTROL

Numerical control, or control by numbers, is not new in the sense that numbers have always been used for accurate tool-table movements. However, numerical control as it is spoken of today is a much more complex control based on electronic computations. These computations are used in giving physical movement, continuous or intermittent, through servomotors to the tool, machine table, or auxiliary functions.

Many other industries have shared with the machine-tool industry in the development of numerical control. Wherever equipment is instructed to perform a function now done by an operator, there is a potential application of numerical control.

Table and tool movements may be made by electrical, hydraulic, or mechanical means, as long as the input signal represents a numerical value.

**Types of Numerical-Control Systems.** The two main forms of numerical control, as applied to machines, are point-to-point positioning and continuous-path control.

POINT-TO-POINT POSITIONING. Point-to-point (PTP), or *discrete*, positioning is a relatively simple method of moving the tool or work on the $x$- and $y$-coordinates to a desired position. This type of control is applicable to a wide variety of machine tools, such as drill presses, boring mills, punch presses, and jig boring machines. PTP positioning, as used in drilling or boring holes, requires a high degree of accuracy as to location of the coordinates, but it places no restriction on the path used in achieving the location.

CONTINUOUS-PATH OR CONTOURING SYSTEM. Contouring is by far the most complicated system of numerical control. In milling a profile of a three-dimensional surface, the entire path must be described by the data input medium. The tool path may be for any shape, whether it is a parabola, arc, sphere, square, or any combination of these in three dimensions. In reality, this system is an advanced stage of the point-to-point system. The controls are such that the tool can move only in a straight line. The straight-line distances can be made so short (0.0005 in.) and so well blended that they will appear as a continuous smooth cut. The systems are compared in Fig. 19.18.

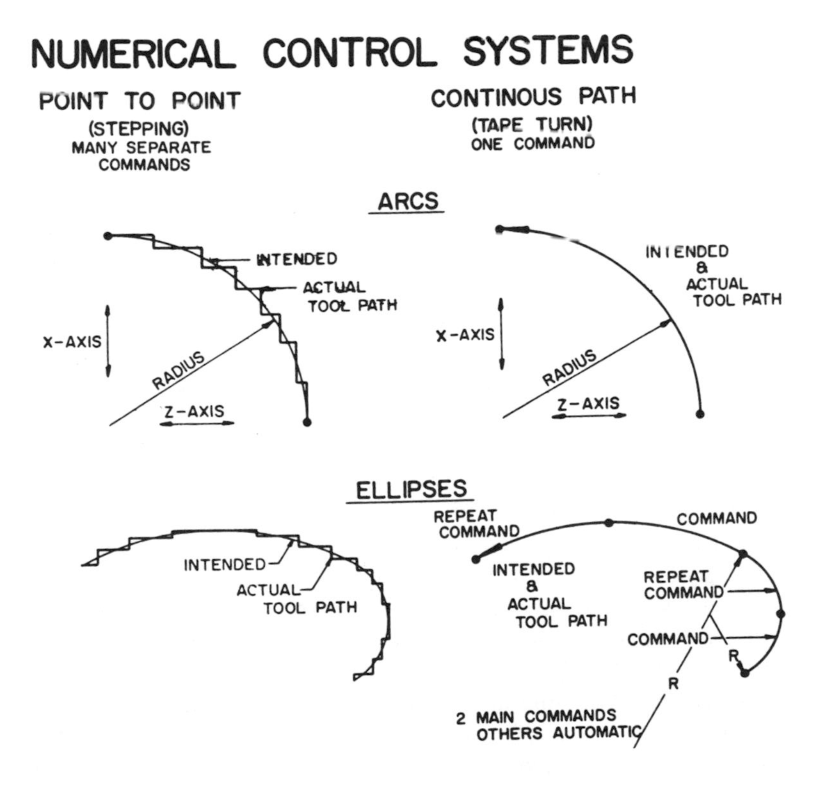


**Fig. 19.18.** The PTP positioning system compared to continous path.

Of the two systems, PTP positioning is the most commonly used. Contouring was first developed in cooperation with the aircraft industry for complex machining jobs (Fig. 19.19), and this is still one of its best applications. As the industry grows and more technical help becomes available, it is likely that tool-and-die shops may become the biggest users.

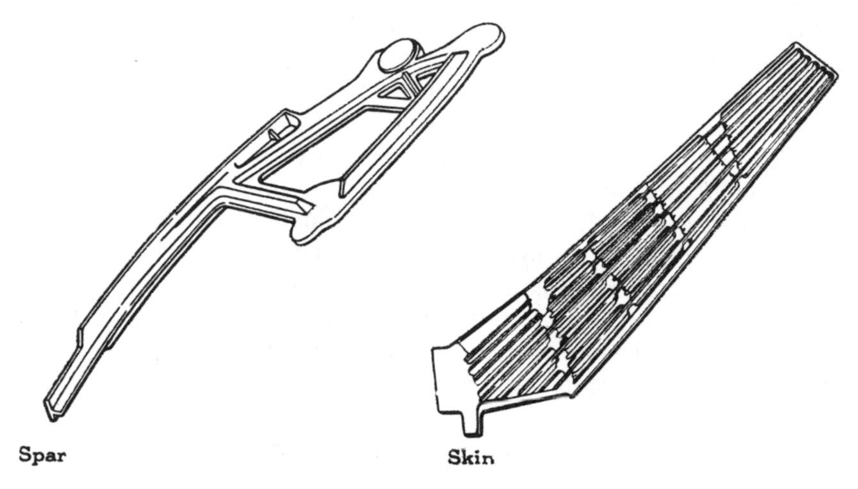


**Fig. 19.19.** Complicated aircraft members milled with continuous-path numerically controlled machine.

Computers are rarely needed to prepare data for positioning systems, but the thousands of points needed for continuous-path control provide a complex problem not practical to solve except with a computer. The fact that a computer could be used to calculate the thousands of physical points needed for any mathematical shape had a marked influence on the development of numerically controlled machines.

The three axes of movement, termed $x$, $y$, and $z$, are normal to each other. More movements are available. There is an axis of rotation about each of the basic axes, making six in all. Numerically controlled machining often calls for extra maneuverability (Fig. 19.20).

**Numerical-Control Programming.** Numerical control may best be understood by following the various steps from the time the design is conceived until the finished part is completed. The steps in continuous-path control concern design, numerical drawings, part programs, process cards, postprocess cards, the computer, final output, and machine controls (Fig. 19.21).

**Programming.** After the method of making the part has been decided, the *part program manuscript* is written. This manuscript describes the operations to be performed in a sort of "pidgin English." It, in turn, is copied by a keypunch operator into a deck of punched cards. Each

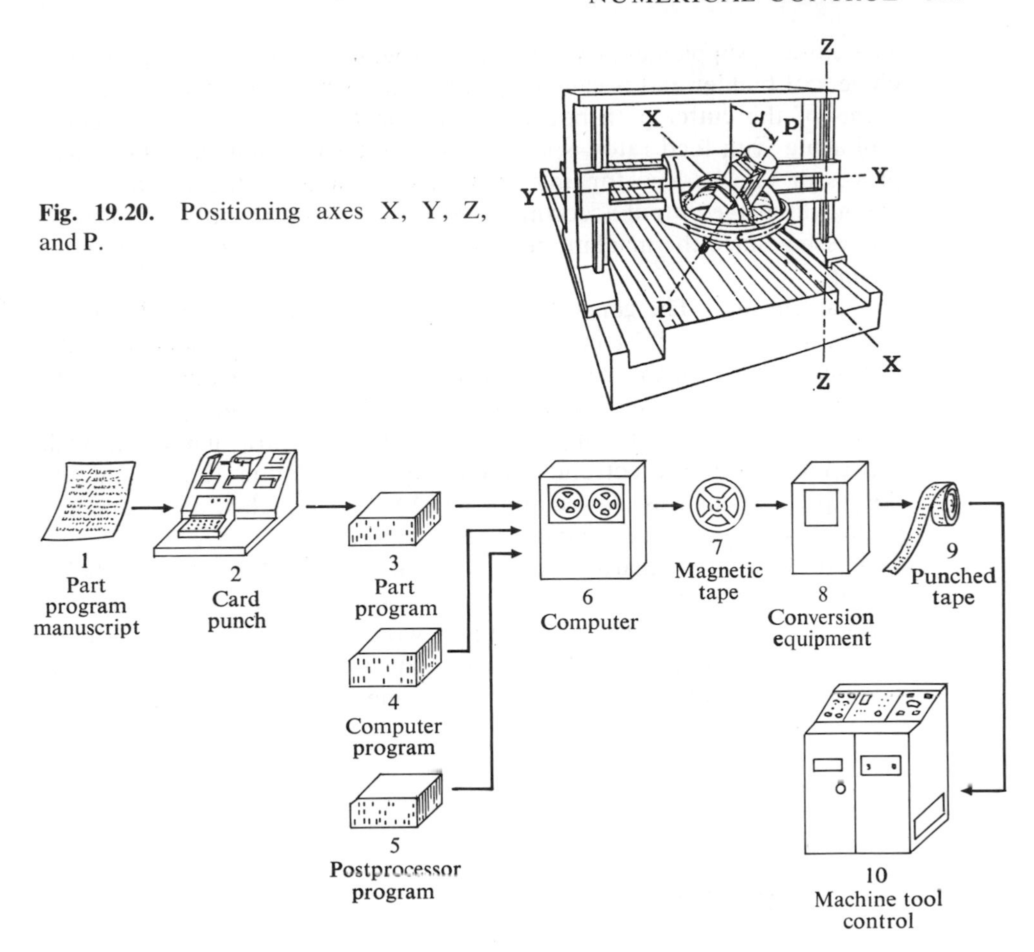


**Fig. 19.20.** Positioning axes X, Y, Z, and P.

**Fig. 19.21.** Steps in computer programming for N/C.

card contains a specific machine-tool instruction. A deck of cards adds up to a *part program.*

Two major elements make up the modern N/C computer tape preparation system: the *general processor* and the *postprocessor.* Often the two are considered a single unit; however, for purposes of clarification, it is best to consider each one separately.

**The General Processor.** The general processor furnished by the computer manufacturer accepts the part program instructions and calculates all the coordinate points that the cutter will follow in producing the part. It also contains diagnostic elements which, upon discovering an error in the part program, stops the machine thus saving costly computing time.

**The Postprocessor.** One machine tool control system may differ from another. It is necessary to convey these differences to the computer.

Therefore a supplementary computer program is furnished by the machine tool builder and is called a postprocessor program. The coordinate points of the cutter path become output in the format of the machine tool along with feed rates, speeds, and auxilliary commands. The postprocessor checks also for any limitations peculiar to the machine tool and modifies the output accordingly. Some machines do not require postprocessing since the machine tool is made to accept the general purpose coding.

**Final Output.** The general processor accepts the programs described and shown in Fig. 19.21 in either magnetic tape form or punched cards. The final output of the computer contains all the commands needed to produce the part in either magnetic or punched tape. Many ways have been devised to store the machine information but now, mainly through the efforts of the Electronics Industries Association (EIA) and the Aerospace Industries Association (AIA), a one-inch-wide, 8-channel perforated tape has been standardized.

A portion of a standard tape is shown in Fig. 19.22. Each set of holes going across the tape is referred to as a *word* which causes a specific action of the machine tool. A group of words that offers a complete instruction for a specific machine movement is a *block*. Blocks are separated by an end-of-block character in channel 8.

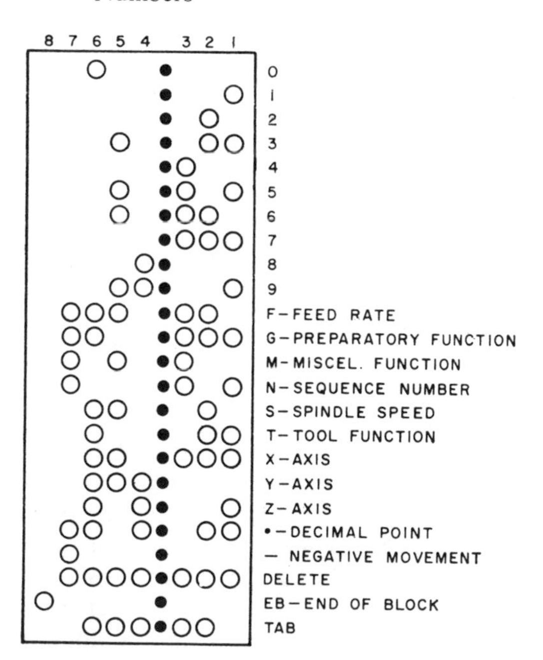


**Fig. 19.22.** A portion of the standard 8-channel one-inch wide punched tape. The address codes are only a portion of those contained in the E.I.A. standard.

Not all computers are made to put out the standardized 8-channel tape; therefore a conversion unit is used after the computer as shown in Fig. 19.21.

**Design and Numerical Drawing.** The engineer who has become familiar with numerical control adapts his designs so as to facilitate machining and handling. After the design has been made, it is given numerical dimensions; that is, all dimensions are in the form of coordinates referenced to a common axis (Fig. 19.23). Much of the conventional dimensioning is omitted.

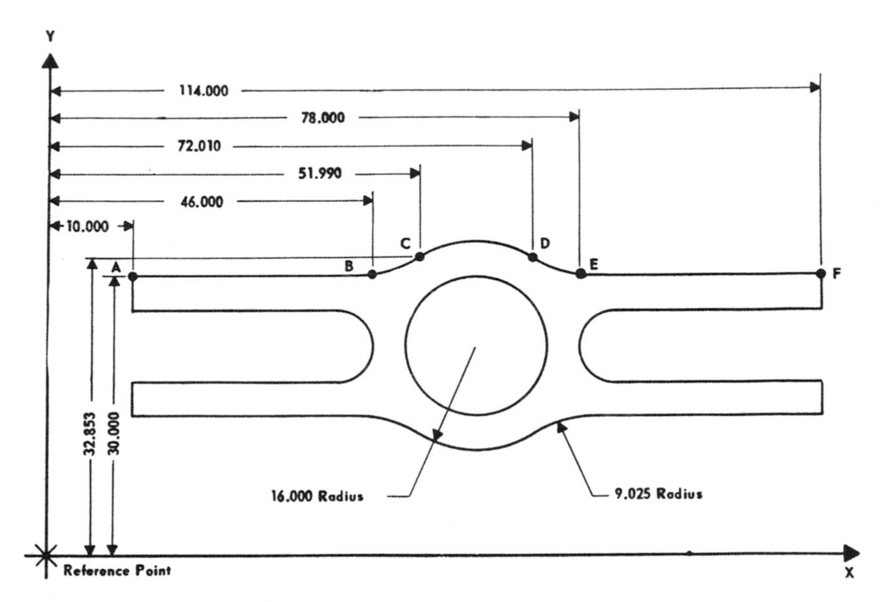


**Fig. 19.23.** Numerical drawing.

After the drawing is complete, it goes to a methods engineer or a methods group. It is at this point that decisions are made as to the best way of making the item. Weaknesses of the design may also be detected, and more suitable materials selected. Cutters are chosen, and speeds and feeds are set for each cut.

Command information on tape consists of binary numbers corresponding to the number of command pulses that must be delivered to each axis in order to drive the cutting tool along a straight line from one point to the next specified point. Other information for full numerical control consists of instructions to the machine, such as direction of motion, start, stop, reverse, rapid traverse, etc.; numerical values, such as dimensions or length of travel and speeds, feeds, number of pieces, etc.; and auxiliary functions, such as tool changing, coolant on or off, chuck open or closed, end-of-cycle signal, operate lubricator, etc.

In semiautomatic systems, signals from the tape can tell the operator when to perform auxiliary functions manually.

**Machine Control Unit.** The tape is mounted in the machine control unit on a photoelectric reader, which senses the holes in the tape. One block of information from the control tape is received by the electronic buffer unit immediately and is stored. The storage unit supplies information to the decoder as required. The coded digital information is changed to command pulses by the decoder. Each pulse commands an increment of travel, which may be as small as 0.0005 in. These pulses serve as input to the servos for the machine lead-screw motors.

**Machine Interpretation of Numbers.** A machine is not able to understand Arabic numbers, so they must be changed to a two-number, or *binary,* system. In this system only two numbers are used, 0 and 1. A circuit is either open (0) or closed (1). The machine has only to sense this difference. So, if a hole is punched for 1 and none for 0, the machine scanning device can quickly obtain the input information.

In order to handle large numbers easily, the figures are spread over four columns in what is known as the *Binary Coded Decimal* (BCD) system. The BCD characters are coded with *b*inary dig*its* (or bits) in the first four tracks or channels. Any decimal digit is then made up in the row by the BCD code. The figure 12.345, for example, is punched on the tape as shown in Fig. 19.24. You will note that numbers three and five are made

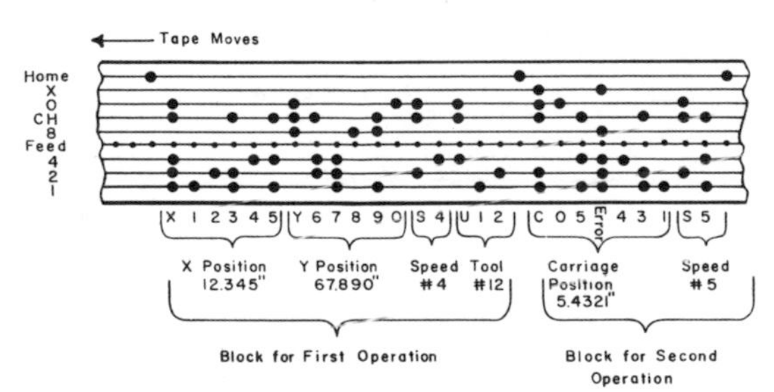


**Fig. 19.24.** A punched-tape program.

by punching holes in more than one channel. Three uses channels one and two, five uses channels four and one. The decimal is given certain positions in the tape block.

Another tape format is the straight binary (SB) system. It is more complicated since it is based on exponents (or powers) of two. It is similar to our decimal system, which uses the powers of 10. For example,

$$214 = (2 \times 10^2) + (1 \times 10^1) + (4 \times 10^0).$$

The numbers are written down in sequence. A zero is used to establish a positional sequence whenever any other multiplier is missing. In the binary system, 214 would be

$$(1 \times 2^7) + (1 \times 2^6) + (0 \times 2^5) + (1 \times 2^4) + (0 \times 2^3) + (1 \times 2^2) + (1 \times 2^1) + (0 \times 2^0).$$

The number is again written by listing the multipliers in front of each expression: 11010110. If the ones are holes and the zeros spaces the machine can readily read this system as *on* or *off*.

The BCD is the newer of the two systems and the most used because it can be programmed without a computer if the calculations are not too complex. It is now standard for PTP positioning and is used in some continuous-path work also. BCD is more easily understood; however, it requires about $3,000 to $6,000 worth of extra equipment in the controls. Some machines are made to operate on either system.

**Manual Position Selectors.** Provision is made, on most numerically controlled machines, for manually feeding the information for the desired coordinates into the system (Fig. 19.25). The manual arrangement consists of a number of 10-position rotary switches, each of which provides one digit of numerical position information. This information is converted within the machine to the binary-coded decimal form and is fed to the servomotors by means of impulses from the decoded punched tape.

**Newer Developments.** A newer system of tape programming being developed is known as APT, or automatic programmed tool. The aim of the developers of this system is to make input information as close to human language as possible.

APT consists of over 30,000 preprogrammed instructions to the computer. Thus it adapts a general-purpose computer to a special-purpose computer for programming machine tools. The computer, when given instructions, can, with lightning speed, find the preprogrammed information and apply it to the job at hand. This is why the process sheet becomes comparatively simple. An example of the process-sheet information is given in Table 19.2.

One disadvantage of the APT system is that it requires a very large computer, which may not be readily available for most companies. As a

**Table 19.2**
**Sample APT Process Sheet Information**

| Part program | Explanation |
|---|---|
| CUTTER/1 | Use 1-in.-diameter cutter |
| TOLER/.005 | Tolerance: 0.005 in. |
| FEDRAT/80 | Feed rate: 80 in. per min |
| HEAD/1 | Use head 1 |
| SPINDL/2400 | Turn on spindle at 2,400 rpm |
| COOLNT/FLOOD | Turn on coolant, flood setting |
| PT1 = POINT/4,5 | Define a reference point, PT1, as the point with coordinates 4,5 |
| FROM/(SETPT = POINT/1,1) | Start the tool from the point called SETPT, which is defined as the point with coordinates 1,1 |

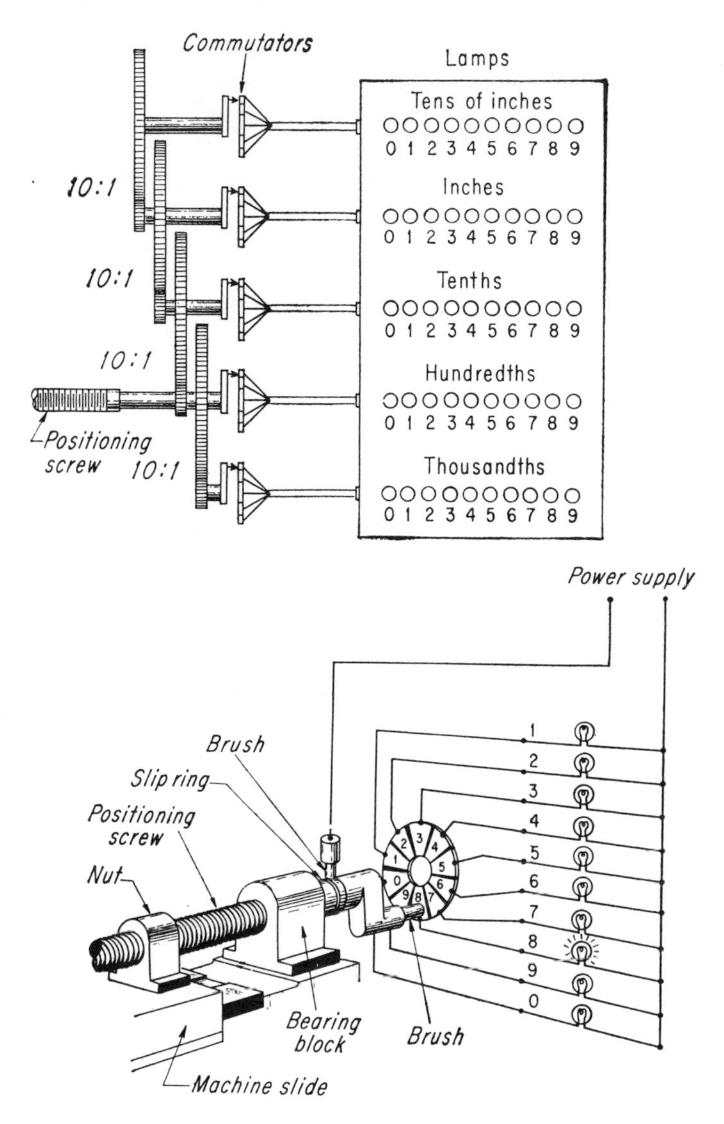


**Fig. 19.25.** An analog-to-digital converter is shown schematically. Any degree of fineness can be provided through the dial settings. Courtesy *American Machinist*.

result, a second program was sponsored by the U.S. Air Force to make simplified programming for smaller computers. The name of this program is ADAPT which is an adaption of APT. It maintains exactly the same part programming as APT.

Another program that utilizes simple manuscript instructions is AUTOPROMPT. It was developed by IBM and is used in conjunction with APT and therefore requires a large computer. Complete surfaces such as hemispheres and cylinders can be machined by simple script instructions.

A PTP program, offered free to IBM users of the 1620 computer and at about $30/hr to others, is AUTOSPOT (Automatic System for Positioning Tools). Simple language can be used to describe the engineering drawing, which is translated into detailed instructions for the machine tool. An example of the machining to be done on a mill and the statement required is shown in Fig. 19.26. A 1-in. groove is to be milled down the part's longitudinal center line, 0.500 in. deep using a 1-in. dia cutter.

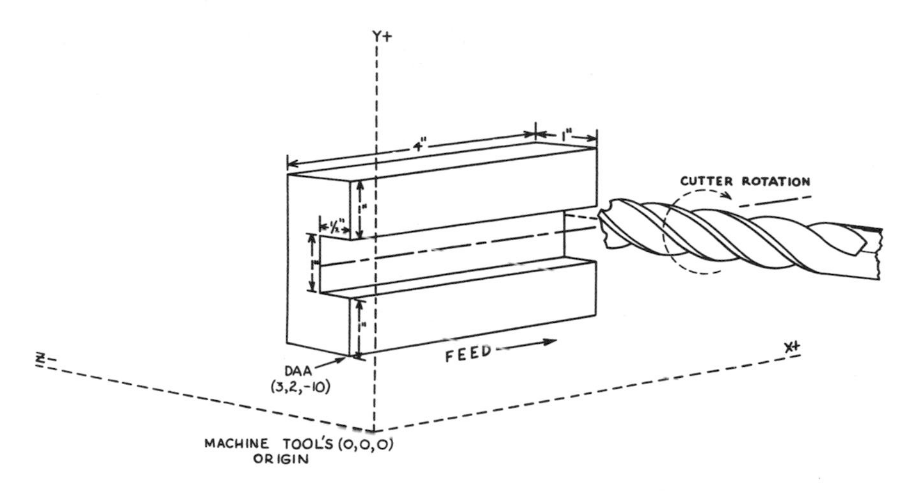


**Fig. 19.26.** An example of an Autospot statement for a simple milling operation: "MILL, 0219/DAA (0.0, 1.5) RTO (4.0)/DP (0.5) $" The statement specifies that with respect to DAA, starting at x = 0.0, y = 1.5, the cutter moves right to (RTO) x = 4.0 at a depth of .500. From any location, a cutter may be moved up to (UTO), right to (RTO), down to (DTO) or left to (LTO) another location. If a rectangular path were to be cut, the machine statement would be: "MILL, 0221/DAA (X,Y) RTO (X) UTO (Y) LTO (X) DTO (Y)/DP ( . ) $" Courtesy IBM.

This simplified programming points the way to the production of parts merely by feeding design information directly into the computer, thereby eliminating even the part drawings. A step in this direction is the numerically controlled drafting machine. With this and the aid of a computer, prints that formerly took many hours to produce can be made in minutes.

Closely associated with numerically controlled drafting is a newer concept known as design augmented by computers (DAC). The hardware of the new system consists of (*a*) an IBM 7094 computer augmented with extra disc and drum storage devices, (*b*) an IBM 7960 special image processing system that allows the computer to read and generate drawings, and (*c*) a graphic console equipped with a viewing screen that serves as a direct two-way communication link between the designer and computer (Fig. 19.27).

**Fig. 19.27.** The position-indicating pencil, often termed *light pencil*, is used to pinpoint some area of the design either for enlargement or for dimensional change. The pencil functions through a conductive screen of thin, transparent tin oxide placed on the glass plate in front of the display tube. When the pencil touches the screen it detects a voltage that is proportional to the distance from the top and left of the screen.

## N/C ADVANTAGES AND ECONOMIC CONSIDERATIONS

The benefits that can be gained from numerical control are many and can be only briefly summarized here as the process affects various phases of manufacturing, tooling, production control, etc.

*Manufacturing:*

(1) Reduction in costs due to less variation in cutting speeds and feeds.
(2) Reduced lead time, from weeks to days.
(3) Improved quality and closer tolerance.
(4) Less scrap and improved quality.

(5) Mirror-image parts can be easily made.

*Tooling:*

(1) Special jigs and fixtures are largely eliminated.

(2) Tool storage is greatly reduced. Storing tapes requires very little room.

*Engineering:*

(1) Increased flexibility of design. Parts thought impractical or impossible are now common. Advanced designs are made.

(2) Finer control of weight and strength. A high degree of repetitive accuracy is obtainable.

(3) Ease of part design change.

*Production Control:*

(1) Scheduling and routing can be done more efficiently since much of the work can be done on N/C machine centers.

(2) Flexibility of machining due to quick-change tooling and simple fixturing.

(3) Increased output due to less time spent in setup and operator error.

(4) Less inventory is required since parts can be more easily produced as needed.

(5) Short runs can be economical.

**Economic Considerations.** Despite all the advantages mentioned, the old question, "How long does it take the machine to pay for itself?" is still as important as ever. Machining centers may cost $150,000 or more whereas some PTP machines may sell for about $15,000. Two examples of PTP N/C savings are shown in Fig. 19.28.

The operating cost for the N/C drill presses used to produce the parts shown is $9.00/hr compared to $7.00/hr for the conventional machine. The savings ranged from $6.50/hr to $22.00/machine-hour. On this basis, the machines could be expected to pay for themselves in 700 to 2,300 hr or in 4 to 13 months of one-shift utilization.

**The Competitive Area for N/C Machines.** N/C machines are not made for high-production work. More specialized machines serve this area. As shown in Fig. 19.29, N/C is best suited for small to medium sized lots.

Computer service centers, some equipment builders, and a few N/C users provide tape programming services for shops that do not have adequate staff or facilities.

**Actuator Body**

| | NC Machine | Conv'l |
|---|---|---|
| Mach. time/pc | 0.225 hr | 0.54 hr |
| Mach. cost/pc | $2.02 | $3.78 |
| Setup time | 2.2 hr | 7.0 hr |
| Annual req., pc | 900 | 900 |
| Lot size, pc | 90 | 90 |

**Total Job Cost**

| | | |
|---|---|---|
| Machining | $1,818 | $3,402 |
| Setup | 198 | 490 |
| Fixtures | 75 | 850 |
| Tape prep. | 66 | — |
| Engrg. modif. | 24 | 75 |
| TOTAL | $2,181 | $4,817 |

**Operations:** mill 3, drill 19, bore 1, c'sink 10, c'drill 8, ream 8 locations.

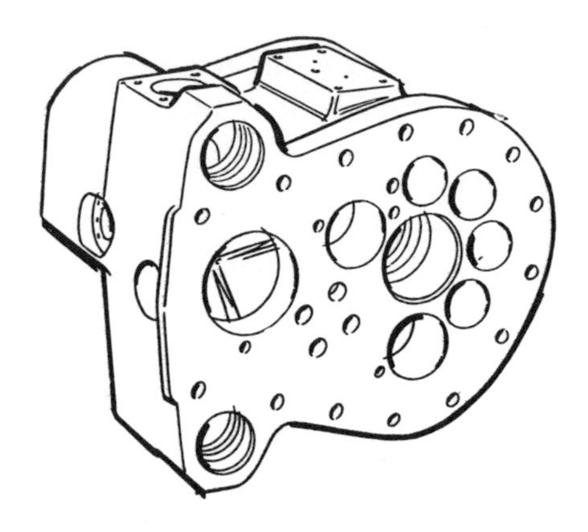

**Servovalve Body**

| | NC Machine | Conv'l |
|---|---|---|
| Mach. time/pc | 0.099 hr | 0.17 hr |
| Mach. cost/pc | $0.89 | $1.20 |
| Setup time | 1.18 hr | 5.20 hr |
| Annual req., pc | 1,000 | 1,000 |
| Lot size, pc | 250 | 250 |

**Total Job Cost**

| | | |
|---|---|---|
| Machining | $ 892 | $1,200 |
| Setup | 42 | 146 |
| Fixtures | 125 | 428 |
| Tape prep. | 21 | — |
| TOTAL | $1,080 | $1,774 |

**Operations:** mill 3, drill 13, tap 1 and countersink 4 locations.

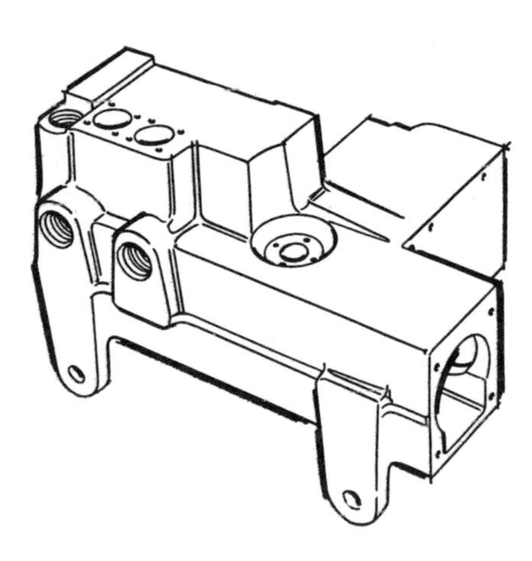

**Fig. 19.28.** A comparison of conventional and N/C machining costs.

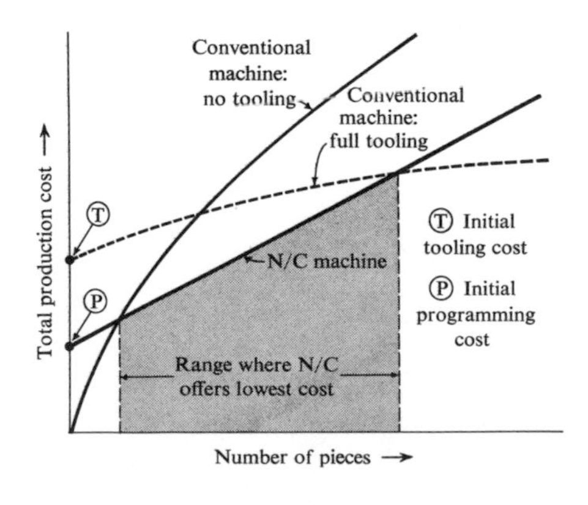


**Fig. 19.29.** Compared to conventional manually operated machines and machines fully tooled for production, N/C shows its best advantage in the small to medium lot area.

## QUESTIONS

**19.1.** What is one of the main events of engineering and science that has accelerated our advancement in recent years?

**19.2.** (*a*) In the air-hydraulic circuit in Fig. 19.2, the cylinder is made to travel at two different speeds. When would it have the slow stroke? (*b*) Explain why.

**19.3.** What is meant by an N/C machining center?

**19.4.** What are some advantages of an integrated line?

**19.5.** Compare a unitized line with a transfer line as to when one should be selected over the other.

**19.6.** What are some of the "building block" units that can be used to help make up an automated line?

**19.7.** What type of line movement would a drilling operation require?

**19.8.** Sketch a section of a vibratory hopper track to show how it would be used to automatically orient a small round-head bolt 1/2 in. long with the heads down.

**19.9.** What is the meaning of point-to-point positioning for a robot?

**19.10.** What are the advantages and disadvantages of the APT programming system?

**19.11.** What is the difference between APT and ADAPT?

**19.12.** How is the designer able to pinpoint an area for enlargement or other dimensional change on the DAC display console?

**19.13.** Explain the functions of the general processor and the post-processor.

**19.14.** What is meant by a tape block?

**19.15.** How is binary information fed to a machine?

**19.16.** Explain the difference between the BCD and the SB system.

**19.17.** Compare the advantages and disadvantages of BCD and SB.

**19.18.** Why does the N/C machine tool fit into the small to medium production area the best?

**19.19.** What are some of the cost cutting advantages of N/C in each of the four categories given?

**19.20.** What makes up a *part program*?

## PROBLEMS

**19.1.** Draw a section of standard punched tape. Show how 15.875 would be punched on it using BCD.

**19.2.** Make a sketch of a panel similar to the one in Fig. 19.25. Show how 15.875 would be arranged.

**19.3.** The caption to Fig. 19.26 has two AUTOSPOT statements. Make a translation of the bottom statement similar to that made for the top. Assume a 3/4-in. dia end mill is being used.

**19.4.** Refer to the comparison between an N/C drill press and a conventional drill press as given in Fig. 19.28. Assume the hourly rate in making the actuator body for machine time and setup is $40/hr on the N/C machine and $10/hr on the conventional machine. Use the formula as given in Chap. 14 to find the break-even cost between these two machines. Use the machine cost just as given in the example.

**19.5.** Find the break-even cost for the Servovalve in Fig. 19.28 if the machine time and setup cost is $50/hr for the N/C machine and $10/hr for the conventional machine.

## REFERENCES

ASTME Collected Papers, Numerical Control, Detroit, 1964.

Childs, J. J., *Principles of Numerical Control,* The Industrial Press, New York, N.Y., 1965.

Emerson, C., "Which Tape Format for N/C?" *American Machinist,* April 15, 1963.

*General Motors Engineering Journal,* Second Quarter, 1965.

*Numerical Control in Manufacturing,* edited by Frank W. Wilson for ASTME, McGraw-Hill Book Co., New York, N.Y., 1963.

Report "N/C – The Time to Act Is Now," *Metalworking,* June 1964.

Thomas, R. A., "The Language of Tape," *American Machinist,* January 6, 1964.

# 20

# NEWER METHODS OF MANUFACTURE

NEW AND EXOTIC MATERIALS often present difficult fabricating problems. These have sparked man's creative genius into providing new concepts of material removal and metal forming. The aircraft and missile industry has had to pioneer some of these new processes because of the demand for minimum weight, high strength, and high-temperature-resistant materials. Competition also serves to spur the search for newer, faster methods that will bring reduced manufacturing costs.

Conventional machining cannot remove metal without pressure and stress. New methods have been developed which are able to remove hard-to-machine metals without touching them with a solid tool. Four of these methods are discussed briefly in this chapter: chemical machining (CHM), electric discharge machining (EDM), electrochemical machining (ECM), electrochemical grinding (ECG), and electrochemical honing (ECH).

## CHEMICAL MACHINING

Chemical machining is a method of removing metal with the aid of chemicals. Basically, it is an etching process that can be divided into chemical blanking and chemical milling.

**Process.** The chemical blanking process consists basically of preparation of a part drawing, metal cleaning, image printing, etching, and resist removal.

The drawing (Fig. 20.1*a*) is carefully and accurately made since the finished part can be no better than the artwork from which it originated. Drawings are sometimes made 100 times actual size with predetermined allowance for the etching undercut.

The finished drawing is checked (*b*) and then reduced photographically on film to the required size.

The metal to be etched must be thoroughly cleaned, usually with the aid of solvents.

After cleaning, the metal is coated with a photosensitive resist and an

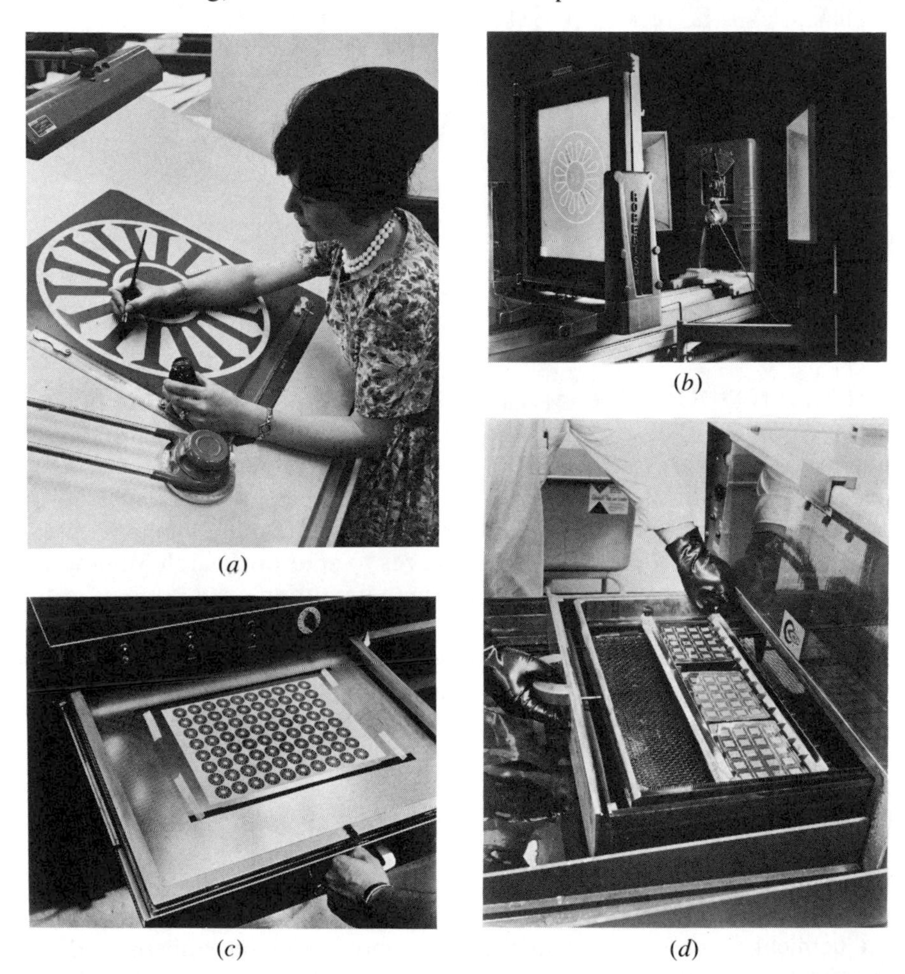

**Fig. 20.1.** The process of chemical blanking. (*a*) Preparation of the drawing. (*b*) Checking the drawing on an optical comparator. (*c*) The metal is coated with a photosensitive resist and the image is transferred photographically to it. (*d*) Spray etching the metal to leave only the finished part. Courtesy Chemcut Corporation.

image of the part to be made is photographically printed (*c*) on one or both sides of the metal surface. By using both sides, the etching time is cut in half. The metal is then developed in a solvent that removes the coating except in the exposed areas, which is now the etchant-resist image of the part.

The metal is then spray etched (*d*) to dissolve all unprotected areas and leave only the finished part, which is washed in a solvent and flushed with water to remove the resist.

The etchant-resist can be placed on the panel by silk screening, but the results are less accurate.

**Chemical Milling.** The process of chemical milling is not limited to photographic or silk-screen transfer. Large parts are machined by spraying a surface with a thin plastic film after it has been thoroughly cleaned. Templates are put on the surface and scribed around with a sharp knife. The cutout areas are removed, and the part is placed in an etchant tank.

The chemical blanking is limited to flat surfaces, but chemical milling can be used on surfaces of irregular geometry. Both large and small parts are shown in Fig. 20.2.

**Advantages.**

(1) Virtually any metal can be used.

(2) Design modifications are easily made by altering the part drawing.

(3) Setup costs are low; maintenance and tooling costs are also low.

(4) Small delicate parts and curved surfaces can be machined that would be extremely difficult if not impossible by any other method.

(5) Applicable for short runs on intricate designs.

**Disadvantages.**

(1) If close accuracy is required, etching all the way through is limited to metal thicknesses of about 0.0625 in.

(2) Surface defects in the metal tend to be accentuated.

**Surface Finish, Accuracy, and Cutting Rate.** Surface finish is dependent on the condition of the original surface up to about 0.015 in. removal. After that, the piece develops its own mat-type finish. The general range is about $30\mu$ to $125\mu$. The accuracy on through cutting is about $\pm 20$ percent of the metal thickness, but $\pm 10$ percent can be obtained in critical areas. The time required in the etchant tank is determined by the depth cut, the acid concentration, and the temperature of the bath. The approximate rate of metal removal is 0.001 cu in./min.

## ELECTRIC DISCHARGE MACHINING

If you have at some time or other observed an arc caused by an accidental short circuit and noticed the pitting done to the surface of the shorted material, you have seen the fundamental principle of electric

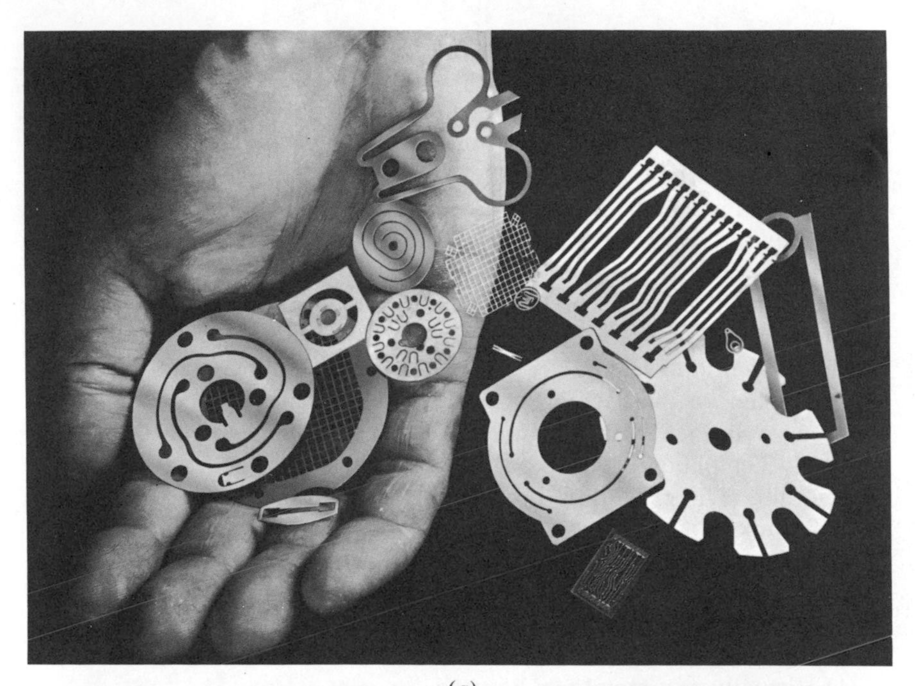

(*a*)

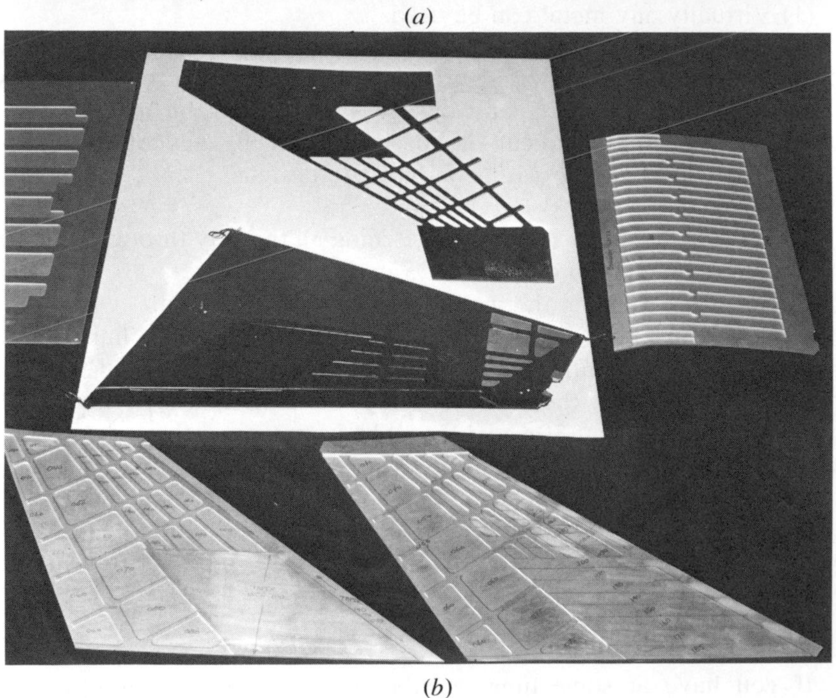

(*b*)

**Fig. 20.2.** Chemically blanked parts (*a*) produced by photographic copying and (*b*) chemically milled parts. Courtesy Chemcut Corporation.

discharge machining (EDM). Of course the EDM process is more complex, but it can be defined as *a controlled source of regulated sparks used to erode away a workpiece to a specific shape* under a dielectric.

**Basic Process Theory.** Although the principles of EDM have been observed ever since the discovery of electricity, it was not until 1943 that the Russian scientists B. R. and N. I. Lazerenko proposed a thermal theory, and then serious work began. The theory stated that electrons jumping from a tool and striking a workpiece cause a sudden rise in the workpiece surface temperature, in the immediate area of the spark impact, which is sufficient to melt and expel a small globule of molten metal (Fig. 20.3). There are billions of electrons in each spark and thousands (10 to 20 thousand) of sparks/sec. It is estimated that at high current levels the temperature on the workpiece surface reaches 10,000°F.

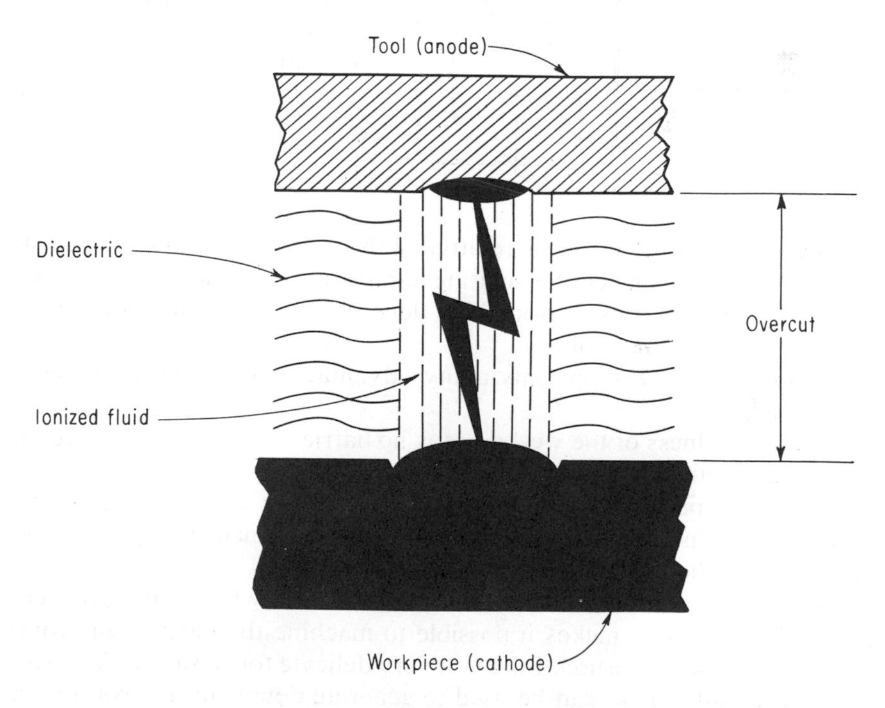


**Fig. 20.3.** When electrical energy ionizes a portion of a dielectric fluid, a spark from the tool erodes a tiny crater on both the tool and the workpiece. Courtesy *American Machinist.*

**Basic Process.** The tool has a positive charge and is called the electrode, and the workpiece is the cathode. The tool feeds down, under control of a servomechanism to the distance of the spark gap or about 0.001 in. At this point, the electrical potential between the work and tool exceeds the strength of the dielectric and sparks flow (Fig. 20.4). A small amount of the dielectric becomes ionized, the chips are washed away, and the power, after a few microseconds, builds up to repeat the process. Chips that are not washed away by the dielectric cause further sparking on the tool, which results in unwanted wear, scrap, and machine downtime. For this reason, a good flow of the dielectric is one of the most essential features of the EDM process.

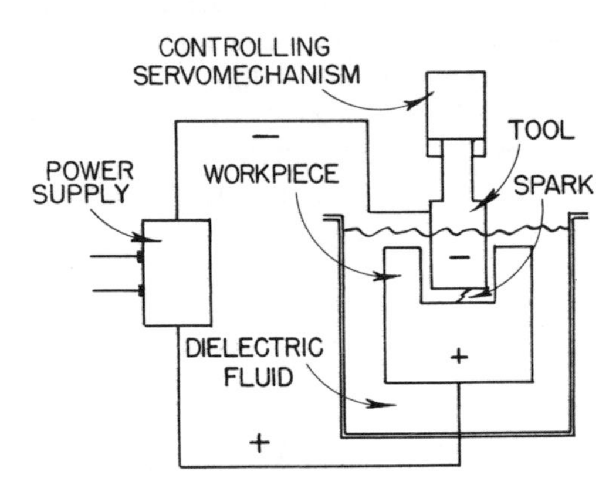


**Fig. 20.4.** A schematic sketch of the main components used in EDM. The servomechanism monitors the gap voltage (shown greatly enlarged) during cutting, moving the tool into or out of the work as required. The power supply changes standard ac current to dc.

The wear of the electrode is relative to the heat and ion flow that attack it. Therefore the higher the melting temperature of the electrode, the less it wears. For this reason, high-density carbon is one of the best overall electrode materials.

**Advantages.** EDM presents many advantages that stem from three basic facts:

(1) The hardness of the workpiece is no barrier. As long as the material can conduct current, it can be machined.

(2) Any shape that can be produced in a tool can be reproduced in the workpiece. Sometimes complicated tooling must be made in segments and fastened together with a material like Eastman 910.

(3) The absence of almost all mechanical force while cutting (except dielectric pressures) makes it possible to machine the most fragile components without distortion. Likewise very delicate tools, such as fine wire only a few mils thick, can be used to generate details in the workpiece.

**Disadvantages.**

(1) The workpiece and tool must be able to conduct electricity.

(2) EDM is slow when compared to conventional methods or even to ECM. Since it is slow, wherever possible the hole, slot, or cavity is roughed out prior to using EDM.

**Surface Finish and Accuracy.** The surface finish is a controlling factor. It determines the amperage, the capacitance, the frequency, and the voltage setting. Heavy metal removal is associated with high amperage, low frequency, high capacitance, and minimum possible gap voltage. The opposite is true for finishing cuts (Fig. 20.5). The best finish that it is economical to produce is about 30 $\mu$.

Carbides, hardened tool steels, cermets, germanium, cobalt, nickel, and titanium, to mention some of the more difficult materials to machine, can be cut to tolerances of 0.0002 in. Closer tolerance requires slower cutting since the chip gap is narrower.

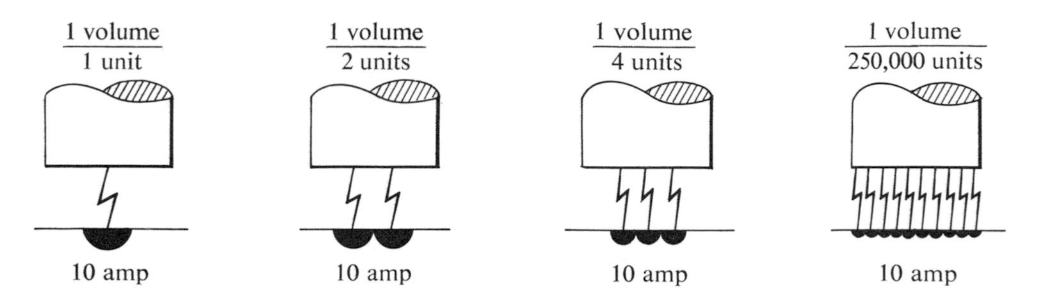


**Fig. 20.5.** With the current maintained at a constant value (in this case, 10 amperes), smaller craters and finer finishes can be produced by increasing the frequency of the discharges. Courtesy Matson Co., a subsidiary of Ingersoll Milling Machine Co.

## ELECTROCHEMICAL MACHINING

A British patent was issued in 1930 covering the electrochemical machining process; however, the development has not been until recent years. The process is similar to electroplating except that the primary interest is in the metal's being removed at the work (anode) rather than being deposited at the cathode. Also current densities are about 5,000 times that used in plating.

**The Process.** Although ECM is compared to an electroplating process, it is much more. The fact that it can be used to produce specific shapes with precise dimensional control qualifies it as a modern machining process. A typical ECM system is shown schematically in Fig. 20.6.

In order to have a basic understanding of the process, it is best to divide the system into its four basic elements: (1) the electrolyte, (2) the electrode, (3) the machining gap, and (4) the current density.

**The Electrolyte.** The electrolyte is a water solution that carries a current, such as a caustic soda or caustic potash solution. The electrical energy starts a chemical reaction in the electrolyte solution which results in the formation of gas at the tool and dissolvement of metal from the workpiece. This reaction causes heat, which is dissipated by a swift flowing stream of electrolyte that also serves to wash the dissolved metal from the area. The amount of electrolyte flow varies with the system, but some systems use as much as 200 gallons/min at 300 lb/sq in. pressure.

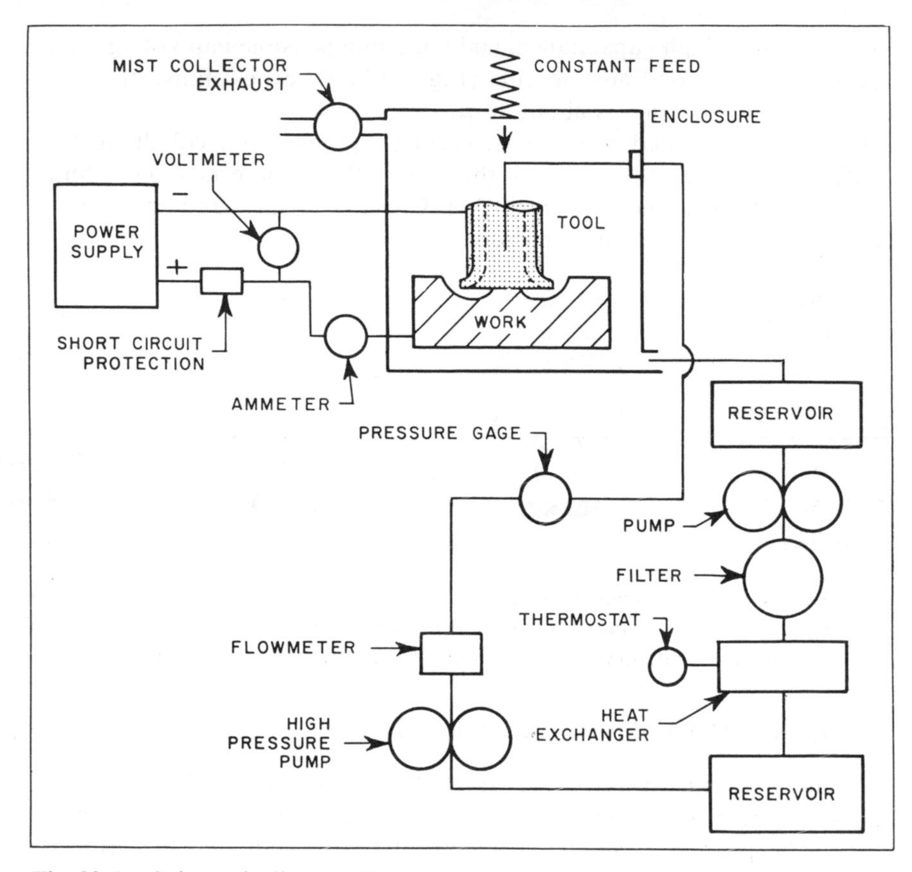


**Fig. 20.6.** Schematic diagram illustrating typical ECM system.

The electrolyte can flow through the tool or from one side to the other under the tool.

**The Electrode.** The electrode must have four characteristics: it must be machinable, be rigid, conduct electricity, and resist corrosion. The three most used materials are: copper, brass, and stainless steel. The outside of the electrode is insulated to prevent machining action on the sides. The tool is somewhat larger on the end (overcut), which permits the electrolyte to have an exhaust route.

**The Machining Gap.** The machining gap is the distance between the work and the tool (Fig. 20.7). This gap is maintained as small as possible, 0.001 to 0.003 in. Any contact of the tool with the work will result in arcing with serious damage to both members. On the other hand, if the gap is permitted to become large, more electrolyte will flow, offering increased resistance for the current with a subsequent reduction in stock removal. A sensitive constant feed drive to the quill is necessary in order to accurately control this gap.

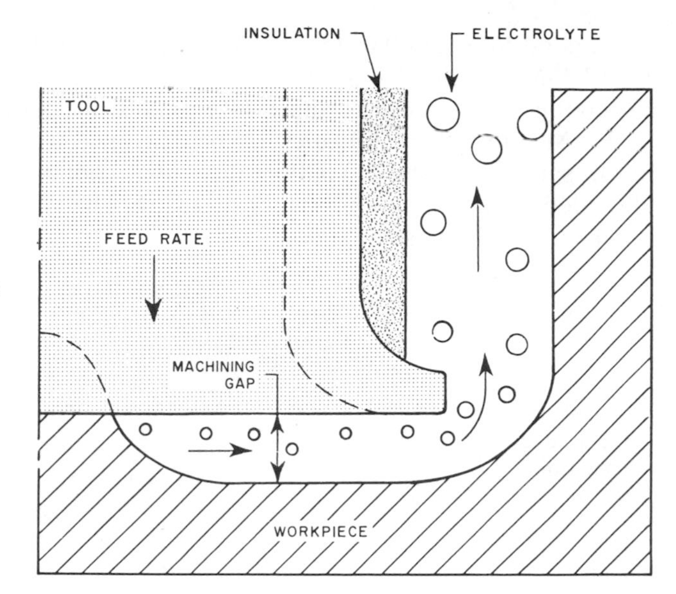


**Fig. 20.7.** The machining gap.

**The Current Density.** The current density refers to the amount of current that is able to pass into a square inch of work area. The metal removal rate is proportional to the current that passes through the machining gap. Current densities as high as 10,000 amp/sq in. have been used.

**Advantages.** The advantages of ECM are similar to those given for EDM with some additions:

(1) The tool does not wear. Once the tool is developed, it can be used indefinitely.

(2) There are no thermal effects on the workpiece.

(3) The faster the stock removal, the better the surface.

(4) Metal removal rates are higher, e.g., the feed rate with similar tooling for "drilling" a 1/32-in. hole in a high speed steel for EDM is 0.015 in./min, whereas with ECM it is 0.060 in./min.

**Disadvantages.**

(1) The tools are more difficult to design and make.

(2) Special fixturing is required for the workpiece to withstand the high electrolyte flow.

(3) The cost of the basic equipment is several times that of EDM.

(4) The most common electrolyte, sodium chloride, is corrosive to equipment, tooling, fixturing, and work material.

**Surface Finish and Accuracy.** The surface finish varies with the type of electrolyte used and the workpiece material. Any defects in the tool, nicks, scratches, or burrs will be produced as mirror images on the work surface. The surface finish produced may be as fine as 25–30 $\mu$.

The overcut per side of the electrode is about 0.005 in., and repeatability is good to within 0.0015 in. An example of the type of work done by both EDM and ECM is shown in Fig. 20.8. The 1/2-in.-deep, 0.050-in.-dia cooling holes on the trailing edge of the cast turbine vane were cut by EDM. More than 100 holes can be machined with proper tooling in a single setup. The small 0.040-in.-dia radial cooling holes visible at the shrouded vane end were produced by ECM. The holes are 3 in. deep and all machined at the same time.

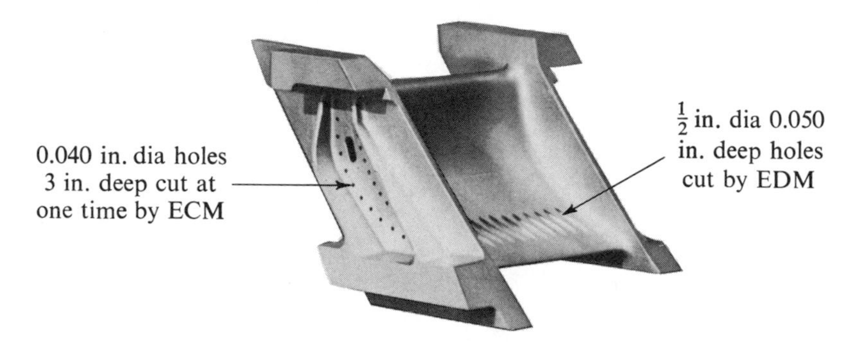


**Fig. 20.8.** A cast turbine vane with trailing edge holes machined by EDM. Holes visible from the end machined by ECM.

## ELECTROCHEMICAL GRINDING

Electrochemical grinding is similar to ECM with an assist from the grinding wheel. The workpiece, which must be metallic, is the anode, and the metal-bonded abrasive wheel is the cathode (Fig. 20.9). Short circuiting is prevented by the protrusion of abrasive particles (diamonds) from the surface of the wheel. An electrolyte (water-base fluid) flows across the surface of the wheel, passing current from the work to the wheel. This causes the anode to dissolve.

**Metal Removal Rate.** The rate at which the metal is removed is directly proportional to the amount of current passing through the elec-

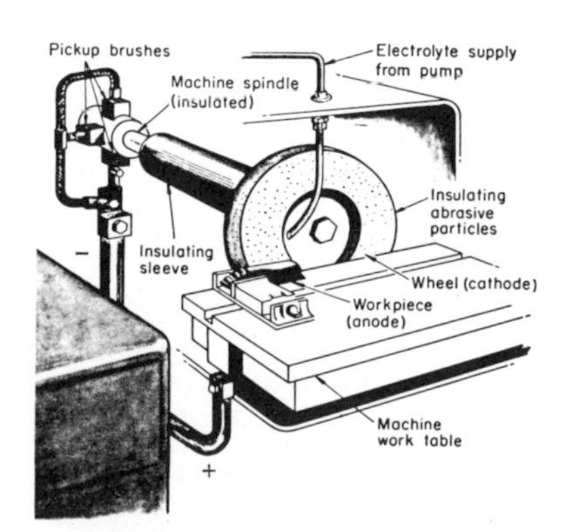


**Fig. 20.9.** Setup for electrolytic grinding. Courtesy Anocut Engineering Co.

trolyte. The current used is determined largely by the area of contact between the wheel and the work. Hardness and tensile strength are not important. As a general rule, about 0.100 cu in. of material can be removed per minute with each 1,000 amp of current.

The diamond particles in the wheel, in addition to being insulators, help to scour the metal that is being cut, making it more receptive to the "unplating" action of the electrolyte. Since cutting, or making chips, is not the main function of the diamonds, the life of the wheel is increased many times.

**Finish and Accuracy.** Carbides, hardened tool steels, cermets, germanium, cobalt, nickel, and titanium, to mention some of the more difficult machining materials, may be cut with tolerances of ±0.0002 in. Cutting time is reduced by about 30 percent over conventional grinding. A 10- to 20-$\mu$ finish can be maintained.

Since this is essentially a "cold" grinding process, there is no minute softening of heat-treated surfaces. Also, very thin parts, such as stainless-steel honeycomb, can be cut without leaving a burr on the foil-thin edge (Fig. 20.10).

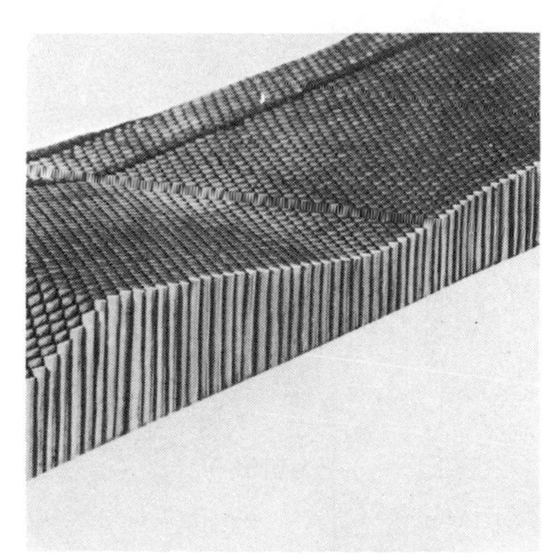

**Fig. 20.10.** Electrolytic grinding performed on stainless-steel honeycomb leaves a burr-free edge. Courtesy Anocut Engineering Co.

## ELECTROCHEMICAL HONING

The same principles as those given in ECG apply to ECH. The advantages claimed over conventional honing are: (1) up to 300 percent increase in abrasive efficiency, (2) more accuracy, (3) stress-free surfaces, and (4) allows slight realignment or position changes of the bores.

In order to retain the benefits of a crosshatch pattern, the current is turned off for the last few strokes.

## NEWER METAL REMOVAL PROCESS SUMMARY

The newer processes of metal removal, as discussed here, are compared and summarized in Table 20.1.

**Table 20-1**
**A Comparison Summary of Some of the Newer Methods of Metal Removal**

| | CHM | EDM | ECM | ECG |
|---|---|---|---|---|
| Tolerances | 0.002 in. thick ± .001 possible .002 in. practical 0.020 in. thick ± .004 possible .010 in. practical 0.060 in. thick ± .006 possible .012 in. practical | Practical .002 – .005 in. Possible .0001 –.0005 in. | Practical ± .005 in. Possible ± 0.0005 in. | Practical ±0.0002 Possible ± 0.0005 in. |
| Surface | Av. rms. aluminum 90 Magnesium 50, Steel 60 Tungsten 50 | .010 in.$^3$/hr. – 30 rms. .5 in.$^3$/hr. – 200 rms. 3.0 in.$^3$/hr. – 400 rms. (Heat affected zone .0001 – .005 in.) | 4 to 50 rms. easily obtained. (No heat affected area.) | Tungsten Carbide 5 – 15 rms. Steel is 15 – 30 No heat affected zone |
| Removal Rate | Penetration rate – .003 in./Min. | .01 – 12 Cu. in./hr. | 1 Cu. in./ Min. for 10,000 amp. unit | 0.01 Cu. in./Min./ 100 amp. |

## ULTRAHIGH-SPEED MACHINING

There has long been a theory in the field of metalworking that materials have a critical impact velocity which, if exceeded, causes instantaneous failure. Less than the critical value merely causes plastic deformation. Within the critical speeds, normal wear would not occur. Some basis for this belief was found in the experiments of Salomon, of Germany, who was issued a patent in 1931 for work in which he plotted speed against temperature. The results indicated that, in trial milling runs using 55,000 sfpm on nonferrous metals, the heat generated followed a peak curve. Both the tool and the work were able to go beyond this peak. From these data, estimated curves for steel and cast iron were made, showing 148,000 and 128,000 sfpm, respectively. The conclusion was that temperatures rise to the critical cutting speed and then decrease as the critical speed is exceeded.

The questions arise: Why must the theory be proved, and will it ever be practical? The incentive to find the answers is the difficulty that is being experienced in machining space-age materials. Aluminum, which can be machined at 15,000 sfpm, is giving way to materials such as titanium and stainless steel, whose cutting speeds may range from 50 to 300 sfpm. If the same amount of machining should be necessary on these harder metals, it would mean a considerable outlay in plants and equipment. This does not mean, should the facts back up the theory, that whole plants would be changed over. It would, however, have a profound influence on all metalworking, and those operations that could receive the most benefit would be done at ultrahigh speeds.

**Equipment.** An early approach consisted of projectiles fired past a stationary cutting tool (Fig. 20.11).

The first projectiles were fired with rifles, using a .30 – '06-caliber Mauser bored out to accept a 0.3-in.-dia slug 2 in. long. The rifle was later changed to a 20-mm cannon which fired a supersonic sled at Edwards Air Force Base. With this equipment, speeds as high as

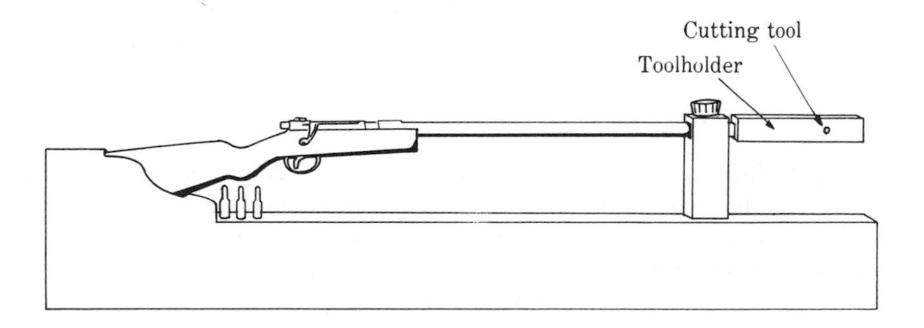


**Fig. 20.11.** Early equipment used to test the theory of ulta high-speed machining.

240,000 sfpm were obtained. Also, tool contact time was increased as cuts 4 ft long were made. Conventional single-point tools were used. These were made from high-speed steel, stellite, and ceramics. It was found that there was a decrease in tool wear at the higher speeds. This may be accounted for, in part, by the difference in chip formation (Fig. 20.12). Conventional metal cutting results in a chip thicker than the depth of the cut, but in ultrahigh-speed machining the reverse was found to be true. The favorable shear angle makes less work for the tool.

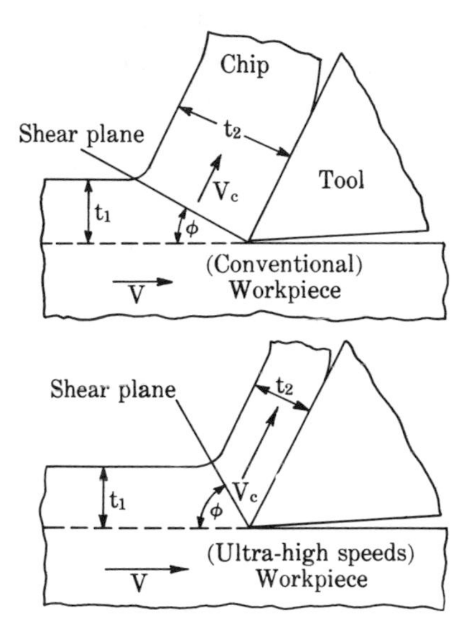


**Fig. 20.12.** Comparison of chip formation between conventional and ultrahigh-speed machining. Note the difference in chip thickness and shear plane. Courtesy *American Machinist.*

**Future Applications.** It is too early to make any critical evaluation of the test results obtained in ultrahigh-speed machining. Further work is being carried on to determine cutting forces under various conditions and to relate these to horsepower requirements. The types of machine design best suited for this work are also being studied. Associated problems are vibration dampening, bearings, drives, operator safety, acceleration, deceleration, and centrifugal and dynamic forces.

## HOT AND COLD MACHINING

**Hot Machining.** Another approach to the difficult task of efficiently machining high-strength, high-temperature-resistant materials is to machine them at elevated temperatures. This is a radical departure from conventional machining, where every effort is directed toward keeping both tool and material temperatures as low as possible.

The theory of hot machining is that the strength of alloy steels is less at elevated temperatures. This reduces the needed cutting force; as the metal acquires a plastic character, it eliminates chip formation in the accepted sense.

Hardened alloy steels can be cut by hot machining. For example, an AISI H4340E steel hardened to 50–52 Rockwell C was cut on a milling machine with a depth of cut of 0.062 in. and a feed of 0.009 in./tooth. A radio frequency heating machine is used to provide localized heating in a hot band just ahead of the cutter and to the desired depth of cut. Ideally, the heat should be raised quickly and confined to the shear zone in which the chip is formed.

Although the tool forces are less, the fact that the work is hot causes the tool to heat up, especially on larger sections, where milling-cutter teeth are engaged for longer periods of time. Jobs done with short tool-contact time have shown that tough metal can be successfully hot-machined. This method has proved particularly effective on metals that tend to work harden rapidly.

**Cold Machining.** At the opposite extreme of machining theory is the method of keeping the metal at a very low temperature, which, in turn, reduces tool-tip temperatures. Ways to accomplish this include surrounding the area with a cold mist (−109°F), using dry-ice coolant (−76°F), or deep freezing the part itself.

Temperatures at the tool-and-work interface may sometimes reach 2,000°F when machining tough, high-alloy materials. Problems arising from this much heat include galling, seizing, work hardening, low-temperature oxidation, and short tool life. Conventional practice has been to reduce the surface foot speed to prolong the tool life but the resulting reduced output has spurred the search for a better way.

Extensive research on subzero machining has been done by Convair for the Air Materiel Command. Some of the results indicate that mist cooling and flooding are superior to having the part cooled. Tool life improved, on some alloys, from 100 to 300 percent over conventional machining. Still further research is necessary; the information gained thus far indicates only that subzero-temperature machining may have merit and needs further study.

## HIGH-ENERGY-RATE FORMING METHODS (HERF)

**Explosive Forming.** Explosive forming is an outgrowth of drop-hammer forming. Several years ago, "trapped rubber" was used to form sheet metal blanks. The method was similar to the Guerin process except that, instead of the slow squeezing action, the rubber was made to strike the part. In this instance the rubber was encased in steel, which made a total free-falling weight of 8,970 lb. The results were so successful, in that there was no springback in the formed parts, that it was decided to extend the process to include much greater forces.

One of the first explosive-forming models used a 12-gauge shotgun shell without the shot. From this model, it was found that relatively high instantaneous forming pressures could be obtained. The 12-gauge shotgun shell developed forces equivalent to 4,500 tons in a conventional press. Figure 20.13 shows how a cartridge explosion drives a ram, forcing trapped hydraulic fluid to bulge the blank held in the forming die.

**Cartridge**

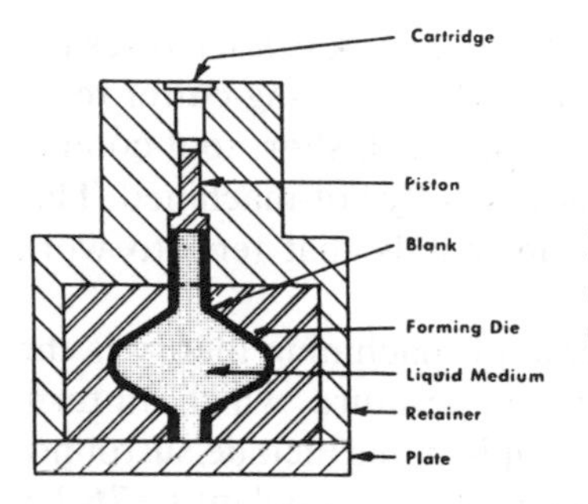


**Explosion drives ram, forcing trapped hydraulic fluid to bulge blank into forming die.**

**Fig. 20.13.** Cartridge explosive used as the energy source for explosive cylinder bulging. Courtesy *Metalworking*.

**Explosive Action.** Explosive action is the application of extremely high pressures – from 100,000 to over 1,000,000 psi – during a very short (microsecond) interval. Metal displacements are 100 times those obtained by conventional presses. During this extremely short ultrarapid stress-loading period, the metal passes through its elastic range into the plastic range, where it assumes a permanent set. It has not yet been determined why metals behave differently under this ultrahigh-stress loading, although it has been known for years that metals can momentarily withstand stresses and undergo elongations that would cause failure by the usual forming methods. How these phenomena relate to the improved metallurgical properties and highly uniform metal distribution in the part formed, without effect on grain structure, is also unknown. It can only be theorized that, under high stress-strain rates associated with explosive forming, the atoms go through a series of slip, break, and heal events very rapidly. Failure occurs when cohesive forces between the atom of the metal are insufficient to cope with continued slipping. This happens when the stress duration has been applied for too long a period. Explosive-forming stress duration is very short, which accounts for its success.

**Equipment.** The basic setup for explosive forming consists of the energy source, transfer medium, blank, die, and some type of retainer.

**Energy Sources.** The energy sources most often used are: high explosives, low explosives, compressed gases, electrohydraulic and electromagnetic forces.

**Types of Explosive Operations.** In explosive forming, large differences in energy requirements and in the resulting behavior of the workpiece will occur depending on whether it is done as a *standoff operation* or a *contact operation.*

In the standoff operation, the energy is released some distance from the workpiece and is propagated mainly in the form of a pressure pulse through some intervening medium.

In contact operation, the energy source, usually an explosive charge, is in intimate contact with the workpiece. High-intensity, transient stress waves are induced into the workpiece.

## HIGH EXPLOSIVE FORMING

**The Process.** The high explosive process elements are shown in Fig. 20.14*a*. It consists of a tank containing a liquid, female die, vacuum system, and explosives with a means of detonation. The liquid serves as a medium for the transfer of the shock waves and gas pressure. A vacuum

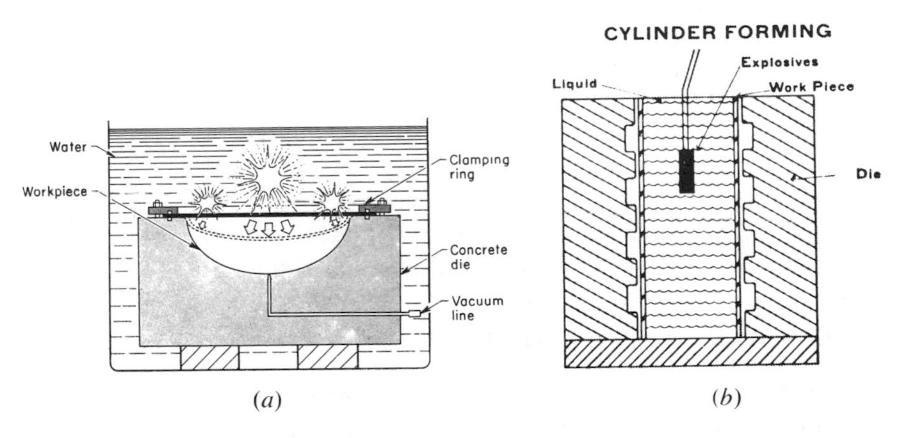


**Fig. 20.14.** Typical setups for under-water forming of a hemispherical part (*a*) and forming or sizing a cylinder (*b*).

pump is required to evacuate the die cavity to reduce forming resistance and eliminate burning of the die side of the material and the die. Temperatures as high as 10,000°F are generated by the compression of any gas entrapped in the die cavity.

**Energy Transfer Media.** The under-liquid detonation of high explosives yields a shock wave and a bubble of hot gaseous detonation products. This shock wave is the major energy for the forming operation.

Pressures range from about 2,000,000 to 4,000,000 psi. The forming time is measured in microseconds. The speed of this conversion front which moves through the explosive is known as "detonation velocity" and ranges from 6,000 to 2,800 fps.

**Tooling.** Some of the most commonly used die materials are: cast steel, Kirksite, epoxy resin, concrete, and laminated fiberglass. Steel is used for long runs, Kirksite, fiberglass, and concrete are restricted to a small number of parts. Epoxy dies are gaining in importance and are replacing steel. They represent lower fabrication costs since they can be produced directly by casting on the master. The smooth epoxy surface can take repeated impact without deterioration, yet design changes are not difficult to make. Less common but very useful at times for large short-run production are dies made of ice.

**Explosive Materials.** Some explosives and their detonation velocities

**Table 20.2**
**Detonation Velocities of Explosives**

| Explosive | Detonation velocity meters/sec |
|---|---|
| TNT | 6800 |
| RDX/TNT (60/40) | 7810 |
| Tetryl | 7500 |
| PETN | 8400 |
| Nitroglycerine | 7800 |

are shown in Table 20.2. These explosives are used in the form of powder sheets, liquid cords, pellets, or cylinders. The sheet type explosive is shown in Fig. 20.15 along with the detonator.

## LOW EXPLOSIVE FORMING

**The Process.** Low explosives such as smokeless or black powders do not explode but, rather, burn or deflagrate at a very rapid rate of several hundred feet per second with a rapid evolution of gas. The pressures produced are considerably lower than the high explosives, being about 40,000 psi (Fig. 20.16). Here the containment helps increase the time during which the pressure is effective and it also helps deliver a greater impulse to the work (Fig. 20.17).

**Applications.** All of the common steels, stainless steels, titanium, and aluminum alloys have been formed explosively. Other applications have been blanking, coining, powder compacting, cutting, embossing, cladding, drawing, expanding, flanging, hardening of metals, joining, sizing, and stretching.

**Fig. 20.15.** A detonator and a sheet of flexible high explosive material. Masking tape should be used to secure an intimate bond between the detonator and the explosive. Courtesy E. I. DuPont De Nemours & Co., Inc.

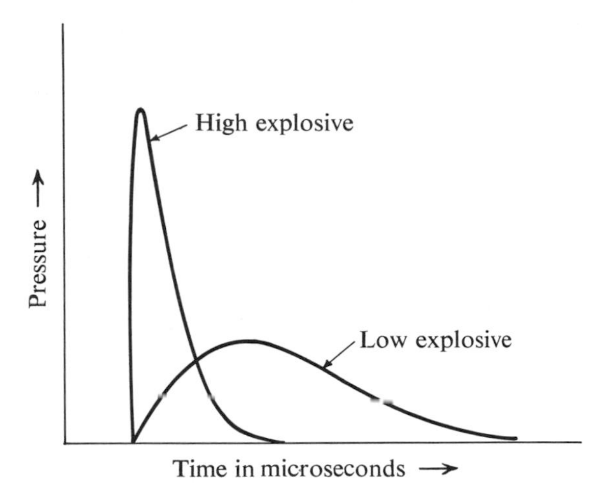


**Fig. 20.16.** A schematic comparison of the pressure time relationship for the two types of reactions for high and low explosives. The energy release is the same, but the rate is different.

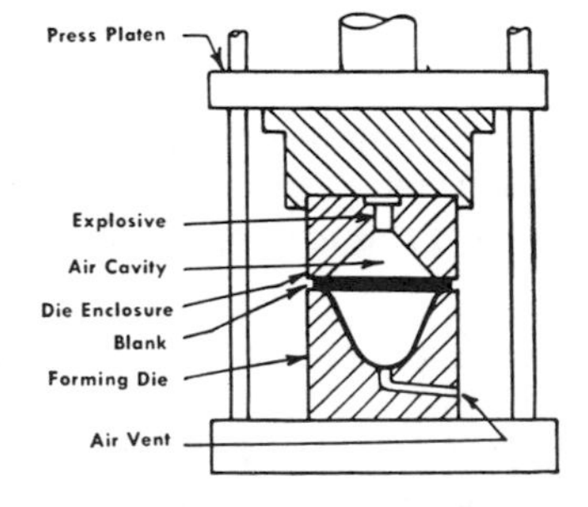


**Fig. 20.17.** Smokeless powder forming. Low explosives are usually used in a closed chamber.

**Accuracy.** The tolerance capabilities of explosive forming is a function of the die design, vacuum drawn part size and configuration, explosive charge, and standoff distance. Normal working tolerances are about ±0.010 in. For example, in forming a 54-in.-dia missile dome of AMS 6434 steel varying in thickness from 0.080 to 0.150 in., the OD tolerances were held to ±0.010 in. and the elliptical contour to ±0.020 in. The material thickness tolerance was ±0.004 in.

**Advantages of Explosive Forming.** The benefits of explosive forming may be summarized briefly as follows:

(1) It is possible to obtain high working energies (up to several million pounds per square inch). This is higher than is economically or physically possible with mechanical or hydraulic presses.

(2) Springback is virtually eliminated. Parts may be formed to finished size. Close tolerances—within ±0.010 in.—are possible. This means better mating between interchangeable parts.

(3) Greater deformation (up to 2.5 times the elongations normal with usual forming methods) can be obtained without failure of the part. Many parts that formerly required multiple draws or spinning steps, with intermediate anneals, can now be formed with a single draw.

(4) Physical properties are improved. The hardness of ductile metals and metals that are not treatable by heat can be increased for greater strength. Yield strength can also be increased.

(5) Economic benefits can be obtained through less expensive capital equipment. Simpler setups require only a well or an underground tank large enough to accommodate the die. Die costs are reduced, since explosive forming requires only one die, usually a female. Die materials can be relatively inexpensive, such as cast epoxy and epoxy-faced dies, Kirksite, and aluminum. Standard tool steels are preferred for long runs.

(6) Excellent repeatability can be obtained where the quantity and placement of the charge are carefully controlled. Even greater repeatability is evidenced by compressed-gas and hydrospark forming.

**Limitations of Explosive Forming.** (1) The formability of brittle metals or metals with 1 to 2 percent elongation is not improved.

(2) Proper selection of the type, quantity, and shape of the explosive charge is an art. Too little energy will not form the part; too much will destroy it. In designing the charge, small models are used, and the results are extrapolated for full-scale workpieces with almost 100 percent accuracy.

(3) Production rate is relatively slow.

## GAS FORMING PROCESSES

**Gas Mixture Process.** The gas mixture process consists of using the energy released by the chemical reaction of a fuel (hydrogen) with an oxidizer (oxygen). The system utilizes separate sources of pressurized fuel, oxidizer, and diluent (usually nitrogen) gases. The combustion chamber where the gases are mixed is part of the forming die (Fig. 20.18).

A glow wire or spark plug is used to ignite the mixture. An exhaust valve is needed to allow removal of the fired gas pressures before disassembling the tool.

When the gases are ignited, a three-phase reaction occurs: (1) the sub-

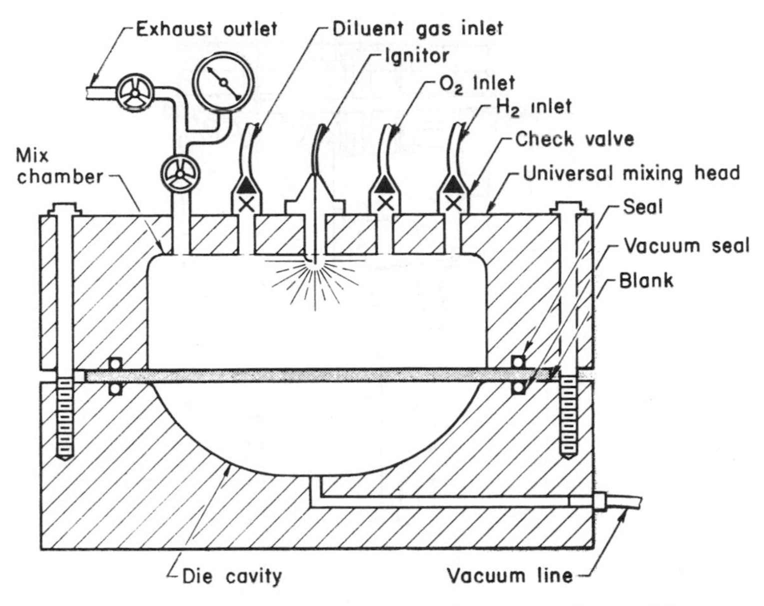


**Fig. 20.18.** Explosive gas forming setup. Courtesy *The Tool Engineer.*

sonic flame front accompanied by changes in pressure, temperature, and volume follow the laws of adiabatic reaction, (2) an extremely turbulent reaction is accompanied by severe shocks, (3) the turbulent reaction is followed by a thin segment of stable shock front moving at supersonic velocity.

**Applications.** The process is particularly suited for rapid cycling and forming of small tubular parts, complex shapes, and large thin sections.

**Pneumatic-Mechanical Forming.** The pneumatic-mechanical forming system utilizes compressed air at high (2,500 lb) pressure to drive a ram at velocities up to 800 ips. One such machine is the Dynapak shown in Fig. 20.19. The punch is mounted on the ram and the die on the bolster plate. The ram derives its energy mostly from adiabatic expansion of compressed gas. The energy level is made infinitely variable between the range of 5 to 100 percent of its rated 550,000 ft-lb. Four 100-hp motors are used to operate the hydraulic jacks which lift the ram off the workpiece and recompress the driving gas.

**Applications.** The pneumatic-mechanical forming machine is used primarily for forging both ferrous and nonferrous materials either hot or cold. The rocket motor body shown in Fig. 20.20 was made from the billet on the left, in one stroke.

Pneumatic-mechanical machines take up less floor space than drop hammers of equal capacity and do not require heavy foundations. The dies used are not as heavy as conventional forging dies so changeover usually take less than one hour.

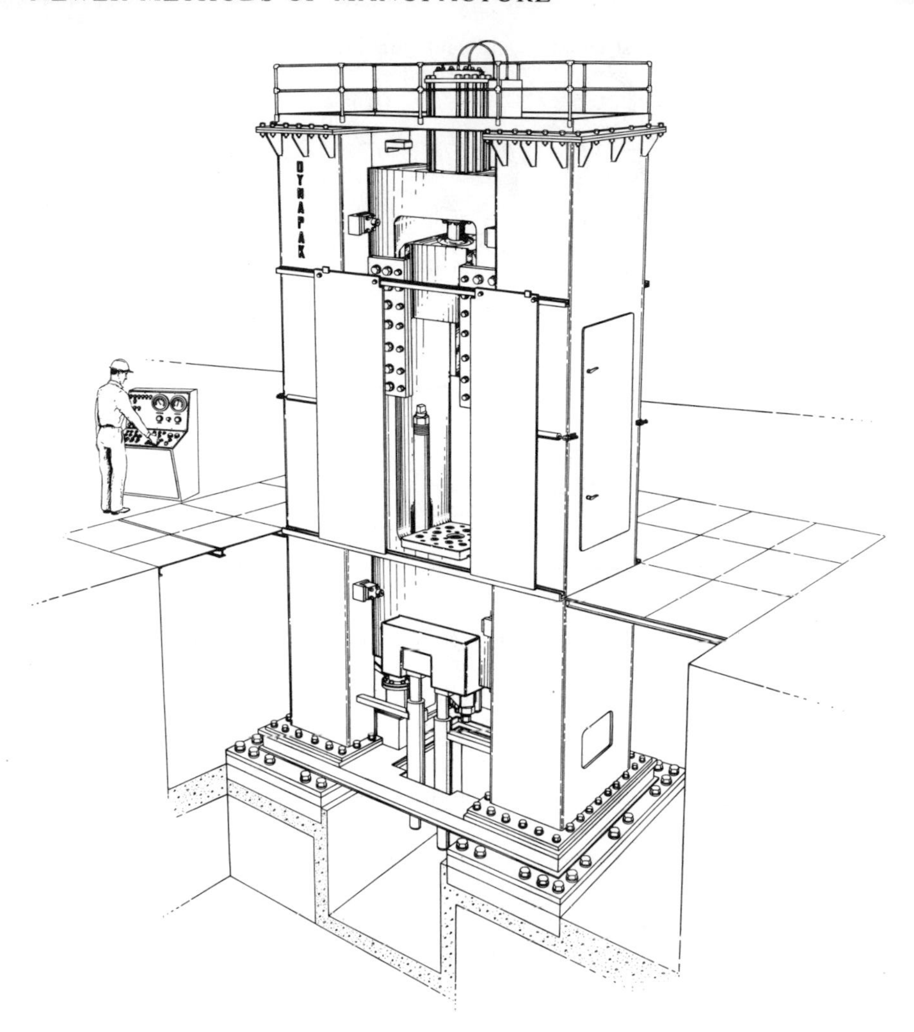


**Fig. 20.19.** A pneumatic mechanical machine used for high energy rate forming, forging, or compacting. Courtesy General Dynamics Corporation.

## ELECTROHYDRAULIC FORMING

**Hydrospark Process.** Two methods are used to create electrohydraulic energy. The first of these is a discharge of an electric spark under water to produce a high velocity shock wave (Fig. 20.21).

A high-voltage power supply is required with capacities for storing the charge, a discharge switch, and a coaxial electrode. The applied force needed can be varied by varying the voltage. Energy increases as the square of the voltage.

**Fig. 20.20.** An example of the type of work performed on a pneumatic-mechanical type press. Courtesy General Dynamics Corporation.

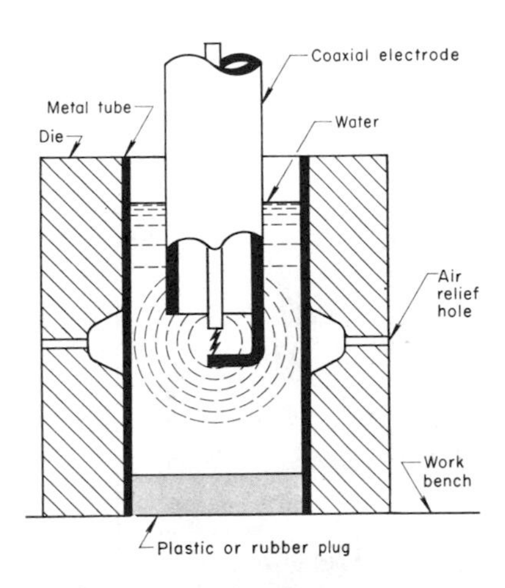


**Fig. 20.21.** Hydrospark forming used to bulge a cylinder. Courtesy *The Tool and Manufacturing Engineer.*

**Exploding Bridge Wire.** The exploding bridge wire (EBW) utilizes the same type of equipment as the hydrospark process except that a wire is used between the two electrodes. The wire vaporizes, and the metal vapor carries current that makes the vapor even hotter. The vapor expands, and the water surrounding it is compressed and forced out radially.

The hydrospark method is faster since there is no need to replace the wire. On the other hand, the wire lends itself to better control. The shock wave can be shaped by bending the wire to conform to the cavity. Also, longer gap lengths can be used for a given amount of energy stored in the capacitor.

**Applications.** The most important applications are for sheet metal forming and for the bulging of tubular parts. Parts have been formed ranging from 3 to 48 in. in dia and material thicknesses from 0.030 to 1.000 in. have been formed.

**Advantages.** The advantages of electrohydraulic forming are greater safety, more precise control, and lower cost. It is estimated that this tool can be constructed at one tenth the cost of a conventional hydraulic press used to do the same work, and it occupies only a fraction of the floor space.

## MAGNETIC FORMING

The newest concept in high-energy forming is magnetic forming. Electric energy in the form of magnetism acts as the forming force. The magnetic pulses last only 6 millionths of a second and exert pressures up to 560,000 psi.

Magnetic pressures of this magnitude require a 1-megagauss field. (The gauss is used to indicate the density of the magnetic flux.)

The commercial magnetic-forming machine shown in Fig. 20.22 is capable of forming pressures up to 50,000 psi in pulses with durations of 10 to 20 millionths of a second. Electrical energy is stored in a capacitor and discharged rapidly through the coil, as shown in the schematic

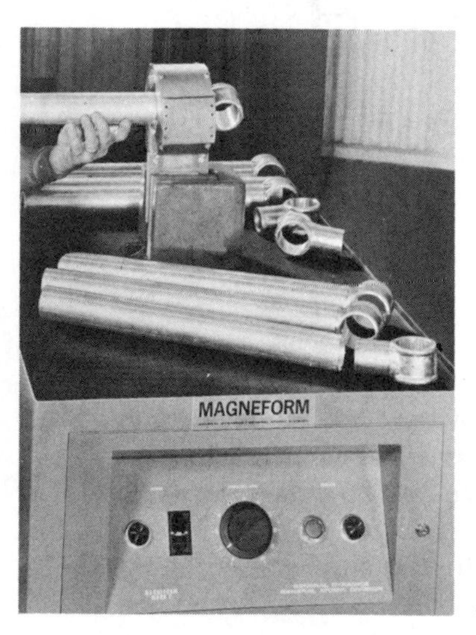


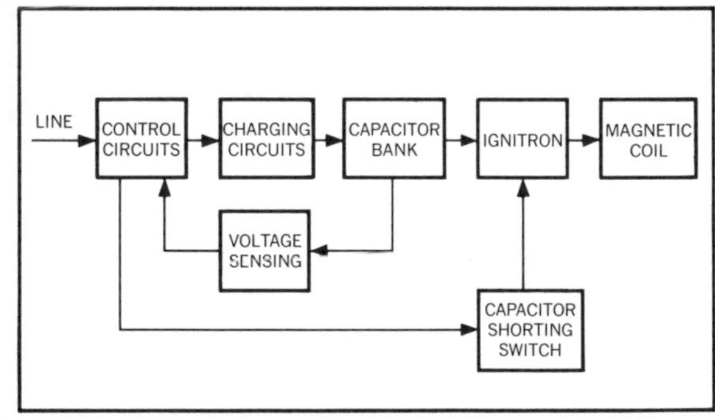


**Fig. 20.22.** The type of machine used for magnetic forming, showing parts being placed in the coil and a schematic diagram of the circuit used. Courtesy General Dynamics Corporation.

diagram. The current is induced from the coil to the conducting workpiece. The induced field in the workpiece interacts with the coil field and produces the necessary forming force.

**Principal Methods.** Magnetic forming can be classified in three basic arrangements—compression forming, expansion forming, and hammer forming.

COMPRESSION FORMING. The principle of compression forming is shown in Fig. 20.23. The tube is formed against a mandrel. The pulsed

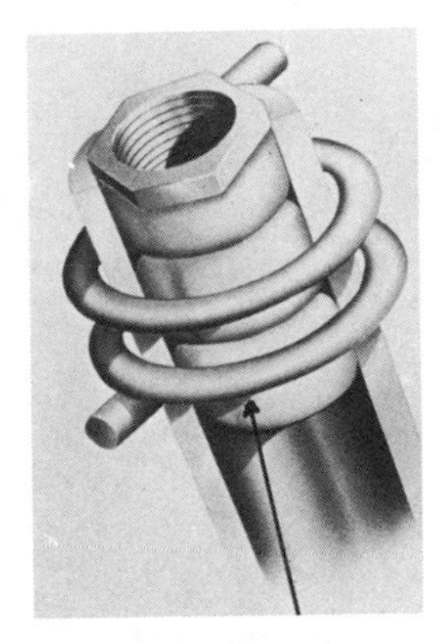

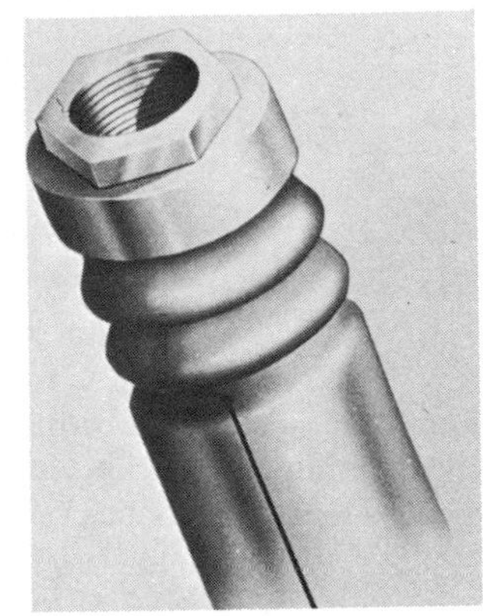

**Fig. 20.23.** Compression forming causes the metal to shrink, compress, or collapse against the mandrel. Courtesy General Atomic Division, General Dynamics Corporation.

forming time is so short that the depth of penetration of the field is small compared to the thickness of the tube. Therefore, there is no magnetic leak to the inside of the tube.

When insulated mandrels are used, the collapse rate is retarded. The magnetic flux must leak through the tube to fill the entire cross-sectional area before a significant back pressure develops.

EXPANSION FORMING. In expansion forming, the work coil is placed inside the material to be formed (Fig. 20.24). A slowly rising magnetic

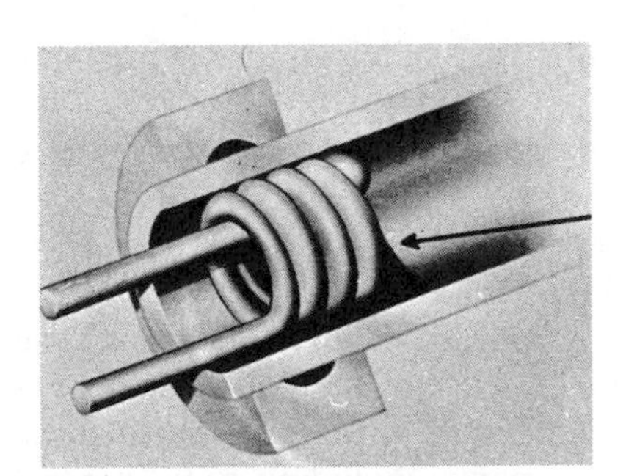

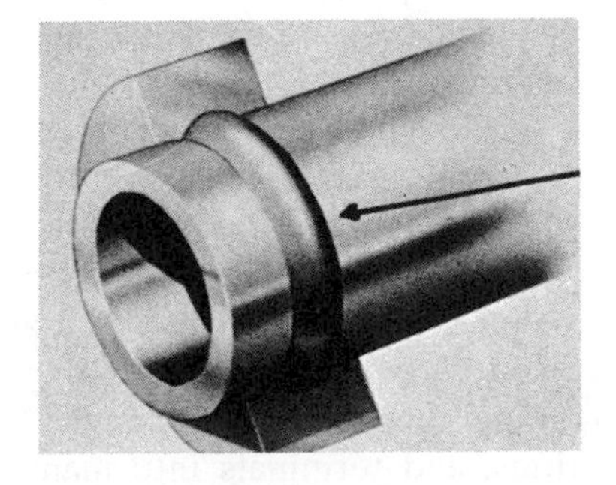

**Fig. 20.24.** Expansion forming is done by placing the coil on the inside and using the forming force to expand the workpiece. Courtesy General Atomic Division, General Dynamics Corporation.

field is established in the coil and in the tube to be expanded. Then the field outside the tube is reduced to zero in less time than was taken for the field penetration. The trapped magnetic field presses the tube outward.

HAMMER FORMING. Flat materials can be formed by placing them between an insulated die and the coil (Fig. 20.25). The same principles used in tube forming apply.

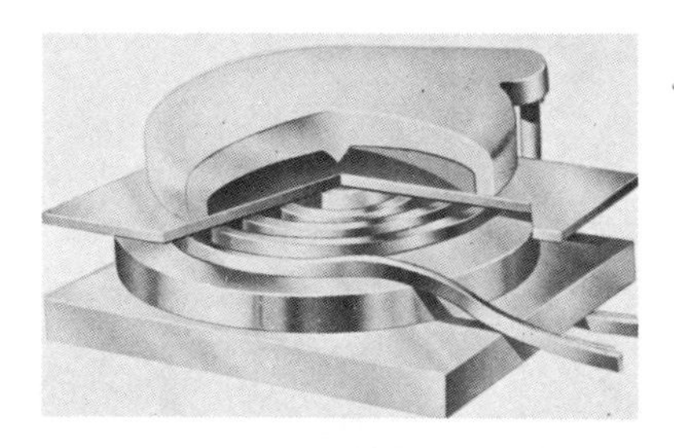

**Fig. 20.25.** Hammer forming can be used to form dimples or blanks or to emboss by pressing the workpiece against a die form. Courtesy General Atomic Division, General Dynamics Corporation.

**Advantages.** This new forming method has several distinct advantages that facilitate production operations.

(1) It does not mar or scratch the work surface and thus eliminates further finishing operations.

(2) The operation can be used to produce quantities—as many as 600 forming operations per hour.

(3) The fact that there is no moving part minimizes maintenance and provides for quiet operation.

(4) The process can be used on refractory metals when covered by an inert atmosphere or a vacuum.

(5) Coils can be easily changed for different applications, minimizing downtime.

(6) The exact amount of force needed can be adjusted and set so there will be no variance from part to part.

(7) Forming operations that are impractical or are inaccessible by other means can often be formed by this method.

**Applications.** Some general applications in use at the present time (Fig. 20.26) are forming tubing into precise and difficult shapes; expanding tubing into bushings, hubs, and split dies; swaging inserts, fittings, and terminals into many diverse parts, including rope, control cables, etc.; and rapid coining, shearing, and blanking.

A summary of high energy rate forming methods is given in Table 20.3.

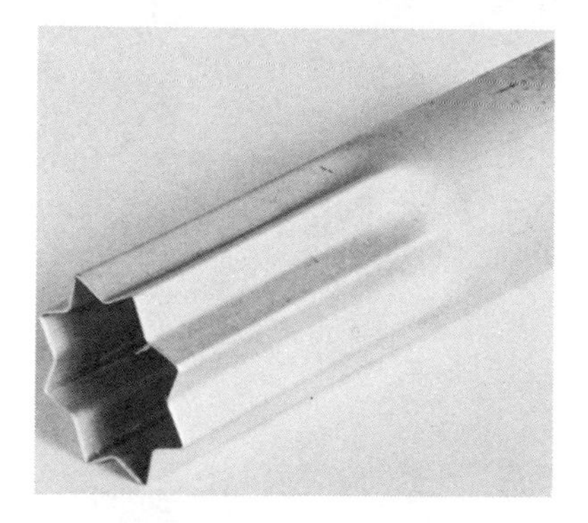

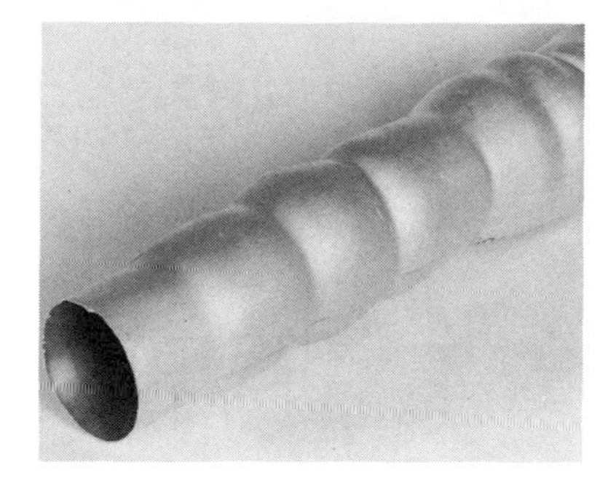

**Fig. 20.26.** Examples of the different types of work performed by magnetic forming. Courtesy General Atomic Division, General Dynamics Corporation.

**Table 20.3**
**Summary of High Energy Rate Forming Methods**

| Characteristics | Explosive forming | Electrohydraulic forming | Electromagnetic forming | Pneumatic-Mechanical forming |
|---|---|---|---|---|
| Types of operation | 1) high explosion<br>2) low explosion<br>3) gas mixture | 1) exploding wire<br>2) spark discharge | 1) compression<br>2) expansion<br>3) hammer | pneumatic-mechanical forging and extrusion |
| | chemical detonation or chemical burning | vaporization of wire or ionization of medium | collapsing magnetic field | quick release valve between high and low pressure gas chambers |
| Energy transfer medium | water, air, sand, and gas | water, air | air | air |
| Part sizes formed | parts 5′–10′ dia, varying thicknesses | parts 3″–4′ dia, 0.03″–1.0″ thick | parts up to 12″ dia, 1/8″ thick | parts up to about 1 1/2′ dia |
| Applications | blanking, coining, powder compacting, embossing, drawing, expanding, bulging, and stretching | bulging, stretching, sizing, coining, embossing, and expanding | swaging, joining, and shrinking | compacting, drawing, extruding, forging, and stretching |
| Tolerances achieved | ±0.001″ times material thickness to ±0.004″ | close tolerances up to ±0.002″ | ±0.001″ tolerances have been obtained | ±0.005″ limits can be obtained |

| | | | | |
|---|---|---|---|---|
| Materials formed | all of the common steels, stainless steels, titanium and its alloys, aluminum alloys, Vascojet 1000 H-11, AMS 6434 | aluminum, stainless steels, copper, titanium, columbium Rene 41 | tungsten, beryllium, columbium, tantalum, and other refractory metals; also, common ferrous and nonferrous materials | refractory metals, high-temp alloys, low-alloy steels, aluminum, alloys, titanium alloys, stainless steels, copper and its alloys |
| Main advantages | 1) close tolerances can be achieved<br>2) elimination of springback<br>3) large part sizes can be formed<br>4) better surface finish | 1) needs less space<br>2) energy released is closely controlled<br>3) cheap tooling<br>4) fast cycle times<br>5) results are repeatable | 1) process suitable for both short and long run work<br>2) close control of magnetic pulse<br>3) reactive materials can be formed<br>4) reproducibility of parts | 1) process speed; elimination of repetitive poundings<br>2) excellent surface finish obtainable<br>3) superior grain flow, high strength of parts formed<br>4) close dimensional control |
| Limitations | 1) production rate is low<br>2) careful handling is necessary | 1) energy rating of the capacitor bank<br>2) amount of energy that can be dumped by triggering device | 1) permanent coils costly for short runs<br>2) capability depends on energy storage capacity | large cycle time |

## QUESTIONS

**20.1.** Explain briefly the difference between CHM and ECM.

**20.2.** What are the principle differences between ECM and EDM?

**20.3.** Why is CHM well adapted to short run production?

**20.4.** Why can good accuracy be obtained on very small parts in CHM?

**20.5.** Explain the theory of EDM.

**20.6.** Where do you think EDM serves its best purpose?

**20.7.** Explain the difference between ECM and ECG.

**20.8.** (*a*) Which of the three processes EDM, ECM, or ECG gives the best metal removal rate? (*b*) surface finish? (*c*) tolerance?

**20.9.** What is the interesting phenomenon associated with ultra high-speed machining?

**20.10.** What is the theory behind hot machining?

**20.11.** What is the theory used to explain the outstanding successes obtained in explosive forming?

**20.12.** What are two less dangerous ways of employing the benefits of HERF?

**20.13.** What is the effect of "standoff" in HERF?

**20.14.** What would happen if no vacuum were put on the die side of the metal before explosive forming?

**20.15.** Compare the rate of conventional forming to HERF in feet.

**20.16.** What is the advantage of gas forming?

**20.17.** What impact velocity can be obtained by the pneumatic-mechanical system?

**20.18.** Explain the difference between EBW and hydrospark forming.

**20.19.** Explain the principle of magnetic forming.

**20.20.** Compare the tolerances obtained by each of the HERF methods.

## PROBLEMS

**20.1.** At what rpm would a 4-in.-dia shaft have to turn to attain the high-speed machining range of 150,000 sfpm?

**20.2.** What is the approximate shear angle of the material shown in (*a*) conventional cutting, (*b*) high-speed machining in Fig. 20.12?

**20.3.** The coefficient of friction can be calculated from the formula

$$\mu = \frac{F_t + F_c \tan \alpha}{F_c + F_t \tan \alpha}$$

where $F_t$ = thrust force; $F_c$ = cutting force; and $\alpha$ = rake angle. Find the coefficient of friction for each cut in Fig. 20.12 with the following conditions: *Tool A:* $F_c$ = 400 and $F_t$ = 150 lb; *Tool B:* $F_c$ = 150 and $F_t$ = 50 lb.

**20.4.** (*a*) Approximately how long will it take to chemically mill an aluminum alloy material that is to have four 1-in. square holes milled to a depth of 1/8 in.? (*b*) What would the surface finish of the bottom of the hole be if the original surface finish were 65 mu?

**20.5.** At maximum capacity, how long would it take to cut a 1/2-in.-dia hole in 1/8-in.-thick tungsten carbide by (*a*) EDM? (*b*) ECM?

**20.6.** What elongation might be expected by explosive forming of corrosion resistant high-alloy steel? See Table 3.6.

## REFERENCES

Cadwell, G., "Spark Forming Goes to Work," *American Machinist,* November 1961.

Degroat, G., "Ultra High Speed Machining," *American Machinist,* February 1960.

*High Velocity Forming of Metals,* American Society of Tool and Manufacturing Engineers, Prentice Hall, Inc., Englewood Cliffs, N.J., 1964.

Parr, J., "Hydro-Spark Forming," *The Tool and Manufacturing Engineer,* March 1960.

Rose, C. N., "Chemical Milling Today," *The Tool and Manufacturing Engineer,* October 1961.

Schwinghamer, R. J., "*Magnetomotive Forming Developments and Magnetohydraulic Forming,*" ASME Paper No. 66–MD–19, May 1966.

Strauss, W. A., Jr., "Photoetched Parts," *Machine Design,* June 4, 1964.

Welky, E. A., and C. T. Syburg, "High Velocity Forging of Conventional Materials," *Machinery,* April 1965.

Young, A. W., "*HERF*orging Hits Its Stride," *Iron Age,* May 20, 1965.

# APPENDIX

**Appendix Table 1**
**General Properties of Thermoplastics**

| Plastic and representative tradenames | Approx specific gravity | Approx resin cost ($/lb) | General properties | Typical applications |
|---|---|---|---|---|
| ABS (Acrylonitrile-butadiene styrene) Abson, Cycolac, Kralastic, Lustran, Boltaron | 1.04 | 0.35 to 0.43 | Good balance of properties—hard surface, high chemical resistance, good electrical properties, abrasion and oil resistance, and impact strength; excellent creep resistance; easily decorated, including chrome plating; will support combustion; can be injection and extrusion molded, vacuum formed, or blow molded. | Instrument panels, telephones, appliance housings, luggage shells, pipe and tubing. |
| ACETAL Delrin, Celcon | 1.42 | 0.65 | Good dimensional stability; same static as dynamic coefficient of friction; good electrical properties, high strength, good impact, and low creep characteristics; easily machined; can be injection or extrusion molded. | Bearings, cams, small gears, knobs, handles, and electric tool housings. |

| | | | | |
|---|---|---|---|---|
| ACRYLIC<br>Acrylite, Lucite, Plexiglas (Implex), Perspec | 1.08 to 1.19 | 0.45 to 0.53 | Outstanding optical properties; excellent mechanical properties for short-term loadings; inert to most chemicals; can be cast, extruded, injection molded, or vacuum formed; good machinability; easily decorated. | Aircraft canopies, windows, signs, pump housings, optical components, photographic lenses. |
| CELLULOSE ACETATE<br>Celanese, Plastacele, Tenite, Vuepac | 1.3 | 0.45 to 0.52 | Easily molded; good dielectric properties and good tensile strength; easily heat sealed, machined, and decorated; no limitations on color; molded grades are self-extinguishing; films are slow burning. | Coil forms, appliance housings, knobs, cutlery and tool handles, blister packaging, toys. |
| CELLULOSE PROPIONATE<br>Forticel, Tenite | 1.2 | 0.62 | Good impact strength, fine surface finish; not recommended for continuous use under high load or high temperature; processed by injection and extrusion but can be blow molded and thermoformed; easily machined or heat sealed; excellent weatherability and is slow burning. | Telephone and radio housings, knobs, bezels, tool handles, steering wheels, vacuum-cleaner attachments. |
| CHLORO-TRIFLUOROETHYLENE<br>(CTFE) Kel-F | 2.09 to 2.14 | 4.70 | Chemically inert; thermally stable between about −400 and +400 F; excellent electrical properties and reasonably good mechanical properties but does creep like other fluorocarbons; | Gaskets, pump and fluid meter parts, diaphragms, valves, seals, linings, packaging of corrosive materials. |

| | | | | |
|---|---|---|---|---|
| | | | good elastic memory; can be compression, transfer, injection, or extrusion worked; difficult to bond. | |
| FLUOROCARBON (TFE, FEP) Teflon, Halon | 2.1 | 3.25 (TFE) 6.60 (FEP) | Outstanding electrical properties and heat and chemical resistance; good mechanical properties; good impact strength; outstanding weatherability; good machinability; nonflammable; FEP is melt-processable; TFE is not. | Nonlubricated bearings, high-temperature electronic parts, gaskets, bushings, chemical-resistant pipe, tubing, and pump parts. |
| PHENOXY Bakelite (by formulation number) | 1.17 to 1.34 | 0.85 | Tough, rigid, strong, high-impact plastic; not to be used over 170 F; can be extruded or injection or blow-molded; resistant to acids and alkalies, but has poor solvent resistance; can be solvent cemented; lowest coefficient of expansion of any unfilled thermoplastic; nonflammable. | Gas and crude-oil pipe, phonograph records, sporting goods, appliance housings, packaging components, high-strength hot-melt adhesive. |
| POLYAMIDE (nylon) Zytel, Catalin, Fosta, Firestone, Plaskin, Spencer | 1.1 | 0.50 to 1.50 | High tensile strength; good impact strength and high abrasion resistance; good chemical and electrical properties; good heat resistance; processed by injection, extrusion, blow molding or compression casting. | Gears, bearings, cams, tubing, rollers, combs, housings, pipe, coil forms. |

| | | | | |
|---|---|---|---|---|
| POLYCARBONATE<br>Lexan, Merlon | 1.2 | 1.15 | Excellent dimensional stability, low water absorption, and excellent creep and impact resistance; good electrical properties and machinability; processed by injection or extrusion molding, casting, blow molding, or thermoforming. | Switch components, tool housings, filter bowls, safety helmets, pump impellers, optical lenses. |
| POLYETHYLENE<br>Alathon, Alkathene, Agilene, Dylan, Oripon, Petrothene | 0.90 to 0.98 | 0.15 to 0.28 | Excellent dielectric properties; low coefficient of friction; good machinability; good to excellent chemical resistance; fair to good weatherability; subject to creep under load; processed by injection molding, extrusion, or blow molding. | Housewares and toys, squeeze bottles, packaging films, pipe, wire and cable insulation. |
| POLYIMIDE<br>Vespel (molded)<br>Kapton (film) | 1.42 | Not available in resin form | Excellent dry film bearing and abrasion characteristics; good thermal stability; excellent dimensional stability; very low creep between −400 to +700 F; outstanding electrical properties; is infusible and nonflammable; cannot be injected or extruded; available as coatings, films, or solid forms. | Sleeve bearings, bearing cages, valves, valve seats, piston rings, rotary compressor vanes, printed-circuit boards, binder in diamond abrasive wheels. |

| | | | | |
|---|---|---|---|---|
| POLYPROPYLENE Catalin, Chevron, Dow, El Rex, Escon, Grace, Marlex, Moplen, Petrothene, Poly-Pro, Pro-Fax, Shell, Tenite | 0.90 | 0.25 | Exceptional toughness, chemical resistance, and electrical properties; has similar properties to polyethylene but offers about 50 deg F higher heat resistance; limitations are creep, difficulty in bonding, and limited low-temperature impact resistance; can be injection, extrusion, or blow molded and thermoformed. | Sump-pump components, hospital and housewares, wire coating, radio and TV housings, packaging, luggage containers, automotive parts. |
| POLYSTYRENE Styrene, Styron, Lustrex, Styrex, Pliolite | 1.04 to 1.09 | 0.18 to 0.36 | Good dimensional stability, low mold shrinkage, low cost, readily extruded or injection molded; excellent electrical insulator; poor weatherability; attacked by oils and most organic solvents. | Appliance housings, refrigerator inner doors, wall tiles, battery cases, toys, displays, electrical insulators, buoys, ornaments. |
| POLYVINYL CHLORIDE Vinyl, PVC, Exon, Geon, Pliovic, Opalon, Ultron, Tygon, Agilide, Marvinol | 1.1 to 1.8 | 0.12 to 0.50 | Good chemical and oil resistance and dimensional stability; good electrical characteristics and abrasion resistance; available in rigid, flexible, and foam forms; does not support combustion; exposure to heat produces degradation. | Chemical piping, valves, signs, phonograph records, plating racks, garden hose, floor tile, wire insulation, footwear. |

**Appendix Table 2**
**General Properties of Thermoset Plastics**

| Plastic and representative tradenames | Approx specific gravity | Approx resin cost ($/lb) | General properties | Typical applications |
|---|---|---|---|---|
| ALKYD<br>Glaskyd, Mesa, Plaskon | 1.9 to 2.3 | 0.35 to 0.75 | Good dimensional stability; hard, stiff, and tough with a good range of physical properties; excellent dielectric properties and high arc resistance; good chemical and heat resistance; properties improved with fillers. | Automotive ignition parts, electronic components, encapsulation of small electronic components. |
| ALLYLIC<br>Acme, Diall, Durez, Rogers, Dapon, Plaskon, Poly Dap | 1.26 to 1.65 | 0.75 to 3.25 | Hard, stiff, and durable, with excellent range of physical properties; excellent electrical properties; excellent chemical, corrosion, fungus, heat, water, steam, and oil resistance; can be compression or transfer molded and is usually filled or reinforced. | Pump impellers, coffee-machine components, radomes, electrical connectors; binder in reinforced structural stock. |
| AMINO (urea, melamine)<br>Cumel, Plaskon, Plenco, Fiberite, Beetle, Gabrite, Sylplast, Melmac, Resimene | 1.5 to 2.0 | 0.19 to 0.34 (urea) 0.42 to 1.05 (melamine) | Rigid, hard, and tough, with a good range of physical properties; good to excellent resistance to heat, water, grease, oil and household chemicals; good abrasion resistance and dimen- | Appliance knobs, dials, piano keys, dinnerware, utensil handles, housings, meter blocks. |

| | | | | |
|---|---|---|---|---|
| | | | sional stability; properties improved with fillers. | |
| EPOXY<br>Hysol, Plaskon, Scotchply, U. S. Polymeric, Fiberite, Durez, Plenco, Devcon-Epolite-Ren Epotuf-Araldite | 1.1 to 1.4 (unfilled)<br>1.6 to 2.06 (filled) | 0.55 to 8.00 | Excellent strength and toughness; good resistance to acids, alkalies, and solvents; excellent adhesion characteristics; low linear shrinkage during set-up; good electrical properties; can be cast, transfer, compression, or plunger molded or foamed. | Usually used for impregnating fiber or cloth lay-ups or for solid castings or potting; molded parts include glass-filled circuit boards, coils, switch components, tools and dies. |
| PHENOLIC<br>Fiberite, Durez, Bakelite, Plenco, Rogers, RCI, Resinox, G. E. Phenolic, Durite, Catalin, Marblette, Mesa | 1.3 to 1.8 | 0.21 to 1.10 | Hard, stiff, and tough, with good range of physical properties; good chemical, heat, and water resistance; good dielectric properties; properties improved with fillers; poor machinability (use carbide-tip tools). | Pump and blower components, electrical switch parts and housings, rotors, gears, handles for tools and utensils. |
| POLYESTER<br>Celanar, Melinex, Mylar, Scotchpak, Scotchpar, Plaskon | 1.05 to 1.4 (unfilled)<br>1.4 to 2.0 (filled) | 0.20 to 0.28 | Can be used up to about 350 F; may be rigid, flexible, or resilient; good environmental resistance. | Usually used as an impregnating resin with cloth or fiber reinforcers, or as a film; used in fiberglass boats, tanks, shower stalls, signs, fan shrouds, fishing rods, archery bows. |

| | | | | |
|---|---|---|---|---|
| SILICONE<br>Dow Corning Silicones | 0.97 to<br>1.13 (resin)<br>1.25 to<br>1.88 (filled) | 3.60 | Exceptional heat resistance to about 700 F; excellent moisture and chemical resistance; good electrical properties and weatherability; can be formed, cast, or compression or transfer molded. | Motor slot wedges, arc and thermal barriers, coil forms, panel boards, heat dams, electrical connectors, terminal strips, potting of electrical components. |
| URETHANE<br>Adiprene, Conathane, Cyanopoline, Elastothane, Estane, Genthane, Mearthane, Multrathane, Roylar, Solithane, Texin | 1.1 to<br>2.0 | 0.80 to<br>1.50 | Exceptional toughness and abrasion and wear resistance; good oil, chemical and solvent resistance; low compression set; can be cast, extruded, calendered, foamed, and injection molded. | Shock absorbers and vibration dampers, check-valve balls, forming dies, rollers, cleats, industrial truck tires, mallets, mattresses, linings for abrasive-slurry pumps. |

# INDEX


Abrasive
  belt machining 519
  machining 505
  materials 493
ABS plastics 641
Acetal 641
Acetylene 234
Acrylic 642
Adaptive control
  machining 401
  welding 279
Adhesives 309
  classification 313
  joint design 315
  principals 311
  strength 314
AISI steels 56 57 58
Alkyd 646
Aluminum 72
  classification 72
  fabricating characteristics 73
Angular measurement 536
Annealing 26
Annealing (*cont.*)
  cyclic 27
  full 27
  spheroidized 26
  stress relief 26
APT 601
Arc spot, welding 203
Arc welding 187
Argon 205
Austenite 23
Autocollimator 545
Automatic
  gaging 556
  screw machine operations 434
  welders 197
Automation 575
  circular indexing 581
  individual machine 576
  multiple machine 582
    building blocks 588
    integrated 582
    transfer 585
    unitized 583

Automation (*cont.*)
  robots 591
  tape control 578
  vibratory hoppers 591
Autopromt 602
Autospot 603

Bainite 26
Bakelite 643
Band saws 486
Barrel finishing 516
Belt polishing 519
Bending 345
Binary system 600
Boring 455
  machines 456
Brasses 76
Brazing 235
Brittleness 41
Broaching 479
  economics 484
  machines 481
  time 483
  tools 479
Buffing 520
Building block units 588

Carbide tools 387
Case hardening 31
  carburizing 32
  cyaniding 33
Castability 55
Cast alloys 387
Cast ferrous metals summary 69
Casting processes review 145
Casting, sand 83
  automation 101
  cores 85 89 90
  design 104
  dry-sand 84
  economics 104 109
  greensand 83

Casting, sand (*cont.*)
  molds 91
  patterns 87 88 116
  process steps 102
  repair 101
  sands 88
  solidification rate 104
  stress concentration 109
  tolerances 113 115
  weight reduction 112
Cast iron 62
  classification 64 65
  ductile 63 68
  gray iron 62
  inoculated irons 64
  malleable 62
Cellulose acetate 642
Centrifugal casting 136
Ceramic tools 388
Chaplets 86
Chemical machining 609
  blanking 609
  milling 611
Chip breakers 385
Chip types 381
Cladding 276
Clearance 550
Clutch-brake control 593
$CO_2$
  core making 90
  welding 200
Cold heading 371
Cold working 341
Comparators
  air 554
  electrical 552
  optical 554
Continuous casting 144
Continuous line 588
Contour cutting 465
Control charts 570
Copper-base alloys 75
  classification 76
  fabricating characteristics 77

Cored wire 195
Cores 85
  core making 89 90
Counterboring 449
Crater wear 389
Creep 49
Critical diameters 219
Critical temperatures of steel 20 22
Crystal structure 12
  lattice imperfections 13
  recovery 22
Cupolas 97
Cutting ratio 395
Cutting tools 382

DAC 603
Depth micrometer 534
Dial indicators 535
Diamond wheels 500
Die casting 138
  draft requirements 141
  metals 141
  timetable 142
  tolerances 142
Dies 346
  classification 349
  materials 347
Dislocations 14
  cold working 17
  edge 14
  mixed 15
  screw 15
  strain 18 23
DQE 223
Drawing 345 369
  dies 348
  forces 369
  ratio 348
Drilling 439
  action 440 444
  economics 454
  feeds 454
Drilling (*cont.*)
  power 455
  speeds 453
  tolerances 447
  torque 455
Drill jigs 442
Drill presses
  gang 451
  multispindle 451
  sensitive 449
  turret 451
  upright 450
Drills
  crankshaft 442
  gun 445
  trepanning 446
Dry-sand casting 84
Ductile iron 65
Ductility 41
Durez 646
Duty cycle 190

Economics
  broaching 484
  casting 104 109
  drilling 454
  forging 162 167
  metal cutting 399
  numerical control 604
  powder metallurgy 320
  surface finish 544
  welding 214
Eddy-current testing 257
Electric discharge machining 611
Electrochemical machining 615
  grinding 618
  honing 619
Electrodes 191
Electroforming 630
Electroslag welding 263
Elongation 36–40
End-quench 32
  hardenability 218

Epoxy 647
Equilibrium diagram 19
Eutectoid 20
Explosive
  forming 623
  welding 275
Extrusion 171
  casting 146
  impact 175 373
  materials 173
  pressure 370
  ratio 172

Fabricating characteristics 54
  aluminum 72
  brasses 76
  cast iron 65
  magnesium 78
  stainless steel 71
  steel 58
  steel casting 60
  titanium 80
Filament winding 303
Fits 551
Flame cutting 243
Flame hardening 34
Flame spraying 240
Fluoroscopy 252
Forging 151
  automatic 154
  comparisons 165
  cost 162 167
  design 163
  drop 153
  force 155
  hand 158
  heavy 158
  materials 161
  press 156
  stress and strain 152
  tolerance 164
  upset 155
Formability 55
Formability (*cont.*)
  aluminum 73
  brasses 77
  magnesium 78
  stainless steel 71
  steel 59
  titanium 80
Forming 341
  design principles 343
  electro-hydraulic 630
  explosive 623
  gas 628
  high energy rate 623
  hydroforming 361
  magnetic 632
  pneumatic-mechanical 629
  roll 366
  rubber and urethane 358
  stretch 362
Frequency distribution 563
Friction
  sawing 487
  welding 274
Frozen-mercury process 132
Furnaces 95

Gage blocks 525
  calibration 527
  classification 526
Gages 548
  automatic 556
  comparator 552
  mechanical 551
  small-hole 535
Gaging
  functions 560
  in-process 559
  postprocess 559
Gamma radiography 252
Gas forming 628
Gas metal arc welding 200
Gas welding 233
Gating of castings 84

Grain size 21
control in welding 179
refinement 24
Gray iron 62
Greensand molding 83
Grinding 493
economics 507
electrochemical 618
operations 501
snag 100 505
surface 503
tool and cutter 505
wheels 495
Guerin process 358
Gun drills 445

Hardenability 27
tests 28 29 30 31
Hardening 27
Hardness conversion chart 43
Hardness testing 41 42 43
Brinell 41
Rockwell 42
Vickers 42
Hard surfacing 238
Heat input 223
Heat treatment 26
High energy rate forming 623
summary chart 636
High-speed machining 621
High-speed tools 387
Hole punching 460
Honing 511
electrochemical 619
Horsepower 394
unit values 396
Hot working 167
drawing 170
extrusion 171
piercing 169
rolling 167
Hydroforming 361

Ideal critical diameter 219
Impact tests 48
Indicating micrometer 531
Indicators, dial 535
Induction hardening 34
Injection molding 291
Inspection methods 561
Inspection, weld 250
eddy current 257
magnetic particle 256
penetrant 258
radiography 251
ultrasonic 253
X-ray 251
Integrated lines 582
Interference microscope 544
Interferometer 530
Investment casting 131
Iron carbon equilibrium diagram 19
Iron-powder electrodes 193
Isothermal transformation 23 25 26 28 29 32 33

Jig boring 459
Jigs 442
Johansson, Carl 525
Jominy, Boegehold Test 28 29 30 31

Lance cutting 250
Lapping 508
Lasers 276
Lathes 406
automatic
cross-slide 417
multispindle 432
screw machines 429
single spindle 427
tracer 416
bench 413
chucks 410

Lathes (*cont.*)
duplicating 413
engine 406
evaluation 413
low production 406
speed 413
turret 419
vertical turret 424
Leak tests 258
Light pencil 604
Linear measurement 530
Lost-wax method 131
Low-hydrogen electrodes 193

Machinability
aluminum 73
brasses 76
cast iron 65
definition 54
magnesium 78
measurement 395
stainless steel 71
steel 58
titanium 80
Machine time
broach 483
drill 455
lathe 436
mill 477
Machining
cold 623
hot 622
ultrahigh-speed 621
Machining center 578
Magnesium 77
classification 77
fabricating characteristics 78
Magnetic forming 632
Magnetic tests 256
Mandrels 411
Martempering 31 33
Martensite 27
Mean 564
Measurement
angular 536
coordinate 547
direct 530
indirect 535
linear 530
metric 525
roundness 539
standard 524
surface 539
Metal-cutting economics 399
Metals of the future 49 50 51
Metric system 525
Metrology 524
Micrometer 530
Micro wire welding 203
Milling
cutters 471
feed 477
methods 475
power 478
production 477
Milling machines 469
column-and-knee 469
combination 470
duplicating 473
fixed bed 471
horizontal and vertical 469
planer 472
Modulus of elasticity 40
Molding equipment 92
Molds
permanent 137
plastic 291
sand 91
shell 121
Monochromatic light 531
Multiple-spindle
automatic lathes 432
drilling heads 452

Natural tolerance 564
New process summary 620

Nitriding 33
Nondestructive testing 250
Normal distribution 563
Normalizing 27
Numerical control 578
  control unit 600
  drafting 603
  drawing 599
  economics 604
  postprocessor 597
  processor 597
  process sheet 601
  programming 596
  systems 594
    continuous path 595
    point to point 595

Optical flats 531
Optical tooling 545
Oxyacetylene cutting 244
Oxygen 233

Parting line 86
  placement 108
Patterns 87 88
  styrofoam 117
Pearlite 20 26
Penetrant inspection 258
Permanent-mold casting 137
Phenolic 647
Photoelasticity 184
Planers 466
Plasma 247
  costs 249
  cutting 248
  spraying 241
  welding 273
Plaster-mold casting 129
Plastics 283
  blow molding 298
  casting 295
  design considerations 306

Plastics (*cont.*)
  elastomers 307
  extrusion 294
  fabricating methods 296
  filament winding 303
  machining 304
  molding 291
  polymerization 284
  properties 286
  thermoplastic
    applications 641
    costs 641
    properties 641
  thermoset
    applications 646
    costs 6464
    properties 646
  types 289
  welding 305
Plexiglas 642
Pneumatic
  circuits 576
  forming 629
Polishing 519
Polyamide 643
Polycarbonate 644
Polyester 647
Polyethylene 644
Polyimide 644
Polymers 284
Polypropylene 645
Polystyrene 645
Polyvinyl chloride 645
Postprocessor 597
Powder metallurgy 320
  design considerations 330
  economics 335
  impregnation 327
  machining 326
  plating 328
  properties 331
  sintering 325
Powder-spray method 241
Presses 352

Pressworking operations 351
Process sheet, N/C 601
Profile recorders 542
Programming, N/C 596
Properties of cast irons 68
Properties of steel chart 69
Proportional limit 37
Protractors 537
Punching 461

Quality control 563
  cost 566
Quenching 28

Radiography 251
Reamers 448
Recrystallization 22
Reduction of area 40
Resistance welding 208
  flash butt 213
  percussion 213
  projection 212
  seam 211
  spot 209
  stud 206
Robot control 591
  point-to-point 593
  continuous path 593
Rockwell hardness test 42
Roll forming 366
Roundness measurement 539

Sampling 562
  plans 568
Sand casting 83
Sawing
  abrasive 489
  band 486
  circular 488
  friction 487
  reciprocation 484
Screw machine operations 434
Screw machines 429
  Brown and Sharpe 429
  Swiss-type 430
Seam welding 211
Shapers 465
Shaw process 134
Shear strength 48 49
Shell-mold casting 121
  blowing 125
  costs 126
  draft 124
  mold ejection 122
  sand recovery 125
  tolerances 128
Shielded-arc electrodes 192
Shielding gas selection 205
Silicone 648
Sine bars 537
Sintering, metal powders 325
Small-hole gage 535
Snag grinding 100 505
Spinning 363
Spot welding 209
Stainless steel 70
  classification 70
  fabricating characteristics 71
Standard deviation 565
Steel, classification 56 57 58
Steel castings 60 61
Strain 37
  true 39
Stress 37
  control in welding 181 183
  engineering 37 38
  true 39
Stretch forming 362
Stud welding 206
Submerged arc welding 197
Superfinishing 512
Super micrometers 532
Surface hardening 34
Surface measurement 539
  optical 544

Surface measurement (*cont.*)
trace type 542
Surface texture 540
arithmetic average 542
by process 545
economics 544
root mean square 542
Swaging 368

Talyrond 539
Teflon 643
Tempering 30
Testing
nondestructive 250 (*see also* Inspection)
tensile 36
Thermal effects of welding 179
Thermal pinch 248
Thermoplastics properties 641
summary chart 641–645
Thermoset plastics properties 646
summary chart 646–648
Titanium 79
classification 79
fabricating characteristics 80
Tolerance 549
natural 564
Tool boards 584
Tool forces 394
Tool life curves 393
Toolmaker's microscope 554
Tool steels 60
Tools, cutting 380
geometry 382
life 389
materials 386
signatures 385
Torch cutting 244
Toughness 43 48
Transfer machines 585
Transformation, isothermal 23
Trepanning 446
Turning 406
Turret lathes 419
Twinning 16

Unitized lines 583
Urea 646
Urethane 648
Ultrasonics 253
inspection 253
welding 271

Vernier calipers 533
Vernier height-gage 534
Vibratory finishing 517
Vibratory hopper 590
Vickers hardness test 42

Wheelon process 360
Weldability 55
aluminum 74
brasses 77
cast iron 66
magnesium 78
stainless steel 71
steel 59
titanium 80
Welding 178
arc 187
arc spot 203
automatic 197
cooling rate 219
corrosion control 182
cracks 182
current characteristics 189
current sources 187
diffusion 273
duty cycle 190
economics 214
electrodes 191
electrogas 266
electron-beam 266
electroslag 263

Welding (*cont.*)
explosive 275
feedback control 279
flashbutt 213
flux-cored 195
friction 274
gas 233
gas metal arc 200
gas tungsten arc (TIG) 205
hard surfacing 238
heat flow 217
ideal critical diameter 219
laser 278
micro-wire 203
new processes 263
percussion 213
plasma 273
positioning 196
principles 178
processes chart 185
projection 212
pulsed current 204
resistance 208
seam 211
shielding gases 205
spot 209
stress control 181 183 184
stud 206
submerged arc 197 264
thermal effects 179
ultrasonic 271
weight of deposit 215
Whiskers 49 50 51
Work hardening 341

X-and-R charts 570
X-ray 251

Yield point 38
Yield strength 38 39